Rossiter W. Raymond

# Statistics of Mines and Mining

Salzwasser

Rossiter W. Raymond

# Statistics of Mines and Mining

1. Auflage | ISBN: 978-3-84605-505-2

Erscheinungsort: Frankfurt, Deutschland

Erscheinungsjahr: 2020

Salzwasser Verlag GmbH

Reprint of the original, first published in 1875.

43D CONGRESS, 2d Session. } HOUSE OF REPRESENTATIVES. { Ex. Doc. No. 177.

# STATISTICS

OF

# MINES AND MINING

IN

THE STATES AND TERRITORIES WEST OF THE ROCKY MOUNTAINS;

BEING

THE SEVENTH ANNUAL REPORT

OF

ROSSITER W. RAYMOND,
UNITED STATES COMMISSIONER OF MINING STATISTICS.

WASHINGTON:
GOVERNMENT PRINTING OFFICE.
1875.

# LETTER

FROM

# THE SECRETARY OF THE TREASURY,

TRANSMITTING

*A report on the Statistics of Mines and Mining in the States and Territories west of the Rocky Mountains.*

---

FEBRUARY 23, 1875.—Referred to the Committee on Mines and Mining and ordered to be printed.

---

TREASURY DEPARTMENT,
*Washington, D. C., February* 19, 1875.

SIR: I have the honor to transmit herewith the report of Rossiter W. Raymond, Commissioner of Mining Statistics, for the year ending December 31, 1874.

Very respectfully, your obedient servant,

B. H. BRISTOW,
*Secretary of the Treasury.*

Hon. JAMES G. BLAINE,
*Speaker of the House of Representatives.*

# CONTENTS.

# INTRODUCTORY.

WASHINGTON, *February* 19, 1875.

SIR: The preparation of the report which I have the honor to submit herewith has been attended by some unusual difficulties in addition to those which beset, under the most favorable circumstances, the collection of trustworthy information from an immense and sparsely-settled region concerning the condition of a constantly shifting industry. The principal trouble has arisen, during the past season, from the lack of means, under the reduced appropriation placed at my disposal, for employing resident correspondents who could furnish me with complete and recent information concerning special localities. This obstacle has been measurably overcome by the cordial co-operation of many gentlemen upon whose services I had no other claim than that which their own personal friendship or public spirit inspired; and I feel justified in believing that the present report will be found not inferior to its predecessors as a trustworthy review of the condition of mining and of metallurgical science throughout the States and Territories to which it refers. I regret to say that, failing to obtain the information promised me from New Mexico, I have been obliged to omit the chapter on that Territory—an omission which I hope to make good next season. Wyoming has been similarly treated, but for a different reason—because the explorations now on foot in that Territory will supersede what information I could at present give with reference to the Black Hills, while from the Sweetwater district there is nothing to report.

The product of the precious metals in 1874 may be estimated, from the best attainable sources of information, as follows:

| 1874. | | |
|---|---|---|
| Arizona | | $487,000 |
| California | | 20,300,531 |
| Colorado | | 5,188,510 |
| Idaho | | 1,880,004 |
| Montana | | 3,844,722 |
| Nevada | | 35,452,233 |
| New Mexico | | 500,000 |
| Oregon | $609,070 | |
| Washington | 154,535 | |
| | | 763,605 |
| Utah | | 3,911,601 |
| Wyoming and other sources | | 100,000 |
| | | 72,428,206 |

For purposes of comparison, the following exhibit of the estimated product of former years is subjoined:

| States and Territories. | 1869. | 1870. | 1871. | 1872. | 1873. | 1874. |
|---|---|---|---|---|---|---|
| Arizona | $1,000,000 | $800,000 | $800,000 | $625,000 | $500,000 | $487,000 |
| California | 22,500,000 | 25,000,000 | 20,000,000 | 19,049,098 | 18,025,722 | 20,300,531 |
| Colorado | *4,000,000 | 3,675,000 | 4,663,000 | 4,661,465 | 4,020,263 | 5,188,510 |
| Idaho | 7,000,000 | 6,000,000 | 5,000,000 | 2,695,870 | 2,500,000 | 1,880,004 |
| Montana | 9,000,000 | 9,100,000 | 8,050,000 | 6,068,339 | 5,178,047 | 3,844,722 |
| Nevada | 14,000,000 | 16,000,000 | 22,500,000 | 25,548,801 | 35,254,507 | 35,452,233 |
| New Mexico | 500,000 | 500,000 | 500,000 | 500,000 | 500,000 | 500,000 |
| Oregon and Washington | 3,000,000 | 3,000,000 | 2,500,000 | 2,000,000 | 1,585,784 | 763,605 |
| Wyoming | ............ | 100,000 | 100,000 | 100,000 | 50,000 | ............ |
| Utah | ............ | 1,300,000 | 2,300,000 | 2,445,284 | 3,778,200 | 3,911,601 |
| Other sources | †500,000 | 525,000 | 250,000 | 250,000 | 250,000 | *100,000 |
| Total | 61,500,000 | 66,000,000 | 66,663,000 | 63,943,857 | 71,642,523 | 72,428,206 |

* Including Wyoming. † Including Utah.

The great event of the year was, of course, the extraordinary development of the large *bonanza* in the Comstock lode. I hope to obtain, with the aid of Mr. C. A. Luckhardt, before the present report is finally printed, a summary and analysis* of the operations on that lode, enabling me to give a very clear account of the nature of the recent discoveries.

Of equal economical importance with new discoveries of valuable ores, though of a less sensational character, are the steady improvements in the art of metallurgy, by which it is becoming possible every year to treat with profit a larger proportion of ores extracted from our mines. I have continued to give much attention to this subject, being ably supported by my assistant, Mr Eilers, with whom it is a specialty; and the efforts which I have put forth, through a series of years, to furnish useful records and suggestions to American metallurgists are, I am glad to know, not unfruitful. A class of men has gradually, but surely, come into the management of important works in this country to whom such information is welcome, because they know how to make use of it; and in return for such a service they freely put at the disposal of him who renders it the results of their own experience, as the means of still greater usefulness. The pages of my reports bear very gratifying evidence of this reciprocity of feeling—a condition indispensable to rapid and thorough advances in any art involving the application of science.

Very respectfully, your obedient servant,

R. W. RAYMOND,

*United States Commissioner of Mining Statistics.*

Hon. B. H. BRISTOW,

*Secretary of the Treasury.*

* Since received. (See chapter on Nevada, under the head of "Comstock lode.")

# PART I.

## CONDITION OF THE MINING INDUSTRY.

# CHAPTER I.

## CALIFORNIA.

Mr. Skidmore, with his usual industry and discriminating judgment, has collected, by correspondence and personal observation, so large a mass of valuable information concerning the condition and prospects of the mining industry in this State that I feel scarcely justified in increasing the length of the present chapter by any introductory remarks. A few brief observations of a general character and references to special information under the headings of the different counties must suffice. The ditches and gravel-mines of California were more extensively operated, for a longer season, and with a larger supply of water, during the past year than ever before. Large undertakings which have been years in progress are now commencing to bear fruit. Among the important mining-ditches completed or advanced toward completion during 1874 may be named the Amador Canal, in Amador County; North Fork and Maxwell Ditches, in Plumas County; the El Dorado Canal, in El Dorado County, and the Milton Ditch, in Nevada County. Detailed accounts of these undertakings and of drift-mining enterprises will be found in the following pages. The gold-quartz-mining industry of the State presents little change. The following table, carefully prepared by Mr. Skidmore, gives the particulars of those mines in California which have produced more than $100,000 each during the the year, so far as the information could be obtained.

*List of quartz-mines in California producing over $100,000.*

| Mines. | Location. | Total number of tons worked. | Average yield per ton. | Total bullion product. | Number of stamps employed. | Cost of mining per ton. | Cost of milling per ton. | Company or custom mill. | Miners' wages per day. | Number of miners employed. | Remarks. |
|---|---|---|---|---|---|---|---|---|---|---|---|
| Idaho | Nevada County | 28,801 | $23 40 | $664,811 | 35 | $6 00 | $2 22 | Company | $3 00 | 175 | Weight of stamps, 850 pounds. |
| Eureka | do | 8,130 | 25 00 | 205,780 | 30 | 10 25 | 2 61 | do | 3 00 | 90 | Do. |
| Empire | do | 11,000 | 16 75 | 187,000 | 20 | 8 00 | 1 75 | do | 3 00 | 80 | Weight of stamps, 900 pounds. |
| Providence | do | 7,200 | *12 00 | ........ | 20 | 2 00 | 75 | do | 3 00 | 42 | Weight of stamps, 750 pounds. |
| Black Bear | Siskiyou County | 10,949 | 19 23 | 211,797 | 32 | 7 75 | 1 15 | do | 3 00 | 100 | |
| Plumas Eureka | Plumas County | 25,021 | 10 00 | 243,950 | 40 | 5 20 | 1 08 | do | 3 00 | 150 | |
| Sierra Buttes | Sierra County | 53,959 | 8 50 | 470,608 | 96 | 3 64 | 83 | do | 2 75 | 250 | |
| Sumner | Kern County | 5,000 | 40 00 | 200,000 | 16 | 2 50 | 2 00 | do | 3 50 | 30 | This company has erected a new 80-stamp mill. |
| Hite's | Mariposa County | 4,000 | 40 00 | 160,000 | 20 | 5 00 | 3 20 | do | 2 50 | 17 | The stamps at mill weigh 500 pounds. |
| Amador Consolidated | Amador County | 22,098 | 10 84 | †259,971 | 40 | 8 03 | 1 04 | do | 3 00 | ........ | The high cost of mining is due to dead work. |
| Keystone | do | 25,146 | 18 00 | 452,507 | 40 | 5 12 | 2 04 | do | 3 00 | 30 | Weight of stamps, 700 pounds. |
| Klamath | Siskiyou County | 15,385 | 13 00 | 200,000 | 32 | 3 50 | 1 50 | do | 3 00 | 75 | Weight of stamps, 600 pounds. |
| Chariot | San Diego | 2,500 | 55 00 | 138,864 | 10 | 10 00 | 4 00 | do | 3 00 | 35 | Weight of stamps, 800 pounds. |

* Free gold. There is 10 per cent. of sulphurets in the ore, which are very rich in gold, and are treated by chlorination. Total product not given, but it has been over $100,000.

† In this sum is included $20,254 from 219½ tons of sulphurets, at $92.27.

The product of gold and silver in California is given by Mr. Valentine, the superintendent of the express business of Wells, Fargo & Co., as follows:

| | |
|---|---|
| Gold dust and bullion by express | $16,015,568 |
| Gold dust and bullion by other conveyance | 1,601,556 |
| Silver bullion by express | 967,857 |
| Ores and base bullion by freight | 1,715,550 |
| Total | 20,300,531 |

In silver-mining there has been considerable excitement, constituting, perhaps, the special sensation of the year, concerning the Panamint district, in Inyo County, a description of which is given in another place.

The mining and reduction of quicksilver-ores have been stimulated to great activity by the high price of the metal, and there is little doubt that, although many of the quicksilver-claims will be abandoned as prices fall, yet some, in which during the period of prosperity valuable deposits have been developed, will continue with profitable production. Considerable information concerning them is given elsewhere in the present report. At this point I will quote an interesting summary of the subject from the San Francisco Mining and Scientific Press of January 23, 1875:

Quicksilver has become a metal of even more importance than ever of late, on account of its scarcity and consequent high price. The quicksilver-excitement, as well as its price, may be said to have kept up to the highest pitch during the whole year. The value of the metal, the accessibility of the localities where it is found, the broad area of country where it is likely to be met with, and the value of a good mine are all incentives which combine to keep up the excitement. It is confined to no particular locality in the State, and all classes of people have turned prospectors to find it. Honest farmers have dropped the plow and taken to the pick, and even the female part of the population in some places have obtained prospectors' outfits and scoured the hills in search of the precious metal. It has been selling at $1.55 per pound for some months past, and is likely to go higher before it falls lower. Our mines must be actively worked to meet the extraordinary demands which the immense mills and heavy mining operations at present conducted require. Only a few of the new mines are yielding quantities of any moment, though many of them are being steadily developed. A great many small claims are being rapidly developed which will in the future add to the production. It is impossible to state the number of quicksilver-claims in California, but after some difficulty we have been able to arrange the following list of *producing* mines, and add to it a list of a number of others which are likely to become good mines when further developed. We give in this list also the number of retorts and furnaces in use at the producing mines, together with the name of the style of furnace:

PRODUCING QUICKSILVER-MINES IN CALIFORNIA.

Almaden mine (Santa Clara County) has six "Almaden" furnaces; building one more. Mr. Randol, the superintendent, informs us that the product of the mine in 1874 was 9,084 flasks.

Redington mine (Napa County) has two Knox & Osborne and two Almaden furnaces running. Building two more Knox & Osborne furnaces. The mine is supposed to yield about 1,000 flasks per month, exceeding the product of the Almaden, and producing more than any other mine in California.

California Borax Company's mines (Lake County) has two Luckhardt furnaces, a Knox & Osborne furnace, a Wallbridge furnace, and five large retorts. Will erect another Knox & Osborne furnace in the spring.

Great Western (Lake County) has one 10-ton Luckhardt furnace; also one known as the Great Western or Green furnace—30 tons.

American mine (Lake County) has a 10-stamp mill for crushing the ore, twelve retorts, a Luckhardt furnace, a Wallbridge furnace, and a Perschbacker furnace.

New Idria mine (Fresno County) has two Almaden furnaces and a Maxwell furnace.

Monterey mine (near New Idria) has a small Almaden furnace.

Guadalupe mine (Santa Clara County) has two small Almaden furnaces and building a Maxwell furnace.

Enriquita, (Santa Clara County,) hoisting ore, which is reduced at the Almaden mine.
North Almaden, (Santa Clara County,) building a Neate furnace.
Cerro Bonito (Fresno County) has a Knox & Osborne furnace.
California mine (on line of Napa and Yolo Counties) has a Knox & Osborne furnace.
Oceanic (San Luis Obispo County) building a Louis Janin furnace.
Keystone, (San Luis Obispo County,) just completed a furnace.
Buena Vista (San Luis Obispo County) has a very small Almaden furnace.
Sunderland (San Luis Obispo County) has a Luckhardt furnace.
Manhattan (Napa County) has a Knox & Osborne furnace.
Phœnix (Napa County) has a Knox & Osborne furnace.
Ætna (Napa County) has a Knox & Osborne furnace.
Ida Clayton and Yellow Jacket (Napa County) have a Knox & Osborne furnace between them.
Abbott mine (Lake County) has a Knox & Osborne furnace.
Buckeye (Colusa County) has a Knox & Osborne furnace.
Rattlesnake (Sonoma County) has a Luckhardt furnace and retort.
Annie Belcher (Sonoma County) has a Knox & Osborne furnace, just completed.
Geyser (Sonoma County) has a Knox & Osborne furnace just completed.
Cloverdale (Sonoma County) has a Knox & Osborne furnace just completed.
Josephine, (San Luis Obispo County,) small 8-ton furnace.
Livermore (Sonoma County) has a modified form of the Knox & Osborne furnace.
Sonoma (Sonoma County) has a Luckhardt furnace.
Missouri (Sonoma County) has five retorts.
Oakland (Sonoma County) has five retorts.
Excelsior, (Sonoma County,) just completed a Winterburn furnace similar to the Green or Great Western.
Saint John's (Solano County) has a Neate furnace.
Kearsarge mine (Lake County) has retorts, but no furnace.
Eastern, (Sonoma County,) erecting a Wallbridge furnace and will put up a Knox & Osborne.
Western or Mount Jackson (Sonoma County) has one small Almaden furnace built by Winterburn.
Washington mine (Napa County) has an Almaden furnace with a Colt condenser.
Oakville mine has three Luckhardt furnaces; no work done during 1874.
Amarillo (Sonoma County) has one retort; commenced running in December.
Comstock mine (Santa Clara County) has one retort.
Elgin mine (Calusa County) has one retort.
Stayton mines (San Benito and Merced Counties; this is a group of twelve mines consolidated in one company) have one retort running, and will erect a Knox & Osborne furnace in the spring.

The following list of quicksilver-claims give promise of being mines when developed, and will have furnaces or other reduction-works this year; the list is not supposed to embrace one-half of the good prospects which will be so improved:

The Los Prietos claim, Santa Barbara County; Jeff Davis, San Luis Obispo County; Todas Santos, San Luis Obispo County; Pine Mountain, San Luis Obispo County; Quien Sabe, San Luis Obispo County; Amador, El Dorado County; Kentuck, Sonoma County; Socrates, Sonoma County; is well opened and ready for furnace. Flag-staff, Sonoma County; Mercury, Sonoma County; Wall Street, Lake County; Columbia, Lake County; London, Lake County; El Madre, Napa County; Georgia, Sonoma County; Cedar, Napa County; Montezuma, Colusa County; Empire, Colusa County; Cochrane, San Luis Obispo County; Live Oak, Sonoma County; Hercules, Sonoma County; Emma, Sonoma County; Illinois, Sonoma County; Peerless, Napa County; Thompson, Sonoma County; Central, Napa County; American, Lake County; Yosemite, Napa County; Bacon mines, Sonoma County; Pilot Knob, Lake County; Alice Cross; Brandt mine, Sonoma County; Lyttle, Trinity County; Boston, Trinity County; Edith, Sonoma County; Empire mine, Mendocino County; Gibson & Phillips, San Luis Obispo County.

There are many other claims being worked, which have yet assumed no prominence, and which are not mentioned above. We are unable to give any figures of the quicksilver-product of the year. Most of the metal is consumed at the mines in the different mining States and Territories, the Comstock mines using the largest amount of any one section. The item of quicksilver, at its present prices, is an important one to the mining interests. How much we have consumed, it is difficult to tell. It all goes out of the State, and none of it ever returns. When our mills lose from half a pound to a pound and a half for every ton they work, it does not take long to use several tons of mercury. By the statistics of the Central Pacific Railroad Company, we see that 432,635 pounds of quicksilver were shipped from this city as through freight, (which means that it went out of the State,) and 36,444 pounds were shipped from San José. As local freight they carried 47,007 pounds. Our exports by sea from this city from 1859 to 1874, inclusive, were as follows:

QUICKSILVER-EXPORTS.

| Year. | Flasks. | Value. |
|---|---|---|
| 1859 | 3,367 | $126,262 |
| 1860 | 8,962 | 318,320 |
| 1861 | 35,218 | 1,112,654 |
| 1862 | 35,707 | 1,169,197 |
| 1863 | 26,060 | 966,748 |
| 1864 | 37,252 | 1,527,963 |
| 1865 | 41,256 | 1,733,283 |
| 1866 | 30,789 | 1,082,940 |
| 1867 | 28,824 | 929,726 |
| 1868 | 43,507 | 1,330,054 |
| 1869 | 23,365 | 747,671 |
| 1870 | 12,959 | 602,051 |
| 1871 | 11,244 | 862,125 |
| 1872 | 14,721 | 875,414 |
| 1873 | 6,169 | 462,495 |
| 1874 | 5,388 | 539,525 |
| Totals | 365,788 | 14,226,441 |

The following table shows the destination of the quicksilver shipped by sea:

| | Flasks. | Value. |
|---|---|---|
| New York | 2,502 | $253,300 |
| Central America | 347 | 54,472 |
| China | 1,150 | 94,500 |
| Japan | 83 | 11,930 |
| British Columbia | 2 | 220 |
| Australia | 50 | 7,500 |
| Mexico | 1,146 | 119,654 |
| New Zealand | 31 | 3,834 |
| Totals | 5,504 | 498,330 |

Quicksilver-mining is destined to become even more important to California than it has ever been before, for we have the only mines of the kind in the United States.

The following compact statement concerning the quicksilver bearing and producing territory of California has been prepared from notes of personal observations furnished by C. A. Luckhardt, esq., of San Francisco. This belt, extending over three hundred miles in a general north and south direction, is occupied by very massive beds of slate-courses, accompanied by gabbro, siliceous and calcareous rocks, and often broken by syenite and dikes of trachytic rocks, which cap it over in places for miles. In and near these slates quicksilver is found wherever they come to the surface, usually either native or as cinnabar. The so-called liver-ore (*Lebererz*) and the selenide of mercury occur only as varieties. The black oxide, of which so much has been said, has never come to Mr. Luckhardt's notice.

These slates extend from San Luis Obispo County to the north of Sonoma County, and re-appear again in Trinity County. They consist principally of talcose slates, carrying siliceous and calcareous slates subordinately, and are in places over one-half mile in width. Accompanying them at the south are calcareous rocks of various ages, and occasionally magnesian rocks, while near the center and at the north are found sandstones and serpentine. The latter seems frequently to have interpenetrated or replaced the slates, causing much irregularity in the outward appearance of the hills, and also an apparently distinctly recognizable disconnection of the slate-belt.

This belt near its southern terminus, in San Luis Obispo County, shows the slates—that is, where their character is still recognizable—standing vertical or dipping from 60° to 80° north and northwest. Farther north the same dip is discernible until Napa County is reached and Mount Saint Helena, where they lie more nearly horizontal, while the north and northwestern portions of the belt show them for miles to dip as much as 40° to the south and southeast. Thus they form an irregularly-shaped oblong basin, of which Mount Saint Helena and a portion of Sonoma County appear to be the center.

The interpenetration of the slates and their accompanying rocks has greatly facilitated the metamorphism of the former, which has in many places advanced to such a high grade as to destroy entirely the characteristics of the several rocks. To this cause chiefly Mr. Luckhardt attributes the occurrence of the quicksilver in so many apparently different rocks.

Traversing, and in some instances running parallel with, the slates—that is to say, imbedded in them—are found calcareous rocks, (compact limestone, calcareous slate, spar, &c.,) and also siliceous rocks, (siliceous slate, quartzite, &c.,) inclosing a stratum of talcose slate, bearing some evidences of motion on the planes of contact; and this has been often mistaken as a sure proof that fissure-veins exist, although a great deal of irregularity appears, and an unequal distribution of the ore through what would be called the vein-matter. The walls are frequently as rich in ore as the vein itself, and it is probable that the idea of a fissure is erroneous, and that, although the slate incased as above described differs very much in appearance from the slates which form the main belt itself, it is nevertheless part and parcel of the same, only more highly metamorphosed.

Quicksilver-ore occurs throughout the above-described territory as impregnations in the calcareous and siliceous rocks and as beds or deposits in the slates, the bodies of the latter class being in some instances so firmly inclosed in the rocks accompanying them as to have the appearance of veins. This distinction between impregnations and beds is, in this case, not really founded in nature, since all the deposits may be conceived as the result of impregnation by sublimation or infiltration. What are here called impregnations in a narrower sense are of comparatively lower grade in richness, and of doubtful profit (or profit with small margin, dependent on the market) in working. The locations and mines situated on or near this belt of country are very numerous, (the stimulus of the high price of quicksilver and the increase of consumption having caused quicksilver-mining to be carried on vigorously, in part for speculation, in part for actual production,) and hundreds of them have been opened within the past eighteen months. Many adventurers have been successful in finding deposits, but most of them are working on impregnations, and these will cease operations as soon as the price of quicksilver declines. Although a large amount of work has been done on all these mines, principally in San Luis Obispo County, Santa Clara County, Solano County, Napa County, Sonoma County, Lake County, Mendocino County, and Trinity County, but few of them have attained a sufficient depth to permit definite conclusions whether the quicksilver will occur in veins or in beds or deposits, and which of the various rocks now carrying quicksilver will become in depth the true gangue or metal-bearing matrix. It is very probable that the quicksilver will be found to be deposited with more regularity in depth than appears to be the case in this ore-bearing belt near the surface. It is reasonable to suppose that the mercurial ores, penetrating by sublimation or otherwise all clefts, openings, fissures, porous rocks, &c., would impregnate also those rocks which, though not now porous, have evidently been subjected to metamorphism, and therefore the true quicksilver-bearing rock (if any exists at depths within the reach of mining) will have to be determined hereafter. For example, in the Idria and St. Anna mines, in the Austrian province of Carniola, where argillaceous slate exists as overlying rock, and where the quicksilver penetrates sandstone, slate, lime-breccia, and bituminous slate, with gypsum and compact limestone near the surface, the bituminous slate became at a depth of 320 feet the only mercury-bearing rock. Also, at the Almaden in Spain, where near the surface the quicksilver was thought to exist only as an impregnation in quartzite, and a dark slate imbedded in graywacke, it was found at 800 feet depth in quartz 67 feet in width, divided only by narrow ribbons of slate, (also metal-bearing,) and eventually, at 1,100 feet depth, quartz was the only metal-bearing rock.

Now all along this belt of slate, quicksilver is found, in some localities in metamorphosed limestone, having a brecciated appearance, again in talcose slate, in quartzite, in serpentine rocks, in sandstone, in siliceous tufa near the thermal springs, (in which the country abounds,) and near extinct craters in a species of ferriferous lava, in bituminous slates, &c. But, as far as developments permit investigation, the various slates, from a blackish bituminous to the compact talcose and siliceous slate, constitute the metal-bearing rocks, and of these varieties the talcose slate predominates.

Explorations have shown that in this territory the ore-impregnations, wherever found of considerable dimensions, stand in connection with ore beds or deposits. As examples, the New Almaden, New Idria, St. John, Great Western, Phœnix, and other mines may be cited. In some cases both the impregnations and deposits (to follow the imperfect distinction here used for convenience) seem to have been formed simultaneously, *i. e.*, the processes by which the materials which formed the ore-beds were deposited affected the surrounding rocks to some extent, removing and depositing minerals where the character of the rocks permitted such changes. In such cases the distance which divides the impregnations from the true ore-bodies cannot be great, and their respective percentage in quicksilver may pass from one to the other so gradually as not to permit a sharp distinction. But it is otherwise if the impregnation is of a later date. Owing to the volatility of the quicksilver it may have penetrated (long after the original deposits were formed) so far into the surrounding territory and in such small proportions as to form outlying impregnations of no practical value to the miner, either in exploration or in exploitation. These impregnations occur mostly in greenstone, calcareous and siliceous rocks, but not often in the metamorphosed slates. They bear from traces to one-half per cent. of mercury. In some localities hills of various rocks free from quicksilver are found to be capped with calcareous or siliceous rocks, bearing metal, over a large territory, and carrying narrow seams and nests filled with ore. From these large bowlders have been detached, and have fallen into the ravines on the slopes of the hills with their *débris*. Such bowlders are mistaken for outcroppings of rich veins, and locations have been made on them in all possible directions for many miles along this slate-belt. These impregnations are generally found in the neighborhood where greenstone-dikes have penetrated the syenite rocks of the mountains, and present large masses, of very irregular dimensions. There is no line of separation visible between the ore-bearing and ore-barren rock. The quicksilver occurs in them principally in very thin flakes of cinnabar, and where crevices are found they contain native quicksilver. Iron, as oxide, not as sulphuret, appears to be the only accompanying metal.

The beds or deposits are found generally in the slates, and the higher the state of metamorphism in them, as a general rule, the more abundant seems to be the ore. The most compact bodies have been met with at or near the line of contact between the slates and immediately overlying rock, which is mostly greenstone and sandstone. They vary from 1 to 40 feet in width, carrying from 2 to 35 per cent. of quicksilver as cinnabar; but close to the surface, in those places where magnesian rocks are found, native quicksilver also occurs. They follow in their dip and strike, with few exceptions, the general position of the main slate-belt, but in different localities they conform in their inclination and course to the local direction of the slates. Throughout the quicksilver-territory the metamorphosed and compact talcose slates inclose the most metal. Siliceous slate, greenstone, and sandstone occur as intruding and accompanying rocks, but their percentage of metal is generally below that of the talcose slate. Calcareous breccia, met with near the surface and imbedded in the slate, also carries metal, sometimes to the extent of 10 per cent. The only exception to this rule is the St. John Mine, at Vallejo, where the sandstone carries more than half of the ore; and the slates, when this is the case, indicate that they have parted with their mercury, which passed into the porous sandstone. These slates are accompanied by gabbro* and sandstone and their modifications. Mr. Luckhardt says he has traced them over hills and valleys for more than two hundred miles, and has found their outcrop invariably mercury-bearing. The ores of these beds or deposits occur as cinnabar, seldom as native quicksilver. Liver-ore is likewise rare. The ores occur in seams, nests, and pockets, and also diffused, varying in thickness from an inch to 30 feet of solid ore. The only accompanying metals are iron and antimony as sulphurets, the latter occurring only sparingly and in the southern portion of the slate-belt. The modifications of these near the surface have colored the ores.

From this description it may be seen how difficult it is to decide what an impregnation may lead to. It is not a certain indication of the existence of an ore body or bodies below its outcrop until the latter has actually been exposed; and if that occurs, it is more the accidental stumbling upon it by the miner than the logic of the geologist which leads to its discovery. Now, these impregnations are found all over the wide territory above described, whence the many locations and mines. Furthermore, it appears that it is impossible to affirm with any degree of certainty (in the absence of facts to substantiate the theory) that it is in fissure-veins the mercury is found.

Suffice it to say, that the appearance of cinnabar over so large a territory has led to the investment of much capital, and work has been carried on in many mines very vigorously and with very good results; that is, wherever perseverance and energy were shown by the miner and capitalist, they have met with success. There is as much likelihood of finding good, compact ore-bodies in one region as in another, all along this slate-belt; but economical success depends in a great measure upon the proper method to be adopted in prospecting and also in the reduction of the ores. Many of the mines now at work will have to reduce low-grade ores until they have been fairly opened.

The reduction of low-grade ores has been made easy. There are sev-

* True gabbro is a granitic granular aggregate of labradorite or saussurite, with diallage or smaragdite, frequently containing also crystals of olivine. I have never personally examined the rock so called in the text, and I do not know whether this term is applied to it with strict mineralogical accuracy. Gabbro is usually found among the older crystalline rocks; but it is known to occur between tertiary strata. It is often associated with serpentine, which, in such cases, is supposed to a product of its metamorphosis.—R. W. R.

eral devices for furnaces, all claiming superiority. Mr. Luckhardt is himself interested as an inventor in one of them, but he frankly declares that it is very doubtful in his mind if any furnace produces above 80 per cent. of the quicksilver contained in the ore; he believes, even, that 65 to 70 per cent. is more correct. However, even at this rate, ore containing 0.5 per cent. of quicksilver can be made to yield a profit. At the Sonoma mines, in Sonoma County, Mr. Luckhardt worked at a profit 12 tons daily of *débris* containing 0.5 per cent. of quicksilver, the market-price being $85 per 100 pounds. He says the cost of beneficiation in the Knox furnace does not exceed $2 per ton, and in the Luckhardt furnace even less. At $4.50 per cord of wood, the cost for years in the latter furnace has not exceeded $1.35 per ton of 0.5 per cent. ore, Chinese labor being employed.

The following is a general description of the leading mines, commencing in the south and going north:

Mercury has been found as far south as San Luis Obispo County, where in 1874 two or three new mines were opened, which are yielding very fairly. The most massive ore is found in compact talcose slate and hornstone and greenstone. The cinnabar is accompanied with much iron-pyrites, and is distinctly crystallized. Very little vermilion is met with, which, judging from other mines further north, speaks better for a continuance of the ore and the probable encountering of ore-beds, than if the quicksilver exists as vermilion, and so, sparingly deposited. The ores vary in tenor from 1 to 4 per cent. on the average. Magnesian rocks predominate. We have here the Guadalupe, New Idria, Oceanic, and Clear Creek mines, which produce now about 600 flasks per month. There are many more, the owners of some of which are contemplating the erection of furnaces, and no doubt much quicksilver will be produced in this region.

Going northward, we find impregnations as far up as Redwood City, but nothing noteworthy has been done in that direction. To the northwest, however, we have the New Almaden, which has not yielded for two years past as well as formerly. At times during the year 1874 it produced only 700 flasks per month. The main reserves of the mine are in the low-grade ores left from former days, pillars left standing, and waste-heaps which contain thousands of tons of 2 to 3 per cent. For years nothing poorer than 4 per cent. was worked; and even now it is a question whether, with the old method of construction of furnaces, &c., it will be practicable to reduce 2 per cent. ore profitably. It is reported that the bottom of the mine looks discouraging, although the past five months have developed a good ore-body. The ores here are remarkably free from iron-pyrites, and have occurred in very large masses of extremely rich ore in serpentine rock and quartzite.

Further northward, the slates apparently disappear. At all events, no explorations have been made for miles till we come to the bay east of San Francisco, where quicksilver has been found back of Oakland, but not developed. The same is true of the territory northward, till we reach San Pablo Bay, where sandstone caps the slate-belt, and where the Saint John and other mines east of the city of Vallejo bid fair to give a good yield. The St. John is the most prominent. It has produced from 75 to 350 flasks per month for more than two years and a half. The undoubted value of this mine gives some cause for surprise that capital is not more largely invested in the neighborhood. Quicksilver is found both north and south of it. The ores are compact and entirely free from sulphurets of iron, containing much vermilion. Talcose slate and sandstone, or quartzite, are the gangue of the deposits.

Going farther northwestward, quicksilver is found as impregnations, and nothing of note has been done for twenty-five miles until we reach Upper Napa Valley, where the slates protrude, massive, and can be followed for sixty miles—in places half a mile in width. The ores occur in impregnations of immense size, and at several places solid beds have been encountered; for example, in Napa, Sonoma, and Lake Counties. Enumerating from south to north the most prominent locations, (among a labyrinth of others,) we have the Oakville, Summit, St. Helena, Great Western, American, Rattlesnake, Oakland, Geyser, Kentuck, Bacon, Missouri, Sonoma, &c., on the west, and the Redington, Knoxville, Washington, and others on the east. Many of these produce from 40 to 300 flasks per month; the Redington produces about 700 flasks. The Pope Valley mines produced altogether, during the year, about 100 flasks per month.

Further north and northeast, impregnations are found; but little work has been done until we come to Trinity County, where several districts have been opened. The facilities for working in these rugged mountains are very small, and the severity of the winters has somewhat retarded developments; but Trinity County will produce probably 75 flasks per month for a year to come. The ores, their occurrence and surroundings, are the same as in Napa, Sonoma, and Lake Counties. All the work done, so far, is near the surface, and the compact slates have not been encountered yet. The Trinity ores, as produced for beneficiation, average 1¾ to 2 per cent. of quicksilver. Further north some locations have been made, but they must be for the present insignificant, since nothing is heard from them.

This constitutes the quicksilver-territory. The main bulk of the ores obtained, outside a few mines, such as the Almaden, Idria, Western, Oceanic, &c., contains 1½ to 2 per cent. At times the miner will stumble into a more compact body and the ore will run up to 4 per cent., but there are few mines here producing on the average 4 per cent. ore.

There is one region, apart from this slate-belt, in Lake County, on the shores of Clear Lake, where quicksilver is found disseminated in the ashes and lava and the subsequently-formed calcareous and siliceous tufas of the sunken crater now occupied by the lake. A company has purchased about 100 acres of land, in which are found deposits of sulphur, having basalt as an underlying rock. Over this basalt lies a very recent sandstone, and over this extends, for several acres, a partially sedimentary and eruptive product, carrying iron-pyrites very abundantly, with massive deposits of sulphur. Above this, again, is found a bed of tufa, with oxides of iron, upon which lie volcanic ashes and *débris* of tufa, mixed with bituminous matter.

Through this hill ooze thermal springs with large quantities of carbonic and sulphurous acid, and waters carrying sulphates of alumina and soda, with borate of lime, soda, and free boracic acid; and throughout this whole mass, 16 to 20 feet in thickness, cinnabar is found in seams in sandstone which overlies the basalt, and also in seams in the amorphous sulphur, and in flakes as vermilion, in proportions sufficient to give an average tenor of from 1 or 2 to 4 per cent. These several layers are plainly discernible, but are deposited irregularly, according to the crests of the hill, so that at times the basalt, in bowlders, overlies the sandstone. The decomposition of the pyrites and solfataric action still exist to a great degree; tufa* is still forming; and sublimation of sulphur goes on. Imbedded in the layers are found masses of chalcedony

* The tufa here alluded to by Mr. Luckhardt is, as I understand it, the calcareous and siliceous deposit or *sinter* from the thermal springs.—R. W. R.

and obsidian. The quicksilver exists as cinnabar and also in metallic form, the latter only in small quantities.

According to a full report made by Mr. Luckhardt upon this property, the upper strata, down to the sandstone, a depth vertically of 15 to 45 feet, extending over the hill, about 600 feet by 800, (as far as quicksilver has been found,) contain, as nearly as could be estimated, 1,110,000 pounds. Mr. Luckhardt says that the future of the property depends upon the continuation of the sandstone. That is, if it is found that the basalt cuts it off entirely, then he does not think quicksilver will be found below it; but if the basalt now found is only an intrusion and has sandstone below it, then he thinks the property may yield immensely. Appearances are in favor of the former supposition. The basalt does not carry quicksilver.

### ALPINE COUNTY.

I had long intended to make a personal visit to Alpine County and study for myself the development of its mineral resources, and this plan, often postponed, I had determined to carry out in the early part of last summer. I had been so often assured by those interested in the county that the Alpine County mineral-belt was only a southern continuation of the geological occurrences of Mount Davidson and vicinity that I thought it important to study this question.

But being again prevented by the pressure of duties in other quarters, I requested my deputy, Mr. A. Eilers, to visit the locality and furnish me with detailed notes of his observations. The result of his investigation is embodied below.

The two principal mining-towns around which operations are now carried on are Silver Mountain and Monitor.

The rocks in this part of Alpine County are eruptive, comprising feldspathic and augitic porphyries, with ground-mass of various colors, (mostly dark-gray and brown,) basalt, (sometimes of magnificently columnar structure,) and white and light-gray volcanic tufas, containing small round particles of quartz. The tufas especially are very prominent, wide belts of them coursing through the country, which naturally first strike the eye by their color and the serrated character of the outcrops, protruding in long lines through the masses of partly decomposed and disintegrated material covering the tufas. These dikes are not dissimilar to the outcrops of immense veins, like the Comstock, and in this way the impression seems to have been created that they are actually the continuation of that vein, especially since, in the vicinity of Monitor, irregular ore-bodies have been found in one of these dikes. So far, however, although the rocks of Alpine County are no doubt closely related to those of the Washoe country, there is no evidence of the existence of a direct continuation of the Comstock in this direction, nor is the character of the ores so far found in Alpine the same as that of the Comstock ores.

The town of Silver Mountain, in Alpine County, is situated about sixty miles southwest of Carson, the capital of Nevada, a station on the Virginia City and Truckee Railroad, and on the eastern slope of the Sierra Nevada range. The town-site occupies a flat in the upper part of the valley of Silver Creek, a tributary of the West Fork of the Carson River. The Exchequer mine is situated two miles north of the town, at the head of Scandinavian Cañon, and the company's mill is about two miles below the town, on Silver Creek. The mouth of the hoisting-shaft at the mine is about 1,300 feet above the mill, a difference in altitude which would permit the advantageous use of a wire tram-way.

The vein, as shown by the underground workings, strikes north and south and dips east. The dip varies slightly at different depths. In the incline it is 76° 53′ down to 60 feet below the surface, (which is here 100 feet above the hoisting-works, at the mouth of the vertical shaft.) From the depth of 60 feet to that of 115 feet the dip is but 53° 30′. From the latter depth to the present bottom of the incline it is 64° 45′; and this angle is shown also in the 100-foot level north from the vertical shaft, which level is 200 feet below the surface at the mouth of the incline, and 60 feet below the bottom of the incline. The angle of 64° 45′ is therefore likely to be the permanent dip, though the depth attained is not great enough to render a further change very improbable.

The vein is from 3 to 7 feet wide, and undoubtedly occupies a true fissure. Along the walls are well-defined clay selvages, or "gouges." The vein-matter is quartz and clay, often much broken up; but where the ore-chutes occur it is solid quartz. Most of the work in drifting, &c., on the vein can be done with pick and gad, but the ore must usually be extracted with the help of blasting. This state of things facilitates considerably the proper exploration of the mine, making dead-work light.

The quartz carries dark and light ruby silver, (pyrargyrite and proustite,) with sufficient iron-pyrites to render practicable a good chloridizing-roasting. In one spot in the mine, namely, in the main tunnel, 700 feet from the intersection of the cross-cut with the vein, occurs, besides the ruby silver, stibnite, ($Sb_2 S_3$,) carrying $40 per ton in silver. The ruby ore, as selected for the mill from the north stope in the 100-foot level, contains, according to the "pulp-assays," (assays of samples taken at intervals from the batteries, mixed, and sampled again,) from $51 to $66 silver per ton. The assays of specimens yielded from $12 to $200 and $300 per ton. A lot of 2½ tons worked at the mill yielded $189.60, or $75.84 per ton.

The underground workings and explorations are the following:

1. The main tunnel, driven from the east 126 feet across the country-rock to the vein, and thence northward along the vein 19 feet, where the top of a body of ore was found, the shape of which has been determined by driving levels upon it right and left and by stoping upwards, sinking an incline on it, and driving levels at frequent intervals from this incline. At the tunnel-level the body was 60 feet long horizontally, tapering toward the surface almost to a point, and increasing in length as increased depth was attained. It pitches southward in the vein. Some of the richest ore has been stoped out, but most of it is still standing, and the bottom of the incline is still in ore. The outline of this body would be represented by two irregular lines diverging downward. In this respect the indications of the incline and levels connected with the main tunnel are confirmed by the 100-foot level from the vertical hoisting-shaft south of the incline, which will be presently described.

The main tunnel has been continued along the vein northward, after passing through the ore-body already mentioned. It is now 811 feet long. Recently it struck low-grade ore (ruby silver and stibnite) 18 inches in thickness, in which it has continued as far as it has advanced up to the present time.

2. The vertical hoisting-shaft: This is located in the east or hanging wall of the vein, which it intersects at the depth of 200 feet. At the depth of 100 feet a cross-cut 25 feet long was driven to the vein, where it intersected a body of ore. This was partially explored by drifting and stoping. The latter work was carried a little below the level; but this procedure was not long practicable, on account of the dfficulty from water and the trouble of hoisting the ore from the stope. It was there-

fore wisely abandoned until the ground could be exploited by upward stopes from the 200-foot level. This ore-body was 50 feet long in the bottom of the drift, and 20 inches wide. It consisted of quartz carrying ruby silver. In the stope above it came almost to a point, and in the stope below lengthened with equal rapidity, so that 15 feet below the 100-foot level it had increased from 50 feet in length to 60 feet. The north edge pitches northward in the vein.

The 100-foot level referred to was continued northward on the vein after passing through this body. At 220 feet from the cross-cut connecting with the hoisting-shaft this level entered a body of ore, which made its appearance first in the bottom. The good ore was found to extend 55 feet in the bottom of the drift and 40 feet in the top. It was partly stoped out. This is undoubtedly the same body as that already developed by the incline and its connections. The dip of the vein (64° 45′ east) and the southerly dip of the ore-body on the vein are the same in the 100-foot level as in the incline-workings. To connect the incline with the 100-foot level, the former must be sunk 60 feet and the latter extended 30 feet. The total length of the incline would then be 200 feet to the 100-foot level from the hoisting-shaft, (in other words, the floor of the hoisting-works is situated nearly 100 feet lower than the tunnel-level.)

It will be seen from the foregoing description that, besides the ore-body in which the northern end of the main tunnel now stands, (which seems to be large, but furnishes at that level low-grade ore, and of which nothing is known at lower levels as yet,) two channels of ore have been developed in this vein, pitching in opposite directions, rapidly approaching each other, and widening or lengthening on the vein as they descend. The inference is obvious and almost irresistible that these are the upper points only of an irregular and much larger body, the main portion of which lies below. It is likely that the 200-foot level from the hoisting-shaft will find these two bodies already united, affording extensive ground for stoping.

Besides the vein already described, which was originally called the Buckeye No. 2, and on which the Exchequer Company owns 4,000 feet, the property includes 2,000 feet on the Accacia and 600 feet on the Fremont and Saugatuck.

The Accacia crops out below the Buckeye, having a similar easterly dip, but striking northwest and southeast. The two veins, therefore, converge toward the north; and in the main tunnel of the Buckeye, 700 feet from the mouth, a cross-cut of 15 feet eastward cut a vein carrying stibnite and ruby silver, and containing $40 silver per ton, which is supposed to be the Accacia. This supposition is confirmed by the croppings of both veins, which are prominent at many points.

Samples from the croppings of the Accacia assay $70 in silver per ton. The vein has not been explored. A tunnel has been driven to strike it, starting from a point 760 feet on the mountain-side (or 120 feet vertically) below the floor of the Buckeye hoisting-works. This tunnel has been run 195 feet in country-rock, and will strike the vein in about 15 feet more. By drifting along the vein, and cross-cutting at one or two points to the Buckeye, the ventilation and drainage of both mines will be greatly facilitated.

The total expenditures of the company to this time, distributed through four years and six months, and including the work in the mine, the original cost of mill-property, an 8-stamp wet-crushing mill, saw-mill, hoisting-works, and buildings, amount to $125,964, of which $21,336 constitutes the present cash indebtedness of the company. The property which the company has to show for its expenditures comprises—

1. Mine-claims:

| | Feet. |
|---|---|
| Buckeye No. 2 | 4,000 |
| Accacia | 2,000 |
| Fremont and Saugatuck | 600 |
| Total | 6,600 |

2. Land:

| | Acres. |
|---|---|
| Excellent woodland on the top of the ridge | 160 |
| Woodland in the valley and on hill-sides around the works. | 600 |
| Total | 760 |

3. At Buckeye hoisting-works: A Bacon's hoisting-apparatus, with two 15 horse-power upright boilers, engine, two drums, and hemp ropes, (the engine is adequate for the hoisting and pumping down to 500 feet;) a Blake pump, No. 3, capable of throwing 3,380 gallons per hour. There is not much water in the mine. The shaft is in two compartments, each 5 feet 4 inches by 4 feet 6 inches. Tubs are used for hoisting, for which cages should be substituted when regular stoping is done. The shaft-house is roomy and substantial, and there is a large boarding-house at the mine.

4. At the mill: Eight stamps; four Hepburn pans, taking one ton per charge; two 8-foot settlers, with siphon attachment; one Knox pan for cleaning up; excellent reservoirs for tailings, slimes, and concentrations. Concentration is performed by two Hendy concentrators. The mill is admirably arranged in terraces, and is a good mill, on the Washoe plan. The machinery is from the Fulton foundery, at Virginia City. The engine was built by Burdon, of New York. Dimensions of cylinder, 3 feet by 12 inches; nominal capacity, about 40 horse-power. To employ chloridizing-roasting, which will undoubtedly have to be done, the mill will have to be altered so as to crush dry, and a roasting-apparatus added. If the supply of ore is large and steady, the Stetefeldt furnace (used at Reno) is the one to be recommended. For smaller operations, the Brückner cylinder, used successfully in Colorado and New Mexico, may be employed. The cost of this mill, over and above the purchase of the mill-site and old mill, with two reverberatories, has been $17,198.

5. Saw-mill: An excellent saw-mill, capable of sawing 10,000 feet of lumber daily. The lumber can be furnished by the company at a cost of $12 per thousand, and sold at $20.

6. Other buildings: Boarding-house at the mill; assay-office; manager's house, stables, blacksmith-shop—all complete and in good order.

Wages are still high in this locality, ranging as follows: Miners, $4 per day; engineers, $4; carmen, $3. Wood can be delivered at the mill or mine for $3 per cord.

When Mr. Eilers was present at the Exchequer Mill, Mr. L. Chalmers, the manager of the company, had sent 50 tons of Exchequer ore to the Monitor and Northwestern Mill, for the purpose of having it worked by chloridizing-roasting and amalgamation. This mill used the McGlew furnace, a shaft-furnace provided with blast at three different heights, with a reveberatory attached. *En passant* I may remark that, in the construction of this furnace, the principles of roasting have not been kept in view. As was to be expected, the furnace did very poor work, and Mr. Chalmers's experiment ended in disappointment, as will be seen from the following:

Mr. Chalmers sent 50 tons of Exchequer ore, the battery-samples of which, taken every half hour by Mr. Chalmers's foreman, assayed:

| | |
|---|---|
| Gold | $8 24 |
| Silver | 61 92 |
| Add for salt in sample* | 2 80 |
| | 72 96 |

He sent also 8 tons of low-grade IXL ore, yielding, by samples taken in the same way, $29.50 per ton.

The two lots should have yielded, at 90 per cent., a gross return of $3,495.60, while the actual return made by the mill was only $653.34, or not quite 17 per cent. The chlorination-tests, made frequently by Mr. Chalmers, ran from 17 to 25 per cent. only, except what came from the dust-chambers, which gave 82 per cent.

Since Mr. Eilers's presence at the mine little work has been done, except that the connection between the 100-foot level of the engine-shaft with the main tunnel has been made. The cause was the temporary exhaustion of the working-capital, which is, however, likely to be supplied.

In the IXL mine, which is only a short distance below the Exchequer, and probably on the Accacia vein, very little work has been done during the year. Mr. Chalmers, who is also superintendent of this mine, reports in regard to it in December:

The north drift from the 200-foot level, though in 208 feet from the new engine-shaft, has not yet tapped the ore-chute which made this mine famous in 1861, and created an excitement which resulted in the erection of the town and the organization and establishment of the mining-district of Silver Mountain.

Not only have we not yet reached this ore-body, as to the existence of which there can be no doubt, for it yielded to the superficial scratching of those primeval days, from a space only 40 feet long by 22 feet high, ore which milled at Silver Creek, now the Exchequer Mill, over $50,000; not only have we not reached this ore-body, but our new works have not yet extended into the original IXL location, which was 93 feet from the present face of the north drift, the conformation of the ground requiring, and economical working demanding, that the engine-shaft should be sunk on the Buckeye portion of the ground.

Two hundred feet more of driving, at a cost of from $12 to $14 per foot, will bring us under the perpendicular of this ore-body, and should it dip north, as I think it does, a few feet more, into a mass of ore which, even at this shallow depth, will redeem some of my prognostications as to the great value of this lode when thoroughly opened up.

Your hoisting-works are in excellent condition, requiring nothing but the addition of a second drum to facilitate the working of, and raising from, two or more levels simultaneously.

Everything has been arranged for economical and efficient working to a depth of 500 or 600 feet.

The engine is a double 8-inch cylinder, by Bacon, of New-York, geared to one winding-drum, with brake-lever and reversing-action mounted on a solid cast-iron bed, very compact and strong. Steam is supplied to both engine and pumps by a strong 40 horse-power horizontal tubular boiler, set in brick in a separate building. Pumping is done by a No. 5 Blake pump, steam-cylinder 7½, water-cylinder 4½, stroke 10 inches; capacity from 8,000 to 10,250 gallons per hour. The main building measures 50 by 30, and 20 feet from the ground-sill to wall-plate, strengthened with 8 by 8 timbers, and strongly trussed and braced in the roof, which has a high pitch, on account of snow. The carpenter's shop occupies a space at right angles to the hoisting-floor, and measures 40 by 24. The blacksmith's shop stands at the mouth of the main tunnel, and about 140 feet northwest of the hoisting-works. The engine-shaft is 212 feet deep, substantially timbered in two divisions, 5 by 4½ in the clear, and strongly bratticed and lined.

The town of Monitor is situated six miles northeasterly from the Exchequer Mill, on a small eastern affluent of the East Fork of Carson River. In its vicinity are a number of mines, such as the Tarshish, Silver Glance, Globe, &c., which have all been described in former reports. Most of them are located on a dike of white tufa, which, in parts, becomes

* That is to say, there being 4 per cent. of salt with the roasted ore, the assays from the battery are increased by 4 per cent. to show the actual value of the ore minus the salt. The change of weight of the original ore by roasting is difficult to ascertain, and probably insignificant, the mass being quartz.—R. W. R.

very siliceous. In 1874 only the Silver Glance, owned by the Monitor and Northwestern Company, was worked. The principal work on this mine is a tunnel 400 feet long, driven in from a few feet above the level of the creek, in tufa. From here a cross-cut 300 feet long is run at right angles to the tunnel, and from this a shaft 150 feet deep has been sunk. In the latter a small body of ore, consisting of galena, blende, iron-pyrites, copper-pyrites, and fahlore, had just been found when Mr. Eilers was at the mine. The body was 3 feet thick, and exposed for a length of only 15 feet. The ore was reported to assay from $180 to $200 per ton, and was to be worked in the McGlew furnace and mill mentioned above. It is situated a short distance below the junction of Monitor Creek and the Carson. There are ten stamps, (dry-crushing,) four common amalgamating-pans, one Knox pan, two settlers, and one Eagle or McGlew furnace, the latter being a 16-foot stack, with three fire-places, one above the other at intervals of about 2 feet, under the grate of each of which a blast-pipe is located. The side of the furnace in which the fire-places are is an incline, on which the ore slides down, being blown against the other wall as soon as it comes before the blast. The draught is downward, carrying the ore into a kind of reverberatory, where it is stirred by hand. The dust passes into condensation-chambers connected with a low chimney. The Tarshish mine was expected to be worked soon again.

In the vicinity of Monitor, and north of it, in Mogul district, is the Morning Star mine, which, though not worked at present, I wish to mention here as the only locality I have yet found in the West where very solid enargite occurs in masses. It assays $90 in silver and $10 in gold, and, being an arsenical-copper ore, it has so far not been worked to advantage, though this might easily be done by smelting.

*Description of leading mines, Alpine County, California, 1874.*

| Name. | Location. | Owners. | Length. | Course. | Dip. | Length pay-zone. | Average width. |
|---|---|---|---|---|---|---|---|
| | | | Feet. | | | Feet. | |
| Exchequer... | Silver Mountain.. | Exchequer Gold and Silver Mining Company, London. | 6,600 | N. 10° W. | 76° 53′ E.. | 390 | 20 inches.. |
| Imperial..... | Monitor .......... | Imperial Company of London. | 10,000 | N. and S. | Not known | ...... | ............ |
| IXL......... | Silver Mountain.. | IXL Gold and Silver Mining Company, London. | 3,010 | N. 14° W. | 78° 30′ E.. | 560 | 12 inches, vein 4 feet. |

| Name. | Country-rock. | Character vein-matter. | Tunnel or shaft. | Length of tunnel. | Depth on vein in tunnel. | Depth working-shaft. | Number levels opened. | Total length drifts. | Cost hoisting-works. |
|---|---|---|---|---|---|---|---|---|---|
| | | | | Feet. | Feet. | Feet. | | | |
| Exchequer.. | Porphyry | Quartz and quartzose ore, ruby, silver. | Both.. | 811 | From 75 to 350. | 200 | 3 | 985 feet besides main tunnel. | $7,055 32 |
| Imperial .... | Porphyry | Not known . | Tunnel | 1,400 | From 600 to 6,207. | ...... | .... | ............... | .......... |
| IXL......... | Porphyry | Quartz and ruby. | Both.. | 900 | 100 to 500 | 200 | 3 | 1,040 feet besides main tunnel. | 10,168 19½ |

*Operations of leading mines of Alpine County, California*, 1874.

*Silver Mountain district.*—Name of mine, Exchequer; owners, Exchequer Gold and Silver Mining Company, London; numberof miners employed, 16; miners' wages per day, $4; cost of sinking per foot, $30; cost of drifting per foot, from $12 to $14; cost of stoping per ton, $2; cost of mining per ton extracted, $3; cost of milling per ton, $6.25; company working its own mill; number of tons extracted and worked, 245½; average yield per ton, $17.05 net; total bullion-product, $4,177.57.
Reported by Lewis Chalmers, manager.

*Statement of quartz-mills of Alpine County, California*, 1874.

*Silver Mountain district.*—Name of mill, Exchequer; owners, Exchequer Mining Company, of London; kind of power and amount, 42 horse-power engine and boiler; number of stamps, 8; weight of stamps, 605 pounds; number of drops per minute, 85; height of drop, 9 inches; number of pans, 4 Hepburn; number of concentrators, 2 Hendy; 23 8-foot settlers; crush wet and work raw; cost of mill, $37,198, including saw-mill and 1 timber-ranch; capacity per 24 hours, 16 tons; cost of treatment per ton, $6.25; tons crushed during year, 100; 1 run showed that the ore could not be treated without roasting.
Reported by Lewis Chalmers, manager.

## INYO COUNTY.

For an account of the operations for the year in this county (except Panamint district) I am indebted to Mr. William Crapo, of Cerro Gordo, whose intelligence and good judgment sufficiently appear in the statements he has furnished. Mr. Crapo desires that acknowledgment be made to Mr. S. W. Cowles, of Cartago, for assistance in preparing an estimate of the bullion-product, and says: "Although Mr. Cowles's figures do not wholly agree with the estimates from other sources, I unhesitatingly accept them as correct, not only on account of his superior facilities for obtaining accurate data, but also because his statements are confirmed by what I know to have been the production of ore."

Mr. Crapo's report to me is dated in November, 1874. I have added some points of later date.

The mining-operations carried on in Inyo County during the past year have been hindered by mismanagement and legal complications and doubts or struggles about titles in the older districts, while rank speculation has characterized the newer ones. The extension of the time for working claims located before the mining-act of 1872 has worked injury in the older districts. In Cerro Gordo district especially is this felt. The first discoverers and locators of claims in this district, taking pains to cover the country with "wildcat locations," and then settling down to the practice of law, have thus far made it lucrative if not honorable to themselves, but expensive to the mining-interests, as capitalists investing in this district know to their cost.

Notwithstanding all these drawbacks, the outlook of mining in this county is brighter now than at any time in its past history. The projected Independence and Los Angeles Railroad, if completed as at present designed, will prove the most powerful restorative to the mining-industries of this region that could well be applied.

Among the projected enterprises completed within the last year is the Cottonwood flume, situated on the eastern slope of the Sierra Nevadas, west of Owens Lake. Projected in 1872 by Sherman Stevens, it was not brought to a successful completion until the spring of 1874. This flume is about six miles long; its head being situated on Cottonwood Creek, at an altitude of 9,000 feet above the sea. The Stevens saw-mill is situated at the upper end of the flume, to which the logs are hauled and converted into lumber and mining-timber. These are sent down the flume to a point near Owens Lake, where they are sold at $45 per thousand. Around the head of this creek, at an altitude of from 9,000 feet to 12,000 feet, are many square miles of scrubby forest, consisting principally of yellow pine, spruce, tamarack, and white fir. The principal drawbacks to this as a lumbering region are the want of suitable timber and the extreme difficulty in getting to mill what there is, owing to the ruggedness of the country. The deep snows and cold weather in winter preclude work in that season.

The completion in May, 1874, of the Cerro Gordo Water-Works, belonging to the Cerro Gordo Water and Mining Company, a Los Angeles incorporation, marks an era in the development of the district. Water, heretofore so scarce and only obtainable at from 7 to 10 cents per gallon, is now supplied to the mines, furnaces, and citizens at a cost of from 1½ cents to 4 cents per gallon.

This enterprise, inaugurated in 1873, was only brought to a successful issue in May, 1874. The water is brought from the Miller Spring, ten and one-half miles northwest of Cerro Gordo, being raised 1,870 feet to the summit of the Inyos by three Hooker steam-pumps, placed 2,800 feet apart, each pump overcoming a vertical lift of 620 feet.

The discharge-pipes are composed of 3-inch boiler-flues fastened together by screw-collars. From the summit the water runs through thirteen and one-half miles of 4-inch No. 18 and No. 16 sheet-iron pipe, having a fall of 950 feet, to the town of Cerro Gordo. The sheet-iron pipe was made in Los Angeles, in lengths of 12 feet, coated inside and outside with asphaltum, and in this condition shipped to the ground. In laying the pipe the joints were fastened together by a riveted sheet-iron collar, made of No. 16 iron, 3 inches wide, being daubed with melted asphaltum and slipped over the joints and held in place by the adhesion of the asphaltum. As originally constructed, a portion of this pipe was intended to withstand a pressure of 650 feet in crossing a sag in the mountain, thus reducing the length to about nine miles; but, after completion, the water being turned on, the joint-fastening was found to be entirely inadequate, the pressure being sufficient to force all the water out at the joints. The expedient of leading the joints was next undertaken, but with no better success; and the engineer in charge not understanding the efficacy of an inside collar, as usually applied in such cases, was obliged to take up the pipe and run it around the mountain, at an increased expense to the company of $26,000, making the total cost exceed $74,000. Since completion it works indifferently well, fully two-thirds of the water which passes through the pumps arriving at Cerro Gordo. As an investment it is estimated that it will pay back the original outlay in two years.

The Potosi tunnel is an enterprise recently inaugurated under the direction of Mr. John Simpson, a mining-engineer. Its point of beginning is at the lower town of Cerro Gordo, whence it is designed to pass under the town and tap the mineral-zone extending along the upper edge of the town. Its total length is to be 4,000 feet, and it is intended to cut the Union vein at a depth of 1,040 feet. (Another estimate, perhaps nearer the truth, is 663 feet.) It is designed to furnish an outlet for the easier working of the mine. The tunnel is to be of sufficient width to admit a double track. The Burleigh drill is to be used in its construction, and it is expected to reach the Union mine in sixteen months. The estimated cost is about $100,000.

The Cerro Gordo metalliferous belt occupies a width of about four miles, the summit of the Inyo Mountains being about the middle. The Inyos are a comparatively narrow ridge, rising somewhat steeply above Owens Lake on the west side, and being much more precipitous on the east side. The metalliferous zone is composed of Silurian (?) limestone, tilted at a high angle, with a sharp westerly dip; the general course of the planes of stratification being north and south. East of the summit, considerable masses of intrusive feldspathic porphyry irregularly alternate with the limestone, while on the west side a few narrow patches of porphyry and argillaceous slate are to be seen. The limestone exhibits all stages of metamorphism, from the dark fossiliferous to the purest marble. The principal mines are situated at an altitude above the sea of about 8,500 feet, and 4,700 feet above Owens Lake, five and one-half miles distant. The ore-deposits of the district, containing lead and silver ores, are collectively fissure-veins.* Their breadth is from a mere seam to 120 feet or more. They may be divided, according to their strike and the nature of the matrix filling them, into two principal classes:

1. Vertical veins, striking north and south, and inclining to follow the planes of stratification of the country-rock, with a matrix composed of fragments of limestone, calc-spar, a little gypsum, (satin spar,) and heavy spar, galena, and iron-pyrites. From the decomposition of these have been formed immense quantities of cerussite, anglesite, a little minium, massicote, wulfenite, and, in some of the mines, pyromorphite. Geodes are of frequent occurrence, being generally incrusted with crystals of cerussite and calcite. This class of veins are very irregular in dip and strike. Their width is from 2 feet to 120 feet.

2. Veins having a general northwest and southeast strike, (conforming in strike to the axis of the mountain-range,) with a varying dip of from 20° southwest to 20° northeast. The gangue is quartz, with ores of silver and copper, (dyscrasite, freieslebenite, pyrargyrite, proustite, stephanite, native silver, malachite, azurite, iron-pyrites, chalcopyrite, erythrite, beiberite, tetrahedrite.) Gold is not found in considerable quantities, except in the neighborhood of the porphyry dikes, where it runs as high as $76 per ton. The veins of this class are strong and well defined, ranging in width from a mere seam to 3 feet. Vertical grooved friction-surfaces, or slickensides, are of frequent occurrence, and may be said to be characteristic of this class, while, in crossing the lead-veins, they fault the latter.

The greater part of the bullion-production of this district for the last year has been derived from the working of two mines, or one mine and a spur, the Union or San Felipe and the Omega. The former is owned and worked by V. Beaudry, of Cerro Gordo, and Messrs. Belshaw & Judson, of San Francisco. The Omega is owned and worked by the Owens Lake Silver-Lead Company, whose reduction-works are located at Swan-

* This is Mr. Crapo's classification. Having never personally examined the mines, I can pass no judgment upon it. A description by Mr. Eilers will be found in my Fifth Report, (rendered February 8, 1873,) on pages 17-22.—R. W. R.

sea, near Owens Lake. The war of monopoly carried on for the last four years between these two companies finally culminated, early in 1873, in a suit of ejectment brought against the Union Company ostensibly by the San Felipe Company, but really by the Swansea Company, (the stock of the former being owned by the latter company.) The case came up for trial in July, 1873, and resulted in a verdict for the plaintiffs. It was claimed by the Swansea Company that the Union was a "jumped" title, being located on the San Felipe vein, which was the older location. The Union Company attempted to show that the San Felipe was located on a "silver-vein" cutting diagonally across the "lead-vein." On the trial it appeared to the jury that the discovery-shaft of the San Felipe was sunk at the point of intersection of the two veins; and a verdict was rendered giving both veins to the San Felipe or Swansea Company. The case was appealed to the Supreme Court, where it still hangs. Meanwhile, the Union Company (under a nominal bond) has been vigorously at work robbing the mine of the rich ores, leaving all ores assaying less than 25 per cent. in lead in the mine, or putting them over the dump as waste.

As a specimen of mine-engineering, the Union mine is by no means a model. But few timbers are used; consequently accidents are of frequent occurrence from the falling of large flakes of ore and rock. The mine is developed to a vertical depth of about 430 feet, the various inclines and shafts requiring about 700 feet to reach this depth. At present the mine is worked through the Freiberg and Bullion tunnels, which are about 650 feet and 400 feet in length, respectively, and both cut the vein at a vertical depth of 160 feet. Placed underground, at the end of the Freiberg tunnel, is a 16-horse engine, used to raise the ore to the level of the tunnel, where it is dumped into a chute, thence run into a car and transported to the dump. Here it is sorted by hand, thence hauled to the furnaces, 150 yards distant, in wagons; requiring thirteen men to dispose of the ore from the top of the shaft to the furnaces. The mine employs twenty miners, and raises 60 tons of ore in 24 hours. The cost of mining is estimated at $7 per ton. With better arrangements, the cost ought not to exceed $1.50 per ton delivered at the furnace.

This mine is noted for its large quantities of compact anglesite. Immense masses are found, weighing several tons, and generally exhibiting a concentric or banded structure throughout. When concentric, it generally contains a kernel of galena.

One-ninth of the ore raised is composed of galena, compact anglesite, and cerussite; the remainder is gray and ferruginous carbonates.

From December 1, 1873, to November 1, 1874, this mine has produced 12,171 tons of ore, of an average assay-value of 47 per cent. lead and 87 ounces in silver per ton.

The ores are reduced in two small furnaces, one owned by V. Beaudry and the other by Belshaw & Judson, the ore being divided equally between the two. The exact amount of ore smelted during the year is not easily arrived at, since there were large quantities on the dumps at the furnaces at the beginning of the year, and there are at present about 5,000 tons at Belshaw's furnace, though none at Beaudry's.

In the latter part of 1873, Beaudry's furnace was remodeled to conform to the more modern method of smelting. Its present dimensions are: At the level of the slag-spout, 5 by 3 feet; at the level of the tuyeres, 34 by 40 inches; above the bosh, 4 by 4.3 feet. The whole height above the tuyeres is 9 feet. There are five tuyeres of 3.5-inch nozzle. A No. 6 Root blower furnishes the blast.

*Charges in twenty-four hours.*

| | |
|---|---|
| Charcoal, 19,750 pounds | $7\frac{14}{16}$ tons. |
| Ores, consisting of galena, sulphates, and carbonates | 25 tons. |
| Slag | 2 tons. |
| Quartzose-silver ores, (when procurable) | 1.75 tons. |

*Cost for twenty-four hours.*

| | |
|---|---|
| Charcoal, 19,750 pounds | $261 25 |
| Labor: | |
| 3 smelters, at $4 | 12 00 |
| 3 helpers, at $4 | 12 00 |
| 3 chargers, at $4 | 12 00 |
| 2 engineers, at $4 | 8 00 |
| 2 foremen, at $8 | 16 00 |
| 12 roustabouts, at $4 | 48 00 |
| Cost of water per day | 25 00 |
| Blacksmithing, fire-clay, repairing furnace, wear and tear of machinery, oil, &c | 75 00 |
| Total cost per day, (excluding cost of ore) | 469 25 |

The above equally applies to Belshaw's furnace,* with the exception of the shape of the furnace, which is round.

These furnaces, when in good running order, turn out on an average 215 lead-bars, weighing 85 pounds each, or $9\frac{1375}{2000}$ tons, in twenty-four hours.

The Omega mine, owned by the Swansea Company, after producing 481 tons of ore, closed down in the early part of summer. Mismanagement and financial and legal difficulties are the presumed cause.

The Santa Maria, Jefferson, and Carman lead-mines, formerly producing much ore, have been entirely idle during the past year.

The Diaz mine, discovered and opened during the past year, has been exploited to a depth of 98 feet, and gives promise of becoming a valuable mine in the future. It is a claim of 1,500 linear feet; has a north and south strike; is 3 feet wide on the surface, increasing to a width of 8 feet at a depth of 98 feet. On the surface it dips 64° east, but becomes nearly vertical at a depth of 98 feet. It has produced 130 tons of ore, mostly sulphur-yellow oxides, (minium and massicote,) with small quantities of galena. Green lead-ore (mimetite) is found in small quantities. The average assay-value is 65 per cent. lead, with from 78 to 288 ounces in silver. All the ores hitherto extracted have been sold to Beaudry's furnace, for $9.50 per ton. The mine is now idle on account of conflicting "wildcat" claims and small profits.

Of the many silver-veins formerly worked, but one, the Raya, has produced more than a few tons of ore. This is owned by the Union Company, and is variously known as the San Felipe, Raya, and Mohawk. The ore is of low grade, its principal value being to form a silicate of lead in smelting the lead-ores. The product has been about 130 tons, of a gross value of $25 per ton.

The San Ignacio, Belmont, Widdekind, San Lucas, Buena Suerta,

* By a comparison of the figures with those given in page 355 of my Report of 1873 (for 1872) for the works of Belshaw & Judson, it will be seen that much more fuel is employed than is there stated.—R. W. R.

Friendship, and several smaller veins of less note, have each produced a few tons of silver-ore, the whole of them together probably not exceeding 300 tons. They are at present all idle.

The prices paid by the "furnace combination" for silver-ores are $10 per ton for third class, $20 for second class, and $30 for first class. The third class assays 60 ounces or less; the second class, from 60 ounces to 180 ounces; and the first class from 180 ounces to 560 or more in silver per ton of 2,000 pounds. The majority of the richer ores contain from $10 to $76 in gold per ton. In custom-working, $1.15 per ounce is allowed for silver and nothing for the gold, while $50 per ton is charged for working the ores. Ores containing less than 100 ounces in silver per ton usually bring the mine-owners in debt to the furnaces. All the silver-mines worked during the past two years (with a single exception) *have become indebted to the furnaces on the sale of ores.* This accounts for their being closed down. The owners, having neither singly nor in the aggregate sufficient means to erect reduction-works, prefer awaiting the advent of outside capital before attempting to realize a profit from the working of their mines.

The bullion-production of the district from December 1, 1873, to November 1, 1874, amounts to—

| From the works of— | No. of tons. | No. ounces in silver, per ton. | Total value in silver, at $1.2929. | Total value in lead, at 6c. | Aggregate value. |
|---|---|---|---|---|---|
| Belshaw & Judson | 2,593 690-2000 | 134 | $449,353 16 | $309,939 48 | $759,292 64 |
| V. Beaudry | 2,381 491-2000 | 140 | 431,064 28 | 284,528 73 | 715,593 01 |
| Owens Lake Silver Lead Company | 120 487-2000 | 140 | 21,808 85 | 14,395 26 | 36,204 12 |

Total production, $5,095\frac{668}{2000}$ tons, of an aggregate value of $1,511,089.77 The production of the Union mine for the month of November, 1874, will not be less than 1,200 tons of ore. The bullion from Belshaw and Beaudry's furnaces will be about 250 tons each. The total silver-product for twelve months was, therefore, about $1,000,000; lead, say, $680,000. Correct data showing the average assay-value of the lead-ores are not obtainable, for the reason that no record is kept, and the loss in smelting (which is very great) not being known, it cannot be accuratly calculated.

The average of a considerable number of blow-pipe assays gives 47 per cent. lead and 87 ounces in silver per ton of 2,000 pounds. There is a much greater loss in the smelting-process than the proprietors are willing to acknowledge.

The average value of the bullion in the above table is deduced from the statements of the owners.*

*Waucobia district* is situated on the east side of the Inyo Mountains, about forty miles north of Cerro Gordo. It is claimed to be a continuation of

* At the close of 1874, a correspondent of the Mining and Scientific Press of San Francisco writes as follows to that paper concerning the Cerro Gordo mines:

"There are two furnaces, each about thirty tons ore capacity, now in constant operation—one owned by M. W. Belshaw and the other by V. Beaudry. The ore used by these furnaces is supplied almost entirely by the Union mine and the Ignacio silver mine. The base bullion produced by each furnace is about twelve and a half tons per day, and is worth about $115 in silver and $90 in lead per ton. The lead will pay expenses of mining, smelting, and transportation to San Francisco, where its cost for refining is about $25 per ton, leaving a net profit of about $90 per ton. The Santa Maria and Omega mines and the smelting-works owned by the Owens Lake Silver and Lead Company are now idle, but are expected to start again soon. These mines are looking well, and ought to yield as much as the Union.

"To give some idea of the profits made against the capital involved here since the

the Cerro Gordo mineral belt. The principal owners of this district, Messrs. Brady and Reddy, are not actively developing it at present.

*Kearsarge district* is situated high up on the eastern slope of the Sierra Nevada, opposite Fort Independence. A little prospecting has been going on during the past summer, and some considerable bodies of low-grade ore are in sight. The mines consist of the Kearsarge, Silver Sprout, Lamb, and Virginia Consolidated. A 10-stamp mill is connected with the mines—idle.

*Russ district* contains the Eclipse mine, owned by an English company. Prospecting has been going on since last June, but no considerable bodies of ore have yet been found. The ores are galena, silver, and gold in layers, forming a combed structure. A mill and furnace are connected with the mine—both idle.

*Coso district* contains a half score of Mexicans, who work, with their usual dilatoriness, a few narrow gold-veins, producing about $1,200 per annum.

*Panamint district.*—The following data concerning this district are taken from a manuscript report by Mr. C. A. Stetefeldt, of San Francisco:

The district is located on the west slope of the Panamint range of mountains, in Inyo County. The road by which the mines are at present reached starts from Indian Wells, situated on the east slope of the Sierra Nevada, not far from Walker's Pass, a station on the road from Los Angeles to Cerro Gordo, one hundred and sixty-five miles distant from Los Angeles. Panamint is northeast of Indian Wells a distance of seventy miles by the present wagon-road, which has to cross three valleys and three low ranges of mountains between Indian Wells and Panamint.

The Panamint range has a general course of north 20° west and south 20° east, and is, both in altitude and extent, the most prominent one in this neighborhood, its highest peaks reaching an elevation of about 10,000 feet above the level of the sea; hence it attracts the clouds from the surrounding country, and is subject to heavy rains and cloud-bursts. On its east slope is Death Valley, which, according to Williamson's observations, is considerably below the level of the sea. Observations taken on the road from Indian Wells show that the descent of the valleys east of the Sierra Nevada is gradual, and terminates finally in Death Valley. In the immediate vicinity of the highest peaks, on the west slope of the range, are the cañons in which Panamint district is located. Of these, Surprise Cañon is the most interesting and prominent,

---

smelting-works were first started, I subjoin the following table, which I think is nearly correct:

" *Capital involved in furnaces, mines, bed-rock tunnels, &c., each year.*

| " Year. | Amount. | Total. | Bullion produced, tons. | Net profits at $90 per ton. |
|---|---|---|---|---|
| " 1869 | $15,000 | $15,000 | 1,000 | $90,000 |
| " 1870 | 15,000 | 30,000 | 1,500 | 135,000 |
| " 1871 | 30,000 | 60,000 | 2,500 | 225,000 |
| " 1872 | 15,000 | 75,000 | 4,000 | 360,000 |
| " 1873 | 10,000 | 85,000 | 5,000 | 450,000 |
| " 1874 | 15,000 | 100,000 | 6,000 | 540,000 |
| " Total | | | | 1,800,000 " |

These statements are exaggerated as to the amount of bullion per furnace daily. I give them without indorsing them.—R. W. R.

containing the best mines of the district and the town-site of Panamint. South of Surprise Cañon is Happy Cañon, and to the north Narboe Cañon. Panamint Range is of considerable width in this vicinity, the distance from its summit to Panamint Valley being eight and one-half miles, and to Death Valley, as I am informed, eighteen miles. Surprise Cañon has a general course of north 70° east, bending, however, to the northeast at its upper end. In ascending this remarkable cañon, one is surprised, indeed; such steep, bold, and barren mountains, intersected by deep gulches; such a variety of rocks; such grand traces of the work of the unfettered elements! One can only compare it with a chart of the moon, and conceive that such must have been the aspect of the whole of our earth in its earliest state. Neither grass nor soil cover these corroded mountain-sides, and only a few huge cactus have fastened their roots to the rocks. The bottom of the cañon is formed of coarse gravel and great bowlders, some of the latter having been washed down from the very summit of the range. Unusually heavy storms have, again, from time to time, torn this young and gradually-formed conglomerate and swept it farther onward, cutting perpendicular channels 10 and 12 feet deep.

Several springs make their appearance, the first one two miles from the mouth of the cañon. The cañon here becomes narrower, but more friendly. Luxuriant vines cover the rocks; wherever the soil is moist there are copses of willows, and occasionally a small pine-tree springs from a cleft in the rocks. Six miles from the mouth of the cañon we reach the town-site of Panamint, laid out on most dangerous ground, in the immediate neighborhood of the mines, and about 6,000 feet above the level of the sea, or about 4,700 feet above Panamint Valley. Here the cañon is again wider, and two miles farther up it is barred by the summit, dividing it into a north and east fork.

The gulches which intersect the mountain-sides at oblique angles with Surprise Cañon, commencing below the town-site of Panamint, are named, respectively, on the north slope, Woodpecker, Jacob's Wonder, Stewart's Wonder, Sourdough; on the south slope, Cannon, Marvel, Little Chief, Stern. The ridges on either side of Surprise Cañon, near its end, rise to an altitude of about 7,500 to 8,700 feet above the level of the sea. These gulches are well timbered, the trees being nut-pine and cedar.

Although the district lies in a barren country, and is rather difficult of access, yet its natural advantages in regard to wood, water, and salt are excellent. The geological features of the Panamint Range are exceedingly grand and interesting, and to do full justice to the subject would require a long study. Mr. Stetefeldt gives, therefore, only such outlines as have an important bearing upon the formation of the mineral deposits, and for this purpose confines himself to the geology of Surprise Cañon alone.

Surprise Cañon, from its mouth to the summit of Panamint Range, presents a succession of sedimentary metamorphic rocks, elevated, disturbed, and transformed by a series of eruptive rocks at various intervals. For two miles from the mouth of the cañon we find a formation of dark-colored mica slates, alternating with quartzitic slates, with little patches of limestone here and there on the top of the ridges. In proceeding farther, we find this slate traversed by dikes of greisen. Greisen is a rock of very rare occurrence, and consists of white crystalline quartz and white lithia mica (lepidolite). Its eruptive nature, which has been doubted by some geologists, becomes here very evident. This rock becomes soon predominant, and forms an immense mass, estimated to be about half a mile in width. Here, owing to the extreme hardness of the rock, the cañon is very narrow, in places just wide enough for the road. The greisen itself is again perforated by dikes of diorite, and where the greisen ends we find a more extensive eruption of diorite, which incloses a small formation of crystalline limestone. Then we meet a formation of light-colored, highly metamorphized quartzitic slates, (with mica,) which have altogether lost their stratification, traversed by dikes of diorite. This is followed by white crystalline limestone of about 1,000 to 1,500 feet in thickness, also containing diorite dikes. Then, again, dark-colored metamorphic quartzitic slates, with mica and hornblende slates, cut by diorite and diabase dikes, occur for a mile and a half. These carry more or less limestone on the mountain ridges.

By this time we have reached the last two miles of the cañon, which here intersects that part of the formation which interests us most, namely, the mineral belt. The latter consists mainly of a bluish crystalline limestone, alternating with dark-colored limestone and calcareous quartzitic and mica slates. Finally, the chain is closed by an enormous eruption of porphyritic trachyte.

In reviewing this grand formation, we arrive at the following conclusions in regard to the geology of Surprise Cañon. We recognize two great centers of eruption and elevation. The first and older is by the greisen, which upheaved and broke the sedimentary rocks, followed by eruptions of diorite, which did their share in metamorphizing the sedimentary rocks. The second and more recent one by the porphyritic trachyte, which elevated the whole range to its present height, and undoubtedly was the cause of the formation of the mineral belt which leans against this rock.

It is thus evident that Surprise Cañon is of eruptive origin; that is, a fissure riven in the mountain. The numerous dikes of diorite which traverse the sedimentary rocks are rarely found to intersect both sides of the cañon, but are mostly found either on one side or the other, which would be unaccountable, if the channel of the cañon had been formed by the action of water.

The mineral belt of Panamint district extends from west to east two to two and a half miles, and from north to south about five miles. On the north slope of Surprise Cañon, where the formation is more regular, it consists of the following succession of rocks: Hornblende slate; crystalline bluish limestone; dark-colored limestone, calcareous slate, quartzitic slate, and mica slate; dike of diorite or diabase; white marble; dark slates; limestone; and then the enormous eruption of the trachyte. This rock is composed of large crystals of sanidine in a flesh-colored feldspathic ground mass, inclosing occasionally crystals of hornblende and dark mica.

While in nearly all the limestone districts of Southeast Nevada, as, for instance, in Cortez, Mineral Hill, Eureka, Reveille, and White Pine, the ore occurs in the form of irregular aggregations, no matter if they are defined or not to a certain extent by some wall, or in the form of impregnations in the limestone—as is often the case in White Pine—Panamint district carries its ore in the much-desired shape of veins, or, to use the common pleonastic expression, in true fissure-veins. I pronounce them veins, for the following reasons:

1. Their croppings can be traced without interruption for long distances.
2. There is a well-defined system of parallel fissures.

3. These fissures not only traverse the limestone, but continue through the slate.

4. They intersect the country-rock independent of the strike and dip of its stratification.

5. The ore is entirely confined to the gangue between the walls of the veins. In no instance did Mr. Stetefeldt find even a trace of ore in the limestone. Those who are familiar with the above-named districts in Nevada will remember how the limestone teemed with little croppings and bunches of ore of most irregular shape.

We recognize in Panamint district one main system of veins, which strikes east of north with a very steep dip west of north, to which belong nearly all the prominent ledges, and then a subordinate system, the veins of which strike west of north and dip east of north. Examples of the first system are the Wonder, Marvel, and Hudson River; of the second, the Esperanza and War Eagle. In comparing the strike of the different veins of the main system, there seems to be at first sight great confusion and irregularity; but these disappear at once, as soon as we take into consideration their relative position in the mineral belt. The fissures of the main system do not run in straight lines, but in curves. At the extreme west end of the belt, they have a course of about north 22° east. Proceeding farther east, they swing more and more east of north, reaching nearly due east, and finally sway back to a northeast course. Keeping this in mind, we ought to be rather astonished at the great regularity of the system.

From the preceding statements it will be seen that the main system of veins runs nearly parallel with the course of Surprise Cañon. Hence the gulches which intersect the slopes of the mountain-sides must also cut the veins. In examining these points of intersection, we find that the veins are more or less disturbed and broken. It is apparent, also, that these gulches owe their existence to eruptive forces. On the south slope of the cañon we find also fissures with crushed limestone—for instance, near the Esperanza—causing a break in this vein.

It has been remarked before that the veins are not confined to the limestone, but also continue through the slate. When they enter the slate, however, we find them generally split up into several branches, which condition is best illustrated at the east end of Stewart's Wonder.

Another peculiarity of the district is the occurrence of branch veins, which dip very flat, are sometimes much more developed, and carry richer ore than the main vein. This case is well illustrated in Jacob's Wonder.

There is scarcely a mining-district where more continuous and bolder croppings are found than in Panamint. In mounting the ridges on either side of Surprise Cañon, the whole system of veins lies spread before the observer like a huge map. These croppings of compact quartz vary in width from a few inches to 20 feet, and even more, and, being less destructible than the country-rock, often project, like cliffs, over 10 feet above the ground. They carry rich ore in continuance over long distances, but are also barren for long distances.

The gangue of the veins is a hard compact quartz, laminated in thick layers parallel to the walls. The only exception to this rule was found in the Sunrise, situated at the extreme east end of the mineral belt. Here a great part of the quartz is spongy, the cavities often filled with crystals of quartz. In this mine also calc-spar is found in the gangue, which has not been observed anywhere else in the district.

The predominant silver-bearing mineral which gives the ore its value is a rich silver fahlore. It appears entirely undecomposed very close to the surface. In many places, however, we find it changed to stetefeldtite—that is, a combination of antimoniates of copper, lead, iron, and zinc, with sulphuret of silver and water. Where this mineral makes its appearance it is accompanied by the blue and green carbonates of copper. Much less frequently we find silver-glance, chloride of silver, and native silver. These latter minerals are predominant only in one mine—the Sunrise. Of the base minerals we find blende and galena.

As a general rule, the silver-bearing minerals are not finely disseminated through the quartz, but form visible grains, from the size of a pin-head to a large pea, and occasionally much larger. The ore occurs also in pay-streaks parallel to the walls of the vein. Both these characteristics make it very easy to assort the pay-ore from the low-grade ore and barren quartz. The ore in the Sunrise is the most prominent exception to this rule, as the spongy quartz contains the silver-bearing minerals finely impregnated.

The ores of Panamint are of very high grade, as the following assays for silver show:

| | Per ton. |
|---|---|
| Pure fahlore from Stewart's Wonder | $919 57 |
| First-class ore from Stewart's Wonder | 215 19 |
| Average of pay-streak from incline of Stewart's Wonder | 83 24 |
| Selected first-class ore from Jacob's Wonder | 348 11 |
| Low-grade ore from Jacob's Wonder | 12 56 |
| Rich ore from croppings of Wyoming | 609 47 |
| Average of pay-streak from Little Chief | 84 82 |
| Ore with blende and galena from Little Chief | 59 69 |
| Ore from pay-streak from Hemlock | 197 91 |
| Ore from pay-streak from Harrison | 152 33 |
| Average of pay-streak from upper cut of Hudson River | 80 10 |

It may be safe to intimate the value of ore from the pay-streaks of these mines to be from $75 to $100 per ton. But much higher grades could be selected should this be desirable.

In reviewing these assays, and estimating the percentage of fahlore contained in the different samples, Mr. Stetefeldt concludes that the mineral belt carries much richer fahlore to the south than to the north. Indeed, the ore found in Narboe Cañon, north of Surprise Cañon, becomes more base, showing more lead-minerals and a fahlore richer in copper, but poorer in silver.

In connection with the above-stated facts, is also to be remarked that the slate seems to have considerable influence upon the distribution of the ore in the veins. All the extensive ore-chimneys are found in the limestone, and wherever a vein runs deep into the slate the croppings soon become barren. There are some veins cropping in the slate alone which show no trace of ore.

The croppings of the veins show ore-chimneys of great extent, measuring on the prominent veins from 300 to 600 feet in length. From observations made in the gulches which intersect the veins, one is led to believe that the dip of these ore-chimneys will be east of north. Another reason which confirms this opinion is that the limestone dips in that direction.

It is to be regretted that so little has been done in regard to the development of the mines in depth. And it is to be regretted still more that the most important developments show nothing favorable, and this has evidently deterred the prospectors from continuing their explorations. These explorations have in some cases been started at the wrong place, in others on inferior mines, and in others with lack of judgment.

In considering the magnificent geological formation of Surprise Cañon, the regularity of the grand system of parallel veins, the uniform character of the silver-bearing mineral in this system, the extensive ore-chimneys exposed on the surface, the presence of an eruptive rock—which is highly favorable to the formation of mineral deposits—in close proximity to the mineral belt, Mr. Stetefeldt is confident that these veins will continue in depth, and also that the ore will continue in depth, and very likely of similar richness to that found at the surface.

The following two are among the principal and most characteristic veins of the district.

The remarkable vein called the Wonder has been located on the north slope of Surprise Cañon, for a distance of 5,250 feet, and its croppings, intersecting Woodpecker, Jacob's Wonder, and Stewart's Wonder gulches, can be traced nearly 5,000 feet. The most valuable of the claims are Jacob's Wonder and Stewart's Wonder, which are separated by the Challenge claim of 750 feet. The croppings of Jacob's Wonder striking nearly west-east, with a dip to the north of 80°, run over the ridge between Woodpecker and Jacob's gulch, intersect the latter, and extend to the summit of the ridge between Jacob's and Stewart's gulch. In Jacob's gulch (425 feet above Rain's camp) we observe a slight break in the vein. To the west, the croppings extend up the steep mountain to its summit, a height of 500 feet above the gulch, the main vein being joined near the summit by a south branch vein 15 to 20 feet in width. These croppings vary considerably in width, and are mostly barren. The ore-chimney lies east of the gulch. Here we find the croppings of the main vein to continue east in a width of from 4 to 6 feet, falling, however, in places considerably below these dimensions. South of the main vein is a branch vein from 10 to 12 feet wide, which dips 35° north. In the gulch both croppings come together, and form an enormous mass of quartz, 20 to 25 feet wide. This branch does not continue west of the gulch, but can be traced east nearly to the Challenge claim. On account of the topography of the hill, the croppings of the branch vein diverge to the east from the croppings of the main vein. At the same time they diminish in width to about 3 feet. Both croppings show good ore for about 500 feet. The branch vein, however, shows much more ore than the main vein, and contains pay-streaks several feet wide. No work has been done on the main ledge. On the branch (475 feet above Rain's camp) an incline 55 feet long has been sunk to the foot-wall of the main ledge, but without penetrating further. This incline shows barren quartz at its bottom. But the Mexicans who did the work have done it without sense. Instead of going down on the foot-wall they ran the incline through the quartz near the hanging wall, and after a while left it altogether and pushed the roof of the incline in imestone, probably to get easier work. A few cuts west of the incline show very rich ore. There is another break in Stewart's gulch, and the disturbance here is much greater than in Jacob's gulch, (same altitude as Rain's camp.) No croppings appear in the gulch to the west opposite the Stewart's Wonder croppings; they are found again, however, some distance to the north, in the Challenge. Stewart's Wonder claim (1,500 feet) runs northeast from the gulch, the vein taking here a northeast course. The croppings show very prominent, and from 4 to 7 feet wide, nearly to the top of the ridge, where an incline has been sunk. In the gulch the quartz is perfectly barren, but higher up an ore-chimney commences, which I estimate to be 350 to 500 feet long. Near the incline (530 feet above Rain's camp) to the northeast, dark-colored limestone and slate commence, and the vein splits into three distinct branches, which carry for some distance very good ore, but become barren in the slate. The incline, about 20 feet deep, sunk on one of these branches, shows near the surface a ledge 5 to 6 feet wide, dipping northwest, with a pay-streak 2 feet wide, the rest being barren quartz. Lower down a wedge of limestone separates the pay-streak from the barren quartz, and at the same time the pay-streak widens considerably over 3 feet.

Next to the Wonder, the Marvel, located on the south slope of Surprise Cañon, is the most prominent vein of the district, and, if we consider the Surprise to be on the same vein, the Marvel is nearly as extensive in length as the Wonder. Marvel gulch is the deepest and longest of the intersections on the south slope of Surprise Cañon. The formation here is less regular, and the disturbances greater, than inother gulches.

The croppings of the Marvel exposed in the gulch are huge and exceedingly barren. Near the summit to the east I measured them and found a width of 20 feet. On the summit, (1,700 feet above Rain's camp,) however, they split up into a number of thin stringers intersected by barren little cross-veins. Descending a short distance to the head of Little Chief gulch, the Wyoming claim commences. Here we find very rich croppings, (1,600 feet above Rain's camp,) from 5 to 6 feet wide. These croppings continue ore-bearing to the northeast, but only in a width of 2 to 3 feet, over the next ridge, which divides Little Chief from Stern's gulch. The length of this ore-chimney is estimated to be from 300 to 350 feet. A short distance below the ridge, (1,600 feet above Rain's camp,) at the northeast end of the ore-chimney, a tunnel about 100 feet long has been run in a southwest direction on the ledge. This tunnel shows for about 20 feet from its mouth a streak of rich ore, 6 to 8 inches wide. In progressing farther southwest the vein widens to 3 feet. During this whole distance the quartz is mostly barren. Above the tunnel are bold croppings. In the vicinity of the tunnel the limestone is dark-colored, and to the northeast we find slate. This accounts for the barren nature of the vein in the tunnel. The Wyoming croppings appear again on the ridge on the other side of Stern gulch, but are barren and split up.

From the topographical description of Surprise Cañon, it will be evident that a splendid opportunity is offered to open the mines by tunnels. Apart from other advantages, this would save the expense of constructing expensive tramways and roads for the conveyance of the ore and supplies.

The Panamint ore is admirably suited for amalgamation, but it must first be roasted.

In conclusion, Mr. Stetefeldt remarks that it is rarely the good fortune of a mining engineer to form so favorable a judgment of an entirely undeveloped district as he feels justified in expressing in regard to the mines of Panamint district.

*Operations of leading mines of Inyo County, California,* 1874.

*Cerro Gordo district.*—Name of mine, Union and Guadalupe; owners, M. W. Belshaw & Co., and V. Beaudry; number of miners employed, 65; miners' wages per day, $4; cost of sinking per foot, from $2 to $25; cost of drifting per foot, from $2 to $25; cost of stoping per ton, about $1; cost of mining per ton extracted, from $6 to $7; cost of smelting per ton, $18; company or custom mill, company smelting-furnace; number of tons extracted and worked, 50 daily; average yield per ton, 30 per cent. silver-lead bullion; percentage sulphurets of lead, 20; total bullion-product, 15 tons daily.

Reported by J. L. Porter, superintendent.

### THE SOUTHERN COUNTIES.

The mining-counties of Southern California are Kern, San Diego, San Bernardino, and San Luis Obispo, the first four possessing districts in which vein-mining on ledges of gold-bearing quartz forms the principal mining-interest, and the latter possessing promising quicksilver-mines of recent discovery, and still in an undeveloped condition. The counties of San Diego, Los Angeles, and San Bernardino all contain placers, but owing to the want of water none are worked on the extensive system pursued in the central portion of the State. The discovery of gold was made in Los Angeles County, near the mission of San Fernando, as early as 1812, and in 1828 gold was shipped from the port of San Diego. It was, however, the policy of the *padres* who controlled the missions to suppress the knowledge of the existence of gold, and it was not till Marshall's discovery at Coloma, in 1847, that the knowledge became public.

*Kern County.*—Feeling convinced, by the accounts reaching me from various quarters, that the mining-industry of this county, long subjected to great depression, would find in its quartz-lodes a new basis of prosperity, I desired to visit the region in person, and was able in the summer of 1874 to satisfy this wish for the first time, though under circumstances of too great haste to permit a thorough survey of more than one or two localities. Since my visit the railroad extending southward from Lathrop has been finished to Bakersfield, which is the point of departure for Havilah, the county-seat of Kern County, forty-five miles distant. The road constituting the main route via Havilah to Kernville, and, I believe, to Panamint and other districts east of the Sierra, ascends at first gradually out of the great basin in which Bakersfield lies—the southern termination, in fact, of the vast interior plain of California. In the distance may be distinguished the sheeny surfaces of Tulare and Kern Lakes—immense puddles, apparently, without defined banks. Beyond them is the blue outline of the Coast range, and to the east, in impressive neighborhood, rise the summits of the Sierra, less lovely with forests and snowy crests than the peaks that look down upon the Hetch-Hetchy or the Yosemite, but, on the whole, not deficient in grandeur. Southward the two great ranges draw together, forming a bar of purple mountains across the plain. Through one of the clefts in this barrier runs the road to Los Angeles. The road bears southeastward, climbing toward the Sierra and leaving behind the broad yellow valley with the green oasis of Bakersfield and the verdant lines of willow and cottonwood that mark the distant rivers.

The mountains improve somewhat on nearer acquaintance. True, the foot-hills are barren and brown, but as we advance we come occasionally on cosy little ranches nestled by the mountain-streams, and at last we ascend into regions of oak opening and pine forest. Far glimpses of the precipitous cañon of Kern River give hints of a grandeur of scenery confirmed by views obtained from many a grade and divide on the way.

Havilah is a pleasant town, once prosperous, and now suffering a sad, I trust only a temporary, relapse; but there are good quartz-mining districts in the mountains around the place, and it is not impossible that Havilah, in the general revival of industry which seems likely in this county, may assume its ancient importance. Its location, on the highway to Kernville and the new districts beyond the range, is a point in its favor.

One of the best mines now actively worked in the county is the Bright

Star, at Piute, about eighteen miles from Havilah. It is owned by the Bahten Brothers, who extracted from it during the year ending June 30, 1874, about $110,000. The vein is said to be narrow but rich, carrying quartz stained with copper and heavily charged with arsenical pyrites. It courses north-northeast and south-southwest, and dips south about 85° in a country-rock of slate. The quartz yields in the mill by battery-amalgamation from $15 to $75 per ton. The mine is opened to the depth of 300 feet, and employs thirty men, at wages of $100 per month. The mill contains 10 stamps, of 750 pounds each, and is not believed to effect a complete extraction. The arsenical pyrites, assaying up to $700 per ton, are partially caught and concentrated by two Hendy concentrators, after which they are roasted and amalgamated—with what metallurgical success I did not learn. It is too cold in this district to wash blankets, the altitude being 9,000 feet, or 2,000 feet above Havilah. I believe the same cause hinders the steady working of the mines, though, with a mine as far advanced underground as the Bright Star, and with proper foresight and outlay in the accumulation of supplies, the climate will not cause serious interruption.

About eighteen miles beyond Havilah, on the main road, is the town of Kernville, once known to pioneers as Whiskey Flat. It is picturesquely situated by the rushing stream of Kern River, in the banks and bars of which, and along numerous small mountain-gulches tributary to it, placer-mining was rife and profitable in that period, a score of years ago, fondly called "early times" by the fast-living population of the Pacific coast. Only a few diggings, on a small scale, remain to bear witness to the golden traditions of the past. The present hope of Kernville lies in quartz, and particularly in a mammoth vein known as the Sumner or Big Blue.

This vein has been traced for some two miles in the hills on the north side of Kern River, and, after much negotiation and some litigation, the ownership of 11,300 feet of claims upon it is now vested beyond dispute in the Sumner Company, the principal stockholders of which are Senator J. P. Jones and Messrs. Burke and Strong. The two latter reside at the mine.

The patents of the Sumner Company cover on this vein the following claims:

| | Feet. |
|---|---|
| Beyond the north extension of the Sumner | 3,000 |
| The north extension of the Sumner | 1,200 |
| The Sumner | 1,200 |
| From the Sumner to the Big Blue | 1,800 |
| The Big Blue | 2,600 |
| The Nelly Dent | 1,500 |
| Giving a total of | 11,300 |

These figures are not taken from the patents themselves, but I believe they are correct. The company owns, in addition, six small cross-veins or feeders, which have been worked to some extent by former possessors, and found very rich, at least in spots.

The value of the main lode lies in its great size and the vast amount of good milling-rock which it can furnish. The principal mine upon it at present is the Sumner. In this mine I found the lode underground to be 84 feet between walls. The hanging wall is granite, the foot-wall is slate, between which and the vein lies a zone of clay 15 feet in thickness. Of the 84 feet of quartz, 42 feet, in two zones or pay-streaks of 35 feet and 7 feet, respectively, consist of pay-rock. The narrower zone is the richer, but the average yield of the whole material in the two is about $18. At the other end of the vein, on the extension known as the Nelly Dent, the vein is 200 feet wide between walls, but carries wholly low-grade ore, yielding, say, up to $10 or $12 per ton.

The Sumner is opened by an adit and cross-cuts 146 feet below the highest surface-croppings. Two shafts are being continued below the tunnel to open a new level below. The richest ore in the upper level has been removed, so far as the ground is open, but a large amount of lower grade is still standing, and there are some 15,000 tons of third-class ore ($12 to $15 per ton) lying on the stulls in the mine. This can be treated with profit when the new mill is completed, as will be seen by estimates given below.

The mills are situated on the bank of Kern River, about half a mile from the mine. Hauling, by a good road and on a down grade, costs but a few cents per ton. The old mill is the only one running, as the new one is not yet finished. The old Sumner Mill contains 16 stamps, of 650 pounds, dropping 7 to 8 inches 85 times per minute, and crushing 20 tons of quartz, or 1.19 tons per horse-power, daily—a moderate efficiency. The discharge is very high. For screens, wire-cloth, 50 meshes to the inch, is employed, and greatly preferred to punched Russia iron. The mortars are cast for a fore-and-aft discharge, but the rear opening is closed with a plank plated with copper. This catches a good deal of amalgam, and can be very easily removed and cleaned. On the other hand, the decrease thus caused in the area of discharge may possibly be the reason that the capacity of the mill is rather low. The wire screens in front are not over six inches high, and are set about nine inches above the dies. This again hinders the maximum discharge, though it is favorable to the amalgamation within the battery. The stamps are fed by hand.

The quartz is bluish and charged with arsenical pyrites. The sulphurets are separated in Hendy concentrators, of which there are two, receiving the battery-pulp from the sixteen stamps. This number is not sufficient. The concentrators have to work too much material in a given time; and the use of two other concentrators below these two, for a second treatment of the tailings of the latter, does not relieve the case. In fact, the second concentration has but little effect. Neither the bulk nor the rate of work having been substantially changed, the greater part of the tailings escaping from the first machines passes through the second, to be thrown irrevocably into the river. One Varney pan and settler work the sulphurets, which are said to yield $65 per ton.

The water-power, which is abundant and convenient, is furnished by a flume, giving a fall of 28 feet. It is utilized by a 41-inch turbine with central discharge.

Close by the old mill, the site and foundations of the new one were, at the time of my visit, the scene of busy activity. This mill (by the present time far advanced toward completion) is to be the largest, and in many respects the best, in the State. The batteries are to contain 80 stamps, standing in one continuous row—five stamps to each battery; the weight to be 750 pounds and the maximum speed 90 drops per minute—a combination which promises the highest efficiency, if adequate discharge is provided. The machinery is furnished by Booth & Co., of San Francisco, the leading manufacturers of the country in this line. I had the opportunity to inspect some of it at the railway-station of Delano, and found it to be, as might have been expected, of the most improved patterns and substantial workmanship. The cost of the machinery for such a mill would probably be about $30,000 coin at San

Francisco, and, adding the expense of freights, grading, foundations, and buildings, I infer that the mill will cost from $80,000 to $100,000.

The process is to be the same as that now followed in the old mill, except that there is to be a Hendy concentrator to every battery of five stamps, and probably no second concentration. The tailings from the Hendy concentrator will be allowed, so far as I could learn, to run off into Kern River and be lost. This is the only great mistake which I observed in the plan of working. It is partly explained by the confined space on the river-bank available for a mill-site, but that difficulty is one which need not have been considered insuperable. It is true that a larger area between the mill and the river would afford a more convenient opportunity to save tailings or treat them more thoroughly before allowing them to escape; and it is also true that, as the mill is now located, the rise of the river would be likely to sweep away accumulations of tailings. But I cannot believe that these considerations dictated the plan of throwing the tailings away. It is more probable that the intention was from the beginning to rush through the mill large quantities of quartz, at small cost for working, get out what gold could be obtained by simple and easy means, and get rid of the rest as soon as possible. The old Benton Mills, on the Mariposa grant, of which Ashburner, I believe, once said that they worked more cheaply and more wastefully than any others in the State, were planted on the banks of the Merced precisely as the mills of the Sumner Company are planted on the banks of Kern River. There, as here, the tailings were incontinently hurried down-stream, out of the way; but there it could at least be said that the ore usually treated was not full of sulphurets, while here it is rich arsenical pyrites that inevitably escapes from the hasty operation of the Hendy machines. Sooner or later this company will save its tailings, either in bulk or in some partly-concentrated form, to the tangible advantage of its exchequer.

The power of the new mill is to be supplied by the ample flume now in use, and will be utilized by a Leffel's American double turbine, 48 inches in diameter, which will furnish a little over 200 horse-power. The power developed by the fall of the stamps will be about 109 horse-power, and the remainder of the motive-power will be available for overcoming the friction of gearing, driving-pans, settlers, and concentrators, &c., with an ample margin for contingencies.

Mr. Charles Strong, (well known as the manager of the Gould and Curry mine, on the Comstock lode, in its days of glory,) one of the resident owners of this property, calculates, from the data acquired by actual experience here, that $3 per ton will cover all the expenses of mining, hauling, and milling in the new mill. The paper-estimate, without allowance for unforeseen expenses, brings the cost as low as $2.50 per ton. This is surprising, but neither incredible nor unprecedented. It must be remembered that the cost of mining is very small in a vein so enormous and at depths so trifling, while the treatment of the quartz in a mill run by water, provided with every labor-saving improvement, and expressly intended to work on the large scale, would naturally be unusually cheap. Similar results have been obtained in Australia; and low as Mr. Strong's estimate appears to be, I see no ground for doubting it. Of course it does not include interest on capital or general expenses; but, on the other hand, these items will be much smaller per ton of ore treated in a mill crushing at least 100 tons per day than in a smaller establishment. In a word, the Sumner Company is making the most of every natural advantage offered by the property itself, and adding the very great advantage of operating on a large scale, with adequate means.

The old 16-stamp mill is said to have earned between $100,000 and $200,000 profits, which have been swallowed up in purchases, improvements, and litigation. The new mill is paid for by an assessment, which is scarcely more than a commercial advance on the value of ore already mined.

*San Diego County.*—This is the most southerly county of the State. Ledges of gold-bearing quartz were discovered and worked on a small scale, on the Wolfskill ranch, thirty-five miles northwest of San Diego, between 1860 and 1864, but the yield is not supposed to have been greater than $100,000. It was not until the discovery of the Washington ledge, in the Julian district, on the 22d of February, 1870, that this could be considered a successful mining-region. Soon after the discovery of gold the owners of the Cuyamaca ranch attempted to float their grant over the mining-district, and for four years contested the claim, until the case was finally disposed of by the Land Department, and decided in favor of the miners in July last. This question of title has been the means of keeping out all foreign capital. The first discoverers were inexperienced, and yet so rich were the mines that they have continued to be developed. The history of the mining-section is peculiar, from the fact that the development has been done entirely from the products of the mines themselves, and bullion to a very large amount taken out. With the settlement of the Cuyamaca case the introduction of capital commenced, and, with the experience gained by those now there, we may look for a largely increased production of the precious metals from this county. At this time there are more than twenty well-defined gold-mines, of which eleven have proved to be of great value, and there are not less than seventy-five stamps running in the Julian and Banner districts.

The Chariot is the leading mine of the county. It has been described in former reports. The old shaft on the large vein is now 150 feet deep. The ore in this part of the mine was not very good; a drift was run to the eastward, to the "Joker" ledge, where very rich rock was struck. The ledge is very small, being only 6 or 8 inches wide. The new shaft on the main ledge is about 180 feet deep; it is 600 feet north of the old one. In this new shaft there is said to be ore as rich as that in the Joker, but the ledge is from 6 to 10 feet wide. There are hoisting-works at both shafts. The mill belonging to this company is four miles from the mine, in San Felipe Cañon. It is a 10-stamp mill of improved pattern, with copper plates and five of Hendy's concentrators.

The Ready Relief mine has a first-class 10-stamp mill, with a 25-horse-power engine. This mine is situated in a peculiarly advantageous position for working. It is on the face of a bluff, and tunnels are run on the ledge, no hoisting being necessary. One tunnel is in 150 feet, one 200 feet and another 150 feet, all on the ledge. There is a shaft from the surface to the end of the lower tunnel, 110 feet deep, which is used as an ore-chute. A small tram-way was built, about 60 feet long, from the mouth of the lower tunnel, and from it the ore is dumped about 30 feet, alongside of the mill. The ledge is said to be 8 feet in width, and to average $50 per ton.

Within the past year a new district has been discovered, which is described by the local press as follows:

The Bladen gold and silver ledge is situated near the summit of the highest foot-hills of the Coast Range Mountains, and is distant about eighty miles from San Diego and fifteen miles northeast of Temecula. The mining-district borders on the San Jacinto

Plains, at the base of the San Jacinto and Coahuila Mountains. It is twenty-five miles from "Old Grayback," the highest peak of the San Bernardino Mountains. The ledge has been discovered at different points each side of the Bladen mine for a distance of about fourteen miles, and bears a northwesterly and southeasterly direction, following the range of hills in as straight a line as may be in the direction of the Julian mines on the south and the San Bernardino Mountains on the north. Water is abundant at the mines, and is said to run in the San Jacinto Creek during the year. Oak timber is plentiful enough in the district for fuel, and will supply quartz-mills for many years to come; while in the mountains, from fifteen to twenty miles distant, abundance of fine redwood and cedar timber of the best quality is found. The Bladen ledge has been opened up in six or eight different places by shafts sunk only a few feet in depth. The deepest shaft is about 50 feet, in which the ledge is solid the whole distance between the wall-rocks, and is over 4 feet in width. In all of the other claims the ledge averages over 6 feet in width from the surface down, and in some of them much more than this width. The altitude of the locality is about 500 feet below that of the Julian district, which is about sixty-five miles distant. The character of the country immediately surrounding the Bladen district is more hilly and broken than that around Julian.

The San Diego Union of January 7, 1875, says:

The bullion shipment from the mining-section of this county in 1874 comes short of what was expected at the beginning of the year. The uncertainty as to the decision in the grant contest, which was not ended until recently, prevented the development of mines, and a number of the best ore-producing mines were engaged for several months in sinking on their leads. The total bullion shipment from our quartz-mines is estimated at $193,369. There has also been shipped hence of gold-dust, chiefly from the Lower California placers, $37,984; making the aggregate shipment of bullion from the city during the year $231,353.

*San Bernardino County* is the largest county in the State, having an area of 23,472 square miles. The population is less than 5,000. The San Bernardino range of mountains divides the county into two parts, differing from each other in topography, climate, and nature of soil. The eastern part lies within the Great Basin, or valley of the Colorado, and may be said to be worthless agriculturally. The western part extends from the San Bernardino range to the Los Angeles line, including within its boundaries the beautiful, fertile valley of San Bernardino. The western slope of the county contains an area of about two thousand square miles, nearly all of which is highly fertile. The San Bernardino is divided from the San Gabriel range by the Cajon Pass. East of the San Bernardino range the vast area of country extending to the Colorado is a barren waste, for the most part uninhabited, and almost uninhabitable.

Holcomb Valley and Bear Valley, constituting one mining-district, are situated near the junction of the Coast range with the Sierra Nevada. Holcomb is a small valley, about 8,000 feet above the sea-level, varying from one and a half to one and three-quarters miles in width by about four in length. The entire area included in the valley does not probably exceed ten square miles. Underlying this surface are the placer-deposits, and in the adjacent hills are found the ledges of ore-bearing rock. There is abundance of pine timber and a good supply of water for milling and sluicing purposes. The mines are accessible by a good wagon-road running through the mountains. The placer-mines in Holcomb Valley proper have been worked since 1860. There is very little drift or alluvium, the gold having all come from ledges near by. Men have also been working for some time on narrow but very rich "stringers," which have paid well. There is one quartz-mill in the valley, but placer-mining is principally carried on there. The ledges in Bear Valley are very large. One ledge is about 60 feet wide, and crops out for over two and one-half miles from 10 to 60 feet high. No work of any consequence is being done anywhere on this ledge, except upon the property of the Bear River Gold-Mining Company, which is now running a tunnel in parallel with the vein at a place where the mountain breaks off into the valley. The intention is to cross-cut to the vein. This tunnel is in now about 100 feet. The property comprises two claims—the Moonlight and Rainbow—of 1,500 feet each. The quartz-mines of Holcomb Valley are in a granite formation, the veins varying from 6 inches to 6 feet in width, comprising veins bearing free gold and iron-pyrites, with a slight proportion of lead and copper. Mining in this section may be said to be at its very commencement. Some of the mines were worked ten or fifteen years ago, but were long since abandoned, and it is only within the last year that work has been resumed.

The Holcomb Valley Company controls the principal quartz-mines of this section. This company has a quartz-mill which reduces about 15 tons of ore per day. The shaft has been sunk to the depth of 200 feet, through ore, with side drifts running to the westward from the shaft, and covering a block of ore 170 feet long by 80 feet in height. The ledge is said to average from 18 inches to 2 feet in thickness, and to pay in the neighborhood of $40 to the ton.

*San Luis Obispo County* lies between Monterey and Santa Barbara Counties, on the north and south, having the Pacific Ocean on the west and Kern County on the east. The topography of the country is broken and the coast-line bold. The occurrence of hot springs and evidences of recent solfataric action, similar to the quicksilver region north of San Francisco Bay, attracted the attention of prospectors early in 1874, and resulted in the discovery and opening of a large number of quicksilver-mines, of which two or three at least give evidence of permanence, and will materially contribute to the product of this metal in 1875.

Among these is the Oceanic mine, a consolidation of three locations, situated near the head of Santa Rosa Valley, about five miles from the village of Cambria, and seven miles from the Pacific Ocean. The matrix of the ore is composed of a coarse sandstone and a conglomerate interstratified. This conglomerate is composed principally of shale and quartz-pebbles, cemented together by silica. The lode can be traced superficially in length east and west from 3,000 to 4,000 feet. The main deposit, so far as known, is where the three claims join, the principal work having been done there. The greatest width of the vein is 100 feet. There are two main tunnels penetrating the hill. The ore does not lie in spots, with here and there a specimen in sight, but there seems to be a vast body.

Mr. J. Galloway, who visited the mine in December, 1874, says:

The mine is located about 100 feet above the level of the ocean, and the first tunnel or lower level is about 400 feet above where the reducing-works are being erected. The ledge runs east and west, and varies from a vertical about 15°, dipping toward the north. The matrix is a soft, friable sandsone, interstratified with conglomerate; the country-rock is highly metamorphosed and the vein-matter remains unchanged. The cap-rock is a close, hard rock, between shale and sandstone, and this cap or wall rock, on the south, is impregnated with cinnabar from 5 to 10 feet before you cut the lode. The north side, or hanging wall, is talc and siliceous schist, and is more broken than the south. The vein-matter is from 12 to 20 feet wide, wherever cut through, in a number of places. It shows a uniformity of width and unbroken length for a distance of 312 feet, rarely met with in any vein-matter, and, still more strange, it shows a uniformity of richness seldom found in cinnabar-mines.

The location could scarcely be more convenient. The road is nearly on a level until you reach the hacienda. From the reducing-works to the mine is about one-half mile, and gives an elevation sufficient to place the working above and beyond the effects of water.

It is only about five months since this mine was bought by the present company, and not more than three and a half months since they commenced work upon the ledge, and yet they have accomplished a good deal. Five tunnels have already been driven into the ledge, thus prospecting it in five different points, four of them on different levels. The lower level is a tunnel, which cuts the ledge at a distance of 187 feet from its mouth, and has a drift 25 feet west, run in ore, now filled with rich cinnabar, and one east on the ledge 16 feet. The vein-matter is 13 feet wide at this point. In this drift is found much native quicksilver. It is also colored with cinnabar. The next tunnel is the Ravine, which is in 50 feet on a level with the North tunnel, and is driven for the purpose of bringing the ore from it to where the tram-work is erected, where the friction-reel for lowering the metal to the dump-house is situated. This tunnel exposes as fine metal as any in the lode. The North tunnel is in 125 feet, starts nearly west of the ledge, and strikes the lode at 30 feet, and is 95 feet in ore. This tunnel tests the ledge for nearly 100 feet. The main drift is in 177 feet, and between the North tunnel and Main is a space of about 40 feet. This drift is in ore all the way, showing the vein-matter to be of nearly uniform width and richness. The Cedarberg tunnel is 65 feet above the main drift, and is farther east or back upon the ledge than any of the other tunnels. The same showing, both as to the width and richness of metal, is here made. The upper cut or tunnel is now into the cap-rock, which shows a richer impregnation of cinnabar than any cap-rock they have yet cut through. The cut is back upon the ledge 553 feet, and at an altitude, measured vertically, of 250 feet from the lower level; the dip of the ledge will make a small amount more. Thus the area of the lode already prospected, allowing the ledge to be but 10 feet wide, and allowing for the angle of the hill, is 28,000 tons of ore. In addition to this there is much of what is called the cap-rock that will pay well for reducing, and might be termed good furnace-ore.

Mr. Galloway made a series of tests with a view of ascertaining the percentage of the ore, and estimates the entire mass of available working ore in sight at 5 per cent., equal to 100 pounds of quicksilver, or, at present prices, $150 per ton. This estimate does not appear extravagant, when we consider that the product of the New Almaden for ten years varied between 5 to 10 per cent. per annum. The cost of extraction and reduction is estimated by Mr. G. at $5 per ton. He says:

The cost of excavating is small. The rock is soft, easily drilled, and throws well. It could be gadded, but would break up too much. The ore is very porous, being, as I stated, a sandstone, and will allow the quicksilver to escape readily. Some of it will crumble down, and will have to be made into adobes or bricks, and it will require for this purpose 20 per cent. of clay, perhaps, to be added. This will add to the cost of handling and kneading into brick, but this is labor that any one can perform. Wood is distant from the mine some miles, but the Oceanic Company have bought 200 acres of good timber-land, and will for a long time have their own wood at cost of cutting and hauling for five miles. They have their own water for the condensers.

The works, furnaces, &c., are of the most substantial kind. The furnace is 15 by 14 feet square at the bottom, and 27 feet high. It is built of good, substantial cut-stone and fire-bricks, except the upper part, which is of fire-brick exclusively, surrounded with an iron jacket of boiler-iron. It is of a capacity of 20 tons per diem, and can be run to a higher figure in adobes—say 36. It has four fires, and thus the heat is kept evenly under the ore. There are two openings for the reception of metal in the top, four-inch holes, and then the fumes start off through an 18-inch connection-pipe into the soot-chamber and condensers. The soot-chamber is a solid, square-cut stone structure, into which the smoke and vapor goes, and, passing down and under a brick arch, enters into the iron condensers. These are on the plan of the Idria mine in Austria. These condensers are cylinders 20 inches in diameter and 10 feet in length, of cast-iron 1 inch in thickness, and weight 2,200 pounds each. They lie horizontal, with a depth of 8 inches, and have connecting-pipes 20 inches in diameter at opposite sides and alternate ends, so that the vapor and smoke pass in at the end of the first, is forced to the other end and into the next, and goes zigzag through them clear to the other end. There are 20 of these cylinders, 10 feet each, and the connecting-pipes 2¼ feet, making the circuit through which the vapor passes 260 feet, where a Knox blower gives draught to the whole affair by drawing out the smoke and driving it up a wooden flue, which is 500 feet long, running up on the hill-side. This will carry away the smoke and catch the quicksilver that may escape the iron condensers. These iron condensers are to be kept wet by a series of sprinkling whirligigs, that are turned by hydraulic pressure, and will keep them cool. This water has a fall of 100 feet; the condensers an area of 1,600 feet inside surface. The quicksilver is run out of small tubes in the ends of the condensers into a trough, that declines to the center into a reservoir, which from it is flasked, weighed, and sent to market.

The deposit has a general east and west course, tending northerly at its western extremity, and dipping toward the north. The width of the best class of ore varies from 10 to 20 feet. The greatest depth attained, 175 feet.

Dr. R. A. Cochrane, the superintendent, who has given much attention to the geological formation, says:

The presence of sand and pebbles demonstrates that it is an aqueous formation produced by currents. The mode of occurrence is through fumes. There are always thermal springs found in connection with cinnabar. Each particular pebble in this conglomerate appears covered with cinnabar. Small bowlders—the hardest kind of flint-rock—are found also. All around this flint-rock, on the surface, cinnabar is deposited, it not having been able to force its way into this hard substance. On breaking this mass of silica no indication of cinnabar is seen, showing that the fumes could not impregnate it. Several recently extinct hot springs are found in the neighborhood. There are other evidences of recent origin than those referred to. From the top of the chimney to where they are now working, they find disseminated all through the mass iron-pyrites, which shows not the least indication of decomposition. In other mines, as, for instance, the Pine Mountain region, evidently older in formation, the surface is covered with red oxide of iron, produced by decomposition of iron-pyrites. In the Oceanic this iron-pyrites is sometimes in cubical and regular masses, having a firm adiated structure, and even experts have mistaken it for copper or gold.

Among the early locations made in San Simeon district is the lode known as the Keystone, one mile further south, and still lower down. Cross & Co., of San Francisco, agents of English capitalists largely engaged in mining on this coast, purchased this claim (1,200 feet) one year ago, and have gone to work in earnest to prove its richness. A shaft has been sunk to a depth of 80 feet on an incline with the ledge, with satisfactory results. A tunnel 250 feet long taps this shaft. There are 500 feet of drifts in the mine in different parts of the ledge. Twenty miners are employed, and the force will soon be increased. The company has a furnace in course of construction, the capacity of which will be 12 tons per day. Large quantities of good ore are also on the dumps at this mine.

All these mines lie due east of San Simeon Bay a distance of about seven miles. The altitude is 2,000 feet. Recently discoveries have been made at the headwaters of Santa Rosa Creek, six miles east of Cambria, which attract considerable attention just now. It is the same character of mineral as above described. The facilities for working these mines are unusually favorable. There is plenty of wood and water close by, and they are all accessible.

*Description of leading mines of Kern County, California*, 1874.

*Cove mining-district.*—Name of mine, Sumner; owners, Sumner Gold and Silver Mining Company; length of location, 1,200, 1,800—3,000 feet; course, northerly and southerly; dip, west.; length of pay-zone, 700 feet; average width, 30 to 40 feet; country-rock, granite; character of vein-matter, bluish quartz, with arsenical pyrites; worked by both tunnel and shaft; depth on vein in tunnel, 146 feet; depth of working-shaft, same; number of levels opened, one; total length of drifts, 1,300 feet; cost of hoisting-works, now just up, $100,000.

*Operations of leading mines of Kern County, California*, 1874.

*Cove mining-district.*—Name of mine, Sumner; owners, Sumner Gold and Silver Mining Company; number of miners employed, 30; miners' wages, $3.50 per day; cost of sinking, $30 per foot; cost of drifting, $10 per foot; cost of stoping, $2 per ton; cost of mining, $2.50 per ton extracted; cost of milling, $2 per ton; company uses its own mill; number of tons extracted and worked, 5,000; average yield, $40 per ton; percentage of sulphurets, 1 per cent.; total bullion-product, $200,000.

*Statement of quartz-mills of Kern County, California,* 1874.

*Cove mining-district.*—Name of mill, Sumner; owners, Sumner Gold and Silver Mining Company; kind of power and amount, water—no limit; number of stamps to mill, 16; weight of stamps, 700 pounds; number of drops per minute, 90; number of pans, 1; number of concentrators, 4; cost of mill, $25,000; capacity per 24 hours, 16 tons; cost of treatment, $2 per ton; tons crushed during the year, 5,000; method of treating sulphurets, by pan.

*Description of leading mines of San Diego County, California,* 1874.

*Banner district.*—Name of mine, Chariot; owners, Chariot Mill and Mining Company; length of location, 1,200 feet; course, N. 42′ E.; dip, 65°; length of pay-zone, 120 feet; average width, 6 feet; country-rock, slate west, granite east; character of vein-matter, quartz, with clay "gouge;" worked by shaft; depth of working-shaft, 260 feet; number of levels opened, 2; total length of drifts, 500 feet; cost of hoisting works, $5,000.
Reported by F. L. Perkins, superintendent.

*Operations of leading mines of San Diego County, California,* 1874.

*Banner mining-district.*—Name of mine, Chariot; owners, Chariot Mill and Mining Company; number of miners employed, 35; miners' wages per day, $3; cost of sinking per foot, $60; cost of drifting per foot, $20; cost of stoping per ton, $6; cost of mining per ton extracted, $10; cost of milling per ton, $4; company uses its own mill; number of tons extracted and worked, 2,500; average yield per ton, $55; percentage sulphurets, ½ one per cent.; total bullion-product for 1874, $138,864.69.
Reported by F. L. Perkins, superintendent.

*Statement of quartz-mills of San Diego County, California,* 1874.

*Banner district.*—Name of mill, Chariot Mill; owners, Chariot Mill and Mining Company; kind of power, and amount, steam; number of stamps, 10; weight of stamps, 800 pounds; number of drops per minute, 90; height of drop, 6 inches; number of concentrators, 5; cost of mill, $20,000; capacity per 24 hours, 18 tons; cost of treatment per ton, $4; tons crushed during year, 2,500; method of treating sulphurets, selling them.
Reported by F. L. Perkins, superintendent.

*Description of leading mines, San Luis Obispo County, California,* 1874.

*San Simeon district.*—Name of mine, Keystone Quicksilver; owners, Cross & Co.; length of location, 1,200 feet; course, north and south; dip, east 50°; length of pay-zone, 150 feet; average width, 3 feet; country-rock, serpentine; character of vein-matter, sandstone; worked by both tunnel and shaft; length of tunnels, 450 feet; depth on vein in tunnel, 185 feet; depth of working-shaft, 80 feet; number of levels opened, 5; total length of drifts, 710 feet; cost of works, $30,000.
Reported by Cross & Co.

## MARIPOSA COUNTY.

Mariposa County is bounded on the north by Tuolumne, south by Fresno, and west by Merced, and extends from the edge of the San Joaquin Valley to the summit of the Sierra range. Its length is about sixty-five miles from east to west, and its width, from north to south, about thirty miles. Merced, its principal river, has its source in the mountain-fastnesses above the Yosemite Valley, in the regions of perpetual snow, and then flows over towering precipices and through deep and precipitous cañons until it reaches the western boundary of the county. Nearly the whole of Mariposa is extremely rugged and mountainous, and there is comparatively little land suitable for agricultural purposes. Quartz-mining is the principal business, and there are numerous lodes that have been profitably worked for many years.

The "Las Mariposas" grant, a large tract of land ceded by the Mexican government before California was acquired by the United States, and which has since become the property of the Mariposa Land and Mining Company, lies within the limits of this county. The Mariposa grant covers an area of forty-eight thousand acres of land, including some of the richest mineral-lands of the county, and within its limits are the towns of Bear Valley, Princeton, and Mariposa.

I am indebted to Mr. Charles Schofield and Mr. G. E. Webber, of Hornitos, for the following review of mining-operations in this county for the past year:

The early mining history of this county is well known. It has stood among the first counties of the State, but its rich placers have been almost entirely worked out, and quartz-mining, for some years past, has not been prosecuted with much vigor. This inactivity has been caused partly by the failure of many of the early enterprises, on account of the loose and careless way in which quartz-operations have been carried on and the want of proper machinery, and partly on account of the stoppage of work on the Mariposa estate.

There are at present but two or three first-class mines regularly worked in this county. These are the Hite mine, on the South Fork of the Merced; the Hasloe mine, on the north side of the main Merced River, and the Washington mine. I shall limit this report to the latter and other mines of less importance in the same district.

*The Washington mine* is located in the Quartzburgh or Hornitos district, about one mile and a half north of the latter town, and about a half mile west of the Hornitos and Bear Valley stage-road. The surrounding landscape is a succession of low grass-covered hills, with occasional patches of oak timber, though the most of this has been cut down and used as fire-wood, leaving only the stumps and occasionally a tree to show where a forest formerly stood. The country-rock is red sandstone, occasionally mixed with slate and traversed by veins of green trap. Veins of gold-bearing quartz are quite numerous, and crop out from nearly all the hills, most of which have been some time or other claimed and prospected, though very few are being worked at present.

The Washington mine was located in 1850, and, with the exception of about three years, from 1864 to 1867-'68, it has been worked nearly ever since, though with varied success and several changes of ownership. The present company came into possession in 1867, and built the mill now running, which is the third one, which has been built in connection with the mine, the first having six stamps, the second ten, and the present twenty. The latter is now nearly worn out, and a new one, with forty stamps, is already projected to replace it. It is thought that the large body of rich ore now in sight will fully justify the erection and furnish constant employment for a mill of that size for many years to come. The mill was idle last winter and spring, while the miners were engaged in sinking, drifting, and timbering. The mine was only run about four months during the year, or from the 20th of June to the 10th of October, and at the beginning of November was awaiting the erection of new more powerful hoisting-works, which were expected to be ready for use in two or three weeks.

The vein where it was then being worked is in some places 16 feet in width, solid ore, which pays on an average about $25 gold per ton. This body was almost untouched from the 400-foot to the 600-foot levels, and no one knows how much deeper.

For the pump and hoisting-works a boiler 4 feet in diameter and 26 feet in length, with a good 60-horse-power engine, has been erected, with another boiler of smaller dimensions as a reserve. A reversible engine is used for hoisting. It is estimated that these works will be sufficiently powerful to hoist water and rock from a depth of 2,000 feet. The rock is drawn in a car up a tram-way to the crusher in the mill.

Attached to the rear of the mill is one of Schofield's patent double-rigged concentrators, which receives the pulp directly from the battery without handling, and does the concentrating for the whole twenty stamps, turning off about one ton of clean sulphurets per day, worth $90 per ton, at a cost of $3 per day—the wages of two Chinamen—no power being required.

The sulphurets, when thus cleaned, are worked by chlorination, works for that purpose having been provided by the company, at a cost of about $6,000, consisting of a roasting-furnace with three compartments, one above another, each of the necessary capacity for roasting one ton at a charge; and, as the time for keeping the charge in each separate chamber is eight hours, three tons per day are usually roasted, with the consumption of one and a half cords of wood per day, at $6 per cord. The whole works, consisting of the mill, hoisting-works, and chlorination, consume about nine cords of wood per day, at $6 per cord. For the chlorination, two tubs of three tons capacity each are used, which are sufficient to keep pace with the furnace.

As all the work of roasting and chlorinating is now done by Chinamen, without the services of an overseer, the whole cost does not exceed $9 per ton.

Next to the Washington in point of present importance is the Italian mine, located about three-fourths of a mile east of the town, and owned by Signor Campodonica. The mine is opened to the depth of 150 feet by shaft, and shows a pay-shoot 120 feet in length, averaging about 3 feet in thickness.

A new steam-mill, of five 750-pound stamp and 12-horse-power engine, was erected about fifteen months ago, and has been running on about half time ever since. During this period it has crushed about 2,000 tons of rock, averaging $12 per ton in free gold, and containing 3 per cent. sulphurets, which will work $90 per ton. Only four men are employed in the mine, one in the mill, and two Chinamen to hoist and spawl the rock.

The Pool Mining Company, recently incorporated, has commenced operations upon a group of mines, five in number, located on Pool's ranch, about one mile east of the Italian mine. These mines have been worked occasionally during the past twenty years, sometimes by arrastras, and at other times by hauling the rock to custom-mills, the yield having ranged all the way from $5 to $100 per ton. One hundred tons worked during the present year turned out one ounce ($13) per ton. Samples recently assayed, from one of the ledges about 2¼ feet thick, indicated $70 per ton; and the sulphurets, of which there is 10 per cent. in the rock, assay $110 per ton. The company is about erecting on the last-mentioned ledge an 8-stamp mill, which will also be convenient to all the ledges, being not more than one-fourth of a mile from the most distant. It will be driven from six to nine months of the year by water from the Bear Creek ditch, in which the company owns an interest, and which it intends to improve with several large reservoirs, so as to keep water in reserve. By this means, it is estimated that rock can be crushed at an expense of less than $1.50 per ton, and the sulphurets concentrated at an additional expense of $3 for each ton of the same.

The No. 9 mine is located about one mile northeast from the Pool mines, and on the same water-ditch. It is owned by Major Hardwick and Frank Thorn. This vein strikes nearly east and west, and dips about 30° north. It has been worked by shaft to the depth of 125 feet, and shows a pay-shoot about 6 feet thick and 100 feet long. About 2,000 tons of this rock was crushed by Dan Jones, at the Gaines Mill, two years since, and yielded $7.50 per ton. But during the winter of 1873-'74 Major Hardwick erected a mill with four stamps, propelled by a 33-foot over-shot water-wheel and crushed about 200 tons, which averaged $9 per ton. The whole expense of mining, hauling, and crushing the rock does not exceed $5 per ton.

The Gaines mine is located at the base of Bear Mountain, about four miles northeast of Hornitas. The vein is a large one, and is represented to have been quite rich near the surface, but to have soon run into a sulphuret rock which does not contain much free gold; and, as the mine was very wet, it was soon abandoned, and has not been actively worked for the last eight or ten years. Eight tons of sulphurets from this mine, worked at the Washington chlorination-works, about two years since, yielded $1,200 to $1,500 per ton. There is a good 10-stamp steam-mill near this mine, belonging to the same company, which is now doing custom-work when any such work is to be had. The charge for working is $4.50 and $5 per ton, according to quantity, but the mill is not kept employed more than one-fourth of the time. There is now some prospect of work being resumed upon the company's mine.

The old Does mine now belongs to Mr. A. H. Brooks, who purchased it of McCullough three years ago for $13,000. It is located about one mile and a half southeasterly from Hornitas. Its strike is north and south, and its dip 45° east. It has been worked by shaft 110 feet in depth, and develops a ledge 10 feet thick all the way down, and at least 12 at the bottom, of solid vein-matter. About 400 tons of rock from this mine, worked by Brooks & McCullough, about two years since, paid from $10 to $12 per ton, free gold, besides 15 per cent. sulphurets, 28 tons of which, cleaned by Schofield's concentrator, and worked at the Washington works last spring, yielded $23.28 per ton. There is no mill belonging to this mine, nor even a good mill-site nearer than one mile from it.

The foregoing are about all the mines of any known value that are worked at present in this district, except a few "pocket-mines" about Hornitos, which are occasionally worked, mostly by Mexicans, who sometimes make a "strike." Their good luck is generally of short duration. They feast and frolic till their money is gone, and then hunt for more; but their days of fasting and toil between "strikes" are often longer than their periods of success.

Quartz-mining generally in this county is beginning to look up, and whoever reports upon it one year from now will probably record quite an improved state of affairs.

The principal mines on the Las Mariposas grant were described in the Commissioner's report for 1869. Since 1870 but little has been done on them. Mr. Edmund C. Burr, the manager of the Mariposa Land and Mining Company, writes me as follows respecting the operations of the company:

The principal object of the company now is the prosecution of the river tunnel, commencing at a point on the Merced River near the Ophir Mills, and about 50 feet higher. The course of the tunnel is nearly southeast, running between the serpentine and blue slates, and following a course parallel to the observed line of the outcrops.

The projected length of the drift will be 17,500 feet. What has been accomplished thus far amounts to 664 feet, up to 625 feet being only a drift of 7 feet by 7 feet 6 inches. From 625 feet on it will be 10 feet by 7 feet 6 inches, the clear thus allowing the use of two drilling-machines side by side.

The progress is necessarily slow at present, owing to the fact that we attack the slates edgewise, a great deal of the force of the powder b eing wasted or lost. By the use of two of the Burleigh drills we hope to make 150 feet per month. As yet work has not been carried on long enough to furnish us sufficient data by which to make an accurate estimate.

As remarked, Burleigh mechanical drills are used, driven by compressed air. Ample power is furnished by the company's dam on the Merced River.

The first object of the tunnel is to tap the Pine Tree vein about 750 feet below the present workings. At that time the tunnel will be in 6,350 feet, but, as there are numerous rich outcrops along the whole line from 900 feet on, it is hoped that we may be taking out ore long before we reach the Pine Tree. When the drift is in as far as the Josephine mine it will be 2,000 feet below the outcrops.

It is now too early to say anything about proposed work on other parts of the estate, but I hope to be able to give you some information in time for your next report.

Since last June no work has been done on the estate except placer-washing by Chinese and Italians, and it is almost impossible to obtain information from them.

The accompanying section will show the relations of the tunnel now in progress to the leading mines at that end of the Mariposa estate.

Mariposa, like others of the mining-counties, has made but little progress for the last few years, yet it possesses many resources that will eventually be developed, and its resources are sufficient to support a large population.

The following statistics are from official sources, and show something of the present condition of mining in Mariposa County:

| | |
|---|---|
| Quartz-mills | 33 |
| Tons of ore crushed | 7,000 |
| Mining-ditches | 5 |
| Miles in length | 25 |

Full returns of the operations of the leading quartz-mines will be found elsewhere. The following account of the mine at Hite's Cove is the result of a personal visit made by me in June, 1874:

From Mariposa a road of eighteen or twenty miles leads over the hills and along the Chowchilla Valley to this picturesque locality on the South Merced.

A striking picture is presented from the summit of the last hill on the road—a deep, precipitous, romantic gorge, through which tumbles the South Merced; side-gorges emptying into this, and filling the landscape

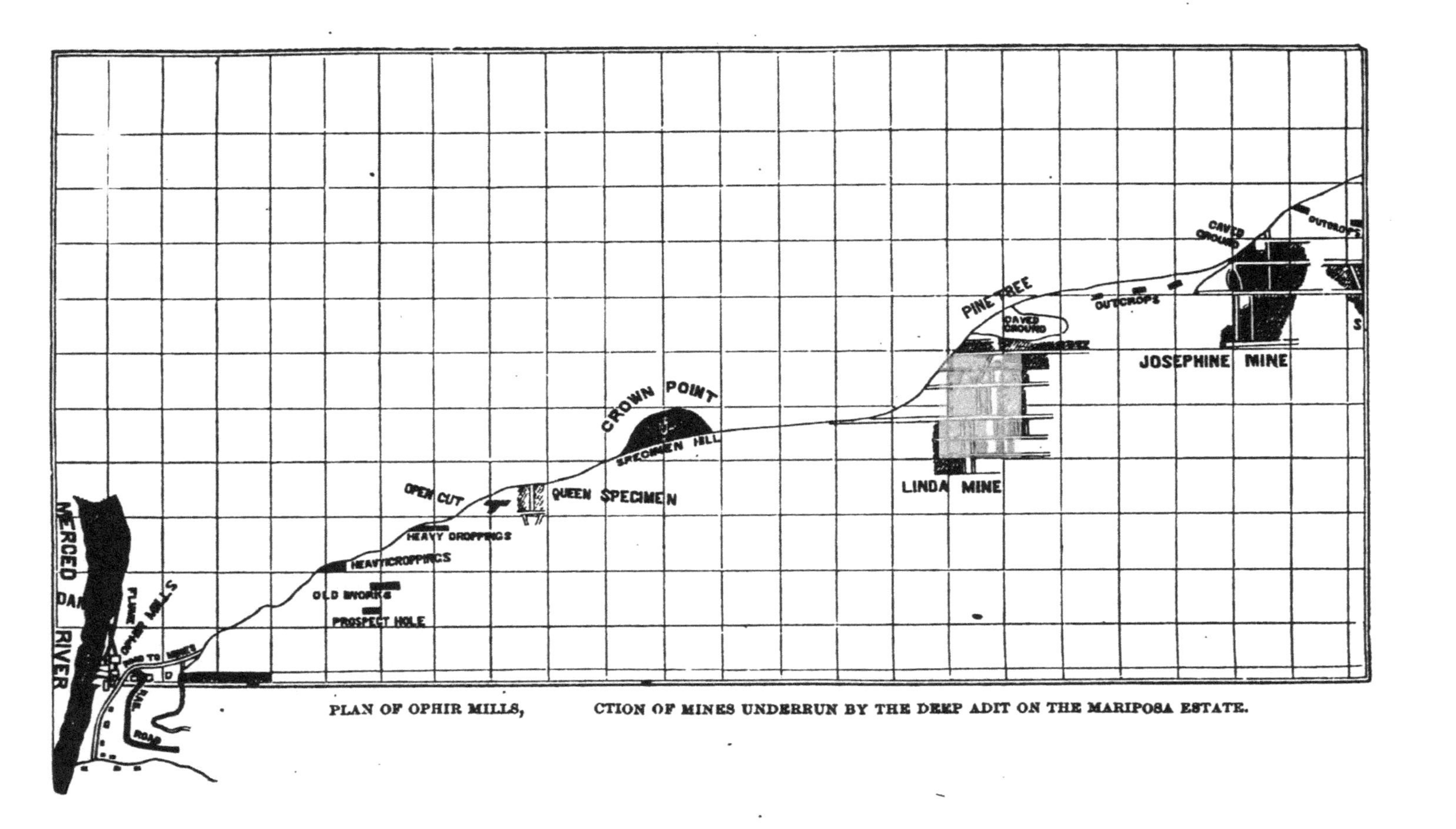


PLAN OF OPHIR MILLS, CTION OF MINES UNDERRUN BY THE DEEP ADIT ON THE MARIPOSA ESTATE.

with infinite variety of surface, clad in forest, chaparral, or bold, outcropping rocks; opposite a steep, sharp, mountain-ridge, separating the South Merced from the main river; beyond that, first a deep cañon, and then other ranges and summits. The cañon is felt rather than seen. One instinctively conceives, from the dim blue haze that hangs over it, and from the configuration of the mountains eastward, that it must lead up to something grand. A safe presentiment; for, by taking yonder slender trail over the divide into the cañon, and following up the stream a dozen miles, one may ride under the shadow of a famous rock known as El Capitan, and guarding the entrance of Yosemite. In fact, this is one of the best trails into the Great Valley, though impracticable for wheels, and at present not employed to any extent by tourists.

On the face of the divide already mentioned, may be seen the indications of a successful and actively prosecuted mining enterprise—dumps, both large and fresh; cars running out of a tunnel, down a steep tramway, to a stamp-mill, and tailings from the mill staining the current of the stream. But not all these are visible from the top of the hill on the Mariposa road. The buildings and the mill are too deep in the gorge to be seen; and the distance is too great for discerning the details, even of such operations as would be otherwise visible. For the South Merced flows some 2,500 feet below the hill-top, and the distance in air-line obliquely across to the mines cannot be less than a mile. To reach the mine after having seen it requires a journey of five or six miles down a steep, zigzag trail, cut in the side of the hill. All the supplies for the settlement and mine at Hite's Cove are carried down this trail. The wagon-road ends a mile or two below the summit of the hill, at a storehouse, where everything is unloaded till required. We passed a gang of Chinamen laboriously rolling over and over, down the trail, a heavy iron-casting for the mill.

Arrived at the bottom of the tedious hill, we gallop for a mile along the banks of the river, and at last, turning a projecting point, come suddenly upon the settlement, situated at the narrowest part of the rocky gorge, and plastered, as it were, against the mountain-side. A slender suspension-bridge leads over to the cluster of buildings, comprising boarding-house, store, stamp-mill, a dwelling or two, blacksmith's shop, &c. The houses stand on narrow terraces, wherever they can get room; and above them all the blue dump of the mines, a thousand feet higher than the river, hangs like a horn of plenty inverted over the settlement. This is indeed scarcely a figure of speech. The mine is the sole source of existence of a settlement in this wild, secluded hole in the mountains. Mine, town, and everything belong to Mr. John R. Hite and his partner; and for about eleven years, almost unheard of by the outside world, they have been steadily and profitably moving. Not being a stock company, and having no motive for advertising the value of the property, they have made no appearance in California street, or Wall street, or Lombard street, and at this day few Californians, even, are aware of the existence at Hite's Cove of one of the most remunerative mines of the State.

The characteristic features of the Hite's Cove mine may be summed up as follows: Vein, 4 to 18 feet wide; course, northwest and southeast, nearly; dip, 80° northeast; country-rock, slate; quartz, white and bluish, with black seams, and heavily charged in places with arsenical pyrites. The gold is fine and seldom visible. Average yield in mill, the year round, $35 per ton. Length of claim, 1,500 feet; no mines on extensions. Within the claim, three pay-chutes, between which the vein pinches. The largest is on the west end, and extends horizontally

300 feet. Vertical depth, unknown. Another, at the east end, is 250 feet in horizontal; and between these two there is a third, of smaller dimensions. In depth, the limits of these bodies have not been reached, though they are opened through a vertical distance of 600 feet.

The mountain in which this vein occurs has a general course not far from that of the vein. The outcrop is perhaps a thousand feet above the river. The entries to the mine, besides the discovery-shaft, are two cross-tunnels, cutting through the foot-wall. The upper tunnel is 210 feet long, and the lower one, 500 feet below, is 1,400 feet long. The latter has been recently completed, and will be the working-entry. The body of reserves is nearly 600 feet in height, measured on the vein. As to the quality, both tunnels and the wings and shafts in the vein corroborate the prolonged experience of the mill, in promising a maintenance for years to come of the uniformly profitable character of the quartz. It contains an average proportion of 2 to 3 per cent. of arsenical pyrites. In pieces which are more heavily impregnated, or massively charged with sulphurets, the contents of gold are high—running up sometimes into the hundreds. The complete natural drainage, easy transportation to mill, and large size of the lode, render mining cheap. Last year the mining-expense was $2.45 per ton of rock worked. Reckoning as much more for milling, (an excessive estimate, since the mill is run with a turbine-wheel,) we have the total running-expense, to which, fortunately, there are no heavy office-expenses, law-fees, salaries, &c., to be added. Mr. Hite manages the mine himself. Miners' wages, $50 and $55 coin per month, with board.

The mill, close to the river, contains 20 stamps, weighing 500 pounds each, and dropping 9 inches 65 to 70 times per minute. The capacity is 20 tons per day, or 1.33 tons daily, per horse-power exerted by the fall of the stamps—a good average degree of efficiency. Quicksilver is used in the battery and on a small apron (copper-plated) in front. After this, the pulp is treated in Wheeler & Randall Excelsior pans, of which there are three to the 20 stamps, and settled in one large Varney pan, which is palpably insufficient for the work it has to do. There ought to be at least two settlers.

A Peer & Lanquist concentrator, for concentrating the sulphurets after panning, is on trial in the mill. It is a cast-iron, ellipsoidal, concave shaking-table, evidently intended to imitate the motions of the horn-spoon and pan of the placer-miner. These hand-implements are mechanically not satisfactory, but imperfect and slow in operation. Their excellence consists in their cheapness, portability, and rudeness. Rudeness may be a virtue in a tool under some circumstances. Thus a jack-knife is rude compared with a chisel, gouge, or auger; that is, rude for each special purpose. But its rudeness leaves room for varied application and the control of the skilled hand. The thing to carry in one's pocket is a jack-knife. And so a horn-spoon or pan, which leaves everything to the skill of the operator, varying his motions to the nicest shades of necessity, is a very good thing to be joined to human intelligence for the performance of work. But I would as soon think of building a machine with an arm and a knife to whittle bed-posts, instead of a lathe to turn them, as to imitate in rigid mechanism the free swing and shake of the pan, or the delicate and variable indications of the spoon. Moreover, these two contrivances concentrate without previous concentration or sizing, which ought never to be done with a machine. The reduction of all material to uniform grades of fineness is an indispensable prerequisite to successful concentration, as might be shown with learned equations and demonstrations, if the principle were not so well known. There is nothing in the results obtained thus far at the Hite's Cove mill, from the concentration referred to, to invalidate the considerations just set forth. But the trial was not yet over at the time of my visit.

A Blake crusher prepares the rock for the stamps, to which it is delivered by a Stanford self-feeder, (described and figured in my report rendered in 1873.) Mr. Hite says it works very well, securing no special saving of labor, but an increased capacity of one-fifth for the mill. The loss of quicksilver in this mill is insignificant—about three flasks per annum. The greater part is used in pans and settler.

*Description of leading mines, Mariposa County, California, 1874.*

| Name. | Location. | Owners. | Length of location. | Course. | Dip. | Length pay-zone. | Average width. | Country-rock. | Character of vein-matter. | Worked by tunnel or shaft. | Length of tunnel. | Depth on vein in tunnel. | Depth working-shaft. | No. levels opened. | Total length of drifts. | Cost of hoisting-works. |
|---|---|---|---|---|---|---|---|---|---|---|---|---|---|---|---|---|
| | | | Ft. | | | Ft. | Ft. | | | | Ft. | Ft. | Ft. | | Ft. | |
| Washington | Hornitos | Incorporation | 3,000 | N. W. and S. E. | S. | 500 | 15 | Slate and sandstone. | Quartz and slate. | Two shafts | | | 650 | 5 | 500 | $10,000 |
| Hite | Hite's Cove | J. R. Hite and Chas. Maine. | 1,000 | N. W. and S. E. | 80° | 800 | 5 to 6 | Talcose slate | Gold-bearing quartz. | Tunnel | 1,450 | 825 | | 6 | 3,000 to 3,600. | |
| Theresa | Coulterville | Incorporation | 1,500 | S. E. | 40° | | 2 | do | | Incline | | 200 | 200 | 1 | | 10,000 |
| Bandereta | do | do | 1,500 | S. E. | 80° | | 2 | do | | Tunnel and incline. | 1,400 | 1,400 | 100 | 3 | 3,000 | |
| Martin & Walling. | do | do | 1,500 | N. and S. | 45° | | 4 | Slate | Slate and quartz. | Incline | | 150 | 150 | 2 | 600 | 10,000 |
| Hasloe | do | do | 1,500 | S. E. | 40° | 600 | 2 | Talcose slate | | do | | 400 | 450 | 8 | 2,000 | 12,000 |

*Operations of leading mines, Mariposa County, California, 1874.*

| Name. | Location. | Owners. | Number of miners employed. | Miners' wages per day. | Cost of sinking, per foot. | Cost of drifting, per foot. | Cost of stoping, per ton. | Cost of mining, per ton extracted. | Cost of milling, per ton. | Company or custom mill. | Number of tons extracted and worked. | Average yield per ton. | Percentage of sulphurets. | Total bullion-product. |
|---|---|---|---|---|---|---|---|---|---|---|---|---|---|---|
| Washington | Hornitas | Incorporation | 30 | $1.75 to $3.50 | $17 50 | $7 00 | $1 50 | $5 00 | $2 50 | Company | 1,700 | $24 27 | 8 | $41,259 00 |
| Hite | Hite's Cove | J. R. Hite and Chas. Maine. | 17 | 2.00 to 2.50 | $6 to $15 | $7 to $8 | 1 00 | 5 00 | 3 20 | Company | 4,000 | 40 00 | ½ to 1 | 160,000 00 |

*Statement of quartz-mills, Mariposa County, California, 1874.*

| Name. | Location. | Owners. | Kind of power, and amount. | Number of stamps. | Weight of stamps. | Number of drops per minute. | Height of drop. | Number of pans. | Number of concentrators. | Cost of mill. | Capacity per 24 hours. | Cost of treatment, per ton. | Tons crushed during the year. | Method of treating sulphurets. |
|---|---|---|---|---|---|---|---|---|---|---|---|---|---|---|
| | | | | | *Lbs.* | | | | | | *Tons.* | | | |
| Bandereta | Coulterville | Incorporation | Water, 25 horse-power | 10 | 450 | 60 | 6 inches | 2 | ...... | $7,000 | 7 | $3 50 | 1,200 | |
| Hasloe | do | do | Steam | 10 | ...... | 60 | 6 inches | ...... | ...... | 5,000 | 8 | 5 00 | 1,143 | |
| Hite's | Hite's Cove | Hite & Co | Water, 130 horse-power | 20 | 510 | 62 | 7 inches | 5 | 4 | 25,000 | 18 | 3 20 | 4,000 | Oxidation. |
| Washington | Hornitas | Incorporation | Steam, 70 horse-power | 20 | 750 | 70 | 8 inches | 1 | 2 | 20,000 | 25 | 7 50 | 1,700 | Chlorination. |

### TUOLUMNE COUNTY.

Tuolumne County is bounded on the north by Calaveras, and on the south by Mariposa, and extends from Stanislaus County on the west to the summit of the Sierra on the east. The Tuolumne River, rising in the Sierras, in the region of perpetual snow, flows through the central portion of this county. The Stanislaus River, which separates this county from Calaveras, has two branches, which rise in the higher Sierras, and have their whole length in this county, and these rivers, with the Tuolumne and its branches, all flow through deep and precipitous cañons, abounding in grand and picturesque scenery. This county, like Calaveras, has been noted for its rich placer-mines, and it still contains great undeveloped mineral resources.

One of the most striking and remarkable geological formations in the State is Table Mountain, which extends for nearly thirty miles, and is composed of masses of basaltic lava, with nearly vertical sides and flat on the top, the summit being at an elevation of about 2,000 feet above the Stanislaus River, near which it runs for a considerable distance. In this mountain there are rich deposits of gold-bearing gravel, some of which are still profitably worked. Some of the gulches and creeks of this county have been immensely rich. At present, however, there is but comparatively little placer-mining done. Considerable attention is paid to quartz-mining, and many rich lodes have been discovered; and when this interest is more fully developed, there is good reason to believe that Tuolumne will again take a prominent place as a producer of the precious metals.

The county at one time contained several towns of importance, but most of them give little evidence of their former prosperity. Where were once thriving and populous mining-camps, in some instances hardly a building is left to mark the spot, and even the soil has been washed away to secure the mineral wealth which it contained.

Sonora, the county-seat, has, however, been more fortunate, and it still remains a beautiful and prosperous mountain town, and a place of considerable business. Columbia, at one time its rival, and equal in population and wealth, has rapidly gone to decay. Fine brick buildings have been torn down, in order that the ground upon which they stood might be mined, and this practice has continued until a large portion of the town, with its public streets, building-sites, and fine gardens and ornamental grounds, has been washed away, leaving in their places the unsightly *débris* and uneven surface of the worked-out mining-claims.

Other towns once noted throughout the mining-districts of the State have shared a similar fate, and none in the county have wholly escaped.

Tuolumne County has 24 quartz-mills, crushing from 10,000 to 15,000 tons per annum, and four mining-ditches, aggregating one hundred miles in length, and supplying 2,500 miners' inches of water (about 70,000,000 gallons daily.)

In common with Mariposa on the south and Calaveras on the north, this county has two systems of quartz-ledges, one occurring in the granite in the higher portion of the county, the other in the slates. The latter is generally known as the mother lode. The representative mine of the county, in granite formation, has been the Confidence, owned by Jesse Halladay and others, and situated thirteen miles northwest of Sonora. The company's claim consists of 2,500 linear feet, of which only the northern portion has been explored. I have been permitted to make the following extracts from a report by Mr. Ashburner, M. E.:

The Confidence vein has a direction about northwest and southeast. It lies very flat, dipping to the east at an angle of 24°. The country-rock is granite. The main shaft is sunk on the dip of the vein 665 feet, or about 270 feet vertical depth.

Five levels have been opened. Nos. 1 and 2 have a length of about 400 feet each. Nos. 3 and 4 have been driven from 300 to 350 feet north of the shaft. No. 5, the lowest level, has been driven north 638 feet and south 60 feet. The shaft has been sunk 115 feet below this level.

The various drifts showed the following width of ledge: In No. 1 a large body of pay-ore was opened, the vein averaging 3 feet in thickness north of the shaft, and 8 feet in thickness south of the shaft. No. 2 developed a large body of rich quartz, averaging 7 feet in thickness. The shoot here had a length of 70 feet, and the yield was about $31 per ton. South of the shaft the vein was 13 feet thick in places, and the yield varied from $5 to $15 per ton.

The No. 3 level was run on a vein 12 feet wide; it developed a fine body of quartz, averaging 10 feet in thickness up to the No. 2 level for a length of 85 feet. The quartz between these two levels is said to have paid at the rate of $47.25 per ton. The end of this shoot is about 150 feet from the shaft where the quartz pinches out. At this point the drift bends to the west and continues in barren ground 29 feet to the foot-wall, which it follows in a northerly direction on an irregular bunchy body of quartz of good quality. The end of the drift is 300 feet from the shaft following its continually changing course, or about 250 feet in a straight line following the general direction of the vein. At this point the vein has pinched out; the foot-wall of granite is visible, but the upper or hanging wall is a hard barren rock, very different from the eastern country.

On the No. 4 level the first 200 feet of the drift passes through quartz of a lower grade than that above, paying about $7 per ton, and the vein varies in width from 3 to 18 feet. From this point there is a pinch for 50 feet; then there is a shoot of good ore, paying about $20 per ton for 90 feet; then for 35 feet the drift passes through barren matter to its end. This shoot of quartz was about 20 inches thick, and lies next the foot-wall; above it is the same barren rock which is found in the No. 3 level, and also, as will be stated hereafter, in the level below. Above this level all the quartz in the shoot has been removed; below work was continued by underhand stoping until the yield fell off to $5 and $8 per ton, which left little or no margin for profit, and the stopes were abandoned, in hopes of meeting with the shoot in the fifth level.

Level No. 5 yielded some rich rock, a small quantity milling as high as $75 per ton, but the vein was irregular, narrow, and uncertain where rich and of low grade where it was wide.

The work of developing this mine was commenced in January, 1868, and continued until 1873, when work was suspended, owing to the rich shoot of ore above described, which had paid largely in the upper levels, becoming poor in depth, and at that time it was not deemed advisable to incur the expense of prospecting new ground.

The first work done consisted in clearing out and opening the adit level which had been driven many years previously by the former owners. A shaft was sunk on the vein to a depth of 50 feet. A level was opened at this point and the vein drifted north 150 feet, and south the same distance. The present hoisting and pumping works were erected in September, 1868, and the work of prospecting was actively continued until the results warranted the erection of a mill, which was begun in June, 1869, and commenced running September 1 of the same year. This mill consisted at first of 20 stamps and 1 arrastra 9 foot in diameter. These were increased in March, 1870, to 30 stamps, and 2 more arrastras were added in June of the same year. Subsequently, during the summer of 1871, 10 more stamps were added, making the mill, as it now stands, to consist of 40 stamps and 3 arrastras, with a crushing and amalgamating capacity of from 40 to 50 tons daily, depending upon the fineness of the screens employed. There is also a large-size rock-breaker, of a capacity of 50 tons during twelve hours running. This machinery is driven by a 60-horse-power engine. At the mouth of the shaft there is an engine of about 50 horse-power, for driving the pumps and hoisting-works. The quartz, after being hoisted to the surface, is passed over a short tramway and dumped upon the mill-floor, so that it requires no superfluous handling.

Between September 1, 1869, when the mill began crushing, to October 25, 1873, when work was suspended, about 32,000 tons of quartz were treated. This quartz varied in yield during periods of several months from $9 or $10 per ton to $33. A large quantity was treated which yielded still higher, and several runs made which paid nearly or quite $50 per ton. The gross amount of bullion produced was $522,222.52, which would give for the average $16.32 per ton. The detailed statement of the receipts is as follows, the figures having been taken from the books of the company:

*Detailed statement of the receipts.*

| Date. | Quartz worked. | Yield. | Yield per ton. |
|---|---|---|---|
| | *Tons.* | | |
| From September 1, 1869, to June 1, 1870 | Uncertain. | $136,904 90 | ........ |
| From June 1, 1870, to June 28, 1871 | 6,856 | 107,314 14 | $15 65 |
| From January 28, 1871, to June 6, 1871 | 3,439 | 113,854 95 | 33 10 |
| From June 6, 1871, to March 26, 1873 | 14,843 | 140,172 46 | 9 44 |
| | 25,138 | 498,246 45 | |
| From arrastras, between June 1, 1870, and March 26, 1873 | | 19,238 08 | |
| | | 517,484 53 | |
| From March 26 to October 25, 1873 | | 4,737 99 | |
| Total | | 522,222 52 | |

The actual cost of mining and milling the quartz while the mill was in operation was $7.64 per ton. The total expenses of all kinds were, from January, 1868, when work first commenced, to October 25, 1873, $356,187.69. These expenses include the cost of the property originally, the cost of the mill and hoisting-works, the work of opening the mine, as well as all other expenses.

EXPENSES.

| | |
|---|---|
| General expense account | $110,043 45 |
| Mill and machinery account | 32,320 14 |
| Fuel | 24,417 62 |
| Sundries | 8,789 64 |
| Mill-labor | 40,971 58 |
| Mine-labor | 139,645 26 |
| Total | 356,187 69 |

The general expense and mill and machinery accounts, aggregating $142,363.59, comprise the cost of mine and mill supplies, salaries of superintendent and book-keeper, wages of several surface-men, express-charges, keeping of horses, and all other incidental expenses, such as surveying and taxes. Also included in these accounts are several items not properly chargeable to the working of the property. These are principally the original cost of the property, the cost of the mill and hoisting-works, as well as of all other buildings connected with the plant. All these expenses could not have amounted to less than $60,000 or $70,000. This sum, or say $65,000, should therefore be deducted from the expenses given above, in order to arrive at the profits realized exclusive of original cost, &c. We should have then:

| | | |
|---|---|---|
| Receipts from bullion as given above | | $522,222 52 |
| Total expenses as given above | $356,187 69 | |
| Deduct | 65,000 00 | |
| | | 291,187 69 |
| Profit | | 231,034 83 |

Mr. Ashburner further says:

Of the 2,500 feet of ground embraced by this property, only about 300 have been really proved, most of which with exceedingly satisfactory results. No work to speak of has been done upon the Independence, and none at all upon the Jesse. Although the fifth level, which is the longest, did not turn out favorably, I do not yet consider the work performed there as demonstrating definitely the valuelessness of that portion of the mine.

Bearing in mind what precedes, namely, that this drift runs next the foot-wall, the occurrence of the irregular masses of quartz found there presents no new features different from what I am informed were met with in the upper works in what is known as the foot-wall vein, for, accepting the conclusion that there are two distinct veins, it appears clear that in many portions of the mine what now remains above the vein, and is called the hanging wall, is not the real hanging wall, and should not be so regarded until after further search and drifts have been run toward the east into the country-granite. This hanging wall, so called, is a fine-grained homogeneous rock, greasy to the touch, and probably containing magnesia, which very likely forms, in connection with the quartz, the filling of the fissure in the granite, for in many places it appears intruded into the mass of the vein, leaving quartz on both sides. The granite of both the eastern and western country is generally similar in composition, though perhaps that on the under or foot side of the vein may contain hornblende, which renders it darker and harder, while that of the upper or hanging wall is so far much softer and pervious to water. In the fourth level, where I had an opportunity of examining this granite, I found it very soft, even friable, and much stained by oxide of iron. All the water in the mine is said to come from and through the eastern country, showing this rock to be generally permeable, while the filling of the fissure, being more compact, is not.

Drift-mining on the old channels underlying the basaltic capping of Table Mountain still continues, and, in the main, with excellent results. This class of mining has been fully described, under heading of Sierra County, in present report. The Alpha Company, near Jamestown, has demonstrated the existence of two channels under the mountain; these are known as the "front" and "back" channels, and are separated by a high rim. That on the east side is the most ancient, and is known as the Caldwell or Saratoga Channel, and carries heavy, black, coarse gold, wherever it has been struck in Table Mountain. The west channel is of later formation, and carries finer and brighter gold; no black gold is found in this, and no pipe-clay, as is the case in the other. Miners are positive there are two distinct channels, and that both have been found at the head of the mountain, at the Buckeye, and also at the Humbug or Blue Gravel claims, several miles apart, would seem to prove the fact.

*Description of leading mines of Tuolumne County, California*, 1874.

*Eighteen Miles Southeast of Sonora.*—Name of mine, Buchanan; owners, Tulloch & Gashwiler; length of location, 1,650 feet; course, southwest and northeast; dip, 70° east; length of pay-zone, 200 feet; average width, 6 feet; country-rock, slate; character of vein-matter, sulphurets and free gold; worked by tunnel, 400 feet on vein; depth on vein in tunnel, 130 feet; depth of winze below tunnel, 100 feet; 1 level opened below tunnel; total length of working-drifts, 100 feet.

Reported by J. Tulloch.

*Operations of leading mines of Tuolumne County, California*, 1874.

*Eighteen Miles Southeast of Sonora.*—Name of mine, Buchanan; owners, Tulloch & Gashwiler; number of miners employed, 6; miners' wages per day, \$3; cost of mining per ton extracted, \$4; cost of milling per ton, \$2; company works its own mill; number of tons extracted and worked, 300 tons per month; average yield per ton, \$20; percentage sulphurets, 2 per cent.

Reported by J. Tulloch.

*Statement of quartz-mills, Tuolumne County, California*, 1874.

| Name. | Location. | Owners. | Kind of power. | Stamps. | Weight of stamps. |
|---|---|---|---|---|---|
| | | | | *No.* | *Lbs.* |
| Buchanan | S. E. Sonora | Tulloch & Gashwiler | Steam | 8 | 600 |
| Spring Gulch | | Spring Gulch Mining Company. | Water | 10 | 650 |

*Statement of quartz-mills, Tuolumne County, California*, 1874—Continued.

| Name. | Drops per minute. | Height of drop. | Pans. | Concentrators. | Cost of mill. | Capacity per twenty-four hours. | Cost of treatment per ton. | Crushed during year. | Method of treating sulphurets. |
|---|---|---|---|---|---|---|---|---|---|
| | *No.* | *Inch.* | *No.* | *No.* | *Dolls.* | *Tons.* | *Dolls.* | *Tons.* | |
| Buchanan | 80 | 6 | ...... | ...... | 10,000 | 10 | 2 | *1,600 | Pan and arrastra. |
| Spring Gulch | 75 | 8 | ...... | ...... | 7,500 | 10 | 1 | 2,880 | Not treated. |

* Started in September.

## CALAVERAS COUNTY.

Calaveras County derives its name from the Calaveras River, which runs through its central portion. It is bounded by Amador on the north—and separated from that county by the Mokelumne River—and by Tuolumne County on the south, its boundary on that side being the Stanislaus River. It has an average width of about twenty miles, and extends, from the edge of the San Joaquin Valley on the west to the summit of the Sierra Nevadas on the east, a distance of about fifty miles. In everything that relates to its topography, soil, climate, and natural productions, it is very similar in character to Amador County. Bear Mountain, a rocky wooded range, about 2,000 feet high, runs across the county from north to south, dividing it into two sections, the western or lower section being composed of foot-hills and low, rolling-prairie lands, while the eastern or upper section is more rugged and broken. West of the Bear Mountain range are extensive copper-veins, some valuable quartz-lodes, mines of iron and manganese, and quarries of slate of an excellent quality. The hills of this portion of the county are sparsely covered with oak trees, but it has no timber suitable for other purposes than for fuel. This region is also naturally arid. That portion of the county east of the Bear Mountain range has contained some of the richest placer-diggings ever discovered in the State, and this kind of mining is still extensively carried on in several localities, hydraulic washing having taken the place of the original and primitive ways of prosecuting this work. There are also many very valuable quartz-mines, and considerable attention is now being paid to the development of this important interest. There are inexhaustible quarries of marble and limestone, and also valuable deposits of brown hemitite iron-ore in close proximity thereto. In the more elevated portions of the county are extensive pine, cedar, and redwood forests. In this timber-belt are situated the celebrated groves of the *Sequoia gigantea*, called, before their genus was determined, *Washingtonia* by patriotic Americans, *Wellingtonia* by English tourists, and known to mankind in general as the Big Trees of California.

The most important towns of this county are San Andreas and Mokelumne Hill. San Andreas, the county-seat, is located near the center of the county, in what was formerly a rich mineral district. There is still some mining done in this vicinity, both in placers and in quartz, but the business of the town is much less than formerly. Mokelumne Hill is situated in the northern part of the county, and was at one time one of the most prosperous in the southern mines of California. In this vicinity were some fabulously-rich placer-diggings, and there are still hydraulic claims that are being worked with profit. There are also several valuable quartz-mines in this district. This town is very romantically located, in a rugged and broken region, adjacent to the Mokelumne River, and notwithstanding the decline in business resulting from the exhaustion of the placer-mines, it is still one of the most beautiful towns in the mountain counties of California.

Copperopolis, located on the copper-belt, which passes through the western portion of the county, was at one time one of the most important towns in the county, being supported by the copper-mines that during several years were extensively and profitably worked at this place. These mines are not exhausted, but the great cost of transporting the ores to the seaboard makes it unprofitable to work them at present. Murphy's, situated in the southern part of the county, was one of the early mining-camps of this region, and in its vicinity were formerly extensive and rich placer-diggings, which have, however, been mostly exhausted. This town is located on a table-land that extends for several miles at an elevation of some 2,000 feet above the Stanislaus River. On this table-land are also situated the former prosperous mining-towns of Vallecito and Douglass Camp. This district was formerly very populous, but, like all the placer-mining regions, has declined, and these towns retain few signs of their former prosperity. In the vicinity of Vallecito hydraulic mining is still being extensively carried on, and there is yet considerable ground that can be profitably worked by this process.

There are in the limits of the county twenty-two mining-ditches, aggregating four hundred and ninety miles in length. The amount of water used per day will approximate 4,000 inches. The county contains thirty-four quartz-mills, crushing from 30,000 to 40,000 tons annually.

Calaveras County contains several hundred quartz-mines, but not more than ten or fifteen have been worked with profit during the past year. The prominent districts are West Point, Sheep Ranch, and San Bruno. The large quartz-vein or series of veins known as the Mother lode traverses the county from north to south, but, with the exception of the Paloma or Gwin mine, situated near Mokelumne Hill, and which has attained a depth of 1,000 feet, there are no marked successes to record in working this belt in Calaveras County. Several mines on the mother lode, at Angel's Camp, have temporarily suspended operations.

Quartz-mining is actively carried on at West Point and vicinity, where the Sanderson mine has a promising 2-foot ledge developed to a depth of 400 feet. The Zacatecas and Prussian Hill mines, in the same disirict, have developed strong veins. The formation here is granite.

The Sheep Ranch mine, owned by Wallace & Ferguson, has produced during the fiscal year ending October 31, 1874, the sum of $61,000. The expense of working the mine and milling the ore ranged from $800 to $1,200 per month during that period. Taking $1,000 per month as the average expense, the net profit of the mine for the year was $49,000. The ledge is not worked upon an extensive scale, it being only necessary to take out rock fast enough to supply a 5-stamp mill, that being the capacity of the battery that has done all the crushing.

The Glencoe district is situated ten miles northeast from Mokelumne Hill, on the east of the south fork of the Mokelumne River, south of West Point district. The district has numerous veins of gold-bearing quartz, mostly running in an easterly and westerly course, in a slate formation. Facilities for mining are here excellent, there being an abundance of wood and water, good roads, and a genial climate.

The Good Hope mine, which has the deepest shaft in the district, was formerly worked by Mexicans, with the aid of arrastras. In 1872 it was purchased by the present owner, C. J. Garland, who erected substantial steam hoisting-works, and sunk a shaft 270 feet. The former owners had run a tunnel 300 feet, which struck the vein at a depth of 130 feet, where the ledge had an average width of 3 feet. The pay-shoot was found to be 125 feet in length, and the ore averaged $18 per ton. From the surface to 40 feet below the tunnel-level the vein stood nearly vertical, inclining somewhat to the south. Forty feet below the tunnel the vein was broken and inclined north. At 260 feet the vein came in regular, with a slight dip to south; fissure from 3 to 4 feet; pay-vein 1 foot, and widening as sinking progressed. Work was suspended in mine in December, 1873. In connection with the mine is a substantial 18-stamp steam-mill, furnished with four of Cochrane's automaton self-feeders—an excellent machine, feeding all kinds of ores, wet or dry, with great regularity, adding much to capacity of stamps and saving materially in labor, wear, and breakage of shoes and dies. Since work was suspended in the Good Hope mine the mill has been employed part of the time in crushing ore from other mines in the district.

The San Bruno mine is situated on the same vein east of the Good Hope. It has been worked through a tunnel to the depth of about 250 feet. They have several shoots of ore, though the present company have taken ore only from one. The ore is of high grade, milling from $25 to $60 per ton. The San Bruno mine is owned by Charles Honchner, William Sigler, and Keys. The company is at present putting steam hoisting-works in the tunnel for the purpose of working the men at greater depth.

The San Bruno, like the Good Hope and other prominent mines of the district, is an east and west location, in slate formation. It was worked by Mexicans before it came into the possession of the present owners, but no great depth was attained by them—their lowest workings being 150 feet. The principal chimney or ore-shoot has a length of 150 feet, and varies in thickness from 1 to 3 feet. The mine is opened by a series of tunnels run on the ledge and a connecting engine-shaft. About 422 tons of ore have been crushed during the present year, yielding $13,700.

The Monte Christo mine, owned by Albright, Rochenbach & Co., is situated south of the San Bruno. Average width of vein, 2 feet; average pay to the ton, $30. The pay-shoot, which at the surface was only 6 feet in length, increased to 52 feet at the 125-foot level.

The Barney mine, owned by Skinner & Co., is situated just east of Mosquito Gulch, a few rods above Good Hope mill. Vein worked by tunnel to a depth of about 80 feet. Vein from 1 to 4 feet. First pay-shoot 50 feet in length. On high grade, paying from $20 to $60 per ton. Mine at present leased to Abbus & Co., who have extended tunnel and struck another shoot of ore that prospects well.

The Bismarck mine is owned by Michler & Trask. Two shoots of ore, each 20 feet long. Mine worked through shaft to depth of 90 feet. First shoot of ore 15 feet from shaft; barren between shoots, which are 25 feet apart. Width of vein from 10 to 16 inches. On high grade, first shoot milling from $100 to $125 per ton, second shoot from $60 to $80. Steam hoisting-works small, and inadequate for the further working of mine.

The Mosquito mine, formerly Valentine, once the most famous mine in the county, is now owned by Bancroft, Knight & Co. First worked by Mexicans, some twenty years ago, who ran 100 mule-arrastras for a long time, supporting a large number of miners. A 15-stamp mill was erected on this mine and run with success for a time, but the pay-shoot was lost and operations suspended until the property passed into the hands of the present owners, who have run a tunnel which strikes the ledge at a depth of 240 feet. This is a large vein, varying in width from 8 to 40 feet, with a chimney exposed on surface to a length of 200 feet. It is reasonable to suppose that energetic prospecting of the ground on the tunnel-level would discover the lost shoot of ore so clearly defined above.

The Woodcock, owned by Fairchild, Wellets & Smith, is worked by a tunnel 200 feet below surface. The vein is from 1 to 6 feet in width, and the ore at present of low grade. There are many other mines in the district which have not been worked this year.

At Mokelumne Hill several hydraulic claims are being successfully worked, and at Central Hill or vicinity some very rich gravel-mines have recently been struck, that are being worked by tunnels and shafts, and paying largely. The ground is all secured under the United States mining-law, and is owned by few individuals, in lots from 40 to 160 acres, thus excluding the many from any advantages of the discovery, except to work for daily wages.

A company has been formed in San Francisco for the purpose of working a group of mines on Carson Hill, in this county, forming a portion of the Mother lode. These mines were worked several years since, and noted for their yield of ores of telluride of gold, but, the ore proving refractory by the ordinary mill process, the project was abandoned in 1865.

The situation of the mines is advantageous; Carson Hill, on the summit, is 2,400 feet above the sea-level. The Stanislaus River, at Robinson's ferry, 800 feet; a difference of 1,600. By running a tunnel from near the river lengthwise on the Stanislaus vein, the work of ore-extraction will be easy, and it is expected that the ore extracted in running this tunnel will pay the expenses. To strike the lower works of the Stanislaus, the tunnel would have to run about 650 feet, and would then drain the mines. The declivity of the hill being sharp, the veins will be reached on the eastern side by new tunnels at a depth of 500 or 600 feet, and reduction-works erected near by. Three mill-sites, with the old Coyote mill, belong to the property. The Stanislaus lode forms the middle branch of the Mother lode, of which the extreme end on the summit of Carson Hill is the famous Carson mine, only the Melones Point Rock, of 1,275 feet in extent, being between the Mineral Mountain and the Morgan. This last mine produced in 1850-'51, by arrastras and hand-mortars, over one million and a half dollars, as reported by Ross Browne in 1868. In the Stanislaus lode the rich deposits were found in chimneys and in the small quartz-feeders which follow the line of slate formation, and at the junction of these with the principal lead. The compact quartz varies from 4 to 6 feet in thickness, and contains rich sulphurets. The free gold which was found on the surface was soon replaced by tellurets of gold and silver and auriferous iron-pyrites. The ores are known to be exceedingly rich, but only from 5 to 10 per cent. were saved by the common mill process. Mr. Küstel recommends that the ore be crushed, concentrated, and roasted in furnaces. The expense of treatment is estimated at from $12 to $15 per ton. The ore is exceedingly rich, a sample of 50 bags, weighing 5,710 pounds, having yielded by chemical treatment $3,555.94.

The Garibaldi mine is located two miles below Robinson's Ferry, on the Stanislaus River, and is owned by L. J. Lewis and the estate of J. P. Rogers, M. D., of San Francisco. They have employed from ten to fifteen white men, with some Chinese. Wages range from $3 down to

$1.50 per day. The work done on the mine is one shaft, 47 feet, which is to connect with a drift in ore for ventilation. A tunnel 152 feet is run, which crosscuts the ledge where the vein of pay-ore is 13 feet wide. It contains "ribbon-rock" of the same character as that in the Amador mine. There is a drift on the vein 70 feet long, showing a large body of milling-ore lying between slate and greenstone. Openings on the surface have been made in several places to trace the ledge, but all permanent work will be done through the tunnel and by levels run on the vein on the north end of the mine, where backs can be had over 600 feet high, between the lower level (which is above water-line) and the top of the hill. The vein dips at 45°, and runs north 47° west. The owners claim 3,000 feet in length. There are 160 acres of gravel-land lying between the vein and the river, which will pay well when a ditch, two miles long, is built to receive the 5,000 inches of water that can be had free, even in the dryest season. This will be sufficient for running 200 stamps. Cost of mining and milling the ore is not expected to exceed $2.50 per ton. Average yield per ton, about $10, though ore has been milled yielding as high as $28 per ton. There is 2 per cent. of sulphurets in the ore, which, after concentration, are worth about $60 per ton. For the foregoing figures I am indebted to Mr. L. J. Lewis, one of the proprietors.

*Description of leading mines, Calaveras County, California, 1874.*

| Name. | Location. | Owners. | Length of location. | Course. | Dip. | Length pay-zone. | Average width. | Country-rock. | Character of vein-matter. | Worked by tunnel or shaft. | Length of tunnel. | Depth on vein in tunnel. | Depth of working-shaft. | No. levels opened. | Total length of drifts. | Cost of hoisting-works. |
|---|---|---|---|---|---|---|---|---|---|---|---|---|---|---|---|---|
| | | | Ft. | | | Ft. | Ft. | | | | Ft. | Ft. | Ft. | | Ft. | |
| Sanderson Gold | Railroad Flat | Sanderson Gold-Mining Company. | 1,500 | N. and S. | 65° | .... | 1½ | .................. | Gold-bearing | Shaft and drift. | .... | .... | 420 | 4 | 500 | $10,000 |
| Union | Lower Calaveritas | John Rathgeb | 1,400 | N. W. and S. E. | E. | 300 | 4 | East, granite; west, slate. | Gold-bearing quartz. | Shafts | .... | .... | 170 | 2 | 75 | 1,500 |
| San Bruno | Glencoe | Siegler, Hoerchner & Co. | 1,000 | E. and W. | N. | 160 | 2 | Slate, granite | Quartz | Tunnel | 560 | 340 | .... | 2 | .... | ...... |
| Good Hope | ....do | C. J. Garland | 1,200 | E. and W. | S. | 125 | 2 | Talcose slate | Quartz, clay on walls. | Shaft and tunnel. | 400 | 130 | 270 | 2 | 400 | 6,000 |
| Valentine | West Point | Henry & Son | 1,500 | N. and S. | W. | 300 | .... | Granite | Quartz | Shaft | .... | 120 | 120 | .. | 100 | ...... |
| Thorpe | Forman's | M. Thorpe & Sons | 1,000 | N. W. and S. E. | 38° | .... | 6 | .................. | Sulphuret, talcose slate, and quartz. | ....do | .... | .... | 75 | 1 | 200 | 800 |

*Operations of leading mines, Calaveras County, California, 1874.*

| Name. | Location. | Owners. | Number of miners employed. | Miners' wages per day. | Cost of sinking per foot. | Cost of drifting per foot. | Cost of stoping per ton. | Cost of mining per ton extracted. | Cost of milling per ton. | Company or custom mill. | Number of tons extracted and worked. | Average yield per ton. | Percentage of sulphurets. | Total bullion-product. |
|---|---|---|---|---|---|---|---|---|---|---|---|---|---|---|
| Sanderson Gold | Railroad Flat | Sanderson Gold-Mining Company | 12 | $3 50 | $20 00 | $15 00 | ........ | ........ | ........ | Custom. | 100 | $50 00 | ........ | $5,000 00 |
| Union | Lower Calaveritas | John Rathgeb | 4 | 2 50 | 15 00 | 10 00 | ........ | $3 00 | $2 25 | Company | 300 | 15 00 | None. | ........ |
| San Bruno | Glencoe | Siegler, Hoerchner & Co | 7 | 3 00 | ...... | 5 00 | $2 00 | 4 00 | 3 75 | Custom. | 700 | 28 00 | ........ | 17,600 00 |
| Maddra | West Point | William Henry & Son | 4 | 3 00 | 8 00 | 3 00 | 2 50 | ........ | 5 00 | Custom. | 100 | 30 00 | 10 | 3,000 00 |
| Thorpe | Forman's | M. Thorpe & Sons | 6 | { 2 50<br>3 00 } | 10 00 | 8 00 | 3 50 | 3 87 | 2 00 | Company | 400 | 7 00 | 8 or 10 | 2,815 25 |

*Statement of quartz-mills, Calaveras County, California, 1874.*

| Name. | Location. | Owners. | Kind of power and amount. | Number of stamps. | Weight of stamps. | Number of drops per minute. | Height of drop. | Number of pans. | Number of concentrators. | Cost of mill. | Capacity per 24 hours. | Cost of treatment per ton. | Tons crushed during the year. | Method of treating sulphurets. |
|---|---|---|---|---|---|---|---|---|---|---|---|---|---|---|
| | | | | | *Lbs.* | | *In.* | | | | *Tons.* | | | |
| Rathgeb | Lower Calaveritas | John Rathgeb | Water | 10 | 700 | 65 | 6 | 3 | ...... | $5,000 | 15 | $2 25 | 300 | None. |
| Good Hope | Glencoe | C. J. Garland | Steam, 30 horse-power | 18 | 650 | 80 | 9 | ...... | ...... | 15,000 | 36 | 1 25 | 2,150 | None saved. |
| Do | ...do | Garland & Denison | Steam, 60 horse-power | 18 | 750 | 75 | 8 | ...... | ...... | 18,600 | 22 | ...... | ...... | |
| Thorpe | Forman's | M. Thorpe & Sons | Water, 6 horse-power | 5 | 550 | 72 | 7 | ...... | ...... | 4,000 | 5 | 2 00 | 400 | |

## AMADOR COUNTY.

This county is situated east of the San Joaquin Valley, and is the northernmost of the tier of mountain counties on the eastern side of the San Joaquin basin. It extends from the edge of the valley proper to the summit of the Sierra range, and has a total area of about six hundred square miles. In its western portion there is considerable good arable and grazing land. Jackson and Ione Valleys are particularly fertile. Amador, however, excels in its mineral wealth, and its rich quartz and placer gold-mines, veins of copper, quarries of marble and slate, and extensive deposits of coal make it one of the richest counties in California. The leading interest at the present time is quartz-mining, some of the most valuable lodes on the Pacific coast being located here. Some of these mines have been profitably worked for twenty years and are still valuable. There are still valuable placer-mines which can be profitably worked since the completion of a large mining-canal, recently constructed to lead the waters of the Mokelumne River into the mining-districts. Several rich veins of copper-ore have been discovered; and one has been profitably worked for several years.

One of the prospective sources of wealth possessed by this county is a deposit of lignite coal, which is believed to extend through the western portion of the county. The deposit has been found in different localities, and in places it is very extensive. Where excavations have been made to test its existence and extent, the vein has been found ranging from 4 to 10 feet in thickness, and in places it is fully one-half mile in width. From the developments already made, it is evident that a large supply of cheap fuel will be easily obtained. This coal is well adapted for manufacturing purposes, and has been successfully used for several years at a steam flour-mill at Ione City, and more recently at the quartz-mills of Sutter Creek.

The principal towns of Amador are Jackson, Sutter Creek, Volcano, and Ione City. Jackson, the county-seat, was during the era of placer-mining a place of considerable importance. It has a history similar to most of the mining-towns of California, having passed through eras of prosperity and adversity. The exhaustion of the placer-mines, and the consequent withdrawal from the locality of the large population that had been supported by that industry, seriously damaged the business-prospects of the town; but the development of the rich quartz-mines in the vicinity and the horticultural and agricultural resources of the surrounding country will eventually cause this place to recover something of its former prosperity. In the immediate vicinity of Jackson are many valuable quartz-mines that are profitably worked. Sutter Creek is a town of considerable importance, principally supported by the rich and extensive quartz-mines in the vicinity. Here are some of the finest and most complete quartz mills in California, and the annual yield of the precious metal is constantly increasing. Volcano is one of the once prosperous mining-towns, in the center of a valuable mineral district, and there is still considerable mining done in that vicinity.

There are nine mining-ditches in the county, aggregating one hundred and forty-eight miles in length, and supplying 12,000 miners' inches. The county has sixteen quartz-mills, crushing from 80,000 to 100,000 tons annually. From 6,000 to 8,000 tons of coal are mined annually. The coal-product will be largely increased on the completion of the narrow-gauge road from Stockton to Ione Valley.

The Amador Canal is forty-six miles long, 8 feet wide on top, 5 feet on bottom, and 4 feet deep. It receives permanent water from perennial lakes among the snowy heights of the Sierra Nevada. One of its leading purposes is to water the broad sweep of undulating lands, which, for want of irrigation, have not attracted settlement. This important work was fully described in the report of 1874.

The copper-ores of Amador are worthy of notice. The richest copper found in this State is in the section of the great copper-belt that runs north and south through the whole length of Amador and Calaveras. Developments, stimulated during the term of high prices, warrant this assumption. The Newton copper-mine was opened to a depth of 280 feet. The vein is remarkable for regularity. It is 6 feet wide, and it averages 16 to 20 per cent. of metal. Two feet of the ore is rich enough to pay for wagoning forty miles to Stockton, and thence shipping by river to tide-water.

The Amador Consolidated continues to be one of the leading mines, though the operations of 1874 were not brilliant in results, as may be seen by the following extracts from the report of Mr. John A. Steinberger, the superintendent, which, however, covers two months more than a year, namely, the period of fourteen months ending April 1, 1875:*

During this period, 22,098 tons of quartz have been extracted from mine and milled, the average yield of which is shown in the secretary's report. Nineteen hundred and seventy-three feet of drift has been driven during the past fourteen months, to wit:

| | Feet. |
|---|---|
| 800-foot level | 432 |
| Panama level | 250 |
| Colton level | 130 |
| London level—north, 310; south, 265 | 575 |
| Lower level, (not named) | 100 |
| Badger shaft, (500-foot level,) north 150; south, 236 | 386 |
| Badger shaft, (400-foot level,) north | 100 |

The face of 800-foot level is now 460 feet north of North shaft. North 150 feet from shaft the rock gave out; or, in other words, we reached the north end of chute of rock. We then drove through barren ground 150 feet, following the hanging-wall, (which was very regular and solid,) when we struck rock varying in thickness from 1 to 3 feet. This body of rock lasted in length 30 feet. We then drove through barren ground some 30 feet and again struck rock averaging in thickness 2¼ feet and in length 20 feet.

The face of this drift now shows some quartz, mixed with slate, lying next the hanging-wall. I should judge from present indications that we are near another body of rock. We are still pushing this drift ahead. On this level, both north and south of shaft, we have opened out a large body of bowlder-rock, which we have been and are still working, though in quality it is low-grade rock.

We have opened a new drift under the old Panama level (north) and driven it 250 feet to bowlder-vein, the old Panama drift being in such bad condition that we deem it much cheaper to open a new level than it would be to retimber the old one.

The vein in face of this drift is 14 feet in thickness, and in quality about seven-dollar rock.

No work has been done during the past fourteen months on Green, De Lask, or Latham levels, on the bowlder-vein north of shaft. The rock being low-grade, it was deemed best for the present to leave the rock in place. The Colton level, south of shaft, has been stoped out. North of shaft there is still one stope which we are working.

The north drift has not been driven ahead to strike the bowlder-vein owing to the poorness of the bowlder-rock in the levels above. This level when started looked very promising, but the good rock continued but a few feet south of shaft.

The average of rock both north and south of shaft was poor on this level.

A cross-cut opposite the shaft on London level was driven 100 feet to the east; the ground passed through was slate and granite, alternate, and most of it very hard. Nothing was found in this cross-cut.

The London level (100 feet below the Colton level) has been a decided improvement (south of shaft) over the Colton level, the rock being better and vein larger. We are now stoping out the rock both north and south of shaft on this level.

The face of south drift is now 265 feet from shaft. The first 60 feet driven from shaft the vein was much broken and mixed with slate. At this point the vein became solid and so continued for some 100 feet, when it again showed confusion, being mixed with

* This document was received in April, several weeks after the Commissioner's report was sent to Congress. It was found practicable to introduce it before the sheets went to press.—R. W. R.

slate and granite, and has so continued to near its present face, when the south end of chute was reached.

The north drift on this level has been driven 310 feet. The first 150 feet driven the ground was very much confused and badly broken, showing no rock. At this point the rock came in next the hanging-wall, and has continued, varying in size from 1 to 10 feet, to within 20 feet from present face of drift—the last 20 feet showing no rock. The quality of rock in this drift is not near so good as that in the south drift.

The North shaft has been sunk (half size) below the London level 220 feet. The first 100 feet we had the vein varying in size from 2 to 5 feet in thickness.

The last 120 feet of sinking-ground showed much confusion, hanging-wall being much broken and confused and vein small and much mixed with slate and granite. Two hundred feet below the London level we started a drift south; the first 15 feet we passed through the same character of ground shown in shaft. We then struck the vein, 6 feet in thickness, though having no regular hanging-wall. This rock continued 65 feet, when we struck an offset in vein of 5 feet. To the south of this offset the vein is much mixed. At the offset there is a seam of clay running east at a right angle with course of vein of a foot in thickness. For the present, we have stopped the driving of drift, and are following this seam; we are now in 60 feet. The ground passed through is more like foot than hanging wall; it is soft slate, much mixed with quartz. The true hanging-wall must be to the east, and I am in hopes of finding rock when we reach the true wall. It is very evident from the change on this level over those above that the vein at this offset has been thrown to the east. If so, we will strike it in this cross-cut.

We have as yet started no drift north on this level. This level is 1,885 feet below the surface.

We have driven a prospecting-drift south from Badger shaft (500 feet from surface) 236 feet; struck several small bodies of rock, but nothing lasting.

Also, have driven north 400 and 500 feet from surface into the old works, and found a large body of bowlder-rock, which had been left standing, upon which we have been working for many months past, and will last for many months to come.

In my last annual report I spoke of the condition of North shaft under the Green level as being in very bad condition, and that retimbering would soon be necessary. We have managed to keep that part of shaft in running condition by patching it up from time to time, being most anxious to prospect the bottom of mine.

As soon as we thoroughly develop our bottom level we should at once commence retimbering that portion of shaft under Green level.

During the past year we have done considerable repairing on Middle and Badger shafts, both of which are now in very fair running order.

From Colton level to bottom level (being 300 feet) the rock is in place, excepting five stopes having been partially worked on London level.

The Amador Canal, for several years in course of construction, was finished and water turned into it last November.

As yet they have not been able to give regular supplies of water, owing to a number of breaks in canal from time to time. This will be a source of annoyance for some time, or until the canal becomes well puddled.

We are now making preparations to use water from this canal. The wheel is being built and pipe being made. Within the next six weeks we will be running the mill by water.

From the secretary's report for the same period I extract the following statement:

RECEIPTS.

| | | |
|---|---|---|
| Cash on hand with John A. Steinberger February 2, 1874 | $1,225 01 | |
| Cash on hand February 2, 1874 | 2,939 55 | |
| Total | | $4,164 56 |
| Bullion-account: | | |
| Proceeds of 22,098 tons of ore, (average per ton, $10.84 $\frac{793}{1000}$) | | 239,717 35 |
| Sulphuret-account: | | |
| Proceeds of 219½ tons of sulphurets, (average per ton, $92.27 $\frac{344}{1000}$) | | 20,254 02 |
| Premium-account: | | |
| Premium received on bullion | | 1,410 63 |
| Mine-account: | | |
| Sale of old machinery, iron, rope, lamps, &c | | 4,283 37 |
| Real-estate account: | | |
| Rent of house at Sutter Creek | | 185 00 |
| Total | | 270,014 93 |

DISBURSEMENTS.

| Item | Amount | Subtotal | Total |
|---|---|---|---|
| **Mine-account:** | | | |
| Drifting and stoping, labor, as per roll | $63,398 90 | | |
| Sinking North shaft, labor, as per roll | 4,542 11 | | |
| Sinking air-way, labor, as per roll | 250 00 | | |
| Badger shaft and general repairs, labor, as per roll | 9,604 01 | | |
| Surface-men, labor, as per roll | 32,566 65 | | |
| Sundry labor, as per roll | 25 00 | | |
| Total labor | | $110,386 67 | |
| **Supplies:** | | | |
| 3,603½ cords wood | $21,651 12 | | |
| Coal | 26 76 | | |
| Charcoal | 948 67 | | |
| Lumber | 1,656 34 | | |
| Timbers | 10,958 80 | | |
| Spiling | 2,761 95 | | |
| Hardware | 5,141 90 | | |
| Steam-pump | 518 50 | | |
| Wire-rope | 4,221 36 | | |
| Rope | 9,582 62 | | |
| Powder and fuse | 5,187 86 | | |
| Oil | 1,539 75 | | |
| Candles | 1,860 25 | | |
| Tar | 67 50 | | |
| Soap | 69 00 | | |
| Canvas | 25 50 | | |
| Oil-cloth | 12 25 | | |
| Water | 250 00 | | |
| Sundries | 75 65 | | |
| Total supplies | | 66,555 78 | |
| Total disbursements at mine | | | $176,942 45 |
| **Eureka Mill account:** | | | |
| Labor, as per rolls | | $17,002 72 | |
| **Supplies:** | | | |
| Hardware | $542 46 | | |
| Castings | 6,011 17 | | |
| Quicksilver | 1,466 40 | | |
| Water | 480 00 | | |
| Coal | 3,049 42 | | |
| 112½ cords wood | 618 75 | | |
| Oil | 735 05 | | |
| Candles | 527 25 | | |
| Tar | 30 00 | | |
| Soap | 6 00 | | |
| Lumber | 244 51 | | |
| Timber | 13 75 | | |
| Shingles | 6 65 | | |
| Sluicing | 131 87 | | |
| Canvas | 10 38 | | |
| Sundries | 49 27 | | |
| Total supplies | | 13,922 93 | |
| Total disbursements at Eureka Mill | | | 30,925 65 |
| **Rose Mill account:** | | | |
| Labor, as per rolls, (watchman) | | | 150 00 |
| **Freight and team account:** | | | |
| On coin | $289 50 | | |
| On bullion | 660 50 | | |
| On supplies | 2,228 04 | | |
| Total freight | | $3,178 04 | |

| | | | |
|---|---|---|---|
| Harness and repairs | $148 75 | | |
| Horseshoeing | 101 11 | | |
| Wagon-repairs | 84 62 | | |
| Feed for teams | 1,275 38 | | |
| Use of team | 52 50 | | |
| Total team expense | | $1,662 36 | |
| Total freight and team disbursements | | | $4,840 40 |
| Sulphuret-account: | | | |
| Charges for working 219¼ tons | | | 3,891 97 |
| General expense accounts: | | | |
| San Francisco expense | | $1,090 97 | |
| Sutter Creek expense | | 870 65 | |
| Salaries paid secretary, superintendent, and clerk | | 9,380 00 | |
| Interest on overdrafts | | 304 36 | |
| Insurance on mill, hoisting-works, &c | | 1,605 00 | |
| Assay of bullion | | 340 84 | |
| Law-expense | | 15 50 | |
| Total | | | 13,607 32 |
| Dividend-account: | | | |
| Amount of dividends paid to stockholders | | | 30,000 00 |
| Tax-account: | | | |
| State and county tax on mine, real estate, and office-furniture for the year 1874-'75 | | | 4,322 34 |
| Cash-account: | | | |
| Cash on hand with John A. Steinberger April 1, 1875 | | $355 47 | |
| Cash on hand April 1, 1875 | | 4,979 33 | |
| Total | | | 5,334 80 |
| | | | 270 014 93 |

ASSETS, APRIL 1, 1875.

| | | |
|---|---|---|
| Real estate: | | |
| Eureka Mill, 40 stamps | $30,000 | |
| Rose Mill, 16 stamps | 5,000 | |
| Badger Mill, 16 stamps | 1,000 | |
| Hoisting-works at North shaft | 50,000 | |
| Hoisting-works at Middle shaft | 15,000 | |
| Hoisting-works at Badger shaft | 15,000 | |
| California engine and hoisting works | 4,000 | |
| Stores, dwellings, tram-way, magazine, &c | 18,600 | |
| | | $138,600 |
| Personal property: | | |
| Timber, lumber, cord-wood, coal, charcoal, powder, tools, supplies, &c. | $14,086 | |
| 3 steam-pumps | 2,500 | |
| Wire and Manila ropes and 3 shaft-cages | 10,400 | |
| 5 horses and mules, feed, harness, wagons, cars | 4,375 | |
| Buckets for water and rock; furniture and safe | 2,200 | |
| 25 tons of sulphurets | 2,500 | |
| | | 36,061 |
| Total real and personal property | | 174,661 |

The Oneida, which may be ranked as one of the most valuable mines in the county, is presenting a flattering appearance. New and substantial hoisting-works have been completed during 1874. The shaft has reached the depth of 1,000 feet, and the mine has been thoroughly proven to the lowest level. The chimney increases as greater depths are reached, and at this time presents an unbroken ledge of rich pay-ore, 500 feet in length. The rock now being taken from the mine will average $20 per ton.

*Description of leading mines, Amador County, California, 1874.*

| Name. | Location. | Owners. | Length of location. | Course. | Dip. | Length pay-zone. | Average width. | Country-rock. | Character vein-matter. | Worked by tunnel or shaft. | Length of tunnel. | Depth on vein in tunnel. | Depth working-shaft. | No. levels opened. | Total length drifts. | Cost hoisting-works. |
|---|---|---|---|---|---|---|---|---|---|---|---|---|---|---|---|---|
| | | | Feet. | | | Feet. | Feet. | | | | Feet. | Feet. | Feet. | | Feet. | |
| Gold Mountain .... | Lower Rancheria. | Gold Mountain Mining Company. | 1,600 | N. 20° E. | 45° W. | 200 | 7 | Slate and metamorphic slate. | Free quartz, arsenical and iron sulphurets. | Tunnel | 400 | 200 | 100, (air.) | Tunnel runs on ledge. | ...... | ........ |
| Keystone Consolidated. | Amador..... | Keystone Consolidated Mining Company. | 1,800 | N. and S. | 45° E. | 500 | 7 to 15 | Greenstone.. | Quartz and slate. | Shaft.. | ...... | ...... | 550 | 5 | ...... | $15,000 |
| Original Amador... | Amador City | English Corporation. | 1,400 | N. and S. | 45° E. | ...... | ........ | Slate and greenstone. | Quartz...... | Shaft.. | ...... | ...... | 800, 300, 220 | 12 | 2,630 | 27,000 |

*Operations of leading mines, Amador County, California, 1874.*

| Name. | Location. | Owners. | Number miners employed. | Miners' wages per day. | Cost sinking per foot. | Cost drifting per foot. | Cost stoping per ton. | Cost mining per ton extracted. | Cost milling per ton. | Company or custom mill. | No. tons extracted and worked. | Average yield per ton. | Percentage sulphurets. | Total bullion-product. |
|---|---|---|---|---|---|---|---|---|---|---|---|---|---|---|
| Gold Mountain .... | Lower Rancheria.. | Gold Mountain Mining Company. | 6 | $3 | ...... | ...... | ...... | $1 50 | $3 00 | .......... | 700 | ...... | ...... | .......... |
| Keystone Consolidated. | Amador .......... | Keystone Consolidated Mining Company. | 75 | $2 50 and $3 | $10 | $6 | ...... | 5 12 | 2 04 | Company . | 25,146 | $18 00 | 2 | $452,506 84 |
| Consolidated Amador. | Sutter Creek...... | Consolidated Amador Mining Company. | ...... | .......... | ...... | ...... | ...... | ...... | ...... | .......... | 22,098 | 10 85 | 1 | *259,971 37 |

* Including $20,254.02 from 219¼ tons of sulphurets, at $92.27 per ton; for 14 months from February 2, 1874.

The discovery of lignites in this and other counties on the foot-hills of the Sierra Nevadas has attracted the attention of the public during the past two years. Dr. J. G. Cooper, formerly one of the corps of Professor Whitney on the State geological survey, and who, in that capacity, had made examinations of more than one hundred localities in which coal-strata had been found, recently read a paper on the subject before the California Academy of Sciences, of which I reproduce an abstract, as the subject is one of growing interest in this State.

Although the unscientific sneer at geological facts and fossils as not practically useful, they are really the only reliable guide in determining the age and probable value of coal-deposits. The true coal of the Carboniferous rock in other countries was formed from tree-ferns, algæ, and other plants of low organization.

None such has been found on this coast, and from the fact that ours contained remains of coniferous and dicotyledonous trees, geologists had long considered it all as lignite, but practically it was as good as much of the older coal, at least that of Vancouver's Island, Bellingham Bay, Coos Bay, and Mount Diablo. The most northern localities mentioned had been determined beyond doubt by the fossils as of Cretaceous age, but there is still some doubt as to those of California, which may be partly or entirely above the Cretaceous strata, like the Rocky Mountain coal, which is generally considered Eocene.

This, however, does not affect the value of fossil-evidence, as all the species of both these formations are extinct, and any coal found associated with fossils of living species must be of later date.

No paying beds of coal have been found anywhere of later date than these. It does not follow, however, that because a stratum is Cretaceous it will pay. Numerous strata in that formation in the Coast range are too thin to pay, though of pretty good quality. None will pay if less than two feet thick, and in most places a thickness of four feet is necessary, if the coal is no better nor more accessible than that of Mount Diablo. Much of the Cretaceous strata is also so metamorphosed that the coal has been ruined by infiltration of iron and silica, with other minerals, the surrounding sandstones being converted into jasper or serpentine.

The fossil shells found in connection with this coal show that it was formed by accumulation of trees, &c., in shallow bays, at the mouths of rivers in fresh or brackish water, and therefore along the shores of older continents or large islands. Often these deposits have been sunk afterward, and strata with marine shells have accumulated above them to a great depth, when all would be again raised above the sea. In the Coast range, Cretaceous coal-strata exist, above which Miocene-Tertiary strata, full of shells of living kinds, were deposited to a thickness of 1,000 feet, but afterward removed sufficiently to show the coal beneath.

The beds of undoubted Tertiary age are numerous in the Coast range, and usually show the vegetable structure so plainly as to be recognized as lignites by everybody, besides differing from coal in a more or less brown tint. Some lignites may pay for working, for local use especially, as they do in some parts of Europe.

Nearly all of that in the Coast range is, however, in either too thin beds or too full of sulphur and other impurities. In a few places it has been purified and hardened so as to resemble anthracite, apparently by the action of subterranean heat, when the strata are in contact with igneous rock beneath them.

The lignite-beds of Ione Valley, Amador County, and Lincoln, Placer County, appear to be of one age. The former is described by Professor Whitney, in the State Geology, Vol. I, as being very soft material, approaching peat, and useful only for local consumption. It forms a bed seven feet thick, occupying several small basins in the foot-hills, apparently the beds of former lakes. Numerous fossil plants are found in it, and are considered by him to prove its Pliocene-Tertiary age. The large deposits found near Lincoln, at a much lower elevation, show that this Pliocene lignite probably occupies large portions of the Sacramento and San Joaquin Valleys, where marine Pliocene fossils have long been known to exist, as well as fresh-water and terrestrial fossils, which occupied it successively as the country rose above the level of the sea. Much of this coal was, no doubt, formed in lakes, which in filling up left the present marshes.

Pliocene coal is also found in the Coast range, but nowhere in paying quantity. Strata from an inch to a foot thick may be seen by any one visiting Long Beach, south of Lake Merced, where the Pliocene strata, full of marine fossils, (which prove their age by the large proportion of living species,) are uplifted with a dip of 30° to 40° to the northeast.

In an article in the Proceedings of the California Academy, Vol. IV, p. 244, Amos Bowman described and figured this Pliocene formation as "terraces," most of which Doctor Cooper thinks exist only in imagination. As seen from east of the bay, the top of the ridge as this point appears tolerably level, but the strata along the beach are plainly inclined 30° to 40°, and were so described in California Geology, Vol. I.

At a distance of ten to twenty miles many such "terraces" may be seen along the ridges around the bay, but none of these ridges are really terraced in the upper strata, which are everywhere highly inclined. True Pliocene terraces do exist at low levels around the bay and in Livermore Valley, containing fossil remains of land-animals. Doctor Cooper was investigating these when the survey was suspended in 1873. The marine terraces described by Professor Davidson, in Vol. V, part 1, do not extend within the mouth of the bay, or very near it.

There is a fresh-water deposit in the basin of San Pablo Creek, containing thin beds of good lignite, full of fresh-water shells, indicating a lake-deposit of probably the Miocene age.

The strata have been very much disturbed by volcanic action in all the places where Doctor Cooper has examined them, and are not likely, therefore, to be profitable. Indications of the effects of the great volcanic convulsions about the end of the Pliocene epoch, which destroyed the then existing tropical fauna and the flora of California, (as described by Professor Whitney,) are to be seen in the coast strata of all the counties so far explored north of the bay, as well as in the gravel-terraces containing the remains of the tropical animals and plants, so far known chiefly by the collections of Dr. L. G. Yates, of Centerville, (described by Doctor Leidy in a recent publication of the "United States Geological Survey of the Territories.")

*Statement of quartz-mills, Amador County, California,* 1874.

*Amador district.*—Name of mill, Keystone Consolidated Mining Company; owners, Keystone Consolidated Mining Company; kind of power and amount, 96 horse-power, steam; number of stamps, 40; weight of stamps, 700 pounds; number of drops per minute, 84; height of drop, 8 inches; number of concentrators, 22; cost of mill, $50,000; capacity per 24 hours, 90 tons; cost of treatment per ton, $2.04; tons crushed during year, 25,146; method of treating sulphurets, chlorination.

## EL DORADO COUNTY.

The history and physical characteristics of this county may be considered as typical of a large portion of the central mining region of California, which, after a period of decadence, is now beginning to feel the beneficial influence of capital and labor, as manifested in the construction of extensive mining-canals, the introduction of water, and the opening of long-abandoned mines. In 1850 El Dorado County had a population of 20,000; in 1860, 20,500; and in 1870, only 10,300. The same proportion holds true in Amador, Calaveras, and Placer Counties. It is estimated that El Dorado County must have produced one-tenth of the $600,000,000 of gold produced by California before 1860, or $60,000,000, of which Georgetown divide produced half. They have had until recently no water on Georgetown divide to sluice systematically. The veins and seam-belts have not been sufficiently understood to be worked discriminatingly.

Among the projects of magnitude for the introduction of water in this county, I may enumerate the operations of the California Water Company, on the Georgetown divide; the El Dorado Water and Deep Gravel Mining Company, on the Placerville divide; and the Mount Gregory Water and Mining Company, in the higher portions of the county. The California Water Company's canal will supply the country between the Middle and South Forks of the American River, while the El Dorado Company will perform a like service for the country south of the South Fork.

The main ditch of the El Dorado Company, commencing near Sportsman's Hall, on the Placerville and Carson road, and ending at the confluence of Alpine Creek with the South Fork of the American River, will be twenty-six miles in length. It will be about 5 feet wide at the bottom, 13 feet wide at the top, and 6 feet deep. It runs through a dense forest of pines nearly all the way, over a rough country, across deep cañons, over precipitous bluffs, along the sides of almost bottomless ravines. Numerous tunnels and flumes will be required, but the work is pushed vigorously forward, and will probably be completed before the end of the season of 1875. When this is done the company will be in a position to command at least 10,000 inches of water for the washing of the auriferous beds. And this water will then be available for the irrigation of a large section of country now consumed by the long summer droughts. The estimated cost of this enterprise is $400,000.

On the Georgetown divide, the California Water Company, a San Francisco-company, comprising many well-known mining-capitalists, (among them Mr. J. T. Pierce, of Smartsville,) has constructed and acquired by purchase a vast system of ditches and an extensive area of valuable auriferous territory. The main water-supply is a system of lakes, known as the Rubicon basin, situated at a high altitude in the eastern portion of the county.

Running lengthwise (northwest and southeast) in the heart of the Sierras for a distance of fifteen or twenty miles, the Rubicon River basin holds several hundred square miles of snow, 10 to 30 feet deep, the melting of which begins in April or May in the bottom of the valleys, and recedes to higher altitudes later in the season. The entire basin of the Rubicon, as well as that of the Little South Fork, is glaciated and dotted with innumerable lakes of glacial origin. Some of these are of great extent and very deep. Inclosed within lateral and terminal moraines, consisting of a narrow rim of loose material easily dug, these lakes are natural dams or reservoirs, capable of standing an enormous pressure, and in most cases of being raised by a slight artificial reconstruction of the eroded outlet to a reservoir-capacity greatly increased above their present natural capacity.

This water will be brought to the mining-region by a system of ditches, flumes, tunnels, and iron pipes, having an aggregate length of three hundred and ten miles and a capacity of 16,470 inches, miners' measurement.

Every portion of the divide is covered by ditches of the California Water Company. Besides the main ditches for summer supply, the old Stone ditch and the El Dorado ditch leading up into the high Sierra, there is a large number of distributing-ditches, connected with numerous subordinate ranch-ditches, which were constructed to take advantage of local streams for mining-purposes, yet which may all be considered, and are equally useful, as agricultural ditches in the summer season. The original survey and construction was the work of the predecessors of the California Water Company.

The California Water Company sells water for mining-purposes in the following districts: Georgetown, Georgia Slide, Pilot Hill, Crane's Gulch, Mount Gregory, Volcanoeville, Tipton Hill, Spanish Dry Diggings, Greenwood, Saint Lawrenceville, Kelsey's, Rich Flat, Centerville, Wild Goose, and along the South Fork of the American for agricultural purposes.

Through the courtesy of the company, I am permitted to make the following extracts from Mr. Amos Bowman's report on their property:

In gravel-mining, 800 inches at 100 feet "head," working for ten hours = 800 ten-foot cubes of water = 800,000 cubic feet, weighing 24,880 tons, (without adding thereto the pressure arising from the "head" employed,) will move, through ordinary sluice-grades of 8 to 12 inches to the box, 3,000 cubic yards of loosened gravel, or 2,000 cubic yards of ordinary uncemented bank-gravel; say an average of 2,500 cubic yards, weighing 8,300 tons, or $(\frac{24880}{8347}) = \frac{1}{3}$ of the weight of water employed.

Reckoned by inches, the amount of gravel moved = *three times* as many cubic yards as there are miners' inches employed.

*Weight of water.*—A cubic foot of water, at 62° Fahrenheit, weighs 62.321 pounds One thousand cubic feet weigh, accordingly, 62,321 pounds, or 31.160 tons.
The weight of gravel, sand, rock, &c., is exhibited in the following table:

| Material. | Pounds in a cubic foot. | Pounds in a cubic yard. | Specific gravity, water being equal to 1. |
|---|---|---|---|
| 1. Clay | 120 | 4,800 | 1.92 |
| 2. Sand, dry | 88.6 | ........ | 1.42 |
| 3. Sand, wet | 118 | ........ | 1.9 |
| 4. Trap-rock | 170 | 4,500 | 2.72 |
| 5. Basalt | 187.3 | 5,060 | 3. |
| 6. Quartz | 165 | 4,450 | 2.65 |
| 7. Shale | 162 | 4,370 | 2.6 |
| 8. Slate, (clay) | 180 | 4,800 | 2.9 |
| 9. Decomposed shale, (estimated) | 100 | 2,700 | 1.8 |

Water is measured by the California Water Company by the customary square-inch aperture under a pressure of 6 inches, making 1 inch equal to 94.7 cubic feet per hour. In other localities, the pressure used is 10 inches, making 109.1 cubic feet per hour, as calculated. The average of the miner's inch in California is, then, about 100 cubic feet per hour, or 1,000 cubic feet per day of ten hours. The average price is 10 cents per 1,000 cubic feet, equal to a cube or tank of 10 feet, measured either way.

For purposes of comparison with quantities elsewhere, Mr. Bowman suggests that the pressure or gauge of the water-agent should be so regulated, in general, as to deliver the average of one hundred cubic feet per hour.
He gives the following table for comparison:

*The standard miners' inch.**

| Pressure from surface to top or middle of orifice, (varying.) | Miners' inch. | In cubic feet, (each 6.23 gallons.) Per second. | Per minute. | Per hour. | Per 24 hours. | Authority. |
|---|---|---|---|---|---|---|
| Six-inch pressure | 1 | .039 | 2.33 | 140 | 3,360 | Hittell. |
| Do | 1 | .026 | 1.57 | 94.7 | 2,274 | Carpenter. |
| Do | 38 | 1 | 60 | 3,600 | 86,400 | Do. |
| Do | 1000 | 26¼ | 1580 | 94,700 | 2,274,000 | Do. |
| Ten-inch pressure | 1 | .03 | 1.8 | 109.1 | 2,618 | Do. |
| Six to ten inch pressure | 1 | .027 | 1.6 | 100 | 2,400 | Standard experimental miners' inch. |
| Do | 10 | .27 | 16 | 1,000 | 24,000 | |
| Do | 100 | 2.7 | 166 | 10,000 | 240,000 | |
| Do | 1000 | 27 | 1666 | 100,000 | 2,400,000 | |

* The usual acceptation of the miner's inch is that given by Hittell in this table. It may be calculated by the formula of Haswell, $\frac{2}{3} b \sqrt{2g} (h' \sqrt{h'} - h \sqrt{h}) C = V$; $b$ being the breadth, $h'$ the distance from the sill to the surface, and $h$ the distance from the top of the opening to the surface, in feet, while C is the coefficient of discharge, assumed at .750, and V the volume in cubic feet per second. Thus:

$$\frac{2}{3} \times \frac{1}{12} \sqrt{2g} \left( \frac{7}{12}\sqrt{\frac{7}{12}} - \frac{6}{12}\sqrt{\frac{6}{12}} \right) \times .750 = .031, \text{ nearly.}$$

The coefficient of discharge is perhaps too large.—R. W. R.

The following is the basis on which Mr. Bowman made his estimates of the quantity, in miners' inches, of water observed flowing in streams:

The breadth, depth, and velocity of the stream, in feet per minute, (as traveled by a chip,) were estimated by the eye. The sectional area being reduced to square feet and decimals thereof, we have multiple × 60 = the cubic feet per hour. Divided by 100, or moving the decimal point two places to the left, = the miners' inches. Or, observe 6 seconds, and distance × area × 6 = miners' inches.

The completion of the California Water Company's canal will place this company, with all its distributing-ditches, in a position commanding a larger area of mining, agricultural, and timber lands than any other corporation of the kind in California, or probably in the United States.

For the following description of El Dorado County and the nature of its varied mining-interests, I am principally indebted to a report on the Georgetown divide, which has been kindly placed at my disposal by the above company.

*Topography.*—This portion of the western slope of the Sierra Nevada does not differ materially from the characteristics of the range elsewhere. Situated nearly opposite, in the line of drainage of the mountain-streams, to the outlet of Sacramento Valley through Carquinez Straits, the eastern end of the county lies in a region showing, in the summit culminations and lakes, more strikingly than any other in the range, the tendency of the Sierra Nevada to split into northerly and northwesterly trends of mountains, characteristic of the entire Pacific slope west of the Rocky Mountains. The northerly trends are peculiar to the great plateau of Nevada and Utah, which extends across into California as far as Mount Shasta, embracing the greater portion of Lassen and Siskiyou Counties.

At the point where Georgetown divide joins the summit, there are three summit-ranges, Tell's Mountain range, the Main or Western summit, and the Eastern summit, or Washoe range. The two western are the highest, being nearly equal in height. But the most westerly range carries the most snow. Its summer stores of water are never exhausted. Between the snowy Tell's Mountain range and the summit runs the Rubicon River, a stream very large in midsummer and autumn, constituting the principal basin of drainage of the melting snows of late summer.

From a general altitude of 8,000 feet, at its junction with the crest of the Sierra, Georgetown divide (like every other divide of the range) sinks gradually and with great regularity, in fifty miles of horizontal distance, to an altitude of 175 feet at the margin of Sacramento Valley plain, near Folsom. Being nearest to the outlet of the valley, (Carquinez Straits,) the rivers of this portion of the Sierra are more deeply eroded, in proportion to the altitude of the range opposite, than elsewhere.

East of Loon Lake basin the divide assumes an Alpine character. The surface changes from glacial *débris* overlying the slates to perfectly bare, polished, glaciated rocks. The forests re-appear only in the higher and less glaciated summits, on ridges where the soil was not removed by the ice-beds, on account of their higher altitude than the glacial levels; or in valleys, or moraine promontories of glacial detritus.

Taking in the western slope at a glance, there are ten parallel swells or corrugations of uplift to the west of, and, in general, parallel with the main summit, with transverse ridges having a northeasterly and southwesterly trend, and here the strikes of the slates are correspondingly altered.

*Erosion.*—The gorges of the Sierra measure, at mid-slope, 3,000 feet deep. At Forest Hill, the cañon of the Middle Fork is below the town.

2,500 feet; below the top of the hill, 2,800 feet. At Deadwood, the cañon is 1,600 feet deep. At El Dorado Cañon, the river is 2,800 feet below the bluffs. The angle of slope in the latter cañon is nearly 45°. The upper edges of the walls are only three-quarters of a mile apart. Probably the average angle of slope is not far from 30°. The streams are mere gutters at the bottom, 100 or 200 feet wide. Under such conditions, when winter torrents rise to 20, 40, and 50 feet above the usual level, flowing at the rate of six or eight miles per hour, carrying huge grinding rocks along with the water at the bottom, one can easily understand how these rivers were capable of executing the titanic work of erosion we still find them engaged upon—under the condition of free eroding grades varying between 50 and 175 feet to the mile.

Independently of erosion, the slates of the divide, which are generally bare of volcanic or detrital matter, have maintained a certain average outline of surface, about ten miles in width, between the two great cañons of the Middle and South Forks, remarkably regular on top, in consideration of the extent of the erosive action to which the country has been subjected.

Otter Creek, Pilot Creek, Little South Fork, and the Rubicon, on the north; Greenwood, Dutch, Rock, and Silver Creeks, on the south, are the principal lateral erosions. Yet they have scarcely been able to give a mountain character to the divide beyond the immediate vicinity of the two great cañons that bound the divide. One hundred and fifty square miles of undulating country, below 2,500 feet altitude, are only here and there intersected by an abrupt branch of the principal cañons. And the country above Georgetown embraces many succeeding areas of fifty square miles in extent, comparatively flat, or diversified by knolls and ridges that seldom rise over 500 feet.

*Seasons.*—From May to August there is no rain-fall worthy of mention, or capable of measurement, on the western slope of the Sierras—nothing beyond a sprinkling. The dry season is at its height in August and September. In the lower zone the rainy season, from November to May, corresponds to what in the higher zones is a similar season of very heavy and continuous rains. The streams in the cañons then become terrific. The suddenness of the mountain-floods is such that the water rises in the narrow cañons in a few days from 20 to 30 feet. Spring and autumn become definite seasons only in the high Sierra.

The rain-zones on the flanks of the Sierra correspond to the forest-zones. The foot-hill belt shares the climate of Sacramento Valley, which also grows spreading oak. Snow never falls, or only as a nine-years' wonder. No rain falls between May and November. The light-forest zone has rain later in the summer—a shorter dry season. Snow lies, in winter, for several days at a time. The heavy-forest zone has snow lying all winter through. The summit-zone, *per contra*, has snow lying all through summer on the higher points, or even lower down; and so late into the summer as to stunt vegetation which is not favorably situated as to soil and sunshine. Snow at the summit lies from 10 to 40 feet deep.

The annual rain-fall at Georgetown, (2,500 feet above the sea-level,) as observed by Mr. McKusick, is from 40 to 47 inches. At Placerville the observations indicate an inch or two less each season.

The winds are ordinarily from the valley in the day-time and from the mountains at night. The foot-hill zone, like the valley of the Sacramento, is subject in summer to parching, often scorching, north winds. In the heavy-timber zone, winter blasts from the east are sometimes very strong. In the summit-zone, summer thunder-storms, accompanied by light rains, come from the north; while the ordinary daily winds are, as in the lower altitudes, westerly and southwesterly breezes. In the foot-hill zone the ice barely forms on winter nights. In summer the thermometer keeps near 100° for many hours in the day. In the summit-region, *per contra*, the ice forms at Ward's Valley, opposite Lake Tahoe, on midsummer nights.

At Georgetown, near mid-slope, the thermometer has been observed by Mr. McKusick for several years—the result showing an average, from readings taken at 9 a. m. and 3 p. m. in winter, of 50° to 65°; sometimes down as low as 40°, but not often; in summer, at the same hours, from 78° to 90°, occasionally as high as 100°.

*Water-supply.*—In his report on the property of the California Water Company, Mr. Bowman describes a series of natural reservoirs, usually of glacial origin, situated in the drainage basins of perpetual snow. He says:

The depth of the annual fall of snow in this region is, by gauge-measurement, 18 feet; or, reduced to water, 6 feet over the entire area. And the "snow-line," or contour of altitude at which the sun's rays, through the long dry season of summer, fail to bring this quantity of snow to the liquid state before the next season's snows pile on, may be set down at 7,500 feet above the sea. In some years, however, the snow remains lying below 7,000 feet; the snow-line oscillating in periods of about ten years.

The ice-beds of the cold period of the Post-pliocene, known as the glacial period, associated with the phenomena of the lake-reservoirs, plainly extended much further down. The lowest point at which glacial gravel is observed in this divide was below Forney's, about 5,000 feet above the sea; though at Bear Valley, near Emigrant Gap, on the Central Pacific Railroad, a glacier reached down to 4,000 feet above the sea. At very numerously-repeated points, within a range of 2,500 feet of altitude, and over an area extending twenty-five miles west of the summit, Nature has laid out, for the great mining region and valleys of the foot-hills and plains of California, a noble system of reservoirs, into which an abundant precipitation pours during months in the driest summer when it never rains. How admirably the glacial valleys and lakes of the region are adapted to a doubled and quadrupled catchment, with trifling labor when the natural dams have been partially broken, can scarcely be realized by persons not familiar with these stupendous works (as they may be called) of design. From ten to twenty miles of ditching connects them with the region where gold has been concentrating for ages, both on and under the surface; and thence downward, upon the western slope, the vine flourishes and the orange blooms.

*Seam-mining.*—Hydraulic mining on Georgetown divide is confined chiefly to the seam-diggings. These consist of decomposed or slightly metamorphosed slates and shales, trending in belts in the strike of the country-rock, as represented at several points on the map. The country-rock has become so soft as to be easily removed in many places with the pipe, but in other localities this can only be done to advantage with the aid of blasting. Harder spots are met with, it is true, which are removed with little difficulty without blasting, as the rock crumbles into the sluices, and is carried away with the aid of large quantities of water and an unusually high sluice-grade, say 12 to 18 inches to the box, (12 feet.) In mines of a character so irregular as the seam, pocket, and the lens-shaped quartz-mines of this region, no average yield can be arrived at or stated. Even the yield in particular cases, during a limited period, is difficult to obtain. No "run" is like another.

In underground operations, of course, there can be no difference in this, from ordinary quartz-mining, as to the method of working. Nor is there any difference of principle between seam-hydraulicking, so long as it continues above ground, and ordinary gravel-mining. The conditions of seam-mining differ from gravel-mining in this:

1. That the miner cannot proceed to wash away the whole hill indiscriminately; for he would only be washing away barren country in one case, while in another the fine gold, or the nugget-bowlders, would be swept wholesale through the sluices.

2. That the "pay" does not run along the surface of the earth horizontally, like the gravel-deposits, but continues vertically, in a narrow pay-channel of quartz-seams, related to some well-defined fissure or wall, which sometimes cuts off all seams on the one side, and always pitches at a steep angle.

3. If the "pay" is followed under the ground, it is not always closely confined within two perfect walls, but often disseminated in a space of from 20 to 50 feet on the one side or on the other of the main fissures. It is generally in association with a series of lenticular masses of quartz, lying or crossing parallel to each other, and having the same dip; and in the form of pay-chimneys, located where some other system of courses of quartz veinlets or porphyry crosses the former. Although these courses are continuous in threads, the tendency to form lenticular masses makes pockets of quartz at the crossings, and the gold-deposits are accordingly in the form of sheets, chimneys, or pockets.

In order to discover and extract these sheets, chimneys, and pockets, wherever found, only one rule of mining applies, viz, to follow the deposit whithersoever it leads. If near the surface, and the ground is decomposed, or the pay-deposits are numerously and widely distributed, it is a very economical method to remove the entire hill with water, which does the sorting and separating in the act of moving. As soon as the deposit is beyond the reach of the water the pay must be followed down by shafts in the usual way, and prospecting-levels along the strike of the belt, connected with prospecting-drifts right and left, at right angles to the lenticular masses sought. These must be systematically run ahead, in order to discover the pay wherever it has been interrupted. Often these lenses measure only a few feet each way; no less frequently they measure 40 feet in length and depth, and a few feet in thickness. Doubtless there are plenty of seams in the country which will develop into something like regularity and certainty in the nature of these deposits. As soon as these seams are thoroughly understood, mining may be pursued permanently in them with profit.

The question whether these seams continue in depth, or unite into a single vein, becomes one of great importance. It matters little whether the pay is found in a solid quartz-vein or in lenticular masses. The question is whether it is continuous and regular in depth, and sufficiently confined, or concentrated in character, to justify following it with shafts, levels, and drifts. As this is a question which only the local conditions of mining and the character and richness of the seams themselves can solve, the best solution which can be given is to furnish a particular description of the character of the several deposits in the mines visited. The geological sections observed answer it, so far as erosion to 1,000 feet depth is able to testify.

The means, methods, and costs of extraction in general vary greatly, always according to the nature of the deposit which is exploited. New conditions, new necessities, and the application of new principles are constantly revolutionizing mining. Hydraulic-mining heads of from 400 to 1,000 feet pressure, vertical, are not at this day uncommon. The principle of hydraulicking veins—of tearing down and transporting the bed-rock slate, with its contents—has been applied on Georgetown divide, in violence to all preconceived notions. The conception is as thoroughly practical as it is original and bold. It is applicable only in the mountains, of course, where there is the advantage of an abundant grade. To the miners of Georgia slide is due the merit of inaugurating and of carrying on successfully, upon a large scale, this novel method of vein-mining during seventeen years past.

Georgia slide for a long time constituted the only "seam-diggings" in the country. The mines were discovered from Georgia flat, near the bed of Cañon Creek, where a portion of the hill had slidden down from the seam-belt. They consist of thin seams in a country of metamorphic porphyry. The pay, as found, is regular and easily followed. The method of working is by hydraulicking, combined with shafting and drifting, wherever the local deposit is unusually rich. Subsequently the side seams and the entire country-rock thus opened up are piped down as far as there is any outlet-grade.

The Parsons claim has been worked in this manner for twelve years. Before this character and method of mining were understood, it had been abandoned by the original owners as worthless.

This branch of hydraulic mining has paid well at the French Hill mine, Greenwood; at the Davis claim, Spanish Dry Diggings; at the Saint Lawrence mine, Greenwood; and it has paid steadily for twenty years at Georgia slide, where there is grade enough left to continue mining by this process for many years to come.

The pay is found on the side of the quartz, away from the sand-streaks. Ordinarily the principal pay is found at the junction of the two systems of veins in pockets.

A section taken from a point in the Beatty claim, looking north, illustrates the peculiar character of the quartz-veins. The quartz may there be seen to run for a while along the strike of the slates, and then jump along irregular bendings to another parallel stratum or bench.

The Parsons claim now pays, according to one of its owners, $700 per month, clear of all expenses, to three men employed. They have taken out as much as $1,800 or $1,900 a month, clear of all expenses. The latter amounted to $100 per month for powder, fuse, &c., and $1.50 a day for 10 inches of water, at 12½ cents an inch a day, (twenty-four hours,) the company owning a small ditch. Probably Georgia slide has yielded by this system of working $500,000.

The Beatty claim yielded, according to Mr. Barklage, $1,000 in one month, clear of all expenses, to a one-eighth interest; making a total of $8,000 per month clear of all expenses. It has been worked with constant profit for seventeen years, but at what rate is unknown.

*Vicinity of Georgetown.*—On the Georgetown ridge we find a group of mines situated nearly on the strike of the slates at Kelsey's. The principal workings are at Empire and Manhattan Cañons. The point between Empire and Manhattan ravines is quartz-seamed, the slates striking north 15° west. It is mined by the hydraulic method at two places—the Castile and the Hart mines.

At the Castile mine there is a fissure or ore-channel similar to that of the Nagler mine, at Greenwood, having two 3-foot veins of decomposed material, separated by 3½ feet of slate. It is hydraulicked off 100 by 70 feet, and 18 feet in depth, the sluices draining east-northeast into Empire Creek, which is distant about 16 chains east.

The Hart mine has on the surface a seam formation about 80 feet wide, which has been hydraulicked out longitudinally twice that distance to a width of 50 feet. At a depth of 95 feet, explored by shaft, the seams come together in a nearly solid mass of quartz over 8 feet wide. It is several hundred yards north-northeast of the Castile mine, and has had washed off about 175 by 50 feet, and 40 feet in depth, of scarcely altered slate.

The Crane's gulch, or Whitesides mine, is also in unaltered slates, in a seam-belt which shows several strong parallel veins running through the middle of the mine in the usual direction. Owing to the course of the gulch, hydraulicking has been done crosswise of the belt. The pit trends in a southeasterly and northwesterly direction. About 150 by 250 feet and 70 feet in depth has been washed out, which yielded $100,000.

The Swift & Bennett mine, immediately south of Georgetown, is situated in a narrow, decomposed belt. It has recently paid largely, prospects varying from 25 cents to $1 per pan.

The Blasdel mine, on Dark Cañon, is on a seam-belt having all the general characteristics of other seam-belts on the divide. There are two hydraulic pits open on the north side of the hill, 175 feet apart. Between these, and extending over the hill for 2,000 feet north and south, prospecting shafts and cuts have been dug at intervals, demonstrating the existence of pay through the entire zone. There are two main veins or seam-zones, each about 8 feet wide. The most westerly vein or zone shows a series of "sand-streaks" running east and west, dipping south 24°. The westerly decomposed quartz-vein is 8 inches wide on the top of the hill; the easterly one, 10 inches. Both dip toward the east, with the slates nearly vertical.

The Maddox mine, on the southerly slope of Little Bald Hill, is situated in a region of great metamorphism, the effects of which are observable to the summit of Bald Hill. A great variety of minerals are here found: crystallized gold, hornblende, asbestos, actinolite, serpentine, talc, &c. The porphyry and vein courses, in crossing each other, form rich pockets or chimneys.

*Neighborhood of Greenwood.*—Beginning at the north, and going southward along the strike of the slates, on the top of the hill west of Greenwood we find, first, the Spanish mine. This shows the usual characteristics of parallelism to the slates, the vein standing nearly vertical, and being intersected by minor cross-seams. There are two principal payseams, running northerly and southerly, and embracing a seam-belt about 100 feet wide. All the smaller seams and side stringers carry gold. A space of 400 feet by 24 feet, and 4 feet in depth, has been hydraulicked off, yielding $13,600. The pay has been followed farther down in a shaft 80 feet deep, and explored in tunnels and drifts for 20 or 30 feet in either direction. The vein at the bottom of the shaft is 2½ feet wide. The dip of the eastern vein worked appears to be toward the southwest 50°; but there are indications elsewhere of conformity of dip with the general dip of the country-rock.

The French or Nagler claim is worked in a vein-system in the strike of the slates; course, south 37° east; dip, northeast 75°. The width of the seam-belt here is about 200 feet. On the west side of the principal vein lies 100 feet of porphyry, strikingly different in color and mineralogical constituents from the rest; and on the east side, decomposed slate-rock, lithologically probably much the same as the "porphyry." There are found isolated portions of the same porphyry in different stages of consolidation, from hard, blue diorite, through all stages of hardness to red and brown loam. An area of about two thousand square yards is hydraulicked off to an average depth of about 30 feet. Gross yield of mining operations, about $100,000. The main vein is 4 feet wide, and consists of decomposed quartz and clay "gouge." A cross-vein of solid quartz, running at right angles to the former, starts east from the middle of the claim, and bends around to the southward. It is probably one of the same system of intersecting veins, the course and dip of which are observed repeated in the Fenton and

Saint Lawrence seam-mines, at this point intersecting the main French claim vein, and enriching it at the point of junction. A shaft has been sunk on this, 24 feet deep, at a point 40 feet from the main vein, to a depth of 30 feet. It has proven not only the continuance of a considerable vein of quartz, 2 feet wide at the bottom, not counting the stringers, but of a chimney of good pay. The course of this vein is north 68° east for the first 40 feet from the vein; thence it is followed by a tunnel, 26 feet in length, running south 85° east. The dip alters in that distance. Starting out at 70° toward the south, it is as steep as 80° or 90° near the surface on the east face of the bank of the claim; while in the tunnel, 26 feet into the bank, its dip is, on the level of the sluice, 80° toward the south. The dip of the vein at the bottom of the shaft is the same as at its mouth at the tunnel. Hence it is a twisting cross-vein at the point where it runs into the main vein, and its true or general course can be better judged by identifying it with other cross-veins of the same system, found, as stated, at a greater distance. This intersecting vein does not cross the main vein. It is cut off by the strongly-marked porphyry to the west on the foot-wall of the main vein. Though strongly developed on the one side, not a sign of the vein can be found on the other.

At the Saint Lawrence seam-mine an immense excavation has been washed out in unusually soft ground, with 100 inches of water, at a cost of ten or twelve dollars per day. Ten months' time did the work. There was paid for this water only $8,000. Four men were employed, and the yield was $23,000. The hill to the west of the Saint Lawrence, a continuation of Greenwood range, is at this point not constituted of greenstone porphyry entire, but of slates, in part unaltered, in part considerably metamorphosed, and only assuming the form of greenstone porphyry in nests. The local cause of the existence of this hill makes itself prominently known to the eye in the form of a great longitudinal quartz-vein which is situated at the apex. It is in the strike of the slates, and forms the west or foot wall of the seam series. The entire hill-top is covered with quartz-croppings. This ledge, associated with greenstone porphyry, forms cones all along the range to its termination at the south. About the middle of the Saint Lawrence mine there is a vein which is considered by Mr. Nagler to be the same as the main vein in the French claim. It is about 18 inches wide, and has a very perfect foot-wall. A shaft was sunk down alongside this, 38 feet deep, in solid slate. Mr. Nagler says he has traced this vein over the hill all the way to the French claim. From the main vein of the Saint Lawrence another series of seams runs off, striking to the southeast. The seam-belt is on the east or hanging wall in this case, and is about 125 feet in width.

The French claim, at Greenwood, has paid from $20,000 to $30,000 to the water company alone. The rate has been as high as $120 a week, but the average would probably be about $80 a week, running two-thirds of the time. The total yield, from the best information obtained, is about $100,000.

*Spanish Dry Diggings.*—The Grit claim shows a pit about 50 by 150 feet, and 60 feet in depth; and is also in the strike of the slates at their junction with a soap-rock belt. The pay-belt is about 50 feet wide. As at the Cedarberg, (quartz-mine,) the slates are on the east and the soap-rock on the west. This body of soap-rock runs to a point to the southward, in the town, at a distance of about 200 feet from the Grit claim. At right angles to the strike, going east, in a distance of about 250 yards, some red-spotted slates set in, which continue around the south end of the soap-rock, and cut it off in that direction. At the Grit there are no seams of any size, nor any signs of quartz noticeable above a depth of 40 feet. At about 40 feet below the surface there was some hard quartz, which occurred in several repeated swells or lenses, about 3 feet wide and 30 feet long, perhaps 100 feet deep. The pay ran along the quartz, but was not in the quartz. Captain Swift, of Georgetown, owned and worked in the Grit claim for seven years prior to 1867. He reports that they got out $100,000 during that time. The principal gold was taken out in 1852. Mr. Waun, of Spanish Dry Diggings, estimates that they took out of the Grit claim, altogether, about $300,000. In December, 1874, a 7-pound nugget was taken from this ground, which is now worked by the hydraulic method.

At Rocky Chucky, we find a number of small parallel stringers of quartz 3 to 5 inches in width. This is all that remains of the seam-belt at this point. The pay is in a soft, red dirt, intermixed with partially decomposed broken quartz in veinlets.

The Waun mine, formerly known as the Taylor & Rice, is on a parallel belt, half a mile west of the Grit belt. Here are four series of seams in a pay-zone of 50 feet in width, running also in the strikes of the slates and dipping to the east about 80°. A pit of 125 feet in depth has been hydraulicked into the steep hillside. Two of the pay-seams are worked from the pit into the hill a width of 3 feet. About 100 feet farther south (the river here running north) a tunnel has been run in at right angles to the seams, and one of the worked seams has been intersected and followed afresh, with highly remunerative results. It was paying richly at the time of Mr. Bowman's visit. The method of working is by tunnels, cross-levels, and stopes. The pay was found in a series of little parallel veinlets running longitudinally, also in an intersecting series of veinlets of a uniform easterly and westerly course. The claim has yielded about $60,000.

The French Hill mine, at Spanish Dry Diggings, is situated in about the same line of strike as the Waun mine, half a mile south of the latter. The most fantastic forms of quartz-deposit were observed in this mine. The accompanying section shows the character of the deposit.

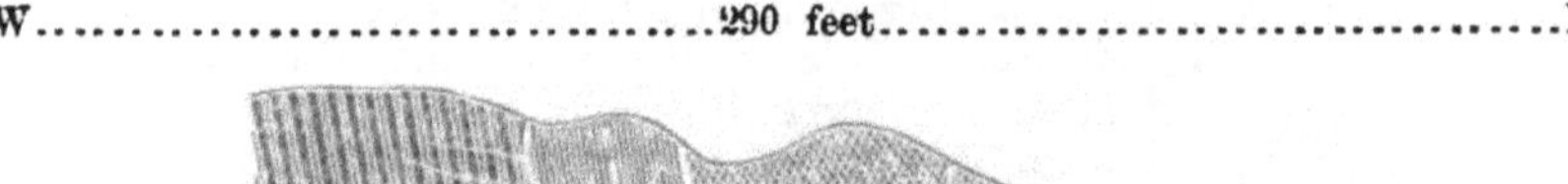


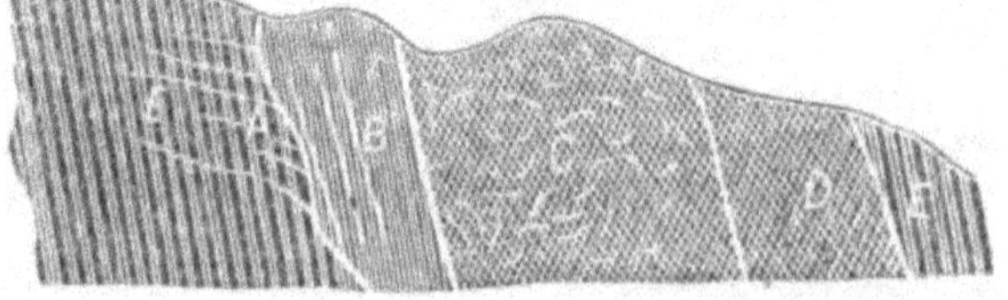

SECTION OF FRENCH HILL MINE, AT SPANISH DRY DIGGINGS.

*A*—Seams of quartz running off into the country-rock and forming rich pockets.
*B*—Quartz-vein in parallel slate-bands, 8 feet; comparatively barren.
*C*—Yellow and grey "porphyry," 70 feet.
*D*—Soap-rock, 100 feet.
*E*—Slate.

*Quartz-mining near Placerville.*—It is evident that the soft, decomposed and metamorphosed slates and porphyry trend in belts showing parallelism and the same general longitudinal direction. That they are in spots, and not necessarily connected, is a reasonable conclusion under the circumstances. On this divide, as on Georgetown divide, the quartz is disposed to pinch out in lenticular forms, and chimneys, like those on Georgetown divide, and the ledges stand nearly vertical.

The Mitchell mine is nearest the town, and the Epley, Pacific, and Harmon mines are probably on a continuous vein. The Mitchell and Poverty Point mines may be a very little farther east in the same general zone of parallel veins. The Mitchell mine has a knob of greenstone to the westward, forming the top of the hill south of, and in the rear of, the main street of Placerville. Ochery pockets, in connection with lenticular quartz, occur in it.

The German company has run a tunnel from the east side into Quartz Hill, about 100 feet, and intersected an 18-inch quartz-vein, which is succeeded on the west by a white "soap-rock." The strongest central body of quartz represented in the section, at the German mine, is considered by the miners as the real mother-vein of the southern counties—along the west side of which are situated the Epley, Pacific, Harmon, Sheppard, and Gross mines, and on the east side the Mitchell, &c.

A section across the hill, given herewith, shows the structure of Quartz Hill.

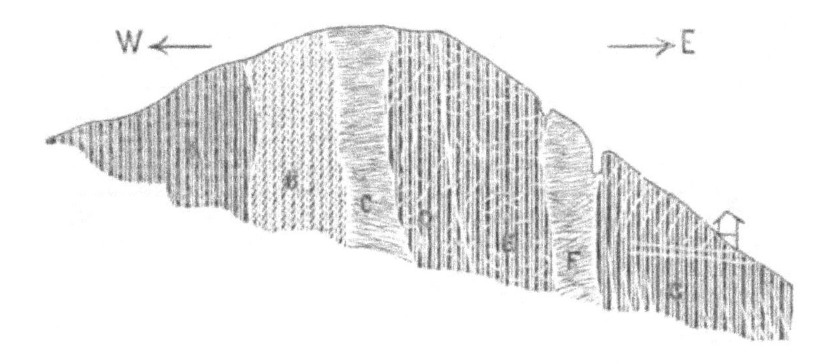


SECTION ACROSS QUARTZ HILL AT THE GERMAN MINE, NEAR PLACERVILLE.

*A*—Soft slates, perfectly laminated and undecomposed.

*B*—Decomposed slate and soap-rock zone, fifty yards wide, everywhere showing conformability to the dip of the slates. A few quartz-veins of three or four inches thickness pass through it, striking N. E.

*C*—Quartz-ledge, thirty-five feet wide, called the mother lode.

*D*—Thirty feet of a yellow ferruginous crisp quartzose rock, showing lines of slaty structure.

*E*—Soft, soapy slates.

*F*—Dioritic-trap masses.

*G*—Soft finely-laminated slates.

At the Pacific mine, a little farther south, there are two well-marked ledges on the surface, 50 or 60 feet apart, accompanied by hydrous magnesian minerals and country-rock, decomposed and metamorphosed after the usual character of the seam-belts.

At the Fiske mine, one-quarter of a mile farther north, the worked vein strikes north 22½° west, and stands nearly vertical; on the surface, apparently dipping 80° west. A section across the same hill, at this point, shows a repetition of the structure of the German mine, from which the continuousness of identical geological features may be observed.

The Hodge & Lemon mine is an extension of the Fiske, forming the extreme north end of Quartz Hill. Seam-mining has been carried on here to such an extent as to lay open half a mile of the principal seam or pay-vein of the district running in the strike of the slates. For several hundred yards an open cut exists, from 30 to 40 feet deep, with continuations in shafts, levels, and drifts under ground. There are cross-veins on the Hodge & Lemon ground dipping north 40°. Where these strike the foot-wall or east wall of the porphyry, the pay is rich. They do not cut into the east wall of the vein. Porphyry is the local designation given to the vein-matter in the fissure. It is about four feet wide at the end of the hill.

The Gross mine, on Big Cañon, one-half a mile northwest of the Fiske mine, shows a ledge 2 inches to 6 feet wide, bearing rock containing $160 to the ton. It has been prospected for a distance of 1,000 feet.

The mother lode of Quartz Hill appears to strike toward the little flat just west of the road on the top of hill, at Kelsey's, and to be as near as possible in the same line of strike as the Doncaster and Saint Lawrence quartz-mines. On the stage-road from Kelsey's to Big Cañon Bridge, Mr. Bowman observed, in this zone, a section showing the character of its continuation in depth 1,000 feet lower than the hill at Kelsey's, and about 1,100 feet lower than the section at the German mine. No very heavy ledge is visible on the north side of the river, where the road descends. The quartz-veins observed in the cañon of the South Fork resembled those of Kelsey's and Saint Lawrenceville.

The Saint Lawrence quartz-mine has been mined to a depth of 600 feet, where there is a fine vein. It follows, in general, the strikes of the slates, and establishes for itself the character of a true fissure-vein by its variation in detail from the bedding of the slates, cutting across them diagonally occasionally, but following the strike in the main. The lenticular masses of quartz described in this mine, along with the associated stringers or seams running off into the slates on one side, are so characteristic of both the seam-belts and quartz-mines of Georgetown divide in general, that the character of the Saint Lawrence may be referred to as typical, geologically, of the veins and seam-belts on the divide. As they follow in most cases some well-marked fissure, yet not always precisely with the bedding of the slates, with accompanying metamorphism and branching of seams and stringers peculiar to the vein-chemistry of the slates, Mr. Bowman considers them as true fissure-veins. An examination of the vein fully develops the fact that the quartz occurs in lenticular forms, not continuing for any great distance, but invariably found to come in again in similar bodies lying along the main fissure. At a depth of 550 feet in the main shaft the quartz had entirely given out, and clay took its place, following an exceedingly well-marked foot-wall. On the 500-foot level, at a distance of 160 feet north of the shaft, the quartz gave out, and there was in its place a five-foot gouge, showing slickenslides on both walls. Before giving out in this direction, the quartz jumped several times from the hanging-wall to the foot-wall, and *vice versa.* Everywhere else throughout the mine slickenslides were noted. The foot-wall is everywhere the strongest throughout the mine. The hanging-wall is occasionally, to all appearance, repeated two or three times in parallel slickenslides, seen in the gouge on the hanging-wall, which run off at regular angles and inclose bodies of slate, separated by little lenses and stringers of quartz. For example, on the hanging-wall, the gouge is full of these seams; and there is no distinct line of demarkation between the vein and the slates. The vein-material appears to branch off into the slates. Wherever the quartz in these side stringers pinches out, the fissure along which it was formed runs uniformly down to the foot-wall. At this point the lens of quartz is on the increase as you go down, attaining its greatest width about 30 feet lower; after which it wedges out gradually as far as the 500-foot level, as represented. The underground foreman states that they have never exposed any regular hanging-wall, so far as he has known. There are two chutes of quartz below the 300-foot level. Above the 300-foot level they run together into one. Where the quartz runs out on the north, the characteristic of "gouge" coming in to take its place is everywhere the same, and precisely as observed at the north extremity of the 400-foot level.

The north and east base of Saint Lawrence Hill is strewn with vein-bowlders from local veins and stringers on both sides of Cincinnati ravine, the recipient of the decomposed croppings. In the early days of placer-mining this ravine paid largely. In this vicinity the Doncaster mine is worked to a depth of 100 feet on a vein of decomposed quartz.

The Taylor quartz-mine, situated south of the old Georgetown road, has been opened to a depth of 400 feet, and worked north and south on the vein for a distance of nearly 100 feet either way. The ledge may be distinctly traced for a distance of one-half a mile. There is a fine vein at the bottom. A thick gouge is in the hanging-wall, into which the shaft has been sunk to a depth of 400 feet, for the purpose of fully testing the character of the mine. The works show conclusively the repetition of lenticular masses of quartz. The quartz lies along a well-defined fissure, the gouges continuing along the fissures wherever the quartz gives out. The quartz pinched out 100 feet south of the shaft, on the 100-foot level. It narrowed in a similar manner on the same level north of the shaft. The ore-pile at the time of Mr. Bowman's visit contained plenty of rock in which gold was visible. Parallel to the Taylor on the east we find several heavy ledges, but carrying comparatively barren quartz.

The Cedarberg was formerly considered a seam-mine. It was discovered by Cedarberg and his partners by following traces of gold from the cañon up on the hillside to the vicinity of the ledge. After a long search in its immediate vicinity, failing to strike anything tangible, and coming to the end of their financial resources, they asked for and were allowed by the California Water Company a supply of water for a day's prospecting, by washing away the surface-dirt. Their flume broke down and the water then did the work in its own way, disclosing a ledge of 2 or 3 inches in width. This was followed down to a depth of about 25 feet, and from it and several small adjacent veinlets, parallel to the first, there was taken out, from ninety cubic yards, $45,000 in a few months. The vein had been worked, at the time of Mr. Bowman's visit, to a depth of 200 feet, where it was 2 feet wide, and consisted of solid blue and white quartz. On the 100-foot level the vein-specimens are less solid, and are considerably intermixed with slate The quartz runs in chimneys, measuring from 50 to 80 feet horizontally, and vertically to unknown depth. Similar chimneys are found in the Taylor and Saint Lawrence mines. The sulphurets follow the blue quartz, while the gold is in the white and pure quartz. Sulphurets are scarce. At the surface the gold was in flakes and sheets. At 200 feet depth the character changed to crystalline, and irregular. The course of the vein is about north 10° west, following the strike of the slates at this point. The vein is along the boundary of a belt of greenish soap-rock, which also follows the strike of the slates. The average assay of the Cedarberg rock, at the time of Mr. Bowman's visit, was from $45 to $62 a ton. There is a 10-stamp mill, which crushes 10 tons a day, at a total cost, for fuel, labor, and running, of $20 per day, or at the rate of $2 per ton. (November, 1873.) Cost of mill, at usual prices, $6,000; actual cost to the company (having been put together second-hand) was only $1,700. The mill has a 15-horse-power boiler and engine. Total employés in mill and mine, 26; wages, $3 per day.

The Sliger mine, in this vicinity, has been noted for its yield of "specimen-rock." The quartz contains granular gold, and is valuable to jewelers, being well adapted to the manufacture of ornaments. The productive portion of the vein lies on the foot-wall, (metamorphic slate.) The east country is talcose slate. The mine is well opened by a shaft 200 feet in depth, and a connecting tunnel 375 feet in length.

The Woodside, near Georgetown, also a "specimen-mine," has produced from $30,000 to $50,000. The vein is 2 to 3 feet in width, and occurs in slate. It has been opened to a depth of 100 feet. The mine has not been worked for several years. The pay was in a chimney north of the shaft, which pitched north 30 feet in 110 feet, and was in the form of spangles or grains of fine gold, with occasionally a chunk of gold. Five or six inches of gold would hang together. The mine was re-opened in December, 1874.

*Volcanoville.*—The McKusick ledge, near Volcanoville, runs with the slates, locally at least, north 20° west, and dips southwest 75° or 80°. It is 2½ or 3 feet wide; in some places 4 or 5 feet; in others again only 1 foot. A shaft has been sunk on the ledge to a depth of 90 feet, and a tunnel run in 100 feet lower from the southeast. At the point of intersection by the tunnel the ledge was only 1 foot wide; but on drifting along it 78 feet toward the north it widened to 5 feet; while in drifting the same distance toward the south it decreased to nothing, though the walls remained perfect, and from 2½ to 3 feet apart, filled with gouge. The foot-wall is the best defined. The rock pays from $13 to $14 a ton.

The Trench ledge, or Yellow Jacket, several hundred yards north-northwest of the McKusick ledge, on Quartz Cañon, runs north 5° west, dipping east 45°, along a metamorphic, trappean belt, which accompanies it on the west, with serpentine beyond, and is from 1 foot to 6 feet in width, varying. On the east are slates. This mine was made famous, in the early days of quartz-mining, from the circumstance of a sheet of gold having been found on it, lying very nearly horizontal. When worked, it paid $70 a ton. Many small veins and strings of quartz run off from the main ledge. The gold was generally found in isolated nests and bunches, of extraordinary richness.

*Geographical relations of seam and vein mines of Placerville divide.*—On the south side of the South Fork, in the vicinity of Placerville, there is a continuation of the seams and vein-belts of Georgetown divide, with all their characteristics and peculiarities. A decomposed belt in the same strike of the slates as Kelsey's crosses the main street of Placerville, at the court-house. The most noted seam-mining locality in the vicinity of Placerville is where Fisk, Sanders & Gilbert took out large sums of money at the north end of Quartz Hill, about a mile from town. The general seam-zone on both divides, consisting of metamorphic or decomposed matter in the neighborhood of extensive quartz-veins, continues in the same strike to the Amador mine on Sutter Creek, from which point the "mother lode" of the middle mining-counties is plainly traceable in a further general continuation as far as Mariposa County.

Mr. Burlingham, superintendent of the Taylor quartz-mine, an experienced miner, who has been over the country, and has especially observed the strike and continuation of the slates and seams, thinks that the identical belt on which Placerville is located continues across the

North Fork of the Mokelumne, at King's Fork Junction, and the South Fork of the Mokelumne, at Bacon's Bridge, continuing thence through Plymouth to the Amador mine. While the identical continuation is not a matter so easily made out, nor, indeed, very probable, the continuation of quartz-ledges and of fissures and decomposed "porphyry" belts, with seam-deposits in places, of the character of Georgetown divide, situated in the same general strike of the slates, and in the same trend of vein-formation and of chemical concentration of gold—in short, geologically identical rather than physically continuous—is a question admitting of no further doubt. It is assuming too much, however, to undertake to trace anywhere, for more than two or three miles, a perfect unbroken continuation of the identical veins, or seams, or "porphyry" belts on Georgetown or Placerville divide. The general system, the geological position, and the chemical conditions of concentration and of precipitation of gold in connection with vein-formation are the same, which is all that can be said, and they are in continuation of the mother lode.

"So far as my observation goes," says Mr. Bowman, "the trap or greenstone 'porphyry' accompanying the seam-belts was not continuous in the form of an 'eruptive' dike, which could be traced for any distance. There are occasional combs of metamorphism which have given shape to the hills, rising up in the form of undenuded crests.

"Mr. Burlingham says he has observed that at the Taylor quartz-mine he has trap to the west of the vein in which he is working; while at Placerville it is on the east, and at the crossing of the North Fork of the Mokelumne there is trap on the west again, which continues thence by way of Negro Hill and east of Nashville, between the two forks of the Mokelumne, always on the west side of the seam-zone, as far as the Amador mine.

"Mr. Derby, of Isabeltown, says he has observed that on the Placerville divide there is a porphyry streak, east of what he believes to be a continuation of the mother lode, adjoining the Hodge, Lemon, and Fisk mines, on the north end of Quartz Hill. It is, according to his observation, about 50 feet distant from these mines east, and runs parallel, the porphyry itself being 2 feet thick. The mother lode at this point he measures as 28 feet wide.

"Mr. Roda, late superintendent of the Saint Lawrence quartz-mine, agrees with Messrs. Burlingham and Derby in the opinion that the Placerville and Dutch Creek veins, which strike through the Pacific Mine Hill, Quartz Hill, Poverty Point, Kelsey's, Saint Lawrenceville, &c., constitute a continuation of the identical mother lode of Calaveras County. When it comes to the connections in detail, however, everybody disagrees, because there is no such connection. Like the 'blue lead' of the ancient-river system, it generally passes through the identical ground which is owned by the miner whose judgment is passed upon it."

*Gravel and placer mines.*—The principal gravel-mines of Georgetown divide are situated at Negro Hill and Massachusetts Flat, upon the extreme point where the forks of the American River unite; Wild Goose Flat and other plateaus or "benches," at an elevation of 75 to 100 feet above the present river-bars; Centerville and Five-Cent Hill, below Greenwood; Buffalo, Mameluke, New York, Jones's, Boulder, Gravel, Bald, Cement, Tipton, and Buckeye Hills; Mount Calvary, Mount Gregory, and Kentucky Flat, embracing a region extending from east to west a distance of thirty miles, and varying in width from three to six miles.

Alternating with the gravel-beds, throughout nearly this entire section, are found the "seam" mines. These run from north to south, and have been traced from the North or Middle Fork of the American to the South Fork across the entire divide. They are simply minute quartz-veins in metamorphic talcose slates, the surface of which having undergone various stages of decomposition, the gold has been left free, while to a certain depth the rock is susceptible to attack by the ordinary method of hydraulic mining. Though "seam" diggings are generally classed as placers, they must eventually be worked as are all other mines of rock in place, as they will certainly become hard after a point is reached below all surface disturbances and away from atmospheric influences.

Where these "seams" have been cut by ravines, the latter have been found remarkably rich, the periodical rains having supplied water by which a natural process of sluicing was accomplished, carrying off extraneous matter and leaving the gold behind. Thus, in the vicinity of these larger ravines, many villages sprang up, as Georgetown, Greenwood, Spanish Dry Diggings, Johntown, Kelsey, Georgia Slide, and other towns of lesser size and note, once populous, and distinguished for their heavy product of gold. In this belt or zone of quartz-seams have been found ledges of considerable magnitude, and upon which costly and complete reduction-works have been built. A number of these have been described above.

Mount Gregory ridge is the largest continuous gravel-deposit upon Georgetown divide. It has a general elevation of 1,500 feet above the Middle Fork, and a heavy gravel-deposit of from 25 to 300 feet deep, and six or seven miles in length, running east and west. Average breadth of gravel nearly a mile. The mines tail into Middle Fork of the American River on the north, and into Otter Creek and Missiouri Cañon on the south. Only the edges were worked in early times, yet there was once enough mining done on this ridge to support very large and active populations at Mount Gregory and Volcanoville. Mount Gregory ridge proper is flanked on the north by the Middle Fork of the American, and on the south by Missouri Cañon to its junction with Otter Creek. Toward the center of the ridge the gravel-deposit is overlaid with a heavy stratum of volcanic cement. Formerly large sums of money were made here in the most primitive methods, by sluicing. On the south slope of the ridge the surface was washed away until the heavy deposit was reached, and after that drifting was resorted to. Upon the northern or Middle Fork side of the ridge little has been done in the way of mining. The hill has been pierced on both sides by numerous tunnels, and demonstrated to contain rich deposits of gold. The cause of this immense bed of auriferous drift remaining so long unworked has been lack of water. Until quite recently no ditch with a capacity of more than 150 inches has delivered water to that point, and then there was only one, which for the past four years has been allowed to get out of repair, while the water was diverted elsewhere. In the spring of 1872 the Mount Gregory Water and Mining Company began the construction of a canal from Pilot Creek to this ridge. That canal, fifteen miles long, is now completed and has a capacity of 3,500 inches of water. The distributing-reservoir of the company is located in the central portion of the ridge, has a strong, substantial embankment, overflows an area of about five acres, and is capable, when filled, of supplying, without inflow, 1,200 inches of water for a run of ten hours. Pilot Creek, for five months in the year, affords water to the full capacity of the ditch. During the remainder of the season the water varies from 200 inches upward. An extension of the canal is begun, which taps the Rubicon River, or South Fork of the Middle Fork, at a distance of twenty-eight miles, and, when finished, will afford a constant flow of water to the full capacity of the canal, making it one of the best water-franchises of the State. The Mount Gregory Water and Mining Company has also quite recently opened two extensive hydraulic mines, one tailing into Missouri Cañon on the south side, known as the Bowman & Worthingham claim, with sluices 5 feet wide, under-currents 15 feet wide, iron pipe 22 inches in diameter, and flowing 1,000 inches of water through one of Craig's largest-sized "little giants," with an 8-inch nozzle, night and day during the flush-water season, against a bank of gravel 100 to 125 feet deep. Upon the northern or Middle Fork side the same company has opened and is hydraulicking the Bitters claim, which shows a splendid face, about 100 feet deep, of fine-looking gravel. The estimated yield of these mines is 20 cents per inch for the water used upon them for each day of working. The Mount Gregory Water and Mining Company owns more than 1,000 acres of this gravel-ridge, and has a frontage upon each side of a mile and a half.

Since the Mount Gregory Company began work upon the ridge, the California Water Company has made purchases of ground there, and is now enlarging the old ditch and fitting up one or two hydraulics. The future yield of gold of Mount Gregory ridge must be large, since the gravel-strata will produce from 10 cents to $2 per cubic yard; nearly every pan of dirt showing the color of gold, while an occasional large nugget is found.

Buckeye Hill is an isolated piece of ground several hundred acres in area, situated upon the extremity of a lateral ridge running southwestward from Mount Gregory ridge below Volcanoville. The gravel at this point has been drifted out in many places to 20 feet thickness. The material was taken out in the summer time and washed in the winter time. The gravel-deposit is about 150 feet deep, and will eventually be hydraulicked.

Further down the ridge are other grand deposits of greater or less extent, which have been slightly worked under disadvantageous circumstances, and with variable results. The ridge ends at the junction of Otter Creek with the Middle Fork of the American, near Ford's Bar.

A series of drift-deposits, which blend with the Mount Gregory channel near its upper or eastern extremity, and having a total length of four miles southward, can be followed to Tipton Hill, though separated by the branches of Otter Creek. These include all the diggings between Kentucky Flat and Tipton Hill. Geologically this series does not differ from the gravels already referred to. Barometrical observations, made in 1871, prove that the stream which formerly flowed in Tipton Hill channel emptied toward the north and west; consequently that it was tributary to ancient Otter Cañon Creek.

Bell's diggings are situated at the extreme head of Missouri Cañon. They have been prospected by a tunnel and many shafts, and found to contain gold in paying quantities. This deposit is supposed to have some connection with the Kentucky Flat gravel at its southern extremity, mentioned below.

Kentucky Flat is an extensive drift-channel, having an average width of about three-quarters of a mile. Its gravels are blended with neighboring deposits for a distance of four miles in a northerly and southerly direction, mingling with the Mount Gregory gravels. The extreme southern end of the hill at this mine has been washed off. The gravel-deposit is 10 feet deep, but farther north bed-rock declines, and the gravel thickens to probably 100 feet. A tunnel pierces the gravel for 1,000 feet. The deposit is here only a quarter of a mile wide; both north and south of it the area of country covered with gravel widens out.

Bitters & Bowman have worked on the south bank of Missouri Cañon at the north end of the Kentucky Flat channel. Where the gravel-deposit is intersected by Missouri Cañon, it seems to trend toward the west, along the southern base of Mount Gregory ridge, and to form a distinct deposit parallel to that of Mount Gregory. By actual survey, the rim-rock of Bitters & Bowman's claim has been found to be 100 feet lower than at Kentucky Flat.

At Kelly's diggings the gravel shows a depth from the surface to bed-rock of not more than 6 feet on the rim, while the depth in the center of the channel is unknown, never having been reached, though shafts 80 feet deep have been sunk. It appears to be the extreme eastern end of the auriferous zone below Tunnel Hill. The deposit extends southerly to the North Branch of Otter Creek, and toward the north and west blends into the Mount Gregory ridge. The area washed is over 45 acres. The gold was pretty generally diffused through the gravel from top to bottom, and the mine, according to Mr. Kelly's statement, paid from fifteen to twenty dollars a day to the hand. The channel is one-quarter of a mile in width. This deposit continues for about two miles from south to north.

Tipton Hill shows, at Schlein's diggings, the most extensive workings in the whole section of country near the base of Tunnel Hill. It is the southern end of a body of gravel of which the Bitters & Bowman claim forms the northern terminus. On the channel there is a shaft, in volcanic mud and gravel, 120 feet deep, which has not reached bed-rock. Upon its eastern side there is a tunnel 1,100 feet long opening the ground, with an outlet into Rock Creek.

Jackass Hill lies at the head of Otter Creek, near Kentucky Flat, and is a portion of the channel running thence to Tipton Hill.

*Parallel to Mount Gregory ridge.*—Across Otter Creek is the Darling ranch, Bald Hill, and Jones's Hill ridge, embracing an extent of country from five to seven miles in length, and an average width of perhaps half a mile.

Fort Hill is situated west of Darling's. On it there are many gravel-claims, where drifting has been carried on to some extent. The deposit is about one-eighth of a mile wide and two miles in length, trending northerly and southerly.

Boulder Hill is an extensive and deep deposit, favorably situated for rapid hydraulic working, being on a ridge between two deep gorges.

At Darling's ranch, west of Boulder Hill, occurs a large gravel-ridge, which has been explored by numerous shafts and tunnels, proving the existence of gold in paying quantities. It can be opened from Cañon Creek or from Otter Creek.

Bald Hill has a comb of talcose slate and other metamorphosed rock, which deflected the course of the ancient channel, standing across it at right angles. Opposite to it modern denudation has carried away all signs and remnants of the ancient channel.

Harrison Hill is a continuous gravel-ridge, containing a deep deposit, extending east and west.

Cement Hill, situated farther west on the same ridge, opposite Georgia Slide, is three-quarters of a mile long by less than half a mile wide. Many years ago it was pierced by tunnels and the bottom stratum of gravel was extracted, yielding immense sums of money. The auriferous earth was found under a thick deposit of fine clay, 30 feet thick, in which whole trees were embedded.

Bottle Hill diggings occupy an area of about half a mile square, and have been celebrated for their yield. The North Star, Saint Louis, Cuyahoga, Gravoy, and Hopewell tunnels, each extensive works, have pierced the Bottle Hill channel from both sides, and the greater portion of the bottom stratum has been extracted by drifting. The channel is very deep, and will probably pay throughout for hydraulicking.

Mount Calvary, west of Bottle Hill, embraces about 200 acres of good paying gravel.

Gravel Hill, joining Mount Calvary, is also extensively worked by drifts, and is yielding considerable gold.

Jones's Hill is divided by a gulch called Jones's Cañon. That portion of the deposit on the north side consists of a heavy bed of gravel, while that upon the south side consists of seam-diggings. The gravel is deep, and has been drifted out to a great extent. Area, one-half to three-quarters of a mile. The Messrs. Barklage are here hydraulicking, using 15-inch pipe, with a large head of water, and obtaining excellent results. The total product of Jones's Hill is estimated at from half a million to a million dollars.

The only ditch delivering water upon this ridge—which, as above stated, is parallel and south and west of Mount Gregory ridge—is that of the Barklage Brothers, which has a capacity of 200 or more inches, is thirteen miles long, and takes its water from Otter Creek. It is an excellent ditch of its capacity, and has a constant water-supply.

Gopher Hill is situated on a divide between two branches of Cañon Creek, south of and parallel with the Darling Ranch ridge, having a favorable hydraulic opening on the north into a precipitous cañon. The gravel-deposit extends north and south a distance of about a mile. The eastern end is supposed to connect with the Kentucky Flat and Tipton Hill channel. A tunnel on the north and several shafts in the hill have demonstrated the value of the ground. New York Hill lies on the same ridge to the westward, and has yielded considerable gold.

Mameluke Hill and Buffalo Hill are worthy of especial notice. Mameluke Hill was carefully examined in 1854 by Professor Blake, having been drifted out to a large extent prior to that time. There is in the hill a gravel and sedimentary deposit 200 feet thick. Its character proves it to be of local origin. The Mameluke Company drove in a tunnel 800 feet long. A section of the hill showed the following strata, in descending order: Gray argillaceous beds, with volcanic matter, (cement,) 40 feet; auriferous gravel, 8 feet; gray argillaceous beds, with volcanic matter, 60 feet; auriferous gravel and clay on bed-rock.

Cañon Creek flows along the northern base of the above-mentioned four gravel-deposits, and has proved exceedingly rich, as have also its many tributaries. Spanish Dry Diggings is supplied partially with the waters of this stream, diverted by a capacious ditch, owned by Hunter, Simpers & Co.

Pilot Hill is the site of some gravel-mining in an ancient channel of local origin. Before 1860 these mines supported a large and prosperous population. The gravel is from 20 to 30 feet thick, and contains many angular quartz-bowlders and nuggets of gold. The principal ground has been purchased by the California Water Company. On the easterly slope of Pilot Hill there is a belt of seam-diggings resembling those of Greenwood, which have been worked to considerable extent by "coyoteying" shafts and levels and arrastra-milling in true Spanish style. The ground has been surfaced off over forty yards square.

Wild Goose and Massachusetts Flats and Negro Hill have each contributed vast sums of gold to the coffers of the world, and are yet yielding annually small amounts, and will afford employment to a limited number of men for many years to come, but the heavier gravel-deposits toward the Sierra will, in the future, be the attractive feature in the mining-interest of the Georgetown divide.

The gravel-belt in the southern portion of El Dorado County, of which Placerville is the center, is estimated to be ten miles in length and from one to one and a half miles in width. The deepest deposits are from 100 to 150 feet, frequently covered by lava. Twenty companies were at work within the limits of this tract, by the drifting process, in 1874.

I extract from the Mountain Democrat the following curious account of mining-operations within the city limits of Placerville. The same class of "city mining" has been carried on at Sonora, Columbia, and other decaying mining-towns:

There are very few town-lots in Placerville proper but have been mined out and filled in once, twice, or oftener, but within the past week we have noticed quite a revival in this line. Chinese companies pay an agreed sum for the privilege of mining out a town-lot, leaving the buildings thereon intact by underpinning and propping up, and after the gold is all washed out the lots are filled up by turning in the water and depositing the sediment from other mines above. For these mining-privileges in town-lots, prices varying from $250 to $1,000 are paid. Negotiations are pending for a lot 107 by 167 feet at the lower end of Main street, $2,000 being the present owner's price. If the code and the law-officers of the county would permit it, there are well-posted old residents, principally business men, who would give a handsome sum for the privilege of mining out Main street from Jones's corner to the Central House, Coloma street, from the upper corner of our office to Main street, and Sacramento street from the corner of Main to a short distance above Dunn's blacksmith-shop. This would include the width of the streets for a distance of about 300 yards. Responsible parties have offered $10,000 for the privilege of mining out this ground, obligating themselves to leave the streets in an improved condition, with a large and substantial sewer the whole distance, which would much improve the adjacent property. From results obtained in digging cisterns and otherwise, those best qualified to judge are confident that not less than $100,000 could be made in thus mining out the portions of streets above indicated.

*Placerville divide.*—Heavy masses of pipe-clay on the north rim of the Pliocene South Fork at Placerville, near the head of the west branch of Cedar ravine, show plainly that the principal ancient channel of the divide was further to the south. On crossing the ridge on the line of the stage-road to Shingle Springs, the looming banks of Coon Hill hydraulic diggings testify to the central location of the ancient valley, as well as of the chief concentration of gold-bearing material in this vicinity. Mr. Bishop, superintendent of the minging-company at Coon Hill, says, on the authority of the company's books and other reliable accounts, that twenty-five acres of gravel removed from here by hydraulicking have yielded a total of $25,000,000.

Touching the direction of this channel, Mr. W. A. Goodyear, of the State geological survey, says: "It is extremely probable that a deep continuous channel, known here as the blue lead, extends from White Rock in a general southerly direction beneath Dirty Flat and under the two intervening ridges to the extreme south end of Smith's Flat. But whether this channel from there on continues its general southerly course, coming out on the Weaver Creek side at or in the vicinity of the old 'Try Again' tunnel, or whether it makes a sharp bend to the southward in Prospect Flat, is a question impossible to answer with certainty until developments have been pushed farther under ground."

Mr. Bowman concludes his report on the gravel-ridges of the Placerville divide as follows: "Without attempting to trace the course in detail of this ancient river of Placerville divide, I may say that I could look over the entire country from Grey Eagle Mountain and Robb's and Tell's Mountains, and was on the ancient channel of Placerville at two points, Sportsman's Hall and Placerville, and I have no hesitation in stating it as a fact that the topography of the Pliocene period is preserved in the volcanic outflows, showing that the ancient South Fork cannot have varied much in its general course from that of the South Fork of the present day."

*Quartz-mines.*—The principal quartz-mines situated in the general zone above described are—

Between the South Fork of the American and the North Fork of the Cosumnes: The Grosch, Drew, Harmon, Sheppard, Pacific, Epley, Miller, Snyder, and numerous others.

South of the North Fork of the Cosumnes: The Lucky, Baldwin, Bacon, &c. The Havilah, at Nashville, one mile west of the Baldwin, was, according to Burlingham, the first quartz-mine worked in this State.

South of the South Fork of the Cosumnes: The Enterprise, Philadelphia, Alpine, Hooper, and others, at Plymouth; and the Keystone, Lincoln, Mahoney, Consolidated Amador, Oneida, Kennedy, and others, near Sutter Creek, in Amador County.

*Costs of materials, freight, labor, &c., in El Dorado County.*—Lumber, $12 to $17 per thousand at the mill. Freight, from Auburn to Georgetown, 50 cents per hundred; from Sacramento to Georgetown, $4.50. Wages, $2.50 per day; $30 to $40 per month and found; miners, above ground, $2.50; under ground, $3. Excavating, by Chinese labor, under contract, has been done at the rate of $2.50 per yard of ditch 3 feet deep, 6 feet on top, and 4 feet at the bottom. Water is sold at 10 cents per inch per day of ten or eleven hours; 20 cents per inch for twenty-four hours.

*Description of leading mines, El Dorado County, California,* 1874.

*Garden Valley district.*—Name of mine, Taylor mine; owners, Taylor Mill and Mining Company; length of location, 1,200 feet; course, northeast, southwest; dip 70° east; length pay-zone, 400 feet; average width, 3 feet; country-rock, slate; character vein-matter, ribbon-quartz; worked by shaft; depth of working-shaft, 408 feet; number of levels opened, 4; total length of drifts, 223 feet; cost of hoisting-works, $8,000.

Reported by William T. Gibbs, superintendent and one of the owners.

*Operations of leading mines, El Dorado County, California,* 1874.

*Garden Valley district.*—Name of mine, Taylor Mill and Mining Company; owners, Taylor Mill and Mining Company; number of miners employed, 20; miners' wages per day, $3; cost of sinking per foot, $30; cost of drifting per foot, $6.50 to $15; cost of stoping per ton, $3 to $5; cost of mining per ton extracted, $7.50; cost of milling per ton, $2.45; company works its own mill; average yield per ton, $13.75; percentage sulphurets, 1; total bullion-product, $15,000.

Reported by William T. Gibbs.

*Statement of quartz-mills, El Dorado County, California,* 1874.

*Garden Valley district.*—Name of mill, Taylor Mill; owners, Taylor Mill and Mining Company; kind of power and amount, steam, 40-horse

engine; number of stamps, 10; weight of stamps, 700 pounds; number of drops per minute, 87; height of drop, 7 inches; cost of mill, $15,000; capacity per 24 hours, 12 tons; tons crushed during the year, 300
Reported by William T. Gibbs.

## PLACER COUNTY.

Placer County lies between Nevada County on the north and El Dorado County on the south, and extends westward from the line of Nevada to the Sacramento Valley. The county is deeply eroded by three of the principal mountain-streams of the State, Bear River on the north, the North Fork of the American, running through the central portion of the county, and the Middle Fork of the American, which forms its southern boundary. The course of these rivers is from east to west, nearly at right angles to the system of dead rivers which traverses the county. By means of the great erosions of the modern rivers, which have cut gorges from 1,500 to 2,500 feet in depth, the ancient rivers have been exposed, permitting the great system of hydraulic operations for which this county is celebrated. The difference of level between the bed-rock of the ancient rivers and the modern streams is from 1,000 to 1,200 feet.

The principal hydraulic-mining districts of the county are situated in the vicinity of Dutch Flat and Gold Run, between Bear River and the North Fork of the American. These districts are on the same channel, and their interests are so closely allied and their characteristics so similar that they may be considered as one. The characteristic of this position of the blue lead has hitherto been the certainty and uniformity of pay of the surface-dirt. Within the past two years, as the natural outlet of the country became impeded by loss of grade and the accumulated *débris* and tailings resulting from ten years of hydraulic washings of the upper benches, extensive works were projected for opening the bottom of the channel, which, as has been demonstrated by the operations of the Cedar Creek claim of Dutch Flat, and of the Gold Run Hydraulic Company, and Indiana Hill Cement Company of Gold Run, is second to none in the State in extent and richness.

The general features of these districts have been so fully described in my reports for 1871, '72, '73, and '74, that I shall here notice only the most important works now in progress.

Hydraulic mining commenced at Dutch Flat, about the year 1856, and at Gold Run ten years later. The surface-product of these districts for a number of years after the introduction of the hydraulic process was not large, but the dirt was easily moved and yielded fair profits, while the natural grade of the country was open. With the exhaustion of the upper benches came a reaction, and for a few years mining was pursued on a comparatively small scale. In 1868, Mr. James Taeff, of Dutch Flat, sunk a prospecting-shaft in the ground now owned by the Dutch Flat Blue Gravel Company, and demonstrated the richness of the bottom of this portion of the channel. This shaft was sunk a distance of 240 feet through gravel, which steadily increased in richness until the well-known "blue-lead dirt" was exposed. This exploration proved an incentive to the large operations now in progress in these districts.

At Gold Run the yield of the upper benches was even less than at Dutch Flat. In 1871, Professor Pettee, then connected with the State Geological Survey of California, estimated that 43,000,000 cubic yards of dirt had been washed by the hydraulic process, which had yielded $2,074,356, or, approximately, 5 cents per cubic yard. Up to that period, however, only the upper stratum had been washed. As succeeding benches were removed the ground yielded larger returns. Here, as at Dutch Flat, prospecting-shafts revealed the rich nature of the bottom, and the discovery prompted the extensive works of the Gold Run Ditch and Mining Company now in progress. The "'49-'50," a prospecting-shaft, sunk from the bottom of a ravine intersecting the trend of the channel, had at a very early period proved the nature of the bottom gravel; but, the shaft having been filled for many years, the development made had become to some extent a matter of tradition, when in 1873 and 1874 the Gold Run Hydraulic Mining Company sunk a shaft on the Sherman & Cedar ground, belonging to the company, and demonstrated the existence of a bottom stratum as rich as that of the most favored districts of the State. This shaft reached bed-rock at the depth of 181 feet. The bottom-dirt exposed is said to have been "prospected" at the rate of $2 per pan, which would be equivalent to $130 per car-load of 16 cubic feet, or $232 per cubic yard.* While there is no special reason for doubting the accuracy of this "prospect," which was officially announced by the superintendent to the London owners, it will doubtless be the opinion of all competent judges that the shaft must have struck on an exceptionally rich spot, and that the test should not be made the standard of calculation of the value of the bottom. Probably the results obtained in the neighboring ground of the Indiana Hill Blue Gravel Mining Company afford a better basis of estimating the value of the blue-lead stratum of the district.

*Gold Run district.*—The Indiana Hill Blue Gravel Mining Company, of Gold Run, owns a claim which was worked from 1854 to 1864, at intervals, as a placer. In 1864 an 8-stamp mill was erected for the purpose of crushing the cemented gold-bearing gravel exposed by the company's tunnel. In April 1872 the company acquired by purchase the Warren and Remington ground, a tract of 9 to 10 acres adjacent to their original claim, and incorporated under their present name. The total area owned by the company is about 40 acres.

The claim is situated at the extreme southern limit of Gold Run district, at a point where the ancient river has been cut by the erosion of the North Fork of the American, which here passes through a gorge nearly 2,500 feet in depth, or 1,500 feet lower than the bed of the ancient river. The continuation of the Dutch Flat and Gold Run channel may be observed on the east side of the North Fork of the American, where it has been mined at Monona Flat, Independence Hill, and Iowa Hill. Owing to this favorable situation of the company's claim, no rim-rock tunnel was necessary; and the channel was followed northerly on the bed-rock by means of a main tunnel 1,600 feet in length, with branches or gangways toward the west, where the richest ground was discovered. The company employs 25 men, at $3 per day, who extract from forty to fifty car-loads per twenty-four hours, according to the nature of the ground. They are now carrying forward a "breast" 110 feet in width, and taking out from six to seven feet in depth of gravel, inclusive of the "bed-rock picking." The gravel is run out from the tunnel and breasts in cars and dumped in the mill near the mouth of the tunnel. The mill has eight stamps of 500 pounds, dropping 76 times per minute; height of drop, 8 inches. The motive-power is water. Seventy-five inches, with a head or pressure of 80 feet, are projected against a hurdy-gurdy wheel 10 feet in diameter. The crushing capacity of the mill depends upon the quality of the cement, which varies in respect to hardness in different portions of the mine.

* A "miners' pan" contains about 400 cubic inches of loose dirt.

The greatest amount crushed in twenty-four hours was 90 car-loads; the lowest amount, 35 loads; the average capacity of the mill is estimated at 42 car-loads per twenty-four hours. A car-load at this claim weighs 1,600 pounds and contains 19½ cubic feet. The company owns a ditch and water-right, supplying from 100 to 200 inches of water during the mining season, which is here nine months of the year.

The following tabular statement will show the operations of the Indiana Hill Blue Gravel Company from the date of incorporation, April, 1872, to the close of the mining season of 1874:

| Mining season. | Number car-loads. | Yield per load. | Total yield. | Total cost mining and milling. | Net profit. |
|---|---|---|---|---|---|
| 1872 | 3,680 | $5 27 | $19,410 97 | $7,360 00 | $12,050 97 |
| 1872–1873 | 6,300 | 4 00 | 25,200 00 | 13,175 00 | 12,025 00 |
| 1873–1874 | 10,017 | 3 08 | 30,811 50 | 20,034 00 | 10,777 50 |
| Total | 19,997 | .......... | 75,422 47 | 40,569 00 | 34,853 47 |

RECAPITULATION.

| | |
|---|---|
| Average yield per car-load, (19½ cubic feet) | $3 77 |
| Average yield per ton of 2,000 pounds | 4 71 |
| Average yield per cubic yard | 5 29 |
| Average cost per car-load, (19½ cubic feet) | $2 03 |
| Average cost per ton of 2,000 pounds | 2 54 |
| Average cost per cubic yard | 2 90 |
| Profit per car-load | $1 74 |
| Profit per cubic yard | 2 39 |

The product of the Indiana Hill Blue Gravel Company's ground from the commencement of mining-operations to the date of incorporation of the present company (April, 1872) is estimated at $125,000. If we add the product of the company from 1872 to 1874, inclusive, we obtain a total of $200,000, from one-twentieth of the area of the claim. The tabular statement above given includes only the yield up to September 12, 1874, the close of the fiscal year. The company resumed operations in November, and opened ground of unparalleled richness. In one instance $1,000 worth of gold was obtained from two car-loads of dirt. It will be observed that while the average yield per cubic yard has been double that of the Bald Mountain Company, of Sierra County,* the percentage of profit has been less. In the Bald Mountain claim the profit was 50 per cent., and in the Indiana Hill 46 per cent. The former company washes its dirt through 6,000 feet of sluices; the latter crushes in an 8-stamp mill. The Indiana Hill is the only notable successful enterprise in operation in California of mining gravel by the crushing or mill process.

The Gold Run Ditch and Mining Company, a local corporation, owns eight tracts of mining-ground, aggregating 328 acres, together with valuable water-rights and an extensive system of ditches, twenty-eight

*See, under heading of Sierra County, description of BaldMountain claim.

miles in length, taking water from Bear River and the South Yuba The company was incorporated in 1870, with a capital stock of $905,000. It has invested $220,000 in mining-ground and $80,000 in ditches and water-rights. Up to the present time the surface-gravel only has been worked, to a depth of from 100 to 175 feet, the lower stratum of over 200 feet in thickness not being workable for want of fall to carry off the tailings. The company is now engaged in the prosecution of an enterprise that cannot fail to have a very beneficial influence on the mining-interests of this section. For several years the yield of gold has been declining, and the surface-beds are rapidly being exhausted. The Gold Run Company has enough ground to keep its men employed for about two years on the present available grade. The richest beds of gravel, fully 200 feet in depth, are yet untouched. To procure the necessary fall and enable it to work these rich beds, the Gold Run Company is running a tunnel through the rim-rock, at a depth of 600 feet below the crest of the ridge. The lower end of this tunnel opens into Cañon Creek, while the head will open directly under the great deposits of blue gravel and furnish means of working them to the bed-rock. The main tunnel is 10 by 12 feet, and is intended to be 2,200 feet long, when it will tap the old 49-50 shaft, which is sunk on the channel of the blue lead to a depth of about 200 feet in fine hydraulic gravel. Four hundred and fifty feet from the mouth of the tunnel their branch for the Indiana Hill claim starts in. This branch is 8 by 8 feet, and is now in over 1,000 feet. It will be run 200 feet longer, when their shaft or incline will be raised. This tunnel was commenced in July, 1873, and has been run through very hard rock in about sixteen months, (actual working time perhaps not more than fourteen months.) The work has all been done with the Burleigh drill.

R. H. Brown, secretary of the Gold Run Ditch and Mining Company, writes, under date of June 5, 1874, to Parks & Lacy, agents of the Burleigh drill in this city, as follows:

We ran two of the Burleigh drills in our tunnel, twenty-four days during the month of May, and made 126¼ feet of tunnel, through hard rock, at an expense as follows:

| | |
|---|---|
| XX Hercules powder | $1,064 43 |
| Fuse and caps | 15 00 |
| Piping, air and water | 129 25 |
| Wood for boiler | 175 00 |
| Coal | 16 38 |
| Steel and iron | 50 00 |
| Oil for lights and machinery | 78 00 |
| Lumber tracking | 25 00 |
| Labor | 1,400 00 |
| | 2,953 06 |

The first week in June we made 31 feet of tunnel; the second week struck the most difficult rock to drill we have ever had. However, we made 26 feet of tunnel last week. We count on making 100 feet a month, and I believe we will average 125 feet until the branch is completed. We are running our rock out by hand. By using mule or steam power I think the expense could have been reduced materially this month. We have worked the car-men eight hours, and have not been delayed with rocks in the way of the drillers.

The main tunnel, which is 12 feet wide by 9 feet in height, has been driven into the mountain a distance of 600 feet, and when completed will be 2,200 feet in length. At a distance of 454 feet from the lower end a branch tunnel leaves the main cut. The branch is 8 feet square, and will be about 1,000 feet long. The whole force is now at work on the branch, which will be completed by the 1st of January, 1875, when washing will begin. The branch is run for the purpose of reaching claims not tapped by the main tunnel. As soon as it is finished work will be resumed on the main cut, and it will be put through in about two years, or by the time the surface-gravel belonging to the company is exhausted. The Burleigh drill is driven by compressed air. The work is now progressing (November, 1874) at the rate of 150 feet per month. Twenty-five men are employed, and about $3,500 disbursed monthly for labor and materials. The total cost of the tunnel is estimated at $125,000.

The Placer Argus, in its mining review, says:

> The importance of this enterprise is manifest when we consider that the ground, which by its aid alone can be worked, is far richer than the best of the gravel that has already been washed, and that there is enough of it to employ all the water that can be had for the next century. Shafts have been sunk 200 feet to the bed-rock, and good pay found the whole distance. Under the influence of this great work, which is to unlock the hitherto inaccessible riches of the district, every kind of business is brightening up, and the future prospect of the towns dependent on the mines are proportionately improved.

The Gold Run Hydraulic Mining Company, of London, England, owns the ground formerly known as the Cedar and Sherman claims, lying immediately adjacent to the properties of the Gold Run Ditch Company and Indiana Hill Company, above described. The company's claims are 2,400 feet in length by 600 feet in width. During the water-season two of Hoskin's giants are used for piping. For convenience of handling in front, and to gain a more concentrated stream, these pipes were made, by special order, about 4½ feet longer than those of ordinary construction, being 12 feet in full. Including waste-water, 840 inches are used daily. Water-rates in this district are 12½ cents per inch per twenty-four hours. The sluices of this company are 1,500 feet long and 4 feet wide; water, 4 to 5 inches deep; grade, 7 to 7½ inches in 12 feet. During the present year, a prospecting-shaft 181 feet in depth was sunk on this claim, disclosing a remarkably rich bank of gravel, which will be available for washing on the completion of the branch tunnel now being run from the Gold Run Ditch Company's tunnel to intersect the prospecting-shaft. As nearly 200 feet in depth of dirt has been removed by the hydraulic operations of ten years on this ground, the original depth is shown by this shaft to have been 380 feet from surface to bed-rock.

In addition to the companies above named, the following are engaged in extensive hydraulic operations in Gold Run district: Hoskins & Bro. are working a claim of 70 acres. They employ fifteen men and use 600 inches of water. The Fish-hawk Company has a claim of 40 acres, uses 550 inches of water, and employs ten men. Sachs & Co.'s claim contains 100 acres. They use 350 inches of water, and employ six men. O. Harkness uses 500 inches of water, employs seven men, and works a claim of 40 acres.

*Tailings.*—The removal and utilization of tailings—the *debris* of hydraulic or gravel ground, once washed and deprived of a portion of its auriferous contents—has for several years past attracted the attention of the miners of California. In all the streams forming the outlets of the gravel-mining districts tailings have accumulated to great depth, often "backing up" to such an extent as to destroy the necessary grade for the sluices and flumes of the miner, and causing the suspension of hydraulic mining until long and expensive tunnels can be run through the bed-rock, having their outlets below the accumulated *débris* of the district.

The contents of these tailings should bear a relative proportion to the product of the ground which forms their source of supply, and this proportion is governed by the natural advantages and mechanical appliances possessed by the owner of the mining-ground for saving his gold and amalgam. Where the miner possesses the advantages of great fall from his claim to its natural outlet, permitting the construction of numerous under-currents and "dumps" or "falls" on long lines of sluices, (as, for instance, at the American and other claims, on the ridge between North San Juan and French Corral, in Nevada County,) the tailings which eventually escape his sluices should be poor; while, on the other hand, in districts not possessing these natural advantages, (as at Gold Run,) we may reasonably expect to find rich tailings, since the disintegrated matter released by the water passes beyond the miner's control and on the property of others, bearing a considerable proportion of its valuable contents, which are deposited in the natural outlet, or caught in the flumes and under-currents placed therein by the owner of the creek-bed.

In the investigation of the value of the tailings deposited from year to year in the beds of streams forming the natural outlet of a hydraulic-mining district, we meet with many difficulties, owing to the absence of any data other than the results obtained by the owners of such claims; and these vary widely, depending on the original contents of the ground removed and the appliances, natural or artificial, of the mining-ground which formed their source.

The gold-product per cubic yard of gravel washed in the hydraulic-mining region of California has for several years been the subject of investigation, and it is an accepted result of numerous calculations and measurements that the quantity of gold obtained from the auriferous ground washed by the hydraulic system has varied in different districts from 5 cents to 25 cents per cubic yard. Where the results of washing large tracts, from surface to bed-rock, have been greater than 25 cents or less than 10 cents per cubic yard, they have been considered exceptional.

The auriferous contents of an area of tailings derived from the washings of a district not possessing the natural advantages of grade and fall should be greater than that of a like area "in place," since the tailings are the product of the concentration by specific gravity of the valuable particles of the washed ground, viz, the gold and quicksilver carried over and through the miner's flume and the resulting amalgam, all of which are deposited in the bed of the stream which forms the ultimate receptacle of the tailings, excepting, of course, the proportion caught in the miner's sluice-boxes, which depends on his appliances, natural or artificial, for their detention. Thus a cubic yard of tailings may represent the residual product of many cubic yards of standing ground.

The expense of running a tailing-claim, after the first outlay in the construction of the necessary flumes and under-currents, is less than that of any other class of mining. The principal items are, lumber for repairs of flumes, iron bars for under-currents, labor in cleaning up and watching against sluice-robbing. Claims of this class are "cleaned up" once a year, after the close of the mining-season. The blocks with which the flume is lined are taken up, the amalgam is gathered, and the blocks are replaced for the run of the next season.

The Moody & Kinder claim, of Gold Run, is of the class known in the hydraulic-mining counties of California as a "tailing" or "fluming" claim, and its value as a mining-property depends on the quantity and richness of the tailings which have accumulated or are passing over the flumes and under-currents placed on the bed of the creek for the purpose of saving the gold, quicksilver, and amalgam which escapes the miner on the hydraulic ground above. Moody & Kinder own 5,000 linear feet of the bottom and sides of Cañon Creek, which is the natural outlet of one-half of Gold Run district, comprising 400 to 500 acres of hydraulic ground. This creek rises in the high divide between the waters of Bear River and the North Fork of the American, pursues a southerly course, and empties into the great gorge of the American River, below Gold Run district. On the west it is intersected by three ravines, which form its principal feeders from the gold-washings of the district. Above the intersection of Gold Run ravine, which is about the center of the district, the creek has a light grade; and here the tailings resulting from ten years' hydraulic washings have accumulated to the depth of from 30 to 40 feet in the center of the channel, and the width of from 50 to 60 feet. Below Gold Run ravine the grade becomes steeper, with occasional falls, and hence there is a diminished depth of tailings. The claim terminates at a high dam near the mouth of the Gold Run tunnel. That portion of the claim above the intersection of Gosling ravine contains about 20,000 cubic yards for each section of 100 yards in length.

Taking the line of the Central Pacific Railroad as the plain of the surface of the country before the commencement of hydraulic washings, we find the difference of elevation at the intersection of Gosling ravine with Cañon Creek to be 300 feet; and this would represent the average height of the hydraulic banks removed during the past ten years, over an area of several hundred acres, which have found their outlet in Cañon Creek. From the head of the first system of flumes to the lower end of the claim the average grade of the creek-bed is about 300 feet to the mile.

The estimates of the value of accumulated tailings differ so widely, that we shall give, without comment, the product of the runs of the past four years over Cañon Creek.

The gold-saving appliances of the Moody & Kinder claim consist of ground-sluices, and flumes, and under-currents, constructed on the most approved methods at present in use, combined with a skillful application of the natural advantages resulting from the occurrence of rapids and falls in the lower end of the claim. Below Gosling ravine, the most northerly of the three tributaries of Cañon Creek, they have constructed a flume 2,400 feet in length, 16 feet in width, and 7 feet in height, divided by a central partition, and terminating in a double under-current at the head of the falls or rapids of the stream. Below this point, several shorter sets of under-currents and flumes have been placed, where the nature of the ground would permit.

The following statement from the books of Moody & Kinder will show approximately the returns from their tailing-claim for the past four years. During one of these years the privilege of cleaning up was sold; and we give the price obtained by M. & K., having no means of ascertaining the amount realized by the purchaser:

| | |
|---|---|
| 1871 the clean up and sale realized | $30,000 |
| 1872, clean up | 34,000 |
| 1873, clean up | 24,250 |
| 1874, clean up and sale | 25,290 |
| Total | 113,540 |

The expenses for four years were $18,000.

*Dutch Flat district.*—This district adjoins Gold Run district on the north, the Central Pacific Railroad running on the line of the boundary of the two districts. The leading claims of the district have been repeatedly described in former reports, particularly in those of 1871 and 1874. On the Cedar Creek claim an English company is engaged in hydraulic mining on an extensive scale. It owns 200 acres of gravel-ground, opened by a bed-rock tunnel 3,000 feet in length, and 8 by 8 feet in dimensions, together with sixty miles of ditches, carrying 6,000 inches of water. The enterprise may be classed among the most promising in this class of mining in the State.

At Dutch Flat giant-powder is used to break up the bowlders found in the hydraulic banks so that the fragments will pass through the sluices. A system of "top-blasting" is practiced, followed by breaking open giant cartridges and pouring the powder on to the rock. This is sometimes built over with clay to hold it in place or give greater effect to the blast. A fuse and cap explode the charge. Huge bowlders which cannot be sledged, nor blasted (without drilling) by black powder, are thus quickly reduced, pitched into the sluices, and entirely got rid of. Pipe-clay, the terror of many mining-claims in early days, is now disposed of by blasting with giant-powder. After the blocks of clay are rolled down from the face of the diggings by the powerful hydraulic pipes, a "clay-augur" is sent with ease into the center of each. This is followed by one-quarter or more of a cartridge of giant-powder, with fuse attached. When all is prepared the miner fires in succession perhaps twenty shots by aid of a hot iron and rod. This disintegrates the cement so that it is practicable to reduce it by action of the water, and prevent solid particles of clay from robbing the sluices, as formerly, by carrying off gold-dust imbedded in its sticky surface.

This district was the first to introduce the great improvements in hydraulic pipes and nozzles which have since been so generally adopted through the State.

The Little Giant, figured and described in former reports, having been first introduced, is the machine in most general use. It is especially adapted for heavy pressure, is simple and durable. Some of these machines are in use under a pressure of 500 feet, and in one instance over 535 feet, and there has been no accident from breakage or difficulty in working. They are made of five sizes, varying from 7 to 15 inches inlet, and adapted for works any size up to 8½ inches.

The Clipper has been but recently introduced, and differs from the Giant in having but one working-joint instead of two, there being no center-bolt, and in requiring no balance. It turns easily, and, as in the Giant, the discharge-pipe remains in any position in which it is placed.

*The Forest Hill divide* embraces the country lying between the North and Middle Forks of the American River. Although a country of extensive mining resources, it does not possess any general system of ditches yielding a summer supply of water; and hence the mining-industry holds a place far below its true merits. The deficiency of water has been, however, to some extent relieved by the construction of the Iowa Hill Canal, (noticed elsewhere in this report,) which supplies the country lying adjacent to the North Fork and between the North Fork and Shirt-Tail Cañon, but does not extend its branches to the Middle Fork gravel-deposits of Michigan Bluff, Bath, Forest Hill, and Todd's Valley. The deep depressions of the Middle Fork on the one side and Shirt-Tail Cañon on the other—each nearly 2,000 feet lower than the summit of the ridge—present a formidable obstacle to the introduction of water on the south slope of the divide.

Forest Hill divide, like so many other mining-regions of the Sierra

Nevada, became almost depopulated after the first historical era of gold-mining in California. It is, nevertheless, one of the most productive mining-districts of the State. Within half a mile of the town of Forest Hill it has been estimated that from $5,000,000 to $10,000,000 have been taken out. The mines began to decline in 1858, at the time of the Frazier River excitement. Prior to that, drifting was the customary method of working, and the lower strata of concentrated gravel yielded enormously.

Mr. Bowman observes that in places on the Forest Hill channel gold has been found in crystals along with quartz-crystals, the angles of which only were broken off. This probably cannot have been situated on any central wash of the ancient channel. The gravels of Forest Hill divide constituted the ancient Middle and North Forks of the Middle Fork of the American.

The following is the reported total yield of leading claims on this divide: The Dardanelles mine with a frontage of 1,000 feet, $2,000,000; the Jenny Lind, with a frontage of 450 feet, over $1,000,000; the New Jersey, with a frontage of 600 feet, and from an area of 500 by 400 feet, $850,000; Deidesheimer, $650,000; the Independence, $450,000; the Fast, Rough and Ready, Gore, and sundry others, each about $250,000; and many claims from $50,000 to $200,000 each. The Gore claim had a frontage of 100 feet and a depth of 200 feet in the hill. About $40,000 from the Gore claim was derived from a basin 380 feet square. The Independence yielded $10,000 from a space 20 by 20 feet.

At Todd's Valley, Pond & Co. now pretty nearly monopolize all the mining-ground. The town itself is comparatively depopulated. The Blue Gravel Range Company occupies the ground of the old "Dardenelles," with Spring Garden ranch and the Powell claim, embracing nearly two miles on the old channel. On the Powell claim a tunnel was run in 750 feet long, but it being found too high to drift the ground to the bed-rock, an incline was sunk from the end 120 feet, and a chamber was excavated, where hoisting and pumping engines and a chimney-shaft were put in. The Mountain Company has a tunnel in 2,600 feet, and a shaft down from the end 60 feet. It was reputed to be netting, several years ago, $1,000 per month. All the claims on the Forest Hill ridge extend through a big channel to the Devil's Cañon, on the north. The tunnels were generally run in too high.

Yankee Jim's, another depopulated, historical town, has yielded its millions, and employed many thousands of miners for ten or twelve years.

Bath had one hundred inhabitants left in 1870. The pay is in hard cement, which has to be thoroughly slacked before washing. Messrs. Rousch & Grinnel, of this place, work their claim by the hydraulic process during the brief water-season, and the results obtained by them this season demonstrate their ground to be equal in richness to any in the divide, if not in the State, probably yielding from 30 to 40 cents per cubic yard of dirt removed.

At Michigan Bluffs there is in sight an unusually large quantity of concentrated pure quartz-gravel, which, in the opinion of Prof. W. P. Blake, is of local origin; though the fact that so little country-rock is intermixed with a large quantity of thoroughly-washed quartz-bowlders would seem to imply that the material had traveled a considerable distance. In the deposits of lighter material there is observable a diagonal stratification caused by varying currents, the evidence being in favor of a local northeasterly and southwesterly course of the channel, like that of the North Fork of the Middle Fork and its branches. The large bowlders in other portions of the diggings testify to the presence, at the time of formation, of a very large quantity of water and a violent transporting-power. One of the claims yielded $48,000 in five months, to nine men, employed night and day, at an expense of $13,000, leaving a profit of $35,000. The usual yield was formerly from six to eight dollars a day of ten hours' work to the man; in some claims from $20 to $30. A four-pound nugget was taken out of these mines in early times, showing that a portion of the gold, at least, was probably of local origin.

Messrs. Wheeler & Breeze, owners of the Paragon claim, situated near Bath, are engaged in hydraulic mining on an extensive scale. This ground was in former years worked by the mill or crushing process, owing to the hardness of the bottom. Several runs were made, with a yeld of $200 per day; but, the ground not proving uniformly rich, this system of mining was abandoned here, as it has been elsewhere in the State, with two or three exceptions, and the hydraulic method substituted. The hardness of the bottom renders blasting necessary. During the early part of the season an extensive blast was fired in this claim, with successful results. Seven hundred kegs of powder, containing 17,500 pounds, were fired by electrical apparatus. The powder was placed in the drifts in iron kegs. The drifting necessary to put off this blast cost $300, and powder, materials, &c., cost $2,700 more. As the owners estimate that from the effects of this single blast they will be able to clean up about $200,000, the outlay of $3,000 was extremely small in view of the benefits derived from it.

In preparing the ground for this blast, drifts were run as shown in the following diagram:

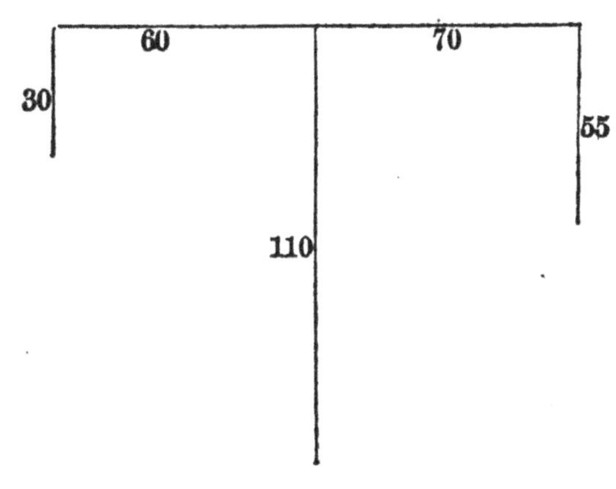


The drift from the face of the bank was 110 feet long. From the end of this a drift was run one way 70 feet and the other way 60 feet. From the end of the 70-foot drift an L was run toward the face of the bank 55 feet; from the end of the 60-foot drift a short drift 30 feet long was also run back toward the face of the bank. In the drifts to the right 400 kegs of powder were placed, and the opposite side, 300 kegs. The mouth of the drift was tamped for 75 feet from the near end, and the T-drifts were tamped 10 feet each way. Plenty of space was left in the L-drifts for the expansion of gases generated by the explosion of powder. The tunnel was 4½ feet high and 5 feet wide, and the bank was 140 feet high. The electrical apparatus by which the blast was fired was 450 feet from the mouth of the tunnel. The whole length of wire was about 1,500 feet. There was 25 feet of tamping—10 feet on each side and that in the width of the tunnel—between the two L-drifts, not counting the 75 feet between that and the outer air. If this blast had been fired by means of fuses, and one side had gone off before the other, much less execution would have been done. Being fired by electricity, the discharge was simultaneous, and was exerted on both sides of the tamping at the same instant, so that the explosive force of the powder was exerted in the proper direction only.

The expensive and slow system of crushing cement in stamp-mills has been superseded, wherever water-facilities will admit, by bank-blasting and hydraulic washing.

Above Bath the country is covered with heavy layers of volcanic *débris*, and mining is carried on by the "drifting" system.

The Hazard mine, now worked by a San Francisco company, is situated on the Volcano Cañon, about one and a half miles from Michigan Bluff. It was located in 1858, by Forrest, Allen, and others, who sunk a shaft 150 feet deep, and found good prospects. On account of the insufficiency of their machinery, they were unable to work to any advantage, the water being in the way. They sold the claim in 1872 to the present company, who commenced operations by putting in an engine of 25 horse-power, hoisting-works, and a 6-inch Cornish pump of 4 feet stroke. They sunk the shaft a few feet deeper and ran a tunnel on the channel 900 feet to the east. The company is working from twelve to sixteen men. They own one hundred and fifty acres of mining-ground, by purchase and location, and have the benefit of two miles of cañon for under-currents, flumes, &c. The bed of pay-gravel is from 6 inches to 6 feet in thickness, the balance of the ground being hard cement. They have already expended about $40,000 in developing the mine.

The Iowa Hill Canal, taking water from the natural streams and basins of the high Sierra, between the North and Middle Forks of the American River, delivers it on the rich extent of mining-ground of which Iowa Hill is the center. From Bird's Flat to King's Hill, two miles below Elizabethtown, on the south side of Indian Cañon, is three miles; from Smiley's store, through Iowa Hill Town, to the second Sugar Loaf, on the north side of Indian Cañon, is three miles, containing, together, on a fair estimate, about 3,000 acres of ground, ranging from 40 to 150 feet deep, which has been proved, by actual but limited washing, to be rich from the surface to the bed-rock. Then there are Wisconsin and Prospect Hills, Sucker and Grizzly Flats, with other places lying intermediate of these points, known to contain gold in paying quantity, all of which are now being covered by branch ditches from the main canal. Over and above all this territory already developed, there is on the line of the main canal a region containing a gravel-deposit, of which nothing more is known than that it carries gold strongly and can be worked to a great advantage, stretching from Iowa Hill ten miles to Damascus on one side, and some fourteen miles to the Forks House on the other. Even beyond this point there is an immense field for successful hydraulic mining.

The main canal is 5 feet on the bottom, 4½ feet deep, with a capacity of carrying 7,000 inches of water. The first large reservoir is about eight miles from the town of Iowa Hill. This contains 48 feet of water and covers 100 acres of ground. From this central reservoir the main canal proceeds past the Forks House to Tadpole, New York, Sailor, and Long, tapping all in turn, to the North Fork of the American River, making a total distance of forty miles. The total cost, with reservoirs, will be $500,000, of which $100,000 has already been expended. To insure a constant supply of water to the capacity of the ditch throughout the year, the company have several excellent reservoir-sites in different places, namely: one at Sailor's Cañon, 25 acres; one at Big Cañon, 25 acres; and several at the head of the ditch, which probably will cover 500 acres more. In the event of these not being sufficient, a tunnel will be run through a gravel-ridge 2,500 feet long, to connect with the waters of the Middle Fork of the American River. The country in the immediate vicinity of Iowa Hill has only had water for a brief season of the year, and, under extreme disadvantages, has yielded from $150,000 to $200,000 per season.

*Ophir and Auburn districts.*—The mines of these districts are all near the line of junction between the slates and the foot-hill granites bordering Sacramento Valley. The best-paying mines are situated near the line of contact, which has the same course in general as the strike of the slates. Auburn is in slate, Ophir in granite. The junction occurs very nearly half way between the Auburn and Ophir mines. Further down the mountain-slope the country is of granite, as far as there is any rock visible, to Sacramento Valley. Both the granite and slate, in this region, are very hard at 100 feet depth. On the surface they have, in some places, weathered soft to the depth of 20 feet.

The veins or seams in the district all have the same course, parallel to the strike of the slates—north-northwest and south-southeast. The quartz and vein-material generally, including the ore, is pockety in character, resembling, in that respect, all the pay-mines described as worked in Georgetown and Placerville divides. Lenticular masses continue for some distance, and then pinch out. There is always a fissure, however, that continues, leaving no doubt that they are true fissure-veins. The quartz is from 6 inches to 4 feet wide. The pay is in the form of flakes and sheets. Most of the ledges are impregnated with sulphurets of iron, which assay never less than a few dollars to the ton. The country-rock adjacent is considerably impregnated with sulphurets. The yield varies greatly. In the case of a replevin suit, thirty sacks of ore out of the Ophir mine contained $2,300. A good deal of the rock is worth several hundred dollars to the ton. A rich pocket was struck in the seam-mine, however, which was almost all gold. With a hand-mortar a man might frequently make $50 a day out of the vein-material.

The principal mines of Ophir and Auburn districts are the Saint Patrick, the Bellevue, the Green, the Julian, the Ophir, the Saint Lawrence, and the Booth. The three first-named have attained a depth of from 250 to 300 feet, and are the deepest in the district. All of these mines, except the Booth, which is a new discovery, have been noticed in detail in former reports.

In the granite-belt near Ophir the surface of the country is intersected in all directions with veins and seams of quartz, often of a "pockety" character, and having but little regularity in mode of occurrence or in pay. These veins are worked, in the summer months, by means of hand-windlasses or horse-whims, by prospectors, who manage to realize sufficient to support them through the rainy season, when this method of working is not practicable, and occasionally to make a "strike" which yields them a few thousand dollars. These periodical "finds" are the incentive to this class of mining, rather than the legitimate profits of the average of the quartz-veins. The district is well provided with mills, which charge from $3 to $4 per ton for crushing. The cost of hauling rarely exceeds $1 per ton. The cost of mining in this manner is from $2 to $3 per ton, which leaves a living profit on $10 rock. The Saint Patrick, Bellevue, Julian, and Saint Lawrence, however, in the same granite-belt, have defined ledges and powerful hoisting-works; but it must be admitted that deep mining in this district has not proved uniformly profitable. To the eastward, in the metamorphic-slate belt, the ledges are stronger and the pay more regular, the Green and the Booth being representatives of mines in that formation.

Placer County contains numerous ledges of quartz; but outside of Auburn and Ophir districts but little progress has been made in their development, and this not of an encouraging nature. A notable exception, however, is the Rising Sun mine, near Colfax, which has for several years been under the skillful and prudent management of the Messrs. Coleman, of Grass Valley, Nevada County. This mine has yielded over $300,000, of which $100,000 has been profit. Heretofore the mine has been worked on a small scale, there being but a single battery of five stamps and a single shaft. A new shaft has been sunk about 100 feet east of the old one. It is down 500 feet, in good-paying ore, the ledge being fully two feet in thickness. The shaft is 12 feet by 4½, giving a double hoisting-shaft and a compartment for the pump and ladders. A new building, large and well designed, has been erected, in which a new engine, in addition to the two now in use, will be erected, furnishing ample power for hoisting, pumping, and driving the stamps. A battery of 10 stamps is likewise being erected. The entire cost of the improvements this season will be about $30,000. The ore taken out is uniformly rich. There are no "specimens," the gold being diffused quite evenly through the rock; but the whole ledge pays well for working, and, with the amount now in sight, gives assurance of paying well for an indefinite length of time.

The Manter mine, situated in the foot-hills near Lincoln, has been extensively developed; but the ores have so far proved refractory, and but a small percentage of gold was saved. This mine has recently been sold to an English company. There is perhaps no subject of more importance to the mining-interest of California than the economical treatment of gold-bearing ores, and it is a fact worthy of consideration that, after twenty years' experience in the business of quartz-mining, it is conceded that, with a few exceptions, from one-third to one-fourth of the assay-value of the ores now being worked, amounting to several millions of dollars annually, passes off in the slums, and is irretrievably lost. This proposition is clearly supported by the results of the experiments of Mr. G. F. Deetkin, of Grass Valley, Nevada County, as described in his article on gold-bearing ores, in my report of 1873. Yet, during this period, immense sums of money have been expended in machinery and inventions, having for their object the introduction of some system or process by which our gold-bearing ores could be worked up to a higher percentage.

Among the practical miners who have devoted time, money, and a close observation to the causes of this loss of gold is Col. I. M. Taylor, of San Francisco, formerly superintendent of the Manter mine, near Lincoln, Placer County, who has during the past year closely investigated this subject, and arrived at the conclusion that the principal loss of gold is to be traced to the imperfect methods of concentration and amalgamation.

Colonel Taylor gives the results of his observations and experiments in the following article:

CONCENTRATION.—I commenced this business believing, as many other theorists have done, that all gold-ores could be concentrated on coming from the battery by machinery without handling, and that gold-sulphurets could be treated successfully only by the chlorine process. I continued in this belief for many years, spending time and money, and accomplishing nothing. Some eight months since, in connection with moneyed men of San Francisco, I purchased a mine which had the reputation of producing ore of a very refractory character. This ore assayed $30 per ton, but not more than one-fourth of it was sufficiently free to admit of its being amalgamated in the battery. I erected a five-stamp mill, and tried various methods for concentrating the sulphurets. The best result obtained was 25 per cent., which, together with the free gold, formed only 50 per cent. of the assay-value of the ore. At this juncture I abandoned everything with the word patent on it, and, going back to first principles, constructed an old-fashioned Cornish buddle, and sized the ore in two sizes, using two pointed boxes, after the plan adopted by the most improved mills in Grass Valley; all the materials held in suspension by the water were allowed to pass over the second box and go to waste. I found by concentrating the two sizes separately in the buddle that I could get about 8 per cent. more than when they were concentrated together. In this way 10 per cent. more was saved than by any other plan yet tried.

The tailings as they came from the buddle were assayed, and found to contain 10 per cent., leaving 30 per cent. unaccounted for. A tank was then constructed 12 by 12, with a partition in the center, and the slum that ran over the second box was allowed to pass into the one and out of the other, giving it plenty of time to settle. In this way one-fifth of all the ore crushed was settled in the tanks, the contents of which assayed 23 per cent., being at the rate of about $6 per ton of ore, making a saving of an assay of $48 per day with an 8-ton mill. Deducting from this 10 per cent. for loss in concentrating, 25 per cent. for working, and $1 per ton for cost of concentrating, resulted in a net profit of $24 per day, or $720 per month, to the mill. A barrel holding 60 gallons was placed under the stream of water from the tank, and when full was left twenty-four hours to settle, a little alum having been added. The top was then carefully poured off, when the sediment was found to contain about 1 per cent. of the ore, which was held in suspension by the water after it had become comparatively clear; 8 per cent. could not be accounted for. It could easily have been wasted in the battery, or more than an average might have been got in sampling the mine. This latter was hardly possible, however, as great care was taken to insure a fair sampling by drilling through the ledge in various places. This result did not surprise me in the least, having long been aware that a large percentage from most mills had been lost in this way. The question was how to concentrate these tailings up to a higher grade, they not being rich enough to pay for chlorinating. Various methods for accomplishing this were tried. The best result from the round buddle, using an ordinary broom for sweeping, was 50 per cent. A buddle was then constructed on a larger scale, and with much less grade than the one already in use. A piece of common mill-blanket was put on the arm for sweeping, and a small stream of water turned on; this proved a success, as shown by assay, 12 tons having been reduced to 1 ton, at a cost of 75 cents per ton. There was still a loss of 10 per cent. Various tests were made in order to determine what grade of sulphuret ore would pay to concentrate. Some 20 tons of coarse tailings had accumulated from the buddle, which, after testing, proved to be worth $2.25 per ton. This was reduced, at a cost of $7.50, to 500 pounds, which had an assay-value of $30. Deducting from this 25 per cent. for loss in working, and $7.50 for labor of concentrating, left a net profit of $14. I now became satisfied that no machine yet invented can concentrate the majority of ores to more than 50 per cent. of their assay-value without their having first been sized and settled in tanks. Assuming that ore requires settling before it can be concentrated up to a high percentage, it is only a waste of time and money to attempt its concentration before settling, as the cost is the same whether it be high or low grade. As a consequence, any machine that fails to take out more than half the value of the ore is of no practical use. All ores must be sized in three different sizes before they can be properly concentrated. Lead-sulphurets are nine and one-half times heavier than water, and five times heavier than quartz. Common iron or copper sulphurets are seven times heavier than water, and three times heavier than quartz. The coarse pulp and sulphurets capable of passing through an ordinary No. 6 mill-screen are, perhaps, on an average fifty times coarser than those found in the slum-ores. They should, therefore, be concentrated separately, otherwise a current of water sufficient to carry off the coarse pulp will also carry off the sulphurets, notwithstanding the latter are from three to five times heavier than the pulp, which, being composed of quartz about fifty times more bulky than the sulphurets, exposes a corresponding surface to the action of the water.

To obviate this trouble, we must equalize the tailings, bringing the sulphurets and the pulp to the same size, then equalize the water to correspond with the fineness of the pulp, and a current that will carry off the latter will leave the heavier sulphurets behind. No man experienced in milling will ever spend a dollar trying to concentrate in violation of these rules or natural laws. Any of the quick-motioned concentrators now in use will separate a large percentage of the coarse sulphurets from the coarse sand, but at the same time they will hold the fine sulphurets, which are of the most value, in suspension so long as the water continues in motion. The round convex buddle is the best equalizer in use, it being fed around the center-post, which is about 12 inches in diameter. As the water recedes from the center it spreads, and consequently decreases in force. If the current of water be strong enough to start the finest sulphurets from the head of the buddle, it will become so diminished before reaching half the distance from the center to the circumference that the sulphurets are left behind. The concave or center-discharging buddle is fed on the outer rim and discharged in the center, consequently the water increases in force toward the center; hence fine sulphurets leaving the circumference of the buddle will be carried toward the center with the pulp by the increasing force of water. In the tin, lead, and copper mines of England, where concentration has been carried to a higher state of perfection than anywhere else, they have long since discarded the center-discharging buddle, and use only the convex. Many mining superintendents contend that their ores are not rich enough to justify handling and concentrating in round buddles, but I am of opinion that all ores below permanent water-level will pay to concentrate if they will pay to work at all. * * * It has yet to be practically proved that any air-concentrator will successfully concentrate the great mass of gold-ores. After the concentration was perfected, 200 tons of ore were run through the mill, and further concentrated to 15 tons, which were shown by assay to contain 91 per cent. of the gold found in the ore after being settled in the tanks, and before it was concentrated. This second concentration cost 63¼ cents per ton.

AMALGAMATION.—The business of concentrating gold-bearing ores, though not without its difficulties, is yet simple and inexpensive compared with that of their amalgamation. The former is mechanical, the material requiring to be handled according to its specific gravity. Any machinery or method of handling that will answer for one ore will do also for another, provided the weight, bulk for bulk, be the same; and where this is not the case, all that is necessary is to so equalize the air or water as to correspond with the weight of the pulp. In the amalgamation of ores, however, we have to deal with chemical as well as mechanical laws and agents, rendering the business much more costly and complicated. The treatment that will answer here for one class of ore fails when applied to another, owing to the presence of different minerals, or to the same minerals being present in different proportions, causing chemical combinations in endless variety. Although the desulphurizing and amalgamation of gold-sulphurets has for many years been extensively experimented upon in California, no plan has been brought into general use whereby refractory ores can be satisfactorily treated or low-grade sulphurets worked with profit, the only tolerably successful method for reducing sulphurets being by the chlorine process—one that is attended with too much expense to answer for low-grade material. Some amalgamators contend that gold, being found in a metallic state, only requires grinding to a certain degree of fineness to admit of amalgamation. This rule, however, will not hold good with a majority of ores, if, indeed, it will with any. However fine the gold may be, it is more or less coated with sulphur, iron, or other base metals, from which friction fails to free it. There is no limit to the divisibility of gold. A certain percentage of gold in all sulphurets must be submitted to the action of either fire or chemicals before it can be freed from these coatings. Either will answer, the question being which is the most economical. Nitric acid desulphurizes very effectually, but is too expensive for general use, though on high-grade ores, which are difficult to treat, it can sometimes be employed to advantage.

In the course of my experiments the plan was tried of grinding from four to twelve hours in a Hepburn roller-pan and amalgamating in a wooden-bottomed settler, the best result obtained from clean sulphuret being 40 per cent. This ore under the same treatment yielded from 50 to 60 per cent. without concentrating, the quartz assisting in freeing and brightening the gold during the process of grinding. This ore was also experimented with in a Varney pan, grinding from twelve to twenty-four hours, with but little better results. A reverberatory furnace on a small scale was then constructed with a view to desulphurizing and amalgamating in pans. No difficulty was found in desulphurizing, but with a reverberatory furnace it was found impossible to oxidize the iron and copper. All gold-ores carry more or less iron, while many contain sulphurets of lead and copper, all of which must be thoroughly oxidized before the gold and silver can be amalgamated in pans. If the iron and copper are not oxidized before being ground in the pans, they will pulverize into a fine powder, resembling emery, which, on coming in contact with the quicksilver, changes it from negative to positive, in which condition it has no affinity for gold. Any portion of the quicksilver that fails to flower and rise to the surface of the water will become coated over with a black scum from the iron and copper, and no satisfactory results can be obtained. Thoroughly oxidize the ore, however, and the iron and copper will dissolve and be held in suspension in the water, leaving the gold free to be taken up by the quicksilver. Any failure in getting a good result in amalgamation when the ore is in this condition is owing to mechanical causes, and not to the chemical condition of the ore. The gold in some ores being very fine, it is difficult to bring all the particles in contact with the quicksilver, even if it be free. The best result obtained from ores imperfectly oxidized was from 50 to 60 per cent. of the assay-value, which could only be obtained by using salt in grinding. This method possesses no economy over the chlorine process, notwithstanding the ore, after being roasted and amalgamated, can be reconcentrated in the slum-buddle, and all that remains in it can be brought up to a higher grade than before roasting. Ores, I found, concentrated much better after roasting than before, everything in them losing specific gravity but gold and silver. The difficulty of oxidizing ore in a reverberatory oven consists in not being able to furnish it with a plentiful supply of oxygen. In order to do this the ore must come in contact with air. The oxygen contained in the air taken in through the furnace is destroyed by heat before it reaches the ore. Again, the air cannot get at the ore so long as it lies in bulk on the floor of the oven, no matter how much it is stirred. It must be varied and allowed to fall through the air after it has attained a proper degree of heat, the air being supplied in some other way than through the grate.

A furnace constructed on the principle of the Stetefeldt, which would continue raising and lowering the ore through the air after it had reached a proper degree of heat, would thoroughly oxidize and desulphurize any gold it might contain in from one to three hours. An instance establishing this fact might be given: 25 pounds of ore was placed on a small sheet-iron furnace constructed for the purpose, and when at a red heat was raised and allowed to fall through the air for three hours, after which it was ground and amalgamated in a small Varney pan. By this treatment it was made to yield from 70 to 80 per cent. of the gold it contained.

Without indorsing in all points Colonel Taylor's conclusions, I give the foregoing article in full, because everything in the nature of intelligent experiment, free from preconceived notions or the hallucinations of inventors, is extremely valuable in this field.

*Description of leading mines, Placer County, California, 1874.*

| Name. | Location. | Owners. | Length. | Course. | Dip. | Length of pay-zone. | Average width. | Country-rock. | Character of vein-matter. | Tunnel or shaft. | Length of tunnel. | Depth on vein in tunnel. | Depth of working-shaft. | No. of levels opened. | Total length of drifts. | Cost of hoisting-works. |
|---|---|---|---|---|---|---|---|---|---|---|---|---|---|---|---|---|
| | | | *Feet.* | | | | | | | | *Feet* | *Feet.* | *Feet.* | | *Feet.* | |
| Crandall ........ | Auburn ........ | J. R. Crandall ........ | 1,400 | 12° S. of E. | S. .... | .... | 4 feet. .... | Slate ........ | Quartz, sulphurets. | Shaft .... | 300 | 55 | 115 | 1 | 50 | $1,000 |
| Orleans ........ | Auburn ........ | Booth Gold-Mining Company. | 3,000 | E. and W. | S. 66° | 350 | 3½ feet .... | Slate, metamorphic. | Iron and galena sulphurets; quartz, auriferous. | Shaft .... | .... | ...... | 160 | 2 | 300 | 100 |
| Bellevue ........ | Ophir ........ | Bellevue Mining Company. | 2,000 | E. and W. | S. 75° | 100 | 15 inches .. | Slate ........ | Sulphurets. ....... | Shaft .... | .... | ...... | 280 | 6 | 1,413 | 2,000 |
| Crater ........ | Ophir ........ | St. Patrick Gold-Mining Company. | 700 | E. and W. | S. 40° | 300 | 15 inches .. | Granite .... | Sulphurets. ....... | 3 shafts | .... | ...... | 1,490<br>1,200<br>390 | 8 | 1,950 | 4,000 |

*Operations of leading mines, Placer County, California, 1874.*

| Name. | Location. | Owners. | No. of miners employed. | Miners' wages per day. | Cost of sinking per foot. | Cost of drifting per ton. | Cost of stoping per ton. | Cost of mining per ton extracted. | Cost of milling per ton. | Company or custom mill. | No. of tons extracted and worked. | Average yield per ton. | Percentage of sulphurets. | Total bullion-product. |
|---|---|---|---|---|---|---|---|---|---|---|---|---|---|---|
| Crandall ........ | Auburn ........ | J. R. Crandall ........ | None at present. | $3 | $20 per 6 f. by 8 f. | .......... | ...... | ...... | $3 to $5 | Custom ... | 208 | $4 85 to $22 | 5.31 | .......... |
| Orleans ........ | Auburn ........ | Booth Gold-Mining Company. | 10 | 3 | $10 00 | $5 00 | $1 00 | $2 40 | $3 00 | Custom ... | 2,008 | $23 00 | 5.00 | $46,540 20 |
| Bellevue ........ | Ophir ........ | Bellevue Mining Company. | 20 | 3 | 50 cents per cub. ft. | $6 per linear foot. | 12 00 | 9 00 | 5 00 | Custom ... | 600 | 18 00 | 1.00 | 10,800 00 |
| Crater ........ | Ophir ........ | St. Patrick Gold-Mining Company. | 35 | 3 | 50 cents per cub. ft. | $7 per linear foot. | 12 00 | 9 00 | 4 00 | Company . | 1,586 | 22 00 | 1.00 | 34,000 00 |

*Statement of quartz-mills, Placer County, California*, 1874.

*Ophir district.*—Name of mill, St. Patrick Mill; owners, St. Patrick Gold-Mining Company; kind of power and amount, steam, 40-horse; number of stamps, 15; weight of stamps, 800 pounds; number of drops per minute, 65; height of drop, 9 inches; number of pans, 2; number of concentrators, 6; cost of mill, $10,000; capacity per twenty-four hours, 20 tons; cost of treatment per ton, $3.50 to $4; tons crushed during year, 4,140.

Reported by L. W. Greenwell, superintendent.

## NEVADA COUNTY.

This county, which may be considered as the leading mining-county of the State, in respect to both the area and richness of its hydraulic ground and its numerous profitable ledges of gold-bearing quartz, is situated in the center of the mining-region of California, having Yuba and Sierra Counties on the north and Placer and Yuba Counties on the south and west, the eastern limit of the county extending to the boundary-line of the State of Nevada. The eastern or more elevated portion of the county, on the western water-shed of the Sierra Nevada, consists of high ridges lying between the headwaters of the Middle and South Yubas and Bear River. This portion of the county is crossed in a general northeast and southwest direction by the ancient-river system popularly known as the Blue Lead, which has for years been profitably worked in Sierra County on the north and Placer County on the south.

Within the limits of Nevada County, however, are found the works of greatest magnitude in this class of mining, among which may be enumerated those of the North Bloomfield and Milton Companies, which were fully described in my report for 1874.

The North Bloomfield Company owns more than fifteen hundred acres of auriferous gravel, situated on the ridge between the Middle and South Yuba Rivers. The depth of ground, as demonstrated by numerous prospecting-shafts, is from 200 to 300 feet; and the quality of the gravel, ascertained by numerous and careful tests, is equal to the average of the most favored hydraulic regions of the State. The lower stratum, being of the character known as "blue gravel," gave large results in coarse gold. It is estimated that it will take fifty years to wash this large tract of ground, which has its outlet, by means of a tunnel, into Humbug Creek, and thence into the South Yuba. During the greater portion of the year 1874 the company employed from five hundred to six hundred men on the tunnel and ditches.

It has taken this company eight years of steady work to perfect its operations. During this time it has acquired not only all the vast estate at Bloomfield, but also one-half of the large property of the Milton Mining and Water Company and one-half of the property of the Union Gravel Mining Company, so that it now virtually owns and controls over seven miles of auriferous channel. It has also constructed two enormous reservoirs, away up near the summit of the Sierras, and built, including the Milton Company's, over 100 miles of canals or ditches, to bring the water to the gravel. These canals and reservoirs have cost over one million of dollars. The Bloomfield Company alone has expended over two millions of dollars. The Union has also expended quite a large sum, and the Milton has expended about one and a half millions, making a total expenditure by this group of companies of over three and a half millions of dollars, without interest, and yet the stockholders in the three corporations do not exceed thirty in number. It will take the Milton Company some two years yet to complete all its deep tunnels. Two of them are now complete, another is about half finished. The Union Company completed its tunnel in July, 1874. When all the works of the three companies are completed they will have six deep tunnels, of an aggregate length of over 20,000 feet. The combined water-supply of the companies is 100,000,000 gallons of water per day throughout the year.

The North Bloomfield Company's tunnel was commenced April 25, 1872, and was prosecuted without interruption until its completion, on November 15, 1874. Eight shafts, averaging 200 feet in depth, were sunk to expedite the work. A diamond drill was used during the last year from the tunnel-mouth, but all the other fifteen faces have been driven principally with single-hand drills and giant powder. The tunnel was run from a point on Humbug Creek, (a tributary of the South Yuba River,) under the creek and 200 feet below it, 7,874 feet, to a point under the great blue-gravel channel at North Bloomfield. The shafts were sunk at such points in the cañon as the location would admit, averaging 1,000 feet apart, and the tunnel was driven straight from the tunnel-mouth to shaft 1, where a deflection was made to reach shaft 2, and there is an angle in tunnel-line at each shaft. The rock through which this tunnel has been driven is metamorphic slate. The power used at all the shafts for hoisting, pumping, and driving air, as also for driving the diamond drill, was water-power, under a pressure varying from 300 to 550 feet. This tunnel alone cost $550,000, and the company commenced mining through it in January, 1875. The upper end of this long and expensive work terminates at a point under the great gravel-channel, and 200 feet below the surface. To this point a vertical shaft was sunk from the surface, passing through a great depth of gravel. It is at this shaft where the hydraulic process commences. The gravel, falling down this shaft, is washed 8,000 feet through the tunnel, over gold-saving flumes, into the creek at the mouth of the tunnel, and then 4,000 feet further, over the creek-bed and under-currents or gold-saving appliances, to the South Yuba, when it is swept away down the mighty cañon into the valley below. There is an angle in the general course of the tunnel at each shaft, as they were sunk in the deep cañon at favorable points, and the tunnel was driven from shaft to shaft, making seven angles in the tunnel. An examination of the accompanying table will show that the engineering work was exceedingly accurate. The country through which this tunnel runs is very much broken, and the cañon under which the tunnel was constructed is sinuous, with very high and steep banks, making it very difficult to run correct preliminary surveys. Yet, as will be seen by an examination of the table, the maximum errors, as between the surveys on the surface and the surveys of the work completed, are as follows: In alignment, $\frac{9}{16}$ of an inch; in levels, $\frac{5}{16}$ of an inch; in distance, $1\frac{1}{4}$ inches. Had the tunnel been constructed upon a straight line for its entire length, these variations would not be considered as worthy of notice even with the very nicest work, but with a tunnel 7 by 8 feet, with small shafts only about 5 or 6 feet square, and with seven angles in the tunnel, such accuracy is extraordinary. The work was constructed under the control and supervision of Mr. Hamilton Smith, jr., the engineer and superintendent of the company.

The only general delay of the work was in December, 1873, caused by blocking of ditches, conveying water for power, and involving a delay of from sixteen to thirty days and an increased cost of $3,000.

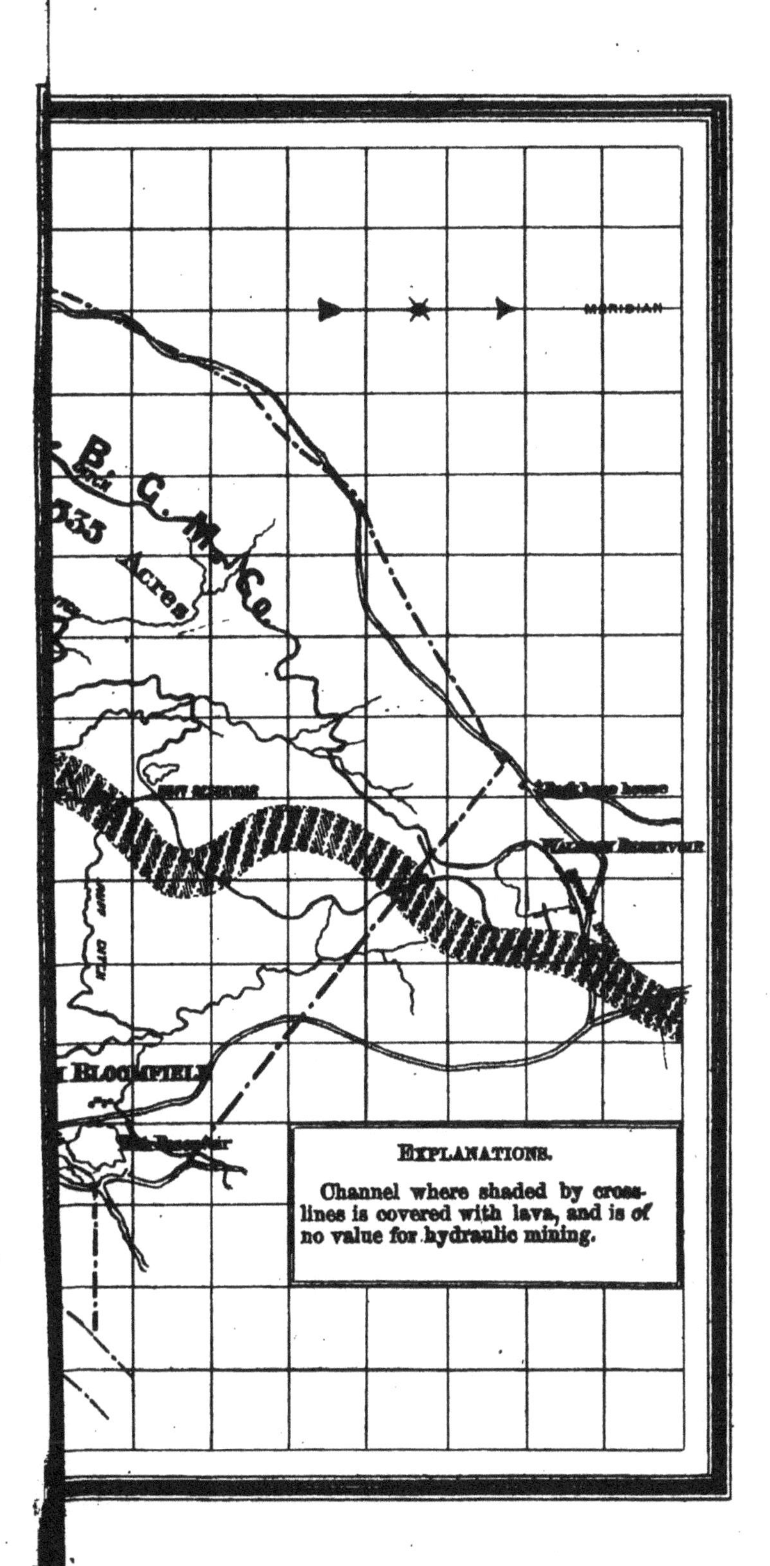
MERIDIAN
C. M. Co.
533
Acres
EXPLANATIONS.
Channel where shaded by cross-lines is covered with lava, and is of no value for hydraulic mining.

The only breakage of machinery worth mention was that of spur-wheel at No. 1, (probably caused by carelessness of brakesman,) which required ten days for repairs.

Water for power was conveyed through a sheet-iron main (15 inches to 7 inches diameter) and branches, (7 inches,) aggregating 9.960 feet in length, and discharged against "hurdy-gurdy" wheels 17 to 21 feet diameter, under pressures varying from 285 to 549 feet. Water-pipe was put together "stove-pipe" fashion, and gave little trouble by leaking or otherwise. The charge for water used was merely the cost to the company of running it from its upper reservoir to the tunnel-pipe. The headings employed three shifts of two men each per day, working under contract, the miners paying for explosives and lights and the company sharpening tools and taking rock from headings. Shafts were sunk by day's work, by three shifts of three men each. In addition to the credit of $9,732.97, there is machinery, &c., now remaining which cost in the city shops nearly $33,000. From this source, perhaps, $12,500 can be realized, thus reducing the net cost of the tunnel to, say, $486,500, or $61.63 cost per linear foot.

*Tabular statement for entire work of Bloomfield tunnel,*

| Working. | Distances completed. | Water pumped, (maximum.) | | Cost for period. | | | | | |
|---|---|---|---|---|---|---|---|---|---|
| | | | | Water-power. | | | Labor. | Materials. | |
| | | Pumps. | Cubic feet per hour. | Miner's inches. | Rate. | Cost. | Amount. | Explosives. | Lubricants and lights. |
| | *Feet.* | | | | *Cents.* | | | | |
| Mouth | *1,177.4 | ...... | ...... | 11,400 | 3.64 | $414 96 | $11,844 54<br>10,369 50 | $1,415 27<br>8,128 87 | $219 50<br>398 65 |
| Shaft 1.—Lower heading. | 398.5 | 1 | ...... | 5,730 | 3.61 | 206 85 | 8,798 79 | 597 69 | 172 59 |
| Shaft | 210 | 2 | 1,210 | 6,400 | 3.61 | 231 04 | 13,497 10 | 627 30 | 300 78 |
| Upper heading. | 432.4 | ...... | ...... | 4,840 | 3.61 | 174 72 | 12,301 49 | 777 78 | 276 29 |
| Shaft 2.—Lower heading. | 497.2 | 2 | ...... | 7,960 | 3.61 | 287 36 | 15,092 87 | 1,044 30 | 332 03 |
| Shaft | 200 | 1 | 470 | 3,200 | 3.61 | 115 52 | 11,250 70 | 1,202 80 | 179 75 |
| Upper heading | 478.9 | ...... | ...... | 5,930 | 3.61 | 214 07 | 15,353 68 | 918 00 | 262 98 |
| Shaft 3.—Lower heading. | 427.9 | 1 | ...... | 6,913 | 3.61 | 249 56 | 11,672 00 | 1,028 07 | 224 37 |
| Shaft | 200 | 1 | 625 | 3,300 | 3.61 | 119 13 | 8,147 37 | 648 35 | 190 75 |
| Upper heading | 516.6 | ...... | ...... | 5,343 | 3.61 | 192 88 | 13,594 17 | 1,038 37 | 213 84 |
| Shaft 4.—Lower heading | 389.5 | 1 | ...... | 6,913 | 3.61 | 249 56 | 11,227 90 | 705 42 | 182 72 |
| Shaft | 200.2 | 2 | 1,750 | 7,200 | 3.61 | 259 92 | 13,939 17 | 790 20 | 290 21 |
| Upper heading. | 393.3 | ...... | ...... | 7,300 | 3.61 | 263 85 | 11,596 70 | 823 18 | 208 72 |
| Shaft 5.—Lower heading | 495.9 | 1 | ..... | 8,770 | 3.61 | 316 60 | 16,142 12 | 943 78 | 325 05 |
| Shaft | 186 | 1 | 584 | 3,600 | 3.61 | 120 96 | 10,637 25 | 708 18 | 162 45 |
| Upper heading | 501.8 | ...... | ...... | 6,267 | 3.61 | 226 24 | 14,211 48 | ,906 27 | 297 79 |
| Shaft 6.—Lower heading | 402 | 1 | ...... | 6,630 | 3.61 | 239 34 | 12,486 07 | 682 40 | 231 09 |
| Shaft | 209 | 1 | 358 | 2,700 | 3.61 | 97 47 | 8,068 52 | 618 40 | 205 00 |
| Upper heading. | 550.8 | ...... | ...... | 5,200 | 3.61 | 187 72 | 17,182 25 | 646 76 | 240 72 |
| Shaft 7.—Lower heading. | 411.2 | 1 | ...... | 6,480 | 3.61 | 233 93 | 13,213 03 | 958 57 | 294 94 |
| Shaft | 189 | 1 | 356 | 2,500 | 3.61 | 90 25 | 8,937 38 | 1,187 60 | 169 90 |

* 546.5 feet hand-work; 630.9 feet diamond-drill work.

*driven by North Bloomfield Gravel-Mining Company.*

| Cost for period. | | | | Total cost of entire works. | | Time occupied by work. | | | Contract-prices of heading. |
|---|---|---|---|---|---|---|---|---|---|
| Materials. | | | | | | | | | |
| Diamonds and steel. | Lumber, timbers. | Coal, tools, and sundries. | Total. | Sum. | Per foot. | Commenced. | Finished. | Days. | |
| $213 67 | $375 00 | $625 50 | $2,848 94 | $14,693 48 | $26 89 | Apr. 25, 1872 | Sept. 13, 1873 | 506 | $21 00<br>26 00<br>26 75<br>30 00 |
| †700 00 | 75 63 | ‡2,488 01 | 11,791 16 | 22,508 62 | 35 77 | Sept. 16, 1873 | Aug. 15, 1874 | 333 | |
| 118 95 | 48 00 | 167 01 | 1,104 24 | 10,109 88 | 30 78 | July 14, 1873 | July 11, 1874 | 362 | 22 00<br>20 00<br>23 00<br>21 00 |
| 54 96 | 856 00 | 1,122 39 | 2,961 43 | 16,689 57 | 79 47 | May 24, 1872 | July 12, 1873 | 414 | |
| 118 95 | 48 00 | 182 50 | 1,403 52 | 13,879 73 | 32 10 | July 14, 1873 | Nov. 17, 1874 | 491 | 23 00<br>24 00<br>24 00 |
| 137 96 | 44 00 | 248 68 | 1,806 97 | 17,187 20 | 34 57 | Mar. 17, 1873 | Nov. 14, 1874 | 607 | 26 00<br>25 00<br>24 00<br>23 00<br>25 50 |
| 69 19 | 342 00 | 721 99 | 2,515 73 | 13,881 95 | 69 41 | May 28, 1872 | Mar. 15, 1873 | 291 | |
| 137 95 | 44 00 | 250 66 | 1,613 59 | 17,181 34 | 35 88 | Mar. 17, 1873 | Oct. 15, 1874 | 577 | 35 00<br>32 00<br>25 00<br>32 00<br>22 00 |
| 118 89 | 44 00 | 219 77 | 1,635 10 | 13,556 66 | 31 68 | Apr. 22, 1873 | Aug. 5, 1874 | 470 | 26 00<br>25 00<br>26 00<br>21 00<br>23 00 |
| 79 48 | 485 00 | 634 77 | 2,038 35 | 10,304 85 | 51 52 | May 25, 1872 | Apr. 19, 1873 | 320 | |
| 118 90 | 44 00 | 219 80 | 1,634 91 | 15,421 96 | 29 85 | Apr. 22, 1873 | Aug. 1, 1874 | 466 | 25 00<br>22 00<br>22 00 |
| 124 49 | 48 00 | 165 98 | 1,226 61 | 12,704 07 | 32 62 | Sept. 1, 1873 | Sept. 20, 1874 | 384 | 26 00<br>24 00 |
| 64 05 | 458 00 | 895 92 | 2,498 38 | 16,697 47 | 83 40 | June 7, 1872 | Aug. 30, 1873 | 449 | |
| 124 50 | 48 00 | 161 09 | 1,365 49 | 13,226 04 | 33 63 | Sept. 1, 1873 | Nov. 7, 1874 | 432 | 25 00<br>24 00 |
| 145 27 | 57 00 | 314 43 | 1,788 53 | 18,247 25 | 36 80 | Feb. 10, 1873 | Nov. 7, 1874 | 635 | 25 00<br>28 50<br>28 00<br>25 50<br>25 50 |
| 65 42 | 342 00 | 392 05 | 1,870 10 | 12,637 31 | 67 94 | May 26, 1872 | Feb. 8, 1873 | 258 | |
| 145 30 | 57 00 | 318 85 | 1,795 21 | 16,162 93 | 32 20 | Feb. 10, 1873 | Oct. 4, 1874 | 601 | 20 00<br>22 00<br>25 00<br>24 00<br>23 50 |
| 133 38 | 41 00 | 262 05 | 1,349 92 | 14,075 33 | 35 01 | Jan. 20, 1873 | May 16, 1874 | 481 | 20 00<br>22 50<br>27 00<br>25 00 |
| 73 23 | 420 00 | 672 69 | 1,989 32 | 10,155 31 | 48 59 | May 30, 1872 | Jan. 18, 1873 | 233 | |
| 133 38 | 41 00 | 271 28 | 1,333 14 | 18,703 11 | 33 96 | Jan. 20, 1873 | May 18, 1874 | 483 | 25 00<br>30 00<br>27 00<br>25 00 |
| 139 38 | 51 00 | 208 28 | 1,652 17 | 17,099 13 | 41 58 | Jan. 18, 1873 | Aug. 29, 1874 | 588 | 25 00<br>31 00<br>27 50 |
| 79 27 | 330 00 | 664 01 | 2,430 78 | 11,458 41 | 60 63 | May 26, 1872 | Jan. 16, 1873 | 235 | |

† Diamonds. ‡ Repairs, &c.

*Tabular statement for entire work of*

| Working. | Distances completed. | Water pumped, (maximum.) | | Cost for period. | | | | | |
|---|---|---|---|---|---|---|---|---|---|
| | | | | Water-power. | | | Labor. | Materials. | |
| | | Pumps. | Cubic feet per hour. | Miner's inches. | Rate. | Cost. | Amount. | Explosives. | Lubricants and lights. |
| | *Feet.* | | | | *Cents.* | | | | |
| Shaft 7.—Upper heading | 414. 6 | ...... | ...... | 5, 440 | 3. 61 | $196 38 | $14, 705 52 | $930 79 | $307 05 |
| Shaft 8.—Lower heading | 456. 4 | ...... | ...... | 10, 900 | 3. 61 | 393 49 | 20, 144 59 | 1, 537 98 | 461 01 |
| Shaft.......... | 182. 7 | 1 | 423 | 3, 400 | 3. 61 | 122 74 | 7, 620 77 | 290 00 | 216 50 |
| Shafts ............ | 1, 576. 9 | 10 | 5, 785 | 32, 300 | ...... | 1, 166 03 | 82, 098 26 | 6, 072 83 | 1, 715 34 |
| Headings .......... | 7, 874. 4 | 8 | ...... | 112, 025 | ...... | 4, 047 51 | 231, 929 70 | 23,083 50 | 4, 649 34 |

*Machinery for 8 sets of works, including water and*

Total cost of shafts ..........................................................................

Total cost of headings........................................................................

Foundery-work.—Cost in San Francisco and Nevada City, $30,597.96; freight to Bloomfield, $1,723.09.
Erection, track, &c.—Timbers and lumber, $7.389.71; track and cars, $3,397.26; labor, $18,958.46; hard
Water-power pipes, gates, &c.—Cost in San Francisco, $12,113.62; freight, $793.57; labor, and lumber

Drill.—Cost of drill, pipe, and fixtures, $3,373.40; labor in fitting, $1,875.43; lumber, $327.36; supplies,
Diamonds.—Purchased 238 carats, $1,914.96; less diminution in weight and value by use, $700..........

*General*

General work and roads.—Foreman, machinist, pump-men, track-layers, general smithing, fitting up
general use, feed for teams, fire-wood, and general supplies, $5,590.23..............................
Surveys and administration.—Salary of company's superintendent and accountant (part) and assistants,

*Deduc*

Materials sold and transferred.—Pipe and pumps sold, $1,037.27; water pipe and gates, (part,) T-iron

Total ..........................................................................................

*Bloomfield tunnel, &c.*—Continued.

| Cost for period. | | | | Total cost of entire works. | | Time occupied by work. | | | Contract-prices of heading. |
|---|---|---|---|---|---|---|---|---|---|
| Materials. | | | | | | | | | |
| Diamonds and steel. | Lumber, timbers. | Coal, tools, and sundries. | Total. | Sum. | Per foot. | Commenced. | Finished. | Days. | |
| $134 96 | $51 00 | $267 54 | $1,691 34 | $16,593 24 | $40 02 | Jan. 18, 1873 | Aug. 21, 1874 | 580 | $25 00<br>30 00<br>28 00<br>36 00<br>27 00<br>30 00<br>28 00 |
| 318 63 | 62 00 | 504 70 | 2,944 32 | 23,482 40 | 51 45 | Feb. 8, 1873 | Aug. 21, 1874 | 559 | 30 00<br>30 00<br>28 00<br>30 00<br>33 50 |
| 53 36 | 652 73 | 697 79 | 1,910 38 | 9,653 89 | 52 84 | July 8, 1872 | Feb. 6, 1873 | 213 | |
| 538 96 | 3,885 73 | 6,001 61 | 18,214 47 | 101,478 76 | 64 35 | | | | |
| 3,064 56 | 1,178 63 | 6,939 13 | 38,915 16 | 274,892 37 | 34 91 | | | | |

*air pipes, tracks, cars, &c., and diamond drills.*

| | | | |
|---|---|---|---|
| .......... | | | $101,478 76 |
| .......... | | | 274,892 37 |
| .......... | $32,321 05 | | |
| ware, nails, iron, &c., $4,149.05 .......... | 33,894 48 | | |
| in laying, $1,724.25 .......... | 14,631 44 | | |
| | | $80,846 97 | |
| $61.50 .......... | 5,637 69 | | |
| .......... | 1,214 96 | | |
| | | 6,852 65 | |

*expenses.*

| | | | |
|---|---|---|---|
| pumps, &c., $19,689.02; roads, $2,122.76; buildings and tools for .......... | 27,402 00 | | |
| taxes, &c .......... | 17,086 38 | | |
| | | 44,488 38 | |
| | | | 132,188 00 |
| | | | 508,559 13 |

*tions.*

| | |
|---|---|
| and cars, and set of machinery at No. 7, $8,695.70 .......... | 9,732 97 |
| .......... | 498,826 16 |

The following is a summary of the progress of this great work:

The first 300 feet of the heading from the tunnel-mouth was in comparatively easy ground, the remainder hard. About one month's time was lost when drill was being put in heading. Four hundred feet of the ground driven by the drill was very hard, the remainder softer, with occasional seams.

Shaft No. 1 was sunk for first 72 feet through very wet and loose material, requiring careful spiling. An attempt was made to puddle shaft from hard rock to surface, (97 feet.) Timbers 6 inches by 8 inches, with a clay puddle-wall 12 inches thick, were used. Puddling proved a failure, as water forced its way through the lower joints. Cost of puddling was about $2,800. Tank then put in, which worked well.

No. 2. About 2½ inches of water pumped from the drift of shaft No. 2 through 450 feet suction-pipe and vertical lift of 23 feet. No delays were encountered in sinking shaft No. 2. The drift passed through a belt of solid black quartz 65 feet thick, the hardest rock found in the tunnel.

No. 3. Fair rock for entire length of this drift. Deducting sixty days' loss of time from fire, gives two hundred and sixty-nine days' time sinking and timbering shaft. Shaft-houses at 3 and 4 were destroyed by incendiary fires just at their completion. About two months' time was lost in replacing works, at a cash cost of $6,000.

Shaft No. 4 followed an open seam for 100 feet down, from which at one time 18 or 19 inches of water flowed. This quantity of water in the shaft delayed work of sinking, and accounts for its large cost. Pumps in this shaft sometimes pumped 28 to 30 inches of water to the surface.

No. 5. This drift made 3 inches of water, pumped through 450 feet of suction-pipe and vertical lift of 25 feet. No delays in sinking shaft, and but little bother from water. Rock generally hard, but with an occasional good slip.

No. 6. Drift through close rock, making but little water. Shaft for first 140 feet was sunk by Chinese labor; rock became harder, and whites were then employed. Drift for about 110 feet required timbering, and being the only part of the tunnel that is timbered.

No. 7. Rock in this drift hard, but with favoring slips. No delays in sinking shaft, which was in hard rock its entire depth. Drift through hard rock.

No. 8. Drift through hard rock; some very hard granite mixed through the country-slate. Shaft was sunk through 110 feet gravel, then through rock. Shaft made about 4½ inches water, which was not "tanked." Delays caused by this water occasioned large cost.

*Statement of errors of connections made in several headings of North Bloomfield Gravel-Mining Company's tunnel—instrumental tests expressed in feet.*

| Headings. | Errors in junction. | | Distance. | | Error. |
|---|---|---|---|---|---|
| | Line. | Level. | Outside measure. | Inside measure. | |
| Mouth to shaft 1 | .000 | +.008 | 1,505.94 | 1,505.95 | +.01 |
| Shaft 1 to shaft 2 | .020 | +.009 | 929.56 | 929.46 | —.10 |
| Shaft 2 to shaft 3 | .005 | —.022 | 906.77 | 906.81 | +.04 |
| Shaft 3 to shaft 4 | .035 | +.008 | 906.08 | 906.03 | —.05 |
| Shaft 4 to shaft 5 | .040 | +.026 | 889.25 | 889.28 | +.03 |
| Shaft 5 to shaft 6 | .010 | —.004 | 903.79 | 903.73 | —.06 |
| Shaft 6 to shaft 7 | .005 | —.012 | 961.97 | 962.01 | +.04 |
| Shaft 7 to shaft 8 | .046 | —.020 | 871.00 | 871.04 | +.04 |
| Total | .161 | —.007 | 7,874.30 | 7,874.31 | —.05 |
| Mean | .020 | | | | |

NOTE.—Maximum error in line, .046 or 9-16 inch. Maximum error in level, .026 or 5-16 inch. Maximum error in distance, .010 or 1¼ inches.

The share-capital of the parent company, the North Bloomfield, is 50,000 shares—45,000 issued—upon which about $30 per share has been paid in assessments, besides borrowing $500,000 from a few of their shareholders and expending $250,000 from bullion taken from their preliminary washings. The Milton Company has a share-capital of 20,000 shares, of which about 15,000 have been issued, and assessments up to $100 per share paid thereon, besides borrowing from a few of their shareholders $200,000 and the expenditure of large amounts derived from bullion taken from their preliminary workings. The Union Company has a capital of 8,000 shares, upon which about $25 per share has been paid and expended, in addition to quite a large amout of bullion taken from their preliminary workings. The North Bloomfield Company owns all its own property, also one-half of the share-capital of the Milton Company, and over one-half of the share-capital of the Union Company.

A.—*Statement of construction of Milton ditch, (Eureka to Milton,) built by the North Bloomfield Gravel-Mining Company in the years* 1872, '73, '74.

| Length. | Chains. | Miles. |
|---|---|---|
| Eureka to South Fork | 563 | 7.04 |
| South Fork to Drop-off | 96 | 1.20 |
| Drop-off to Milton Dam | 884 | 11.17 |
| Total distance | 1,553 | 19.41 or 102,498 feet |

| Fluming. | Twelve-foot boxes. | Linear feet. |
|---|---|---|
| Eureka to South Fork | 961 | 11,536 |
| South Fork to Big Bluffs | 264 | 3,168 |
| Big Bluffs to Milton | 1,113 | 13,352 |
| Total | 2,338 | 28,056 |

The above 2,338 boxes include 56 boxes built in the ditch, and most of which are supported by heavy cribbing. In addition to the above, there are several small branch flumes, one large crossing-flume, and 130 feet of ditch-lining.

A dam was built at South Fork at a cost of, say, $350; old dam at Milton was used.

| Waste-ways. | | |
|---|---|---|
| Eureka to South Fork | 14 wastes, aggregating | 112 feet. |
| South Fork to Big Bluffs | 12 wastes, aggregating | 48 feet. |
| Big Bluffs to Milton | 24 wastes, aggregating | 114 feet. |
| Total | 50 wastes, aggregating | 274 feet. |

The accompanying diagrams, on a scale of $\frac{1}{48}$, give sections through the ditch and flume.

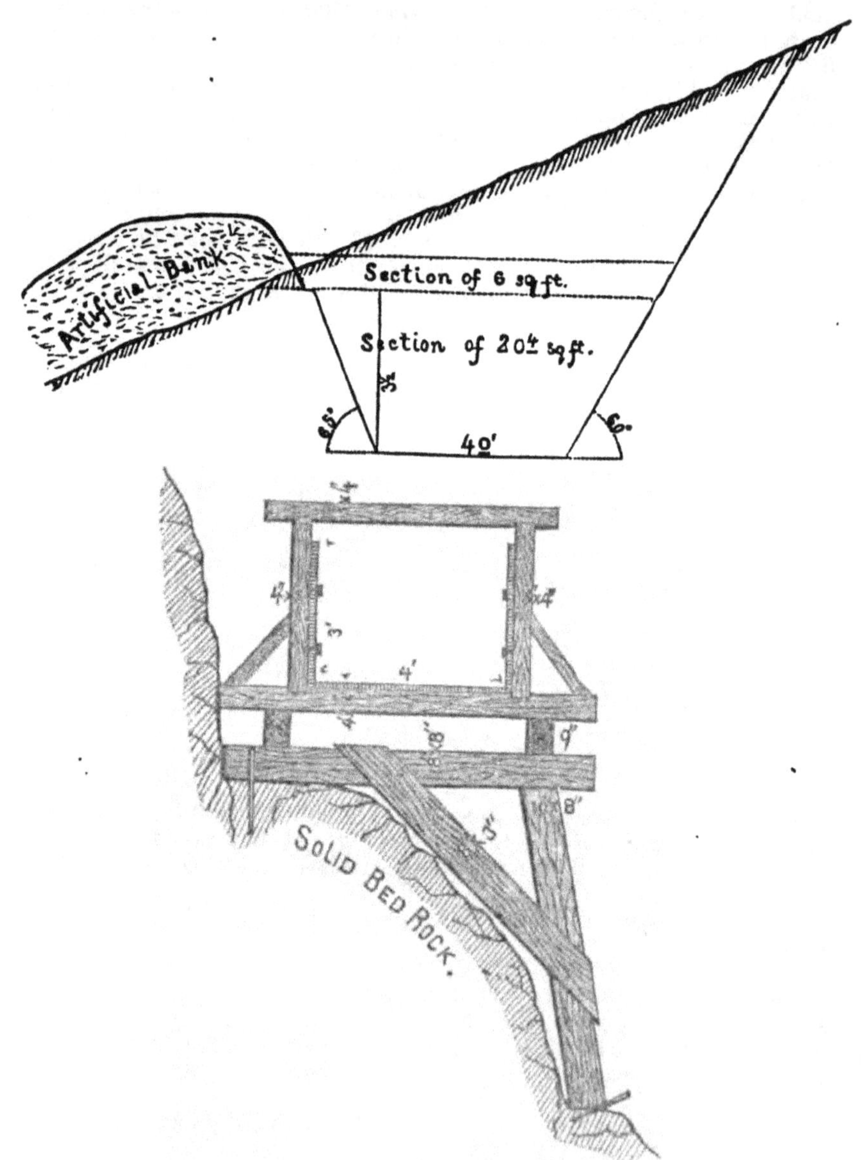


The ditch is graded in from slope-pegs from 6 to 36 inches. The general grade is 19.2 feet per mile. All trees within 15 to 25 feet of the edge of the upper bank are cut. The logs, brush, and leaves from the lower bank (under the artificial bank) were carefully removed. The foundation for the entire width of the flume was generally cut. The sketch shows the method of posting along cliffs, where the foundation was occasionally narrower than the flume. Where flumes connect with the ditch, the posts of the flumes, for a distance of several boxes, are 4 and 4½ feet high, allowing an additional side-plank. The grade of the flume is 32 feet per mile. The planking is 2 inches thick.

B.—*Cost of Milton ditch, from Eureka to Milton,* 19.41 *miles.*

EXCAVATION, ETC.

| | Distance. | Labor. | Explosives. | Tools. | Steel. | Coal. | | |
|---|---|---|---|---|---|---|---|---|
| Ditch .......... | 14.1 miles. | $69, 664 92 | $4, 098 46 | $1, 606 67 | $319 48 | $933 38 | $76,642 91 | |
| Flume-foundation. | 5.3 miles. | 15, 013 40 | 2, 866 72 | 525 50 | 213 00 | 301 11 | 18,919 73 | |
| Clearing line... | 19.4 miles | 3, 582 01 | ............ | 90 00 | ........ | .......... | 3,672 01 | $99, 234 65 |
| | 19.4 miles. | 88, 260 33 | 6, 965 18 | 2, 222 17 | 532 48 | 1, 245 49 | | |

FLUME.

| | Feet. | Feet. | | | |
|---|---|---|---|---|---|
| Lumber, &c., Milton to lower end Big Bluff....... | 1, 083, 434 | | | | |
| Less sold to Milton Company ...................... | 200, 000 | | | | |
| | | 883, 434 | | | |
| Eureka to Big Bluffs, 11,225 boxes ........................... | | 765, 911 | | | |
| Total on hand and used for 2,338 boxes................. | | 1, 649, 345 | $32, 015 28 | | |

NOTE.—Of the above amount of 883,434 feet, it is supposed that there is on hand, say, 130,000 feet, thus leaving 750,000 feet as the amount used for 1,113 boxes from Milton to lower end of bluffs.

| | | | |
|---|---|---|---|
| Timbers cut by hand, stringers, posts, &c ............................ | 1, 301 49 | | |
| Hauling lumber to Milton, Little Poor Man's, &c...................... | 1, 650 00 | | |
| | | 34,966 77 | |

CARPENTERS, ETC.

| Gang. | Boxes 12 feet long. | Labor. | Nails and iron. | Tools. | Totals. | | |
|---|---|---|---|---|---|---|---|
| Young .............. | 1, 145 | $10, 902 81 | $1, 499 57 | $50 00 | $12, 452 38 | | |
| Marriott ........... | 1, 193 | 10, 497 90 | 1, 559 57 | 50 00 | 12, 107 47 | 24,559 85 | |
| | 2, 338 | 21, 400 71 | 3, 059 14 | 100 00 | | | 59, 526 62 |

GENERAL COST.

| | | | |
|---|---|---|---|
| Surveys.—Engineer (who was also foreman) and assistants ........................ | | 4,610 50 | |
| Roads.—South Fork to Bowman's, 3¾ miles ............................ | $1, 200 00 | | |
| South Fork to Little Poor Man's, 2¼ miles..................... | 200 00 | | |
| | | 1,400 00 | |
| Hauling.—Transportation of tools, material, and men.............................. | | 1,450 94 | |
| Boarding.—Loss in boarding laborers, who were charged 75 cents per day......... | | 685 75 | |
| General expense.—Being a portion of North Bloomfield Gravel-Mining Company's cost of management, office, taxes, &c., while ditch was being built ... ........... | | 3,564 63 | |
| | | | 11, 711 82 |

DAMAGES.

| | | | |
|---|---|---|---|
| Eureka Lake Company.—Damage to it by breaking its miner's ditch by blasts.... | | .......... | 1, 635 87 |
| Total cost ............................................................ | | .......... | 172, 108 96 |
| Collected from Milton Company for account extra work ........................ | | .......... | 689 30 |
| Leaving Milton Ditch account (November 10, 1874) on company's books .......... | | .......... | 171, 419 66 |

NOTE.—If the 130,000 feet of lumber supposed to be at Milton is sold for cost, ($20 per thousand,) the total cost of the ditch will be reduced to $169,508.96, or, say, $8,700 per mile. In that event—

COST PER FOOT, ETC.

Ditch: 74,442 ft. long, cost for, say, 117,600 cub. yds., $76,642.91, or 65 cts. per cub. yd., or $1.03 per lin. ft.

Flume: 28,056 ft. long, { cost for excavation, $18,919.73, or 67 cts. per lin. foot. / cost for lumber, labor, &c., $59,526.62, or $2.12 per lin. foot. } $2.79 per lin. ft.

The Omega Mining and Water Company owns a large tract of gravel-ground at Omega Hill, on the south slope of the ridge between Bear River and the South Yuba River. The ground had been worked for sixteen years before it came into the possession of the present owners, and up to that time is estimated to have yielded $1,000,000. Since the

Omega Company acquired the ground, a period of five years, the yield has exceeded $500,000. The yield of the ground for 1874, between April and September, an exceptionably short season, was $95,000. There remains a large extent of ground to be washed away, estimated to be equivalent to twenty seasons' work.

The gravel-mines on the San Juan ridge extend from Snow Point to French Corral, a distance of thirty miles, with occasional spaces of non-auriferous ground. From North San Juan to French Corral, a distance of six miles, the gravel-range is continuous and uninterrupted. This channel is probably the same as the lower channel noticed under the heading of Sierra County, where it has been worked at Brandy City and Camptonville. The gravel and hydraulic claims are located on an immense ancient-river-bed, that traverses the ridge of land embraced between the South and Middle Forks of the main Yuba River, coming from the direction of the Sierra Nevada Mountains, and terminating at French Corral, which is the western extremity of the ridge. At this point it is broken by the deep gorge of the Yuba River, and again makes its appearance at Mooney's Flat, near Smartville, Yuba County, the intermediate portions having been swept away by Deer Creek and the Yuba River.

This ancient channel is in elevation from 400 to 600 feet higher than either of the rivers between which it lies, and its gravel-deposit is from 200 to 1,500 feet in width, and from 100 to 400 feet in depth. Gold is disseminated throughout the entire deposit, but the gravel is always the richest near the bottom or bed rock. The color of the gravel is, in the upper portion of the deposit, red and white, and near the bed-rock deep blue. The blue gravel is generally cemented, and requires blasting before it can be washed. The channel has been cut in places by deep ravines and cañons, and broken by land-slides.

The most important mining-districts between the forks of the Yuba are French Corral, Empire Flat, Kate Hayes Flat, Birchville, Buckeye Hill, Sweetland, Manzanita Hill, San Juan, Badger Hill, Cherokee, Chimney Hill, Columbia Hill, Kennebec Hill, Grizzly Hill, Lake City, Malakoff, Humbug, Relief Hill, Woolsey Flat, Moore's Flat, and Snow Point.

Three water-companies supply the mining-ground of this ridge with water, viz: The North Bloomfield, the Milton, and the Eureka Lake and Water Company. The Bloomfield has a ditch with a capacity of 3,000 inches; the Milton Company one of 2,500 inches, and the Eureka Lake Company three ditches, with an aggregate capacity of 4,000 inches. All these companies own large tracts of mining-ground.

The Milton tunnel, located near French Corral, is intended to open up the Eddy diggings, near French Corral. It is now about 1,200 feet from the lower face. It will be necessary to run it about 1,800 feet farther. The claims into which the tunnel is being run are very rich. The facilities for saving gold when the mine is opened will be unexcelled, as there will be plenty of fall for sluices and undercurrents before the tailings reach the South Yuba River. These diggings are now being worked through a high tunnel which was run several years ago, but the bottom dirt to the bed-rock has never been reached, hence the necessity of running the new tunnel. The company also owns the Empire, Bed-Rock, and Manzanita ground, all situated between North San Juan and French Corral, on which active hydraulic operations are prosecuted.

The Buckeye, near Sweetland, belonging to an English company, has a tunnel 5,000 feet in length, and an outlet on the Middle Yuba. This company used from 500 to 600 inches of water per day during the present season, and this amount will be largely increased in 1875. The mine is said to be paying a handsome profit to the owners, but no particulars could be ascertained, as the principal office is in London.

The Sweetland Creek mines, situated in the immediate vicinity of the Buckeye, is also an English incorporation. Both are under the superintendence of Mr. George D. McLean. The Sweetland Creek was floated in London four years since on a basis of $300,000 capital stock, (a valuation considered too high, from a California stand-point, at that time,) and has since that time paid its owners fourteen dividends, aggregating $126,000, of which $67,000 was paid during the fiscal year ending June, 1874, after a season of limited water-supply.

The American mine, whose claims are situated at Sebastopol, a mile below San Juan, was the second company which introduced hydraulic washing on a large scale. This company has paid dividends with great regularity for the past twenty years. The company's ground has been described in my former reports. Its operations are on a large scale. The main water-pipe is 32 inches in diameter, and furnishes a fall of about 250 feet at the bed-rock. The American has been working without cessation ever since 1852, and its present owners are among the original locators. An old tunnel has been in operation for several years, and has been low enough to wash a very large section of country. Now, however, the tunnel has got so high that they can only wash the top-dirt, and for the past season this is all the washing they have done. About twenty acres of gravel are now ready to be washed as soon as the new tunnel is completed. The new tunnel is now in 3,100 feet, with 700 feet more to run before it will be completed.

The auriferous belt of Nevada County is very wide, and includes several belts of granite. Some of the most profitable quartz-mines of the State are situated in the principal granite-belt, near Nevada City; among these the Providence, Soggs or Nevada, and Wyoming. This granite-belt passes east of Grass Valley, through Nevada City, and extends across the country into Sierra County, taking North San Juan in its course. Immediately east of the granite the slates are highly altered, and contain the Grass Valley group of quartz-mines, among the most noted of which, for present and past yield, are the Idaho, Eureka, Empire, North Star, and Allison Ranch. The two last-named have ceased to be productive, but their position as producers of bullion will probably be filled during the coming year by the New York Hill and the Massachusetts Hill.

The Idaho mine, of Grass Valley, is a "close corporation," owned by a few shareholders, mostly residents of the town, and is worked on strict business principles, without regard to stock quotations. The 35-stamp mill, steam hoisting-works, and machinery are among the best in the State, and the mine is in a prosperous condition. This mine is an example of many others in California, worked by private companies, which, while they materially increase the bullion-product, are rarely heard of except through the annual reports of their officers.

The president and superintendent of the Idaho, Mr. Edward Coleman, says, in his annual report for the year ending December 7, 1874:

During the year we have crushed 28,801½ tons of rock, of which 1,942½ tons came from the 400 level, 1,886¾ tons came from the 500 level, 5,581¼ tons came from the 600 level, 16,433½ tons came from the 700 level, 1,594¼ tons came from the 800 level, and 963 tons from the shaft. This gave a gross yield of 36,169 ounces of gold, $631,190.56; 217 tons of sulphurets, $21,600; specimens, $38.50; gross yield of tailings, $10,989.25; gold from old copper, $692.39—total, $664,811.20—giving an average of $23.40 per ton.

We have driven during the year 1,149 feet of drift. This amount of drift has opened up more ledge than we have worked out, and it may safely be estimated that we now have at least four years' work of pay-ore in sight. All the machinery is in good working order, and the business of the company is satisfactory.

MILL AND MINING.

| | |
|---|---|
| Surface labor | $45,605 36 |
| Underground labor | 123,912 08 |
| Wood and poles | 21,522 37 |
| Powder and fuse | 4,055 08 |
| Hardware | 5,673 47 |
| Lumber | 1,653 02 |
| Candles and oils | 5,220 05 |
| Coal | 3,303 40 |
| Quicksilver | 3,016 25 |
| Foundery | 9,312 52 |
| Drill-steel | 2,239 69 |
| Water | 396 50 |
| Superintendent's salary | 6,000 00 |
| Sundries | 1,752 40 |
| | 233,662 20 |

Average cost per ton, $8.22¼.

SAVING 217 TONS OF SULPHURETS.

| | |
|---|---|
| Labor | $2,788 00 |
| Paid for working 64 tons by chlorination | 1,600 00 |
| Repairs | 48 50 |
| | 4,436 50 |

The balance of sulphurets (153 tons) were sold.

GRINDING TAILINGS WORKED ON PERCENTAGE.

| | |
|---|---|
| Wood and oil | $1,023 50 |
| Foundery | 265 00 |
| | 1,288 50 |

M'DOUGALL'S CONCENTRATOR.

| | |
|---|---|
| Labor, lumber, material | $223 01 |
| Paid McDougall | 70 00 |
| | 293 01 |
| Burleigh drills | 11,567 20 |
| New pumping-works | 25,001 41 |
| Repairing old shaft and fixing pumps | 9,186 86 |
| New steam-pump for 200 level | 1,685 02 |
| Sinking main shaft | 23,375 89 |
| General account | 24,270 99 |

RECAPITULATION.

| | |
|---|---|
| Mill and mining | $233,662 20 |
| Sulphurets account | 4,436 50 |
| Tailing account | 1,288 50 |
| McDougall concentrator | 293 01 |
| Burleigh drill | 11,567 20 |
| New pumping-works | 25,001 41 |
| Repairing old shaft for pump | 9,186 86 |
| New steam-pump for 200 level | 1,685 02 |
| Sinking main shaft | 23,375 89 |
| General account | 24,270 99 |
| Total expense of working mine | 334,767 58 |
| Dividends, $102.50 per share | 317,750 00 |
| | 652,517 58 |

RECEIPTS.

| | |
|---|---|
| Cash on hand | $10,297 10 |
| 36,167 ounces bullion | 631,190 56 |
| Sulphurets worked and sold | 19,243 35 |
| Percentage from tailings | 5,557 13 |
| Pan-rent | 1,288 50 |
| Old copper | 992 89 |
| Lease of surplus water | 400 00 |
| Specimens sold | 38 50 |
| Old rope | 15 00 |
| Total receipts | 669,023 03 |
| Expenditures | 652,517 58 |
| Cash on hand | 16,505 45 |

From the report of the secretary, of same date, (December 7, 1874,) it appears that there are but twenty-two shareholders. The shares, 3,100, correspond to the number of feet in the location. The largest amount of stock held by one person is 575 shares, and the smallest, 10 shares. From an inspection of the list of shareholders it appears that about one-tenth of the stock is owned by practical miners, probably employés of the mine.

I quote the following from the report of the secretary, Mr. George W. Hill:

The monthly receipts of the company from all sources for the fiscal year ending December 1, 1874, are as follows, viz:

| | |
|---|---|
| Receipts from all sources for December, 1873, amount to | $62,518 71 |
| For January, 1874 | 56,137 81 |
| For February, 1874 | 49,328 67 |
| For March, 1874 | 70,628 68 |
| For April, 1874 | 64,013 75 |
| For May, 1874 | 58,314 19 |
| For June, 1874 | 66,610 99 |
| For July, 1874 | 44,815 02 |
| For August, 1874 | 53,080 88 |
| For September, 1874 | 43,227 25 |
| For October, 1874 | 35,411 09 |
| For November, 1874 | 54,638 89 |
| Total receipts for fiscal year ending December 1, 1874 | 658,725 93 |
| Balance on hand December 1, 1873 | 10,297 10 |
| Shows assets, for year, of | 669,023 03 |

The monthly expenditures of the company for the period as above for all purposes amounted as follows, to wit:

| | |
|---|---|
| December, 1873, with dividend No. 54 | $58,967 44 |
| January, 1874, with dividend No. 55 | 65,921 17 |
| February, 1874, with dividend No. 56 | 49,987 12 |
| March, 1874, with dividend No. 57 | 61,652 88 |
| April, 1874, with dividend No. 58 | 60,590 16 |
| May, 1874, with dividend No. 59 | 52,350 61 |
| June, 1874, with dividend No. 60 | 69,885 13 |
| July, 1874, with dividend No. 61 | 44,203 92 |
| August, 1874, with dividend No. 62 | 47,514 01 |
| September, 1874, with dividend No. 63 | 41,247 73 |
| October, 1874, with dividend No. 64 | 41,755 89 |
| November, 1874, with dividend No. 65 | 58,441 53 |
| Total expenses, including dividends, for year | 652,517 59 |
| Assets brought forward | 669,023 03 |
| Balance in treasury December 1, 1874 | 16,505 44 |

In the foregoing monthly expenditures are included 12 dividends declared by the board of trustees for the year herein stated, and at the times and amounts, as follows, to wit:

| Number of division. | Declared. | Per cent. | Amount. |
|---|---|---|---|
| 54 | Jan. 5, 1874 | 10 | $31,000 |
| 55 | Feb. 2, 1874 | 10 | 31,000 |
| 56 | Mar. 2, 1874 | 10 | 31,000 |
| 57 | Apr. 6, 1874 | 12½ | 38,750 |
| 58 | May 4, 1874 | 12½ | 38,750 |
| 59 | June 1, 1874 | 10 | 31,000 |
| 60 | July 6, 1874 | 10 | 31,000 |
| 61 | Aug. 3, 1874 | 7½ | 23,250 |
| 62 | Sept. 7, 1874 | 5 | 15,500 |
| 63 | Oct. 5, 1874 | 5 | 15,500 |
| 64 | Nov. 2, 1874 | 5 | 15,500 |
| 65 | Dec. 7, 1874 | 5 | 15,500 |
| Being for the year 102½ per cent. on the capital stock, and amounting to | | | 317,750 |

The following are the aggregate receipts and expenditures of the company for the last six fiscal years, that being the time in which the mine has paid dividends, and before which prospecting only was done:

| | |
|---|---|
| Receipts from all sources for the fiscal year 1869 | $306,038 75 |
| For 1870 | 183,450 23 |
| For 1871 | 407,301 16 |
| For 1872 | 404,035 52 |
| For 1873 | 1,010,612 20 |
| For 1874 | 669,023 03 |
| Total receipts for 6 years | 2,980,460 89 |

There have been paid out in dividends as follows:

| Years. | No. of dividends. | per cent. | Amount. |
|---|---|---|---|
| 1869 | 11 | 55 | $170,500 |
| 1870 | 7 | 12 | 37,500 |
| 1871 | 12 | 75 | 232,500 |
| 1872 | 11 | 52½ | 162,750 |
| 1873 | 12 | 220 | 682,000 |
| 1874 | 12 | 102½ | 317,750 |

Being 65 dividends, 517 per cent. on the capital stock, amounting to $1,602,700.

The Idaho, like its neighbor, the Eureka, has been noted for the uniformity of its yield rather than for extraordinary richness of rock. Specimen gold-rock in either mine is of rare occurrence. It will be observed by the foregoing report that, out of a total product of $669,023, only $38.50 was realized from the sale of specimens. The average yield per ton was $23.70 from 28,801 tons crushed during the year. The Idaho Mill carries 35 stamps of 850 pounds' weight, dropping 65 blows per minute from a height of 10 inches. The cost of sinking, per foot, has been $120, while drifting, per foot, has cost $15. The cost of stoping, hoisting ore, and milling, altogether, is $8.22½ per ton. The percentage of sulphurets is ¾ of 1 per cent. The length of the pay-zone is 1,300 feet. The average width of the ledge is 3½ feet. The course of the ledge is northeast and southwest, with the dip toward the south. The country-rock is greenstone and metamorphic slate. Depth of the working-shaft is 920 feet, and the number of levels opened is eight. The total length of the drifts is 3,988 feet. The hoisting-works cost $38,689.42. The company own the mill, which is worked by a steam-engine of 90 horse-power. The number of pans in use is 12, and 2 concentrators of sulphurets. Cost of the mill, $24,430.42. Capacity of the mill for every 24 hours, 91 tons. The sulphurets are treated by the chlorination process. The Idaho has been worked for six years. The company is now using five Burleigh drills—one in the shaft and four in the drifts. The average time of drilling in the Idaho quartz is 5 feet of hole in each hour, the hole being 2 inches in diameter. In hand-drilling the average is for three men to put in 6 feet of hole during a shift of 8 hours. It will be seen at once that while three men are drilling 6 feet of hole, with hand-drills and hammers, three men with a Burleigh machine will drill 40 feet of hole.

During the year 1874 the pumping capacity has been increased and the old shaft retimbered, involving large outlays, which have had the effect of placing the mine in an effective condition for the year 1875. The underground work has been pushed ahead with due diligence, and, having in view the importance of keeping the mine well opened, work has been constantly going on in the drifts and in the main shaft.

The superintendent, in his annual report, says:

The shaft is down 75 feet below the 800 level. The ledge is somewhat broken up, thus rendering it necessary to carry a very large shaft, as well as make it more expensive. The quartz seems to be improving, and it is thought it will form a solid ledge before reaching the 900 level. The 800 west level is in 116 feet from the shaft, but little rock has been taken from the backs. This is an average quality of rock. Also the 800 east is of average-grade ore, and the drift is in 109 feet from the shaft. The 700 west drift is in to within 35 feet of the Eureka mine, and a few months more will exhaust the backs. The 700 east drift is in 424 feet from the shaft, and the backs are worked through to the 600 level 233 feet from the shaft. The 600 south drift is in 803 feet from the shaft, or 322¼ feet from the split. The ledge is exhausted in the drift, and it is low-grade ore in the backs. The north branch is in 421 feet from the split, and 901½ feet from the shaft. The ledge has been very small. It is now opening out larger, but it is low-grade ore; however, from indications, it ought to come in better. The rock in those backs is good mill-rock, and it is worked through to the 500 level 341 feet from the shaft. The 500 backs are of an average quality; they are not yet worked through to the 400 at any point. The 400 is exhausted.

In conclusion, Mr. Coleman remarks:

In reviewing our operations for the past year, you will notice that the yield of the rock has not been so much or the dividends so large as those of the previous year. Still it has been good, and the dividends in the aggregate amount to the sum of $317,750 for the year.

The Eureka mine, situated about two miles east of Grass Valley, and adjoining the Idaho on the west, has employed during the year 80 miners, at $3 per day to the man. The cost of sinking, per foot, in exploring, has been about $65, while the cost of drifting has been about $25 per foot The cost of stoping has been about $10.50 per ton of ore. Milling the ore costs $2.61 per ton; the company owning its own mill. The number of tons extracted and worked during the year is 8,130, the average yield of which has been $25 per ton. The percentage of sulphurets in the rock amounts to 1.5. The total bullion-product has been, for the year, about $205,780. The Eureka's location is 1,680 feet, for which the company has a patent. The course of the ledge is nearly northeast and southwest, and the dip is toward the south. The length of the pay-zone is about 1,000 feet, with a ledge of four feet in thickness. The country-rock is metamorphic slate and greenstone. The mine is worked through a shaft, which has a total depth of 1,250 feet. There are eight levels opened, and the total length of drifts is 9,000 feet. The cost of the hoisting-works is $48,000. At the mill a 60 horse-power engine is in use, and the number of stamps is 30, each of which weighs 850 pounds. These are dropped, each, 65 times per minute, and the drop is 10 inches. There are two pans and two sulphuret-concentrators in the mill. The cost of the mill was $30,000, and is capable of crushing 65 tons of ore in twenty-four hours. The sulphurets are treated by the chlorination process. All the stamps of the mill have not been employed during the year. The lower portions of the mine do not show good pay-rock, but explorations which are now going on may result in something good. The Eureka went into operation October 1, 1865, and up to and including the 30th of September, 1874, had taken out bullion to the value of $4,273,148.49. During that time it paid dividends to the amount of $2,054,000. On the 1st of October, 1874, the company had on hand, in cash and value of supplies, the sum of $101,646.73, which will enable them to explore the ledge to a much lower depth than has yet been reached.

Mr. William Watt, the superintendent of the Eureka mine, in his annual report, made September 30, 1874, says:

In Eureka mine proper we have driven 735 feet of drifts, 49 feet of a rise, 191 feet of cross-cuts, and sunk 207 feet of shaft; also, sunk 53 feet of No. 3 shaft on the Roannaise, and 107 feet on Eureka ledge No. 2. We have hoisted 8,207¼ tons of quartz, and crushed 8,130 tons in 303½ running-days, as follows: 87¾ days with 15 stamps, and 216 days with 10 stamps, averaging a little over 2¼ tons per day to each stamp. We have concentrated 73¼ tons of sulphurets and worked 89 tons, and have now on hand 16¼ tons, which I value at $700 net. There are 325 tons of quartz on the surface, and 350 broke in the mine ready for hoisting. The amount of quartz extracted during the past year from the various levels has been 8,207¼ tons, and there are upward of 1,000 tons of ore in sight which will pay a profit.

The superintendent concludes his report by remarking that, notwithstanding a great amount of dead-work and exploration had been carried on during the past year, no favorable results had followed.

The following is from the secretary's report for the fiscal year ending September 30, 1874:

RECEIPTS.

| | | |
|---|---|---|
| By book accounts, October 1, 1873 | $113,656 25 | |
| By bullion account | 205,050 60 | |
| By mine account | 21 00 | |
| By mill account | 723 50 | |
| By wood-ranch | 1,751 03 | |
| By wood account | 14,282 84 | |
| By premiums | 679 82 | |
| By McDougal works | 729 36 | |
| By interest | 4,884 66 | |
| | | $341,779 06 |

DISBURSEMENTS.

| | | |
|---|---|---|
| To dividends | 80,000 00 | |
| To mine account | 83,254 33 | |
| To mine account, prospecting | 14,989 25 | |
| To mill account | 22,461 22 | |
| To sulphuret-concentration | 1,495 00 | |
| To sulphuret-reduction | 480 00 | |
| To construction | 16,539 00 | |
| To Great Western mine | 4,600 06 | |
| To Roannaise mine | 1,115 00 | |
| To Eureka No. 2 mine | 1,203 50 | |
| To wood account | 8,804 27 | |
| To wood-ranch | 883 78 | |
| To McDougal works | 325 00 | |
| To bullion expenses | 804 78 | |
| To general expenses | 6,712 48 | |
| To discount account | 21 03 | |
| To book accounts, September 30, 1874 | 98,140 36 | |
| | | 341,779 06 |

### ASSETS AND LIABILITIES.

| | | |
|---|---|---|
| **Assets:** | | |
| Available on hand, September 30, 1874. | | |
| Cash balance | $98,140 36 | |
| 16½ tons of sulphurets, estimated value | 700 00 | |
| 350 tons of ore broke in mine, (cost, $9) | 3,150 00 | |
| 325 tons of ore on surface, (cost, $10) | 3,250 00 | |
| 134¾ cords of wood | 606 37 | |
| Supplies at mill | 1,000 00 | |
| Supplies at mine | 1,200 00 | |
| | | $108,046 73 |
| Real estate: | | |
| Mill, estimated value | 30,000 00 | |
| Mine improvements and buildings, estimated | 20,000 00 | |
| McDougal works, estimated | 2,000 00 | |
| Wood-ranch, 160 acres | 800 00 | |
| | | 52,800 00 |
| | | 160,846 73 |
| Liabilities, none. | | |

### MINE STATEMENT.

| | | |
|---|---|---|
| October 1, 1873.—Ore on surface, tons | | 247½ |
| Ore hoisted during the year— | | |
| From fourth level, tons | 40 | |
| From fifth level, tons | 159 | |
| From sixth level, tons | 8,008½ | |
| | | 8,207½ |
| | | 8,455 |
| September 30, 1874.—Ore on surface, tons | | 325 |
| Worked at the company's mill during the year | | 8,130 |
| September 30, 1874.—Ore on surface, tons | 325 | |
| Ore broke in mine, tons | 350 | |
| Ore reduced during the year, tons | 8,130 | |
| | | 8,805 |
| October 1, 1873.—Ore on surface, tons | 247½ | |
| Ore broke in mine, tons | 250 | |
| | | 497½ |
| Ore mined during year, tons | | 8,307½ |

### SULPHURET STATEMENT.

| | | |
|---|---|---|
| October 1, 1873.—Number of tons on hand | 32 | |
| Number of tons concentrated during the year | 73½ | |
| | 105½ | |
| September 30, 1874.—Number of tons on hand | 16½ | |
| | | 89 |
| Number of tons worked during the year | 12 | |
| Number of tons sold | 77 | |
| | | 89 |

### ORE STATEMENT.

| | | |
|---|---|---|
| 8,130 tons of ore worked by mill process yielded | | $196,839 62 |
| Also sulphurets as follows: | | |
| 12 tons worked by chlorination yielded | $1,884 48 | |
| 77 tons sold yielded | 6,326 50 | |
| | 8,210 98 | |
| Less 32 tons on hand October 1, 1873—estimated value | 3,500 00 | |
| | 4,710 98 | |
| Add 16½ tons on hand September 30, 1874—estimated value | 700 00 | |
| | | 5,410 98 |
| Add results of McDougal works | | 729 36 |
| | | 202,979 96 |

Or an average of $24.97 per ton.
Average yield of sulphurets, $100.37 per ton.

BULLION STATEMENT.

As reduced by mill process:

Average fineness, .848—equal to $17.52$\frac{7}{100}$ per ounce.

As reduced by chlorination process:

Average fineness, .992—equal to $20.50$\frac{17}{100}$ per ounce.

Returns from McDougal works:

Average fineness, .794—equal to $16.41$\frac{33}{100}$ per ounce.

Return of bullion:

| | | |
|---|---|---|
| 11,233.93 ounces, at $17.52$\frac{7}{100}$ | $196,809 09 | |
| 91.85 ounces, at $20.50$\frac{17}{100}$ | 1,883 08 | |
| 44.37 ounces, at $16.41$\frac{33}{100}$ | 728 26 | |
| 1.89 ounces assay chips and grains | 33 03 | |
| Sulphurets sold | 6,326 50 | |
| | | $205,779 96 |

Weight of bullion:

| | | |
|---|---|---|
| Before assaying: | | |
| Ounces | | 11,373.33 |
| After assaying: | | |
| Face of bars, ounces | 11,369.15 | |
| Assay chips and gains | 1.89 | |
| Loss | 2.29 | |
| | | 11,373.33 |

COST OF MINING.

| | | |
|---|---|---|
| Supplies on hand October 1, 1873 | $1,500 00 | |
| Paid for sundry supplies and labor during the year | 83,254 33 | |
| | | $84,754 33 |
| Deduct: | | |
| Supplies on hand September 30, 1874 | 1,200 00 | |
| Merchandise sold during the year | 21 00 | |
| | | 1,221 00 |
| Cost of mining 8,307¼ tons | | 83,533 33 |

Or an average of $10.06 per ton.

COST OF MILLING.

| | | |
|---|---|---|
| Supplies on hand October 1, 1873 | $500 00 | |
| Paid for sundry supplies and labor during the year | 22,461 22 | |
| | | $22,961 22 |
| Deduct: | | |
| Supplies on hand September 30, 1874 | 1,000 00 | |
| For custom-work done and supplies sold | 723 50 | |
| | | 1,723 50 |
| Cost of milling 8,130 tons | | 21,237 72 |

Or an average of $2.61 per ton.

COST OF CONCENTRATING SULPHURETS.

| | |
|---|---|
| Number of tons concentrated during the year | 73½ |
| Cost of concentrating: | |
| Paid for pay-rolls | $1,495 00 |

Or an average of $20.32 per ton.

STATEMENT OF PROFITS FOR THE YEAR ENDING SEPTEMBER 30, 1874.

Receipts:

| | | |
|---|---|---|
| From bullion | $205,050 60 | |
| Amount of sulphurets belonging to last year | 3,500 00 | |
| | | $201,550 60 |
| From sulphurets on hand | | 700 00 |
| From other receipts | | 6,815 06 |
| | | 209,065 66 |

| | | | |
|---|---|---|---|
| Cost of same: | | | |
| Supplies on hand October 1, 1873 | | $8,281 25 | |
| Paid for supplies and labor during the year, (including $21,857.81 paid for prospecting or dead-work) | | 123,325 29 | |
| Paid all other expenses | | 7,517 26 | |
| | | 139,123 80 | |
| Off for supplies on hand September 30, 1874 | | 2,806 37 | $136,317 43 |
| Net profits | | | 72,748 23 |

DISTRIBUTION OF PROFITS.

| | | | |
|---|---|---|---|
| Paid dividends, $4 per share | | | $80,000 00 |
| Paid for construction | | | 16,539 00 |
| | | | 96,539 00 |
| On hand October 1, 1873: | | | |
| Balance cash | | $117,156 25 | |
| Balance supplies | | 8,281 25 | |
| | | 125,437 50 | |
| On hand September 30, 1874: | | | |
| Balance cash | $98,140 36 | | |
| Balance supplies and sulphurets | 3,506 37 | 101,646 73 | |
| Amount of former balance of cash and supplies reduced during the year | | | 23,790 77 |
| | | | 72,748 23 |

*Statement showing the receipts and disbursements of the company from the date of its going into operation, October* 1, 1865, *to date.*

| | | | |
|---|---|---|---|
| Receipts: | | | |
| By bullion taken out | | | $4,273,148 49 |
| By other receipts | | | 35,390 38 |
| | | | 4,308,538 87 |
| Disbursements: | | | |
| To sundry titles; to paid on the purchase of mine; for Whiting ground or square location; purchase of Mobile and Roannaise mines and perfecting titles | | $301,906 50 | |
| To construction | | 158,383 71 | |
| To dividends | | 2,054,000 00 | |
| To mining, milling, and all other expenses | | 1,692,601 93 | |
| | | 4,206,892 14 | |
| On hand September 30, 1874: | | | |
| Balance of cash | $98,140 36 | | |
| Balance of supplies | 3,506 37 | | |
| | | 101,646 73 | |
| | | | 4,308,538 87 |

*Statement of profits from October* 1, 1865, *to September* 30, 1874.

| | |
|---|---|
| Receipts: | |
| From bullion | $4,273,148 49 |
| From other receipts | 35,390 38 |
| | 4,308,538 87 |
| Cost of same: | |
| Paid for mining, milling, and other expenses | 1,692,601 93 |
| Net profits | 2,615,936 94 |

Distribution of profits:

| | | | |
|---|---|---|---|
| Paid for sundry titles | | $301,906 50 | |
| Paid for construction | | 158,383 71 | |
| Paid for dividends, $102.70 per share | | 2,054,000 00 | |
| | | 2,514,290 21 | |
| Balance of cash | $98,140 36 | | |
| Balance of supplies | 3,506 37 | | |
| | | 101,646 73 | |
| | | | $2,615,936 94 |

The following review of the operations of other mines in Grass Valley district for the year 1874 has been prepared by Mr. Rufus Shoemaker, of the Grass Valley Union, and Mr. Frank J. Beckett, the latter having compiled the tabular statements for this report:

The Omaha mine is situated south of Grass Valley, on Wolf Creek, and about three miles distant from the town. It is owned by an incorporated company, which has its principal place of business at Sacramento. Sixteen miners are at present employed, and $3 per day per miner are paid. The cost of sinking so far per foot has been $25. The cost of drifting per foot has been $10, and the cost of stoping per ton has been $6. The cost of extracting ore per ton has been about $10, and the cost of milling the ore per ton has been $3.50. The milling is done at a custom-mill. The number of tons which have been worked is 100, and gave an average yield of $21 per ton. The percentage of sulphurets is about 1.5. The total bullion-product has been $2,500. The length of the location is 1,400 feet, and the course of the ledge is north and south, with a dip to the west at an angle of 32°. The length of the pay-zone, as far as explored, is 170 feet, and the vein has an average thickness of 15 inches. The country-rock is serpentine. The work is done through a shaft which is 260 feet deep. The ledge in the bottom of the shaft is fully 3 feet thick, and shows free gold in great quantities, besides good sulphurets and general good quality of rock. The walls of the ledge are well defined and smooth. Two levels have been opened, and these are, together, of the length of 255 feet. The hoisting-works are run by water-power, and cost $1,500.

The Pittsburgh mine is situated about four miles southwest from this place. There are 12 miners employed, at $3 per day each. The cost of sinking per foot has been $7; the cost of drifting, $3 per foot; while stoping has cost $3 per ton. The cost per ton for extracting ore has been $6, and the cost of milling the same has been $4 per ton. The crushing is done at a custom-mill. About 100 tons of ore have been extracted and worked, and this gave an average yield of $75 per ton, or a total of $7,500. The percentage of sulphurets is very large. The location consists of 1,500 feet, and the course of the vein is northeast and southwest, with a dip to the northwest at an angle of 60°. The length of the pay-zone, as far as is known, is 400 feet, with an average thickness of 30 inches. The country-rock is slate. The mine is worked through a shaft which is now 104 feet deep. Two levels have been opened, of the total length of 400 feet. The hoisting and pumping are done at present by horse-power.

The Empire mine is one of the oldest quartz-mines in the district, and is, we believe, the oldest of any now being worked. The mine employs 80 miners, at $3 per day as wages. The cost of sinking per foot is $12; cost of drifting per foot, $8; and the cost of stamping per ton $5. The cost of extracting ore per ton is $8, and milling costs $1.75 per ton. The company own the mill. The number of tons taken out and worked during the year is 11,000, and the average yield has been

$16.75 per ton. The percentage of sulphurets has been 2½. The total bullion-product has been $187,000. The length of location is 2,800 feet; course of ledge, north and south, with the dip toward the west. The pay-zone of this mine has a length of over 1,000 feet, and an average thickness of 15 inches. The Empire is worked through a shaft which has a depth of 1,200 feet. There are 12 levels opened. Total length of drifts, 7,900 feet. Cost of hoisting-works, $40,000. Steam-power is used, and the mill has 20 stamps, which weigh 900 pounds each. Each stamp drops 72 times in a minute, and the height of the drop is 9 inches. Number of pans, 4; number of concentrators, 10. The cost of the mill was $40,000, and its capacity is 40 tons for every twenty-four hours. The sulphurets are treated by the chlorination process. The Empire is owned by an incorporated company, whose principal place of business is San Francisco.

The New York Hill mine, situated about two and a half miles south of Grass Valley, employs 45 miners. The cost of mining and milling is given at $19.50 per ton, but as this is excessive, we presume it includes a great deal of dead-work. The number of tons worked by custom-mill during the year has been about 500, and the average yield has been not less than $50 per ton, making a total bullion-yield of $25,000. The location is about 3,000 feet in length, and the ledge runs northeast and southwest, dipping to the northeast. The length of pay-zone, so far as explored, is 1,400 feet, with an average thickness of 2½ feet. The country-rock consists of greenstone and slate. The ledge is worked through a tunnel, which has now a total length of 750 feet. There are two drifts opened from the tunnel. There are 400 feet of backs above the tunnel-workings. On the ledge there is a shaft, not used at present, but which is available at any time for working, of the depth of 700 feet. The total length of drifts in the mine is 500 feet.

Two miles west of Nevada City, and situated in the granite-belt, is found a group of quartz-mines, which have been worked at intervals for the past fifteen years, with many alternations of success and adversity, but all of which have within the past two years entered on a career of prosperity. Within a radius of less than one-fourth of a mile from sixty to eighty stamps are engaged in crushing quartz on the Providence, Nevada, California, and Wyoming.

I visited the Providence July 3. It is about 1¼ miles from Nevada City, on the banks of Deer Creek, and comprises a claim of 3,100 feet on a vein running north and south. The claim extends under the creek to the north side, where the Nevada occupies an extension. The incline was at the time of my visit 585 feet deep, on the dip of the vein, 38° to 43° east. At 87 feet it is intersected by the Green tunnel, an adit 140 feet long; at 200, 300, 400, and 500 feet there are levels. On the 400 and 500 foot levels the drifts at the time of my visit were 50 feet south and 70 feet north, 120 feet being at those levels the horizontal dimensions of the pay-chute. The country-rock in the mine is granite; but 600 feet south of the incline slate appears as foot-wall, and will probably be found still farther south to constitute the hanging-wall also. The vein-matter is white and blue quartz, seamed and ribboned, carrying bunches, zones, and streaks of sulphurets. The width of the vein down to the 400-foot level is 10 to 12 feet. Just above that level a foot-wall splice or additional zone appears, enlarging the total width to from 27 to 30 feet, of which perhaps 20 feet is pay-ore. The foot-wall zone is the richer. There is a body of it about 12 feet thick, between the 400 and 500 foot levels, comprising perhaps 12,000 tons, and expected to yield in mill about $18, besides containing, say, 6 per cent. of sulphurets, worth $150 per ton. The other reserves of the mine were also large at the time of my visit; and I learn that the opening of the 600-foot level since that time has greatly added to its visible resources, having exposed an 80-foot vein of increased average richness. The operations of the mine and mill during 1874 will be found on page 12.

The mill and chlorination-works are crowded on the steep bank close by the mine. The former contains 20 stamps, weighing 750 pounds each, dropping 8 inches, 75 times per minute, and crushing 38 tons in twenty-four hours. Amalgamation takes place in the battery and upon aprons in front, from which the pulp is conveyed to launders and buddles for the separation of the sulphurets. These comprise about 6 per cent. of the ore, and consist of iron pyrites, with arsenical pyrites, galena, &c. The product of sulphurets is about 45 tons monthly. They are treated by Plattner's process of chlorination in works having 2½ tons daily capacity. Mr. Lüdemann, the metallurgist in charge, hoped to be able to roast the sulphurets with salt in such a way as to get chloride of silver, and, after chlorinating and leaching out the gold, to obtain the silver by a new lixiviation with hyposulphite of soda. The practicability of such a treatment appears doubtful. Whether the silver in the sulphurets is sufficient to pay for a special extraction I am not informed. The presence of galena is encouraging as regards the probability of a considerable silver value, and embarrassing as regards successful chlorination.

Work is being prosecuted on the copper-mine at Spenceville, in this county, under the superintendency of Mr. G. F. Deetkin, with every prospect of success. The shaft is down 100 feet, and the ledge at that depth is 70 feet in width. The rock is richly impregnated with native copper. The ore is taken out and roasted in a large furnace, after which it is turned into three large vats, upon which a stream of cold water is turned, and the copper, in a state of solution, is then conducted from the vats into a large cylinder of about 12 feet in diameter. In this is placed old or refuse iron, for which the copper has an affinity. The cylinder is made to revolve rapidly by steam, by which means the copper is collected on the iron. The superintendent thinks the process of separating copper from the ore in which it is contained is no longer a matter of experiment. There are many other ledges in the vicinity equally as rich, and are awaiting the success of working this one.

There are within the limits of Nevada County sixty mills for the reduction of gold-bearing quartz and cement, located as follows: Grass Valley Township, 24; Nevada, 14, (besides several that are at present idle;) Eureka, 6; Washington, 4; Little York, 2; Bridgeport, 7; Bloomfield, 1; Rough and Ready, 1; Meadow Lake, 1. There are seven metallurgical works located, four in Grass Valley Township, two in Nevada Township, and one in Meadow Lake Township. The total cost of these, in round numbers, is in the neighborhood of $1,250,000. Forty-two of these mills are run by steam and twenty-five by water power.

*Rain-fall at Nevada City and Sacramento.*

| Period. | Nevada City. | Sacramento. |
|---|---|---|
| | *Inches.* | *Inches.* |
| October, 1870, to June, 1871, inclusive | 39.23 | 8.47 |
| October, 1871, to June, 1872, inclusive | 78.22 | 24.05 |
| October, 1872, to June, 1873, inclusive | 38.70 | 14.20 |
| October, 1873, to June, 1874, inclusive | 62.91 | 21.89 |

*Description of leading mines, Nevada County, California*, 1874.

| Name. | Location. | Owners. | Length of location. | Course. | Dip. | Length of pay-zone. | Average width. | Country-rock. | Character vein-matter. | Worked by tunnel or shaft. | Length of tunnel. | Depth on vein in tunnel. | Depth of working-shaft. | Number of levels opened. | Total length of drifts. | Cost of hoisting-works. |
|---|---|---|---|---|---|---|---|---|---|---|---|---|---|---|---|---|
| | | | *Ft.* | | | *Ft.* | | | | | | | *Ft.* | | *Ft.* | |
| Eureka | Grass Valley | Incorporation | 1,680 | E. and W. | S | 1,000 | 4 ft | Clay, slate, and serpentine. | Quartz. | Shaft | | | 1,250 | 8 | 9,000 | $48,000 |
| Empire | do | do | 2,800 | N. and S | W | 1,300 | 15 in. | Slate | do | do | | | 1,250 | 12 | 7,900 | 40,000 |
| Omaha | do | do | 1,400 | N. and S | 32° W | 170 | 15 in. | Granite and serpentine. | do | do | | | 236 | 2 | 255 | 1,500 |
| New York Hill. | do | do | 3,000 | N. W. and S. E. | N. E | (*) | 2½ ft. | Greenstone and slate. | do | Tunnel | 750 | 400 | 700 | 2 from tunnel. | 500 | 20,000 |
| Pittsburgh | do | Sanford & McCook Bros. | 1,500 | N. E. and S. W. | 60° N. W. | 400 | 30 in. | Slate | do | Shaft | | | 104 | 2 | 400 | Horse-power. |
| Idaho | do | Corporation | 3,100 | E. and W. | S | 1,300 | 3½ ft. | Serpentine | do | do | | | 920 | 4 | 3,988 | 24,430 42 |
| Wyoming | Nevada | Incorporation | 2,200 | N. and W. | E | | 18 in. | Slate | do | Both | 1,500 and 1,100. | 450 | 250 | 2 from incline. | 3,000 | 3,500 |
| Providence | do | Providence Mining Company. | 3,100 | N. E. and S. W. | 50° E | 900 | 50 in. | Granite and slate | Sulphureted quartz. | Shaft | | | 700 | 6 | 1,200 | 9,000 |

* One 400 feet, other unexplored.

*Operations of leading mines, Nevada County, California*, 1874.

| Name. | Location. | Owners. | No. of miners employed. | Miners' wages per day. | Cost of sinking per foot. | Cost of drifting per foot. | Cost of stoping per ton. | Cost of mining per ton extracted. | Cost of milling per ton. | Company or custom mill. | Number of tons extracted and worked. | Average yield per ton. | Percentage of sulphurets. | Total bullion-product. |
|---|---|---|---|---|---|---|---|---|---|---|---|---|---|---|
| Eureka | Grass Valley | Incorporation | 90 | $3 00 | $65 | $25 | ........ | $10 25 | $2 61 | Company | 8, 130 | $25 00 | 1.5 | $205, 780 00 |
| Omaha | do | do | 16 | 3 00 | 25 | 10 | $6 00 | 10 00 | 3 50 | Custom | 100 | 21 00 | 1.5 | 2, 500 00 |
| New York Hill | do | do | 45 | 2 50 | ...... | 15 | ........ | 15 00 | 4 50 | do | 500 | 50 00 | 3 | 25, 000 00 |
| Pittsburgh | do | Sanford & McCook Bros | 12 | 3 00 | 7 | 3 | 3 00 | 6 00 | 4 00 | do | 100 | 65 00 | 10 | 7, 500 00 |
| Empire | do | Incorporation | 80 | 3 00 | 12 | 8 | 5 00 | 8 00 | 1 75 | Company | 11, 000 | 16 75 | 2.5 | 187, 000 00 |
| Wyoming Gold | Nevada | do | 55 | 3 00 | 15 | $5 to 8 | 1 50 | 3 00 | 1 25 | do | *750 | 18 00 | 3 | *12, 000 00 |
| Providence | do | Providence Mining Company. | 42 | 3 00 | 15 | 6 | 1 25 | 2 00 | 75 | do | 7, 200 | †12 00 | 10 | |
| Idaho | Grass Valley | Corporation | 175 | 3 00 | 120 | 15 | | $8 22½ | | do | 28, 801¼ | 23 40 | ¾ of 1 | 60[illegible], 811 20 |

* Per month. † Free gold.

*Statement of quartz-mills, Nevada County, California*, 1874.

| Name. | Location. | Owners. | Kind of power and amount. | No. of stamps. | Weight of stamps. | No. of drops per minute. | Height of drop. | Number of pans. | Number of concentrators. | Cost of mill. | Capacity per 24 hours, tons. | Cost of treatment per ton. | Tons crushed during year. | Method of treating sulphurets. |
|---|---|---|---|---|---|---|---|---|---|---|---|---|---|---|
| | | | | | *Lbs.* | | | | | | | | | |
| Eureka | Grass Valley | Incorporation | Steam, 60 horse | 30 | 350 | 65 | 10 in | 2 | 2 | $30, 000 00 | 65 | $2 61 | 8, 130 | Chlorination. |
| Empire | do | do | Steam | 20 | 900 | 72 | 9 in | 4 | 10 | 40, 000 00 | 40 | 1 75 | 11, 000 | Do. |
| Idaho | do | Corporation | Steam, 90 horse | 35 | 350 | 65 | 10 in | 12 | 2 | 38, 689 42 | 91 | ........ | 28, 801½ | Do. |
| Wyoming | Nevada | Incorporation | Water, hurdy and overshot. | 16 | 800 | 75 | 10 in | 2 | 6 | 8, 000 00 | 30 | 1 25 | 8, 000 | Do. |
| Providence | do | Providence Mining Company. | Water | 20 | 750 | 68 | 10 in | 1 | 8 | 20, 000 00 | 35 | 75 | 7, 200 | Do. |

## YUBA COUNTY.

Yuba County is situated in the foot-hill country on the east side of the northern portion of the Sacramento Valley, and is bounded on the north by Butte County; east by the counties of Nevada and Sierra; south by Placer County, the western portion of the county covering the agricultural and grazing lands of the Sacramento Valley. It is about fifty-seven miles in length by eighteen miles in width. The mountainous portion of the country is drained by the Yuba River and its tributaries. The principal mining-interest of the county is the deep gravel deposits, worked by the hydraulic process, at Sucker Flat and Smartville.

It is scarcely necessary for me to repeat in this place the general observations made in former reports on the nature and importance of this peculiarly American process, the product of native ingenuity and enterprise, acting upon natural conditions such as no other country has afforded in equal degree. The dimensions and situation of the deep-lying placers or gravel and cement beds, the character of the climate, furnishing a season of abundant water, followed by a dry season, and the configuration of the surface, permitting the discharge of tailings, are circumstances which, while they operate as hinderances to placer-mining of the ordinary type, greatly favor the extended operations of hydraulic mining. And in no part of California has this method been pursued with greater boldness or developed with greater skill than in Yuba County.

The placer-diggings of Timbuctoo and Sucker Flat were discovered in December, 1849, and were worked by rockers and small sluices on the surface in the winter season only until hydraulic washing was introduced in 1865 and 1866, no claim using more than 40 to 50 inches prior to that time. In early days the claims were all small, (100 by 120 feet.) Four or five owners would work their claim together, sharing the profits equally. Gradually the claims were consolidated and more ground was located. The owners were fewer in number and the magnitude of mining-operations constantly increased as the surface of the ravines was worked out. At first, where the banks were low, the bottom was worked out with picks and the top allowed to fall in, and then a stream of water was turned on and the dirt was run off through the sluice-boxes. This method, which was at once wasteful and perilous to life, continued till about 1860, when Mr. J. P. Pierce introduced a method of blowing up the banks with powder. It was also discovered that the gravel in the hills could be more profitably worked by using more water and large sluices.

Prior to 1855 the workings in Timbuctoo and Sucker Flat were similar, but as the ground was developed it was discovered that in Timbuctoo proper the gravel was lying on the bed-rock, (or country-rock,) while toward Sucker Flat it was within a rim of the bed-rock of the channel. At Sucker Flat, as the surface was worked off and the gravel in the banks was exposed, it soon became evident to Mr. Pierce that more extensive preparations were required. The main beds of gravel here were harder and required longer flumes in order to extract the gold. The rim-rock, showing on both sides, was lowest on the side toward the Yuba, where a dumping-space of 300 vertical feet between it and the Yuba afforded much better hydraulic-mining facilities than Big Ravine, at Timbuctoo. All that portion of the gravel above the edges of the rim-rock, (denominated "top-gravel,") was worked through flumes intersecting the rim by short bed-rock cuts. As the upper layer began to be worked out, shafts were sunk, and the fact was developed that a large body of the famous "blue gravel" lay in the old channel between the rims of Sucker Flat and Smartville. It also became evident that in order to work this lower lead and to extract the gold, a tunnel must be run through the rim at sufficient depth to allow the gravel at the bottom of the ancient channel to be run off.

The Blue Gravel Company was the first to run such a tunnel, 1,200 feet in length. It was commenced in 1855 and finished in 1864. During a large part of this time work was suspended for lack of funds. As soon as this tunnel was completed the mine yielded very largely. Adjoining mine-holders were stimulated to run similar tunnels. To the westward, the Pittsburgh, as now consolidated, owning about 1,000 feet on the channel; next the Rose Bar, owning 2,000 feet; next the Pactolus, owning 1,000 feet, have since completed deep bed-rock tunnels. To the eastward, the Blue Point, owning about 1,100 feet; the Smartville Consolidated, owning 1,200 feet, and lastly, the Enterprise, owning 1,300 feet, have all run long and costly bed-rock tunnels. Thence the channels pass through the hills to Mooney Flat, where is in progress the last and perhaps the most important of these bed-rock tunnel undertakings, the Mooney Flat tunnel.

The Blue Gravel claim is one of the most famous gravel-claims in the State. It was incorporated by Mr. McGaumry and others in 1855. The company commenced running a tunnel the same year, an undertaking which was regarded by most of the miners as hazardous, but with the most hopeful anticipations by the men who incorporated it, and who had ascertained by the sinking of shafts that there was a heavy deposit of "blue gravel" below the cuts, on the rim-rock. The total cost was $80,000, all of which, except $10,000, was paid from the profits of the upper lead, worked through a deep-cut, and a short tunnel of about 600 feet still higher on the rim-rock. Through this short tunnel about 10 acres of ground were worked off to a depth of 80 feet. The first clean-up of the claim through the first 1,400-foot tunnel cleared off all the indebtedness of the company, under which it had staggered for nine years. In May, 1864, the first dividend was declared. The head of water ordinarily used was 500 inches. Between this time and January, 1869, the sum of $643,000 was disbursed in dividends.

During the year 1868 it became evident that part of the lower lead above the 1,400-foot tunnel was nearly exhausted, and that there was a still lower stratum of gravel which could not be worked through this tunnel. Accordingly the company stopped declaring dividends and began another tunnel 65 feet lower. Work was vigorously prosecuted on this tunnel, and it was finished in July, 1872, at a cost of $75,000. During its construction the company worked the banks of the ancient river southward of the main channel. The flume for the 1,400-foot tunnel, through which this outside bank has been washed, being over the main channel, it was not abandoned until February of the present year, (1874.) At the present time the Blue Gravel Company is running through the lower tunnel, using a head of 800 inches of water. The claim is now so well opened that there need be no more delays before all the ground left on both the upper and lower leads will be worked off. While it is not expected that any large dividend will be declared from the beds of gravel left standing south of the main lead, yet they will always pay more than water-expenses, and the lower lead will be richly profitable until it is worked out.

After the Blue Gravel Company had declared several hundred thousand dollars in dividends, the Blue Point Company began a tunnel 2,200 feet long, which was finished in three or four years by working at the lower end and from two shafts at the same time. This tunnel cost $146,000, or $63 per foot. It was finished in 1872; and since then the Blue Point

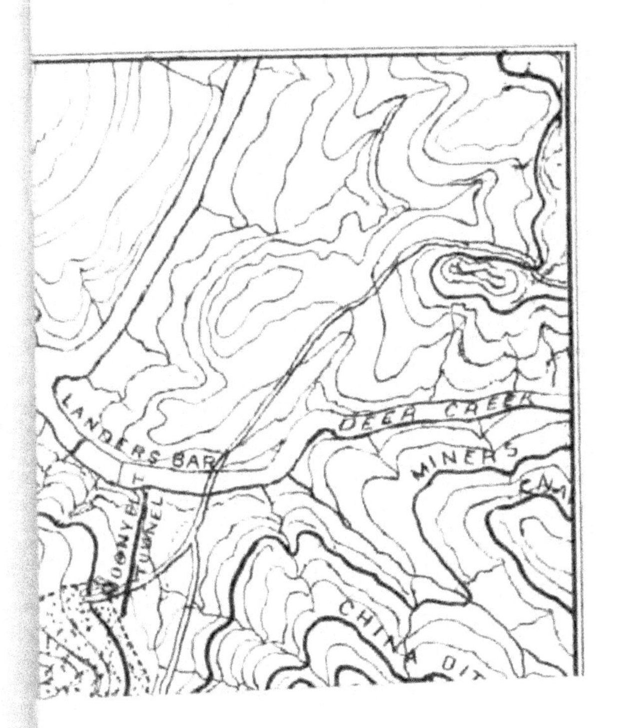
LANDERS BAR
DEER CREEK
MINERS
TUNNEL

Company has run off the lower lead about 3½ acres of gravel of an average depth of 60 feet, equivalent to 320,000 cubic yards, from which $210,000 of profit has been divided in two years.

The productiveness of this region in gold from the hydraulic mines of hill gravel of Yuba Basin has been greater than that of any other mining region of California. The country extending from Timbuctoo to Mooney Flat (three miles) is generally conceded to have been the richest in actual returns from hill-gravels, not only on the Yuba, but of all hydraulic-mining districts in California up to the present time.

The placer-mining of 1849, which was confined to the present-river beds, lasted but a few years before the entire population had transferred its sphere of operations to the hill-gravels or deep placers, the sources which had enriched the fabulous "bars."

Messrs. Pierce and McGanny, of the Excelsior Mining and Water Company, of Smartville, have kindly placed at my disposal an able and comprehensive report on this region by Mr. Amos Bowman, from which I make the following extracts:

The mining-properties of Sucker Flat and Timbuctoo are situated at the ancient debouchure of the Pliocene Yuba River into what is now the Sacramento Valley—the modern Yuba River being one of the principal affluents, from the western slope of the Sierra Nevadas, of the Sacramento River, which drains the great valley of Northern California. The Pliocene Sea, as traced by the State geological survey, left its marks on the hills below the gravel-ridges, at Swiss Bar Knot, a point two hundred miles from and 400 feet above the Pacific Ocean.

The property of the Excelsior Water and Mining Company consists of water-rights, ditches, mines, lands, and appurtenant property, as follows: About one hundred and ten miles of ditches, of which about sixty miles are in use. The whole mining-district extending from Nevada City to Sacramento Valley, and bounded by the Yuba and Bear Rivers, is supplied with water by the ditches of this company. The expense of taking care of the ditches is very light compared with ditches located higher in the mountains, for the reason that the ground at this lower altitude rarely freezes and no snow lies in winter. As no fluming is requisite from Deer Creek to Smartville, the banks of the ditches have settled, and in the process of time become covered by a permanent growth of grass and bushes, so that a serious break, even during the severest storms, is almost an impossibility.

The South Yuba ditch is thirty-five miles long, conveying water from the South Yuba River to the mining-region of Smartville and the foot-hills of the Sacramento Valley. Size: 8 feet wide on top, 4 feet on the bottom, and 4 feet deep. The Booyer ditch, fifteen miles in length, conveys water from Deer Creek to the same locality. It has nearly the same dimensions as the South Yuba ditch.

The water sold by the Excelsior Canal Company is obtained, however, not only from the South Yuba River and Deer Creek, (a tributary of the Yuba,) but from Squirrel Creek, a tributary of Deer Creek.

The grade of the ditches is 10 feet to the mile. The general depth of the fluming-current is 2¼ feet. The South Yuba ditch is carried across Deer Creek at a narrow cañon by a wire suspension-flume, and across Squirrel Creek by a truss-flume, with a span of 60 feet, both being out of all danger from the highest freshets known. The length of this ditch, from Deer Creek to Smartville, is about twelve miles.

The South Yuba ditch runs a continuous stream of 1,200 inches, miner's measure,* and by aid of the large reservoir near Smartville, to hold the water at night, it affords a salable quantity of 2,800 inches per day of ten hours.

As early as the spring of 1852 ditches had been dug to run water to Rough and Ready, Newtown, and Smartville and vicinity by different companies. All of these water rights and ditches are now owned by the Excelsior Canal Company. The capacity of these ditches is about equal to that of the South Yuba ditch for supplying water at Timbuctoo and Smartville, and a reservoir is attached sufficiently large to hold the water running at night. The above ditches, taken together, were estimated in 1868 to supply 5,000 miner's inches for eight months in the year, 3,500 inches for two months, and 3,000 inches for the remaining two months in the year. The water can all be sold at 10 cents per inch. At the present time the ditches supply 5,000 inches a day during six months, 4,000 inches during one month, and 3,000 inches during five months; or an

* The most common miner's inch is measured under a pressure of 6 inches, but the miner's inch *as here measured* is the amount of water an inch square will discharge under a pressure of 10 inches. The usual "head" of water sold to any one company is 500 inches, which are measured in a box having an aperture at the discharge-end of 4 × 125 inches, under a pressure of 9 inches from the center of discharge.

average of over 4,000 inches a day during the year. The first cost of the above-mentioned ditches was over $500,000.

*Pipe.*—Between Smartville and Timbuctoo the company has about 6,000 feet of iron pipe (made of boiler-iron) 40 inches in diameter, and 2,500 feet of 20-inch pipe, also of boiler-iron, all of which is coated with asphaltum. Smaller lengths of pipe, too numerous to mention, are scattered all over the mining-banks. The cost of the large main iron pipe was $40,000.

*Dams and reservoirs.*—The bank or dam of the main or Union reservoir is 700 feet long, 30 feet deep in the middle, 5 feet wide on top, and 100 feet wide at the bottom, and lined with a riprap stone wall. The Boyer and a dozen other smaller reservoirs scattered over the country, and owned, but rarely used, by the company, are similarly constructed. The principal dams, namely, those in the beds of the South Yuba and of Deer Creek, are of timbered framing bolted firmly into the bed-rock and loaded with stone rubble.

The permanent improvements made by the Excelsior Canal and Mining Company amount in cost to more than $1,000,000, Messrs. Pierce and McGanny having spent for repairs alone over $100,000 since 1869. The mines purchased by them, including the Kentuck, Pennsylvania, Greenhorn, Live Yankee, Smartville Consolidated, Rose's Bar, &c., cost $200,000 to $300,000, exclusive of the improvements mentioned.

In the history of this district, the Blue Gravel Company, the parent of the present water company, was the first to *recognize* the situation by running the necessary deep-bed-rock tunnels and thereupon working on the wholesale hydraulic scale.

*Yield of gold.*—The yield of gold from the ancient rivers may be subdivided under four heads: 1. From the bed of the present Yuba River, adjacent to its intersection with the ancient river. Wherever the present-river system intersected the ancient system, the bars yielded their greatest product. 2. From the creeks and ravines tributary to the present-river system, leading to and intersecting the ancient-river hills. 3. From surfacing to a depth of from 1 to 5 feet over the gravel-hills. 4. From the deep placers or hill-gravels of the Lower Pliocene Yuba, by the process of hydraulic mining.

The result in general of the deep operations to which river, ravine, and surface mining were obviously only introductory may be summed up in the brief statement that everything of that kind in this district has paid dividends.

The aggregate yield of the hill or deep-placer mines between Timbuctoo and Mooney Flat is generally estimated at from eight to ten million dollars in twenty years of mining. Mr. Carpenter, one of the oldest and best-informed miners, estimates that about one-third of the gravel originally in place has been moved to produce this sum.

Before the Blue Gravel Company commenced work, there were several million dollars of gold taken out at Timbuctoo. The Blue Gravel Company had taken out $1,700,000. This is shown by the books, bullion-receipts, and vouchers of the company since the present management had control of it. About $100,000 was taken out before that time. A very large portion of the ground—from one-fourth to one-half—yielded, according to the detailed statements given of particular mines, 50 cents per cubic yard. The blue gravel of the deeper portion of the channel has yielded here and elsewhere, in placers, upwards of $3 per cubic yard by crushing in cement-mills. At French Corral the cost of crushing in mills was $1.10 per cubic yard out of a yield of $3.50, leaving a net profit of $2.40 per cubic yard, (Eddy.) Similar results were obtained in the Babb cement-mill, but not for any extensive body of ground. The average yield of the gravel-banks cannot be correctly arrived at for business purposes, except by working-tests made on large masses of ground.

Even locally on the channel the average richness varies when taken on the same level. The basis of the yield of the Michigan, given below, as estimated by Mr. McAllis, was about the average of the gravel lying north of Independence Hill, and west of the end of the pipe near Sand Hill. To the north and west of that point the ground was not so rich; it may be estimated at half the average yield of the Michigan. This general estimate holds good from the surface to the bottom of the gravel. General averages easterly of Independence Hill are given in connection with the yield of the Rose's Bar mines.

*River and bar mining.*—Mr. Jeffries states that at Rose's Bar, in the summer of 1851, there were two or three large companies, the Excelsior, the Patch, and the Go Easy. At Lander's Bar, there were the Go Easy, the Ohio, the National, the Irish Wing-Dam, and three or four others, employing forty or fifty men each. At Cordeway Bar, at Barton's Bar, and at Park's Bar similar operations were in progress. The bars at these places paid very richly, being mined at low water. At Park's Bar the pay-ground was three hundred yards wide. The Ohio Company took out $100,000 per season, for two or three seasons. The bar-gravel contained sometimes $100 to the pan in scale gold. The claims were 100 feet by 300 feet.

On the bars there were usually employed five or six men to one quicksilver-machine—two getting dirt and three running it. It was 3 feet deep from the bar to the bed-rock. They took out 40 ounces of amalgam a day to the machine; value, at $10 to the ounce, $400 per day. The average yield of all the bar companies was about 20 ounces a day. At Rose's Bar there were three different companies. Some individuals (Lyman

King and Joe Taylor are mentioned) made $30,000 a season. A man with a rocker would make $1,000 to $1,500 per season. Numerous companies made $75,000 a season in the rivers.

*Ravine-mining.*—At the mouth of Rabb Ravine, near Timbuctoo, the yield used to be $100 per day to the man.

In Timbuctoo Ravine, (Big Ravine,) Mr. McAllis, the owner since 1859, estimates that the amount taken out up to 1869, below the bridge where the Marysville road crossed the ravine, was $400,000. Above the bridge, the distance is twice as great along the gravel-deposit, and probably about $200,000 was taken out in the same time. Between 1859 and 1869, Mr. McAllis himself took out $200,000 of the $400,000 mentioned above, which should, however, be credited rather to the hill from which the material was dumped into the ravine.

Mr. Carpenter, for many years secretary of the Excelsior Water Company, says that $10 was the wages in 1849 and 1850. There were twenty men at Timbuctoo, who made an average total of $400 a day without any proper mining-facilities in water or sluices. He estimated $20,000 as the yield of the ravines, &c., at Timbuctoo for the winter of 1849-'50. Mining on a large scale only began when the water was introduced in 1852. Mr. Carpenter estimates the yield of 1850-'51 as not largely in advance of the first winter. From 1851 to 1853, in the Rabb Ravine, I have no estimate.

Between Sand Hill and Squaw Creek, on the north side of the channel, are numerous ravines which yielded richly. On the Cement Mine Ravine, Andrew Morrison used to make $200 a day; and two ounces a day to the man was very common. The Pennsylvania and the Ridge Ravines did not pay much. Probably the total yield of the Cement Ravine was $10,000. Squaw Creek was the first ravine of importance draining north and lying east of Sand Hill. The distance mined on it was half a mile, though it cut into channel-gravel for only a few hundred feet. Mr. Carpenter says that on Squaw Creek and Sucker Flat Ravines in 1852, 1853, and 1854 there were seventy-five men employed—about twenty-five men on Squaw Creek and fifty men in Sucker Flat—the number being about the same for three winters. They used the rockers and quicksilver-machine. Prior to 1851 they had only rockers and "toms." They averaged from $30 a day upward. It was not till after 1854 that ditch-water was brought in.

Mr. Jeffries says that in 1851 and 1852 Squaw Creek Ravine was mined, together with about 66 feet on each side. There were five or six sluices in it, and five men to the sluce, making thirty men in all. On the banks there were about one hundred men engaged in drifting, cutting pillars, and undermining without any pressure of water. They earned from $12 to $15 a day, sometimes $40. The surface-mining was only spade-deep. The sags in the hill were the richest places. According to Mr. Jeffries, on Squaw Creek the companies made runs of eighteen days, in which three or four men usually netted from $5,000 to $6,000. Sucker Flat Ravine was about equal in yield to Squaw Creek.

*Hill-surface mining.*—Mr. McAllis says it was common for miners engaged in surface-mining anywhere over Independence Hill or Sand Hill to make from $50 to $100 a day to the hand. Nine-tenths of the area of the gravel-country below Squaw Creek was "surfaced." Mr. McAllis estimates that $10 a day to the man would be low enough as the average yield in "surfacing," and that $15 a day would be more nearly correct. Miners engaged in this kind of work always had plenty of money; some of them made $30,000 or $40,000, and then went away. Taking the year 1855 as an average for the region below Squaw Creek, there were engaged in surfacing twenty men, who worked half of the time through the year. Estimating the average yield per man at only $8 per day from 1850 to 1860, the gold taken out by these men would amount to $288,000.

*Deep-placer mining.*—Beginning at the west end of the district, the Haywill, the Warren or Bullard, and the Davis mines together yielded $200,000 to $250,000. The Burgoyne, the Gallagher, and the Chase yielded together $70,000. The Antone yielded $300,000. The bed-rock of all these claims has been worked over three or four times. The Union yielded $259,585. This mine had a frontage on the channel of 300 feet. The Michigan, with the same dimensions, yielded at the rate of 50 cents per cubic yard, and the total yield was $350,000. The Hyde claim yielded $90,000, and the Babb claim $270,000.

A group of mines, known as the Live Yankee, the Plam, the Savage, the Marpile, and the Marllove, were very productive. In these the lower "lead" has not yet been worked. Dividends were paid from the working of the upper lead. The ground now owned by the Pactolus Company has yielded $300,000.

The Rose's Bar Consolidated embraces a large number of separately-worked and originally independent mines, the yield of which can be arrived at as a whole, pretty nearly, by a comparison with that of the Boston Company, worked by Joseph Taylor, and centrally situated.

The Rose's Bar Company is at present using 800 inches of water and employing six men, at a cost of $140 a day, and producing from $200 to $250 a day, leaving a net profit of $50 to $100 a day. This ground is worked solely for the purpose of getting at the lower lead, and is the poorest ground that has been worked in this district, yielding at the rate of from 8 to 10 cents per cubic yard. The yield of the ground from 1869 to 1873 was $256,147. The Cement Company, adjoining, yielded, up to 1869, $200,000.

The Boston Company took out $90,000 from an area of 300 feet square, the height of the bank being 40 feet in front and 130 feet in the hill, or an average of 70 feet, making 38¼ cents per cubic yard. The expenses were about half the yield.

The above instances will suffice to show the average yield of ground worked by the hydraulic method. It would be useless to give details of all the companies working this ground for the past ten years. We will notice, however, some of the principal ones.

The total yield of the Blue Gravel since its incorporation is $1,560,000, of which more than $1,000,000 has been profit.

The Blue Point mine paid since the completion of the new tunnel, within eighteen months, according to the statement of Mr. McGanny, one of the stockholders, $31,000 in dividends to one-seventh of the property. There were seven clean-ups, amounting in all to $374,000. The mine has been worked since 1854. The total yield from the surface down, according to Mr. McGanny's estimate, was about $90,000. A portion of this mine has been worked to the bed-rock. The expenses have been unusually large —estimated at about three-quarters of the total yield.

The Smartville Consolidated, being the last mine worked on the channel, has yielded since 1869, according to Mr. McGanny, the representative of half the stock, dividends as follows: In 1869, $16,000; in 1870, $36,000; in 1871, $15,500; in 1872, $16,000; total, $83,500. From January 1, 1869, to March 12, 1870, the total yield, according to Mr. McGanny, was $131,000 in four clean-ups, of $11,000, $30,000, $41,000, and $49,000. The expenses were about $37,000.

According to Mr. Carpenter, the total yield of the Smartville since the consolidation was about $300,000, and prior to that, while owned by Peterson and others, about $100,000; making a total of $400,000.

Mr. Thurston estimates that ordinarily about two-fifths of the yield in this portion of the channel goes to dividends, as everything is in favorable condition for hydraulic operations. In the Blue Gravel the expenses were about 45 per cent., though often in twenty days' run $50,000 was taken out, at a cost of only $10,000, after the mine had been opened.

At Mooney Flat the yield from the hill gravels was, according to Mr. Middleton and others, about $16,000.

*Extent of ground.*—The average of the Excelsior Company's mining-ground, lying in eight separate bodies, is approximately as follows:

| | Acres. |
|---|---|
| 1. The Blue Gravel mine | 70 |
| 2. The Rose's Bar mine | 65 |
| 3. Essner, Greenhorn, and Live Yankee | 13 |
| 4. Pennsylvania, Michigan, &c | 31 |
| 5. King claims | 1 |
| 6. Smartville Consolidated, (half of 30 acres) | 15 |
| 7. Enterprise, (⅛ of 18¾ acres) | 2¼ |
| 8. Mooney Flat, (⅔ of 189 acres) | 126 |
| Total | 323¼ |

The company also owns interests in various claims at Timbuctoo, Smartville, and Sucker Flat, and several valuable tailing-rights, together with the appurtenant tunnels, flumes, reservoirs, ditches, and other accessaries of mining, and 660 acres of agricultural lands, with buildings, offices, &c.

*Gravel removed and left standing, corresponding to value.*—The general estimate of gravel removed and left standing bears the relation of 1 to 3. Touching the ground owned or partially owned by the Excelsior Company, a preliminary estimate furnishes the following proportions of workable ground remaining:

| Claim. | Original area. | Area remaining. |
|---|---|---|
| | *Acres.* | *Acres.* |
| 1. Blue Gravel | 70 | 35 |
| 2. Rose's Bar | 65 | 32 |
| 3. Essner, Greenhorn, and Live Yankee | 13 | 2 |
| 4. Pennsylvania, Michigan, &c | 31 | 3 |
| 5. King | 1 | 1 |
| 6. Smartville Consolidated | 15 | 10 |
| 7. Enterprise | 3 | 3 |
| 8. Mooney Flat | 126 | 124 |

*Years' run and probable yield.*—The period of run requisite for the water to remove the gravel at hand or owned by the company, taking the past ten years as the rate of washing, cannot be less than fifty years. The rate of washing is 2,500 cubic yards per day to 800 inches of water. As both Messrs. McAllis and O'Brien move gravel (of very different degrees of hardness) at the rate of 2,000 to 3,000 cubic yards per day of ten hours, an average ten hours' execution can be set at 2,000 cubic yards with safety. Five days will move a cubic or solid chain, (1 chain deep,) or 10,000 cubic yards. Taking as a moderate average ditch-delivery at 4,000 inches, or 5 heads like the above, the execution to the above is at the rate of 1 cubic chain or 10,000 cubic yards per day; in a ten days' run an acre, one chain deep, or 100,000 cubic yards.

Taking the moderate average of 2,000 cubic yards of gravel per day of ten hours as moved by 300 inches of water, five days would move 10,000 cubic yards or a cubic or solid chain one chain deep.

The total ground originally in place between Timbuctoo and Mooney Flat comprised 125,000,000 cubic yards.

Amount removed 25,000,000 cubic yards, yielding, as shown by known returns, $6,000,000.

Probable amount thus accounted for, at least $4,000,000.

At $6,000,000, (the known yield,) the rate of yield per cubic yard would be 24 cents.

To apply this to the 100,000,000 cubic yards of gravel remaining might give too favorable a promise, in view of the fact that much of the Mooney Flat Hill gravel is higher relatively than was the bulk of the foregoing.

The report concludes as follows:

It would be futile to calculate how near the theoretical times of running off this gravel it would be possible to arrive at, even by the most energetic management, in a district which has been opened up for advantageous working like this one. Certain it is that the clean-ups and other necessary interruptions would extend the time of running off to three or four times the ten thousand days.

The shortest possible time, then, in which it would be possible to take out the $12,000,000 to $30,000,000 remaining would be from nine to twelve years. How much is to be done within that time will depend entirely upon the policy and management of the company in charge of this property.

I should add that the foregoing extracts have been condensed, by omissions and by running different parts together, in such a way that while the sense is (I believe) correctly given, the author of the report is not strictly responsible for it in this form. Some further statements of recent date may be here appended relative to the operations of the year 1874.

During the past season all the mining-lands of Mooney Flat have been consolidated into one company, known as the Deer Creek Mining Company. The whole tract embraces nearly 500 acres, ranging in depth (from bed-rock to surface) from 70 feet to 550 feet, the latter being the depth at the summit of the hill dividing Mooney Flat and Sucker Flat. During the summer several shafts were sunk, which are said to show the gravel to be of the same kind and quality as that of Sucker Flat and Smartville, worked for many years past with such marked success.

The Deer Creek Mining Company, having secured this range of gold-bearing gravel, is now engaged in running a bed-rock tunnel for the purpose of opening this gravel range. The tunnel is to be constructed so as to permit a flume 5 or 6 feet in width to be placed in it, which will require some 1,500 inches of water under a ten-inch pressure to operate it. With a flume of this description, on a grade of 6 inches to 12 feet, many thousand cubic yards of gravel can be washed in a day, realizing profits from gravel that has heretofore been considered worthless. This tunnel is now completed some 800 feet, at a cost of about $40,000. More than two years have been consumed in the work, and there are about 300 feet of rock yet to be tunneled before the gravel will be reached. This is expected to be accomplished in 1875.

The Blue Gravel Company, of Sucker Flat, has been, during the past year, operating through its lower tunnel, which enables it to work to the bed-rock the ancient-river channel that passes through this vicinity, yielding better returns than the upper strata, which were worked some years since. The blue gravel is, with one exception, the only company which has worked this ancient channel to the bed-rock, and this company has worked only a few hundred feet of it. The exception is the Union Company, on the east, which has just cleaned up a short run of thirty days' washing of the lower strata to the bed-rock, yielding $73,000 in gold bars.

West from the Blue Gravel Company is the Rose's Bar Mining Company, which has been some five years running a bed-rock tunnel, which is now nearly completed. It is expected that the company will be able to open its mine to the bed-rock during the spring of 1875. This tunnel is run on a grade of 6 inches to 12 feet, and will be fitted up with a flume capable of running 1,200 inches of water. The upper strata were worked off some years since, and shafts sunk for prospecting indicate that this mine will yield as well as any mine on the ancient channel.

West from the Rose's Bar Company's mine is that of the Pactolus Mining Company, which, in 1873, completed a bed-rock tunnel 1,100 feet in length, at a cost of $50,000. During the summer of 1874 it was engaged in opening the mine from the tunnel. The opening clean-up realized about $800 per day's working. The company now has pipes and apparatus for washing down into its pit, which give additional facilities for working, and will no doubt make the returns much larger. It uses 800 inches of water, and there is cause to regret now that the flumes were not made wide enough to run one-third more water.

### BUTTE COUNTY.

The physical characteristics of this county have been fully described in former reports. The principal mining-interest of the county is centered in its extensive hydraulic banks, which have been opened on a large scale near Cherokee Flat. The Spring Valley Canal and Mining Company is here conducting an enterprise second in magnitude to none in the State. This company has acquired, by purchase, 10,719 acres of mineral and agricultural land, at a cost of $240,527, together with valuable water-rights, and has expended, in the necessary flumes, ditches, and reservoirs, to utilize the water-privileges, over $1,000,000. The assets of this company at date of their annual meeting, July 1874, amounted to $4,344,496, and the actual liabilities to $143,869. The company disbursed during the fiscal year $150,000 in dividends. The average depth of pay-gravel in the company's ground is 300 feet, and it will require from twenty to thirty years of hydraulic washing to accomplish its removal.

During the year there has been constructed a continuous line of sluices 3,500 feet in length and 6 feet in width. About 400 feet of tunnel has been run. The water used by this company is brought by two ditches, sixty miles in length, from Butte Creek, and from the headwaters of the West Branch of the Feather River. The ditches are 6 feet wide at bottom and 8 feet wide on top. They are 4 feet deep, and run a constant stream of 2,200 inches of water. They have on the line of their ditch about four miles of iron pipe 30 inches in diameter. One section of this pipe conducts the water across the West Branch of Feather River. It is laid in the form of an inverted siphon, and has a vertical depression of 856 feet. The diagram opposite page 392 of my report of 1873 will give an idea of the position of the pipe. The receiving-arm has a head of 180 feet vertical pressure. The length of the inverted siphon is two and a half miles, and the pipe is 30 inches in diameter. There are ten miles of sluices, varying from 4 to 6 feet in width, and 23 under-currents from 10 to 40 feet in width. For the year ending July, 1874, the sum of $476,112 in gold was washed out and shipped. They employed 160 hands all the year round, and expended $125,000 during the same time, of which $85,534 was for labor. The quicksilver alone used by the company for the year cost $13,309. For iron pipe they paid out $8,839.

The following statement of assets and liabilities and items of current expenses is taken from the annual report of the secretary, Mr. L. Glass:

ASSETS AND LIABILITIES.

| | | | |
|---|---|---|---|
| Assets: | | | |
| Property account | | | $4,028,811 50 |
| Ranches purchased | | $240,527 48 | |
| Ranch interest per Central Pacific Railroad | | 534 08 | |
| Ranch expenses, dam, &c | | 6,428 44 | |
| Total | | | 247,490 00 |
| Claims purchased | | | 28,308 79 |
| Permanent improvement: | | | |
| Permanent expense account | | 4,139 09 | |
| Permanent labor account | | 1,718 51 | |
| Concow dam of 1873 | | 4,603 09 | |
| Ditch and pipe of 1873 | | 4,065 01 | |
| Dewey ditch | | 1,132 33 | |
| Iron pipe | | 8,839 62 | |
| | | | 24,497 65 |
| Cash on hand | | | 2,079 23 |
| Quicksilver on hand | | | 13,309 51 |
| | | | 4,344,496 68 |
| Liabilities: | | | |
| Capital stock | | | $4,000,000 00 |
| Gold bars | $476,112 23 | | |
| Other sources | 350 00 | | |
| Found in flume | 280 00 | | |
| Gross receipts | | $476,742 23 | |
| Less current expense | 17,184 47 | | |
| Less current labor | 85,534 47 | | |
| Less teaming account | 2,579 38 | | |
| Less merchandise account | 14,487 02 | | |
| Less lumber account | 4,282 45 | | |
| Less interest account | 1,235 21 | | |
| Less litigant account | 1,810 00 | | |
| | | 127,113 00 | |
| Balance, actual profit | | 349,629 23 | |
| Less dividends made | | 150,000 00 | |
| Residue | | | 199,629 23 |
| Total actual liabilities | | | 143,869 54 |
| Balance on current accounts | | | 997 91 |
| | | | 4,344,496 68 |

| | | |
|---|---|---|
| Current expense: | | |
| For office expenses, injured hands, traveling expenses of superintendent, coal-oil, recording, repairs of pipes, guns, ammunition, surveying, subscription to roads and other purposes, crucibles and expenses of melting-room, tolls, maps, repair of wagons, provisions for ditch-hands, damages to house-lots, and repair of bridges, aggregate | $4,230 26 | |
| For hay, barley, harness, and harness-repairs | 1,492 19 | |
| Taxes | 10,269 25 | |
| Freights, railroad, by team, and express charges, including gold bars | 1,192 77 | |
| Total | | $17,184 47 |

| Merchandise: | | |
|---|---|---|
| For black powder used during the year | $9,367 06 | |
| For giant powder used during the year | 1,746 25 | |
| For drill-steel, hammers, nails, blacksmith's supplies of iron, coal, &c., blasting wire and fuses, grizzly bars and tools | 3,373 71 | |
| Total | | $14,487 02 |
| | | 31,671 49 |

A peculiar feature in this claim is the fact that diamonds are found in the washings; most of them, however, by the primitive method of rocking. One diamond worth $250 was cut in Boston in 1864, and last year several were tested in Amsterdam and Paris, and pronounced diamonds of the first water. Professor Silliman has examined these sands carefully, and enumerates the mineralogy of the Cherokee washings as yielding gold, platinum, iridosmine, diamonds, zircon, topaz, quartz in several varieties, chromite, magnetite, limonite, rutile, pyrites, garnets, epidote, and almadine. One of the diamonds found weighed 2¼ carats.

### SIERRA COUNTY.

Sierra County is a rugged, mountainous region, furrowed by deep cañons and gorges and cut down from east to west to great depths by the eroding influences of the North and Middle Yuba Rivers and their numerous intersecting tributaries. The lowest portion of the county has an altitude above sea-level of 2,000 feet; many of its peaks and ranges rise to heights varying from 5,500 to 8,000 feet.

The geological features are similar to those of the central mining-region of California; viz, on the eastern border, granite, forming the summits of the main range of the Sierra Nevadas; westward, basaltic lavas and volcanic breccia overlying the slates, succeeded by slates in various degrees of metamorphism and belts of serpentine as we descend toward the western line of the county.

The principal source of wealth of the county is mining, which is carried on extensively in three branches, quartz, hydraulic, and "drift" diggings. The river-bars, once noted for their great yield, are now nearly exhausted. Having ceased to pay the demands of white labor, this class of mining has been abandoned to the patient and plodding Chinese, who are still engaged on the banks and bars of the Yuba in washing ground which has been worked three or four times by white labor. Their system of operations is to turn the rivers from their beds by means of long flumes and run the dirt through boxes. By this method they are enabled to realize Chinese wages—from $1.50 to $2 per day—and sometimes to make a "strike" on ground overlooked by their predecessors.

Quartz-mining in this county is prosecuted on an extensive scale, as will appear by the returns of the Sierra Buttes mine and other mines, the statements of which may be found elsewhere in this report.

Hydraulic mining is carried on in the westerly portions of the county and at points on the Slate Creek Basin, also at Brandy City. At these points the auriferous gravel is not deeply covered by the volcanic material, which flowed over the eastern or higher portions of the county. This class of mining has been fully described in the report for 1873, by C. W. Hendel, United States deputy mineral surveyor for the county.

One of the peculiar and interesting features of the mining-interest of Sierra County is its "drift-claims." This class of mines consists of locations on the ancient-river beds covered by the great volcanic outpouring which succeeded the Pliocene period. The work of extraction pursued is by means of tunnels, generally run through the side or "rim-rock" at sufficient depths to tap and drain the gutter or channel. The main tunnel is carried on "up stream," and gangways are run toward either bank or rim, the gravel being run out by cars and washed in sluice-boxes, and the worked-out ground being allowed to cave and fill up as the main tunnel, which is securely protected by timbers, progresses. The method of mining in drift-claims is well illustrated by the accompanying diagram of the underground workings of the Bald Mountain Company, at Forest City. The depth of gravel which it is found will pay by this system is from 3 to 5 feet, including from 3 to 6 inches of bed-rock, which is picked down and washed with the gravel. The lower stratum, including this bed-rock, is invariably the richest.

This branch of mining, formerly pursued with great success in various portions of the county, had of late years fallen into decadence, until it was revived, in 1872 and 1873, by the remarkable results following the opening of the Bald Mountain claim of Forest City. Before the development of the mine, in 1872, Forest City, once a lively and prosperous town, had degenerated into an almost abandoned mining-camp. Houses and lots were to be had for the payment of taxes, and in one instance a well-furnished house, with a lot on which it stood, sold for $75. The results of the first clean-up of the Bald Mountain, however, changed, for the time at least, the destinies of the town, and now it has resumed all its former prosperity, and given a strong impetus to surrounding camps. As the Bald Mountain may be considered a representative claim of the class known as drift-diggings, it deserves a description in detail.

Three distinct and well-defined channels or Pliocene rivers cross the county in a northerly and southerly direction, and have been cut by the modern rivers at right angles. The principal rivers are the North and Middle Yuba, their course being east and west. The bars of these rivers and their accumulation of gold are due to the disintegration of those portions of the ancient rivers which were crossed and eroded by them. The two more easterly of these channels have been covered to great depths, at some places as much as 1,000 feet, by an immense outpouring of volcanic *débris* and lava. The western channel is on the border of the lava-flow, and is well exposed at Brandy City, Camptonville, and other places, where hydraulic mining has been extensively and profitably pursued. It is on the central channel, known throughout three counties of the State (Placer, Nevada, and Sierra) as the Great Blue Lead, that the Bald Mountain Company's ground is located.

The course of this ancient river, supposed by many eminent geologists to have been the principal stream of the Pliocene period, was nearly north and south, on a grade of from 70 to 100 feet to the mile. It has been distinctly traced from Plumas County, on the north of Sierra County, through Sierra and into Nevada County, a distance of twenty-five miles. Its bed has an altitude at Forest City of 4,350 feet above sea-level, being from 1,500 to 1,600 feet higher than the beds of the modern rivers, (the North and Middle Yuba,) which have crossed it from east to west.

The principal places where this channel has been worked are Monte Christo, City of Six, Rock Creek, Forest City, Wet Ravine, Alleghany, Chip's Flat, and Minnesota. Here it has been cut by the Middle Yuba, and makes its re-appearance south of that stream at Snow Point, in Nevada County. This ancient river has been distinctly followed at many of these points by long connecting-tunnels. The workings of Forest City and Alleghany were so connected several years since, and it was possible to enter a tunnel at Alleghany and emerge at Forest City, a distance of more than 6,000 feet. Also between Rock Creek and City of Six, a distance of one mile, similar underground connection was made.

At Forest City, the principal claim up to the period of the development of the Bald Mountain Company's ground was the Live Yankee, 2,600 feet in length on the channel, most of which is now worked out. This ground yielded during the nine years in which active mining was carried on (from 1855 to 1863, inclusive) the sum of $698,534, of which $370,166 were expended in opening the claim and current mining expenses during the whole period of the nine years, while the dividends disbursed during the period were $328,368. For several years thereafter, and until the ultimate exhaustion of the company's ground, the dividends averaged $10,000 per annum. For a period of seven consecutive years after the opening of this mine, and preceding the first symptoms of exhaustion, the dividends averaged $49,191 per annum. The total product of the ground may be safely estimated, from the date of opening the mine to the suspension of operations, at not less than $1,000,000, of which probably one-half was disbursed in dividends; and this was during a period of high-priced labor and in the pioneer days of drift-mining.

At about the time of the decline in prosperity of the Live Yankee, (1867-'68,) Mr. M. Redding, a practical miner of Forest City, who had worked in the Live Yankee, observing an easterly trend of channel in that ground, formed the opinion that the main channel of the Blue Lead diverged from its north and south course and passed northeast toward and under a high mountain capped with basaltic lava, lying east of Forest Hill. Acting on this theory, he acquired by location and purchase the present ground of the Bald Mountain Company. He was unable to impress his neighbors with the correctness of his views, and entirely unsuccessful in interesting San Francisco capitalists in the project of developing this ground. Finally, after much perseverance and the display of remarkable tenacity of purpose, he succeeded in forming a company, consisting of practical men, whose only capital was their labor, and vigorous operations were commenced in the fall of 1869. A shaft was sunk at a point considerably east of what was popularly considered the course of the Blue Lead. The shaft, 3 feet by 7 feet in dimensions, passed through alternate strata of surface-loam, volcanic cement, pipe-clay, sand, gravel, &c., until at a depth of about 260 feet the long-sought blue-colored pay-stratum was reached, and bed-rock found at a depth of 269 feet. By a singular coincidence, the particular portion of the channel tapped by the shaft was in the poorest ground in the mine. The results of the prospect were barely encouraging for further development, and one or two of the owners parted with their interests at a loss. Had the shaft been sunk 10 feet in any direction from its position, it would infallibly, as proved by the subsequent working of the mine, have struck gravel of remarkable richness. In fact, the forty-ounce nugget found in the company's boxes in 1873 is supposed to have come from this portion of the ground. However, Mr. Redding and his associates were satisfied with the character of the gravel, and determined on running a tunnel to connect with the shaft and open the ground. This tunnel was commenced in June, 1870, and prosecuted, with periods of interruption, to its completion by connection with the prospect-shaft in April, 1872, a period of twenty-two months. The distance was 1,800 feet. The ground was easy to work, excepting a rim of serpentine, which was passed through about 400 feet from the mouth of the tunnel. The work was accomplished by manual labor, single-hand drills, and giant powder, and cost, including the prospect-shaft, about $20,000, being $1,000 to each original interest. This represents the outlay of the company on the mining venture. No further assessments were levied, but something over $200,000 in dividends were disbursed between April, 1872, and January 1, 1875.

The following statement from the company's books at the close of the fiscal year ending June 30, 1874, will show the receipts and dividends up to that time, the amount of ground worked, and the comparative yield:

*Bald Mountain Company.*

| | |
|---|---|
| Taken out from July 1, 1872, to July 1, 1873 | $131,780 71 |
| Dividend | 54,000 00 |
| Taken out from July 1, 1873, to July 1, 1874 | $196,571 67 |
| Dividend | 110,000 00 |
| Total receipts to July 1, 1874 | $328,352 38 |
| Dividend to July 1, 1874 | 164,000 00 |
| Total car-loads extracted to July 1, 1874 | 115,950 |
| Paid per car-load, (1 cubic yard) | $2.76 |
| Paid per square foot of ground | $1.09 |
| Amount of square feet worked | 292,200 |
| Per cent. of dividend on gross amount taken out | 50 |
| Superficial feet contained in channel | 7,232,000 |
| Value, at $1.09 per square foot | $7,882,880 |

The first "clean-up," consisting of dirt taken from the tunnel, yielded only $132. From this time the returns steadily increased, in June of the same year amounting to $1,000, and subsequently to four, five, six, and seven thousand dollars, until, with the progressing development of the mine, the present remarkable yield has been reached, and attention has been thereby called to this long-neglected branch of mining. On April 25, 1874, two years after the first modest return of $132, the result of the clean-up was $13,733.24, and one week thereafter, (May 1,) $14,694.76. As will be seen by the foregoing statement, the total yield up to July 1, 1874, was $328,352.38. From July 1 to September 30, 1874, there was taken out only $16,726.84. This was owing to a scarcity of water, the dirt being allowed to accumulate in the company's yards. The actual bullion-product, therefore, at the time of Mr. Skidmore's visit, (October, 1874,) had been $345,079.22, from a piece of ground about 1,000 feet in length by 500 feet in width, exclusive of the unworked ground within this area, and not including the value of the gravel-piles in the company's yards. The company owns 8,000 linear feet of unworked ground on the channel not shown in the annexed diagrams.

Mr. Edwin Stone, secretary of the Bald Mountain Company, says that the company uses from 4,000 to 6,000 pounds giant powder per year, from 2,000 to 3,000 pounds steel, and from 300 to 400 boxes of candles. The company commenced in November, 1874, the washing of the gravel which had accumulated in its yards during the dry season. Eight thousand car-loads of gravel from the small dump yielded 2,400 ounces of gold-dust, and 16,000 car-loads remained to be cleaned up before the close of the year. At that time they were working ninety men, and had room for about twenty more if the supply of water should continue. The quality of gold taken out is what is denominated as "coarse," the largest piece yet found weighing 65 ounces. The superintendent, H. Wallis, an experienced gravel-miner, informs us that in his experience of twenty years he has never seen a channel that paid so uniformly. The further into the mountain they go the richer becomes the gravel, and the dividends for 1874 will probably reach $10,000 to the share, or $200,000 in the aggregate.*

The system pursued in opening the ground is well illustrated by the preceding plan of the company's workings, made October 1, 1874. The main adit or tunnel was then 3,000 feet in length. Only 1,900 feet are shown in the plan. The distance from mouth of tunnel to shaft is 1,800 feet. The course of the tunnel is north 20° east. This brought the end of the tunnel on rising rim-rock, as shown in drifts Nos. 8, 10, and 14. It seems that the channel is crossed diagonally. Explorations, by means of drifts to the east of the main adit, (see drifts 9, 11, and 13,) demonstrated the gutter or center of the channel to be in this direction, and to follow a course indicated by the words "branch main tunnel." This branch will be connected with the main adit at drift No. 5, near the shaft, and will form the principal tunnel for opening the ground. The course of the stream is northeast by east, with a tendency to swing still further to the eastward. This fact, taken in connection with explorations in the grounds of the North Fork Company, indicates that the Bald Mountain Company has a channel tributary to the main Blue Lead, the point of its intersection being Forest City Basin. While this theory is not yet accepted by the owners, there are strong probabilities of its correctness. The high rim-rock observed on the west side of the main tunnel and the existence of gravel on the ground west of the North Fork of Oregon Creek are strong evidences of the intersection of two channels at this point. The only question of interest is, which is the main channel?

The term "drifting," as applied to this class of operations, relates to the mode of extracting the auriferous gravel by means of tunnels and gangways, or galleries, and washing the dirt in sluices. This system is rendered necessary in consequence of the capping volcanic matter overlying the ancient channels, and rendering hydraulic operations impracticable. In hydraulic operations, the entire face of the bank is removed by the use of heavy streams of water, under great pressure, thrown against the banks, by means of pipes and nozzles. In "drifting claims," only the lower stratum of gravel, lying on the bed-rock, is mined and washed. The average depth of pay-ground, when mined in this manner, is about 3 feet.

The pay-roll of the Bald Mountain Company contains generally from 90 to 100 names. Of these, 85 are miners, receiving $3 per day. Thirty cars are employed in running out the gravel, which is dumped in the yards at the mouth of the tunnel, and thence into the flume. This flume passes from the mouth of the tunnel down the banks of Oregon Creek, which is the ultimate receptacle of the tailings. The flume is 6,600 feet in length, (one and a quarter miles,) 17 inches in width by 18 inches in height, paved with blocks 16 by 16 inches, and 4 inches thick. The gold, being as a general rule coarse, settles in the interstices of the blocks. No

*The results of the operations of this company may be compared with those of the Indiana Hill Company, of Placer County, which is engaged in the same class of mining—the Bald Mountain Company using sluices for washing, and the Indiana Hill Company using stamps for crushing the gravel.

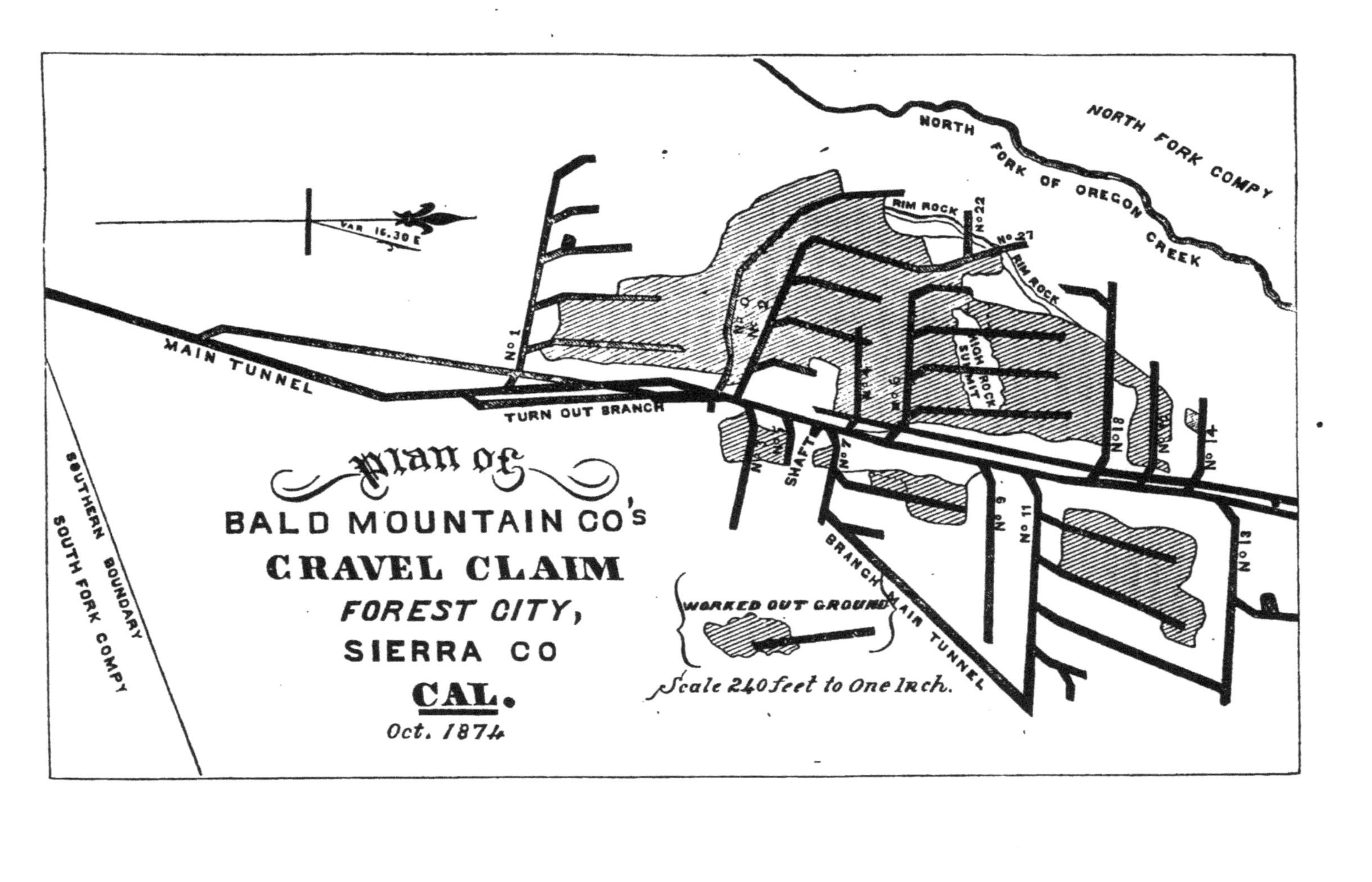
Plan of
BALD MOUNTAIN CO'S
CRAVEL CLAIM
FOREST CITY,
SIERRA CO
CAL.
Oct. 1874
NORTH FORK COMPY
NORTH FORK OF OREGON CREEK
RIM ROCK
No 22
No 27
RIM ROCK
HIGH ROCK SUMMIT
No 1
No 18
No 14
No 7
SHAFT
No 9
No 11
No 13
MAIN TUNNEL
TURN OUT BRANCH
BRANCH MAIN TUNNEL
WORKED OUT GROUND
Scale 240 feet to One Inch.
VAR 16.30 E
SOUTHERN BOUNDARY
SOUTH FORK COMPY

quicksilver is used in this class of mining. The usual head of water of the company is from 100 to 150 inches. The flume is cleaned up in sections about twice a month; and a general clean-up is made twice a year, when the worn-out blocks are turned over or replaced. A set of blocks will last about ten months, but during this time they are turned over. The company also owns one mile of the bed of Oregon Creek, below the end of the flume. The flume is built on a grade of 6 inches to 12 feet, equivalent to a fall of 225 feet for the whole distance of one and a quarter miles.

The main tunnel is in dimensions 6 feet 3 inches on the bottom by 3 feet 8 inches on top and 6 feet in height. It is substantially timbered and laid with iron track. The cars are 4½ feet long, 2 feet wide, and 2½ feet high, containing one cubic yard of loose dirt, which is equivalent to about half a yard of ground in place; therefore, the result of $2.76 per car-load, obtained by this company from April 1872, to July 1, 1874, would be equivalent to $5.52 per cubic yard of standing ground.

The method of stating the yield of gravel varies in different localities. In some counties the standard is the "car-load"—generally 16 cubic feet, or a ton—but as there is no arbitrary size of a car, and the dimensions and capacity depend on the size of the tunnel, this term alone affords no standard of comparison. Again, the yield is sometimes expressed as so much to the superficial yard of channel-ground, but here we meet with the difficulty that the height of ground worked is not uniform, being at some places 3 feet and at others from 4 to 5 feet. The only unvarying standard of comparison is the cubic yard; but unfortunately this is not in general use among miners, and necessarily our means of comparison are limited. It is, however, well understood that gravel should pay, under ordinary circumstances, from 50 cents to $1 per car-load of 16 cubic feet, to be considered as profitable for drifting. We may therefore fix the minimum remunerative yield at from 85 cents to $1 per cubic yard of broken ground.

It will be observed that a distinction should be made between gravel "in place" and gravel broken down to the condition of loose dirt. It is apparent that a cubic yard of ground, as it stands in its position in the mine, where it is so compressed as to require the use of drills and gunpowder for its extraction, cannot be contained in its loose or broken condition within the limits of a car or box of the dimensions of one cubic yard. Careful observation and comparison have demonstrated that in a majority of cases a cubic yard of standing ground will when broken down fill two cars of the capacity of one cubic yard each. The measurement of the ground worked in the Bald Mountain compared with the number of car-loads extracted confirms this estimate.

Mr. Charles Hendel, United States deputy mineral-surveyor for Sierra County, has furnished the following interesting and authentic statement of the yield of mines on the Slate Creek Basin channels, in the northern portion of the county:

At Howland Flat and Potosi the—

| | | | | Pay per sq. yd. |
|---|---|---|---|---|
| Down East Company took from | 287,000 | sq. ft., 4½ ft. high, | $346,000 00.... | $10 80 |
| Union Company took from .... | 1,300,000 | sq. ft., 4½ ft. high, | 1,187,284 74.... | 8 10 |
| Hawkeye Company took from . | 348,000 | sq. ft., 4½ ft. high, | 365,819 40.... | 9 45 |
| Pittsburgh Company took from | 430,000 | sq. ft., 4½ ft. high, | 352,549 81.... | 7 38 |
| Total .................. | 2,365,000 | .................. | 2,251,653 95 | |

which gives an average pay of 95 cents per square foot of ground 4½ feet high, or $5.70 per cubic yard of gravel, (drift,) which it is claimed was mined at a cost of 47 cents per square foot, leaving a clear profit of 48 cents per square foot.

At Grass Flat, in the Pioneer Company's ground, opened as a drift mine, but by its development proving to be a hydraulic mine, the following results have been obtained:

In excavating their branch tunnel, most of it 15 feet above bed-rock, 3,000 square feet of tunnel, or 475 cubic yards of gravel, yielded $384.52; which produces the following result, viz, $2.30 $\frac{72}{100}$ per linear yard of tunnel, $1.15 $\frac{36}{100}$ per square yard of tunnel, and 80 $\frac{95}{100}$ cents per cubic yard of gravel.

The main tunnel of the Pioneer was run 1,186 feet in gravel, from 10 to 20 feet above bed-rock, by which 6,616 square feet of ground was worked, producing 883.37 cubic yards of gravel, which yielded $1,400.53, corresponding to the following results:

| | |
|---|---|
| Per linear yard of tunnel | $3 56 |
| Per square yard of tunnel | 1 91 |
| Per cubic yard of gravel | 1 59 |

The above claims, it will be observed, were worked on the hydraulic principle, and for this method, by which the whole surface of the bank is removed, the yield has been exceptionally high.

The success attending the development of the Bald Mountain claim has attracted the attention of owners of ground in the vicinity, and several claims hitherto abandoned, or worked on a small scale, are now being opened by means of tunnels, in anticipation of tapping the same channel. The most important of these are the Ruby on the north, and the Rising Star on the south.

The Ruby Company owns a tract of ground 5,000 feet in length by 4,400 feet in width, approximately 550 acres, situated on the north side of Bald Mountain and adjoining the Bald Mountain Company's ground. The channel heretofore spoken of as the Blue Lead runs through the company's ground in a general north and south direction. It has been exposed near their north line by means of tunnel and shaft, and work is now being prosecuted at that point with encouraging results, although under many disadvantages, owing to the influx of water. The company has projected and commenced a rim-rock or bed-rock tunnel from the head of Little Rock Creek, which will open its ground near the south line. This tunnel will be 2,000 feet in length, and its completion will enable the company to drain and work the entire extent of its ground on an up-grade, instead of working down-stream, as at the Rock Creek opening. Capt. William Irelan, the superintendent, informed me that at the Rock Creek tunnel the yield was about $2.20 per cubic yard of loose dirt. It is contemplated that the new tunnel can be finished, by means of the Burleigh drill, during the year 1875, and at a cost not exceeding $30,000. A portion of this cost will be borne by the old workings, which have been placed in a position to contribute largely during the year 1875.

The construction of long bed-rock tunnels at points where, as in this claim, the existence of the channel has been demonstrated, relieves this class of mining of its risks. The width of the Blue Lead channel varies from 300 to 600 feet. The portion which will pay by the drifting system is rarely narrower than 180 feet, and often wider than 300. As we have seen, it is in the Bald Mountain 500 feet wide. If we take 200 feet as a fair average of the width of the pay-gravel, and 4½ feet as an average of height of the profitable stratum, we have 165,353 cubic yards of standing ground to 5,000 feet of location, equal to 230,706 car-loads of loose dirt. At $2.10 per car-load, which is somewhat under the mark for this district, we should have a gross yield of something over half a million dollars to 5,000 linear feet of channel. The usual ratio of profit is over 50 per cent.

Southeast of the Bald Mountain claim, and about one mile west of Alleghany, the Rising Star Company (Messrs. Hanly, Crane, and others) is engaged in running a tunnel under the ridge, with the expectation of striking the continuation of the Bald Mountain channel. This tunnel, which will be 2,000 feet in length, is being rapidly carried forward by means of a Burleigh drill driven by compressed air.

The South Fork Company has run a tunnel between 1,200 and 1,300 feet, opening some good prospects, and expects to reach the channel early in 1875. The North Fork Company is now in over 2,500 feet. The Red Jacket, formerly Young America, owned by Heintzen, Nelson & Co., have already commenced their tunnel, which will be extended until it drains the whole ridge back to the Highland and Masonic shaft. This is one of the best located claims in Forest City, lying between Crane, Hanley & Co.'s and the Live Yankee, cornering on the Highland and Masonic. Next in importance is the Wisconsin, on the northwest side of Forest City ridge. The tunnel here, 1,500 feet in length, was commenced in 1854, and has cost up to the present time over $36,000. It is too high to work the ground, but the company contemplates running another.

*Operations of leading mines, Sierra County, California*, 1874.

*Sierra City district.*—Name of mine, Sierra Buttes; owners, English corporation; number of miners employed, 250; miners' wages, $50 per month and board; cost of sinking per foot, $20 to $40; cost of drifting per foot, $6 to $15; cost of stoping per ton, $1.94½; cost of mining per ton extracted, $3.63½; cost of milling per ton, 83 cents; number of tons extracted and worked, 53,959; average yield per ton, $8.50; percentage sulphurets, 1; total bullion-product, $470,608.41.

Reported by William Letts Oliver.

*Statement of quartz-mills, Sierra County, California*, 1874.

| Location and name. | Owners. | Kind of power. | No. of stamps. | Weight of stamps, lbs. | No. drops per minute. | Height of drop, in's. | No. of pans. | No. of concentrators. | Cost of mill. | Capacity per twenty-four hours, tons. | Cost of treatment per ton. | Tons crushed during year. | Method of treating sulphurets. |
|---|---|---|---|---|---|---|---|---|---|---|---|---|---|
| SIERRA CITY. | | | | | | | | | | | | | |
| Hanks | English corporation. | Water. | 30 | 750 | 80 | 8 to 10 | 1 | 2 | $50,000 | 65 | $0 68 | 22,160 | Amalgamation. |
| Beis | do | do | 16 | 650 | 76 | 8 | .. | .. | 15,000 | 28 | 88½ | 9,606 | None. |
| Hitchcock | do | do | 40 | 700 | 68 | 8 | 2 | 2 | 55,000 | 65 | 86½ | 21,768 | Amalgamation. |

## PLUMAS COUNTY.

This county, situated north of Sierra, contains the head of the system of ancient channels or Pliocene rivers which form the basis of gravel and hydraulic mining in California. The surface of the county is extremely rough and mountainous. The main crest of the Sierra Nevada, rising to an altitude of 6,000 feet, with culminating peaks of 7,000 to 9,000 feet, runs near the western line of the county. Owing to the rugged nature of this portion of the county and the difficulty of building roads, Plumas has only imperfect communication with other portions of the State, and its mineral resources have until recently been overlooked or but partially developed. Within the past year, however, attention has been attracted to the vast extent of auriferous gravel found in the county, and several operations of great magnitude and promise have been inaugurated. Among these may be instanced that of the North Fork Mining Company.

A party of Sacramento capitalists, having purchased the mines near the North Fork of Feather River, and lying about six miles south of Big Meadows, known as the Dutch Hill, Barker Hill, Cummings Hill, and Owl Hill mines, incorporated under the name of the North Fork Mining Company, with a capital of $500,000, for the purpose of working their ground. These mines had long been known as among the richest in Plumas County, but, situated as they were on an isolated point at a great altitude, it was deemed impracticable to bring water upon the ground; but this company, knowing that the quality of the ground would justify an immense outlay, had surveys made, and upon those surveys has constructed over twenty miles of ditches and laid a line of pipe with a capacity of 2,000 inches of water.

The following description of the North Fork Company's ground is condensed from the Plumas National:

The mining-ground consists of the claims known as Cummings Hill, Dutch Hill, Barker Hill, and Owl Hill. These mines have been worked more or less during the last twenty years, but the facilities at the command of the owners of these claims were so poor that little was accomplished. Sufficient work was done, however, to prove the value of the ground in the different claims. Lying contiguous to and in such a position that one ditch would cover the ground in each of the claims, the North Fork Company purchased the whole ground from the different owners, and in February, 1874, made application to the United States Government for a patent to the same, giving the claims thus consolidated the name of the North Fork placer-mine, and claiming in the application 371$\frac{17}{100}$ acres. Since the application for patent the company has taken up and bought claims adjoining the mine to the amount of nearly 500 acres, and has now, in all, about 800 acres of mining-land in one body, lying from 200 to 500 feet below the ditch.

The first of these claims that is reached by the ditch is Cummings Hill. This is the highest ground within the boundaries of the North Fork mine, some of the gravel being 100 feet above the line of the ditch. This, however, will be worked by water pumped from the ditch. For this purpose the pump and engine now at Ohio Valley are to be used. The pump is capable of raising 100 inches of water to the height of 200 feet. The gravel at Cummings Hill is very rich. Until the water was brought to the ground by the Ohio Creek ditch the claim was worked by drifting, and the dirt hauled in carts to water caught in a reservoir, and fed by a small spring affording only a few inches of water. Even by this slow process as high as $1,000 per day has been made, and $110 has been washed from a single load of dirt, hauled to the water in a hand-cart. By this mode of working, very little ground, compared with the immense body in the hill, has been worked. Considerable gravel taken from Cummings Hill, and not considered sufficiently rich to pay for carting and washing, has been dumped at the face of the works, and will now be washed.

Dutch Hill is the next claim on the line of the ditch. A considerable amount of this claim can be worked by piping, but it is more properly a drifting-claim. Three bed-rock tunnels have been run on this ground, one of which goes into the hill from the south, and dumps into Barker Ravine; the other two enter the hill from the west, and dump into Owl Creek. The last tunnel run, one of those dumping into Owl Creek, is the main tunnel of the mine. It is run in a distance of 700 feet, and has been run by the North Fork Company since they came into possession of the mine. The work on this tunnel was commenced in 1873, and was run a considerable distance by hand-drilling, the miners making about 2 feet per day in bed-rock; but one of Wood's steam-drills has been used during the past summer, which averaged 5 feet per day. The motive-power used was Ericsson's air-condensing engine. The main drifting-ground in the

Dutch Hill claim is about 1,400 feet wide, and 10 feet deep between the lava-cap and the bed-rock. Where the gravel has been reached and worked, it has proved exceedingly rich, paying as high as an ounce per day to the hand. This gravel is capped with lava, and the tunnel passes through solid bed-rock for its entire length, tapping the gravel at what is supposed to be the lowest point in the channel.

Barker Hill, to the east of Dutch Hill, is the last point reached by the ditch. This is a fine body of gravel, similar to that in the Badger, Gopher, and Hungarian Hill mines, near Quincy. This ground is easily worked, and is very rich. On the greater part of Barker Hill a pressure from the ditch of 200 feet and over can be used. The old Ohio Creek ditch reached this claim, and while water lasted in it—which was only about 200 inches for a month or so in the spring—the claim was worked. Last spring the present proprietors laid a flume up Barker Ravine to bed-rock in this claim, and, using a Little Giant machine, worked off considerable ground, with such good results that they feel confident of cleaning up $1,000 per day with proper machinery and the water which will be rendered available from the new ditch. Besides the ground within the boundaries of the old Barker Hill claims, about two hundred acres of a gravel-flat lies below it, which the company has already located or otherwise secured, and which is now, with this new water-privilege, rendered very valuable.

The North Fork Company will obtain its supply of water from Rice Creek. This stream heads under the south side of Lassen's Butte, and is continuously fed from melting snows, and is the extreme westerly branch of the North Fork of Feather River. It contains at its very lowest stage about 3,000 inches of water, (miner's measurement,) thus affording a never-failing supply.

Surveys demonstrated only one practicable route for the water, viz, crossing the ridge into Tehama County, thence carrying the ditch on the south side of the point or ridge dividing Butte Creek from the Big Meadows. This could only be done by means of a tunnel 1,150 feet long, starting sufficiently low to tap all the waters of Rice Creek Meadows. The contract for building this tunnel was let for $8 per foot, the work to be completed by the 1st of December, 1874. To finish this work within the time specified, it was necessary to sink two shafts on the line of the tunnel, thus affording a chance for six gangs of men to work at the same time, one set of men at each end and two sets in each shaft, working in opposite directions. Even with this force it required the best management on the part of the contractor to finish 1,150 feet within the prescribed time. The tunnel is now completed. It is 6½ feet in height, 5 feet wide, has a grade from one end to the other of 40 inches, and is capable of carrying all the waters of Rice Creek in ordinary seasons.

The water, after passing through this tunnel, will descend about 100 feet, and empty into a lake, in Tehama County, known as Tule Lake, which, in an ordinary stage of water, covers an area of about 160 acres, and is about 10 feet deep in the deepest part. This lake the company proposes to use as a reservoir for the ditch. The water is conducted from Tule Lake through a cut in solid lava 660 feet long, 7 feet wide, and 11 feet deep at the deepest point. From this cut the water is conveyed in a flume built on a solid lava foundation, a distance of 924 feet, where it is taken up by the ditch, which extends a distance of fifteen miles to the head of the pipe. The first six miles of this section of the ditch is on the western slope of the divide-ridge between the waters of the North Fork and West Branch of Feather River, the summit of which constitutes the county-line between Plumas and Tehama, and which ceases to be a defined ridge at what is known as Deer Creek Pass, on the divide between Butte and Deer Creeks. At six miles the ditch crosses the ridge into Plumas County, and follows the side-hill to a point on the Stover Mountain, overlooking the Big Meadows. With the exception of about two and a half miles at the head of this section, the line is granite rock, which rendered blasting necessary on nearly the whole line. On this work one of Wood's steam-drills was used, which proved a decided improvement on hand-work, doing the work of six men and sinking a hole in hard granite rocks at the rate of 22 inches in six minutes. On this section of the ditch both white men and Chinamen were employed, the Chinamen digging and the white men blasting and trimming up the ditch.

The line of pipe connecting with the ditch is a few feet over eight miles in length, and heads at the Stover Mountain, passing in an almost direct line to the mountain just north of the upper end of Butte Valley. It has a pressure of 150 feet, and at a distance of five miles from the head has a depression of 500 feet, measured from a grade-line. The pipe is made from the best Pennsylvania iron, and was manufactured in Pittsburgh for this particular work, and over 600 tons have been used in its construction.

From the discharge of the North Fork Mining Company's pipe a ditch is carried a distance of six miles on the face of the ridge sloping toward Butte Valley. This ditch is 6 feet wide on the top, and has a uniform depth of 3 feet. At six miles from the discharge of the pipe the ridge dividing Clear Creek from Butte Valley is crossed by means of a tunnel, 835 feet in length, and 200 feet below the apex of the hill. Hard bed-rock was encountered almost from the very commencement, necessitating blasting nearly the whole length of the tunnel. From the tunnel an old ditch, known as the

Ohio Creek ditch, is reached by a new ditch one and one-fourth miles in length. Ohio Creek ditch, which runs to the Dutch Hill mines, has been enlarged to the capacity of the new ditch leading to it.

To complete the twenty-five and one-fourth miles of ditch and eight miles of pipe, a large force of men has been employed during the whole summer of 1874.

Mr. Morris Smith, of Spanish Ranch, writes as follows concerning mining in his vicinity:

The Golden Enterprise Fluming and Mining Company is laying in Spanish Creek a flume 10 feet wide, to tap the great gold-basin at the foot of Spanish Peak. This basin contains an area of six miles square, a great portion of which is auriferous ground.

The mines upon the more elevated portions of this basin are now being successfully worked.

The Plumas Water and Mining Company worked upon Badger Hill for one hundred days last summer, with 200 feet hydraulic pressure and 1,000 inches water, miner's measure, and cleaned up $23,000. They have the bed-rock and much of the bottom-gravel to work in yet. This mine is worked through a bed-rock tunnel, with a 5-foot flume, upon a grade of 10 inches to 12 feet length. The depth of the hill in the center is 160 feet; the gravel is about 50 feet in depth, the remaining ground being streaked with pipe-clay and fine gravel. This hill contains 37½ acres.

Gopher Hill has about the same depth as Badger, contains 140 acres, and is owned by five different parties. This hill has been poorly supplied with water. Thomas Haycock worked one hundred days this year, with 100 feet pressure and 300 inches water, through a 24-inch flume upon a grade of 14 inches to 12 feet, and cleaned up $25,000. Several of the locations upon this hill are being drifted, and yield from $3 to $30 to the day's work. The Home Stake claims, upon this hill, owned by Morris Smith and others, have produced from drifting in one-half acre of ground $13,000. The number of acres in this location is 15½.

The general course of the bed-rock in this region is north about 40° west. There are many claims worked by individual miners, with varied success; but all appear to be satisfied with their locations.

The Maxwell Ditch and Mining Company owns a large tract of ground on Spanish Creek, six miles from Quincy. Work was commenced on its ditch in 1872. The water is taken from Spanish Creek. The flume, 5 feet in width and about the same in depth, is 900 feet in length from the head to where the water enters the ditch. It was an expensive piece of work, as in many places large points of rock had to be blasted out of the way to make room for the foundation, and in others it is upheld by frame timbers 30 or 40 feet in height. Lumber and timbers for the flume were "shot" down the almost vertical face of the mountain over a quarter of a mile. The ditch is 3 feet wide on the bottom, 3 feet deep in solid ground, and carries 2,000 inches of water. At the crossing of Black Hawk Creek the water is led into a section of 27-inch iron pipe, 420 feet long, and having a depression and elevation of 75 feet. About half a mile below the crossing of Black Hawk the company has two "giants" at work, making an opening in the gravel with good prospects. The water lasts the whole season. The ditch will be extended down the river as fast as required, and it is well known that numerous bars and gravel-beds under the ditch-line will pay handsomely for working in this way.

At La Porte, in the southern portion of the county, near the line of Sierra County, extensive and profitable hydraulic operations have been carried on for the past twenty years by Messrs. Conly & Gowell, and it is estimated that at least fifteen years of mining will be required to exhaust the ground now opened. About 6 acres have been washed during the present year, yielding, approximately, $100,000. Conly & Gowell have a flume 6,000 feet in length, on a light grade. They use two "monitors," running through each pipe 700 inches of water, besides a large stream turned on during the water-season for the purpose of running off the tailings. The bank is from 50 to 80 feet in depth, and the pay so uniform that the clean-up of the season can be calculated before operations are commenced. One of the peculiar features of the claim is the heavy deposit of auriferous black sand, which is concentrated and sold in San Francisco, whence it is shipped to Swansea, Wales. These concentrated sands have been sold at rates as high as $560 per ton.

The Argentine mine, owned by Messrs. Heath & Judson, is situated in a district through which run innumerable stringers and ledges of quartz, all carrying more or less gold. The bed-rock is talcose slate, carrying gold-bearing sulphurets which assay from $50 to $500 to the ton, while to crush the whole indiscriminately will pay from $5 to $20. A great portion of the quartz is decomposed, from which the gold readily frees itself upon being washed, and it was this fact which induced the owners to fit up hydraulic diggings. Their flume at the head is 1,200 feet long, dumping into a steep cañon; thence the tailings run over a rough ground-sluice 700 feet, when they are picked up by a flume and run through an under-current, after which they resume their travels, to be caught up again by another under-current. This operation is repeated four times after leaving the main flume, and the tailings are finally dropped into Greenhorn gulch, about one mile from their starting-point. This system of working the auriferous slates is similar in many respects to that of the seam-diggings of El Dorado County.

Plumas County contains many quartz-ledges; but this branch of mining has not so far been conducted with marked success. The principal quartz-mine of the county is the Plumas Eureka, owned in London, England, the returns from which will be found elsewhere in this report. An extensive ledge of gold-bearing quartz, which has been but partially prospected, traverses the county from northwest to southeast, cropping out boldly on French Ravine, near Rich Bar, on the North Fork of Feather River, and showing a width of from 30 to 40 feet, in a formation of talcose slate. This ledge is now being developed by Mr. A. J. Warner, of Rich Bar, and Mr. B. F. Whiting, of Quincy, and their associates. The completion of a wagon-road from Oroville up the North Fork of Feather River to Sierra Valley will open a rich mineral region in Plumas County now comparatively unknown.

Dr. W. H. Harkness, of Sacramento, has been engaged during the summer of 1874 in making explorations of the high region on the borders of Plumas and Lassen Counties, and has made some discoveries of an interesting character.

The region visited by Doctor Harkness presents a vast field of volcanic disturbance, covering an estimated area of about eight thousand square miles, within which limit he found no less than one hundred extinct craters. He was occupied twenty days in his explorations, accompanied by guides. At one point, on the dividing line of Plumas and Lassen Counties, he discovered a large crater which was surrounded by unquestionable evidences of having been in a state of eruption at a recent period. It had burst forth at one end of a lake, which now covers an area of about three square miles, turning the outlet from its original channel and raising the waters of the lake a considerable height. The volcanic cone rises to about the same height as Vesuvius, and is much steeper. It is covered with pumice and volcanic ash of recent origin, as no signs of disintegration of the lava are apparent. Scattered about the lake, which has been hemmed in and raised, are the stumps of trees, some rising to a height of 40 feet above the surface of the water, and in a comparatively good state of preservation. The overflow of the volcano was evidently not in a molten state entirely, as the trunks of trees are found standing in different portions of the lava-field only partially burned through. In other places the lava-beds are thickly pitted with cavities, varying from 2 to 4 feet or more in diameter. On sounding these cavities the stumps of trees are found at the bottom, the trunks above having been burned away after the lava had cooled and set. At other places, on high land not reached by the lava, the sides of the trees toward the volcano have been burned away by the heat. Nature has since been at work repairing the mischief, and the charred sides of the trees have been partially covered again, the overgrowth uniformly showing about twenty-five annular rings, which affords very conclusive evidence that the burning did not occur much longer than twenty-five years ago. The lava-overflow from this crater covers about one hundred square miles, and bears as fresh an appearance as if it had but recently cooled.

*Description of leading mines, Plumas County, California, 1874.*

| Name. | Location. | Owners. | Length of location. | Course. | Dip. | Length of pay-zone. | Average width. | Country-rock. | Character of vein-matter. | Worked by tunnel or shaft. | Length of tunnel. | Depth on vein in tunnel. | Depth working-shaft. | Number levels opened. | Total length of drifts. |
|---|---|---|---|---|---|---|---|---|---|---|---|---|---|---|---|
| | | | *Feet.* | | | *Feet.* | *Ft.* | | | | *Feet.* | *Feet.* | *Feet.* | | *Feet.* |
| Genesee ....... | Genesee .. | Genesee Valley Mining Company. | ...... | N. W. and S. E. | E. and W. | 250 | 3 | Slate............ | Black decomposed | Tunnel | 315 | 195 | 50 | 2 | 1,500 |
| Baker Mill and Mining Company. | Cherokee . | Company .......... | 1,800 | N. E. and S. W. | E......... | 300 | 5 | Clay, slate, and granite. | Yellow quartz.... | Shaft.. | ...... | ...... | 300 | 6 | 1,750 |

*Operations of leading mines, Plumas County, California, 1874.*

| Name. | Location. | Owners. | Number of miners employed. | Miners' wages per day. | Cost of sinking per foot. | Cost of drifting per foot. | Cost of stoping per ton. | Cost of mining per ton extracted. | Cost of milling per ton. | Company or custom mill. | Number of tons extracted and worked. | Average yield per ton. | Percentage of sulphurets. | Total bullion-product. |
|---|---|---|---|---|---|---|---|---|---|---|---|---|---|---|
| Plumas Eureka.................. | Quartz.... | Company ................ | 150... | *$55 00 | $25 to $40 | $8 to $20 | $2.73 | $5 20¼ | $1 08¼ | Company . | 25,021 | $10 | 2¼ | $243,949 94 |
| Baker Mill and Mining Company. | Cherokee . | Company ................ | 8 to 15 | 2 00 | $13 | $3 to $5 | 50c to $1 | 1 50 | 1 00 | Company . | †25 | 10 | ...... | 85,000 00 |
| Genesee Valley ................. | Genesee .. | Company ................ | 5 to 7. | 2 00 | $3 to $7 | $4 | $1.25 | 1 50 | 1 00 | Company . | †10 to 13 | ...... | ...... | ........... |

*Per month, and board. †Per day.

*Statement of quartz-mills, Plumas County, California, 1874.*

| Name. | Location. | Owners. | Kind of power and amount. | Number of stamps. | Weight of stamps. | Number of drops per minute. | Height of drop. | Number of pans. | Number of concentrators. | Cost of mill. | Capacity per 24 hours. | Cost of treatment per ton. | Tons crushed during year. | Method of treating sulphurets. |
|---|---|---|---|---|---|---|---|---|---|---|---|---|---|---|
| | | | | | *Lbs.* | | *Inches.* | | | | *Tons.* | | | |
| Plumas Eureka | Quartz | Company | Water and steam, 100 horse. | 40 | 750 | 80 | 8 to 10 | 2 | 20 | $110,000 | 90 | $1.08½ | 25,147 | Amalgamation after roasting. |
| Baker Mill and Mining Co. | Cherokee | Company | Steam, 33 horse. | 12 | 650 | 85 | 7 and 8 | ... | ... | 20,000 | 25 | ... | ... | |
| Genesee | Genesee | Company | Water | 5 | 550 | 75 to 90 | 8 | ... | ... | ... | 10 to 15 | $2.50 to $3 | 3,600 | |

## SISKIYOU AND KLAMATH COUNTIES.

In the northern portion of California, and adjoining Oregon, are situated the counties of Siskiyou and Klamath, having an aggregate area of 10,740 square miles and a population of less than 9,000. Mining, both vein and placer, is carried on in the western portion of these counties, near the Del Norte line, where the auriferous slates occur and massive ledges of quartz are found. Professor Whitney, in the first volume of Geology of California, says of this comparatively unknown portion of the State:

> The whole of the region about the junction of the Del Norte, Siskiyou, and Klamath Counties is highly impressive in character from the evidence it exhibits of having undergone an immense denudation. The great body of the mountains has been lifted up to a height of from 6,000 to 8,000 feet, not in one continuous chain, as in the Sierra Nevada, but in a number of irregular and independent masses, and these have been so cut up and furrowed by the action of denuding forces as to leave a labyrinth of sharp ridges and peaks, separated by deep cañons, cut down to a depth often as great as 5,000 or 6,000 feet, a large part of which can only have been formed by the gradual wearing away of the rock which occupied the space between the walls of these stupendous ravines.

Sawyer's Bar, the center of the mining-region, is reached from Scott's Valley over a trail which is closed by snow four months in the year. The want of a wagon-road is severely felt by every resident of this mining-district. The Salmon River, which has yielded immense riches, flows through the town, and is still worked to advantage by a large number of miners. The quartz-interest is but in its infancy, and promises to be as extensive and rich as any other mining-district in the State.

The works of the Black Bear Company are the largest and most extensive in the northern country, and comprise mill, chlorination-works, saw-mill, foundery, &c. The mill has 32 stamps, with a capacity of 48 tons per day. The new battery is erected in a substantial manner. The mortar-beds are made of two pieces of timber, 10 feet long and 30 by 48 inches, and keyed together. These are stood upon end upon a foundation made of fine quartz-dust, which is tamped down with hot irons, thereby making it as hard, and firmer, if possible, than rock. It takes four men a whole day to lay a half an inch depth of this dust. The space around the mortar-blocks is then filled in with rock, packed carefully and closely together by hand, and the result is a battery without the least jar. The 16-stamp battery now in operation is built in a like manner; eight of the stamps are fitted with revolving blanket-troughs. The quartz is delivered at the mill from the mine in wagons, which carry eight tons. These wagons when they arrive at the mill are backed up to the dump, the hind wheels blocked, a block and tackle, suspended from a derrick, hooked on to the fore end of the wagon, which is then hoisted up, and the rock slides out of the hind end in close proximity to the batteries. From here the quartz is fed into the batteries, and crushed by the stamps. The treatment is the usual California process—amalgamation in battery, copper plates, and blanket washings. The blankets are changed and washed about every ten minutes. The blanket washings and tailings are put into a Stevenson pan, which grinds them still finer; from this pan they run into a settler, from which, by constant stirring, the sulphurets are caused to flow out and into two of Hendy's concentrators. After leaving these, they pass to the chlorination-works. In addition to the Hendy's concentrators, there are two large, old-fashioned buddles, which are likewise kept constantly at work concentrating the sulphurets. The company has also a well-appointed foundery, in which are cast the cams, shoes, and dies used in the batteries, grate-bars, and other articles necessary about the mill or mine; also a saw-mill, blacksmith-shop, boarding-house, a brick office for the superintendent, and three or four neat private residences with tasty gardens, store-houses, stables, &c. The batteries, pan, settlers, concentrators, saw-mill, and blower for foundery are driven by a hurdy-gurdy wheel 17 feet in diameter, with 250 feet pressure when there is plenty of water. When the water gives out the motive-power is furnished by a 60-horse-power double engine, the steam for which is generated in a boiler 16 feet long by 54 inches in diameter. The mill-building is large and spacious, being 84 feet wide by 90 feet long. The pay-roll numbers over ninety men, who receive from $50 to $60 per month. The width of the ledge varies from 5 to 45 feet, and the ore yields an average of $25 per ton. New machinery is being erected underground, for the purpose of opening up the chute and extending the lower level to the Yellow Jacket, which yielded $40 to the ton. The ledge extends, and is owned by the company, 5,200 feet, and runs north and south. The present mill is located over two miles from the mine, which entails considerable expense in the transfer of quartz to the mill.

Two miles distant, on the extension of the same ledge, is the Klamath Gold-Quartz Company. The prospects of the Klamath are equal to those of its neighbor. The mill is an exceedingly fine one, containing 32 stamps. The ledge is steadily increasing as it is followed down, both in width, extent, and richness. Should this development continue and its prospects be as good as at present, the milling-capacity (32 stamps) will be doubled next year. The track from the mouth of the tunnel and incline to mill is 3,000 feet long, yet quartz can be carried the whole distance and delivered in mill for 35 cents per ton. Should the milling-power be increased, it is thought that the ore can be delivered for 30 cents per ton. The district is young and its resources are not fully tested.

On the north side of the township, and about two miles distant in a direct line, is the Morning Star, at the head of Jackass gulch, which has been wonderfully rich in gold. This vein is believed to be the source whence the gulch obtained its very valuable deposit.

It is chiefly the great difficulty of obtaining supplies and of getting over the mountain which has prevented the vicinity of Sawyer's Bar from becoming an important gold-quartz-mining district.

Recent accounts from Oregon advise us of the discovery, in Jackson County, on Rogue River, of extensive ledges of gold-bearing quartz, which, from their situation, will probably be found to be the continuation of the system now being explored in Klamath and Siskiyou Counties.

*Description of leading mines, Siskiyou County, California, 1874.*

| Name. | Location. | Owners. | Length of location. | Course. | Dip. | Length pay-zone. | Average width. | Country-rock. | Character vein-matter. | Worked by tunnel or shaft. | Length of tunnel. | Depth on vein in tunnel. | Depth of working-shaft. | Number of levels opened. | Total length of drifts. | Cost of hoisting-works. |
|---|---|---|---|---|---|---|---|---|---|---|---|---|---|---|---|---|
| | | | *Feet.* | | | *Feet.* | *Feet.* | | | | *Feet.* | *Ft.* | *Feet.* | | | |
| Evening Star. | Salmon | Taylor, Rainey & Tonkin. | 3,000 | N. E. & S. W. | 20° E. | 100 | 4 | Black slate | Quartz | Tunnel | 125 | 150 | | 1 | | |
| Klamath | ...do | Company | 4,200 | N. E. & S. W. | 20° E. | 600 | 6 | ...do | Hard quartz. | ...do | 600 and 500. | 600 | | 2 | | |
| Black Bear | ...do | ...do | 5,200 | N. & S. | 76° E. | 1,000 | 2 to 20 | Slate. | Quartz and slate. | Both | 1,600 | 500 | Below tunnel, 180. | 5 | | $3,000 |
| Morning Star. | Jackass. | Kellner & Co | 2,400 | N. & S. | 40° E. | 200 | 2 | Porphyritic slate. | Hard quartz. | Tunnel | 300 | 200 | | 1 | 400 | |

*Operations of leading mines, Siskiyou County, California, 1874.*

| Name. | Location. | Owners. | Number of miners employed. | Miners' wages per day. | Cost of sinking per foot. | Cost of drifting per foot. | Cost of stoping per ton. | Cost of mining per ton extracted. | Cost of milling per ton. | Company or custom mill. | Number of tons extracted and worked. | Average yield per ton. | Percentage of sulphurets. | Total bullion-product. |
|---|---|---|---|---|---|---|---|---|---|---|---|---|---|---|
| Black Bear | Klamath | Corporation | 100 | $55 per month and board | $18 to $30 | $8 to $16 | $3 88 | $7 75 | $1 15 | Company | 10,949 | $19 23 | 3 | $311,797 25 |
| Evening Star. | Salmon | Taylor, Rainey & Tonkin. | 8 | $2.31 and board | | | | | | ...do | | 12 00 | | |
| Klamath | ...do | Company | 20<br>*10 | $60 per month and board; helpers, $50 per month and board; Chinamen, $14 to $30 per month. | | | | 3 50 | 1 50 | ...do | | 13 00 | 2 | 200,000 00 |

* Helpers.

*Statement of quartz-mills, Siskiyou County, California*, 1874.

| Name. | Location. | Owners. | Kind of power and amount. | Number of stamps. | Weight of stamps. | Number of drops per minute. | Height of drop. | Number of pans. | Number of concentrators. | Cost of mill. | Capacity per twenty-four hours. | Cost of treatment per ton. | Tons crushed during the year. | Method of treating sulphurets. |
|---|---|---|---|---|---|---|---|---|---|---|---|---|---|---|
| | | | | | *Lbs.* | | *In.* | | | | *Tons.* | | | |
| Black Bear | Klamath | Company | Water and steam, 60 horse-power. | 32 | 550 | 70 | 9 | 1 | 2 | $50,000 | 48 | $1 15 | 11,126 | Roasting, and then amalgamation in pan. |
| Evening Star | Salmon | Taylor, Rainey & Tonkin | Water | 4 | 750 | 75 | 8 | | | 6,000 | 10 | | | |
| Morning Star | Jackass | Kellner & Co | Water | 8 | 600 | 70 | 8 | | | | 16 | | | |
| Klamath | Salmon | Company | Water and steam | 32 | 600 | 76 | 8 | | | 40,000 | 48 | 1 50 | | |

## TRINITY COUNTY.

This county, situated in the northwestern part of the State, between the head of the Sacramento Valley and the Pacific Ocean, derives its name from Trinity River, a large stream rising in the Scott and Siskiyou ranges, flowing in a circuitous manner through the county, and emptying into the Klamath, through which its waters find their way to the Pacific Ocean. The surface of the country is covered with chains of lofty mountains composed of granite and slates, the sides of which have been eroded in deep gulches and cañons. The Trinity and Salmon Mountains, separating the county from the Upper Sacramento Valley, reach a great elevation, and are covered with snow nearly through the summer. The passes over these mountains have an altitude of over 4,000 feet, rendering access difficult during the winter-months. The valley-system of the county has a mean elevation of about 1,700 feet above sea-level.

From 1851 to 1858, Trinity County had a population of about 4,000 working miners, whose earnings aggregated scarcely $2,000,000 per year. Dr. Henry De Groot, who has traveled extensively in this county during the past year, says the miners now there do not exceed one-third of that number, and yet it is estimated that their gross earnings will this year amount to more than a million dollars, and may even reach a million and a quarter, making yearly average wages of nearly $900 to the man, while it can hardly be said that they work more than half, certainly not over two-thirds, of the time.

The gravel or channel system of Trinity County is but little understood. In fact, so far as observed, the auriferous deposits partake more of the nature of enormous drifts filling pre-existing valleys than of the channel-characteristics of Placer, Nevada, and Sierra Counties. Capt. Geo. H. Atkins, of Weaverville, a miner of extended experience and close observation, informs me that two or more of these ranges of drifts extend across the county in a general east and west direction. One of these has been traced forty miles, and forms a succession of high ridges, occasionally cut by modern streams, and showing a depth of gravel unknown in other portions of the State. It probably at one time reached the Pacific Ocean. Another drift-system occurs further south, but a large portion of this has been broken and swept away by the Trinity River, leaving occasionally large deposits of gravel adjacent to either bank where the river has taken a sharp turn. Of this class is the Atkins & Lowden property, eight or ten miles from Weaverville, noticed at some length elsewhere.

Weaverville, the county-seat, is situated in the Weaverville Basin, which seems to be a depression occurring on the line of one of the drifts above alluded to. The deposit of gravel is here very deep, and has no natural outlet deep enough to be available for other than surface-washings. During the year 1852, while placer-mining was actively carried on at Weaverville, the citizens commenced a prospecting-shaft for the purpose of determining the depth of the gravel and ascertaining whether the bottom could be practically worked. Much to their astonishment, no bed-rock was reached that season, and the work was continued during 1853, and finished in 1854. Mr. Sites, who had charge of the work, informs me that the shaft passed through alternate strata of gravel, sand, and bowlders, and occasionally a layer of talcose material. What was supposed to be the bed-rock was reached at a depth of 660 feet. About three-fourths of the matter passed through was auriferous; but, contrary to expectation, the bottom did not prove proportionately richer than the top. This exploration only proved the impracticability of working the ground lower than the natural outlet of the basin.

One of the most interesting mining enterprises of this section is the Davidson flume. This is a large and costly structure laid down along Weaver Creek for the purpose of running off and discharging into Trinity River the tailings that for more than twenty years have been collecting in Weaver Basin, and have so accumulated as to seriously interfere with mining-operations there. Nearly a mile of this flume has been completed, at a cost of about $30,000, leaving five miles yet to be built. It is 16 feet wide and 12 feet high, constructed of heavy lumber, and anchored with iron bolts into the bed-rock. The winter before last being dry, there was not enough water to carry off any great amount of tailings, and it was feared the work would prove a failure. With the present abundance of water, however, it is accomplishing a great deal of good, relieving many of the overcharged claims, and promising the proprietors a rich harvest of gold. This structure is divided into two compartments, the one 10 and the other 6 feet wide. When there is plenty of water both are kept running at the same time. When the water diminishes, it is all turned into one and the other cleaned up. The final success of this flume should encourage the projectors of these tailing-operations elsewhere in the State to perseverance, as only a bountiful supply of water seems necessary to insure good results; and this can in most cases be readily commanded. If $40,000 or $50,000, and, possibly, $60,000 or $80,000, can be extracted from these tailings in the course of three or four months, worked on such a comparatively limited scale, what might be expected from the steady operations of an enterprise like that designed to rewash the tailings of Bear River and similar localities in the central portion of the State?

The McGillivray property is situated fifteen miles below Weaverville, on the main Trinity. This consists of extensive river-bars, farming-lands highly improved, water-franchises, ditches, &c. The clean-up of the present season is estimated at $75,000, two-thirds of it net profits; and this production could, without any very great expenditure, be nearly doubled. A similar property, belonging to Doctor Ware, located one mile above Weaverville, was disposed of in the spring of 1874, the purchasers being mostly resident miners. A ditch has recently been completed carrying a portion of the water appurtenant to this estate upon Oregon Gulch Mountain, where it is now being used for piping off a deposit of rich gravel owned by the company. This ditch, which is four miles long and carries 2,500 inches of water, was built in three months—a feat that speaks well for the energy of the builders, the weather having been much of the time very unfavorable for out-door operations.

The exhaustion of the surface-placers, bars, and streams of Trinity, about the year 1860, was followed by a period of decadence which lasted for ten years, when the deep gravel-deposits of the county attracted the attention of some of the leading hydraulic miners of Nevada and Placer Counties, who acquired by location and purchase extensive areas of gravel-ground, and proceeded to lay out and construct mining-canals for the purpose of bringing in large quantities of water. Several of these operations are only equaled in magnitude by the great operations in the central counties. Among these is the Oregon Gulch Mining Company's property, a few miles west of Weaverville, which was formerly known as the Ward claim, having been worked by Mr. James Ward during the rainy seasons of several years past with a light head of water, at times not more than 100 inches. The present owners, Messrs. Atkins, Loveridge, Lowden, and others, are now engaged in the construction of a ditch for the purpose of bringing an ample supply of water on the ground. This ditch is 4 feet wide on the bottom, 6½ feet on top, 3 feet deep, and run on a grade of 11.20 feet to the mile. It will carry 2,500 inches of water this season, and after becoming settled will carry 3,000 inches or more. The line is graded 7½ feet wide and the excavation made close into the bank, leaving not less than a foot of solid earth on the outside, which greatly strengthens the ditch and will do much to prevent breaks and other damage. Thus far the ground has been of a character easily worked, much of it such that it could be spaded. Most of the rock is of a rotten, seamy character, easily worked, and but little blasting has therefore to be done. No fluming will be required on the line, which, when completed, will be nearly four miles in length, and all ditch. Near the top of the divide they are engaged in running a cut; from this they will tunnel a short distance, raise a shaft, and sluice out a hole for a reservoir. From this reservoir it is intended to run the water down a ravine and take it up in a pipe. Eleven hundred feet of 15-inch iron pipe will be required to convey the water on the claim, and 250 feet pressure will be used. Much more pressure can be secured if needed, but the bank being easy to cut, 250 feet is considered all-sufficient. A 4-foot flume, of forty boxes, will be put in here, and the dirt dumped and redumped, and run through three 6-foot flumes of four boxes each before it reaches the bed of the gulch several hundred feet below. Under-currents and all other improvements and appliances for saving gold will be used extensively.

A writer in the Trinity Journal, in referring to this claim, says:

The gravel-deposit forming the Oregon Gulch Mountain is not equaled by anything of the kind in the State. The reservoir cut on the top of the hill is in gravel of similar character, and prospects equally as well as that being worked by Mr. Ward, some 900 feet below the summit. The slate bed-rock is more than 100 feet lower than the level on which Ward is working, and the gravel is of the same character and quality. If this bed-rock goes in on a level, as it undoubtedly does, we have here a gravel-deposit of not less than 1,000 feet in depth, which prospects very rich wherever it has been tried. As to the prospects, from 6 to 10 cents is frequently found in a shovelful of the gravel, and it is a rare occurrence to wash a shovelful without obtaining several colors. The hole from which Mr. Ward has taken some $10,000 is not so large but that it could be worked in ten days with a Little Giant and a good head of water. On the bed-rock below, Mr. Dyer has this winter been running a claim, using Ward's water; has already taken $600 from his flume, and will probably get as much more in cleaning the bed-rock. The piece of ground he has sluiced off is about equal to half a day's work with a good rig and plenty of water. It may seem like exaggeration when we say that the amount of ground these men have moved was so small and paid so largely, but such are the facts, known to all who are conversant with the matter. And they did not get a very large proportion of the gold at that. Desiring to know how much fine gold he lost, Mr. Ward last summer washed sixty buckets of his tailings in a rocker, and realized $5, or 8 cents to the bucket. We have never made detailed estimates of the probable amount of pay in this ground, but with a good rigging and plenty of water it cannot fall much below $500 per day, and is liable to go to a thousand, or even higher. It is probable that after this season the company will find it necessary to work this immense deposit in three or four benches, as it will be an utter impossibility to work a 1,000-foot gravel-bank from the bed-rock. As to how this immense bed or channel of auriferous gravel was deposited there, much speculation is indulged in, but it certainly seems that in ages long ago Oregon gulch was the natural outlet of Weaverville Basin.

This enterprise is one of many of a similar character in this portion of the county. Several claims having, like this, an outlet on Oregon gulch, are preparing for opening on an extensive scale in the spring of 1875. With the vast extent of rich gravel in this county, it is somewhat remarkable that this interest should have lain dormant for so many years. It was not until the spring of 1874 that the improved appliances of hydraulic mining, so generally in use in Nevada and Placer Counties, were introduced in Trinity.

The Buckeye Company's mining property is between Stewart's Fork of the Trinity River and Rush Creek, a tributary of the Trinity, eight miles in a northeasterly direction from Weaverville. The company have acquired 960 acres of auriferous gravel on the line of the ancient channel, which once had its outlet on the Pacific Ocean. The ground, forming a high ridge with deep gorges on either side, is favorably situated for hydraulic operations on a large scale.

The deposits of this channel form an extensive plateau along the southern base of the Salmon and Siskiyou Mountains, and average three miles in width, with a length of more than forty miles. The depth of auriferous gravel, from the highest point of the plateau to bed-rock, has been estimated at 1,000 feet and more. Prospecting-shafts sunk at various points have demonstrated the richness of this gravel for a distance of nearly forty miles. The bed-rock was reached only in one place, Weaverville Basin.

Another favorable feature in the deposits of this channel is the absence of lava, cement, and pipe-clay; nor is the gravel strongly cemented, being easy to wash, while the gold is of fine quality.

A mining-canal has been surveyed by the Buckeye Company, and partially completed, for the purpose of bringing on this plateau the waters of Stewart's Fork and its tributaries. This canal will be twenty miles long, and will take up the waters of Stewart's Fork and eleven tributary streams. Mr. W. S. Lowden, United States deputy mineral-surveyor, estimates the quantity of water available during the dry season at 8,610 inches, (miners' measurement,) and states that the water-supply will at no time run below 5,000 inches; while for several months the supply will be largely in excess of the capacity of the ditch, or the probable demand for many years. This canal will supply a mining-country having an area of eighteen miles in length, three miles in width, and an estimated average depth of 500 feet, comprising Buckeye Ridge, Brown's Hill, Musser Hill, and Bolt's Hill.

The Buckeye canal, as surveyed, will be twenty-one miles in length, 8 feet wide on top, 5 feet wide on bottom, and 3½ feet deep, having a capacity of from 2,500 to 3,000 inches. Mr. John Simpson, M. E., has made the following estimate of cost of construction: Earth-excavation, per cubic yard, 15 cents; earth and decomposed rock, 20 cents; earth and detached rock, 50 cents; country-rock, (slate,) $1. The total cost will approximate $60,000. Should it be necessary to utilize all the available water, a lower ditch of like capacity can be constructed for the same amount. Water will be sold at 12½ cents per inch.

Until the completion of the system of canals above described, all of which will probably be finished in 1875, mining cannot be carried on in this county on a scale commensurate with the magnitude and proven richness of its gravel-deposits. The topographical features of the country are unfavorable for the accumulation of heavy masses of snow near the gravel-ridges, in consequence of the steepness of the mountain-sides. It has therefore been necessary to tap the streams near their heads, and by this means, with long lines of ditches, an ample supply of water will be obtained for eight or nine months of the year.

Among the important gravel-mining claims of Trinity County may be noted the operations of Messrs. Atkins & Lowden, at Grass Valley Creek, on the south bank of the Trinity River, four miles south of Weaverville Basin. This company owns about 1,000 acres of land, of which the greater portion is hydraulic-mining ground. Water is taken from Grass Valley Creek by means of two ditches, of an aggregate capacity of 1,600 inches; and a third ditch will be constructed during the year 1875, by means of which all the water of Grass Valley Creek may be brought to the highest point on the gravel-ridge. This will furnish a summer supply of 2,000 inches, and at no season will the supply fall below a working-head. The outlet of the claim is through Little Creek into the Trinity River. This seems to be a detached mass of gravel, which has been left standing in consequence of the Trinity River making a sharp turn at this point. The existence of gravel on benches and points above the present banks of the Trinity, for many miles north and east of this large accumulation, indicates that the Trinity has repeatedly cut through the ancient bed of deposit. The workings of 1874 prove the ground to be as rich as the standard of the central counties, a surface-bench having yielded 7½ cents to the cubic yard. The nature of the bottom is unknown.

The resources of Trinity County are comparatively unknown, and until the past year have been unappreciated. Its mountains are intersected with ledges of gold-bearing quartz, and yet the county possesses but one quartz-mill, a small mill of five stamps. Immense ridges of auriferous gravel, of a proven depth of from 500 to 800 feet, whose richness has been demonstrated by surface-washing with the small heads of water accumulated in the rainy season and by prospecting-shafts, may be traced for miles within its boundaries, and yet up to the present time no work has been done for more than three months in the year. Extensive deposits of cinnabar exist near Trinity Center, from which shipments of quicksilver were made within three months from the period of breaking ground. The mean elevation of the valley-system of this county is less than 2,000 feet; and its soil will produce in profusion all the products necessary for the support of a large population. Its mountains are well wooded and watered, and the climate is mild and equable.

## SONOMA, NAPA, AND LAKE COUNTIES.

The counties of Sonoma, Napa, and Lake, lying between San Francisco Bay and Clear Lake, have been the scene of considerable excitement during the year 1874, consequent upon the discovery and opening of rich mines of silver and quicksilver. In Napa County alone it is estimated that at least $250,000 has been disbursed for labor in the development of the mines, and as much more expended for mining-machinery, furnaces, retorts, &c. Probably the counties of Sonoma and Lake have each derived as much benefit through the development of their mineral resources as Napa County.

A belt of eruptive rock of a porphyritic nature extends from the head of San Fancisco Bay through these three counties to the vicinity of Clear Lake. Mount Saint Helena, in Napa County, which reaches an altitude of about 4,400 feet, is the culminating point of this upheaval. Immediately eastward of this volcanic belt is a region of metamorphic slates, sandstones, and serpentine, very similar to the mercury-bearing rocks of New Almaden and New Idria. As early as 1860, cinnabar was discovered in this range, and many locations were made; but the low price of quicksilver, resulting from the large production of the leading mines of the world, rendered this class of mining hazardous, if not unprofitable. Only one mine, the Redington, at Knoxville, Lake County, now the most productive quicksilver-mine in California, has been worked with any degree of regularity. In 1870, the New Almaden, of Santa Clara County, ceased to be largely productive, and shortly afterward the Almaden of Spain was closed, owing to the domestic troubles prevailing in that country. The rapidly-increasing price of quicksilver, which steadily advanced from 50 cents to $1.50 per pound, proved an incentive to discovery and exploration, and resulted in the opening of many valuable mines, the production of which is elsewhere given in tabular form.

I subjoin brief notices of some of the leading mines.

The Great Western mine is situated on the ridge which separates Napa and Lake Counties, a few miles north of Saint Helena Mountain. The company has a patent for the mine and 1,600 acres of timber-land. A 10-ton furnace and another of 30 tons' capacity, with six iron condensers, were in operation in December, 1874. The mine is producing about 350 flasks of quicksilver a month. Two main tunnels have been run from opposite sides to the center of the hill, where they are connected with a shaft 70 feet deep. From the shaft parallels have been run, and the ore is taken out through the tunnel on the east side of the hill just above the reduction-works. A tunnel has been commenced below the works to run 1,600 feet, the length of the developed portion of the mine. The ore will be taken through this tunnel from above and from side-drifts. The rock is soft; much of it crumbles and is made into adobes. There is 1,500 feet of tunneling altogether, splendidly timbered, exclusive of the prospecting-tunnels, eight in number. The quicksilver-ores here are contained in a formation of a serpentine over 200 feet in width, which has been cross-cut for 40 feet, and found profitable for that distance. The lowest depth attained below the surface is 400 feet.

About three miles north of the Great Western is the American mine. This company uses a 10-stamp mill for crushing the ore, after which it is made into adobes, or bricks, and treated by a retort and condensers, with a capacity of 10 tons (ore) per day.

The Ida Clayton, in the same vicinity, will be ranked among the producing mines in 1875.

The Manhattan Company's mine, near Knoxville, Lake County, has been opened to a depth of 200 feet, but there is an abundance of ore near the surface. The company runs a Knox & Osborn furnace, of 24 tons' capacity, and works 16 to 20 tons daily, employing 65 to 70 men in all. Wages, to miners, $45; helpers, $40; common laborers, $35 per month and board, with cabin-accommodations. These are the standard wages of miners and helpers in the quicksilver-region.

The California Company's works are situated three miles north of Knoxville, near the line of Yolo County. The lower tunnel, through which all the ore from the various tunnels and surface-workings for 300 feet above it is discharged, is about 500 feet long, with a well-ironed car-track. Above this is another tunnel, 600 feet long, with its cross-sections and stopings. Above this is still a third tunnel, between 400 and 500 feet long. And still above are open surface-workings, yielding good mineral, worked to the depth of 40 or 50 feet. The company is using at present one of the best constructed Knox & Osborn furnaces, with ample condensing-chambers, with vapor-draught extending nearly a quarter of a mile away, up the mountain side, thus freeing the whole premises of any danger from hurtful gases or mercury-poison. Capacity of furnace, 24 tons daily. The fine metal-bearing dirt and gravel are concentrated by the aid of water, so that but a small amount passes through the furnace. The amount of metal-yield varies according to quality of ore used; one week gave a yield of 40 tanks of 76½ pounds each, but some weeks run less. The company employs on all the works, including extensive wood and farm operations, 100 men. It erects all necessary dwellings and owns all improvements made on its lands.

The Redington quicksilver-mine, of Knoxville, employs 250 men. There are three Knox & Osborn furnaces, and one more is being erected. These are old-style furnaces, that work about 200 tons per week. The Knox & Osborn furnaces yielded about 600 to 1,000 flasks per month for the past few months. There is plenty of good ore in the mine on the 400-foot level. The company owns the town and miles of the surrounding lands, with a vast store of all the ordinary and extraordinary necessaries, provisions, clothes, medicines, &c., hotel and stable, and shops.

The Commercial Herald, of San Francisco, says of this mine:

The Redington has produced during the year 7,200 flasks, an increase of 3,000 flasks over 1873, averaging about 800 flasks per month during the last half of the current year. It gives promise of a still further increase of product during 1875, having just completed a third new continuous furnace, and having a fourth now under construction. This company will thus, within thirty days, have three furnaces in operation, (each with a capacity of 24 tons daily;) and about April 1, 1875, they will have four such furnaces in operation, and may be expected to produce thereafter about 1,200 flasks per month. The development of their mine has fully kept pace with this increase of their reduction-capacity, and affords a guarantee of an ample supply of ore to keep all their reduction-works busily employed. Besides paying the large outlay for new construction-works, this mine has paid its owners continuous dividends through the year, at the monthly rate of $30 per share, (the capital stock being $1,260,000, in 1,260 shares of $1,000 each,) so that it now fairly stands as the leading quicksilver-mine of the coast.

The main range of the Mayacamas Mountains forms the northeastern boundary-line between Sonoma and Lake Counties. Between Pine and Cobb Mountains, links in the main chain, rises Big Sulphur Creek, flowing through a deep gorge formed by Cobb Mountain and the Hog's-back, a spur running nearly parallel with the main range. In the cañon formed by these two mountains the Geyser Springs are located. Little Sulphur Creek rises in Pine Mountain, flows west of the Hog's-back range around Geyser Peak, and unites with Big Sulphur, isolating, as it were, this immense upheaval, whose highest point, Geyser Peak, is 3,500 feet above the sea-level. On the opposite side of Little Sulphur there is a parallel range of almost equal length, which, to the westward, breaks into irregular hills, terminating finally in the Russian River Plain. We have now Cobb Mountain in the main range for a background, a parallel range known as the Hog's-back or Geyser Mountain, and still another parallel range of almost equal altitude, the trend of each being southeast and northwest. Little Sulphur Creek flows through Pine Flat, which is situated between the two ranges last named. As the mountains approach each other the valley is pinched out, and the creek, leaving the flat, flows through a steep and precipitous cañon. The main quicksilver-lode of the district is supposed to come through Pine Mountain into the Hog's-back or Geyser range, near Pine Flat. The ledge runs along this range in the direction of Geyser Peak, near which it crosses the ridge, running from thence in the direction of Cloverdale. It is said that the outcrop has been traced until it terminates with the coast at a point in Mendocino County.

The Oakland mine is situated in the Devil's Cañon, at the head of a deep gorge in the Geyser range. The location was made by tracing to their source bowlders and "float" rock containing cinnabar, which were found profusely scattered over the cañon. The Oakland Company works a force of forty miners. There are six retorts in operation, producing 100 flasks of metal a month. The main tunnel is run into the hill 113 feet, from which drifts have been run east 36 feet. The company has in operation a crusher, run by steam-power, with a capacity for crushing 100 tons in twenty-four hours. There is also complete 400 feet of railway from the tunnel to the retorts. The ore in the ledge, which is 10 feet wide, is cinnabar. The walls of the ledge are well defined, and the clay lining appears between the ledge and both the upper and lower walls. The ore is richest in the center of the vein.

Adjoining the Oakland mine is the Geyser. Its massive outcrop forms a portion of the northern wall of the cañon. From these two ledges the large amount of float in the bed of the stream originally came. It indicates that there is a large supply in the main lodes from whence these huge cinnabar bowlders were torn.

The Rattlesnake, during the early period of its development, was known as a "native-quicksilver" mine. The metal is not in the form of cinnabar, but adhering to the walls of the tunnels and shafts, and forming little pools in the dirt or earth at the bottom of the tunnels. It presents the same appearance as though one had carried the metal into the tunnels and scattered it broadcast on the sides and roof. This peculiarity is found in the Idria mine of Carniola, Austria.

There are several hundred locations in the counties of Sonoma, Lake, and Napa which are being vigorously prospected, and not less than one hundred furnaces and retorts, varying in capacity from one to ten tons of ore per day, have been erected, a majority of which will run periodically while quicksilver maintains a price over $1 per pound; but the trouble with the majority of these claims is that the mercury-bearing ores are not concentrated in large masses, but diffused and disseminated throughout the country-rock to such an extent as to rarely exceed one-half of one per cent., equivalent to $10 per ton with quicksilver at $1 per pound.

I am under obligation to Dr. A. Blatchly, M. E., of San Francisco, for the following interesting account of the Geyser group of quicksilver-mines situated in Sonoma County:

The first quicksilver-mine discovered in the vicinity of the Geyser Springs was the Pioneer. It was located by a band of prospectors who were hunting for silver in 1859, and found native mercury in the croppings of the vein. This discovery immediately caused a mining excitement, and a number of mining-districts were formed and a multitude of claims were located and recorded. Little or no work was done on many of these claims, while on others a considerable amount of labor was expended; but the low price of quicksilver and want of skill in mining and metallurgy caused all work to cease, and for a number of years the mines were deserted. In about ten years after the first locations were made, the present owners came into possession, formed new mining-districts, and have pushed the development of their mines with skill and energy.

These mines are located in Sonoma, Napa, and Lake Counties, extending in an easterly and westerly direction for fifty or sixty miles, and northerly and southerly for twenty or thirty miles. The boundaries of this district are by no means well defined, and it is probable that future explorations will materially enlarge these limits. The mines so far located are mostly in the northern portions of Sonoma and Napa Counties and in the southern part of Lake County.

The country-rock is metamorphic sandstone, traversed by veins of serpentine, and occasionally by veins of quartz and quartzite, these siliceous veins being generally small and not well defined, with strikes in every direction, while the magnesian veins are large, well defined, and usually strike nearly east and west, with a slight dip to the north.

The cinnabar almost invariably occurs in a siliceous magnesian veinstone, in which the silex greatly predominates. Many of the deposits are net-works of veins of silex, with the interstices filled with silicate of magnesia and cinnabar. These veins are found between the sandstone and the serpentine. In some mines the vein is north of the serpentine, in others south, there being no regularity in this particular in their modes of occurrence. In some veins the cinnabar is found in nodules or deposits, but generally in small veins running across the veinstone, and usually associated with quartz. The native mercury is generally found either in a coarse siliceous sandstone or in clay. In the Flagstaff it occurs in a veinstone which does not differ materially from that which carries cinnabar.

Nearly all of these veins contain iron in considerable amounts, frequently in sufficient quantities to constitute an ore of iron. Gold, silver, and copper are also frequent constituents of these lodes, and occasionally chrome-iron in considerable quantities. But so far as is known, in no instance have the precious metals been sufficiently abundant to pay for the expense of extraction.

Bitumen is found in nearly all of these veins, sometimes a deposit of a gallon or two in one cavity.

Thermal springs are numerous throughout the whole quicksilver-region, and the uniformity of their occurrence leads prospectors to the belief that there is intimate relation between the causes which generate thermal springs or produce deposits of cinnabar, and that where one is found the other is liable to occur in the same vicinity.

*Reduction of the ores.*—For testing the mines, iron retorts are generally employed, but the slowness and costliness of this mode of reduction soon cause it to be abandoned, and where the mine is proved to be valuable, continuous feeding and discharging furnaces take the place of the retorts. These are almost exclusively either of the Knox & Osborn or Luckhardt & Riotte patterns. Their capacity is from ten to twenty tons per day, and the cost of reduction is from $1 to $2 per ton. This low cost of extracting the mercury enables the miners to realize a profit from very poor ores where the expense of mining is not too great. In many of the mines ores that yield over one-fourth of one per cent. of mercury at the present price of quicksilver will pay for working.

The improvements in these furnaces over those formerly employed at the New Almaden are very great, not only on the score of economy in the reduction of the ores, but also in preserving the health of the workmen. By the use of a suction-fan the fumes are drawn into the condensers, and if a leak or opening should be made in the furnace the air would rush in instead of fumes rushing out. By proper precaution no one need ever be salivated unless by some accident.

*Cinnabar-placers.*—In the earth and *débris* below and around the quicksilver-lodes cinnabar is found, like placer-gold or stream-tin. Formerly this was made into adobes, or unburned brick, and these were reduced in the furnaces in the same manner as the other ore. Where this dirt was sufficiently rich, this mode of working did very well, but nearly every mine has large amounts of dirt too poor to pay in this manner. Latterly this material has been concentrated by hydraulic washing and sluicing, and a number of claims too poor to pay otherwise have been rendered profitable.

When we consider the volatile nature of its constituents, the stability of cinnabar is remarkable. When all other sulphides have been decomposed by the action of the elements, it alone remains. Owing to this property, prospectors are enabled to judge with almost unfailing accuracy whether a mountain has any cinnabar-lodes in it by washing a few pans of dirt at its base. If no color is found, it is a waste of time to dig in that locality.

To describe all of the mines in this region that have a promising outlook would be impossible, hence I shall briefly notice a few typical mines, one of each variety, native metal, cinnabar, and hydraulic.

The Socrates, formerly called the Pioneer, is situated in the summit of the mountain-range between the headwaters of Pluto's Creek and Little Sulphur Creek. Here the first discovery of quicksilver north of the bay of San Francisco was made, and more labor has been expended in developing it than on any other mine in this vicinity. Over 1,000 feet of tunnels and drifts have been run, and various inclines and openings on the surface, displaying large bodies of ore. The strike of the vein is northwest and southeast, with scarcely any dip, being nearly vertical. The metal is mostly native, with small amounts of cinnabar, and still smaller quantities of selenide of mercury. In the deepest workings the free metal appears to be more abundant than nearer the surface. Formerly, an expensive furnace was erected for this mine, but after being tested for a short time, it was pronounced a total failure. This was the only attempt to reduce ores on a large scale made by the first locators, and this failure caused all work to cease in these mines for a long time. This mine is located about five miles from the Geyser Springs, on the road to Calistoga.

The Geyser mine is located on the north bank of Devil's Cañon, about two miles south from the Geyser Springs. The strike is southwest and northeast, with a dip to the northward. The vein is large and crops to the surface, forming cliffs 50 or 60 feet high, showing large amounts of ore on the surface. The ore is almost exclusively cinnabar, many specimens being crystallized. A Knox & Osborn furnace is nearly ready to run, and exploration is actively conducted.

The Cloverdale is situated on the north bank of Sulphur Creek, about seven miles from Geyser Springs. The croppings are large, and the veinstone appears to be almost a quartzite, containing more silex than is often found in this region. A furnace is now in operation on this mine, having apparently an abundant supply of ore.

The Yellow Jacket and Ida Clayton, both on the same vein, are about twelve or fourteen miles southeast from the Geyser Springs. The vein in the Yellow Jacket has a strike nearly east and west, with a dip to the north. It is large, nearly 100 feet wide, and soft, so that it can be worked by a hydraulic apparatus, the same as a placer gold-mine. The pipes and sluices are nearly completed, and when fully in operation this will be a very productive mine.

The workmen employed in these mines are generally American, Cornish, and Mexicans, but at the Great Western the employés are mostly Chinamen. The wages range from $1.50 for a Chinaman to $3.50 for a first-class white miner.

The Saint John quicksilver-mine is one of the foremost new enterprises. A system of ridges, evidently a continuation of the Mount Diablo range, makes its appearance north of the Straits of Carquinez and pursues a general northwest course along the eastern edge of Napa Valley toward Mount Saint Helena, with which they are probably connected. On the summits of the rounded and bare hills east of Napa Creek outcrops of sandstone are observed. These are in close conjunction with a belt of volcanic matter, probably the same observed farther north, of which Mount Saint Helena is the prominent landmark. Between these formations we find the Saint John quicksilver-mine, near Vallejo, Solano County, which, for production, extent of ground, and development, may be ranked among the leading mines of the Coast range. The principal characteristics of this mine were described in the report of 1873. The mine is opened by a system of tunnels, seventeen in number, all of which cut the ledge at right angles. The greatest depth attained (September, 1874) was 275 feet. The ledge, which has a northwest and southeast course, conformable to the strike of the sandstones and volcanic belt, has an average width of 6 feet and a well-defined foot-wall. The gangue contains a larger proportion of siliceous matter than the quicksilver-belt of Napa, Sonoma, and Lake Counties. While the vein is narrow, the ores have been of high grade. The production will be found in the tabular statement. The physical characteristics of this mine are quite distinct from those of the group farther north, and there seems to be no connection with that belt.

During the year 1874 considerable progress has been made in the development of a ledge of silver-bearing quartz in Napa County, near Calistoga, known as the Calistoga mine. The vein occurs in a formation of porphyry, has a bold outcrop, and presents all the features of a true fissure. The principal work on this ledge has been confined to the original claim, on which a 10-stamp mill has been erected and was in successful operation during the latter part of the year. Many promising locations are reported to have been made on extensions of the first discovery.

The outcrop of these ledges, for there are ten parallel lines of outcrop, distant from each other about half a mile, is situated on the eastern slope of Saint Helena Mountain, at an elevation of 2,400 feet above sea-level, or 2,000 feet lower than the summit of the mountain. The existence of these ledges was known long prior to their location, as they were crossed by a well-traveled mountain-trail, but the presence of silver was not suspected, it being the popular opinion, based on the statement of some itinerant geologist, that silver and gold bearing veins could not exist in the Coast range. In 1872 a prospector of an investigating turn of mind had some of the rock assayed, and obtained results of from $10 to $20 per ton in silver and gold, the silver slightly preponderating. A system of tunnels was commenced, as shown on the accompanying diagram, and resulted in the opening of a strong vein, from 6 to 10 feet in width, carrying ore which steadily increased in richness as depth was attained. The actual milling results are stated at from $60 to $80 per ton, but the production has not been large, in consequence of the lateness of the season when milling was commenced. Several tunnels have been run on the ledge, and one has been commenced which will tap the ledge at a depth of 400 feet below the outcrop on the hill. Work has been confined to the lower of the two ledges. The upper ledge presents a bold outcrop which may be traced for two

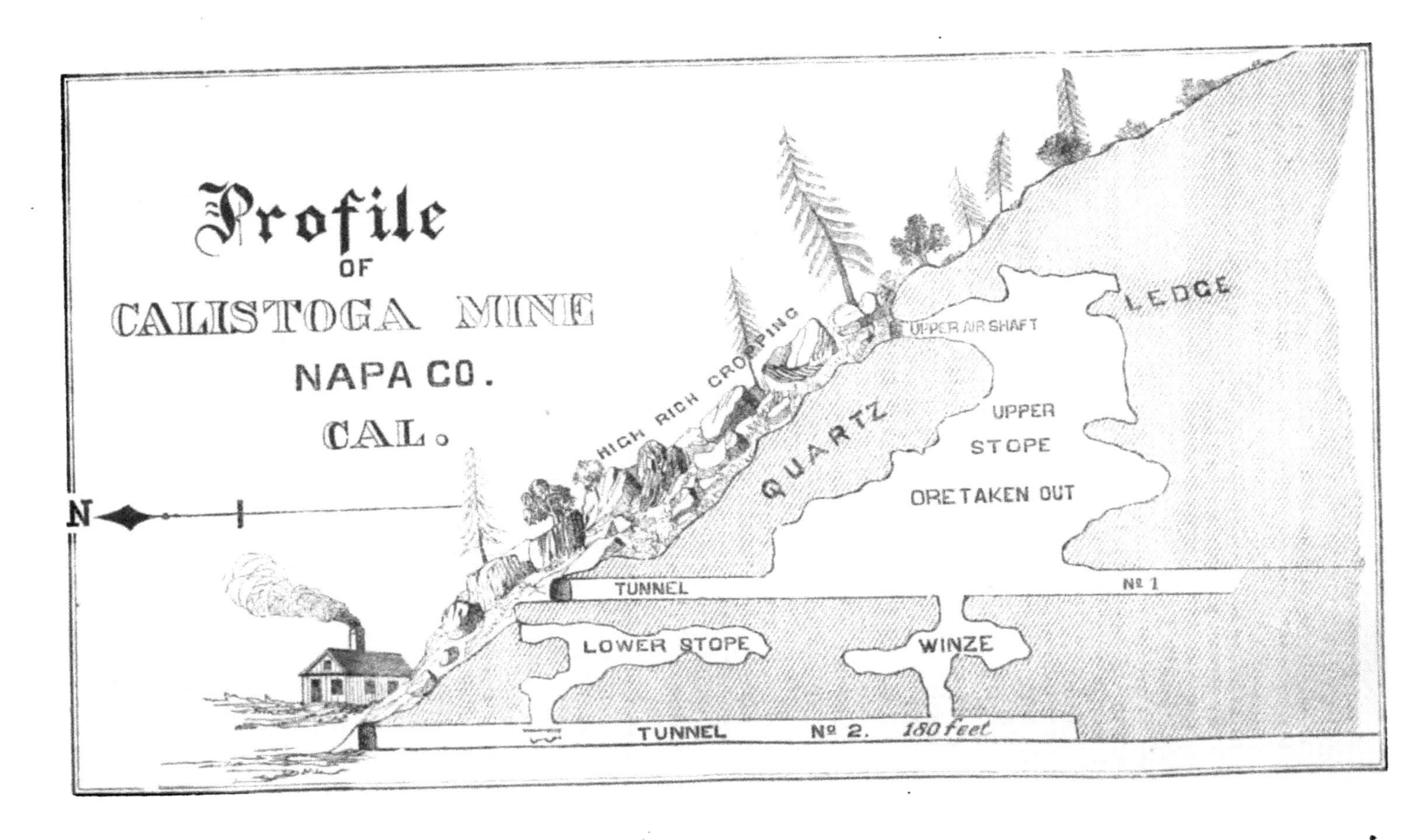
Profile
OF
CALISTOGA MINE
NAPA CO.
CAL.
N
HIGH RICH CROPPING
UPPER AIR SHAFT
LEDGE
QUARTZ
UPPER
STOPE
ORE TAKEN OUT
TUNNEL
Nº 1
LOWER STOPE
WINZE
TUNNEL
Nº 2. 180 feet

miles in a northerly direction. The North Calistoga Company, in this direction, is taking pay-ore from its mine.

In Sonoma, Napa, and Lake, and also in Calaveras and Stanislaus Counties, there have been some very extensive mines of chrome-iron ore discovered, but the demand for it in this country is very limited. Thousands of tons can be laid down in San Francisco for less than $20 per ton.

*Description of leading mines, Napa County, California,* 1874.

*Calistoga mining-district.*—Name of mine, Calistoga; owners, Calistoga Mining Company; length of location, 4,000 feet; course, north 16° east; dip, 71° west; length of pay-zone, 450 feet; average width, 5 feet; country-rock, porphyry; character of vein-matter, quartz; worked by tunnels, (3;) length of tunnels 425, 169, 184 feet; depth on vein in tunnel, 390 feet; total length drifts, 778 feet, (3.)

Reported by A. Badlam.

*Statement of quartz-mills, Napa County, California,* 1874.

*Calistoga mining-district.*—Name of mill, Calistoga; owners, Deane & Co.; kind of power and amount, steam, 60 horse; number of stamps, 10; weight of stamps, 910 pounds; number of drops per minute, 80; height of drop, 12 inches; number of pans, 4; number of concentrators, 1; cost of mill, $25,000; capacity per 24 hours, 18 tons; cost of milling and treatment per ton, $4.25; commenced October 1, 1874; method of treating sulphurets, amalgamation.

Reported by A. Badlam.

## SANTA CLARA COUNTY.

The Quicksilver Mining Company continues to be the leading one, in spite of the great productiveness of the Redington, in Lake County, and the rapid growth of numerous other competitors. From the reports and exhibits prepared for the stockholders' meeting of February 24, 1875, I extract the following statements.

The president, Mr. Daniel Drew, says:

The total product of quicksilver from the mines of the company during the year was 9,084 flasks, the monthly product being as follows:

| | *Flasks. | | Flasks. |
|---|---|---|---|
| January | 675 | July | 775 |
| February | 675 | August | 930 |
| March | 750 | September | 820 |
| April | 735 | October | 857 |
| May | 729 | November | 737 |
| June | 661 | December | 740 |
| Total | | | 9,084 |

Being an average per month of 757 flasks.

The receipts of the company during the year, including quicksilver on hand at current prices, were:

| | |
|---|---|
| From quicksilver | $920,490 72 |
| From ore account, increased | 68,445 42 |
| From rentals and miscellaneous sources | 23,775 53 |
| From interest, (on call-loans, &c.) | 20,566 80 |
| From premium on gold | 40,962 05 |
| Total | 1,074,240 52 |

* A flask of quicksilver is 76½ pounds.

Reference was made in my report of last year to an agreement entered into with Mr. Thomas Bell, of San Francisco, for the sale of the products of the mines, which agreement was considered a fortunate one for the company. The experience of the year has fully justified this belief. The sales have been made at the full market-price, and the returns have been made with promptness. I have no doubt, therefore, but that the stockholders will approve the action of the board in continuing for the coming year the contract with Mr. Bell, the more especially as the slight changes made in the agreement is in the interest of the company. Again, by availing ourselves of the services of Mr. Bell, we have secured a prompt and active home demand, enabling us to get immediate returns, together with the best prices.

The constantly-increasing price of quicksilver during the past year has been of great advantage to our company, and promises to be of still greater advantage the coming year. The highest price obtained during 1873 was $1.20 per pound. The year just passed the average price was about $1.35 per pound, while the price at the close of December was $1.55 per pound. The demand for quicksilver has greatly increased, while the general supply is scarcely augmented. These two facts seem to point to a certain maintenance of the present price and to a probable increase during the present year.

An examination of the report of the manager will show that an amount of "deadwork," so called, has been done and permanent improvements added more than sufficient to maintain the property in as good condition as at the close of the preceding year; and that while the net income has been largely increased, it has not been done by allowing any deterioration in permanent structures, or by omitting any "deadwork" necessary to maintain the mine in its present efficient condition. On the contrary, new bridges have been built, important additions have been made to furnaces, and new and effective machinery, especially air-compressors and rock-drills, has been purchased. In addition to these, the company has purchased during the year a large and valuable tract of woodland, by which purchase they have not only effected a great saving in the present price of fuel, but have solved a problem which was beginning to assume serious importance to the company, in regard to obtaining a full and regu- and regular supply of wood.

The stockholders will note with pleasure the improvement shown in the financial condition of the company. The value of the cash-items, after payment of interest due on bonds January 1, 1875, (the only debt of the company,) is $730,000, all of which amount is applicable to dividends on the stock.

In conclusion, the stockholders can be fully congratulated upon the present excellent condition and prospects of the company.

From the report of the manager, Mr. J. B. Randol, the following is extracted:

| | | |
|---|---|---|
| The earnings during the year ending December 31, 1874, were: | | |
| From 9,084 flasks of quicksilver produced during the year | | $920,490 72 |
| Of this quantity— | | |
| 7,324 flasks were sold through the agency of Thomas Bell, esq., at current market-rates, for account of the company, netting | | $744,490 72 |
| 1,567 flasks were consigned to Thomas Bell, esq., for sale, upon which advances were received to the amount of | | 78,350 00 |
| And it is estimated that the amount to be realized over advances on the 1,567 flasks and the value of 193 flasks at the mine will net | | 97,650 00 |
| | | 920,490 72 |
| From rents and privileges | | 21,804 86 |
| From miscellaneous sources | | 4,871 65 |
| From ore account, increased | | 68,445 42 |
| Total | | 1,015,612 65 |
| The expenses were: | | |
| For mine and hacienda pay-rolls | $389,190 69 | |
| For materials and supplies consumed | 72,743 18 | |
| For miscellaneous property | 1,357 06 | |
| For miscellaneous expenses, including taxes | 30,148 27 | |
| | | 493,439 20 |
| Balance, net earnings for 1874 | | 522,173 45 |

Compared with 1873, the operations of the mine and furnaces present an increase in gross earnings of $188,443.37, an increase in expenses of $94,773.19, and an increase in net earnings of $93,670.18.

The contract for the sale of quicksilver through the agency of Thomas Bell, esq. was renewed, on the 20th of April, for one year, on the same terms as named in my report for 1873, except that it was agreed that, with respect to the local sales and the sales at the domestic subagencies, the price should be determined after consultation with the company's manager, and that interest should be allowed on all sales from the average date of each month in which such sales were realized. On consultation with Mr. Bell, it was decided to require the subagencies at Virginia City, Gold Hill, Austin, &c., to make sales at a price that should, after payment of freight and commission, net an amount equal to the price current in San Francisco. The result was a large increase in the amount of sales at the San Francisco agency, where the price could be more readily regulated and the business centralized.

The market-price of quicksilver in San Francisco for the past year is given in the following table, the several advances in price having been made at the dates named:

| | Per pound. | Per flask of 76½ pounds. |
|---|---|---|
| 1873.—Dec. 31 | $1 20 | $91 80 |
| 1874.—Jan. 22 | 1 25 | 95 62½ |
| April 1 | 1 30 | 99 45 |
| April 20 | 1 35 | 103 27½ |
| Aug. 5 | 1 45 | 110 92½ |
| Oct. 9 | 1 50 | 114 75 |
| Nov. 9 | 1 55 | 118 57½ |
| Dec. 31 | 1 55 | 118 57½ |

As has been the case for several years past, the price was controlled by the ruling rates for quicksilver in the London market, and the steady advance through the year was sustained by an increased consumption attended by a falling off in the production of our own and other old mines, which decrease the many new quicksilver-mines discovered and worked on this coast during the past year have so far failed to make good.

The future for many of these mines is very promising, and while their product for 1875 will probably add to the supply of quicksilver, it will be readily marketed, through the increased demand caused by the late discoveries of large bodies of silver-ore in the Comstock lode and the wants of foreign markets, in all of which the stock is very small.

Many of our men were attracted to these new mines; therefore in outside work, cleaning ore, and picking *terrero* from the dumps, their places have been filled by Chinamen, of whom 80 are now employed at the old mine, 20 are engaged in mining at Enriqueta and San Pedro, and 10 are at general work in the hacienda. Six hundred and two men and boys of all nationalities are now employed at the mines and hacienda, the average for the year having been 567.

For the expenses incurred in the production of ore, the mine pay-rolls present expenditures for labor amounting to $329,012.47, an excess of $53,084.29 over like payments for 1873.

These disbursements were for:

| | |
|---|---|
| Ore paid miners by the carga (300 lbs) | $121,230 27 |
| Ore-cleaning | 10,589 90 |
| Terrero from the dumps | 29,286 25 |
| Tierras from the dumps and mine | 19,556 37 |
| Yardage | 85,377 73 |
| Timbermen and miners by the day | 15,157 61 |
| Tramming | 24,324 38 |
| Foremen | 5,540 00 |
| Laborers and skilled labor | 14,834 92 |
| Transportation by wagons | 3,115 04 |
| Total | 329,012 47 |

The production of mine-ore was 19,139¾ cargas of 300 pounds each. Of this quantity there were produced from the—

| | Cargas. |
|---|---|
| Old mine | 13,079½ |
| San Francisco | 2,810½ |
| Outside mines | 3,249¾ |
| Total | 19,139¾ |

In addition to the ore extracted from the mines, there were gathered from the dump-piles at the old and new planillas 19,380 cargas of terrero, and from the dumps and mines there were taken 79,607 cargas of tierras, making the aggregate production

118,126¾ cargas, or 17,719 tons—a gain of 27,446 cargas over the quantity handled in the preceding year. The ore-product shows an increase of 30 per cent. against an increase of 19 per cent. in the amount of the mine pay-rolls. This increase, as in the previous year, was in the lowest-grade ores, of which the terrero shows an increase of 10,470 cargas and the tierras an increase of 17,431 cargas.

The average price per carga paid to the miners for mine-ore was $6.33, and the cost for cleaning it was 55 cents per carga additional, making the total $6.88.

The cost of terrero averaged $1.51 per carga; the tierras, 25 cents per carga, and the average cost for ores of all grades was $2.78¼ per carga.

In the latter cost are included all the expenditures on the mine pay-rolls, and it presents a lessened cost per carga of 25¼ cents compared with the past year.

The mine-ore was extracted from 35 workings or *labores*, and its total production was only 455¾ cargas less than in 1873.

In the old mine the only body of new ore worked was at the 1,000-foot level, (192 feet below new tunnel.) Ore was developed by this level in March with a yield of 22 cargas, increased to 211 for July, and closing with 249 cargas in December, giving a total product of 1,183 cargas, with good promise for the future.

The principal yield from other points in the old mine was from the Santa Rita West, 2,405 cargas; the Upper mine, 169¼ cargas; the Great Eastern, 1,335¼; the Greeley, 1,156; the Victoria, 1,125¼; and the Ardilla, 1,111 cargas.

In the San Francisco mine the discovery of the "New World," at the lowest level, 597 feet below the apex of Mine Hill, added to its product 1,531 cargas of high-grade ore since last June. The product for December was 348¼ cargas, with a fair prospect for a continued and augmented yield, fifty men being at work extracting ore at the close of the year. The total product of the San Francisco mine was 2,810¼ cargas, or 2,461 cargas more than in 1873.

The outside mines show a product of 3,249¾ cargas, an increase of 1,547½ cargas.

Of the total quantity, the Cora Blanca gave 2,460¼ cargas, and promises to increase in importance. The favorable opinion expressed in regard to this mine in my last report has been well maintained by its product, and all the expenditures made in opening and developing it have been largely repaid.

Work was resumed at Enriqueta in May, and the San Mateo again began to produce ore in June. Its total product was 276 cargas of fair-quality ore, and at the end of the year the ground was of a character to justify further workings and explorations.

The tabular statement of drifting which is appended to this report shows that the running of tunnels and drifts, sinking of shafts, making rises, &c., were carried on at seventy-three points on the company's property, including the old mine and outside works.

The disbursements for labor on this account comprise, for yardage, $85,377.73; for timbermen and miners by the day, $15,157.61; a total of $100,535.34.

Nine thousand six hundred and eighteen feet of tunnels, drifts, cross-cuts, and shafts were driven or sunk for prospecting or communication, at an average cost of $8.87½ per foot.

The sinking of the Randol shaft was resumed in May. It was deepened 135 feet, and well timbered all the way.

After sinking 100 feet, a level, now known as the 1,100-foot level, (it being at that depth below the apex of Mine Hill,) was run from the shaft to cut the vein below the favorable ground in the 1,000-foot level, but its progress was stopped in December, by cutting a large body of water that rose 80 feet in the shaft in forty-eight hours.

This great outflow of water is usually from the vein, and is favorably regarded by the old miners. Its great quantity temporarily stopped the work in the drift and also in the shaft, but the pumps now have it under control, and the sinking of the shaft and work in drift will soon begin again.

On the 31st of December the shaft had reached a depth of 672 feet from grass, equal to 1,102 feet in vertical depth below the original opening of the mine. There had been run from or connected with it five levels, and the cost of the entire work had been included in the ordinary expenses of the mine. It is proposed to sink the shaft 400 feet deeper without delay. This will require a new engine and hoist, for which preparations are already being made.

The Cora Blanca shaft was sunk to and connected with the Deep Gulch tunnel. Its total depth is 181 feet, and it has proved of great service in the explorations carried on in the Cora Blanca mine, in which work a large force of miners have been engaged throughout the year with satisfactory results.

At Enriqueta, the San Mateo shaft was sunk for 87½ feet, partly through metally ground, which will be further explored. Nineteen and a half feet were driven in the Eldridge tunnel in June, but the rock was hard and costly ($20 per foot) and the ground unfavorable; therefore the work was stopped, and drifting renewed in the San Andreas, which was directed to the same point as the Eldridge, was in favorable ground, and could be run at much less cost. Two hundred and eighty-two feet were drifted, and in December a little ore was discovered. The work will be continued until it reaches and explores the ground under the San José labore, which in past days was the richest and largest working in the Enriqueta mine.

In January, 1867, an important work called the Bottom tunnel was started at a point about 1,600 feet distant from and nearly on a level with the furnaces at the hacienda. It was to be driven through favorable outcroppings to reach and traverse the whole length of the Cora Blanca vein. For reasons then deemed good, it was discontinued in June of the same year, and remained idle until the very favorable developments of the Cora Blanca caused us to resume the work last July. Since the latter date it has been driven 143 feet by hand-labor contract; size, 7 by 7 feet.

In November two Burleigh drills were set at work to enlarge the tunnel to 8 by 8 feet. The enlargement was completed and the face advanced 16 feet by the drills at the end of December.

The distance to the ore-ground now worked in the Cora Blanca is 3,000 feet, of which 390 feet are completed, and an advance of at least 4 feet per day is expected to be made with the drills, which so far have worked admirably in every respect.

The tunnel will strike the Cora Blanca mine at a point 450 feet deeper than its present work; and as the ore-bodies are dipping toward the tunnel, it is probable that paying ground will be developed in its course.

In the old mine, at the 1,000-foot level, two Ingersoll drills were placed at work in September; but as the columns for supporting the drills were badly adapted to the purpose, we were compelled to suspend operations with them until new appliances could be had. For this purpose we purchased a column and a carriage commonly used for the Burleigh drill, and with their aid we have greatly increased the efficiency of the Ingersoll drills.

The hacienda pay-rolls show the following disbursements for labor:

| | |
|---|---|
| General account | $44,801 78 |
| Operations of railroad | 3,629 00 |
| Operation of furnaces | 11,747 44 |
| Total | 60,178 22 |

Compared with the previous years, these expenditures present an increase of $18,532.71, or 44¼ per cent.; but it will be noted that the quantity of ore roasted shows an increase of 35.33 per cent., with an addition of 45,553 cargas, or 80.97 per cent. to the ore on hand. This reserve ore was passed over the railroad and handled at the hacienda in addition to the ore roasted, and therefore the charges on pay-rolls for labor were increased for that account. All the expenses at the hacienda were also increased by the large quantity of ore in the form of tierras.

Most of this class of ore requires to be made into adobes, or sun-dried bricks, before it can be roasted. Of these adobes we had on hand December 31, 1873, 380,693. In 1874, 1,214,363 were made. There were roasted 919,193, and there were left on hand 603,836, or about eight months' supply.

The number of furnaces in operation monthly was 5⅓, as in 1873, while the number of charges fired and roasted was an increase of 57 charges.

The quantity of ore of all qualities roasted was 23,454,000 pounds, an increase of 6,123,625 pounds over the work of 1873.

There were consumed in the operations of the furnaces 3,296 cords of wood, at an average cost per charge of $78.30, or a cost per carga of 24 cents; and the cost for each flask of quicksilver produced was $2.09. Coal, charcoal, and coke were also burned, and their cost added to the cost of wood made the cost of fuel per carga of ore roasted 25.65 cents.

The totals of the hacienda pay-rolls and the value of wood burned made the cost of roasting ores per charge $327; per carga, $1.01, and per flask of quicksilver produced, $8.71.

A comparison of these costs with those for 1873 shows an increase, as below, caused by the lower grade of the ore, and consequently the greater quantity worked to produce a less quantity of quicksilver:

| | 1873. | 1874. | |
|---|---|---|---|
| Wood, average cost per charge | $76 64 | $78 30 | $1 66 increase. |
| Wood, average cost per carga | 0 25 | 0 24 | 01 decrease. |
| Wood, average cost per flask | 1 28 | 2 09 | 81 increase. |
| Wood and pay-rolls, cost per charge | 310 75 | 327 00 | 25 25 increase. |
| Wood and pay-rolls, cost per carga | 97 | 1 01 | 04 increase. |
| Wood and pay-rolls, cost per flask | 5 06 | 8 71 | 3 65 increase. |

On the 1st of January, 1874, there were on hand at the furnaces 546¼ cargas of mine-ore, and 55,713 cargas of tierras; a total of 56,259¼ cargas.

During the year there were produced 19,139¾ cargas of mine-ore, 19,380 cargas of terrero, and 79,607 cargas of tierras, which, added to the quantity on hand, made a total of 174,386½ cargas. Of this quantity, there were roasted in the furnaces 19,371 cargas of mine-ore, 18,277 cargas of terrero, and 40,532 cargas of tierras; making the quantity of all ores reduced an aggregate of 78,180 cargas, or 23,454,000 pounds.

These ores gave an average yield of 2$\frac{89}{100}$ per cent., and produced 678,325¼ pounds of quicksilver, equal to 8,867 flasks of 76½ pounds each.

For the first three months of the year the tierras were estimated at 2 per cent., and for the remainder, on account of their poor quality, they were taken at 1¼ per cent.

The percentage of mine-ore and terrero, excluding the product of tierras, was 4$\frac{33}{100}$ per cent., or 3$\frac{57}{100}$ per cent. less than the true percentage of 1873, while the percentage of all ore, 2$\frac{89}{100}$, exhibits a decrease of 1$\frac{33}{100}$ per cent.

The surplus of ores on hand December 31, 1874, was, in cargas, as follows: Mine-ore, 314¾ cargas; terrero, 1,103; and tierras, 100,395; a total of 101,812¾ cargas, a quantity sufficient to supply our present furnaces for one year.

The product of quicksilver was—

| | Flasks. |
|---|---|
| From the furnaces | 8,867 |
| From sluicings and washings | 217 |
| Total | 9,084 |

And its cost, as shown by expenditures for—

| | |
|---|---|
| Pay-rolls, including repairs | $389,190 69 |
| Materials and supplies, including repairs | 72,743 18 |
| Miscellaneous expenses and taxes | 24,091 38 |
| Miscellaneous property | 1,357 06 |
| | 487,382 31 |
| Less ore-account increased | 68,445 42 |
| Amounted to | 418,936 89 |

Average cost per flask, $46.11, or $14.69 more than the average cost per flask in 1873.

During the same period the selling-price of quicksilver increased $26.77½ per flask.

A more economical and improved method of reduction and condensation has been the object of our constant study and experiment. The results, as evidenced by the working of the new iron condensers and those made of wood and glass, are satisfactory, and the improved method of extracting quicksilver from the soot found in the condensers has also worked well.

It is expected that the new iron-clad furnace, to be put in operation in February, will reduce a large quantity of low-grade ore at a very much diminished expense for fuel and labor.

The question of an abundant supply of fuel for the mines and hacienda has engaged my serious attention, and is now happily set at rest by the late purchase of 757 acres of woodland lying within three miles of Mine Hill, and distant one and one-half miles from the Enriqueta mine. Competent judges estimate the quantity of wood to be not less than 40,000 cords, and, with proper management in the cutting and preservation of the timber on these lands, it is certain that we shall have all the fuel that may be required for many years.

*Tabular statement showing number of cargas and tons of ore of all qualities produced from the New Almaden mine in* 1874.

| Months. | Ore, cargas. | Terrero, cargas. | Tierras, cargas. | Total cargas. | Total in tons and pounds. | |
|---|---|---|---|---|---|---|
| January | 1,653½ | 1,075 | 3,828 | 6,556½ | 983 | 950 |
| February | 1,197 | 1,151¼ | 4,850 | 7,198¼ | 1,079 | 1,550 |
| March | 1,607 | 1,118 | 5,610 | 8,335 | 1,250 | 500 |
| April | 1,372 | 1,179 | 5,643 | 8,194 | 1,229 | 200 |
| May | 1,443 | 1,166 | 8,210 | 10,819 | 1,622 | 1,700 |
| June | 1,297 | 938⅓ | 11,038 | 13,273⅓ | 1,991 | 50 |
| July | 1,796¾ | 1,104½ | 8,725 | 11,626¼ | 1,743 | 1,875 |
| August | 1,447¼ | 977¾ | 7,026 | 10,451 | 1,552 | 1,300 |
| September | 1,738½ | 2,391 | 8,169 | 12,298½ | 1,844 | 1,550 |
| October | 1,761 | 2,781 | 9,058 | 13,600 | 2,040 | ...... |
| November | 1,701 | 2,665 | 5,877 | 10,243 | 1,536 | 900 |
| December | 2,125½ | 2,833 | 6,280 | 11,238½ | 1,685 | 1,550 |
| Total | 19,139¾ | 19,380 | 85,214 | 123,733¾ | 18,560 | 125 |

*Tabular statement showing number of cargas and tons of ore of all qualities reduced and flasks of quicksilver produced at New Almaden mine, 1874.*

| Months. | Ore, cargas. | Terrero, cargas. | Tierras, cargas. | Total cargas. | Total in tons and pounds. | | Per cent. average. | True per cent. average. | Number of flasks. |
|---|---|---|---|---|---|---|---|---|---|
| January | 1,598 | 1,075 | 2,313 | 4,986 | 747 | 1,800 | 3.45 | 4.77 | 675 |
| February | 1,235½ | 1,131½ | 2,103 | 4,490 | 763 | 1,000 | 3.83 | 5.44 | 675 |
| March | 1,908 | 918 | 2,843 | 5,669 | 850 | 700 | 3.37 | 4.75 | 750 |
| April | 1,371 | 1,182 | 2,990 | 5,543 | 831 | 900 | 3.11 | 5.00 | 735 |
| May | 1,534 | 1,363 | 2,723 | 5,620 | 843 | ...... | 3.25 | 4.91 | 729 |
| June | 1,139 | 534 | 4,620 | 6,293 | 943 | 1,900 | 2.60 | 5.61 | 661 |
| July | 1,478 | 1,158 | 4,524 | 7,160 | 1,074 | ...... | 2.69 | 4.72 | 775 |
| August | 1,451½ | 968½ | 3,760 | 6,180 | 927 | ...... | 3.79 | 7.34 | 930 |
| September | 2,033 | 1,277 | 3,460 | 6,770 | 1,015 | 1,000 | 2.86 | 4.27 | 820 |
| October | 2,073 | 2,807 | 3,563 | 8,443 | 1,266 | 900 | 2.59 | 3.38 | 857 |
| November | 1,733 | 2,927 | 3,690 | 8,350 | 1,252 | 1,000 | 2.16 | 2.68 | 737 |
| December | 1,817 | 2,916 | 3,943 | 8,676 | 1,301 | 800 | 2.16 | 2.71 | 740 |
| Total | 19,371 | 18,277 | 40,532 | 78,180 | 11,727 | ...... | 2.88 | 4.29 | 9,084 |

*Tabular statement showing product of furnaces for 1874.*

| Months. | No. of charges. | Class and quantity of ore. | | | Total quantity of ore in pounds. | Yield of quicksilver. | | |
|---|---|---|---|---|---|---|---|---|
| | | Grueso. | Granza. | Tierras. | | Per cent. average. | True per cent. of average. | Number of flasks. |
| | | *Pounds.* | *Pounds.* | *Pounds.* | | | | |
| January | 15 | .......... | 802,000 | 694,000 | 1,496,000 | 3.45 | 4.77 | 675 |
| February | 14 | .......... | 716,000 | 631,000 | 1,347,000 | 3.83 | 5.44 | 675 |
| March | 18 | .......... | 848,000 | 853,000 | 1,701,000 | 3.37 | 4.75 | 750 |
| April | 18 | .......... | 766,000 | 897,000 | 1,663,000 | 3.11 | 5.00 | 677 |
| May | 21 | .......... | 869,000 | 817,000 | 1,686,000 | 3.25 | 4.91 | 718 |
| June | 20 | .......... | 502,000 | 1,386,000 | 1,888,000 | 2.60 | 5.61 | 640 |
| July | 22 | .......... | 791,000 | 1,357,000 | 2,148,000 | 2.69 | 4.72 | 755 |
| August | 19 | .......... | 726,000 | 1,128,000 | 1,854,000 | 3.79 | 7.34 | 918 |
| September | 20 | .......... | 993,000 | 1,038,000 | 2,031,000 | 2.86 | 4.27 | 759 |
| October | 25 | .......... | 1,464,000 | 1,069,000 | 2,533,000 | 2.59 | 3.38 | 857 |
| November | 24 | .......... | 1,398,000 | 1,107,000 | 2,505,000 | 2.16 | 2.68 | 707 |
| December | 26 | .......... | 1,419,000 | 1,183,000 | 2,602,000 | 2.16 | 2.71 | 736 |
| Total | .... | .......... | 11,294,000 | 12,160,000 | 23,454,000 | 2.88 | 4.29 | 8,867 |
| From sluicings | | | | | | | | 217 |
| Total product | | | | | | | | 9,084 |

The charges for the improvement-account were $45,318.98, and the following is a summary of the work:

For the mine:

| | |
|---|---|
| Ingersoll drills and compressor, foundations for and setting up machinery, fittings, pipes, &c | $5,492 21 |
| Burleigh drills and compressor, boiler, foundations for and setting up machinery, pipes, fittings, &c | 11,140 00 |
| Boiler, pumping-gear, &c., at Randol shaft | 3,181 64 |
| Boiler and compressor house, smith's shop, and magazine at the Bottom tunnel | 1,144 48 |
| Bridge at Bottom tunnel | 2,104 86 |
| Roads to the same | 272 88 |
| Smith's shop, planilla-shed, chute, and incline at Cora Blanca | 504 14 |
| Road to Deep Gulch tunnel | 493 57 |
| Shaft-house, smith's shop, and magazine at Enriqueta | 477 40 |
| Houses on Mine Hill | 1,656 55 |

For the hacienda:

| | |
|---|---|
| New furnace of 1873 | 1,445 37 |
| Tower at furnaces Nos. 1 and 2 | 215 78 |
| Shed over and near furnaces Nos. 3 and 4 | 3,972 50 |
| Condenser of No. 5 furnace | 4,470 90 |
| Prolongation of flues of furnaces 3, 4, and 6 | 121 85 |
| Iron-clad furnace | 4,307 74 |
| Bridge in furnace-yard | 426 95 |

| | |
|---|---|
| Tool-house and shed in furnace-yard | $350 03 |
| Grading adobe-yard | 445 96 |
| Reservoirs, tanks, flumes, and fences | 1,405 54 |
| Railroad and chutes | 1,137 26 |
| Protection against fire | 549 37 |
| Total | 45,318 98 |

Of this there were expended—

| | |
|---|---|
| For labor | $16,144 55 |
| For materials | 29,174 43 |
| | 45,318 98 |

Mr. F. Fiedler, the foreman of the hacienda, reports as follows:

The following is a statement of improvements and alterations made during the year 1874 for the Quicksilver Mining Company under your orders and directions.

A new bridge 91 feet in length, crossing the Alamitos Creek in the furnace-yard, was constructed to facilitate the transportation of adobes, wood, and other materials necessary for the operation of furnaces Nos. 1 and 2, situated on the opposite side of the creek, and heretofore entirely isolated from the principal part of the reduction-works. This structure was a great necessity, not only in putting all the furnaces in close communication, but also in saving expenses and time, as formerly all materials necessary for these two furnaces had to be carried over a bridge outside of the reduction-yard by a long and circuitous route. Furnaces Nos. 3 and 4, their condensers, and a large space of ground in the rear formerly used for the storage of adobes, are the middle of a block, whose outside lines are formed by Nos. 5 and 6 furnaces, respectively, and were covered by a series of small roofs of all dimensions put up at different times, as necessity required them. These had some time ago become entirely dilapidated, allowing the rain to drift in everywhere, entirely unfitting the free space in the rear of the furnaces for the storing of adobes or other materials liable to be destroyed by water. For this reason these roofs were torn down and replaced by one roof in two joining sections, at a cost of $3,972.50, covering a space of 121 feet in length and 132 feet in breadth, giving ample additional room for storage of adobes, lumber, and other materials.

The mine not yielding ore enough to insure a continuous working of all furnaces with granza, terrero and screenings from tierras had to be substituted to a large amount. These latter, being in smaller particles, require more flues and channels of adobes to obtain their complete combustion and roasting, and therefore a greater amount of adobes had to be made during the summer-months to keep a sufficient supply on hand for the rainy season. The facilities for making and drying adobes in the different yards were limited, and a new yard on the banks of the creek was made by filling in a large space of ground, 150 feet in length and 200 feet in width, with the slag of the furnaces and the soil of the adjacent hills. The house erected for the reception of all working-utensils in use, and serving as store-house for a weekly supply of oils and all sorts of materials required for the carrying on of the general business, was destroyed by fire and replaced by a building nearly twice the former size. This store-house has been put up entirely away from other buildings, and should it burn a second time, will not cause any greater loss than its own destruction. A reservoir has been built in the hills south of the hacienda, of a sufficient capacity to supply water during the summer-months to different yards used for the molding and drying the adobes, these yards having been furnished heretofore only partially with water.

The re-opening of work at the Bottom tunnel necessitated, 1st. The construction of a bridge across the Alamitos Creek, for the transportation of the machinery and boilers, wood, powder, steel for drills, &c. This bridge, put up in a most substantial manner, with a track of T-rail the entire distance, has a length of 415 feet and crosses the creek at a height of 60 feet. 2d. The building of a house for the reception of the boiler and machinery of Burleigh's air-compressor. 3d. The erection of a blacksmith's shop. 4th. The construction of a powder-house. 5th. The building of wagon-roads to deliver all materials at the landing of the bridge.

The increased amount of tierras received from the mine during the past year, all of which have to pass over a screen into a chute to separate the larger rock from the tierras proper, proved the chute erected for that purpose entirely inadequate to perform this work, and a new chute was erected with two compartments, and a screen similar in its construction to the one put up the year previous, requiring the extension of the main railroad-track for a distance of 144 feet, all built on high trestle. Two new sheds were also built for the storage of pipes, iron, and other materials liable to suffer from exposure. A new boiler was placed in position at the Randol shaft and a new pump provided for this place. Many other improvements, appearing small if considered separately by themselves, but representing together a great amount of labor, have been made, to facilitate the operations in the reduction-works and to lessen the expenses.

Four wooden condensers were built for No. 5 furnace, each one being 22 feet in length, 15 feet high, and 8 feet wide, containing an aggregate of 10,650 cubic feet of condensing-space. This furnace was also supplied with a cast-iron condensing-tank of the same pattern as the sheet-iron tank purchased the previous year. The new shaft-furnace, built after the Page patent, not proving as beneficial as anticipated, the discharge of the ore being irregular, was partially torn down and rebuilt into an improved old-style furnace, used now exclusively for the burning of adobes, the capacity of the furnace being 111,000 pounds adobes per charge, or 444,000 pounds per month. The condensing-chambers, being the same formerly used for the original No. 2 furnace, consisting of a solid block of eighteen chambers, were divided into two blocks by cutting out one condensing-chamber, and thereby obtaining a better circulation of air, not only on the outside walls, but also through the lower arches and passage-ways on which these condensers are built. A wooden tower with an upcast and a downcast shaft was erected for Nos. 1 and 2 furnaces, through which the fumes of these two furnaces have to pass after leaving the last condensing-chamber. The downcast shaft is supplied from the top of the tower with a spray of water, which, falling through a series of triangular wooden cross-pieces inserted into the sides of the shaft in alternate opposite directions, is divided into minute particles, carrying along all fumes from the condensers and precipitating all metallic fumes reaching this point.

A continuous furnace, built nearly according to plans of a furnace in operation at the Idria mine in Austria, Europe, is about completed. This furnace being incased in an armor of iron, and having three fire-doors and three discharging-places, will not only insure a complete and even roasting and discharging of the ores, but will also prevent any escape of valuable fumes from the ore-chamber through its walls, besides being of great durability, nothing but slight repairs every two or more years being necessary to keep it in constant operation. Two brick towers and two large brick condensers are now being constructed for this furnace. From these the fumes will pass into No. 5 brick condensers, and then into the four wooden condensers mentioned before. This leaves No. 5 furnace with only two iron condensers, and requires the building of a new series of condensing-chambers for the same, of which two of brick are now completed, and three of wood and glass, being each 12 feet wide, 12 feet long, 20 feet high, with 34 large windows, and divided into four chambers, will be finished by the end of January, 1875. A new flume 1,000 feet in length, of Oregon pine, and leading to a central chimney, whose location has been determined upon, has been commenced. This flue will receive all the fumes from Nos. 3, 4, and 6 furnaces after leaving their present outlet or chimneys, and will be pushed to completion, the weather permitting, with all possible speed. This central chimney will give a better and more even draught to the furnaces, saving all fuel now consumed continually in the draught-fire chimneys. Thirty-one new openings like those reported the previous year have been made in the old brick condensers, and all the old ones have been remodeled and better fitted for the purpose intended, namely, additional help for cooling the mercurial vapors.

The soot deposited in the condensers and collected after each charge contains a large percentage of quicksilver in minute particles. The oil and other fatty substances generated through the combustion of fuel and the roasting of ores adhering to these diminutive globules, prevents their uniting. The quicksilver obtained heretofore by washing the soot in a long row of sluice-boxes formed only a part of the amount contained in this rebellious mass, the residue being partially carried away by the water and the balance saved by returning the soot to the ore-chambers and roasting it again. By the erection of four iron tanks and the application of Wright's process, which was secured for this company, all loss of quicksilver has been avoided, and the soot as obtained from the condensers has proved of no further annoyance.

The secretary, Mr. David Mahany, furnishes the following statement of the business of the Quicksilver Mining Company for 1874:

| | Dr. | | Cr. |
|---|---|---|---|
| To quicksilver and ore on hand December 31, 1873, per last report, and cost of quicksilver and ore produced and mined in 1874 | $663,274 09 | By balance to the credit of income account, December 31, 1873 | $960,437 77 |
| To legal expenses | 1,985 85 | By sales of quicksilver | 893,670 72 |
| To interest on funded debt | 70,000 00 | By rents and privileges | 22,123 26 |
| To taxes | 6,056 89 | By materials and property sold | 1,652 27 |
| To exchange | 2,025 00 | By interest on call-loans, &c | 20,566 80 |
| To convertible-bond stock | 70,000 00 | By premium on gold sold | 40,962 05 |
| To claims adjusted | 53,647 00 | By ore on hand | 159,312 09 |
| To general expenses | 11,372 19 | By quicksilver on hand | 97,650 00 |
| To balance to the credit of income account, January 1, 1875 | 1,318,013 44 | | |
| | 2,196,374 96 | | 2,196,374 96 |

*Balance-sheet, December* 31, 1874.

| | Dr. | | | Cr. |
|---|---|---|---|---|
| Convertible-bond stock | $871,000 00 | Capital stock, (preferred) | $4,291,300 | |
| Real estate and mining property | 11,047,875 60 | Capital stock, (common) | 5,708,700 | |
| Railroads | 79,853 33 | | | $10,000,000 00 |
| Houses and lands | 113,220 00 | Second-mortgage bonds | | 1,000,000 00 |
| Furnaces | 117,500 00 | Income account | | 1,318,013 44 |
| Virginia City property | 3,000 00 | | | |
| Furniture, hacienda, &c | 4,000 00 | | | |
| Machinery and tools | 56,524 61 | | | |
| James B. Randol, manager | 15,364 99 | | | |
| Quicksilver on hand | 97,650 00 | | | |
| Ore account | 159,312 09 | | | |
| Materials and supplies | 85,000 87 | | | |
| Permanent improvements | 16,144 55 | | | |
| E. N. Robinson, tr., (loans on call) | 422,983 19 | | | |
| Miscellaneous property | 12,584 21 | | | |
| Woodlands | 16,000 00 | | | |
| | 12,318,013 44 | | | 12,318,013 44 |

*The following tabular statement of the production of quicksilver at New Almaden for twenty-two years and three months was furnished to the Mining and Scientific Press by Mr. Randol, the manager at New Almaden.*

| Date. | Class and quantity of ore. | | | | From furnaces. | From washings. | Total. | Average amount per month. | Percentage, including all. | Percentage. | True per cent. of ore, excluding tierras and washings. | No. of months. |
|---|---|---|---|---|---|---|---|---|---|---|---|---|
| | Grueso. | Granza. | Tierras. | Total. | | | | | | | | |
| | *Pounds.* | *Pounds.* | *Pounds.* | *Pounds.* | *Flasks.* | *Flasks.* | *Flasks.* | *Flasks.* | | *Tierras.* | | |
| July, 1850, to June, 1851 | ........ | ........ | ........ | 4, 970, 717 | 23, 875 | ........ | 23, 875 | 1, 989½ | 36. 74 | ........ | 36. 74 | 12 |
| July, 1851, to June, 1852 | ........ | ........ | ........ | 4, 643, 290 | 19, 921 | ........ | 19, 921 | 1, 660 | 32. 82 | ........ | 32. 82 | 12 |
| July, 1852, to June, 1853 | ........ | ........ | ........ | 4, 839, 520 | 18, 035 | ........ | 18, 035 | 1, 503 | 28. 50 | ........ | 28. 50 | 12 |
| July, 1853, to June, 1854 | ........ | ........ | ........ | 7, 448, 000 | 26, 325 | ........ | 26, 325 | 2, 193¾ | 27. 03 | ........ | 27. 03 | 12 |
| July, 1854, to June, 1855 | ........ | ........ | ........ | 9, 109, 300 | 31, 860 | ........ | 31, 860 | 2, 655 | 26. 75 | ........ | 26. 75 | 12 |
| July, 1855, to June, 1856 | ........ | ........ | ........ | 10, 355, 200 | 28, 083 | ........ | 28, 083 | 2, 340¼ | 20. 74 | ........ | 20. 74 | 12 |
| July, 1856, to June, 1857 | ........ | ........ | ........ | 10, 299, 900 | 26, 002 | ........ | 26, 002 | 2, 167 | 19. 31 | ........ | 19. 31 | 12 |
| July, 1857, to June, 1858 | ........ | ........ | ........ | 10, 997, 170 | 29, 347 | ........ | 29, 347 | 2, 445¾ | 20. 41 | ........ | 20. 41 | 12 |
| July, 1858, to October, 1858 | ........ | ........ | ........ | 3, 873, 085 | 10, 588 | ........ | 10, 588 | 2, 647 | 20. 91 | ........ | 20. 91 | 4 |
| November, 1858, to January, 1861, (closed by injunction.) | ........ | ........ | ........ | ........ | ........ | ........ | ........ | ........ | ........ | ........ | ........ | ... |
| February, 1861, to January, 1862 | ........ | ........ | ........ | 13, 323, 200 | 32, 402 | 2, 363 | 34, 765 | 2, 897 | 19. 96 | ........ | 18. 64 | 12 |
| February, 1862, to January, 1863 | ........ | ........ | ........ | 15, 281, 400 | 39, 262 | 1, 129 | 40, 391 | 3, 366 | 20. 22 | ........ | 19. 65 | 12 |
| February, 1863, to August, 1863 | ........ | ........ | ........ | 7, 172, 660 | 17, 316 | 2, 248 | 19, 564 | 2, 795 | 20. 86 | ........ | 18. 46 | 7 |
| September, 1863, to October, 1863 | ........ | ........ | ........ | 2, 346, 000 | 4, 820 | 700 | 5, 520 | 2, 700 | 18. 00 | ........ | 15. 67 | 2 |
| November, 1863, to December, 1863 | 54, 800 | 1, 586, 500 | 718, 000 | 2, 359, 300 | 4, 040 | 407 | 4, 447 | 2, 223½ | 18. 65 | 3 | 17. 52 | 2 |
| January, 1864, to December, 1864 | 1, 259, 400 | 18, 730, 300 | 3, 287, 900 | 23, 277, 600 | 42, 176 | 313 | 42, 489 | 3, 540¾ | 13. 96 | 3 | 15. 64 | 12 |
| January, 1865, to December, 1865 | 2, 288, 900 | 25, 749, 000 | 3, 910, 500 | 31, 948, 400 | 47, 078 | 116 | 47, 194 | 3, 933 | 11. 30 | 3 | 12. 42 | 12 |
| January, 1866, to December, 1866 | 1, 506, 000 | 19, 939. 100 | 5, 440, 200 | 26, 885, 300 | 34, 726 | 424 | 35, 150 | 2, 929 | 10. 00 | 3 | 11. 62 | 12 |
| January, 1867, to December, 1867 | 731, 500 | 15, 689, 288 | 9, 603, 145 | 26, 023, 933 | 23, 990 | 471 | 24, 461 | 2, 038¼ | 7. 19 | 3 | 9. 42 | 12 |
| **January, 1868, to December, 1868** | 2, 274, 208 | 14, 566, 600 | 12, 564, 722 | 29, 405, 530 | 25, 577 | 51 | 25, 628 | 2, 135⅔ | 6. 66 | 2 | 10. 12 | 12 |
| **January, 1869, to December, 1869** | 150, 000 | 11, 942, 175 | 13, 366, 000 | 25, 458, 175 | 16, 898 | ........ | 16, 898 | 1, 408 | 5. 07 | 2 | 8. 48 | 12 |
| **January, 1870, to December, 1870** | 30, 000 | 12, 531, 900 | 8, 535, 800 | 21, 097, 700 | 14, 423 | ........ | 14, 423 | 1, 202 | 5. 23 | 2 | 7. 42 | 12 |
| **January, 1871, to December, 1871** | ........ | 13, 661, 700 | 8, 373, 000 | 22, 034, 700 | 18, 563 | 5 | 18, 568 | 1, 547½ | 6. 44 | 2 | 9. 16 | 12 |
| **January, 1872, to December, 1872** | 142, 000 | 12, 777, 000 | 8, 497, 600 | 21, 416, 600 | 18, 391 | 183 | 18, 574 | 1, 548 | 6. 63 | 2 | 9. 57 | 12 |
| **January, 1873, to December, 1873** | ........ | 8, 492, 375 | 8, 838, 000 | 17, 330, 375 | 11, 042 | ........ | 11, 042 | 920 | 4. 87 | 2 | 7. 86 | 12 |
| **January, 1874, to December, 1874** | ........ | 11, 294, 000 | 12, 160, 000 | 23, 454, 000 | 8, 867 | 127 | 9, 804 | 817 | 2. 96 | 1. 62½ | 4. 39 | 12 |
| **Total** | 8, 436, 808 | 166, 959, 438 | 95, 294, 867 | 375, 351, 655 | 573, 607 | 8, 537 | 582, 954 | 2, 183 | 11. 88 | 2. 38 | 14. 80 | 267 |

**Product of Enriqueta from 1860 to 1863, 10,571 flasks.**
**Total product of all the mines on the company's property 598,525 flasks, of 76½ pounds each, or 45,787,162½ pounds.**

## MENDOCINO COUNTY.

This county, situated in the northwestern part of California, has an area of about five thousand square miles, and is bounded on the west by the Pacific Ocean and on the east by the Coast range. This position gives it the benefit of more abundant rains than are enjoyed by the interior plain of the State. Small mountain-torrents from the Coast range traverse the fertile land between the mountains and the sea. Wild oats and clover grow in abundance; sugar-pine, yellow pine, fir, bay, pepper, white-oak, and iron-oak are the principal trees; the madroña and manzanita run wild; vegetation is luxuriant; the climate is healthy and agreeable; and a large portion of the county affords agricultural and grazing land of unsurpassed quality. The population of the county is scanty, but is reported to be rapidly increasing. Ocean communication with San Francisco is maintained by means of sloops and light-draught vessels, such as are able to enter the shallow harbors of this portion of the coast.

An especial interest attaching to this county at the present time is derived from the recent discovery within its borders of a coal-bed of excellent quality. For information concerning it, I am indebted to Mr. Augustus J. Bowie, jr., of San Francisco.

The Mount Vernon coal-mine is situated on both sides of Middle Eel River, (a mountain-torrent,) in townships 21 and 22 north, range 13 west, Mendocino County, California, and lies northwesterly from Ukiah, distant, in an air-line, thirty-eight miles, and about sixty miles by the county-road, which has been constructed from Ukiah City to Round Valley, a Government military station. This road, at its crossing of Middle Eel River, passes within two miles of the Mount Vernon mine, and there is a good trail along the river-side from the road to the property. Noyo is its nearest sea-port. A circuitous road forty-two miles long connects the two places. The projected line of the North Pacific Railroad (a California enterprise) passes within four miles of the mine, crossing at the junction of Middle and South Eel Rivers. The property consists of 7,200 acres of land.

No work of importance has as yet been done to develop the resources of this coal-field. The value of the property can only be prospectively estimated. The coal has been traced six miles. The land is admirably located with reference to the strike and dip of the coal, as shown by its various outcroppings. All the property on which the coal is exposed and the land into which it dips are covered by United States patents.

Commencing at a point about one mile south 35° east from the coal-outcrop in the river, the coal first makes its appearance in a creek on the side of a steep hill, covered by a thin layer of shell limestone, broken up and mixed with decomposed shale and soil colored red with ferruginous matter. Tracing along the line of the croppings north 35° west to the banks of the river, over undulations caused by land-slides, which gradually flatten as the stream is approached, and in which detritus and bowlders of all sizes lie scattered in profusion, an immense body of coal is found, entirely denuded of all its encompassing strata. For nearly 600 feet in length, with a height of 14 feet, a body of coal is exposed, forming an abrupt bluff, over which the water runs, occasionally detaching huge masses of coal from the outcrop and hurling them down the stream.

The strike of the bed, taken from this main outcrop, is north 40° 50′ west, (magnetic;) the dip, 19° north 47½° east. The thickness of the strata as here exposed is about 20 feet. High water prevented the accurate measurement of the bed. Actual measurement, where practicable, showed 14 feet of coal in the stream, and a calculation of the thickness of this bed from the encompassing walls showed it to have been 20 feet through, from 5 to 6 feet having been washed away. These immense croppings extend across the river, a width of 350 feet. The banks on both sides being low, the coal beyond the river gradually enters the detritus and soil, and for a distance of several hundred feet southeast lies only a few feet under ground. A cross-cut made on the surface 400 feet from the river, on the northwest side, is said to have shown a solid stratum of coal 20 feet thick. The coal in the river lies between two strata of gray argillaceous shale. The edges of the upturned strata have slacked and assumed a reddish-brown color. The shale underlying the main coal-bed contains several seams of coal, varying in size from 1 to 8 inches. The exposed thickness at the surface of the shales underlying the coal is about 19 feet, the dip being 21°. The shale is succeeded by a thin bed composed of fossil oyster-shells, some of which are very large. A soft yellow and brown sandstone, partially disintegrated, underlies it.

An examination of the overlying shale on the southeast side of the river, below high-water mark, and a foot below the surface, showed it to be of grayish color, hard, compact, and of such a nature as will form a strong and solid roof, impervious to water, and a great protection for the future working of the mine. Overlying the shale is sandstone, soft and totally disintegrated, and along the banks of the river detritus and metamorphosed rock overlie the sandstone.

On the northwest side of the river there is a steep and precipitous bluff of metamorphosed rock. It is evidently of a later formation than the coal, and is rapidly disintegrating and falling to pieces.

Directly below this bluff, traced up a ravine, (northwest,) the coal shows itself at the surface, covered here and there by black dirt several inches thick. Turning to the west of north, and descending the hill, leaving the bluff to the north, and crossing a small plateau made by a land-slide, another creek is reached, along the sides of which coal-croppings are found. This coal is almost in a direct line with the strike of the coal on the opposite or southeast side of the river, but its altitude is not as great, nor is there any appearance of limestone. The encompassing strata are decomposed shales, sandstone, wash, and soil with a ferruginous tinge. On account of the many land-slides, the edges of the encompassing shales are not visible for any distance after leaving the river. Three-quarters of a mile northwest of the river-croppings, under an abrupt bank, (the sides of which have been cut and washed away,) at an altitude of about 200 feet, a bed of coal 6 feet thick has been exposed. A small shaft was started alongside of this bank, but, after sinking a few feet, work was stopped on account of the water. The height of the bank over the coal is about 75 feet. It is composed of a reddish clay and disintegrated sandstone, 75 feet thick, underlying which is detritus and conglomerate 5 feet thick. Clinging to the face of the conglomerate is the coal.

Three hundred feet above these croppings, about a mile and a half northwest of the river-croppings, coal again shows itself in a mountain creek issuing from an abrupt bank of columnar sandstone about 60 feet high. Black dirt and fragments of coal on all sides, imbedded in a whitish clay, indicate the presence of coal. No defined stratum has here been found; nor has the ground been prospected. Crossing to the north the ridge of sandstone, shale, and detritus, and descending on the opposite side of the hill, a distance of four and a half miles from the croppings in the river, coal again makes its appearance on the side of a creek. A cut was once started on it, but it is now caved in. The neighborhood has always supplied itself with coal from this bed. Northwesterly for half a mile the coal can still be traced.

The coal is hard and bituminous. It burns freely and with a bright flame, gives a good compact coke, and leaves but little ash. A proof of its purity is the circumstance that the water which pours over it from the springs remains pure and drinkable. The coal gives a strong heat,* and will prove an excellent coal for generating steam, producing gas,† and for domestic use.

Although exposed to the action of the air and water, it does not slack to any great extent, but remains remarkably hard and compact. From repeated experiments with it in large quantities, a mere trace of sulphur is perceptible. Annexed is an analysis of the coal, made by Mr. L. Falkenau, State assayer:

| | |
|---|---|
| Specific gravity | 1.282 |
| Volatile combustible substance | 40.20 per cent. |
| Fixed carbon | 49.70 per cent. |
| Moisture | 6.70 per cent. |
| Ashes | 3.00 per cent. |
| Sulphur | 0.40 per cent. |

Amount of gas evolved, 37 cubic feet per 10 pounds avoirdupois of the coal.

The following tabular statement of coal analyses, compiled by Mr. Falkenau, is given by Mr. Bowie, as showing the rank occupied by the Mount Vernon coal:

| Designation. | Specific gravity. | Volatile combustible matter in 100 parts. | Fixed carbon in 100 parts. | Earthy matter or ashes in 100 parts. | Water in 100 parts. |
|---|---|---|---|---|---|
| Lehigh, Pennsylvania | 1.590 | 5.28 | 89.15 | 5.56 | ........ |
| Cumberland, A. and T. | 1.313 | 15.53 | 76.69 | 7.33 | ........ |
| Sydney, New South Wales | 1.338 | 23.81 | 67.57 | 5.49 | ........ |
| Newcastle, England | 1.257 | 35.83 | 57.00 | 5.40 | ........ |
| Sagahalia, Siberia | 1.288 | 35.70 | 56.45 | 6.05 | 1.80 |
| Cannel, Lasmaho, Scotland | 1.228 | 49.34 | 40.97 | 6.34 | 2.00 |
| Mount Vernon, Mendocino, Cal | 1.282 | 40.20 | 49.70 | 3.00 | 6.70 |
| Black Diamond, Mount Diablo, Cal | ........ | 33.89 | 46.84 | 4.85 | 14.69 |
| Bellingham Bay, Washington Territory | ........ | 33.26 | 45.69 | 12.66 | 8.39 |
| Nanaimo, Vancouver's Island | ........ | 32.16 | 46.31 | 18.55 | 2.98 |
| Coos Bay, Oregon | ........ | 32.59 | 41.98 | 5.34 | 20.09 |

From these analyses it is evident that the coal of Mendocino County is a lignite, and there is no reason to doubt that, like the other coals of this variety on our Pacific coast, it is of recent geological age. The comparatively small percentage of water and ash, and the high proportion of fixed and volatile combustible matter, if maintained throughout the bed, will establish the quality of this coal as superior to others found upon the coast. Transportation to San Francisco being, however, at present impracticable, it will be several years, doubtless, before any use is made of this coal. Mr. Bowie intimates that a narrow-gauge railroad to Noyo, in place of the present circuitous wagon-road of forty-two miles, will be necessary, unless the completion of the North Pacific Railroad, now projected, will permit land-transportation at reasonable rates.

* Lead reduced from litharge, 22.89; carbon corresponding to volatile matter, 15.72 units of heat, 54.40.

† Yield of gas per ton of coal, 11,250 cubic feet; illuminating-power of gas, 15.3 candles; carbonic acid, 7 per cent.—London analysis.

## VENTURA COUNTY.

*Description of leading mines, Ventura County, California, 1874.*

| Name. | Location. | Owners. | Length of location. | Course. | Dip. | Length of pay-zone. | Average width. | Country-rock. | Character of vein-matter. | Worked by tunnel or shaft. | Length of tunnel. | Depth on vein in tunnel. | Depth of working-shaft. | Number of levels opened. | Total length of drifts. | Cost of hoisting-works. |
|---|---|---|---|---|---|---|---|---|---|---|---|---|---|---|---|---|
| | | | Feet. | | | Feet. | Feet. | | | | Feet. | Feet. | Feet. | Feet. | Feet. | |
| Castac | Snowy | Treadwell & Sewart | 1,500 | N. 20° E. | E. | 110 | 3½ | Slate | | Tunnel | 60 | 230 | | 2 | 540 | |
| Snowy | do | do | 1,500 | N. 15° W. | E. | 80 | 3 | Slate and granite | | do | 190 | 85 | | 1 | | |

*Operations of leading mines, Ventura County, California, 1874.*

| Name. | Location. | Owners. | Number of miners employed. | Miners' wages per day. | Cost of sinking per foot. | Cost of drifting per foot. | Cost of stoping per ton. | Cost of mining per ton extracted. | Cost of milling per ton. | Company or custom mill. | Number of tons extracted and worked. | Average yield per ton—gold. | Percentage of sulphurets. | Total bullion-product. |
|---|---|---|---|---|---|---|---|---|---|---|---|---|---|---|
| Snowy | Snowy | Treadwell & Sewart | 4 | $2 50 | $10 00 | $6 00 | $0 50 | $0 65 | $0 75 | Custom | 700 | $19 25 | | $13,475 |
| Castac | do | do | 8 | 2 50 | 13 00 | 7 00 | 70 | 90 | 75 | Company | 900 | 20 40 | | 18,360 |

*Statement of quartz mills, Ventura County, California, 1874.*

*Snowy district.*—Name of mill, Castac; Owner, J. B. Treadwell; kind of power and amount, water, 20 horse; number of stamps, 4; weight of stamps, 600 pounds; number of drops per minute, 80; height of drop, 8½ inches; number of pans, 1 (Hepburn;) number of concentrators, 1; cost of mill, $4,000; capacity per 24 hours, 6 tons; cost of treatment per ton. 75 cents; tons crushed during year, 1,640.

Reported by R. G. Sewart.

---

# CHAPTER II.

## NEVADA.

This State now takes the lead in the production of the precious metals. Mr. Valentine, superintendent of Wells, Fargo & Co.'s express, reports the product as follows—I am not able to make any detailed comparison between his figures and those I had gathered from other sources, but there is every reason to believe that his totals are correct:

| | |
|---|---|
| Gold dust and bullion by express | $345, 394 |
| Gold dust and bullion by other conveyances | 34, 539 |
| Silver bullion by express | 30, 954, 602 |
| Ores and base bullion by freight | 4, 117, 698 |
| | 35, 452, 233 |

Accounts of the different districts, so far as they could be obtained during the year, are given below.

*The Comstock lode.*—The following is the summary of productive operations on this lode during 1874:

| | |
|---|---|
| Number of tons milled | 527, 623 |
| Total bullion-product | $22, 400, 783 |

Average yield per ton of leading mines:

| | |
|---|---|
| Consolidated Virginia | $55 |
| Belcher | 54 |
| Crown Point | 39 |
| All other mines | 15 |

The last figure includes the low-grade ores of such leading mines as Chollar, Overmann, &c., and the product of the "outside" mines, situated on supposed branches of the Comstock, south of the Belcher, some in Gold Cañon, others in American Flat.

It is estimated (but I consider the figures probably too high) that $400,000 worth of quicksilver and $9,600,000 gold and silver was lost in the tailings during the year. A considerable portion of this may be recovered by reworking, but it would not be extravagant to say that $4,000,000 has been irrevocably lost.

For a comprehensive account of the extraordinary developments connected with the "Great Bonanza," and of other operations on the Comstock lode, during the year 1874, I am, as usual, chiefly indebted to Mr. C. A. Luckhardt, formerly of Virginia City, but of late years a resident of San Francisco, and interested in the Nevada Metallurgical Works at that place. I should explain that, owing to the diminished appropriation of last year, I was unable to engage Mr. Luckhardt to make a personal examination of these mines until after this report had been rendered, when, Congress having supplemented the appropriation, I could afford to pay the expenses of the necessary journey. As a consequence, a part of the following descriptions refers to a period later than December, 1874, extending, in fact, down to March 31. I have introduced it into the manuscript report while revising the latter for publication.

For facility of description, I will commence, as in former reports, at the north and follow to the south, dividing the lode into three parts, the north, middle, and south.

I. The 12,200 feet of ground comprising the Utah, Sierra Nevada, Union Consolidated, Mexican, Ophir, California Consolidated, Virginia Consolidated, Best & Belcher, Gould & Curry, Savage, Hall & Norcross, and Chollar, has been developed during the past year as follows:

*Utah.*—The ore-body belonging near and at the surface to the Sierra Nevada had been formerly found to extend into the Utah ground, and had been worked by the Sacramento Company to the depth of about 400 feet, showing the same character of ore, varying from $3 to $12 per ton, principally in gold. Since then, the Utah has reached a vertical depth of 500 feet in the new shaft, 600 feet east from the old one. The old shaft reached a vertical depth of 280 feet, and through it the vein was explored at 80 feet, and at 200 feet for 180 feet eastward, and vein-matter (quartz and porphyry) was encountered, 176 feet wide, running north and south, and containing no ore of any consequence. It carried zincblende and an abundance of pyrites. Through the new shaft, at 400 feet depth, the vein was reached by a cross-cut 480 feet west from the shaft, and found to be 167 feet wide, consisting of quartz and porphyry, much broken up and carrying ore in detached seams, which assayed from $5 to $18 per ton. It was bounded on the east by 4 feet width of clay. Explorations were continued westward and the workings were connected with those of the old shaft. The vein, 167 feet wide, was followed southward for 300 feet, and cross-drifts east and west were run for 165 feet through the same quartzose material, which showed a uniform character, while explorations northward for 180 feet disclosed much more quartz than the south mine did. Especially along the eastern clay-seam the quartz carries much iron-pyrites; but it seems to be here still too much divided and intermixed with porphyry to indicate a strong body of ore close at hand. The general course of the vein is south 20° west, and its dip is 40° to 45° east. This vein in the Utah seems to have strength, and the workings of the company will be carried on vigorously in the hope of valuable discoveries. To this end, heavy pumping-machinery is erecting. It is not at all improbable that the quartz and quartzose matter in the vein may increase in quantity as the work progresses in depth, in which case ore in paying quantities may be encountered. The owners of the Sutro mine, immediately adjoining, northwest the Utah, are exploring the ground through a tunnel, now 1,100 feet long, which is expected to cut the vein at 1,300 feet. Farther south and east of this mine locations have been made for several thousand feet. Some are of recent date, some are old titles and claims, at one time abandoned but now revived. Shafts are sinking on every side; some have attained 150 feet vertically, but nothing noteworthy has as yet been demonstrated through any of their workings.

*Sierra Nevada.*—This mine has been worked on the old ore-body, from which during the year 17,708 tons of ore have been extracted, giving an average yield in mill of $5.50 per ton. The location of the mill close by the mine made it possible to work this ore to advantage; but the mill has been idle since January, 1875. There is still a large amount of low-grade ore here, west of the old workings. The old shaft has attained a vertical depth of 770 feet, and is still sinking. At the 700-foot level, work (cross-cut) is progressing eastward, but the vein has not yet been reached. The old upper levels have lately been partially reopened, with the purpose of further exploring the old upper ore-body. The new shaft, which lies 1,450 feet north, 65° east, from the old one, has attained a vertical depth of 650 feet, but as yet no work has been done through it.

*Ophir.*—This mine now comprises 600 feet horizontally on the vein. Last year's work showed, as remarked in my last report, an attained depth of 1,400 feet, with explorations proving the vein-matter to be 300 feet wide, and exposing near the south line the apex of what seemed, at that time, to be a new ore-body. The shaft has been sunk to 1,465 feet vertically, work carried eastward from it for 350 feet, and at this point 250 feet further sunk, making a total depth attained of 1,700 feet. The 1,465 foot level shows the vein-matter to be 300 feet wide in the south mine—*i. e.*, the ground from the company's south line to the shaft, a distance of 450 feet northwest—and two strata of ore have been encountered. This level has been connected with the California mine, besides the connection existing at the 1,300-foot level. The ore much resembles that which was at first found in the Ophir and Mexican mines at a depth of 250 to 400 feet, in the old workings of 1863–'64. It extends in the 1,465-foot level as far north as 160 feet from the south line. It was also encountered in the 1,300-foot level, but lying irregularly, and was followed by a winze, located 140 feet north of the south line, in a northeasterly direction to the 1,600-foot level, where a drift northeast developed ore of the same character. Whether this is the same seam as that in the 1,465-foot level remains to be proved. At the 1,700-foot level the ground was opened southward to the south line near the west wall, and a cross-drift, 140 feet from the south line, was run 120 feet to the east, without finding any ore up to that point. The seams of ore above described have apparently an inclination southward toward the California mine, and probably belong to the body of ore developed in the latter mine, or constitute parallel outliers on the west of it. The ore is of very good quality, carrying traces of zincblende, and seeming identical with the ore found in former times in the Mexican near the surface. No work in the 1,700-foot level has as yet been undertaken toward the north.

*California, Virginia, Best & Belcher.*—The developments described in my report of 1872, in the 1,400 feet of ground situated between the Gould & Curry and the Ophir, and those described in my report of 1873, together with what the Ophir mine developed in the latter year, 40 feet from their south line, led to vigorous operations in 1874; and that ground, which had previously been considered barren, proved to contain a stupendous mass of ore. As all this 1,400 feet of ground is owned and managed by one party, and the workings are all connected, I will describe it in one, although it contains three distinctly-different corporations, viz: next to the Ophir, the California Consolidated Mining Company, having 600 feet; adjoining this, on the south, the Virginia Consolidated; and next to this, and immediately north of Gould & Curry, the Best & Belcher Mining Company.

The Virginia Consolidated shaft, which lies 20 feet south of the California south line, has attained a depth of 1,580 feet vertically, and the Gould & Curry and Ophir shafts are connected with it at the 1,300, 1,400, and 1,500 levels. The 1,167-foot level, which was driven from the Gould

& Curry in the latter part of 1872, developed in 1873 a body of $45 mill-ore 300 feet long, and from 8 to 30 feet wide, in the Virginia Consolidated ground. The 1,400-foot level exposed it also. Here it was about 300 feet in length and 90 feet wide; but the 1,500-foot level has, up to March 31, 1875, proved 700 feet length of ore, in some places 200 feet wide. The Virginia Consolidated is extracting an average of 580 tons of $150 mill-ore per day from between the 1,400 and 1,500-foot levels.

So far as this ground has been explored, up to the date just mentioned, it is impossible to come to any definite conclusion as to the actual dimensions of the ore-body. The first supposition was, that the ore found in the 1,167-foot level, near the Best & Belcher line, belonged to the body which the 1,500-foot levels of the Virginia and California Consolidated have developed for 700 feet in length, and that it was the northern extension of the same ore which was met with in the 1,465-foot level of the Ophir, near the California Consolidated north line. The idea is also prevailing that this supposed one ore-body (for which the Spanish name *bonanza* has been generally adopted) commences at the southern portion of the Virginia Consolidated 1,167-foot level and extends northward, and declines northward through the entire 1,400 feet, and finally sinks in Ophir ground at 1,700 feet depth. A close examination of the 1,500-foot level of the Virginia and California Consolidated, coupled with the difference existing in the character of the ores found south and east, in the center, and north and west, in the ground described, proves this theory erroneous. The Virginia Consolidated shaft penetrates the 1,500-foot level 100 feet east of the west wall of the Comstock, and about 20 feet south of the California Consolidated south line. At 142 feet, directly east from the shaft, ore was first encountered, having a north 48° east course, and dipping 65° east. It was penetrated diagonally to its course for 145 feet, showing ore of $700 average assay-value, and bounded on the east by a clay-seam which I do not consider to be the true "east clay." This cross-drift, 20 feet south of the California Consolidated south line, is called "cross-cut No. 1," and cuts through the ore-body at an angle of 50°, giving, therefore, not exactly a cross-section. The ore here is composed of stephanite and the peculiar "subsulphide" characteristic of the Comstock. Copper-pyrites and argentiferous fahlore occur with it. The metalliferous mineral occurs in seams of a few inches to several feet in thickness, and penetrates the entire mass of quartz for 145 feet. The gangue is quartz, no lime being visible, except very close to the eastern clay, where here and there bunches of calcite are found. At these points the character of the ore also changes a little, and partakes more of the nature of the ore which is found in the southern portion of the Comstock; for instance, in Belcher and Crown Point. One hundred feet directly north from this cut No. 1 is No. 2, which went through quartzose vein-matter for 157 feet, therefore 177 feet east of the shaft, before meeting with the same ore and some thin western clay-partings. It penetrated the ore for 100 feet, when the latter was cut off on the east by feldspathic porphyry. Further explorations will have to be made eastward at this point, as this cannot be the final eastern boundary of the body. The character of the ore is the same as in No. 1 cut. One hundred feet directly north from No. 2 cut is No. 3, which cuts into ore 240 feet east of the shaft. Up to April 2, 1875, this cut had penetrated 25 feet into the ore, which is not of as good quality as that shown by No. 1 and No. 2, and will perhaps not be found as wide as it was in No. 2. No east division has as yet been encountered. One hundred feet north of cross-cut No. 3 is No. 4, 300 feet south of the Ophir south line. It has penetrated, so far, 90 feet east of the main north and south drift, which makes the point reached 110 feet east of the shaft. It is in the same material which constituted the ground west of the ore farther south in cuts 1, 2, and 3; but here, as well as in No. 3, small seams of ore are found. In this cut, at 50 feet east of the main north drift, ore has been found 8 feet wide. It seems to run almost parallel with the ore-body developed in Nos. 1, 2, and 3; but it is 180 feet farther west, carries much more zincblende, and is of much poorer quality. It is probably connected with the western ore-seam of the 1,465-foot level in Ophir, and can hardly be distinguished from the ore found at 360 feet depth in the Ophir workings of former years. Farther north from No. 4 no work has been done eastward from this level; but the main drift, running north, extends into the Ophir ground and connects with the 1,465-foot level of that mine.

Going south from cut No. 1 100 feet is cut No. 2 of the Virginia Consolidated. This encountered the thin western clay-seam of the ore-body about 40 feet east from the main north and south drift, and penetrated for 50 feet through material resembling that through which California Consolidated cut No. 3 passed for 30 feet. It is not a high-grade ore. For 50 feet farther east, here (beyond the 50 feet just mentioned) the ore is a little better, but still retains the character of the ore found in the more northerly cross-cuts. Pressing farther east the quartz changes, and the drift (or rather cross-cut) enters into a different material, quartz predominating, but accompanied by lime, the ore losing more and more its base ores. The cross-cut penetrates 200 feet farther through this ore ($150-ore) until, 350 feet east from the main north and south drift, it meets the east clay, which runs here north 20° east, standing nearly vertical.

Again, 100 feet south of No. 2 cut is No. 3, which encountered the western small clay-seam of the ore 65 feet west of the main north and south drift, and was driven east for 35 feet, where a heavy clay-body was met, which is the east clay of this described ore-body. It runs here north 50° east, and stands nearly vertical. One hundred feet farther south from here is a winze, which connects with the 1,400-foot level, and which shows ore for 60 feet below the 1,400 level of the same character as cross cut No. 3 shows at the 1,500-foot level. The sketch herewith given will afford probably a better notion of these workings than a further detailed description. The following explanations refer to the sketch:

The strike and dip of the east clay at *a*, in cross-cut No. 3 of the Virginia Consolidated, corresponds with that at *b* of cross-cut No. 1, while that at *c* does not. The ore in the eastern portion of cross-cut No. 2, in Virginia Consolidated, is different in character from that found in cross-cut No. 3 of the same mine and that of cross-cuts Nos. 1, 2, and 3 in the California ground, but resembles the ore found in the 1,300-foot level of the Virginia Consolidated 400 feet north of that company's south line. Furthermore, the ore found in cross-cut No. 4 of the California ground and that found in Ophir are identical. Mr. Luckhardt infers that three distinct ore-bodies exist, but they are of too large dimensions, and, in consequence, have been too little explored to permit speculation with any accuracy as to their position and inclination. The northerly and westerly ore (in California No. 4 and in Ophir) seems to be composed of detached ore-seams, sometimes running parallel to each other, which are imbedded in a quartzose porphyritic mass 150 to 250 feet wide. The central body, which has a western clay-seam running very regularly, and separating it from the above, shows a narrowing at cross-cut No. 3 in the California, and the wings sunk below

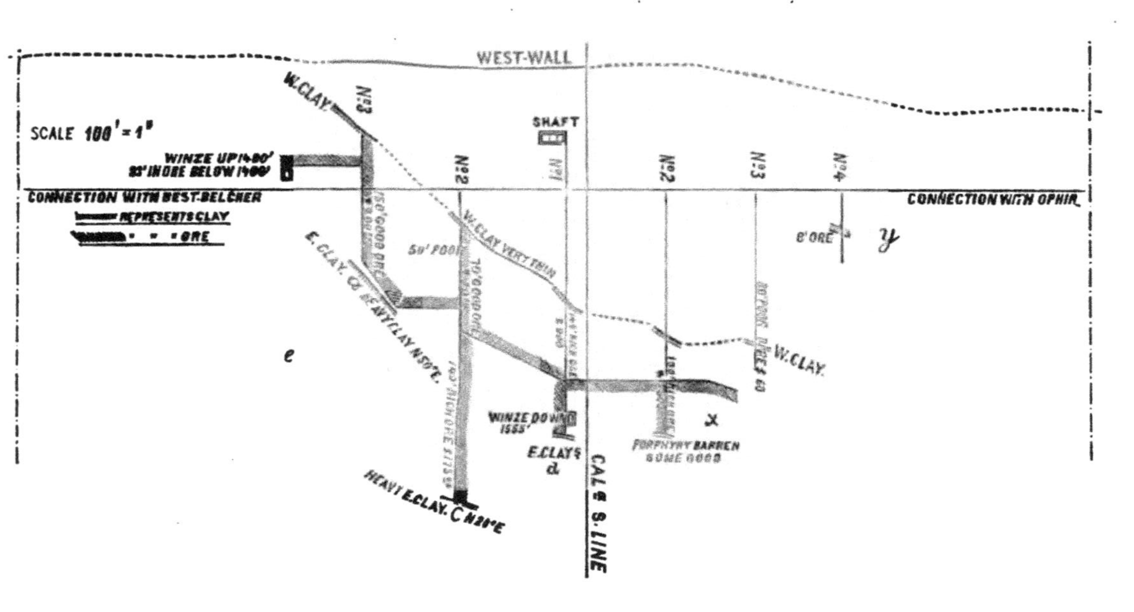


THE GREAT BONANZA—GROUND PLAN AT THE 1,500-FOOT LEVEL OF VIRGINIA AND CALIFORNIA.

the 1,500-foot level prove that the ore descends southward. The clay at *a* resembles the clay at *c*, but also that at *b;* and unfortunately the stopes, &c., at cross-cut No. 2, in Virginia ground, do not allow of an examination whether *a* was connected with *b*. But it is not likely that the clay swings from *a* to *b* and back to *c*, and Mr. Luckhardt does not consider *b* to be the proper eastern clay. The porphyry which is encountered in cross-cut No. 2, in the California ground, to the east of the ore, may be an intrusion, and may have caused the narrowing of the ore in cross-cut No. 3 in the California, and its apparent widening in cross-cut No. 2 in the Virginia. Mr. Luckhardt concluded therefrom that it may be possible that the clay at *a* and *c* is the same, and that the porphyry at No. 2, California, had some influence upon the ore west of it, while south and east of it it had none, or *vice versa*. It is noteworthy that while the feldspathic porphyry, which constitutes almost exclusively all the so-called "horses" of the Comstock lode, is invariably void of all precious metals, there is at *x*, in California cross-cut No. 2, a porphyry-belt close to the ore-body, in which the pyrites (so characteristic of the eastern porphyries) is replaced by silver-ore in places, so that it showed $10 to $20 per ton by assay, though barely to be distinguished from the porphyry farther east in outward appearance. It may be, therefore, that, by pushing considerably eastward the work at *b* and *x* and farther north, the clay *c* will be found beyond this porphyry. This would prove the entire ground for over 900 feet northeast and southwest and for 300 feet southeast and northwest to belong to one ore-body.

The west wall of the Comstock, as indicated by present workings, is supposed to have in this place a nearly north and south direction; and the western boundary of the ore-body must meet it on the south at an angle of 40° to 45°, while the eastern boundary of the ore must meet it to the south at an angle of 65° to 70°. It is possible that this ore-body will be found to flatten against the west wall of the Comstock fissure, run with it, and widen again to the southward. Should this be the case, we would naturally expect at *e* some intrusion different from the vein-matter which has so far been exposed at *e* for 300 feet in depth. But the quartz developed by the 1,200-foot level at the south line of the Virginia Consolidated mine, which carried $40 per ton, is, to all appearances, very massive, as is also the quartz which, on the 1,300-foot level of the Virginia Consolidated, lies 400 feet north of the company's south line and carries fine ore. Considering, also, the 300 feet length of ore 90 feet wide, which the 1,400-foot level of Virginia developed, it appears that the facts will hardly sustain the theory just mentioned; and it seems more probable that a third ore-body, lying *in echelon* with the other two, will be developed to the south and east, appearing in the 1,600-foot level in Best & Belcher. But this difficult question requires further light from connected explorations and careful study.

Contrary to the generally-accepted theory, Mr. Luckhardt is of the opinion that the main bulk of this ore, whether it prove in time to be one or two bodies, will be encountered in the Virginia ground and south of it, rather than in the Consolidated Ophir farther north. The greater amount of explorations in depth to the north, aided perhaps by speculative purposes, has created an impression that the northern ore-body, shown on the sketch in California cross-cut No. 4, and the ore found in the Ophir, prove the northward extension and the northward pitch or inclination of the great *bonanza*. But it is more likely that they belong to a parallel outlying body or group of seams, and that the middle body is not connected with them.

It is impossible to estimate the amount and value of ore standing in the Virginia and California ground. As has been remarked, the ore varies in quality. Extraordinarily rich streaks are met with. In some places for 50 feet in the cross-cuts the ore assays $900, and again sections of 200 feet assay $150. Cross-cut No. 2 in the California, for 80 feet across the ore, assays $412 per ton, and, on the other hand, in some places the ore will not assay above $50; as, for instance, for 50 feet in Virginia Consolidated cross-cut No. 2, near the west clay. Fabulous reports and estimates have been given, varying from $150,000,000 to $1,500,000,000 "in sight." It is reasonable to suppose that the ground from the Best & Belcher north line to the Ophir south line, and in depth from the 1,300 to the 1,550-foot level, will yield, judging from present developments and present yield, ore for two years to come at the present rate of extraction, viz, 450 to 580 tons daily. The quality of the ore is a matter of pure speculation. At present the average is about $150 per ton, but $300 to $500 ore can be extracted at will, by selecting the stopes. Mr. Luckhardt thinks that $160 may possibly be an average of the whole mass referred to; but this is merely a prudent guess. At this rate there would be over $60,000,000 "in sight." But such calculations are at present futile. The width and length of the ground to be exploited necessitates considerable care in working. Experience has shown, in the south portion of the Comstock, that the present mode of timbering, although convenient for wide spaces, is deficient in strength. The caving of timbered spaces has caused the loss of good ore-ground. To obviate this difficulty, the ground is now worked in sections of 50 by 60 feet and timbered in the old way; but every fourth set is then filled up with heavy 10 by 12 and 12 by 12 inch timbers across the whole 60 feet and up to the top of the stopes, all well keyed in, so as to form a solid bulk-head of timber. Narrow passages are left at long intervals, and the rest of the space between the bulkheads is filled with waste from the upper or lower prospective works. It is a very expensive method, but cheaper than masonry would be.

The Consolidated California is not as yet producing ore, and it will be some time before the shaft will be completed. The adjoining mines need the full capacity of their respective shafts to hoist their own ores.

The Virginia Consolidated Company has erected a very fine mill, close to the shaft, capable of reducing 250 tons of ore daily. A tram-way leads to the mill from the shaft.

The following description of this mill is taken from the Virginia Independent:

The site is to the north and east of the hoisting-works, and between the two the descent is such that the ore can be run from the level of the surface of the hoisting-works and reach the mill at its very pinnacle. The ground covered by the mill also descends very rapidly to the east, so that the ore falls naturally to the stamps; from the stamps, by distributing-sluices, the pulp drops to the tanks, from the tanks to the pans, and so on down by a gradual descent to the agitator-room. Some idea of the dimensions of the mill may be gained from the following particulars: The battery-building is 110 feet long by 48 feet wide, the pan-building is 195 feet long by 92 feet wide, the engine-room is 58 feet by 92 feet, and the retort-house 25 feet by 60 feet, the whole covering over 26,000 square feet of ground. Besides these buildings, there is an office, 20 by 30 feet, fronting on G street.

The engine is a compound condensing or low-pressure engine of 600 horse-power, with cylinders 24 by 48 and 48 by 48 inches, respectively. The steam is admitted into the 24 by 48 cylinder and cut off at the half-stroke. It is then taken into the expansion-cylinder, which is 48 by 48, and contains four times the cubical contents of the smaller one, and thus an expansion of eight times is gained. After leaving the expansion-cylinder the steam exhausts into the condenser. This gives the vacuum-power, or the power resulting from the atmospheric pressure at this altitude, in addition to the expansive power of the steam. The engine-shaft is 14 inches in diameter, and carries a band-wheel 18 feet in diameter, weighing 33,000 pounds, which drives the battery part of the works. This shaft itself weighs 15,000 pounds, and the approximate weight of the whole engine is 100,000 pounds, or 50 tons. The foundation of this ponderous piece of machinery contains 450 cubic yards of masonry, laid in cement, and weighs 600 tons. This engine-shaft is coupled to an extension 11 inches in diameter, which drives the amalgamating part of the works.

The boilers which supply this monstrous motive-power with steam consist of four pairs, 54 inches by 16 feet, so arranged as to run a single pair at a time or all together. The smoke-stacks are 42 inches in diameter, and stand 90 feet from the ground. There are, also, in connection with the boilers, two steam-pumps, for feed and fire purposes. To these a water-hose is constantly attached, and ready for use in case of fire. The roof of the engine-room rises to the height of 50 feet; the western earth or embankment wall is 22 feet high, built of hewn trachyte-rock, and, with the other embankment-walls, contains 4,000 perch of stone.

The mill is connected with the hoisting-works by means of a covered trestle-work, 44 feet in height at the mill-end. The same cars used in the mine will be run in trains by mule-power from the mine, 278 feet, to the mill. The trains will consist of from six to ten cars. It will take a car-load every five minutes to supply the demands of the mill. These cars are dumped into chutes, one on each side the center, from which the fine ore sifts into the ore-bins below. The part of the ore first dumped is carried back from the center by chutes, and thus becomes equally distributed into the feed-bins. From the feed-bins the ore is taken by Tulloch's self-feeders and given to the batteries as it is required.

The batteries are driven from a counter-shaft propelled by the large band-wheel below, the connection with which is made by a belt 24 inches wide and 160 feet long. From this counter-shaft the batteries are driven by 14-inch belts 60 feet long. The batteries are run in sets of ten stamps each, and clutches are so arranged as to stop any ten without interfering with the working of the other machinery. The batteries consist of 60 stamps of 800 pounds each. The mortars are so arranged as not to interfere with putting shoes and dies into the batteries, as they discharge at one side, and are two nearly together instead of separately and in the center. From each division of 30 stamps there are distributing-sluices, which convey the pulp into the settling-tanks. From the tanks it is shoveled out upon the platform in front of the pans. There are 16 pans on each side. These are flat-bottomed, 3 feet in diameter, and calculated to hold about two tons each. The pans have steam-bottoms, and are arranged to admit live steam into the pulp or under the bottom. There is a settler to each pair of pans, making 8 on each side. These are nine feet in diameter and 3 feet deep. From the 15 settlers the pulp is discharged into 4 agitators, and these discharge in turn into the tail-sluice, which is placed at the end of all the other appliances. The heavy stuff in the agitators will be cleaned out twice every twenty-four hours, and subjected to further working by 4 more pans and 2 settlers placed in the agitator-room.

The quicksilver-tank is placed in the store-room under the ore-bins, and will come down pipes to the distributing-tanks in the pan-room. From these tanks it will be distributed through pipes to the charging-bowls of the pans. After passing through the pans and settlers, it is discharged from each settler direct into each strainer. A pipe leads from the bottom of each strainer to the main receiving-tank, which is placed below everything and in the center of the pan-room to catch all; and from that it is pumped up into the main distributing-reservoir again. There are two of these reservoirs, one on each side.

The amalgam remains in the strainers, and is taken by an amalgam-car, holding about 20 tons, which runs through the center of the pan-building to the retort-room. The same car takes retorted silver back into the mill-circulation. Each pan and settler can be stopped without interfering with the other machinery. Each row of pans and battery has a traveling-tackle, for convenience of handling the stamps and pan-mullers. The strainers are provided with locks and covers for protecting the amalgam. The retort-room is built of brick, and contains 4 retorts, each of which is calculated to hold 2,000 pounds. These are so arranged that any one or more of them can be run. The flues lead into a brick chimney 50 feet in height. Back of the reduction-works is a cooling-reservoir, about 200 feet square, where the water from the condenser can radiate its heat and attain the required temperature to be used again.

The mill cost, it is said, including the grading, $250,000.

It is the intention of the present manager to erect another and smaller mill, of 25 tons capacity, for the treatment of the first-class ores, which will need roasting.*

A connection with Gould & Curry and Ophir, in the lowest levels of this mine, secures ventilation.

* The ore from this ground of the Virginia and California Consolidated mines has the peculiarity that the lower workings show an increase in the proportion of gold over the ore from the upper levels, (1,300 and 1,400 foot levels.)

Adjoining the Virginia Consolidated to the south is the Best & Belcher, which has been explored within the past year to a small extent through the Gould & Curry workings. It was close to the south line of this mine that, two years ago, the first signs of the now exposed *bonanza* were discovered. It has no separate shaft or hoisting-works of its own. In its ground, at 1,200 feet depth, the above-mentioned quartz-body was found, which will, when well explored, lead undoubtedly to the finding of much good ore. In the 1,500-foot level, at the Gould & Curry north line, a winze was sunk, which penetrated into the west wall and met some ore. Assays run from \$5 to \$25. This work is west of the quartz, and is comparatively dry. It was continued to the 1,700-foot level, and drifts northeast and southeast were run from it, which met some ore, sustaining the opinion above suggested, that the ore from the Virginia will find its deepest point south.

Work is progressing slowly here, and the cross-cutting eastward cannot be continued much farther, as the Gould & Curry has at the 1,700-foot level all the waters of the Virginia and California Consolidated levels down to 1,550 feet to contend with, and fear is entertained that if this quartz is broken into it will flood its works. This flow of water is an indication that the quartz-masses of the two mines named pitch southward, through Best & Belcher, into the Gould & Curry 1,700-foot level.

*Gould & Curry.*—This mine has reached to the 1,700-foot level with its shaft, and a wing sunk near the north line, on joint account with Best & Belcher, has penetrated about 100 feet deeper. The waters coming from the north are so abundant that this 1,700-foot level has been only part of the time accessible. Work eastward and northward progresses slowly. The quartz of the 1,300-foot level looks very well, but developments of importance are expected chiefly from the 1,600 and 1,700 levels, possibly the 1,500. The old workings have yielded no ore this year. In fact the *bonanza* has occupied all minds, and the principal work of the Gould & Curry has been to pump water for the others; but better things may be looked for.

*Savage.*—This mine has attained a vertical depth of 2,200 feet. The shaft is vertical for 1,300 feet, then inclined at 38° east for 160 feet, and was then changed to 45°, continuing to the present bottom. During the past year, the 2,000-foot level was opened from the incline, which lies east of the vein. A cross-cut 100 feet west met with the east clay, which is here from 6 to 12 inches thick. The vein was found to be about 100 feet wide here, and composed of quartz and porphyry. Near the center of it a few small ore-seams were found; the whole mass would assay from \$1 to \$20. The work was carried south on this level, and connected with the Hale & Norcross workings. A cross-drift 25 feet north of the south line showed the vein 120 feet wide. Quartz predominated here over the porphyry. The ore-seams were also found here, but do not contain sufficient ore to pay for extraction. As the vein was wider south than north, or at the incline, a winze was sunk 100 feet north of the Hale & Norcross line, which penetrates down to the 2,200-foot level. It was sunk east of the vein to avoid the water. A drift south for 120 feet has been run from this winze, but all the work is east of the vein.

At the same level the ground north was explored for 600 feet, with cross-drifts every 100 feet, showing about 100 feet of vein-matter. Toward the north, porphyry predominates over the quartz; toward the south the contrary is the case. The 2,200-foot level has been opened so far with a cross cut 175 feet west of the incline, but no east wall has as yet been found.

The 1,900-foot level has been more thoroughly explored. It shows less quartz than the 2,000-foot level. It has a drift to the south line. About 300 feet north from Hale & Norcross a cross-drift showed the vein to be 402 feet wide from the black dike to the east wall, but composed mostly of porphyry. The 1,900 has been connected with the 2,200-foot level east of the vein, and gives very fair ventilation.

The quartz in the 2,000 north and south level and its improvement toward the south, and the fact that it carries some ore, leads to the belief that the 2,200 level in the southern portion of the mine may possibly develop a body of ore. From analogy furnished by the history of the Comstock, this is certainly to be expected.

The company has under construction a splendid piece of machinery west of the present hoisting-works. The engine has two cylinders giving 400 horse-power. The round-steel-wire cable is to be used in preference to the flat, and it is claimed that the plant will be adequate to work to the depth of 4,000 feet. The old 400, 500, and 1,000 foot levels have been overhauled, and passages have been opened for air-connections. The mine produced no ore during the year.

Hale & Norcross has attained a depth of 2,200 feet, but nothing new and noteworthy has been developed. The mine is very closely managed. During the year 17,469 tons of ore was produced, which yielded $295,361.12. The larger part of this ore came from the eleventh and fourteenth levels—but about three-sevenths came from the old second, third, and fourth levels—and to all appearances the mine will continue to yield from the old workings and the tributaries, and "left-ground" of the ore-body of the sixth and seventh levels, possibly 1,200 to 1,500 tons of $1,400 to $1,700 ore per month for some months to come.

Both east and westward of the described ground, numerous old claims and titles have been hunted up within the last year. Surveys have been made for some of them, patents obtained by many, and work has been commenced on more than 40 of them, all represented by an incorporation issuing and dealing in the respective stocks. Probably in the course of time much litigation will arise, when these workings penetrate to a sufficient depth (if the companies do not die a natural death) to meet with ore. In the mean time the old companies which are on the Comstock leave them unmolested. Some of them have quartz and porphyry resembling that of the Comstock, and the "clay-wall" theory may one day be again tenaciously contested. That there are metal-bearing veins existing east of the Comstock is an established fact, and that the eastern boundary of the Comstock is a hard thing to determine is equally true. In the conflict of theories, probably nothing will be left for the contestants but to *connect in ore with one another* or come to compromises.

The companies east of Hale & Norcross have erected large and expensive hoisting-works—as, for instance, the Julia, Senator, &c.—and have penetrated as deep as 1,400 feet, but most of their underground work is done westward from their shafts toward the Comstock. In the Julia some ore-bearing quartz was encountered, but it is claimed to lie in the Chollar Company's ground.

Four miles northeast from the Utah some quartz has been found, in a vein 4 to 6 feet wide, which assayed as high as $80 per ton. It does not at all resemble the Comstock quartz, being much more compact, and carrying very little gold. By next year it will be possible to give more information concerning this deposit. Thus far, nothing noteworthy has been developed.

II. *The middle portion of the vein*, including Chollar, Bullion, Exchequer, Alpha, Imperial, Empire, and the small Gold Hill proper, an aggregate of 1,800 linear feet, gives much fairer hopes for ore this year than the past. Under the stimulus of the eastern developments in Virginia Consolidated, &c., on the north, and Crown Point, &c., on the south, the 1,200 feet of ground, including Bullion and portion of the Chollar which had laid unnoticed and unprospected for a long period, has at length been explored to some extent, and with flattering results.

The Chollar is still at work on its apparently never-to-be-exhausted mass of low-grade ore at and near the surface, and is still producing about 60 tons of $20 mill-ore per day. The fact that the Hale & Norcross workings showed ore led to the former prosecution of explorations exclusively north of the Chollar shaft, the ground south and east being thought barren. Within the past year the shaft has been repaired, and the incline opened to 1,300 feet in depth, and work was started in earnest on the 1,050 and 1,150-foot levels, south of the shaft. The vein is apparently much wider than it proved to the north and above these levels; the labyrinth of clay and argillaceous and porphyritic matter which there filled the vein ceases on the south, and quartz takes the place of these minerals. Here and there spots of ore are encountered, and the company now hopes to find ore in paying quantities—a hope which is strengthened by the finding of quartz in the Julia mine and by the developments in the Bullion described below.

*Bullion.*—Much credit is due to the owners of this mine, who, under adverse circumstances, almost unvarying ill fortune, have never through many years abandoned the explorations for a single day. They have carried on their work, partly through their own shaft, (down to the depth of 1,400 feet, *i. e.*, 800 feet vertical, and then inclined at 45° east to the 1,400 level,) and partly through the Imperial-Empire shaft. In the Bullion shaft, at the 800-foot level, they are 150 feet west of the west clay of the Comstock. Cross-cutting 400 feet east from the shaft, they passed through porphyries and clayey matter, carrying abundance of sulphate of lime. Drifting was done here for 100 feet in a southeast direction, (east of the west clay, and therefore within the vein,) and the small quartz-seams near the shaft were found to unite into 15 feet width of solid quartz, resembling much the quartz-body which the Alpha 500-foot level showed, and also that of the Yellow-Jacket 1,100-foot level, but not to be supposed to belong to either. After 100 feet of drifting farther south, near the west clay, the quartz became much wider, lay much flatter than it had done to the north, and assayed from $10 to $15 per ton, having very little porphyry intermixed. To this quartz reference will be again made below, since it has been supposed to be identical with the quartz of the 1,700-foot level of Imperial, in Bullion ground.

The old workings have been overhauled. The quartz-body which came from the Chollar at the 400-foot level is from 50 to 110 feet wide, carries low-grade ore, ($10 to $12,) ending (barren) at the east wall in the 1,200-foot level, 300 feet south of the Chollar line. This body lay 500 feet northeast from the quartz above described in the 800-foot level. It really never yielded anything worth extracting below the 800-foot level. In the 500 and 600 foot levels this body still exists, bearing low-grade ore; but none has been extracted in that portion, and only lately has work been commenced for its full exploration and exploitation. The Imperial and Empire having finished exploration in its 1,700-foot level, the Bullion Company worked 1,050 feet northeastward from the Imperial shaft in this level. This work lies 700 feet farther east, and 430 feet lower, than the 1,400-foot level of Bullion proper. These 1,050 feet from Imperial followed the "black dike" through the Exchequer ground; and 10 feet north of the Bullion south line a cross-cut east was made,

64 feet in length, before the west clay wall was struck. This 64 feet traversed a quartzose breccia, often found on the Comstock, lying between the west clay and the black dike. It assayed here from $10 to $30 per ton. The same material was found 30 feet wide at the 1,400-foot level of the old Bullion incline, 700 feet west of and 430 feet above this work of the 1,700-foot level, where the quartz-body east of this material, (already referred to as found in the 800-foot level) was followed for 200 feet south and 200 feet north from the bottom of the incline, and found to be from 86 to 125 feet in width, and showing ore in seams and bunches, assaying from $10 to $35 per ton, ending (in ore) 200 feet south of the incline, but apparently gathering strength in the northern portion of this 1,400-foot level, where its northern terminus has not been exposed. Now, in this 1,700-foot level of Imperial, continued into the Bullion ground, the first eastern cross-cut was made, as I have said, 10 feet north from the Bullion south line, cut through this brecciated belt, which in places showed ore-fragments assaying as high as $180. It lies between the "black dike" and the west clay of the vein, and the cross-cut has not passed as yet into the quartz east of it a sufficient distance to permit any conclusion as to what may be expected from it. In the Bullion ground, 90 feet north from this first cross-cut, another cross-cut eastward is now in progress, and has run so far 30 feet in quartz east of the breccia. This quartz shows some very fair ore in seams, assaying as high as $45. It is the same as that developed in the 1,400-foot and 800-foot levels, just east of the west clay. There is, therefore, a considerable piece of ground (700 by 430 feet in Bullion alone) yet unexplored, while above and below the 430 feet of vertical height ore has been exposed. Moreover, the west wall at the 1,400-foot level runs north 5° east, at the 1,700 it runs north and south; the west clay at the 1,400-foot level runs nearly north 6° east, while at the 1,700-foot level it runs in Exchequer north 10° east, and the east clay, which bounds this quartz-body in Imperial and Exchequer, runs north 6° east, but in a distance of 150 feet northward changes in Bullion to north 20° east. If this course continues, it will widen in Bullion the distance (at 200 feet north) between the west and east clays to 125 feet, instead of only 35 feet; and if this quartz maintains its character, which it has every appearance so far of doing, it will make a very formidable body in going north, and will very probably carry a compact ore-mass. I believe there has been no instance in the history of the development of the Comstock where a quartz-body 100 feet wide did not carry a body of ore. As a rule, the wider the quartz the more extensive the ore-body. The ore from the quartz just described, carries two-thirds silver to one-third gold, while the deepest body in Chollar carries three-fifths gold to two-fifths silver, and the upper body two-thirds gold to one-third silver. This quartz, therefore, cannot be taken for a continuation of either. Judging from analogy, it may be expected that explorations in this ground from the 1,100 to the 1,700 foot level will develop large masses of ore; certainly the indications are in the highest degree favorable.

*Imperial.*—This mine has attained a depth of 2,100 feet, an increase of 200 feet in the past year. The 1,850-foot level has been opened this year, and the vein found much wider. The 1,700-foot level showed the vein to be narrow, especially in the northern portion of the mine. The 1,700-foot level in Exchequer showed the east and west clay coming nearly together. For instance, 80 feet north of the Exchequer south line the east and west clays come within 15 feet of one another, but the distance increased, 190 feet north of the Imperial north line, to 30 feet. The 1,850 level was run 75 feet east from the incline, and a drift was pushed south nearly to the Yellow Jacket line. No cross-cuts were made to demonstrate the width of the quartz or of the vein itself. At 100 feet east of the west clay, opposite the incline, the quartz was not very porphyritic. It was followed south 70 feet, where ore of fair quality was encountered, (width not known.) A winze was sunk here 175 feet, at an angle of 42° east, which went through ore in seams for 60 feet below the 1,850 level. This ore was followed for 125 feet in a southwest direction. It is $60 mill-ore—higher in grade, though not as heavy a body, as that found in Bullion. These developments, although much more favorable than any made in Imperial for five years past, do not promise permanent profit, unless the 2,000-foot level should demonstrate the extension of this ore. The 2,000-foot level was run south 20° east and entered the vein, which here carries quartz and porphyry in width not yet determined. Work here is progressing south as fast as possible to connect with the winze coming down from the 1,850-foot level, in order to ventilate both levels. The stifling air (at 126° Fahr.—a temperature which ventilation will of course at once reduce) makes prospecting very expensive.

III. *The southern portion of the Comstock, from the Imperial southward.*—*Yellow Jacket* has attained a depth of 1,840 feet. The vein shows in depth a flatter dip east than above, and in consequence the incline following it pitches 35° east from the 1,600-foot level down, while above the 1,600 level its pitch was 45°. During the past year the 1,740 and 1,840 foot levels were opened. The 1,300 and 1,400 foot levels had developed a body of quartz which carried principally gold, but not in paying quantity. The 1,540-foot level met the east clay 330 feet east from the incline. The vein was found nearly 300 feet wide, and was followed north and south to the boundaries of the claim. Cross-cuts were run at every 200 feet, and the quartz of the 1,300 and 1,400 foot levels was found, but no ore of value. This level will soon (in a few days) be connected with the 1,700-foot level of the Imperial for ventilation, and this will also enable the smaller mines between the Imperial and Yellow Jacket to prospect their ground at that depth. At the 1,740-foot level the vein was reached 220 feet east of the incline, and found to be 350 feet wide; but it carries less quartz than in the upper levels. The east wall is well defined, and dips 40° east. The vein was followed south to the Kentuck line, and north to within 300 feet of the north line, and found to be filled principally with porphyry. The north drift is still advancing, and is now 450 feet north of the shaft. No cross-cuts have been run here. The 1,840-foot level was advanced to date, (April 1, 1875,) 81 feet east from the incline, and 200 feet will have to be run to reach the vein. The mine produces, at present, no ore.

*Kentuck.*—At the 1,540-foot level of the Yellow Jacket, a connection with Crown Point was made through the Kentuck ground. The vein was strong, and carried as much quartz as it did in Yellow Jacket, but no ore was found, except in small straggling seams. At the 1,740-foot level the vein is being prospected, and it is probable that some ore will be found, since the middle ore-seam of the 1,600-foot level of the Crown Point extends north into the Kentuck. (See remarks on Crown Point below.) The mine yielded a small quantity of ore from the 800 and 900 foot levels from the old workings during the past year, and there is low-grade ore yet standing in the mine, which will, some day, be extracted.

*Crown Point.*—This mine reached a depth of 1,700 feet. The incline commences at the 1,000-foot level and sinks at an angle of 36° east. The old ore-body from the 500-foot level has been quite worked out, having ceased entirely at the depth of 980 feet. The newer body, which is now productive, and has paid all the late dividends in Belcher and Crown Point, commenced at the 1,000-foot level, extended at the 1,100-foot level 275 feet north of Belcher line; at the 1,200-foot level, 300 feet north; at the 1,300-foot level, 325 feet north; at the 1,400-foot level, 325 feet north; and has been, up to April, 1875, worked for 350 feet north from Belcher line in the 1,500-foot level. During the past year the mine has produced from these levels 176,745 tons of ore, of an average mill-value of $65 per ton. The 1,600-foot level has advanced 200 feet east, and has reached the quartz of the upper level, and met with what seems to be another body of ore which lies between the east and west ore-bodies of the above levels, only farther north. (It is this body which may be expected to extend into the Kentuck ground.)

The 1,700-foot level has just been opened and the landing-station finished. No prospecting work has yet been done here.

In all these levels, from the 1,000-foot level down, the general formation of the vein holds good for the Belcher also; and a concise general description of it, based upon Mr. Luckhardt's careful personal study, and applicable to both mines, may be of interest. The total width of the vein is not demonstrated. The west wall has a general north and south course, and stands for 600 feet deep wherever exposed at, say, 50° east. The 1,000-foot level showed a comparatively ore-barren quartz-mass, (showing only small spots and seams of ore,) 80 to 100 feet in width, which lay from 20 to 60 feet east from the west wall. In this quartz all the ore-bodies found above this level disappeared, pinching out into thin seams as they came down into it. It is bounded on the east by clay and porphyry; runs nearly parallel with the west wall; has been cut through by the incline about 80 feet above the 1,500-foot level, where it seemed not to be so wide as above. This quartz-body carries little porphyry, and only a very small percentage of carbonate of lime. It dips 48° east. No explorations have been made in it since the 1,000-foot level cut through it. At the 1,100-foot level the quartz-body which has furnished all the ore for the past three years was encountered 275 feet east of the shaft, being divided by porphyry and quartz from the western quartz. Once cut in the 1,100-foot level, it was explored upward, and found to extend above the 1,000-foot level. As depth was attained it widened, and, instead of maintaining a distinct boundary to the east, it became more and more mixed with porphyry. Strange to say, neither the Belcher nor the Crown Point Company has ever had the curiosity to explore the ground east of it at any point farther than 65 feet. This quartz-mass varies in width from 90 to 250 feet, and carries, in places, large quantities of carbonate of lime, especially in its western portion. Explorations so far have proved that it contains two distinct bodies of ore, namely, an eastern body, which has no lime, and a western, which has a good deal. These two bodies unite at the 1,500-foot level of Crown Point, about 350 feet north of the Belcher line, and separate in the Belcher ground, so that about 400 feet south from the Crown Point line the distance between the two is 115 feet, filled with quartz and porphyry. (The porphyry is identical with that which constituted the so-called east porphyry of the Comstock, at and above the 500-foot level.)

The western body kept a steady dip of 45° east, had its widest place at the 1,500-foot level, and carried the ore universally in seams alternating with barren crystalline quartz and calcite. At the 1,500-foot level in Crown Point it had become too poor to work with profit. The eastern body continues deeper than the 1,500-foot level, but lies now (at 80 feet below the 1,500-foot level) at an angle of 27° east, and its course is south 20° east. It has carried its ores more evenly disseminated through the entire quartz-mass, and has been the stronger of the two bodies. Wherever the quartz widens the silver diminishes. The southern portion of it, in Belcher, some 50 feet south of Crown Point, gave, as assayed by Mr. Luckhardt, $45 in gold and $8 in silver per ton. Now these two bodies diverge southward and downward at and below 1,500 feet depth, and the material between them in the north portion, where they are near together, is wholly quartz. In this has been found the third body, above referred to, which may prove a large one, extending, as has been remarked, into Kentuck. The small ore-seams in the west body and the general appearance of the ground lead to the belief that this third body will be larger northward than southward. It is less likely that the porphyries which show themselves in the gap southward will give way to quartz. Moreover, this intermediate quartz to the north is probably the identical quartz-body which the 1,100, 1,300, and 1,400 foot levels of the Yellow Jacket developed. The proportion of the gold and silver in it being the same, is additional proof. If these views are correct, it is highly probable that within the year a fine ore-body will be found between the centers of Yellow Jacket and Crown Point in the quartz which shows itself at both points. The 1,600-foot level of Crown Point, about 365 feet south from the Kentuck line, has very recently encountered ore in the region referred to.

The two bodies already known were explored by five winzes, one being sunk where they join on the north and two in each body farther south. With the exception of considerable solid ground in the east body, the greater part of both has been worked out. The present yield per day is about 500 tons of $30 mill-ore, and there is still standing fully nine months' supply from the lowest points explored by the winzes below the 1,500-foot level to the 1,300-foot level and north and south of the present stopes. The 1,300-foot level showed the ore in the widest place 130 feet, and between the 1,300 and 1,400 are about 125 feet length of ore-ground, standing 40 feet high, averaging 40 feet in width. Below the 1,300-foot level, 350 feet from the Belcher line, the two bodies come together, and at the Belcher line the space between the two is 88 feet, and the west body is poor. At 35 feet north of the Belcher line is the south winze, now 49 feet deep in the east body. The west body has been exhausted at this point. At the 1,400-foot level the south winze is 40 feet north of the Belcher line, and connects with the 1,500 level. The middle winze is 135 feet north of the Belcher line, and the north winze 300 feet north of the Belcher line—all in the east body, and all showing the same quality of ore as above stated, (average $35.) At the 1,400-foot level the ore is 50 feet wide, and is known to extend from the north winze to the south winze, 300 feet in length. The west body, 36 feet wide, is exhausted on this level. Stoping has been commenced 40 feet above the 1,500-foot level north, where the west body gives out in ore—*i. e.*, stops short, without first decreasing gradually in its ore-contents. The west body has two winzes in it between the 1,400 and the 1,500 foot levels, which, however, proved but poor ore, ($20 mill-ore.) At 40 feet north of the Belcher line in this level a cross-drift east was run for 60 feet, but no east wall has been found. At 140 feet farther north another eastward cut was opened, and continued 65 feet east beyond the ore, with the same result. Here the east body is 10 feet wide, dipping very flat eastward. Here the west quartz (away to the west of the west body) has never been explored. Farther north the 1,500-foot level had to pass 60 feet of quartz width before it came to the west quartz body. About 140 feet north of the Belcher line the east cross-cut shows, 60 feet east of the east body, some fine quartz-stringers, and it is surprising that no work further east was done here, as it certainly ought to have been. There is no east wall visible. Below the 1,500-foot level no ore has been extracted. The west body is, so far as explored, valueless, the ores being too poor to pay for working. The east body goes off very flat, and shows signs of weakening. The future success of Crown Point depends much on the explorations and its results now pending in the 1,600 and also the work east in the 1,500 foot level, which ought to be undertaken without fail. The incline bottom is now 367 feet south of the Kentuck line; it traverses the company's ground, commencing at the 1,000-foot level at the Kentuck line. The cars used for hoisting ore and waste lift 7 tons, using round ropes of steel. The dead-weight of the incline-rope which has to be overcome is 17,000 pounds. The company has an extra incline-engine of 134 horse-power, a splendid piece of work. The incline car has brakes attached, and all are made after the pattern of Mr. H. Donelly, of whom I spoke last year, who now is in the Belcher, and showed his good sense there in all the contrivances and appliances for using compressed air.

*Belcher* has attained a depth of 1,580 feet. The shaft is 850 feet vertically, whence begins the incline, 250 feet south of Crown Point. The ore here in Belcher extended up to the 900-foot level, and a considerable quantity of $30 ore is still standing here between the 900 to 1,050 foot level. Below 1,050 feet all the ground has been excavated for 300 feet south from Crown Point, and for 60 to 100 feet in width, (horizontally,) down to the 1,300-foot level. Some low-grade ore is standing south of the stopes; but it is pinching out fast. The heaviest body is the west body of the two above described. At 1,150 feet the ore has been found to extend 380 feet south of the Crown Point line, at least, but it may go 50 feet farther. It yields $25 per ton. At 1,200 feet, 60 feet from the Crown Point line, the east body of the Crown Point is visible, and 50 feet wide.

At 50 feet above the 1,300-foot level the ore is only 15 feet wide, and has been worked for 400 feet south of the north line, where it was at least 60 feet in width, and of excellent quality.

At the 1,300-foot level connections are made with the Overmann, the Crown Point, and through the latter with Yellow Jacket and Imperial. Below the 1,300-foot level from 90 to 100 feet in length south from the Crown Point line has been extracted. The ore is now 12 to 15 feet in width at this point, and will extend 130 to 150 feet (in all) southward from the north line, so that, possibly, 40 or 58 feet of ore-ground is standing here yet. At the 1,400-foot level, 120 feet south of the north line, a winze has been sunk, which shows ore down to the 1,500-foot level. (Gold predominates over silver. Assays, gold, $44.60; silver, $8.10.) There are two more winzes here, (north,) which also connect with the 1,500-foot level, but the ore is not over $35 per ton. At the 1,400-foot level, about 400 feet south of Crown Point, the quartz becomes wider, carrying less silver the farther south it extends, but it also lies much flatter to the east. This quartz, I believe, is the southern portion and terminus of the east body of Crown Point. In all these stopes and below the 1,400 to the 1,500 foot level there are possibly 12 to 14 months of ore-reserves at the rate of 400 tons per day of $30 per ton milling-value. The product now is about 500 tons per day, varying from $27 to $35 per ton. In the ground between the 1,200 and 1,400 foot levels several cross-drifts eastward were run, one of them being 200 feet in length, but no east wall was encountered. Hence, as far as explored, the vein has really a width of over 400 feet, and no east boundary as yet. On the 1,400-foot level the west clay lies 168 feet east of the incline, but the west quartz of Crown Point and Yellow Jacket, above named, into which the upper ore-bodies pinched, does not exist here in Belcher, south. Explorations ought to be carried on eastward much farther than has been up to date. The product of Belcher for the year was about the same as of Crown Point—that is to say, the tonnage. The quality of the ore was better in Belcher. Mr. Donelly, the foreman, has adopted a good principle. The mine has in the west wall a streak of soft porphyritic material, in which the company have sunk a shaft 12 by 6 feet, which is used as an air and a timber-lowering shaft. Through it splendid ventilation is produced. One of these shafts had been constructed before, but was lost by fire from the 1,050 up. It will be sunk west of the incline and all levels will be connected with it.

Burleigh's air-compressor and drill do very good service. The sinking of winzes, hoisting of timbers to the stopes, &c., are very much facilitated by the engines used underground, fed with compressed air from the surface, and it is only by using them that prospecting work can be carried on as extensively as is now the case, in spite of the natural temperature.

*The Overmann.*—This mine has attained a depth of 1,150 feet, of which 200 feet were sunk this year. The ground 500 feet north of the new shaft was explored by a winze, which showed ore in seams of very good quality. The vein was 200 feet wide, and the 900-foot level showed 18 feet width of quartz which was ore-bearing for 200 feet in length. The quartz of the 1,300-foot level of Belcher somewhat resembles this, but the two are probably not the same body. The ore is sparingly distributed. Not enough work has been done to permit any conclusion as to the future developments. The 1,100-foot level has been opened 60 feet west of the shaft, and the ground has been explored for a short distance south and north. Nothing of note has been developed. The ground yields a large bulk of water, which necessitates the erection of new pumping-machinery, and which will retard all work for some months this coming season. South and west of Overmann every inch of ground has been relocated and incorporated, and stock issued. The western quartz-bodies which lie in American Flat have given much employment to miners and capitalists during the past year. But nothing noteworthy has been exposed. Much low-grade ore exists in the western red-quartz body, which courses through Devil's Gate and through American Flat, and some of the companies located thereon are producing from it.

On the south or southwestern branch of the Comstock, toward Devil's Gate, very extensive explorations have been made during the year, and the ground east of Justis and in Justis has disclosed to the east of the so-called "red ledge" (which undoubtedly is the south and southeast outrunners of the body which made Overmann, &c., so famous in former times) a body of white quartz, very similar in its nature, its character of ore, and minerals to that which constituted the west body of the Crown Point 1,200-foot level. In the Justis ground it contains detached ore-seams, in which the ores are barely distinguishable from those of Crown Point, and assay very high. Mr. Luckhardt could not make a close examination, since the development is very recent. The same quartz is 50 to 60 feet wide in the Silver Hill and Dayton mines, farther southwest from the Justis; but it does not carry the ore as yet. Probably during this year much valuable ore-ground will be exposed in this direction. The folly of incorporating questionable titles or ground begins already to show itself between Justis and Woodville. I believe that only a connection in ore between the two mines will settle the ownership. This is the best practical proof, and is not bad, geologically, even.

The Dayton, Silver Hill, Woodville, &c., have extracted much ore within the year from the red-quartz body, and extensive machinery is under erection or has already been erected to search for and thoroughly explore this new white-quartz body. The Silver Hill has attained a depth of 368 feet vertically, and has explored the ground for 1,440 feet northwest and southeast and 360 feet northeast and southwest. The red ledge pinches at 334 feet depth. The quartz (white) is 50 feet wide here, and lies to the south, widening. It shows by assay from $5 to $15 per ton, but is not compact enough here to carry ore. Apparently the red quartz has had, even at this level, some influence upon it, and it is not unlikely that farther east and southeast, possibly a few hundred feet from the present workings, the porphyry will be less abundant and quartz will predominate, when ore in paying quantities may be expected. The red ledge carries an abundance of low-grade ore, which will pay the company's expenses for one or two years to come. It carries principally gold. The company is erecting heavy machinery. The same may be said of the Dayton mine. It has attained 450 feet depth, with explorations of the ground 750 feet north and south by 200 feet east and west. The red ledge shows much fair milling-ore in sight, but begins to give out at or near the 280-foot level. The white quartz sets in at about this depth, to the east of it, and looks very promising. Assays of it show a predominance of the silver over the gold. In Mr. Luckhardt's opinion, which seems to me based on sound local analogy, the increase in bulk of this quartz-body will be attended with the appearance of more compact ore.

*The Sutro tunnel.*—This enterprise has been prosecuted with energy, in the face of the natural obstacles and the bitter opposition from hostile interests, which has constituted an element in its history almost from the beginning. With the merits of the legal and other disputes in which this company is involved I have had nothing to do, and I profess no opinion concerning them. As to the great importance of the tunnel to the development of the Comstock lode, and its thorough and profitable exploitation, I hold the same opinion as I have already often expressed. The discovery of large ore-bodies at great depth cannot fail to encourage the enterprise of a deep tunnel; and the circumstance that many of the shafts are already near or below the tunnel-level does not at all destroy the argument in favor of the scheme, based on the advantages to be expected from drainage, ventilation, reduced hoisting and pumping expenses, safety of miners, and gain of water-power by hydraulic machinery or turbines. I introduce here, while preparing this manuscript for the press, the following account of the tapping of shaft No. 2, from the Virginia Enterprise of March 12, 1875:

The tapping of the water in shaft No. 2, Sutro tunnel, was yesterday successfully accomplished, under the supervision of chief engineer of the works, Carl O. Wedekind, a most accomplished young engineer. He had charge of the central-shaft section of the Hoosac tunnel, and has had much experience in other large works of the kind. He took charge of the tunnel as chief engineer May 1, 1874, and has since been conducting that work. It will be remembered that on the 30th of June, 1874, the men were driven out of shaft No. 2 by the striking of a heavy body of water as they were drifting west from its bottom. The volume of water was too great to be handled by the pumps, and the shaft, 1,040 feet in depth, was filled up to within 100 feet of its top. Work at that point was then suspended until the tunnel-header should be advanced far enough to tap the immense body of water obtained in the shaft. On the 8th instant the regular work of driving the header of the tunnel was suspended, being at a distance of 8,800 feet from the mouth of the tunnel, and 98 feet from the water in shaft

No. 2. A diamond drill was mounted and adjusted by Mr. Wedekind, with the aid of a transit-instrument for the grade, in order to exactly strike the header driven east from shaft No. 2. A double bulk-head, built of 12-inch timbers, was constructed between the drilling-machine and the face of the tunnel, with a quarter-inch boiler-plate slide arranged vertically in its center. Drilling was begun on the 9th instant, at 11 o'clock p. m., the power for the drill being compressed air. It made its way through unusually hard rock at an average speed of 18 inches every 15 minutes, diameter of hole being 2 inches. Excepting one interruption of several hours, caused by the breaking of the diamond-bit, the drilling went on finely, and on March 11, at 2 o'clock a. m., three times three cheers were heard from all when the drill-rod broke through. The water, coming out at the side of the rod under a head of 835 feet, (to which the water had lowered since June,) was white as snow, and quite hard to feel of. Great care was taken in letting back the drill-rod; but after a few feet had come out, the friction between the machine and rod and the grip of every man present was no longer sufficient. It went like a streak of lightning, even not lacking the fiery part, and finally lodged back in the tunnel. The plate in the middle of the bulk-head went down in front of the hole, changing the 120 inches of water coming out into a beautiful white feather-shaped fountain. The idea of striking the workings of No. 2 shaft by a diamond drill through 100 feet of rock, at a distance of one and two-third miles from the mouth of the tunnel, has been doubted by almost every one. Its successful determination establishes here the already well-earned name that Mr. Wedekind has attained by his accurate work on the Hoosac tunnel and other places. In three hours the water had been lowered in the shaft 128 feet.

The following still later report of progress to May 7, from the Scientific Press, is added to this report during publication:

The following report of progress made in the Sutro tunnel, Nevada, for the week ending May 8, was furnished us by Pelham W. Ames, secretary of the company:

| | |
|---|---|
| Number of feet in, May 1 | 9,271 |
| Number of feet driven during week | 71 |
| Distance in, May 8, (feet) | 9,342 |

The work is a conglomerate greenstone base, with angular pieces of trachyte imbedded, so hard and tough as to require repeated charges in same holes before blast would take effect. At the last moment the rock is reported as presenting a more favorable appearance.

The work during the week was as follows: Holes drilled, 372; holes blasted, 405; aggregate depth, 2,303 feet; average depth, 6 $\frac{71}{372}$; powder consumed, 1,099 pounds; exploders consumed, 506.

The water, which had been running in at the face through an open fissure, was left behind on 3d instant, when the fissure took a sudden turn to the south. On the 2d instant 30 inches of water (a new body) were struck, making flow from tunnel 87 inches, but this has since decreased to 71 inches. This is now utilized for cutting a wagon-road through a high, sandy bluff, near Carson River.

Temperature of air at heading 79°; shaft No. 1 east, 76°; west, 70°; shaft No. 2, east, 78°; west, 80°; mouth, 63°. Temperature of water at heading, 79°; shaft No. 1 east, 77°; shaft No. 2 east, 79°; mouth, 75°.

Currents of air pass in at mouth and down shaft No. 1, up shaft No. 2.

## *Report of the Savage Mining Company for the year ending June* 30, 1874.

Mr. A. C. Hamilton, the superintendent, says:

At the date of my last annual report, July 1, 1873, the work of extracting ore from the upper levels of the Savage mine had been discontinued for reasons which were mentioned in that report. It had become quite evident that the future prosperity and value of the mine would depend upon what might be developed by deeper working. I have, therefore, confined our operations during the last twelve months entirely to the sinking of the main-shaft incline and to the opening and prospecting of the ground from the 1,600-foot level downward. This work has been pushed forward with all the force which could be employed to advantage, and it has not been materially retarded at any time by any interruption.

Up to this time we have not been so successful as to reach any body of pay-ore. But as each successive level has been opened and explored, the vein has been found to be continuous, and to preserve its regularity and extent. This fact, taken in connection with the character and appearance of the vein as shown on the 1,900-foot level, is encouraging, and a reasonable expectation may be entertained that the vein will eventually, and possibly before long, make into an ore-body and yield a revenue, as it did in years past in the upper levels of the mine. I am looking with much interest to the development of the 2,000-foot level, which is now being opened, and which we will very soon be prepared to prospect, by running cross-cuts to the west from the east wall of the ledge.

The work accomplished and the progress made during the past year may be briefly summed up as follows: On the 1,600-foot level the cross-drifts which had been commenced have been continued west until they reached what is supposed to be the west wall. On the 1,700-foot level, the main south drift, which had connected with the winze, has been continued 60 feet to the south, or Hale & Norcross line. From this main drift three cross-cuts have been run to the west, one 135 feet, one 280 feet, and one 60 feet in length. From this level two winzes have been sunk, one 75 feet in depth, and the other 240 feet on an incline to the 1,900-foot level. This latter winze reaches the 1,900-foot level at a point 82 feet from the south line. On the 1,900-foot level the main south drift has been run from the shaft incline 295 feet to the south line. From this drift four cross-drifts have been run west, cutting the vein at right angles, one 302 feet, one 50 feet, one 90 feet, and one 85 feet in length. A winze has been sunk from this level—on an incline of 60°—121 feet, reaching the 2,000-foot level at a point 92 feet from the south line. On the 2,000-foot level connection has been made with this winze by a drift run from the shaft-incline—a distance of 210 feet—following the east wall. This drift is now being driven southward to reach the south line.

All the work spoken of above has been done in that portion of the mine which lies on the south side of the shaft-incline. On the 1,600-foot level two cross-drifts have been run in the ground lying north of the shaft-incline, one to the west 125 feet, and the other to the east 245 feet. Below this level (the 1,600-foot) that portion of the ground which is on the north side of the incline-shaft, and which constitutes more than one-half of the entire mine, has not been penetrated in any manner whatever.

The shaft-incline has been sunk 534 feet, from the 1,700-foot station down to a point 50 feet below the 2,000-foot station—all the way in hard blasting-rock—at an inclination of 38¼° from the 1,700-foot station to the 1,900-foot station, and 39¼° from the 1,900-foot station to the bottom.

The excavations made during the year, in linear feet, aggregate as follows:

| | Feet. |
|---|---|
| Shaft-incline | 534 |
| Main drifts | 695 |
| Cross-drifts | 1,689 |
| Winzes | 436 |

To these may be added the large spaces opened in solid rock for the 1,900-foot and 2,000-foot stations.

The power of the pumping and the hoisting engines has been fully equal to the requirements of our deep working, and it is quite sufficient for sinking to and opening another level 100 feet deeper. The large beam-engine, which is used exclusively for pumping, has sufficient power to drain the mine of water to a much greater depth than has yet been reached. But for the purpose of sinking below a 2,100-foot level and hoisting from below that depth, it will be necessary either to have one additional engine at the surface, or to provide some intermediate power placed at some point under ground.

For the purpose of supplying the lower levels of the mine with an ample quantity of fresh air, I am having made and will soon have in place a new wooden air-pipe, 12 inches square in the clear, which will extend from the top of the shaft down to the 2,000-foot station. The air will be forced down this pipe by two air-blowers stationed on the surface.

Everything pertaining to the machinery and to the hoisting-works is in good order, and I have not failed to adopt, from time to time, such improvements which have been suggested by the ingenuity of competent men, who are in charge of the several departments of the mining-works, as would give security against accidents and promote convenience and economy in working the mine.

The putting up of a new office-building during the year has involved an extra expense. The old building, a brick structure, had become so much cracked and warped, by the settling of the ground upon which it stood, that it seemed unsafe to occupy it any longer. I have had it removed, and in its place erected a wooden building set upon a firm foundation and built in a substantial manner. This building serves for office-rooms and for the superintendent's dwelling.

The two mills belonging to the company, the Savage and the Atchison, situated at Washoe, have been for a long time out of use, and they were passing into that state of dilapidation which is incident to property of that kind when it is not kept in use and in repair, and at the same time there was a continual expense for a watchman, for taxes, and for insurance. In accordance with instructions, the buildings and what remained of the machinery and fixtures have been sold at the best price offered. The mill-sites and the water-right still remain the property of the company.

The secretary's report shows the following receipts and disbursements:

RECEIPTS.

| | |
|---|---|
| Cash: | |
| Balance on hand July 10, 1873 | $10,703 59 |
| Assessments: | |
| Nos. 10, 11, 12, 13, and 14 | 400,000 00 |
| Savage Mill: | |
| Machinery and old iron sold | 4,300 00 |
| Atchison Mill: | |
| Machinery and old iron sold | 1,100 00 |
| Virginia and Truckee Railroad Company: | |
| Return freight-charges | 1,545 33 |
| | 417,648 92 |

DISBURSEMENTS.

| | |
|---|---|
| Miners, engineers, &c., and company officers | $212,410 58 |
| Timber and lumber | 23,412 18 |
| Wood and charcoal | 55,299 45 |
| Mining-supplies, hardware, candles, oil, &c., and insurance on hoisting-works | 57,839 76 |
| Virginia and Gold Hill water-works | 9,600 00 |
| Assay-office, net cost | 5,733 87 |
| State, city, and county taxes | 5,180 05 |
| Legal expense | 5,213 70 |
| Freights to Virginia | 868 65 |
| Surveying in mine | 825 00 |
| Horse-feed and vehicles | 1,646 90 |
| Books and stationery | 834 16 |
| Interest on over-drafts at bank | 73 30 |
| Exchange on superintendent's drafts | 1,602 01 |
| Office-rent, porter, &c | 2,641 10 |
| Sundry extraordinary expenses | 7,150 90 |
| Watchman at Savage Mill | 300 00 |
| Profit and loss: | |
| Reclamation on bars | 100 12 |
| Office-fixtures for new office | 592 50 |
| Cash on hand July 11, 1874 | 26,324 69 |
| | 417,648 92 |

ASSETS.

Real estate at Virginia City, Nevada.

E-street works and machinery.

(These works are in equally as good condition and of as much value as they have been at any time during the past five years.)

Mill sites and water-privileges at Washoe.

| | |
|---|---|
| Assay-office, fixtures, and tools—value | $1,000 |
| Office fixtures and furniture | 2,500 |
| Mining-supplies, stores, and assay-materials on hand at mine—cost | 44,500 |
| Cash on hand | 26,324 |

The company has no liabilities.

## *Report of the Chollar Potosi Mining Company for the year ending May 31, 1874.*

Mr. Isaac L. Requa, the superintendent, gives the following summary of the results of work at the company's mine during the year ending May 31, 1874:

Ore exracted........................................................ 32,915 tons.
Ore reduced at mills ................................................ 35,341 tons.

The ore-supplies have been mostly drawn from those sections of the mine known as the Blue Wing and Santa Fé. These old stopes were worked years ago, when such ores as we are now extracting were untouched, owing to their value being too low to pay the milling-rates of those days. At this time we have no other ore-sources that are reliable.

Prospecting has been carried on extensively, nearly ten thousand feet of prospecting drifts, winzes, and inclines having been made. At many points we have discovered quartz in large bodies, carrying very little precious metals. At other sections the material has assayed very largely, while the quantity was limited.

In the last yearly report reference was made to indications at the fourth station, near the north line. Since then that portion of the company's ground has been very thoroughly looked over. I regret to say that the results were not an adequate compensation for the outlay necessary to proper prospecting of that section.

During the past eight months our efforts in seeking ore have been mainly devoted to the southern part of our mine, from the second to the fifth stations, embracing a depth of 700 feet, and 700 feet in length of ground entirely unprospected. At the fifth or lowest level several small bodies of excellent ore have been discovered. Indications seem to justify the conclusion that these small deposits are merely the offshoot of a large and valuable "bonanza" in the immediate neighborhood of present workings. For the purpose of developing the anticipated extensive ore-deposit, operations are ceaselessly carried on at this point. Should success attend the efforts, resort will be had for greater depths to the main incline, which is in good condition and requires only the removal of the water, which can be easily accomplished, to permit of opening a level 300 feet below present fifth station.

The producing portions are affording 90 tons of ore daily, and, so far as can be determined, this quantity will not be lessened for next year, and from present prospects may be largely increased during the twelve months to come.

The main shaft and pumps are in complete working condition. Wire ropes and cages, engines, and all other machinery necessary for full operations at company's mine are in perfect order.

*Condensed statement of cost, production, &c., of Chollar Potosi Mining Company, from June 1, 1873, to May 31, 1874.*

| *Ore-statement:* | *Tons.* | *Lbs.* |
|---|---|---|
| Ore on hand June 1, 1873 | 3,849 | 520 |
| Extracted during year | 32,915 | 1,600 |
| Total | 36,765 | 120 |
| Worked during year | 35,341 | ...... |
| On hand June 1, 1874 | 1,424 | 120 |

| *Cost per ton:* | |
|---|---|
| Extracting ore, repairs, prospecting, dead-work, and incidentals | $6 84 |
| Reduction, including milling, melting, and assaying | 12 44 |
| Total cost per ton | 19 28 |
| Gross yield per ton of ore worked | 17 47 |
| Net loss per ton of ore worked | 1 81 |

| Bullion. | Gold. | Silver. | Total. |
|---|---|---|---|
| Average value of bullion per ounce | $0.61 | $1.25 | $1.86 |
| Average fineness of bullion per ounce | .029.5 | .964 | .993.5 |
| Average proportion of precious metals | 33 | 67 | 100 |

*Condensed statement of cost, production, &c., of Chollar Potosi Mining Co., &c.*—Continued.

| | |
|---|---|
| *Work of assay-office:* | |
| Ounces troy of bullion assayed before melting | 340, 535. 45 |
| Ounces troy of bullion assayed after melting | 331, 965. 55 |
| Average per cent. loss in melting | 2. 5 |
| Number of ore-assays made | 6, 513 |
| Number of bars made | 243 |
| Total weight of bars in pounds avoirdupois | 22, 768 |

| | |
|---|---|
| *Rece'pts and expenses:* | |
| Received from bullion | $617, 391 22 |
| Received from other sources | 2, 538 68 |
| Total | 619, 929 90 |
| Expense by all sources | 684, 023 41 |
| Net loss for year | 64, 093 51 |

*General statement of Chollar Potosi Mining Company, showing net cost of mining, milling, &c., and returns from bullion, &c., for company year from June 1, 1873, to May 31, 1874.*

| Months. | Number of tons worked. | Mine-costs. | | | | |
|---|---|---|---|---|---|---|
| | | Milling ore. | Labor. | Timber. | Wood. | Coal. |
| 1873. | | | | | | |
| Stock on hand June 1 | ........ | ........ | ........ | $17, 053 97 | $3, 210 88 | $1, 280 83 |
| Purchased in June | 1, 000 | $11, 000 00 | $12, 800 00 | 3, 924 80 | ........ | 98 00 |
| Purchased in July | 2, 500 | 27, 500 00 | 12, 286 75 | 4, 728 61 | 198 00 | 170 24 |
| Purchased in August | 3, 771 | 39, 910 00 | 14, 036 25 | 5, 803 51 | 2, 598 75 | 177 84 |
| Purchased in September | 2, 250 | 27, 000 00 | 13, 683 25 | 7, 146 95 | 2, 716 87 | ........ |
| Purchased in October | 2, 280 | 27, 360 00 | 15, 180 25 | 3, 191 55 | 247 50 | 337 05 |
| Purchased in November | 2, 700 | 32, 400 00 | 12, 625 75 | 1, 090 75 | 3, 461 50 | 94 50 |
| Purchased in December | 4, 378 | 55, 914 00 | 14, 119 00 | 361 20 | ........ | ........ |
| 1874. | | | | | | |
| Purchased in January | 3, 802 | 49, 098 00 | 9, 981 75 | 827 30 | ........ | ........ |
| Purchased in February | 2, 800 | 36, 400 00 | 8, 665 00 | 178 26 | ........ | ........ |
| Purchased in March | 4, 450 | 57, 850 00 | 12, 705 87 | 867 65 | ........ | ........ |
| Purchased in April | 2, 975 | 38, 675 00 | 13, 145 50 | 818 42 | ........ | ........ |
| Purchased in May | 2, 435 | 31, 655 00 | 14, 146 25 | 298 60 | ........ | ........ |
| Materials paid for in San Francisco | ........ | ........ | ........ | ........ | ........ | 236 00 |
| Total gross expense | 35, 341 | 434, 762 00 | 153, 375 62 | 46, 291 57 | 12, 433 50 | 2, 394 46 |
| Less materials sold, &c. | ........ | ........ | ........ | ........ | ........ | ........ |
| Less stock on hand June 1, 1874. | ........ | ........ | ........ | 11, 602 44 | 3, 860 58 | ........ |
| Total net expense | 35, 341 | 434, 762 00 | 153, 375 62 | 34, 689 13 | 8, 572 92 | 2, 394 46 |
| Average per ton | ........ | 12 30 | 4 34 | 98 | 24 | 07 |

*General statement of Chollar Potosi Mining Company, showing net cost of mining, milling, &c, and returns from bullion, &c.*—Continued.

| Months. | Mine-costs. | | | | | |
|---|---|---|---|---|---|---|
| | Candles. | Miscellaneous supplies. | Assay-office. | Water. | Taxes. | Ex-change. |
| 1873. | | | | | | |
| Stock on hand June 1 | $870 00 | $2,088 90 | $300 00 | ........ | ........ | ........ |
| Purchased in June | ........ | 916 67 | 500 00 | $550 00 | $407 82 | $146 50 |
| Purchased in July | ........ | 1,924 53 | 539 27 | 275 00 | ........ | 65 00 |
| Purchased in August | ........ | 4,588 57 | 380 00 | 550 00 | 340 30 | 85 00 |
| Purchased in September | 197 50 | 826 65 | 380 00 | 550 00 | 461 25 | 60 00 |
| Purchased in October | 332 75 | 1,848 13 | 619 15 | 550 00 | ........ | 80 00 |
| Purchased in November | 217 75 | 532 07 | 370 00 | 550 00 | 2,943 86 | 80 00 |
| Purchased in December | 111 02 | 871 40 | 600 00 | 550 00 | ........ | 80 00 |
| 1874. | | | | | | |
| Purchased in January | ........ | 1,348 37 | 385 00 | 550 00 | 548 00 | 95 00 |
| Purchased in February | ........ | 1,148 35 | 370 00 | 550 00 | 489 09 | 15 00 |
| Purchased in March | ........ | 933 55 | 380 00 | 550 00 | ........ | 65 00 |
| Purchased in April | ........ | 1,790 72 | 380 00 | 550 00 | ........ | 75 00 |
| Purchased in May | ........ | 1,246 27 | 380 00 | 550 00 | ........ | 65 00 |
| Materials paid for in San Francisco | 1,933 83 | 9,050 04 | 245 96 | ........ | ........ | ........ |
| Total gross expense | 3,662 85 | 29,123 22 | 5,829 98 | 6,325 00 | 5,190 32 | 931 50 |
| Less materials sold, &c | ........ | 1,605 71 | 782 97 | 60 00 | ........ | ........ |
| Less stock on hand June 1, 1874 | 759 58 | 3,024 01 | 300 00 | ........ | ........ | ........ |
| Total net expense | 2,903 27 | 24,493 50 | 4,747 01 | 6,265 00 | 5,190 32 | 931 50 |
| Average per ton | 08 | 69 | 14 | 18 | 14½ | 03 |

| Months. | Mine-costs. | | | Mine-returns. | | |
|---|---|---|---|---|---|---|
| | Legal. | Real estate. | Total. | From bullion. | From sundries. | Total. |
| 1873. | | | | | | |
| Stock on hand June 1 | ........ | ........ | $24,804 58 | ........ | ........ | ........ |
| Purchased in June | $250 00 | ........ | 30,593 79 | $17,025 16 | $168 21 | $17,193 37 |
| Purchased in July | 250 00 | ........ | 47,937 40 | 46,449 75 | ........ | 46,449 75 |
| Purchased in August | 250 00 | ........ | 68,720 22 | 64,479 20 | 60 00 | 64,539 20 |
| Purchased in September | 250 00 | ........ | 53,292 47 | 43,636 70 | ........ | 43,636 70 |
| Purchased in October | 250 00 | ........ | 49,996 38 | 42,902 91 | 371 50 | 43,274 41 |
| Purchased in November | 250 00 | ........ | 54,616 18 | 47,270 66 | 90 00 | 47,360 66 |
| Purchased in December | 250 00 | ........ | 72,857 22 | 72,855 89 | 782 97 | 73,638 86 |
| 1874. | | | | | | |
| Purchased in January | 250 00 | $250 00 | 63,333 42 | 60,028 82 | ........ | 60,028 82 |
| Purchased in February | 250 00 | ........ | 48,065 70 | 47,957 30 | ........ | 47,957 30 |
| Purchased in March | 250 00 | ........ | 73,602 07 | 80,409 24 | 1,066 00 | 81,475 24 |
| Purchased in April | 250 00 | ........ | 55,684 64 | 51,560 53 | ........ | 51,560 53 |
| Purchased in May | 250 00 | ........ | 48,591 12 | 42,815 06 | ........ | 42,815 06 |
| Materials paid for in San Francisco | ........ | ........ | 11,474 83 | ........ | ........ | ........ |
| Total gross expense | 3,000 00 | 250 00 | 703,570 02 | 617,391 22 | 2,538 68 | 619,929 90 |
| Less materials sold, &c | ........ | 90 00 | 2,538 68 | ........ | ........ | ........ |
| Less stock on hand June 1, 1874 | ........ | ........ | 19,546 61 | ........ | ........ | ........ |
| Total net expense | 3,000 00 | 160 00 | 681,484 73 | ........ | ........ | ........ |
| Average per ton | 08 | 00½ | 19 28 | 17 47 | 07 | 17 54 |

*Statement showing the work of company's assay office from June 1, 1873, to May 31, 1874.*

| Months. | No. of bars made. | Numbers of bars. | | Weight of bullion in troy ounces. | | Average per cent. loss in melting. | Gold. | | Silver. | | Total value. | Average value per ounce. | Number of ore-assays made. | Pounds avoirdupois of bullion produced. | Cost of office. |
|---|---|---|---|---|---|---|---|---|---|---|---|---|---|---|---|
| | | From— | To— | Before melting. | After melting. | | Average fineness. | Value. | Average fineness. | Value. | | | | | |
| 1873. | | | | | | | | | | | | | | | |
| June | 6 | 4,136 | 4,141 | 9,535.30 | 9,003.00 | 5.5 | .031 | $5,835 32 | .961 | $11,189 84 | $17,025 16 | $1 96 | 599 | 617 | $800 00 |
| July | 19 | 4,142 | 4,160 | 25,675.10 | 24,880.95 | 3.1 | .030 | 15,472 24 | .963 | 30,977 51 | 46,449 75 | 1 86 | 545 | 1,707 | 539 27 |
| August | 25 | 4,161 | 4,185 | 33,850.90 | 33,232.65 | 1.8 | .032 | 23,200 58 | .961 | 41,278 62 | 64,479 20 | 1 94 | 627 | 2,279 | 380 00 |
| September | 17 | 4,186 | 4,202 | 24,804.10 | 24,127.85 | 2.7 | .027 | 13,458 00 | .967 | 30,178 70 | 43,636 70 | 1 80 | 494 | 1,655 | 380 00 |
| October | 17 | 4,203 | 4,219 | 24,625.50 | 23,941.35 | 2.7 | .026 | 13,017 03 | .966 | 29,885 88 | 42,902 91 | 1 78 | 566 | 1,642 | 619 15 |
| November | 18 | 4,220 | 4,237 | 27,067.65 | 26,439.90 | 2.3 | .026 | 14,444 82 | .960 | 32,825 84 | 47,270 66 | 1 78 | 523 | 1,814 | 370 00 |
| December | 31 | 4,238 | 4,268 | 40,913.10 | 39,881.30 | 2.5 | .028 | 23,006 52 | .967 | 49,849 37 | 72,855 89 | 1 83 | 557 | 2,735 | 600 60 |
| 1874. | | | | | | | | | | | | | | | |
| January | 24 | 4,269 | 4,292 | 33,045.30 | 32,254.45 | 2.4 | .030 | 19,771 20 | .965 | 40,257 62 | 60,028 82 | 1 86 | 344 | 2,212 | 385 00 |
| February | 20 | 4,293 | 4,312 | 26,266.65 | 25,745.75 | 2 | .029 | 15,799 07 | .966 | 32,158 23 | 47,957 30 | 1 86 | 384 | 1,766 | 370 00 |
| March | 30 | 4,313 | 4,342 | 42,818.55 | 41,865.50 | 2.2 | .033 | 28,294 89 | .963 | 52,114 35 | 80,409 24 | 1 92 | 666 | 2,871 | 380 00 |
| April | 20 | 4,343 | 4,362 | 28,030.85 | 27,349.95 | 2.4 | .030.5 | 17,473 41 | .964 | 34,087 12 | 51,560 53 | 1 89 | 592 | 1,876 | 380 00 |
| May | 16 | 4,363 | 4,378 | 23,902.45 | 23,242.90 | 2.8 | .029 | 13,811 30 | .965 | 29,003 76 | 42,815 06 | 1 84 | 606 | 1,594 | 325 96 |
| | 24 | 4,136 | 4,378 | 340,535.45 | 331,965.55 | 2.5 | .029.5 | 203,584 38 | .964 | 413,806 84 | 617,391 22 | 1 86 | 6,513 | 22,768 | *5,529 98 |

| | | |
|---|---|---|
| * Gross cost of assay-office | | $5,529 98 |
| Less amount received from Buckeye Gold and Silver Mining Company for work done | $782 97 | |
| Less bullion-bars made from slag, ashes, dips, &c | 2,726 16 | |
| | | 3,509 13 |
| Leaving actual net cost of office | | 2,020 85 |

The following is the secretary's report:

RECEIPTS.

| | | |
|---|---|---|
| Bullion account: | | |
| Proceeds of bullion sold | | $617,269 86 |
| Superintendent Requa: | | |
| This amount received from him, being last year's balance. | | 1,499 53 |
| Cash account: | | |
| Cash on hand as per last statement | | 133,392 93 |
| | | 752,162 32 |

DISBURSEMENTS.

| | | |
|---|---|---|
| Working ores | | $434,762 00 |
| Labor | | 174,756 78 |
| Timber and lumber | | 29,237 60 |
| Hardware | | 7,585 86 |
| Coal | | 1,140 63 |
| Candles | | 2,814 85 |
| Powder | | 595 50 |
| Oil | | 662 70 |
| Wood | | 9,222 62 |
| Water | | 6,265 00 |
| Assaying | $993 98 | |
| Less received for assaying | 782 97 | |
| | | 211 01 |
| Rent of office | 1,500 00 | |
| Less received | 470 00 | |
| | | 1,030 00 |
| Legal expenses | | 3,400 00 |
| Taxes on real estate in Virginia and ore extracted | | 5,357 97 |
| Stationery | | 1,266 56 |
| Stable | | 2,527 06 |
| Real estate: | | |
| Purchase of lot at Virginia | | 250 00 |
| Freight of bullion | $1,770 40 | |
| Miscellaneous freight | 1,379 58 | |
| | | 3,149 98 |
| Discount on bullion sold | 18,840 86 | |
| Discount on superintendent's drafts | 931 50 | |
| | | 19,772 36 |
| Expense account: | | |
| Incidentals at Virginia office during year | 4,003 37 | |
| Incidentals at San Francisco office during year. | 1,059 05 | |
| Fire-insurance on company's property at Virginia | 1,692 00 | |
| Assessments Nos. 1 and 2 on Comstock Company *vs.* Sutro | 2,100 00 | |
| | | 8,854 42 |
| Machinery and materials account: | | |
| One Rand & Waring compressor | 4,625 00 | |
| Two Babcock fire-extinguishers | 120 00 | |
| Boiler tubes | 1,019 00 | |

| | | | |
|---|---|---|---|
| Gasoline | | $636 40 | |
| Rubber-hose | | 47 50 | |
| Car-wheels | | 392 50 | |
| T-rails | | 632 24 | |
| Sundry bills for materials | | 2,390 25 | |
| | | 9,862 89 | |
| Less material sold, as follows: | | | |
| To Belcher Company, Laflin & Rand battery, &c | $117 13 | | |
| Imperial Company, cage | 35 00 | | |
| Imperial Company, 142 feet pump-column | 994 00 | | |
| Jacket Company, cement | 16 08 | | |
| Virginia Consolidated, pipe | 72 00 | | |
| Virginia and Gold Hill Water Company, iron pipe | 246 00 | | |
| Gould & Curry, iron pipe | 78 00 | | |
| Sundries | 47 50 | | |
| | | 1,605 71 | |
| | | | $8,257 18 |
| Due from Superintendent Requa on account | | | 968 23 |
| Cash on hand | | | 30,074 01 |
| | | | 752,162 32 |

*Report of the Crown Point Mining Company for the year ending May* 1, 1874.

The report of Mr. S. L. Jones, the superintendent, is almost as full as that of last year, and gives a fair history of the mine for the year.

At the date of the last general report of the superintendent the 1,300-foot level was the deepest ore-producing level in the mine. We had, however, cut out a large and commodious station on the 1,400-foot level, and run a drift from it south along the foot-wall a distance of 142 feet, and had driven the main incline down to the 1,500-foot level and made the necessary preparations to cut out the station there.

Since that time the drift on the 1,400-foot level has been continued south to the Belcher line, and has been connected with the 1,300-foot level by three winzes, known respectively as the north, middle, and south winzes, which secure us good ventilation and the necessary facilities for the conveyance of lumber, timbers, &c.

The north winze was started at a point 252 feet due north of the Belcher line, near the east wall, and bears north 33° east, with an average dip of 51°. It shows a continuous ore-body from the 1,300 to the 1,400 foot levels, with many rich spots and streaks, and in some places the ore is considerably mixed with porphyry and barren quartz. The middle winze was started at a point 114 feet due north of the Belcher line, and 40 feet east of the west wall, and bears north 78¼° east, with an average dip of 44°. This winze was started with a dip of 55°, but when it reached the west wall the dip was changed to conform to that of the wall, and the winze was so continued until it reached the 1,400-foot level. The south winze was started on the dividing line between the Crown Point and Belcher mines, and near the east wall. It was sunk vertically 28 feet until it reached the west wall, and thence it followed that wall to the 1,400-foot level. Both the middle and south winzes showed ore of a superior character and quality for about 75 feet down, when they passed into the nearly barren material lying next to the west wall.

The north winze reached the main south drift of the 1,400-foot level at or about the same time that this drift reached the Belcher line, and ventilation having been thus secured, cross-cuts were at once commenced to determine the width and quality of the ore-body. Two cross-cuts have been run on this level. The first or north cross-cut was started at a point in the main south drift 320 feet due north of the Belcher line. After passing through the west wall this drift ran through ore of fair average quality for 26 feet, then through solid porphyry 8 feet, and then into ore of excellent quality for 16 feet, at which point the east wall was found. This last-mentioned 16 feet of ore is unquestionably superior in quality to any body of ore of like size ever developed by any cross-cut in the mine.

From the cross-cut we followed this east ore-body north a distance of 25 feet, all the way in very rich ore, at which point it narrowed down to about 2 feet in width, when work on it was temporarily suspended. We are now running a drift south in the same body, which is already in a distance of 30 feet, and has been from the commencement in ore of superior quality, fully equal to that found in the cross-cut, and assaying from $200 to $600 per ton. In the west ore-body no drift has yet been run north, but, judging from the width and character of the ore in the cross-cut, it is fair to presume that it will extend at least 75 to 100 feet further in that direction. The second or middle cross-cut in this level was started at a point 162 feet north of the Belcher line. In this cross-cut we ran through ore of high grade 40 feet in width, and then encountered solid porphyry, which continued for 88 feet, when the east ore-body was struck. This proved to be of superior quality and fully 40 feet wide. We have commenced breasting out in the east ore-body from the cross-cut, and as the breasts are driven north and south the width and character of the ore-body is maintained.

It will be seen that the north cross-cut developed two bodies of ore, aggregating 42 feet in width, and separated by a belt of porphyry 8 feet wide. The middle cross-cut also exposed two ore-bodies. They show an aggregate width of 80 feet, and the intervening belt of porphyry is 88 feet wide. The distance between the two cross-cuts is 158 feet, but the various seams, stratifications, and other vein peculiarities exposed in both cross-cuts, present convincing evidence that the ore-bodies are continuous, and that the east development in the middle cross-cut is identical with that in the north cross-cut. The fact that the porphyry which separates the two bodies has contracted in the north cross-cut to a belt only 8 feet in width, would seem to strengthen the theory advanced by my predecessor in his last annual report, viz, that the east and west chimneys would eventually unite and form one large and compact ore-body.

On the 1,500-foot level a large and commodious station has been cut out, 45 feet long, 14 feet wide, and 9 feet high. From this station a double drift, 10 by 8 feet, was started, running south 73½° east. At the distance of 250 feet from the place of beginning, and 345 feet due north of the Belcher line, the west wall was struck. In passing through this wall exceedingly rich ore 4 feet in width was found. The drift was further continued for 38 feet in quartz of average low grade, showing frequent spots and streaks of fine ore, when rich ore was again struck and continued for 10 feet, when the wall was found and the drift discontinued. We intend soon to push this drift ahead, in the expectation of finding the east ore-body. The last ore discovered, as above described, far excels in quality any ore ever found in the mine so far north of the Belcher line, the average assays running from $100 to $250 per ton. How much farther it extends can at present only be conjectured, as no drift has yet been run in that direction.

Two winzes are also being driven to connect the 1,400 and 1,500 foot levels. The first or north winze was started in the east ore-body in the north cross-cut of the 1,400-foot level, and dips east at an angle of 37°. This winze is now down to a vertical depth of 84 feet, and for the first 65 feet it continued in ore of very high grade, when for a few feet the ore became badly mixed with nearly barren quartz and porphyry. The bottom of the winze is, however, again in ore of superior quality.

The middle winze was started at a point 140 feet north of the Belcher line and 24 feet east of the west wall, with an average dip of 38°. This winze has attained a vertical depth of 56 feet, and was in ore of fine quality until it reached the west wall, after which the rich ore was confined to the east side of the winze.

On the 1,500-foot level we are now running a double drift, following the course of the foot-wall, which will connect with the north and middle winzes from the 1,400-foot level, and also with the Belcher works. This will secure good ventilation, preparatory to prospecting and opening the level thoroughly. All that is known of this level thus far is what has been developed by the north and middle winzes, and by the first drift above described. So far, however, the developments are more than encouraging, and warrant the belief that in point of richness and extent this level will prove to be equal at least to any of the levels already opened and explored.

We have also during the past year made extensive and important developments in the east ore-body on the 1,200 and 1,300 foot levels. It will be remembered that at the date of the last general report a cross-cut was being run on the sill-floor of the 1,300-foot level, in the expectation of striking the east ore-body, which had already been explored on the third and fifth floors of the same level. In this expectation we were not disappointed, and shortly afterward the cross-cut ran into remarkably rich ore, which continued for 15 feet, when the east wall was struck. From this point a winze was sunk, following what we term the foot-wall of the east ore-body, until it intersected the ore-breasts from the middle cross-cut of the 1,400-foot level, at which last point the east ore-body is fully 40 feet in width.

The first discovery of this east ore-body was made, as you were informed by the report of my predecessor, on the sill-floor of the 1,200-foot level, at which point it was 200 feet in length, measured on the course of the vein. On the sill-floor of the 1,300-foot level it was 250 feet in length. On the sill-floor of the 1,400-foot level it has not yet been traced to its northern extremity, but the north cross-cut on that level shows that it is over 320 feet in length, assuming that it continues south to the Belcher line, as it has been proved to do on the 1,200 and 1,300 foot levels. That it will so continue, I have no doubt. At the time of the discovery of the east ore-body it was generally supposed that it extended but a short distance above the sill-floor of the 1,200-foot level, but since that time we have run drifts on the second, third, fifth, and ninth floors of that level, and in each drift we have found ore worth from $50 to $80 per ton, and averaging 10 feet in width. I think this ore-body will certainly extend up to the 1,100-foot level, and probably much higher. What we have already found will enable the 1,200-foot level to hold out, at the present rate of extraction, one year longer than it otherwise would. On the 1,300-foot level the east ore-body is not at all uniform in width. Where we are working it, between the 1,200 and 1,300 foot levels, it is in some places fully 45 feet wide, and in others only 6 to 15 feet wide.

The foregoing description brings the history of the developments of the mine down to the present time. At all prospecting points work is being vigorously pushed, and I do not doubt that before the next annual report still other important discoveries will be made.

During the year an unusually large amount of dead-work has been necessarily done. On the 600, 700, 800, and 900 foot levels the stations have all been retimbered and put in thorough repair; 300 feet of the main south drift of the 900-foot level have also been retimbered. In fact, these levels were much in need of the thorough overhauling which they have received. On the 1,000, 1,100, and 1,200 foot levels we have in the aggregate retimbered 2,200 feet of drifts, and 400 feet of winzes and timber-chutes. The timber-chute from the 1,100 to the 1,200 foot level is 184 feet long, and 3½ by 5 feet, and has within the year been retimbered four times. We have also made very extensive repairs in the main shaft and incline, and have raised a new compartment from the 1,000-foot level to the surface. This compartment adjoins the north hoisting-compartment on the east, and is intended as a rope-shaft for the new incline hoisting-machinery about to be erected.

Without attempting to enter fully into details, I will state that in the work of prospecting, working, and ventilating the mine, there have been—

| | Feet. |
|---|---|
| Drifts run | 1,200 |
| Cross-cuts run | 500 |
| Raises made | 200 |
| Winzes sunk | 1,250 |
| Drifts retimbered | 2,500 |

The above does not include all the work that has been done in the mine during the year, but it would require too much space to enumerate the many different repairs and improvements. It will be enough to say that I have endeavored to leave nothing undone that was necessary to put the mine in a condition of thorough repair and to facilitate its working.

During the year the following outside improvements have also been made: Two substantial wood-elevators have been erected, by which one man can deliver at the furnaces all the wood required in running the entire machinery of the mine. A carpenter and machine shop, 36 by 100 feet, and two stories high, and a building, 36 by 100 feet, with 19-foot posts, for the incline machinery, have also been erected. In the carpenter-shop we have placed a 12 by 24 inch engine, which at present drives two circular saws. This engine is an old one that we had on hand, but it has been thoroughly repaired, and fully answers the purpose for which it is used.

We have also greatly increased our facilities for contending with fires, should any occur. We have connected the main large reservoir with the mine by entirely new and much stronger iron pipe. The water-company has completed its new ditch from Marlette Lake, and now, instead of having to pump from the mine to keep this reservoir full, it is supplied to overflowing by inexhaustible streams running into it. We have also surrounded the works with hydrants, which will enable us to attack a fire from any point.

In the building erected to receive the new machinery for the incline we have constructed a substantial foundation upon which to place it. This foundation consists of stone and brick masonry, 17 feet in depth from the surface to the bottom, and contains 16 2½-inch anchor-bolts for the pillow-blocks, and 30 2-inch anchor-bolts for the engines. The hoisting-reel belonging to this machinery is the largest on the coast, being 21 feet 6 inches in diameter, with a face of 6 feet 6 inches. The face has a spiral groove of sufficient capacity to wind a 2-inch steel rope to work the mine to a depth of 3,000 feet. The reel is heavy and strong, and will hoist 12 tons of ore at a load. It will be driven by a pair of 20 by 42 link-motion engines, so arranged that they can be run singly or together, as may be required. The two engines are equal to 250 horse-power. The gear and pinion wheels are duplicated, so that in case of an accident these would be very little delay in hoisting.

The following statement will show the economy and advantages of the above-described works over those now in use on the same incline. The old works hoist 2 tons of ore at a load; the car and rope used to hoist the same (figuring from the 1,500-foot level) weigh 11,000 pounds. The car and rope of the new machinery will hoist 12 tons at a load, and will weigh 18,100 pounds. Now, taking the hoisting of 500 tons of ore per day as a basis for calculation, the following result will be shown:

| | |
|---|---|
| To hoist 500 tons of ore by the present works requires 250 trips of the car, which, at 11,000 pounds per trip, aggregate in dead-weight....... | 2,750,000 lbs. |
| To hoist 500 tons of ore the new works would have to make 41⅔ trips, which, at 18,100 pounds per trip, aggregate in dead-weight........... | 754,156 lbs. |
| | 1,995,844 lbs. |

Showing that in hoisting 500 tons of ore the new machinery saves over the old the expense of hoisting 1,995,844 pounds.

To hoist 1,995,844 pounds from the 1,500-foot level requires 2½ cords of wood at $12 per cord, amounting to $30 per day, and for the year to $10,950. The old works wear out six ropes per year, at $3,000 each, aggregating $18,000. One rope, costing $6,000, will last the new works two years. Here is a clear saving in ropes per year of $15,000, and in wood of $10,950.

In addition to the above, a great saving will be made in the more speedy transportation of men and materials to the various levels, and, what is of more importance, it will be attended with much less peril to human life. The cost of this machinery, including the freight and the expense of setting it up, will be about $50,000. It will be in operation by the 10th of July.

The following table will show the amount of ore, fractional tons being omitted, extracted from the various levels during the year:

| | Tons. |
|---|---|
| From the 1,000-foot level........ | 16,576 |
| From the 1,100-foot level........ | 388 |
| From the 1,200-foot level........ | 23,636 |
| From the 1,300-foot level........ | 84,000 |
| From the 1,400-foot level........ | 15,000 |
| | 139,600 |

The actual amount of ore reduced at the mills was 140,132$\frac{1719}{2000}$ tons. Average yield of the ore per ton, $50.96.

There yet remains on each of the above levels a very large amount of ore, and, as will be seen from the foregoing table, the 1,400-foot level is almost intact, while from the 900 and 1,500 foot levels not a pound of ore has yet been extracted. Since the date of the last report the 900-foot level has been more thoroughly prospected, and the ore has been found to be of much better quality than was expected, which, taken in connection with the width and strength of the vein at that point, makes it almost certain that the ore-body will at least extend up to the 800-foot level.

The company's mill has been kept almost constantly running since the last annual report. It has been supplied with an entirely new battery, and everything necessary to keep it in thorough repair has been done. A tank 110 feet in length by 36 feet in width, for the purpose of catching slimes, has also been built within 50 feet of the mill. This tank is divided into 30 compartments, and as about two-thirds of these compartments are constantly full of clear water, suitable for supplying steam fire-engines, it will be of great assistance in case of fire occurring in or about the mill.

For a statement of the cost of extraction and reduction of ores, the average yield per ton, the gross product thereof in bullion, supplies on hand, and all other particulars relating to the financial condition of the company, I refer you to the report of the secretary.

In conclusion, I desire to say that the future of the mine never looked brighter than it does at present. It is true that a very large amount of ore has been extracted from the mine during the year, but at the same time the developments in the mine have been equally great; and, in my judgment, there is more ore in sight in the mine to-day than there was at the time of the last annual report.

The secretary's report contains the following:

RECEIPTS.

| | | |
|---|---|---|
| Bullion.................................... | | $7,417,115 30 |
| Mine expense: | | |
| Lumber sold........................... | $4,807 77 | |
| Wood sold Nevada Mill Company ........ | 1,722 50 | |

| | | |
|---|---|---|
| Kentuck Company, labor, lumber, &c | $2,963 78 | |
| Old iron sold | 304 50 | |
| Pump sold Utah mine | 217 20 | |
| Wire rope sold | 207 50 | |
| Rebate on fire-extinguishers | 100 00 | |
| Rebate on giant powder | 43 20 | |
| Yellow Jacket Company and railroad company | 116 60 | |
| Petaluma Mill, for lime | 4 90 | |
| | | $10,487 95 |
| Rhode Island Mill: | | |
| Slimes sold | 2,316 24 | |
| Quicksilver-flasks sold | 224 40 | |
| Atlas Mill, for soda | 34 50 | |
| | | 2,575 14 |
| Interest: | | |
| Belcher Company, interest on note | | 2,750 00 |
| Balance due San Francisco Refining Works | | 286 54 |
| Cash in treasury May 1, 1873 | | 1,873,891 41 |
| Total | | 9,307,106 25 |

DISBURSEMENTS.

| | | |
|---|---|---|
| Mine expense: | | |
| For labor | $912,039 75 | |
| For lumber and timber | 261,312 64 | |
| For wood | 63,978 00 | |
| For castings, machinery, &c | 50,496 65 | |
| For freight on supplies | 14,532 39 | |
| For candles, oils, hardware, rope, &c | 146,292 43 | |
| | | $1,449,301 86 |
| Mine improvement: | | |
| For two steam-engines, reels, and boilers complete | | 31,200 00 |
| Rhode Island Mill: | | |
| For labor | $45,658 87 | |
| For wood | 28,987 25 | |
| For machinery and castings | 23,003 43 | |
| For water | 8,400 00 | |
| For quicksilver, chemicals, oils, &c | 66,601 60 | |
| | | 172,651 15 |
| Rhode Island Mill improvement: | | |
| For extension of tram-way | 1,001 42 | |
| For material, hauling, &c | 203 50 | |
| | | 1,204 92 |
| Crushing: | | |
| For working 124,084$\frac{1290}{2000}$ tons ore | | 1,488,019 34 |
| Legal expense: | | |
| For attorney's fees | $5,457 00 | |
| For adverse claims and titles | 36,147 72 | |
| | | 41,604 72 |
| General expense: | | |
| For superintendent's salary, Gold Hill office, insurance, &c | | 26,940 85 |

| | | |
|---|---|---|
| San Francisco expense: | | |
| For salaries, office, stationery, &c | | $11,331 14 |
| Taxes: | | |
| For quarterly tax on ores | $79,498 54 | |
| For city and county | 10,598 95 | |
| | | 86,097 49 |
| Discount: | | |
| On drafts | 2,955 32 | |
| On bullion | 109,856 57 | |
| | | 112,811 89 |
| Treasure freight: | | |
| For transportation of bullion | | 30,917 38 |
| Assaying: | | |
| For ore-assays | $10,978 00 | |
| For bullion-assays | 14,035 23 | |
| For coinage and parting bars | 38,397 70 | |
| | | 63,410 93 |
| Assay-office: | | |
| For material and storage | | 48 15 |
| Interest: | | |
| At Gold Hill | | 914 53 |
| Dividends: | | |
| Nos. 31 to 42, inclusive | | 5,300,000 00 |
| Cash: | | |
| In hands of superintendent at Gold Hill | $30,017 51 | |
| In treasury May 1, 1874 | 460,634 39 | |
| | | 490,651 90 |
| Total | | 9,307,106 25 |

ORE STATEMENT.

| Tons. | Lbs. | | |
|---|---|---|---|
| 16,044 | 820 | At Rhode Island Mill, yielded<br>(Average per ton, $53.) | $850,360 61 |
| 37,619 | 1,300 | At Brunswick Mill<br>(Average per ton, $50.79.) | 1,910,658 61 |
| 26,333 | ...... | At Mexican Mill<br>(Average per ton, $51.89.) | 1,366,537 69 |
| 25,479 | 1,740 | At Morgan Mill<br>(Average per ton, $50.23.) | 1,280,062 08 |
| 12,333 | 300 | At Petaluma Mill<br>(Average per ton, $50.64.) | 624,544 12 |
| 10,486 | 550 | At Pioneer Mill<br>(Average per ton, $49.78.) | 522,079 50 |
| 7,490 | 950 | At Atlas Mill | 386,122 08 |

| | | | |
|---|---|---|---|
| | | (Average per ton, $51.98.) | |
| 375 | 1, 100 | At Hoosier State Mill............. | $25, 048 09 |
| | | (Average per ton, $66.70.) | |
| 1, 713 | 900 | At Devil's Gate Mill................ | 75, 828 87 |
| | | (Average per ton, $44.25.) | |
| 1, 804 | 100 | At Sapphire Mill.................. | 84, 636 [illegible]0 |
| | | (Average per ton, $46.90.) | |
| 449 | 350 | At Sherman Mill.................. | 15, 858 21 |
| | | (Average per ton, $35.22.) | |
| 140, 129 | 110 | Total ...................... | 7, 141, 736 46 |

*Average yield per ton for the year.*

There were 140,129$\frac{110}{2000}$ tons of ore worked, yielding $7,141,736.46, making average yield per ton, $50.96½.

BULLION STATEMENT.

Stamped value of bullion, as per assay-certificates received at the office in San Francisco, credited on the books of the company:

| | |
|---|---|
| Value in gold................................ | $3, 107, 325 04 |
| Value in silver.............................. | 4, 018, 935 32 |
| Assay, grains and chips ........................ | 17, 634 97 |
| Reclamation, Belcher Silver-Mining Company ......... | 275, 000 00 |
| Wrecked cars, ores lost, Virginia and Truckee Railroad Company.................................. | 378 84 |
| | 7, 419, 274 87 |

CONTRA.

| | |
|---|---|
| Reclamation on bars computed at standard value ...... | $2, 159 57 |
| | 7, 417, 115 30 |

WEIGHT.

| | Ounces. |
|---|---|
| Amalgam before melting...... | 3, 363, 375. 30 |
| Amalgam after melting....... | 3, 293, 896. 41 |
| | 69, 478. 89—Loss, say 2$\frac{1}{10}$ per cent. |

Tons, 112$\frac{80}{2000}$ = 225,609.11 pounds = 3,293,896.41 ounces troy.

Average value per ounce, $2.16.

COST OF WORKING ORES.

Average cost of working 140,129$\frac{110}{2000}$ tons of ore, $11.85 per ton.

COST OF MINING.

Total cost of mining 140,937$\frac{810}{2000}$ tons of ore, including all cost of running drifts, repairs to shaft and machinery, pumping, &c., $9.24 per ton.

ASSETS.

| | |
|---|---|
| Cash on hand May 1, 1874.............................. | $490, 651 90 |
| Mine improvements, buildings, &c........ ............. | 130, 000 00 |

| | |
|---|---|
| Rhode Island Mill | $80,000 00 |
| Real estate | 15,000 00 |
| Supplies at mine | 28,023 91 |
| Supplies at mill | 14,318 36 |
| Lumber at Clear Creek Flume, 2,600,540 feet | 52,010 80 |
| Lumber at Carson yard, 1,720,743 feet | 36,135 60 |
| Wood at Empire, 2,604 cords | 16,926 00 |
| Ore on hand at mills, $515\frac{440}{2000}$ tons | 22,730 91 |
| Ore on hand at mine, 580 tons | 20,578 40 |
| Sundry book accounts | 3,374 44 |
| | 909,750 32 |

LIABILITIES.

| | |
|---|---|
| Balance due on lumber | $1,542 66 |
| San Francisco Refining Works | 286, 45 |
| | 1,829 11 |

*Product of the Belcher and the Belcher Silver-Mining Company since December, 1865.*

| Dates. | Gross yield. | Totals by years. | No. of tons worked. | Totals by years. | Return per ton. | Pulp-assay. | Per cent. worked. | Mine-assay. |
|---|---|---|---|---|---|---|---|---|
| From 1863 to 1868, inclusive. | $1,474,269 33 | $1,474,269 33 | ........ | 31,725 | ...... | ...... | ... | ... |
| 1869 | 18,313,17 | 18,313 17 | 888 | 888 | ...... | ...... | ... | ... |
| 1870 | 204.252 93 | 204,252 93 | 11,352 | 11,352 | ...... | ...... | ... | ... |
| 1871—August | 50,935 47 | ........ | 680 | ........ | $74 90 | ...... | ... | ... |
| September | 161,103 09 | ........ | 2,009 | ........ | 80 19 | ...... | 74 | $74 |
| October | 313,640 53 | ........ | 4,209 | ........ | 74 52 | ...... | 77 | 119 |
| November | 345,419 28 | ........ | 5,717 | ........ | 60 41 | ...... | 69 | 86 |
| December | 328,036 45 | ........ | 5,853 | ........ | 55 83 | ...... | 61 | 72 |
| | | 1,199,134 87 | | 18,468 | | | | |
| 1872—January | 426,127 51 | ........ | 8,473½ | ........ | 50 29 | ...... | 70 | 72 |
| February | 290,396 57 | ........ | 5,504 | ........ | 52 76 | ...... | 72 | 73 |
| March | 354,681 08 | ........ | 6,721½ | ........ | 52 76 | ...... | 68 | 77 |
| April | 365,915 15 | ........ | 7,542 | ........ | 48 51 | ...... | 65 | 74 |
| May | 429,349 35 | ........ | 7,411 | ........ | 57 93 | ...... | 76 | 69 |
| June | 407,847 53 | ........ | 6,979 | ........ | 58 43 | ...... | 72 | 81 |
| July | 321,421 93 | ........ | 5,000 | ........ | 64 29 | ...... | 75 | 70 |
| August | 251,790 83 | ........ | 5,886 | ........ | 42 77 | ...... | 61 | 45 |
| September | 442,201 16 | ........ | 6,545 | ........ | 67 56 | ...... | 75 | 58 |
| October | 366,326 54 | ........ | 4,690 | ........ | 78 10 | ...... | 73 | ... |
| November | 534,394 63 | ........ | 7,656 | ........ | 69 80 | ...... | 72 | 92 |
| December | 604,206 82 | ........ | 10,787 | ........ | 56 00 | ...... | 69 | 56 |
| | | *4,794,659 10 | | 83,195 | | | | |

* Stamped value of bullion:

| | |
|---|---|
| Value in gold | $3,087,948 56 |
| Value in silver | 1,706,710 54 |
| Total | 4,794,659 10 |
| Number of ounces refined bullion | 1,483,753 |
| Average fineness in gold | 101-thousandths. |
| Average fineness in silver | 890-thousandths. |
| Average value per ton in gold | $37 12 |
| Average value per ton in silver | 20 51 |
| Total | 57 63 |
| Cost of crushing | $12 00 |
| Cost of mining | 9 07 |
| Total cost | 21 07 |

*Product of the Belcher, &c.*—Continued.

| Dates. | Gross yield. | Totals by years. | No. of tons worked. | Totals by years. | Return per ton. | Pulp-assay. | Per cent. worked. | Mine-assay. |
|---|---|---|---|---|---|---|---|---|
| 1873—January | $437, 489 50 | ........ | 8, 501½ | ........ | $51 46 | ........ | 69 | 70 |
| February | 730, 013 80 | ........ | 10, 417 | ........ | 70 07 | ........ | 69 | 71 |
| March | 888, 393 39 | ........ | 12, 008 | ........ | 73 98 | $97 57 | 72 | 75 |
| April | 1, 184, 475 62 | ........ | 15, 981 | ........ | 74 11 | 106 76 | 68 | 128 |
| May | 1, 500, 823 81 | ........ | 18, 965 | ........ | 79 13 | 72 00 | 82 | 127 |
| June | 914, 592 19 | ........ | 13, 074 | ........ | 69 95 | 98 00 | 69 | 107 |
| July | 912, 781 15 | ........ | 13, 685 | ........ | 66 70 | 93 86 | 77 | 100 |
| August | 691, 758 93 | ........ | 10, 483 | ........ | 65 99 | 95 12 | 69 | 100 |
| September | 647, 709 62 | ........ | 9, 876 | ........ | 65 58 | 99 00 | 65 | 103 |
| October | 831, 262 83 | ........ | 12, 825 | ........ | 64 82 | 95 33 | 68 | 97 |
| November | 940, 195 75 | ........ | 13, 702 | ........ | 68 61 | 96 26 | 74 | 89 |
| December | 1, 099, 674 48 | ........ | 15, 147 | ........ | 72 60 | 95 94 | 74 | 106 |
| | | †$10, 779, 171 07 | | 154, 664½ | | | | |
| 1874—January | 1, 067, 674 93 | ........ | 14, 406 | ........ | 74 11 | 95 42 | 78 | 104 |
| February | 984, 041 71 | ........ | 15, 584 | ........ | 63 13 | 77 95 | 82 | 84 |
| March | 895, 561 98 | ........ | 15, 093 | ........ | 59 33 | 67 55 | 84 | 83 |
| April | 988, 985 77 | ........ | 16, 233 | ........ | 60 92 | 75 00 | 80 | 79 |
| May | 796, 691 03 | ........ | 15, 147 | ........ | 52 60 | 69 00 | 73 | 71 |
| June | 667, 138 41 | ........ | 12, 512 | ........ | 53 32 | 75 00 | 68 | 62 |
| July | 695, 055 40 | ........ | 12, 289 | ........ | 56 56 | 65 24 | 85 | 55 |
| August | 975, 613 77 | ........ | 13, 849 | ........ | 70 44 | 88 00 | 89 | 96 |
| September | 619, 014 10 | ........ | 12, 291 | ........ | 50 36 | 68 50 | 72 | 70 |
| October | 622, 903 36 | ........ | 14, 791 | ........ | 42 11 | 57 00 | 72 | 58 |
| November | 457, 452 04 | ........ | 12, 344 | ........ | 37 0[illegible] | 55 32 | 67 | 56 |
| December | 380, 430 60 | ........ | 12, 200 | ........ | 31 18 | 45 23 | 67 | 46 |
| | | ‡9, 150, 563 10 | | 166, 739 | | | | |
| 1875—January | 407, 701 19 | 407, 701 19 | 12, 287 | 12, 2*7 | 33 18 | 43 37 | 76 | 45 |
| Grand total | ........ | 28, 028, 064 76 | ........ | 479, 318½ | ........ | ........ | ........ | ........ |

† Stamped value of bullion:

| | |
|---|---|
| Value in gold | $5, 725, 247 50 |
| Value in silver | 5, 009, 520 51 |
| Assay-grains | 44, 403 06 |
| Total | 10, 779, 171 07 |
| Number of ounces refined bullion | 4, 173, 535. 74 |
| Average fineness in gold | 66½-thousandths. |
| Average fineness in silver | 929-thousandths. |
| Average value per ton in gold | $37 16 |
| Average value per ton in silver | 32 53 |
| Total | 69 69 |
| Cost of crushing | $12 10 |
| Cost of mining | 8 51 |
| Total cost | 20 61 |

‡ Stamped value of bullion:

| | |
|---|---|
| Value in gold | $5, 656, 103 67 |
| Value in silver | 3, 437, 593 05 |
| Assay-grains | 26, 866 38 |
| Total | 9, 150, 563 10 |
| Number of ounces refined bullion | 2, 945, 066. 84 |
| Average fineness in gold | 93½-thousandths. |
| Average fineness in silver | 910-thousandths. |
| Average value per ton in gold | $34 10 |
| Average value per ton in silver | 20 78 |
| Total | 54 88 |
| Cost of crushing | $12 97 |
| Cost of mining | 9 80 |
| Total cost | 22 77 |

*Statement of the assessments levied by the Belcher and the Belcher Silver-Mining Company since December*, 1865.

[Company re-organized and capital stock increased from 1,040 to 10,400 shares November 2, 1868.]

| Assessment number. | When levied. | Amount per share. | Total per share. | Grand total per share. | Amount collected. | Total collected. |
|---|---|---|---|---|---|---|
| 1 | December 18, 1865 | $100 00 | | | $104, 000 00 | |
| 2 | April 10, 1866 | 60 00 | | | 62, 000 00 | |
| 3 | July 21, 1866 | 45 00 | | | 46, 800 00 | |
| 4 | October 17, 1866 | 33 00 | | | 34, 320 00 | |
| 5 | January 1, 1867 | 15 00 | | | 15, 600 00 | |
| 6 | March 25, 1867 | 12 00 | | | 12, 480 00 | |
| 7 | May 30, 1867 | 15 00 | | | 15, 600 00 | |
| 8 | September 21, 1867 | 15 00 | | | 15, 600 00 | |
| 9 | December 27, 1867 | 15 00 | | | 15, 600 00 | |
| 10 | March 13, 1868 | 25 00 | | | 26, 000 00 | |
| 11 | July 13, 1868 | 25 00 | | | 26, 000 00 | |
| 12 | October 6, 1868 | 25 00 | | | 26, 000 00 | |
| 13 | December 31, 1868 | 25 00 | $410 00 | | 26, 000 00 | |
| | | | | | | $426, 400 00 |
| | By increase equal to | | 4 10 | | | |
| 1 | March 15, 1869 | 3 00 | | | 31, 200 00 | |
| 2 | May 4, 1869 | 5 00 | | | 52, 000 00 | |
| 3 | October 21, 1869 | 2 50 | | | 26, 000 00 | |
| 4 | June 8, 1870 | 4 00 | | | 41, 600 00 | |
| 5 | September 6, 1870 | 2 00 | | | 20, 800 00 | |
| 6 | December 2, 1870 | 1 00 | | | 10, 400 00 | |
| 7 | February 16, 1871 | 1 00 | | | 10, 400 00 | |
| 8 | April 14, 1871 | 4 00 | 22 50 | | 41, 600 00 | |
| | | | | | | 234, 000 00 |
| | By increase equal to | | 2 25 | $6 35 | | |
| | Grand total collected | | | | | 660, 400 00 |

*Dividends paid by the Belcher and the Belcher Silver-Mining Company since December*, 1865.

[Original capital stock, 1,040 shares; increased to 10,400 shares November 2, 1868, when the company was re-incorporated as the Belcher Silver-Mining Company; and further increased to 104,000 shares July 26, 1872.]

| Dividend number. | When paid. | Amount per share. | Totals by years. | Dividends. | Totals by years. |
|---|---|---|---|---|---|
| 1 | June 1, 1864 | $21 00 | | $21, 840 00 | |
| 2 | July 1, 1864 | 24 00 | | 24, 960 00 | |
| 3 | August 2, 1864 | 24 00 | | 24, 960 00 | |
| 4 | December 31, 1864 | 24 00 | | 24, 960 00 | |
| | | | | | $96, 720 00 |
| 5 | January 28, 1865 | 51 00 | | 53, 040 00 | |
| 6 | February 25, 1865 | 60 00 | | 62, 400 00 | |
| 7 | March 28, 1865 | 75 00 | | 78, 000 00 | |
| 8 | April 28, 1865 | 75 00 | | 78, 000 00 | |
| 9 | May 30, 1865 | 51 00 | | 53, 040 00 | |
| | | | $405 00 | | 394, 480 00 |
| | By increase equal to | | 4 05 | | |
| 1 | January 10, 1872 | 10 00 | | 104, 000 00 | |
| 2 | February 10, 1872 | 15 00 | | 156, 000 00 | |
| 3 | March 9, 1872 | 15 00 | | 156, 000 00 | |
| 4 | April 10, 1872 | 20 00 | | 208, 000 00 | |
| 5 | May 10, 1872 | 30 00 | | 312, 000 00 | |
| 6 | June 10, 1872 | 30 00 | | 312, 000 00 | |
| 7 | July 10, 1872 | 30 00 | | 312, 000 00 | |
| | Total 150; by new increase equal to | 15 00 | | | |
| 8 | August 10, 1872 | 3 00 | | 312, 000 00 | |
| 9 | September 10, 1872 | 3 00 | | 312, 000 00 | |
| | | | 21 00 | | 2, 184, 000 00 |
| 10 | January 10, 1873 | 3 00 | | 312, 000 00 | |
| 11 | February 10, 1873 | 3 00 | | 312, 000 00 | |
| 12 | March 10, 1873 | 4 00 | | 416, 000 00 | |

*Dividends paid by the Belcher and the Belcher Silver-Mining Company, &c.*—Continued.

| Dividend number. | When levied. | Amount per share. | Totals by years. | Dividends. | Totals by years. |
|---|---|---|---|---|---|
| 13 | April 10, 1873 | $5 00 | .......... | $520,000 00 | |
| 14 | May 10, 1873 | 8 00 | .......... | 832,000 00 | |
| 15 | June 10, 1873 | 10 00 | .......... | 1,040,000 00 | |
| 16 | July 10, 1873 | 8 00 | .......... | 832,000 00 | |
| 17 | August 9, 1873 | 7 00 | .......... | 728,000 00 | |
| 18 | September 10, 1873 | 5 00 | .......... | 520,000 00 | |
| 19 | October 10, 1873 | 3 00 | .......... | 312,000 00 | |
| 20 | November 10, 1873 | 4 00 | .......... | 416,000 00 | |
| 21 | December 10, 1873 | 5 00 | .......... | 520,000 00 | |
| | | | $65 00 | | $6,760,000 00 |
| 22 | January 10, 1874 | 5 00 | .......... | 520,000 00 | |
| 23 | February 10, 1874 | 5 00 | .......... | 520,000 00 | |
| 24 | March 10, 1874 | 5 00 | .......... | 520,000 00 | |
| 25 | April 10, 1874 | 5 00 | .......... | 520,000 00 | |
| 26 | May 9, 1874 | 5 00 | .......... | 520,000 00 | |
| 27 | June 10, 1874 | 5 00 | .......... | 520,000 00 | |
| 28 | July 10, 1874 | 5 00 | .......... | 520,000 00 | |
| 29 | August 10, 1874 | 3 00 | .......... | 312,000 00 | |
| 30 | September 10, 1874 | 4 00 | .......... | 416,000 00 | |
| 31 | October 10, 1874 | 3 00 | .......... | 312,000 00 | |
| 32 | November 10, 1874 | 3 00 | .......... | 312,000 00 | |
| 33 | December 10, 1874 | 3 00 | .......... | 312,000 00 | |
| | | | 51 00 | | 5,304,000 00 |
| 34 | January 11, 1875 | 3 00 | .....3 00 | 312,000 00 | 312,000 00 |
| | Total per share | .......... | 144 05 | | |
| | Total dividends | .......... | .......... | .............. | 14,981,900 00 |

NOTE.—For the foregoing interesting tables I am indebted to H. C. Kibbe, esq., secretary of the company.—R. W. R.

## *Report of the Yellow Jacket Mining Company for the year ending July 1, 1874.*

The following is an abstract of the secretary's and superintendent's reports:

| | |
|---|---|
| Balance on hand July 1, 1873 | $37,427 89 |
| Received from assessment No. 16 | 120,000 00 |
| Received from assessment No. 17 | 120,000 00 |
| Rents and advertising | 1,055 50 |
| From sale of machinery | 4,139 43 |
| From Belcher Mining Company for hoisting | 79,281 37 |
| | 361,904 19 |

CONTRA.

| | |
|---|---|
| Amount of disbursements for year | $321,045 67 |
| Cash on hand | 21,428 55 |
| Stock on hand | 19,429 87 |
| | 361,904 19 |

Prospecting in the mine has been carried on most assiduously, but with poor success. The lode continues very regular in strike and dip, carrying an average width between clay-walls of upward of 300 feet. The men are now drifting for the ledge on the 1,740-foot level, which is 273 feet vertically below the level of the Sutro tunnel. This new level, it is hoped, will carry them below the barren belt which has continued from the 940-foot or last ore-producing level. They hoisted during the past year for the Belcher mine, 50,127 tons ore. The machinery of the mine, both pumping and hoisting, is all in good working condition.

### *Report of the Virginia Consolidated Mining Company for the year ending December* 31, 1874.

**Mr. James G. Fair, the superintendent, reports as follows:**

During the past year 91,167$\frac{1388}{2000}$ tons of ore have been extracted from the several levels of the mine, and 89,783$\frac{1187}{2000}$ tons reduced, producing in bullion $4,979,182.07. There are now in the ore-house and at the mills 2,775$\frac{1638}{2000}$ tons, valued by assay at $222,022.30.

Within the past year the main shaft has been extended to the 1,500-foot level. This level is now partially explored by cross-cuts extending into the ore-body in four different places, each 100 feet apart. The most southerly of these cross-cuts shows a width of ore of 152 feet. The remaining cross-cuts have not yet crossed the ore-body; all having penetrated it over 100 feet and one over 300 feet. The quality of the ore is of very high grade, and far excels in value any ever removed from the Comstock. The quantity now exposed to view is almost fabulous.

On the 1,550-foot level a drift has been run the whole length of the mine, the northern 400 feet of which passes through ore assaying from $200 to $800 per ton. Two cross-cuts have been run on this level, one to the east and one to the west, disclosing a width of ore of over 100 feet, and neither east nor west wall has yet been reached.

Below this 1,550-foot level a double winze has been sunk 110 feet, passing through rich ore, and the bottom of the winze is now in ore of equal value to any yet found.

The greatest quantity of ore extracted has been taken from the 1,200, 1,300, and 1,400 foot levels. Large reserves of ore yet remain on these levels, the northern extent having not yet been reached in any of them.

The production of ore and bullion of the mine could have been doubled during the last four months had we had sufficient milling facilities. This difficulty is now obviated, as the new mill at the shaft of the mine will be in running order within five days, and can crush 260 tons daily.

The quality and quantity of ore developed in the mine the past year far exceed in value those of any mine which has ever come under my knowledge or observation. The main shaft has three hoisting-compartments, with three cages in each, capable of hoisting 1,400 tons daily. The hoisting-engines are all in good condition. The Gould & Curry shaft can be used to hoist 100 tons daily, if desired.

A new shaft of three compartments is being sunk 1,040 feet east of the present one. This shaft will be available for hoisting and ventilation within a year. It is now 82 feet deep.

The stock of timbers and material at the mine is very large, as shown by the tabular statement accompanying.

The large and capacious assay department of the mine has proved a source of economy and convenience, and is capable of melting, barring, and assaying $100,000 bullion daily.

**The secretary reports the receipts and disbursements as follows:**

RECEIPTS.

| | |
|---|---|
| Bank of California: | |
| Balance due last meeting, since paid | $150, 050 95 |
| J. C. Flood, treasurer: | |
| Balance due last meeting, since paid | 125, 000 00 |
| J. G. Fair, superintendent: | |
| Balance due last meeting, since paid | 126, 209 13 |
| Cash on hand: | |
| Balance last meeting | 38 76 |
| Back assessments | 1, 261 25 |
| Rent | 72 00 |
| Bullion-samples | 1, 969 75 |
| Bullion | 4, 979, 182 07 |
| Company's stock: | |
| Being proceeds sale 70 shares stock belonging to company | 6, 978 13 |
| | 5, 390, 762 07 |

DISBURSEMENTS.

| | |
|---|---|
| Wood and timber | $53,398 83 |
| Lumber | 124,097 48 |
| Repairs | 4,699 16 |
| Hauling | 945 10 |
| Freight | 19,382 47 |
| Surveying | 1,285 00 |
| Assaying | 3,295 56 |
| Books and stationery | 310 25 |
| Legal expense | 5,976 50 |
| Water | 5,650 00 |
| Real estate | 9,089 00 |
| Contribution | 205 00 |
| Hoisting | 4,016 94 |
| Construction | 49,536 17 |
| Taxes | 34,869 43 |
| Insurance | 2,473 00 |
| Reduction | 1,236,736 19 |
| Interest and exchange | 3,752 63 |
| Bullion freight | 9,577 50 |
| Bullion reclamation | 25 79 |
| Dividend No. 1 | 323,790 00 |
| Dividend No. 2 | 323,790 00 |
| Dividend No. 3 | 323,790 00 |
| Dividend No. 4 | 324,000 00 |
| Dividend No. 5 | 324.000 00 |
| Dividend No. 6 | 323,998 50 |
| Dividend No. 7 | 323,998 50 |
| Dividend No. 8 | 323,998 50 |
| Dividend No. 9 | 323,998 50 |
| Sutro committee | 11,988 00 |
| C. & C. joint shaft | 8,240 02 |
| Bullion discount | 135,790 81 |
| Advertising | 191 25 |
| Supplies | 84,223 02 |
| Salaries and wages | 573,404 15 |
| Expense | 5,132 59 |
| Bank of California, balance due | 82,450 43 |
| James G. Fair, superintendent, balance due | 4,655 80 |
| Total | 5,390,762 07 |

LANDER COUNTY.

In *Reese River district*, the Manhattan Silver-Mining Company of Nevada has again done the principal work during the year. The company has by degrees secured such a position in this district that it will probably always maintain a more or less pronounced monopoly for the purchase and working of all ores outside of those secured from its own mines. The year has been a prosperous one for the company, as the annexed statement will show. In the management of the company a change has taken place, the majority of the stock having been transferred in the summer to San Francisco ownership. At the mill in Austin many experiments have been made by Mr. A. Trippel, M. E., to whom I am indebted for most of the data given below.

The following is an extract from the financial report of the superintendent for the year 1874:

| Name of mine. | Quantity of ore worked. | | Total value extracted. | Milling expense. | Mining expense. | Total expense. | Net proceeds. |
|---|---|---|---|---|---|---|---|
| | *Tons.* | *Lbs.* | | | | | |
| Oregon | 824 | 754 | $179,693 36 | $27,035 90 | $128,625 21 | $155,661 11 | $24,034 25 |
| North Star | 1,455 | 1123 | 259,305 63 | 47,177 76 | 97,572 13 | 144,749 89 | 114,555 74 |
| South America | 2,252 | 736 | 326,052 20 | 73,750 34 | 81,190 29 | 154,949 63 | 171,102 57 |
| Other mines | 798 | 511 | 142,427 25 | 26,138 98 | 95,417 38 | 121,556 36 | 20,870 89 |
| Custom ores | 1,187 | 958 | | | | | 24,128 85 |
| Total | | | | | | | 354,692 30 |
| Less freights to New York, reclamation, taxes, insurance, interest, losses by flood, fire, cattle, and quicksilver | | | | | | | 108,942 03 |
| Net profit | | | | | | | 245,750 27 |

The following are the mill statistics of the Manhattan Company for the first eight months of the year:

*Mill statistics, Manhattan Mill, January to September, 1874.*

| | January and February. | March. | April. | May. | June. | July. | August. |
|---|---|---|---|---|---|---|---|
| Average amount of ore crushed per hour ... pounds.. | 2,000 | 1,914 | 1,978 | 1,903 | 1,800 | 2,042 | 2,068 |
| Tons per month | 854½ | 595½ | 512½ | 640 | 555½ | 616 | 577 |
| Per cent. of bullion of assay-value produced | 90.21 | 90.84 | 83.71 | 89.95 | 89.80 | 80.18 | 89.55 |
| Per cent. of tailings | 9.45 | 9.15 | 11.68 | 7.66 | 10.3 | 13.7 | 9.3 |
| Monthly average assay of raw ore. | $199.74 | $205.33 | $240.96 | $251.80 | $190.06 | $200.31 | $178.24 |
| Average amount of ore charged in pans ... pounds.. | 1,844 | 2,003 | 1,774 | 1,805 | 1,889 | 1,901 | 1,873 |
| Average feed of ore per minute ... pounds.. | 37.8 | 36.9 | 37.3 | 37 | 38 | 38.2 | 39.4 |
| Average feed of salt per minute ... pounds.. | 3.28 | 3.2 | 3.3 | 3.3 | 3.4 | 3.5 | 3.5 |
| Average percentage of chlorination | 91.7 | 90.8 | 89 | 90.2 | 91.9 | 89 | 89.7 |

*Mill statistics, Manhattan Mill, 1874—expenses per ton of ore.*

| Month. | Labor. | Fuel. | Supplies. | Quicksilver. | Salt. | Office-labor. | Casting. | Hauling. | Total. |
|---|---|---|---|---|---|---|---|---|---|
| January, February | $7 60 | $10 72 | $3 04 | $2 50 | $3 15 | $1 16 | $1 12 | $1 12 | $31 41 |
| March | 9 77 | 11 45 | 1 66 | 2 56 | 3 08 | 1 00 | 82 | 1 18 | 31 52 |
| April | 10 74 | 11 90 | 2 11 | 2 54 | 2 29 | 88 | 93 | 1 17 | 31 75 |
| May | 7 83 | 10 90 | 2 26 | 2 92 | 3 14 | 70 | 98 | 1 12 | 29 85 |
| June | 9 22 | 10 36 | 1 82 | 2 76 | 3 43 | 81 | 84 | 1 19 | 30 45 |
| July | 8 70 | 10 11 | 2 77 | 3 20 | 2 78 | 81 | 1 16 | 1 05 | 30 58 |
| August | 8 50 | 10 80 | 2 38 | 3 28 | 2 90 | 86 | 1 70 | 1 04 | 31 46 |

At the Manhattan Mill, a new battery with 20 heavy stamps was finished in January, and has been working since. As many expected, the greater weight of the stamps did not increase materially the quantity of ore crushed. The present stamps weigh 933 pounds; they make about 90 drops of 9 inches per minute. The average amount of pulp produced (which passes a battery-screen with 1,600 meshes to the square inch) from ore prepared by a Blake's crusher is about 2,000 pounds, or 100 pounds per head of stamp, per hour, not counting stoppings. Considerable improvements were made, the object of which was the stamping of small, extra lots of ore separately, and the feeding of each kind of ore, by means of Standish feeders and proper conveyers, in the desired proportion into the common elevator. From this it is again delivered into a screen, and after being mixed automatically with salt, into the Stetefeldt furnace. The latter has been considerably heightened during the year, and thereby much improved. Up to the 1st September, 4,349 tons, averaging $209 in assay-value, were produced and treated in the mill.

During the first five months of the year the experiments commenced in 1873, and having in view the discovery of a more economical and efficient method of treating the Lander Hill ores, were continued, and developed the fact that the present amalgamation can be superseded with considerable advantage by the humid extraction by means of hyposulphite of soda. It was intended to continue these experiments by verifying their results on a more practical scale, but the transfer of the property to a new organization prevented temporarily the execution of this project. However, the all-important question, how to make the leaner ores available and profitable, has to be solved sooner or later. With the present metallurgical treatment, and the high cost of mining which is unavoidable with narrow though rich ledges in hard rock, only the richer portion of the ledge can be worked; ores of less than $80 assay-value are hardly profitable, those below $60 occasion an absolute loss. Yet the many mines in the possession of this company could undoubtedly produce an immense quantity of lower-grade ores which, under more favorable circumstances, ought to give satisfactory profits.

It was considered that a concentration of the leaner ores, which would insure a product of $300 to $400 value per ton, would be of great advantage, provided that this operation should prove to be metallurgically practicable—*i. e.*, should not entail too great a loss of valuable mineral. On account of the scarcity of water, it was resolved to test the system of dry separation, with Krom's separators; and for that purpose 5 tons of poor ores from three of the principal mines were concentrated in Star City on this machine, with a fair result. Including the value of the dust, (all passing a 10,000-mesh screen,) 83.7 per cent. of the assay-value of the ore was obtained. Considering all the circumstances, this result was held to be satisfactory, and it encouraged the hope that with more suitable arrangements a better result may be obtained. At the end of the year it was decided to put up dressing-works, with a daily capacity of 50 tons. The consequences of the introduction of concentration in many districts of Nevada must appear of great importance, since it will open and render profitable many mines now idle.

The development of the mines of the Manhattan Company has been considerable. Two shafts have been deepened, one to over 700 feet. Hoisting-works and underground pumps have been put in place, the latter giving entire satisfaction; and a large number of feet of levels and cross-cuts has been driven. It is a peculiar feature of the Lander Hill ores that they do not contain a trace of gold, while on Union Hill, south of Austin, several localities produce auriferous ores. At present none of these mines are worked. Another distinctly separate system of veins in Austin carries pyritous ore and galena, which, when dressed, forms a rich and valuable smelting-material. An attempt is now being made to work one of these veins.

There were several accidents in the district, some by explosions of giant powder, some by falling in the shafts, and one in the mill. Altogether four lives were lost. In the summer a cloud-burst broke over the town, and did great damage all around. The Manhattan Mill itself was almost filled with sand, and had a narrow escape from total destruction.

The following is a memorandum kindly furnished me by Mr. Trippel, and giving the results of an examination of the products from Krom's dressing-machinery, the ore being from the South America, one of the Manhattan Company's mines.

Of the crushed, original ore, as per sample sent, there were treated 89 pounds. From this were produced:

| | Per cent. | Pounds. |
|---|---|---|
| Concentrations | 19.2 | 17 |
| Fine dust | 5.2 | 5 |
| Tailings, (two kinds) | 75.6 | 67 |
| Total | 100.0 | 89 |

As stated, there were two kinds of tailings, of which one was designated as containing "particles of ore." The quantity of this kind is said to be so small that by adding it to the other and larger part it would not modify materially the assay-value of the main portion. The subsequent assays showed as follows:

| | |
|---|---|
| Original ore, (value per ton) | $108 37 |
| Concentration, (value per ton) | 416 26 |
| Dust | 92 67 |
| Tailings, (with ore attached) | 43 19 |
| Tailings | 30 62 |

In other words:

| | | | |
|---|---|---|---|
| Value in 89 pounds of ore | | | $4 82 |
| Of which was produced— | | | |
| In concentrations | $3 54 | | |
| In dust | 23 | | |
| | | $3 77 | |
| In tailings | | 1 05 | |
| | | | 4 82 |

| | Per cent. |
|---|---|
| Produced in value in concentration and dust | 78.2 |
| Produced in value in tailings | 21.8 |
| | 100.0 |

In this case, one-fourth (24.4 per cent.) of the original ore would have to be treated further to extract the silver.

The following statement will show the comparative economical results:

At present:

| | |
|---|---|
| 4 tons of ore (at $108.37) yield, at 90 per cent | $390 13 |
| Less for treatment, ($28) | 112 00 |
| Net result | 278 13 |

Krom's dressing:

| | | | |
|---|---|---|---|
| 4 tons of ore, ($108.37,) at 78.2 per cent., yield | | $338 98 | |
| Less dressing expenses | $10 00 | | |
| Treating one ton | 28 00 | | |
| 5 per cent. loss in treating 1 ton | 20 80 | | |
| | | 58 80 | |
| | | | $280 18 |
| Balance in favor of concentration | | | 2 05 |

or about 51 cents per ton of ore, which does not indicate, therefore, any great economy as compared with the present treatment.

In examining the appearance of the tailings, it was easy to observe that their size was *too coarse* to allow a more perfect detachment of the mineral from the rock. To prove this assertion, Mr. Trippel made two assays, one of the finer portion of these same tailings, which was, however, still tolerably coarse, showing already a reduction from $30.60 to $22, and the other of the result of the following experiment: The same tailings (the whole average) were reduced to a coarse sand, and washed in a crude way by hand. A concentrate was thus produced which assayed $83 per ton, and final tailings assaying $14.92 per ton.

Now, there can be no reasonable doubt in the mind of any one acquainted with the dressing of ores in general, that Mr. Krom, by using his system, or by any other suitable system known, and applied properly, ought to have, and could have, produced not only the same but even a better result, *i. e.*, poorer tailings than those obtained by hand-washing, especially as the tailings, on examination, showed no material quantity of chlorides.

The following represents the economical results, on the basis of the tailings, which were produced by Mr. Trippel by a further washing, and yielding $14.92 per ton.

After the additional washing, the weight and values produced from 89 pounds of ore are represented by the following figures:

| | Pounds. | | Per ton ore. |
|---|---|---|---|
| Concentration | 19.5 | $4 09 | $91 82 |
| Dust | 5.0 | 0 23 | 5 38 |
| Tailings | 64.5 | 0 50 | 11 10 |
| | 89.0 | 4 82 | 108 30 |

Each ton of ore yields, therefore, in dust and concentration, $97.20, or 89 per cent. in value; in lost waste, $11.10, or 11 per cent. in value. In weight one ton is produced from four, and has to be subsequently treated. The following comparison with the present system may therefore be made:

At present:

| | |
|---|---|
| 4 tons of ore, (at $108.37,) at 90 per cent., yield | $390 13 |
| Less expenses, ($28) | 112 00 |
| Net return | 278 13 |

Krom's dressing:

| | | |
|---|---|---|
| 4 tons ore, ($108.37,) at 89 per cent., yield | | $385 76 |
| Less expenses for dressing | $10 00 | |
| Treating one ton | 28 00 | |
| Loss in amalgamation, 10 per cent. of $416.26 | 41 62 | |
| | | 79 62 |
| Leaves | | 306 14 |
| The net result after dressing | | $306 14 |
| Net result at present | | 278 13 |
| Balance in favor of dressing | | 28 01 |

Which would be $7 gain for each ton of raw ore.

As remarked before, it is easy to obtain tailings as low as $14.92; but, with due diligence and care, they ought to be below $10, provided that no finely-divided chlorides are present in the ore. Every dollar less in the tailings will increase the gain from dressing the ore, and if the tailings carry even no less than $10, the net gain per ton of raw ore would be close to $11.

A difference of only $7 per ton in the two methods of working the ores is a very desirable and substantial gain in itself; but coupled with the consideration that by adopting this method the consumption of fuel will be diminished fully two-thirds in comparison with the treatment of undressed ore, and that thereby the supply of fuel will be secured for a longer period at present rates, it becomes of the greatest importance.

The most substantial benefit, however, derived from ore-dressing, be it by Krom's or any other method, is the availability of ores for profitable treatment, which now must remain untouched. The truth of this assertion is too manifest and too well proved to need any argument. The system of dressing is old—very old; but lately the old-fashioned routine work has had to give way to a thorough scientific, systematized work with improved apparatus, which has saved numerous mines, and is coming by degrees to be better understood, even in this country. Mr. Trippel says: "I do not claim to be particularly and unconditionally impressed with the advantages of Krom's dry-dressing, but under the local circumstances in most parts of Nevada, it is, in my opinion, either the only system practicable, or at least offers the most advantages."

The following supplement to the foregoing memorandum has also been furnished me by the kindness of Mr. Trippel. The concentrated ore was analyzed, and showed:

| | Per cent. |
|---|---|
| Insoluble matter, rock | 26.92 |
| Sulphur | 23.13 |
| Arsenic | 9.28 |
| Antimony | 8.63 |
| Iron | 18.12 |
| Manganese | 11.10 |
| Copper | 2.16 |
| Lead | 2.75 |
| Silver | 1.43 |
| | 103.72 |

(The surplus is owing to the want of proper apparatus to make a perfect analysis.)

With this ore experiments were made as to the best method of further treatment. One hundred parts were roasted sufficiently to expel most, but not all, of the sulphur. There remained, after roasting, 83.7 parts, and 16.3 parts were therefore lost. The roasted ore was then smelted without any flux, and a regulus (matte) resulted weighing 21.9 parts. This being the result of 100 parts concentrated ore, which assayed before $416.15 per ton, the value of the regulus ought to have been found to be $1,900 per ton (21.9 : 100 :: 416.15 : 1,900.) The actual result was $1,877.93, showing a loss of $22.07. Referred to the ton of unroasted ore, this loss would be as follows:

| | |
|---|---|
| The assay of concentrated ore was | $416 15 |
| Of which was produced in the regulus | 411 06 |
| Loss | 5 09 |

Or 1.22 per cent. of the value of the ore.

This shows that smelting is the cheapest method of treating the ore. The expenses of smelting are not higher than those of amalgamation. The expenses of freighting the regulus to New York, assuming it to be worth $2,000 per ton, are not much higher than the express-charges of the same value in bars; and the loss in smelting is at the outside only one-fifth of that set down in this report for amalgamation. The loss, now calculated, is $5.09; but taking it at $8, it compares well with the loss of $41.62, given above.

Another experiment with 100 parts of concentrated ore was made by roasting the ore first by itself and subsequently with an addition of salt, and amalgamating. So far no satisfactory result has been reached. The losses were by far too high, and unless repetitions give better results, it would be unwise to think of treating the ore by itself by amalgamation.

The New Pacific Mining Company has worked with a small force during the year, but most of the work in the mine was done by tributers. The following account, given by the chairman of the annual meeting, held on October 17, in London, will plainly show the state of affairs. This gentleman had visited the mine late in the summer:

Although there were several points being worked, there were only eleven men at work, the average being eighteen or twenty, and of the eleven men only four were paid by the company, the other seven being tributers. The arrangement with the tributers was that they (as a body) should have 50 per cent. of the net yield of the ore. This system had been successfully worked at Austin by Mr. Pringle. The book-keeper at Austin had given him an estimate, showing that by this system the tributers had done for nothing, in the hope of finding something to repay them for their trouble, what would have cost the company $18,000 if they had done the work. The tributers evinced faith in the mine by continuing to work on day after day, and month after month, and had as yet got but very little. He had seen some of them who had worked for months, and all they had got had been a few bags of ore, but they were content to work on if they only got the bare necessaries of life, trusting entirely to making a strike. He had found that they were burning one cord of wood per day, at a cost of $12, and the quantity of water being pumped was 7,400 gallons; but the machinery was very much strained to do this, and it would be quite inadequate to sink any deeper. The gross weight of ore raised during the year was 129,561 pounds, and the value of the ore was $12,290. Against this, the cost of reducing amounted to $2,009, the discount on the bullion to $641, and $4,790 to the tributers, leaving $4,684 as this company's share of the profits, or £926. Since then, a statement of the last quantity of ore sold had been received, which was for the week previous to the 21st November, showing that six tons of ore had been worked, producing 1,660 pounds of first-class ore, value, $353.44 per ton; 11,359 pounds of second-class ore, value, $98.95 per ton, and four tons on hand not yet assorted. The cost of the mine was $1,500 per month, or a loss of nearly $1,000 a month. There had since been no material alteration in the mine, and therefore the estimated loss per month of nearly $1,000 must be looked forward to, unless the grade of the ore improved or a richer ledge was obtained.

A call for more working-capital was made at the same meeting. Mr. J. D. Pringle, the superintendent, writes to me that during 1875 he expects to work the mine far more energetically than in 1874.

*Kingston district* is located in Kingston Cañon, about twenty miles southeast of Austin. Its entrance is from Smoky Valley, and at its head a part of Bunker Hill Mountain divides it from Big Creek Cañon, which opens into Reese River Valley. The two cañons form thus a line crossing the Toyabe range from east to west. The country-rocks in this district are almost exclusively metamorphic slates, limestone, and here and there some granitic masses. The veins are all between slates and limestone; they carry fahlores, silver-glance, stetefeldtite, with little iron-pyrites, some galena and zincblende, and gold. Two mines have been worked somewhat, both of which are located on the southern slope of the principal peak of Bunker Hill, at an altitude of nearly 9,000 feet. One of them, the Victorine, has produced several hundred tons of ore, which averaged in a lot of 400 tons $54.44 in gold and silver, and in another lot of 800 tons $64.80 in gold and silver, the value of gold being fully one-half of the whole.

Not less than four amalgamating-mills have been built in this cañon, none of which have been in operation since several years ago. The Bunker Hill Mill and another, the name of which has escaped me, have each 5 stamps; the Sterling Mill, built lower down in the cañon, has 20 stamps, and is a first-class mill, with a magnificent water-power, roasting-furnace, and a number of outbuildings. A fourth mill was built at the entrance of the cañon, but its machinery has been removed to White Pine.

The Victorine mine shows a very large ledge, and the ore can undoubtedly be produced cheaply; but the present developments do not warrant a judgment as to its future. The milling of the ores of the vicinity in these mills has not given satisfaction, and could not well do so, on account of the want of a treatment consistent with the character of the ore. It is thought that these mines might fully supply a large mill, but the process employed should be one similar to the Washoe process, as executed in Ely district by the Raymond & Ely and Meadow Valley Companies.

In *Grass Valley*, forty miles northeast of Austin, there are some mineral deposits, which will sooner or later attract more attention than heretofore. The country-rock here is quartzite, shale, and limestone, and the ore-deposits either occur on the contact between the quartzite and the limestone and shale, or very near it in the latter rocks. There are two distinct classes of deposits, the one carrying in a quartz-gangue finely-divided spots of silver-glance, stetefeldtite, ruby silver, fahlore, &c., and the other containing in the same gangue, but in larger veins, often 10 or 12 feet thick, the "base ores," such as iron-pyrites, galena, zincblende, and copper-pyrites. Some shafts, the deepest of which is only 50 feet, have been sunk on these ledges, but nothing has as yet been done to prepare for stopes or to beneficiate the ore. With a view to the latter task, however, some experiments have been made by Mr. Trippel at Austin, who has furnished me with the following memorandum of the result. The object was to see whether the ores could be profitably concentrated, in order, if this should prove possible, to amalgamate or leach the one class and to smelt the other, or perhaps both.

*Concentration of Mr. Frost's ore.*

| | |
|---|---|
| Assay-value of crude ore | $620 45 |
| Concentrations per ton | 1,248 80 |

| | |
|---|---|
| Chloride of silver obtained by leaching, per ton of ore ..... | $18 85 |
| First tailings, assay-value per ton ........................ | 221 48 |
| Concentrations obtained from first tailings, per ton ........ | 1,028 91 |

10.29 ounces of raw ore were taken, and calculations based on the result for 10.29 tons. The sample produced:

| | |
|---|---|
| (1.) 3.88 ounces concentrations—value of 3.88 tons......... | 4,845, 34 |
| (2.) chlorides, by leaching—from 10.29 tons.... | 193 45 |
| (3.) 0.35 ounces second concentration—value of 0.35 ton.... | 360 11 |
| 4.23 ounces produced.............................. | 5,389 90 |
| Assay-value of crude ore, 10.29 tons...................... | 6,383 91 |
| Of which was produced.................................... | 5,398 90 |
| And lost .............................................. | 985 01 |

Of the total value, 84½ per cent. was produced.

The rewashed tailings would represent 6.06 tons, and the ton of tailings contained, therefore, $162.50.

*Tests made with Mr. Williamson's ore.*

This ore is a mixture of sulphurets of copper, iron, zinc, and lead, with quartz. Some of the sulphurets are decomposed, and the ore, in parts, has an ochreous appearance.

Four samples were tested, Nos. 1 and 2 containing principally a lead-zinc sulphide, and Nos. 3 and 4 consisting mainly of copper and iron pyrites in quartz.

| | No. 1. | No. 2. | No. 3. | No. 4. |
|---|---|---|---|---|
| ½-ounce assays for lead....... | 7½ % | $15\frac{4}{10}$ % | $4\frac{6}{10}$ % | 7 % |
| Assays of silver, per ton of the bullion .................. | $125.68 | $122.51 | $342.43 | $131.93 |
| The silver in bullion would represent, per ton of raw ore... | $9.43 | $18.85 | $15.71 | $9.43 |
| Each ton of bullion would require of raw ore, tons............. | $13\frac{3}{10}$ | 6½ | $21\frac{8}{10}$ | $14\frac{3}{10}$ |

A copper assay was made for No. 4. Result, 13 per cent. copper.

The raw ore was then dressed or washed by hand. The results were for silver:

| No. 1. | No. 2. | No. 3. | No. 4. |
|---|---|---|---|
| $15.70 | $128.00 | $62.00 | $69.00 |

*Battle Mountain district*—The mining-camp called Galena is situated about fourteen miles due south from Battle Mountain, a station on the Central Pacific Railroad, in an east-west cañon in the Battle Mountain range. Its easy accessibility by a good road so near the railroad gives its mineral deposits enhanced value. The geological formation in this district is mainly a succession of various clay and chloritic slates, quartzites, and siliceous, dark-gray limestone, the general trend of the strata being north and south, with a dip to the west. There are two principal vein-zones, running and dipping parallel with the stratification, of which the easterly one is on the contact between limestone and quartzite, and the westerly one in quartzites, alternating with the smaller strata of slate. A third lode seems to branch off in a northwest direction, and is running between quartzite and limestone. The mines in Galena which have been most developed are the White and the Shiloh. In these, two shafts have been sunk to a depth of over 200 feet, showing a tolerable regularity of the ore-deposit, which has a steep dip, first to the east, but in depth to the west, and is from 6 to 12 feet wide. The mineral in these mines is argentiferous galena, mixed with ruby silver and fahlore. The galena itself carries, in pure cubical pieces, over $100 in silver per ton, and the mixed mineral, without gangue, often as much as $1,200. The owners of these two mines have also purchased the Macbeth mine, which seems to be either a branch of the former or on the lode above mentioned, running slightly to westward. Both shafts have hoisting-works and pumps, and over 2,000 feet of levels have been driven. At the time of Mr. Trippel's visit, to whom I am indebted for notes on this district, there were not over twenty men at work, producing a quantity of rich smelting-ores for shipping, which undoubtedly has left a considerable monthly surplus over all expenses. The workings were prosecuted within 200 feet of the northern boundary-line of the claim. A large amount of third-class ore has been dumped with the rock, and the accumulation amounts to over 20,000 tons, which by concentration could easily be made profitable. At present a small contrivance called concentrating-works is connected with these mines, but as Mr. Trippel saw it, it had no claim to be called so, and could give no satisfaction. The principal claims on that lode and branches from north to south were Alfred, Enterprise, Macbeth, White, Shiloh, Avalanche, Veritas, Champion, and others, all of these containing argentiferous-lead ores, decomposed down to the water-level. On nearly all of them some work has been done, but the existence of water at a depth of less than 100 feet has prevented any extensive work, the capital invested in most of these mines being very insignificant. Further south, in Copper Cañon, and apparently on the same ore-zone, several claims are worked for copper. An English company is in possession of the Superior mine, which produces large quantities of copper-pyrites, the better quality of which is shipped via San Francisco to Swansea, while the leaner ores await the period of concentration. Aside from this mine, there are several others producing copper-pyrites, and purple copper-ore, namely, the Philadelphia, Trenton, Virgin, Josephine, &c.

The eastern ore-belt is represented mainly by the Trinity and Butte mines, the first of which has been worked to some depth, but in an unsystematical manner. It produces in its upper parts the usual products of decomposed sulphurets of lead, and in depth, galena. The following claims are on this line: Cherokee, Neptune, Little Giant, Buena Vista, and others. None of these seem to have been worked beyond some prospect-work. A small smelting-works with one blast-furnace and an amalgamation-mill have been erected in the lower part of the cañon, of which the first one had hardly been tried, and the latter taken down after some unprofitable work. Nevertheless, both would undoubtedly be in operation if they had been used properly, and the ore been concentrated previous to bringing it to the works.

The mining-facilities in this cañon are excellent. There is sufficient water for dressing-works, and the proposed railroad between Battle Mountain and Austin, which will pass the entrance of the cañon within less than a mile, will increase these facilities considerably.

To the north of Galena a few miles distant, numerous irregular pockets of oxidized copper-ores (malachite, azeorite, and red oxide) are found imbedded in limestone. They were worked slightly in former times, but were not found to be profitable.

## EUREKA COUNTY.

In *Eureka district*, the Richmond Consolidated Company has passed through a very prosperous year, as will appear from the following extract from the official report of the directors for the fiscal year 1873-'74. It was published in the London Mining Journal of November 21, 1874:

The company have continued in quiet possession of the mine since the settlement of the lawsuits with the Eureka Company in June, 1873, and have been enabled during this period to carry on their mining operations without interruption; they have also obtained United States patents for their property, thus placing their title beyond dispute. It will be in the recollection of the shareholders that Mr. Probert returned from his first visit to Eureka in December, 1873, and Mr. Corrigan in January following; and that on January 27 Mr. Probert gave to a meeting of shareholders a most lucid and interesting description of the mine, both with reference to its then present position and probable future prospects, and made numerous suggestions to enable the company to lessen the working expenses, and obtain a larger profit from each ton of ore raised.

The directors, having determined to erect refining-works, prevailed on Mr. Probert to visit some of the smelting and refining establishments of France and Germany, so as to be in a better position to advise with them as to the description most suitable for this company. Immediately after his return, the directors ordered apparatus for refining and desilvering, similar to that in use at the establishment of Messrs. Luce, Fils & Rozan, of Marseilles, where it has been in successful operation for some time, and which system is now in course of adoption at some of the largest establishments in this country. Mr. Probert left England again on June 5 to superintend the erection and starting of these works, and arrived at Eureka on June 28. As Mr. McGee, the late superintendent, was about to leave the service of the company, Mr. Probert, on the urgent request of his fellow-directors, undertook, pending the appointment of another superintendent, to attend personally to the company's affairs at Eureka. The directors are now in negotiation with an experienced manager, whose services they expect to secure almost immediately, so as to be able to relieve Mr. Probert of this part of his present arduous duties.

The expenditure on capital account, in the purchase of new mines, additional buildings and works, defending mine, &c., has exceeded the amount authorized to be raised by shares by £55,285; from this, £7,995 was written off last year out of revenue, and £12,000 this year, leaving the expenditure still in excess £35,289. The directors have appropriated, in further reduction of this amount, the sum received for premium on shares, and for proceeds of shares surrendered by the vendors, amounting in the whole to £11,900, leaving a balance to the debit of capital of £23,389. The net revenue for the year amounts to £86,591, which, with a balance of £19,667 brought forward from last year, makes a total of £106,258. Out of this sum the directors have paid during the year the following dividends: 7*s.* 6*d.* per share on October 13, 1873=£19,911; 10*s.*—5*s.* on January 10, 1874, and 5*s.* on February 28, 1874=£26,904 18*s.*; 5*s.* on May 15=£13,458 4*s.*; 5*s.* on August 14=£13,458 19*s.*; total, 27*s.* 6*d.* per share, or 27¼ per cent. per annum, amounting to £73,733 1*s.*, leaving a balance of £32,525 15*s.* 1*d.* Of this balance, the directors have written £5,000 off "defense of mine account," and £7,000 off "construction account," and they propose to carry to reserve the sum of £15,389 19*s.*, which, with the amount of £8,000 already standing to the credit of that account, will make a total of £23,389—a sum sufficient to meet the present overpaid balance on capital account, or, in other words, to provide out of revenue the whole of the expenditure on capital account in excess of the amount of the authorized share-capital, and to carry forward a balance of £5,135. The directors recommend that a dividend of 5*s.* per share should be forthwith declared, payable on December 14.

The total quantity of Richmond ore smelted during the year has been 28,165 tons, and of purchased ore 1,997 tons, together 30,162 tons, producing 5,130 tons of base bullion, of the estimated value of $1,779,231, (£355,846.) The total value of the base bullion (at assay-value) produced in the year, as reported weekly by cable, amounts to $1,765,000, whereas the actual value amounted to $1,779,231, being $14,231 in excess of the amount reported by cable. In consequence of the unusually severe winter, and the consequent scarcity of charcoal, all the furnaces had to be shut down on February 22. Smelting was not resumed until April 23, when one furnace was started, the three not being in blast until June. In July, further difficulties occurred in the supply of charcoal, which at one time threatened the stoppage of all the furnaces, but, happily, an amicable arrangement was made with the charcoal-burners, and there has been since that time an unlimited supply.

During the latter part of Mr. McGee's management, explorations had not been carried on to any extent. On Mr. Probert's arrival, he put a large force on this work in various directions, and the success attending his efforts has been beyond expectation. The latest explorations of Mr. McGee had laid bare poor irony ore for a length of 80 feet in the deepest workings of the Richmond vein; but fortunately, in working upward and sideways, good ore in large masses was found, and thus what was thought at one time to be the worst point in the mine has proved to be one of the best. Explorations were also made under the Lizette tunnel, where splendid ore was found, and which is developing rapidly. Explorations were also made from the bottom of the McGee, or main shaft, at a depth of 400 feet below the surface of the ground, and here again a fine lode of ore, of exceptionally high grade, was found. The extent and value of the ore opened up by these explorations can hardly be estimated at present, but they are of the utmost importance to the company, and the value of the mine is very greatly increased thereby, the ores discovered being very rich carbonates.

Some inconvenience and loss in smelting has recently taken place from defects in the steam-engine. The mine of late has furnished more ore than could be smelted with the present furnaces; on August 31, 3,802 tons had been accumulated on the dumps ready for smelting. The refining-works will not be ready so soon as was expected, in consequence of the captain of the vessel which was chartered for the purpose declining to take the apparatus, on account of the size of the castings. The first portion did not arrive at New York until September 8. In order to avoid further delay, the directors have arranged to send the remaining portions by steamers, as the buildings are in readiness for the machinery. The proposed railway to Eureka from Palisade Station (on the Central Pacific Railway) is reported to be finished for a distance of fifty miles, where it is proposed to stop for the winter, thus leaving only a length of about thirty miles to be completed.

The refining and desilverization process above referred to is a modification of the Pattinson process, in which the entire work of removing the intermediate products is done by machinery, and the stirring by steam—a very great improvement on the old process. As compared with the desilverization by means of zinc, there is this disadvantage in the Pattinson process, that by its use the contents of silver in the lead-riches cannot be brought higher than from 2.5 to 3 per cent., while, by means of the zinc-process, the lead may be enriched to 10 per cent., and more. Under the conditions prevailing at Eureka, this may, however, be no disadvantage, since a large percentage of litharge may be welcome for the purpose of adding it in the ore-smelting, in order to secure in the charge a higher percentage of lead than the ores themselves furnish. Next year, I hope to be able to give a full description of the new process, which will no doubt be in regular running-order in the course of the coming summer.

The following description of the Richmond mine, as now known after the explorations and developments of the past, and the accompanying information in regard to the operations of the company during the calendar year 1874, I owe to the kindness of the manager, the Rev. Edward Probert:

The Richmond mine is situated on Ruby Hill, which is a spur of Prospect Mountain, and is distant from Eureka about three miles. The vein cropped out on the southwest side of the hill, and it has been followed about 900 feet on the dip, to the northeast or opposite slope, the workings passing below the crest of the hill at a depth of 300 feet or thereabout. On this northeastern slope a shaft has been sunk, (now 500 feet deep,) from which two drifts have been run—one leaving the shaft at a depth of 200 feet, the other at 400 feet, each being about 250 feet long—to intersect the ore-body. This is found to preserve nearly the same course and dip at the greatest depth attained which it possessed at the starting point on the other side of the hill. It has been technically defined as a "pipe-vein," but it is in reality made up of a series of bonanzas of an ellipsoidal shape placed with their longer axes in the direction of the dip. One of these, the third from the croppings, and the largest, extends downward about 300 feet, its greatest lateral extension being about 250 feet, and its maximum thickness (between the walls) about 100 feet. This chamber is invaded at two or three points by projecting buttresses of limestone, diminishing the mass of ore by at least one-third. The lode has nowhere had a lateral extension on what would be termed the "course" of more than 300 feet. To the northwest it "feathers out" completely, while to the southeast it is limited by a bar of red oxide of iron, behind which all traces of the lode disappear. The "country" or containing rock of the lode is metamorphic limestone; the hanging-wall is very irregular and much shattered; the foot-wall is more regular and compact, sometimes running quite flat for a hundred feet or so, and then dipping suddenly for 70 or 80 feet at an angle of 75° to the horizon. The ore itself consists chiefly of an argentiferous carbonate of lead, with nodules of galena interspersed through it, together with a considerable amount of decomposed arsenical pyrites, which is gold-bearing. Large masses of oxide of iron alternate with the richer vein-materials, and constitute what may be termed the "gangue," which imparts to the ore its usual ferruginous or red-earthy appearance. This oxide of iron, which amounts to upward of 40 per cent. of the contents of the vein, is itself more or less rich in gold or silver, and forms, with the other materials, an excellent fluxing mixture. Practically, indeed, the ore has no "gangue," and all that can be done in the way of assorting is to remove some of the poorest-looking lumps of iron-stone and any pieces of limestone which may have fallen in from the hanging-wall. The carbonates of lead vary in color from dirty white to yellow, gray, and dull earthy-red, the latter being due to an admixture of iron oxides. The nodules of galena represent, no doubt, the original condition of the vein-material, the carbonates being evidently the results of decomposition. These nodules of galena are always found imbedded in the carbonates and covered with a crust of sulphate of lead, which seems to be the intermediate condition of the mineral while undergoing the chemical change from galena or sulphuret of lead to carbonate. Some rarer minerals are also found in the mine, such as mimetite, an arseniate of lead combined with a chloride. This is always rich in gold. Also, there are often met with, lining the drusy cavities of the harder vein-materials, beautiful crystals of molybdate of lead, of the usual tubular form and orange color. The mine is perfectly free from water, and at the present depth the moisture in the ore is not greater than it was in the upper workings, the average being about 12 per cent. The ore being for the most part in a friable condition, is easily mined, very little blasting being necessary.

The ore is admirably adapted for smelting in the blast-furnace, requiring no previous fluxing or preparation of any kind. The accompanying analysis was made from the furnace-samples for the month of August, 1873. It will be seen that the percentage of peroxide of iron is very high, and to this are mainly due the simplicity and success of the smelting-operations.

The loss of precious metals and lead during the years 1872 and 1873 exceeded 20 per cent. of the total assay-value of the ore. This year the loss has been reduced to about 12 per cent., mainly through the introduction of a flue, through which the smoke and furnace-gasses pass for a distance of 800 feet before their discharge into the atmosphere. The yield from the first 250 feet of this flue (the portion nearest the furnaces) is between six and eight tons daily when three furnaces are running, the assay-value of the deposit removed from it exceeding that of the ore smelted about 20 per cent. The loss in smelting lead-ores in blast-furnaces is always high; but the large amount of arsenic in the Richmond ore makes the loss higher than the average for other ores. The arsenic volatilizes at a low temperature, carrying with it the precious metals, and filling the atmosphere with a garlicky odor, which is perceptible at a distance of twenty miles.

Our daily consumption of charcoal for three furnaces is about 4,500 bushels, which, at a cost of 30 cents a bushel, amounts to the large sum of $40,000, or thereabouts, monthly. The quantity of ore smelted daily by each furnace is about forty-five tons—sometimes rising to sixty tons, when the percentage of lead in the ore is high. The average percentage of lead may be stated at 23 per cent.—25 to 30 per cent. is exceptional; but with these latter figures the smelting is more expeditious while the consumption of fuel is decreased.

*Analysis of furnace-sample of Richmond ore worked in August,* 1873, *made by F. Claudet, London, assayer to the Bank of England.*

| | | |
|---|---|---|
| Oxide of lead | 26.57= | 24.65 lead. |
| Oxide of copper | .52 | |
| Peroxide of iron | 40.37 | |
| Oxide of zinc | 2.46 | |
| Arsenic acid | 5.82= | 3.79 arsenic. |
| Antimony | Traces. | |
| Sulphuric acid | 2.60= | 1.04 sulphur. |
| Chlorine | Traces. | |
| Silica | 7.08 | |
| Alumina | .77 | |
| Lime | 1.18 | |
| Magnesia | .50 | |
| Water and carbonic acid | 12.60 | |
| | 100.47 | |

Silver, 13 ounces 1 pennyweight per ton of 20 hundred-weight;* gold, 19 pennyweights 14 grains per ton of 20 hundred-weight.

During the years 1873 and 1874, 58,037 tons of ore, averaging about $50 per ton in gold and silver and 24 per cent. of lead, were reduced at the Richmond smelting-works.

The product during the calendar year 1874, from 27,784 tons of ore, was 9,575,900 pounds of work-lead = $4,787\frac{1900}{2000}$ tons.

| | |
|---|---|
| Gold and silver contents, (Mint prices) | $1,263,120 14 |
| Value of lead, at 4 cents per pound | 383,036 00 |
| Total | 1,646,156 14 |

The flue built by the Richmond Company for the purpose of catching the dust mechanically carried off by the blast, and for condensing fumes, which is mentioned above, is a very substantial, and at the same time a very convenient, structure. It will be referred to again in another part of this report, under "metallurgy."

Nearly all the crude lead produced by the Richmond Company during the year has been subjected *in loco* to a refining or calcining process, the object being twofold: first, to remove such impurities (antimony, arsenic, and mechanically admixed speiss and dross) as have heretofore given occasion for serious reductions in the price paid by the desilverizing-works; and secondly, to obtain for addition in the blast-furnaces the resulting oxides, which, of course, contain a very large percentage of oxide of lead. In thus raising the percentage of lead in the blast-furnace charges, it was expected that the loss of gold and silver heretofore suffered in the speiss (see my last year's report) would be materially diminished; and experience has corroborated this view. While the speiss formerly produced contained from $16 to $20, and even $25, of silver and gold, the present product contains rarely more than $12 per ton.

The operations of the Eureka Consolidated Company during the period from December 1, 1873, to October 1, 1874, are described in detail in the following extract from the report of Mr. W. H. Shaw, the superintendent:

On assuming the duties of superintendent here, I found an ore-body on the fourth level and 150 feet west from Lawton shaft; also a body of ore on the first level and 175 feet south from Windsail shaft. These deposits we found to be of considerable extent, furnishing a supply of ore for several months. Being thus provided, we had opportunities for running prospect drifts and winzes in other directions to seek for a supply of ore as a guarantee against future contingencies.

We will first mention the work done in and through the Lawton shaft.

On the first level, the south branch of tunnel had been run 250 feet. Starting from this point, we extended the tunnel in a westerly direction 100 feet for prospecting purposes, when, finding no trace of ore existing here, we discontinued further operations.

From the second level, at a point commencing 100 feet southeast from shaft, we drifted 150 feet, to connect with ore-body and for air.

In the third level we ran the south branch of the tunnel 200 feet for taking out ore; then ran a second cross-cut 300 feet, connecting it by a drift 300 feet long with the main south branch of this level for air.

On the fourth level we continued in the old ore-chamber, located some 150 feet west of the shaft, and extracted large amounts of ore, but finally it became exhausted. From this chamber we followed ore-signs up to the second level, through and into the Kidd ground, finding there a large body of ore, varying in length from 75 to 200 feet, and from 1 to 20 feet in thickness. We extracted ore from this point from about the middle of last December until the 1st of August of the present year, when all signs of ore disappeared. Leaving this chamber we returned to the third level again, and raised 185 feet from drift connecting with south branch of tunnel, cutting through a fine

* This is apparently the ton of 2,240 pounds avoirdupois. In this country the value of silver-ore is usually calculated upon the "short ton" of 2,000 pounds.—R. W. R.

body of ore fully equal to any heretofore found. At the top of this raise we developed a large chamber of ore, on which we have prospected some 75 feet in length by 4 feet in width. The extent or depth of this body we have not ascertained, as we had to drift from this point for air, running 135 feet into the old ore-chamber of the Kidd for that purpose, and are now prepared to explore this ground in every direction to learn its value. On the fourth level we commenced drifting at a point 300 feet from shaft on the foot-wall and running 270 feet, finding a small body of ore of inferior grade, which we sunk down upon, and shall prospect more fully to determine its extent, having some ore still in sight on each side of the winze.

On the fifth level we have run 620 feet of drifts connecting with fourth level by winzes for air. In this level we found a body of ore at a point distant 300 feet from shaft. We ran through this 70 in length by 4 in width, finding a good quality of ore, which we have not explored to any definite extent. Here we rise through to fourth level for a supply of air, cutting through 40 feet of ore on the way. This ore is of an inferior quality, and its extent and character are not yet fully determined. Three winzes have been sunk from this level to connect with the sixth level, the first being 40 feet from shaft and sunk through ore the whole distance, which is, however, of inferior quality and small extent. The second winze is 200 feet from shaft, rising up from the sixth to the fifth level for air, and no ore is in sight. The third winze is 300 feet from shaft and connects with sixth level. This cuts through 25 feet of ore, and is 70 feet north of the ore-body above alluded to. This level is being continued on to connect with the Bell shaft, now distant 175 feet, with good indications of finding ore at any time.

On the sixth level, 300 feet from shaft, we have sunk down 25 feet through ore, which we leave for the present and until we can get a supply of air. The third winze mentioned on the fifth level is being sunk down below the sixth, and we shall continue it down until we find ore; thence we intend drifting toward the winze 40 feet to the south, when the first winze on this level will be completed connecting with this drift.

The sixth level is now in a distance of 500 feet, and will be pushed forward, making for the Bell shaft, 320 feet distant.

In the Lawton shaft proper we are sinking down, having extended the work 40 feet, and will continue down until reaching a depth of 100 feet.

In the Buckeye shaft we found a small body of ore, being a mixture of galena and carbonate of good quality. Work continues on this to learn extent of body.

A body of ore in the Windsail shaft that was being worked when I took charge of the company's business has been alluded to. This chamber is 175 feet south from the shaft and on the first level. The ore was of good quality, and has continued with various changes as to quantity until quite recently, when it pinched out, but came in again with much regularity. At the present we have a small amount of ore in this chamber, and very encouraging prospects of finding new deposits.

From the first level we have run a drift 200 feet from the east branch of tunnel, finding excellent ore, which we hope to find developing into a large body.

On the second level three drifts have been run of 250 feet; also a drift to the south, finding a fair body of ore, which we think connects with ore-body on first level.

From the first and second levels we are now taking out all the ore coming from the Windsail.

On the third level we have run a drift 120 feet from the north branch, finding a small body of ore, now worked out. Tracing from here, we raised 30 feet for prospecting, and abandoned same on account of foul air, but shall resume work as soon as we can get the requisite ventilation, and continue drift to connect with Bell shaft, 200 feet distant.

In Champion ground we have run tunnel No. 1 140 feet east toward Bell shaft, finding ore 75 feet from the mouth of the tunnel, of good quality and unknown extent. This tunnel is being driven forward for prospecting, in the hope of finding ore that extends through the surface-ground.

Tunnel No. 2 we have run 40 feet in a southerly direction, finding ore in small quantities, with good indications for larger bodies.

Tunnel No. 3 we have run 100 feet in a northeasterly direction through ore-seams, with good prospects for finding an ore-body.

Tunnel No. 4 we have run 125 feet east to take out ore from near the surface.

The Champion ground was supposed to be worked out when my administration commenced. Our workings have proved differently, and we hope yet to find such deposits as will justify the expense of further labor. We have already extracted 1,500 tons of good ore, and expect to find more equally as good.

From the foregoing it will be seen that at this time we have ore in larger or smaller quantities in the first, second, third, fourth, fifth, and sixth levels worked from the Lawton shaft; also in the second, third, and fourth levels of the Windsail; also in tunnels one, two, and three of the Champion ground, and in the Buckeye shaft. From these several deposits we expect to glean large supplies of ore for many months to come, though much work has yet to be done in making air-connections and other connections that will give us a safe and easy exit for our ores.

The Bell shaft we commenced with diamond drill, drilling eight holes, 400 feet each in depth, consuming the time from May 15 to July 16. July 18 we commenced blasting, and have worked up to the present time, (with but ten days' loss of time.) We have now reached a depth of 87 feet, and are making good progress.

The timber-yard of the Lawton, which formerly stood at an inconvenient distance across the ravine, has been moved to the immediate vicinity of the works, ground for this purpose having been made by walling up the steep embankment and filling in with waste from the mine, thus giving the foreman complete supervision over every department of labor throughout the works.

On the Windsail ground we have changed the ore-dump from the gulch below to a point higher up and more convenient to the works. By walling up and filling in, we have made a large yard, affording ample space for piling wood, timber, &c.

*Furnaces.*—Since the last annual report we have substituted Root blowers in place of the Sturtevant fan then in use, and find them much more economical, requiring less power of machinery, and affording a blast more powerful, steady, and uniform; in fact, giving better results throughout, and proving highly satisfactory in every respect. These blowers are attached to furnaces Nos. 3, 4, and 5.

We have now three first-class furnaces, viz, Nos. 3, 4, and 5, each of fifty tons' capacity, and capable of greater service under more favorable circumstances. Furnace No. 3 has been running six weeks, and seems in condition now to run that much longer. Furnaces Nos. 4 and 5 are in perfect order for successful smelting, and all three are connected with the dust-arrester. We hope soon to have these furnaces all running to their full capacity.

Furnaces Nos. 1 and 2 we have abandoned, and will use only in cases of emergencies or when we see a way for their utility in some other branch of the reduction of base ores.

In building a dust arrester or flume, it necessitated several improvements. First, in the raising of the top of building some 16 feet and replacing the old roof by a new one of more substantial construction, bracing it thoroughly with heavy timbers to give support to the immense weight of iron beneath, used in the flume, this being suspended from the roof by iron rods running round the flume and extending up through the roof-timbers and securely fastened by heavy nuts. Under the flume is built a plank flooring for car-track, over which is run the car for collecting and conveying the dust outside the building to places convenient for reduction. Underneath this flooring is a large and airy feeding-room, giving plenty of space for the labor required in charging the furnaces with their supply of coal and ore.

The next improvement necessitated was in carrying up the furnaces to a greater height of 20 feet to clear the building and leave room for making the necessary connection to the flume.

These improvements are all completed. The dust-arrester is in place, and little remains to be done now before we shall be able to test the merits of this particular improvement. This flume is most substantially built, being lined throughout with iron. It is considered safe against fire.

The secretary's report for the year ending September 30, 1874, is as follows:

RECEIPTS.

| | | |
|---|---|---|
| From sale of iron | $89 33 | |
| From sale of mule | 74 00 | |
| From sale of lumber | 120 62 | |
| From sale of wood | 16 00 | |
| From sale of slag-pile | 1,000 00 | |
| From sale of 2 bars bullion | 29 90 | |
| From proceeds base bullion refined: 888 tons, product of 1873; 2,941 tons, product of 1874 | 1,286,175 40 | |
| Cash in hands of superintendent October 1, 1873 | 1,759 73 | |
| Over-draft, Bank of California, October 5, 1874 | 60,289 32 | |
| | | $1,349,554 30 |

DISBURSEMENTS.

| | | |
|---|---|---|
| Construction and improvements: | | |
| Cost of boiler for diamond drill | $800 00 | |
| Cost of 3 No. 5 Root blowers, with pulleys | 2,275 00 | |
| Cost of iron, nails, &c., for fume-arrester | 4,644 54 | |
| | | $7,719 54 |
| Mine account: | | |
| For labor | 213,255 97 | |
| For hauling ore | 45,663 88 | |
| For hauling water | 6,821 12 | |
| For timber, lagging, and lumber | 28,883 14 | |
| For wood and coal | 6,801 98 | |
| For tools, iron, steel, and hardware | 5,986 08 | |
| For blacksmithing | 290 25 | |
| For candles, oil, and tallow | 4,253 80 | |
| For castings and foundery-work | 441 93 | |
| For freight on supplies | 2,284 01 | |
| For powder, fuse, and exploders | 3,156 76 | |
| For surveying and recorder's fees | 401 00 | |
| For water-tank | 58 25 | |
| For legal services | 20 00 | |
| For diamonds and bits | 285 44 | |
| | | 318,603 61 |
| Smelting account: | | |
| For labor | 87,887 57 | |
| For charcoal | 250,566 53 | |
| For wood | 3,361 25 | |
| For timber and lumber | 9,056 88 | |
| For hardware, iron, steel, and tools | 9,339 85 | |
| For castings and foundery-work | 4,101 24 | |
| For freight on machinery and supplies | 4,617 90 | |
| For tinsmithing | 315 60 | |
| For blacksmithing | 1,490 00 | |
| For fire-rock, brick, and stone | 5,177 10 | |
| For oil, candles, and tar | 505 14 | |
| For assaying | 925 00 | |
| For paints | 35 12 | |
| For purchase bullion | 300 00 | |
| | | 377,679 18 |
| General expense, Eureka: | | |
| Salary superintendent, employés, and traveling-expenses, &c | 10,816 00 | |
| Ore-tax | 4,713 92 | |
| State and county taxes | 4,203 50 | |
| Repairs and furnishing superintendent's house and office | 1,317 92 | |
| Harness and repairs | 80 88 | |
| Express-charges, franks, and newspapers | 436 20 | |
| Telegraphing, books, stationery, printing, and advertising | 570 75 | |
| Exchange on superintendent's drafts | 487 30 | |
| Barley, oats, hay, and horse-hire | 1,606 80 | |

| | | |
|---|---|---|
| Insurance | $105 00 | |
| Medicines and medical attention to employés injured while in the service of the company | 964 00 | |
| Wood | 99 37 | |
| Assays | 410 00 | |
| Attorney's fees | 1,780 75 | |
| Blacksmithing | 121 00 | |
| Surveying and recorder's fees | 175 50 | |
| Repairing road | 25 00 | |
| Donations | 30 00 | |
| Banners, flags, &c., 4th of July | 90 50 | |
| Locations | 25 00 | |
| Balance for purchase lot | 300 00 | |
| Purchase first bar bullion produced at works | 20 00 | |
| | | $28,379 39 |
| Expense San Francisco: | | |
| Salary of officers and traveling-expenses of president | 6,885 00 | |
| Office-rent and services of porter | 1,185 00 | |
| Books, stationery, and stamps | 228 41 | |
| Printing, advertising, and newspapers | 217 00 | |
| Express-charges and telegraphing | 141 33 | |
| Retaining stock on call-list of board | 100 00 | |
| City, county, and State taxes | 15 25 | |
| Assays | 41 25 | |
| Removing furniture and refitting new offices | 279 99 | |
| Coal and wood | 45 75 | |
| Hack-hire | 5 00 | |
| | | 9,143 98 |
| Interest: | | |
| For interest on over-drafts and notes | | 5,200 68 |
| Freight, refining, &c.: | | |
| For transportation and refining-charges, &c., on base bullion | | 205,882 11 |
| Bills payable: | | |
| For notes becoming due | | 50,000 00 |
| Book accounts: | | |
| For bills carried over from last year | | 1,462 98 |
| Superintendent's drafts: | | |
| For drafts carried over from last year | | 15,613 43 |
| New Champion location: | | |
| For purchase Bateman's interest in Nugget, At Last, and Margaret claims | | 5,000 00 |
| Dividend account: | | |
| For Nos. 11–13 paid stockholders | | 150,000 00 |
| | | 1,174,684 90 |

| | | |
|---|---|---|
| Over-draft, Bank of California, October 1, 1873 | $3,558 17 | |
| Over-draft on bullion shipments October 1, 1873 | 171,310 81 | |
| Cash in hands superintendent October 1, 1874 | 42 | |
| | | $174,869 40 |
| | | 1,349,554 30 |

RECAPITULATION.

Receipts:

| | | |
|---|---|---|
| From sundry sales | $1,329 85 | |
| From product 3,829 tons base bullion refined | 1,286,175 40 | |
| Cash in hands of superintendent October 1, 1873 | 1,759 73 | |
| Over-draft, Bank of California, October 5, 1874 | 60,289 32 | |
| | | $1,349,554 30 |

Disbursements:

| | | |
|---|---|---|
| Construction and improvements | $7,719 54 | |
| For mine account | 318,603 61 | |
| Smelting account | 377,679 18 | |
| For general expense, Eureka | 28,379 39 | |
| For expense, San Francisco | 9,143 98 | |
| Interest | 5,200 68 | |
| Freight, refining, &c | 205,882 11 | |
| Bills payable | 50,000 00 | |
| Book accounts | 1,462 98 | |
| Superintendent's drafts | 15,613 43 | |
| New Champion location | 5,000 00 | |
| Dividends | 150,000 00 | |
| | 1,174,684 90 | |
| Over-draft, Bank of California, October 1, 1873 | 3,558 17 | |
| Over-draft on bullion-shipments October, 1873 | 171,310 81 | |
| Cash in hands superintendent October 1, 1874 | 42 | |
| | | $1,349,554 30 |

RESOURCES AND LIABILITIES.

Resources:

| | | |
|---|---|---|
| W. H. Shaw, superintendent | $0 42 | |
| Supplies at Eureka, per inventory: | | |
| On account mine | 12,779 89 | |
| On account furnaces | 44,334 44 | |
| 218 tons base bullion at refining-works and *en route*, approximate value | 63,250 24 | |
| | | $120,364 99 |

Liabilities:

| | | |
|---|---|---|
| Over-draft, Bank of California | $60,289 32 | |
| Superintendent's drafts, (not presented) | 14,495 85 | |
| Book accounts, (not due) | 1,903 05 | |
| | | $76,688 22 |
| Balance net resources | | 43,676 77 |

COST OF EXTRACTING ORES.

| | |
|---|---|
| Expense of extracting and hauling to furnaces 22,831 tons of ore is | $318,603 61 |
| Supplies on hand October 1, 1873 | 6,962 09 |
| | 325,565 70 |
| Less supplies now on hand, per inventory | 12,779 89 |
| | 312,785 81 |

COST OF SMELTING ORES.

| | |
|---|---|
| Expense of smelting 22,197 tons of ore is | $377,679 18 |
| Supplies on hand October 1, 1873 | 17,552 62 |
| | 395,231 80 |
| Less supplies now on hand, per inventory | 44,334 44 |
| | 350,897 36 |

Or $15.80 per ton.

22,197 tons of ore reduced produce 3,159 tons of base bullion, or, by average, 6.91 tons of ore produce one ton of base bullion, at a cost of .......... $210 02

Transportation and other charges from works to San Francisco, (where now all the company's bullion is being refined,) aggregate per ton about .......... 32 10

W. W. TRAYLOR,
*Secretary.*

During the year 1874 the company worked 23,485 tons of ore, from which were produced 3,365 tons of crude lead, (base bullion.)

| | |
|---|---|
| Value of lead | $269,224 00 |
| Value of gold and silver | 858,243 22 |
| Total | 1,127,467 22 |

The average value of the work-lead in gold and silver was therefore $225.05 per ton.

The Ruby Consolidated Company has principally worked during the year in the Dunderberg and Pleiades mines, which are two adjoining locations on the same vein.

I take occasion here to correct a mistake which was made in a former report, where the Dunderberg was described as an east and west vein. At the time this statement was made the mine had hardly begun to be developed, and the surface-ore seemed, indeed, to indicate an east and west course. But further explorations showed that this ore was only a body detached from the vein proper and enveloped in clay. The vein itself was found a few yards west of the old workings, and the extensive explorations made since show clearly that the vein runs northerly and southerly. It is opened by an incline (58½° west) now 312 feet deep, on which substantial steam hoisting-works were erecting when Mr. Eilers visited the locality in the summer of 1874. In this incline, not very far from the surface, an ore-body was found, which was followed continuously and stoped out to the 170-foot level. Here it divided, one part dipping to the south, the other to the north. Both portions have been encountered again in the 300-foot level, north of the incline, and the development of these bodies both upward and downward was going on at the time above mentioned.

The Pleiades lies lower down on the slope of the hill, (south of Dunderberg,) and is on the same vein. The deepest point in its shaft is 105 feet from the surface. In the 50-foot level of this mine the ore-body was found from 6 to 9 feet wide. At the lowest point the ore is only from 2 to 2½ feet wide. The ore is very similar to that of Dunderberg, being only a little more quartzose. The 120-foot level of the Dunderberg connects with the working of the Pleiades.

The smelting-works of the company have been in operation during only six months of the year. As on former occasions, I am indebted to Mr. O. H. Hahn, M. E., the superintendent of the works, for very full and detailed statements of the campaigns made.

There were smelted during 1874:

| Ore, &c. | Tons gross. | Total gross. | Tons net. | Lead, per cent. | Silver per ton. | Gold per ton. | Total per ton. |
|---|---|---|---|---|---|---|---|
| Dunderberg: | | | | | | | |
| Screenings | 3,06.46 | | | | | | |
| Coarse | 1,969.3 | | | | | | |
| Argillaceous | 26.1 | 5,080 | 4,673.7 | 21.82 | $28 57 | $14 42 | $42 99 |
| Custom ore: | | | | | | | |
| K K | 161.1 | | | | | | |
| Excelsior | 16.7 | | | | | | |
| Eyrie | 25.7 | | | | | | |
| Miner's Dream | 1.6 | | | | | | |
| Adam's Hill Company | 29.6 | 234.7 | 212.3 | 13.98 | 28 87 | 5 97 | 34 84 |
| Total net | ........ | ........ | 4,886 | ........ | ........ | ........ | ........ |

The working-results at the Ruby furnaces during 1874 are given in the following table:

*Working-results at the Ruby Smelting-Works, Eureka, Nev., during the year* 1874.

| Month. | Number of working-days. | Tons of ore, including moisture. | Tons of ore. | Contents of ore in lead. | Contents of ore in silver. | Contents of ore in gold. | Total contents of ore in gold and silver. | Slag used as flux. | Bushels of charcoal purchased. | Bushels of charcoal carried to furnace. | Bushels of charcoal actually used. | Tons of coke used. | Number of bars produced. | Weight of bars. |
|---|---|---|---|---|---|---|---|---|---|---|---|---|---|---|
| | | *Gross.* | *Net.* | *Tons.* | | | | *Tons.* | | | | | | *Pounds.* |
| January | 16.5 | 640.2 | 561.76 | 115.441 | $15,724 87 | $8,161 49 | $23,886 36 | 216.07 | 23,815 | 24,570 | 23,321 | ........ | 2,190 | 195,536 |
| May | 17.0 | 615.6 | 562.27 | 90.100 | 13,284 93 | 9,983 58 | 23,268 51 | 195.69 | 94,961 | 17,946 | 16,847 | 35.34 | 1,601 | 140,474 |
| June | 45.5 | 1,862.3 | 1,750.56 | 325.354 | 43,527 34 | 23,376 94 | 66,904 28 | 472.70 | ........ | 37,587 | 33,438 | 182.00 | 6,707 | 582,284 |
| July | 19.5 | 794.2 | 713.30 | 133.267 | 16,380 70 | 6,513 50 | 22,894 20 | 296.77 | ........ | 26,058 | 24,693 | 16.24 | 2,713 | 236,969 |
| September | 28.0 | 1,049.5 | 964.58 | 269.562 | 35,921 77 | 15,237 04 | 51,158 82 | 444.83 | 52,900 | 39,350 | 37,413 | ........ | 5,256 | 450,380 |
| October | 9.75 | 352.9 | 333.62 | 107.074 | 14,851 60 | 5,366 48 | 20,218 08 | 184.18 | ........ | 12,565 | 12,194 | ........ | 2,074 | 173,774 |
| Total | 136.25 | 5,314.7 | 4,886.09 | 1,049.798 | 139,691 21 | 68,639 03 | 208,330 24 | 1,810.24 | 171,076 | 158,076 | 147,926 | 233.58 | 20,541 | 1,779,417 |
| | ........ | 4,886.09 | ........ | 886.694 | 114,610 52 | 63,593 39 | 178,203 91 | ........ | 147,926 | ........ | ........ | ........ | ........ | ........ |
| Losses | ........ | 428.61 | ........ | 163.104 | 25,080 69 | 5,045 64 | 30,126 33 | ........ | 23,750 | ........ | ........ | ........ | ........ | ........ |
| Losses per cent | ........ | 8.06 | ........ | 15.5 | 17.95 | 7.35 | 14.46 | ........ | 13.8 | ........ | ........ | ........ | ........ | ........ |

*Working-results at the Ruby Smelting-Works, &c.*—Continued.

| Month. | Number of bars in car-load. | Weight of lots. | Contents of car-loads in silver. | Contents of car-loads in gold. | Contents of car-loads in silver. | Contents of car-loads in gold. | Total contents of car-loads in noble metals. |
|---|---|---|---|---|---|---|---|
| | | *Pounds.* | *Ounces.* | *Ounces.* | | | |
| January | 2, 261 | 201, 938 | 9, 765. 33 | 409. 23 | $12, 623 23 | $8, 458 63 | $21, 081 86 |
| May | 1, 382 | 121, 694 | 6, 116. 91 | 272. 97 | 7, 907 40 | 5, 652 51 | 13, 559 91 |
| June | 6, 730 | 584, 100 | 27, 985. 42 | 1, 031. 70 | 36, 163 53 | 21, 536 52 | 57, 700 05 |
| July | 2, 996 | 261, 768 | 12, 000. 08 | 399. 02 | 15, 381 22 | 8, 247 45 | 23, 628 67 |
| September | 5, 149 | 441, 969 | 22, 594. 73 | 672. 21 | 29, 193 35 | 13, 855 00 | 43, 048 35 |
| October | 2, 271 | 190, 660 | 11, 633. 69 | 328. 93 | 15, 038 62 | 6, 799 82 | 21, 838 44 |
| Total | 20, 789 | 1, 802, 129 | 90, 086. 16 | 3, 114. 06 | 116, 307 35 | 64, 549 93 | 180, 857 28 |
| Left over from last year | 158 | 14, 242 | 687. 74 | 30. 24 | 889 06 | 625 08 | 1, 514 14 |
| | 20, 931 | 1, 787, 887 | 89, 398. 42 | 3, 083. 82 | 115, 416 29 | 63, 924 85 | 179, 343 14 |
| Less bullion purchased | 90 | 8, 203 | 624. 79 | 16. 04 | 807 77 | 331 46 | 1, 139 23 |
| | 20, 541 | 1, 779, 684 | 88, 773. 63 | 3, 067. 78 | 114, 610 52 | 63, 593 39 | 178, 203 91 |
| | | Number of tons. | Average silver per ton. | Average gold per ton. | Value of silver per ton. | Value of gold per ton. | Value of gold and silver per ton. |
| | | | *Ounces.* | *Ounces.* | | | |
| Less ounces silver and gold | { 88, 773. 63<br>{ 3, 067. 78 | 889. 842 | 99. 76 | 3. 447 | $128 79 | $71 46 | $200 26 |
| | 91, 841. 41 — | 3. 148 | | | | | |
| | | 886. 694 | | | | | |

Number of furnaces, 2, with 7 tuyeres each.
Number of campaigns, 5, of 27½ days each.
Charcoal per ton of material, 24 bushels.
Coke per ton of material, 0.044 ton.
Charcoal per ton of ore, 32.3 bushels.
Coke per ton of ore, 0.033 ton.
Price of charcoal, 32 cents per bushel.
Price of coke, $60 per ton.
Slag used per ton of ore, 34.06 per cent.

Average assay of ores:

| | |
|---|---|
| Silver | $28 59 |
| Gold | 14 04 |
| Total | 42 63 |
| Lead | 21.4 per cent. |

Average yield of ores:

| | |
|---|---|
| Silver | $23 45 |
| Gold | 13 02 |
| Total | 36 47 |
| Lead | 18.2 per cent. |

The speiss produced was 276.36 tons, or 5.2 per cent. of the gross weight of the ore. The average assay of the speiss for the whole year was:

| | |
|---|---|
| Silver | $11 53 |
| Gold | 10 94 |
| Total | 22 47 |
| Lead | 3.5 per cent. |

The calculated quantity of slag during the year is about 5,960 tons. The average assay of the slag is:

| | |
|---|---|
| Silver | $0. 875 |
| Lead | 0.65 per cent. |

The dust has not been collected during this year. Its quantity, found by calculation based upon the contents of lead in the dust and the losses of that metal, has been 595.7 tons, 11.2 per cent. of the gross weight of the ore treated. The following assays of dust were made at different times during the year:

| | I. | II. |
|---|---|---|
| Lead | 21% | 17.5% |
| Silver | $21 20 | $22 77 |
| Gold | 11 30 | 15 07 |
| | 32 50 | 37 84 |

*Average.*

| | |
|---|---|
| Lead | 19.25 per cent. |
| Silver | $21 98 |
| Gold | 13 18 |
| | 35 16 |

The Hoosac Company has worked during the larger part of the year its one large furnace, the largest in Eureka, and said to have a capacity of over 80 tons daily when in good running order. Ores from the Hoosac mine have been smelted together with purchased ore, the latter principally from the K K mine. The work-lead produced at these works is of inferior quality, as it contains very large amounts of antimony and arsenic. The quantity produced during the year and shipped through Mr. Daniel Meyer, of San Francisco, was 1,150 tons, containing 150 ounces of silver and 3 ounces of gold per ton of 2,000 pounds. Its value may therefore be put down as follows:

| | | |
|---|---|---|
| 1,150 tons of very impure lead, at 3 cents per pound...... | | $69,000 00 |
| 172,500 ounces silver, at $1.25.............. | $215,625 00 | |
| 3,450 ounces gold, at $20.67................ | 71,311 50 | |
| | | 286,936 50 |
| | | 355,936 50 |

The K K Company has worked its mine during the year energetically. Very excellent hoisting-works, made by Booth & Co., in San Francisco, have been erected. The company has sold the larger part of its ore to other companies. In July, however, the Silver West furnace was rented, and smelting was then continued until the end of the year. The production is given in the following statement, obtained through the kindness of Mr. W. S. Keyes, the superintendent:

*Product of K K Consolidated Mining Company's mines for* 1874.

| Months. | Ores sold. | Price. |
|---|---|---|
| | *Pounds.* | |
| May | 2,070,010 | $10,347 04 |
| June | 1,430,456 | 6,618 63 |
| July | 1,065,103 | 2,837 98 |
| August | 2,873,624 | 8,918 09 |
| September | 2,013,314 | 6,655 43 |
| Total | 9,452,507 | 35,377 27 |

| Months. | Ores smelted. | Bullion. |
|---|---|---|
| | *Pounds.* | *Pounds.* |
| July | 3,083,000 | 331,302 |
| August | 1,600,000 | 173,292 |
| September | 3,200,000 | 233,174 |
| October | 2,032,000 | 192,962 |
| November | 3,120,000 | 266,680 |
| December | 3,300,000 | 258,123 |
| Total | 16,335,000 | 1,455,533 |

Value of bullion in gold and silver, $225 to $519 per ton. The average, which is not given by Mr. Keyes, may be assumed as $300 per ton. Product, $727\frac{1533}{2000}$ tons, at $300, $218,330.

The expense of mining and smelting was very nearly $25 per ton of ore.

About twenty-five miles north of Eureka, in the *Diamond district*, Diamond range, the Champion company again made several short campaigns during the fall. But all these attempts seem to have been failures. The ores of the locality are very calcareous and quartzose, and, as I understand, not very abundant, though the Champion vein is about 5 feet thick. The total quantity of lead reported to me as shipped by Mr. Daniel Meyer was only 40 tons. Its value in gold and silver is not stated.

The mines in *Pinto district*, seven miles east of Eureka, have again been taken up by a new English company, so far without any noteworthy results.

*Recapitulation of product of Eureka.*

| Companies. | Lead, tons. | Lead, value. | Gold and silver, value. |
|---|---|---|---|
| Richmond Company | 4,788 | $383,036 | $1,263,120 |
| Eureka Company | 3,365 | 269,224 | 858,243 |
| Ruby Company | 890 | 71,200 | 178,204 |
| Hoosac Company | 1,150 | 69,000 | 286,936 |
| K K Company | 727 | 58,160 | 218,330 |
| Total | 10,920 | 850,620 | 2,804,833 |

This shows a large reduction from the yield of 1873, which was $3,907,401,54.

Total value of production of gold, silver, and lead, $3,655,453.

*Mineral Hill.*—The mines of the English company in Mineral Hill have produced from 40 to 60 tons of ore per week, which assayed from $40 to $55 per ton, and which has been worked by raw amalgamation. The Stetefeldt roasting-furnace has been idle throughout the year. The mines and works seem to have just about paid expenses, and the outlook for the future is not encouraging.

## HUMBOLDT COUNTY.

For the principal part of the report on this county I am again indebted to Mr. D. Van Lennep, M. E., of Unionville.

In taking a general review of the mining-industry of Humboldt County during the year 1874, it must be conceded that there has been progress in more than one direction. A decided one lies in the fact that at present miners have a better chance of finding out the value of their mines by the aid of the reduction-works provided with roasting-furnaces. The confidence in this mode of reduction has much increased. Cases of extravagant and foolish expenditures on mines are now very rarely seen.

No specific progress has been made in the reduction of ores by the discovery of new processes, but more attention and economy are exercised in following the well-known methods of raw amalgamation and of the chloridizing-roasting process. At present there are no smelting-furnaces in the county.

Several new mining-districts have been organized during the year, and the prospects of the development of successful mines are good.

In *Buena Vista district* the Arizona mine, owned by the Arizona Silver-Mining Company, incorporated in the fall of 1873, has been worked during the whole year with a force of from sixty to seventy men. The number of tons of ore extracted during the twelve months is 4,949, of which 4,826 tons were reduced at the mills of the company, and 123 tons shipped for reduction to Reno and Winnemucca. This last ore amounted in value (by pulp-assay) to $27,853.67, or $226.44 per ton.

The milling-ore has averaged from $6 to $10 per ton lower than last year. The prices paid for labor are about the same as heretofore—that is, $4 for miners, $3 to $3.50 for car-men; from $1.50 to $2 is paid to

Chinamen for assorting, cutting sage-brush, and labor in the mills. The same price is paid to Pi-Utes.

The principal work in the mine has been done—

1. On the Stewart part of the mine, mostly above the lower tunnel;
2. In the stopes reached by the most southern incline sunk on the ledge below and east of the Fall tunnel, (it is part of the ground lying between the main east and west tunnels;)
3. In the stopes west of the western tunnel; and
4. Mostly southward on the eastern spur.

The large break which occasioned west of the Fall tunnel the loss of the ledge (ultimately rediscovered by the McDougall shaft) was reached during the year by a drift across the country-rock in the vicinity of the McDougall shaft. This work has permitted the easier extraction of ore out of this part of the mine. A tunnel about 110 feet west of the main tunnel, and about 12 feet higher on the hill, is now being run to reach the works last mentioned, and to further facilitate the extraction of ore and waste, as well as to supply fresh air to the interior western works. The Fall tunnel was not run any farther into the hill this year. The Stewart or eastern main tunnel is within a few feet of reaching the farther incline and drift run east of the Fall tunnel. Prospecting in the mine has been very limited.

In the mills of the company some alterations have been made. A new battery of ten stamps has been placed in the Arizona Mill. The engine and boiler of the tailings-mill were also placed in the Arizona Mill, side by side with the old engine and boiler. The mill has now twenty stamps, five Wheeler and four Varney pans, worked by two boilers and two engines, one engine driving the twenty stamps and the other the rest of the machinery. The silver-mill has undergone no changes, except for the necessary repairs.

The tailings-mill was run about seven months during the year. The capacity of the pans being too small for the poor tailings worked, the mill was torn down, partly used in repairing the other mills and partly sold for old iron.

No dividends have been paid since the incorporation of the company. At the end of the year an assessment of $1 per share was levied, the number of shares amounting to 48,000. The superintendent appointed at the end of 1873 succeeded in extracting from the ore of the Arizona mine, by raw amalgamation, about 10 per cent. more than was obtained before, and increased the fineness of the bullion from .500 to about .920. For certain administrative measures, to which the principal owner of the mine was opposed, he was displaced, and the former administration coming in power, brought back its former manner of amalgamation.

The Henning mine and the Pioneer Mill were leased last summer for a few months to H. R. Logan. The Henning ore being full of sulphurets of the base metals, and not rich enough in silver to extract the necessary amount to meet expenses, and the tailings at the reservoir of the mill being also too poor to be worked, operations at the mill and the mine were discontinued, and both have been idle since. There was an attachment on the property for back salary of the former superintendent, and a suit before the district court for about two years. At the last term of the district court, in the month of December, the case was decided in favor of the attachment, and the property is expected to pass into the hands of new parties, subject to a mortgage on the property, unless differently settled.

The Millionaire, discovered in the fall of 1873, has been worked during the present year steadily by a small force of from four to six men.

The milling-ore, about 30 tons, has been sold and worked at the Arizona Mill, the owners getting about $12 per ton for it. The shipping-ore, about 54½ tons of first and second class ore, was reduced at the Reno and Winnemucca Reduction-Works, producing in gross $6,279.85. In the fall of 1874 the mine changed hands. The discoverers and their partners sold it to Mr. William Beachey, it is said, for $18,000, one-third of this sum being taken in stock of the company formed by the purchaser. The work done on this mine is as follows: From the outcrop and discovery-point, a tunnel about 80 feet in length was run on the ledge, when a break occurred, displacing the vein. The first owners have sunk a small incline on this break, and found some ore with quartz, but abandoned it, their means being not equal to large expenditures in prospecting. They occupied themselves in stoping out the space on both sides of the tunnel, (about 90 feet on the right and about 30 feet on the left,) in some places as far as the break which crosses the ledge diagonally. A large portion of the material extracted was good ore. The ledge is from 1 foot to 18 inches wide. Since the mine has changed hands, the incline has been prosecuted to find the ledge beyond the break, a contract having been taken to run 100 feet. Thus far the ledge is irregular, broken, and carrying little ore.

The North Star was leased by some miners, who took out about 10 tons of shipping-ore. It was found too poor to be remunerative, and the mine was abandoned by them.

In the Peru mine, the water coming in stopped the work in the first part of the year.

Considerable work was performed on the Norman and Occidental, two new claims in the district. They were abandoned for the present for want of means and failure to find a sufficient amount of rich ores.

The Kentuck, a southern extension of the Arizona mine, owned by a Virginia City company, was worked last summer by sinking a double shaft on it, by contract, about 120 feet deep, for the purpose of striking the ledge. Feeders were found and followed to vein-material, with some quartz and little or no ore. Since then the mine has remained idle.

An antimony-ledge was worked profitably during the later part of the summer and in the fall. It is situated about a mile southeast of Unionville, in the foot-hills. It was discovered several years ago, but not worked successfully hitherto. The owner, having had, this year, a market at San Francisco, by virtue of the erection of Starr's smelting-works there, has extracted ore from this ledge, making about a 10-ton shipment every week. The ore extracted is sulphuret of antimony, (gray antimony ore,) not quite as pure as the Bloody Cañon ore. It is found in bodies of several feet width in the vein-matter, which is soft and easily worked out. A tunnel is run on the ledge about 350 feet long, and shafts about 30 feet deep have been sunk in it to find the bodies and extract them.

The following is a statement of monthly bullion shipments from Unionville during 1874 by the express company:

| Month | Amount | Month | Amount |
|---|---|---|---|
| January | $12,916 53 | August | $22,516 62 |
| February | 13,328 56 | September | 10,323 69 |
| March | 11,278 68 | October | 12,230 89 |
| April | 16,257 89 | November | 12,152 47 |
| May | 11,778 82 | December | 11,759 63 |
| June | 14,172 76 | | |
| July | 21,527 17 | Total | 170,243 71 |

The total product of the district may be estimated as follows:

| | |
|---|---|
| Shipping ore from the Arizona, about 80 per cent. of $27,853.67, that is, about | $22,182 93 |
| Shipping ore from Millionaire, 80 per cent. of $6,279.85, that is, about | 5,023 88 |
| Ores of antimony shipped amount to about | 6,000 00 |
| List of bullion above | 170,243 71 |
| Total | 203,450 52 |

*Indian district.*—Here the Eagle mine has attracted most attention. Work was begun in June by the new owners, and prosecuted steadily. The developments on the mine at the time it was purchased by the Oakland Mill and Mining Company consisted of an incline sunk on the course of the ledge at an angle of about 20° and about 30 feet deep, connecting with a shaft north of it. The shaft was a few feet deeper on the ledge than the incline. The company began, by contract, a new shaft on the ledge 10 to 20 feet south of the mouth of the old incline, and carried it to a depth of about 70 feet. Then drifts were run on the ledge in both directions. On the north side the new shaft was connected (by the drift mentioned) with the old shaft by sinking the last some 20 feet deeper. This is about 110 feet from the surface. The new shaft was sunk some 60 feet below the first level, and again drifts were run on each side along the ledge. The shaft had been sunk some 10 feet below the second level at the end of the year. The ledge is from 6 to 12 feet, and probably on the average 10 feet wide, as far as present developments show. At about 20 feet in depth some fine sulphurets of silver, with gold, were found in the new shaft. Sometimes a small bunch of native gold is found in the center of a piece of the silver-mineral. The ore has increased in going down. Assays of the vein-material range from $10 in gold, with but $2 to $3 in silver, to $300 in silver with but about $20 in gold. Native gold is seen in many parts of the ledge. Some choice pieces of mineral have assayed as high as $1,200 and more, in silver.

In October a 15-stamp mill, with four deep Wheeler pans and two settlers, was brought from San Francisco. A substantial building was put up, and the machinery laid on strong frames. The whole was in working order by the 1st of December. Although the machinery is from one of the best founderies on the coast, the cylinder had a flaw, and at the first trial cracked. There being a delay of three to four weeks by this accident, the mill was only started a few days before Christmas. The result of the working is not yet known. The present company was formed in California, in the last days of January, by B. F. Bivens, who purchased the mine. This gentleman has the general management of the business.

North of the Eagle, and on the north side of the cañon, the Comet was discovered last summer. It is about 18 inches wide; the quartz contains much free gold. In November a half interest in the mine was purchased by a company, said to have been organized in the Eastern States, for the purpose of working this and other mines.

Still north of this mine are the Black Hawk and Butte, two mines owned by Indian district prospectors, and worked by them for over a year. These mines have been bonded, and are expected to be sold to the eastern company mentioned above.

The smelting-works at Oreana have been idle all the year. The property was purchased in 1873 by General P. E. Conner, of Salt Lake.

In February, 1874, the property was incorporated in San Francisco, and the stock offered for sale. In the spring, an agent of the company came to the works with his family, bought some charcoal and flux, and it was reported that the works were on the eve of resuming business. Ore, it was said, would be sent from the Salt Lake region. A month or so was spent in idle hope and looking for ore, and the agent then left the country. The works are now in charge of an agent of General Conner. During the fall, a tunnel was run by Chinamen on contract to prospect a mine in Trinity district, connected with these works. This was done in the hope of getting ore for the works, but the ledge had not been reached or found when the contract expired, and the work was abandoned. The owners have applied for a patent for the land on which the works are located, according to the mining-law of 1872 providing for mill-sites.

Mr. Torrey's mill, run by the current of the Humboldt River, about three miles above the Oreana Smelting-Works, was at work all summer and part of the fall on ore obtained from the Jersey mine, in Arabia mining-district.

In *Relief district* the Batavia and Pacific mine and mill were shut down in March, and the property was attached by some members of the company, owing, it was said, to complications arising from informalities in the articles of incorporation. The lawsuit brought by a former superintendent for salary, which was lost and appealed, was decided at Carson City last summer in favor of the company. In December, one of the principal owners came from the East and let a contract to run 100 feet in the tunnel, started to cut the ledge at a depth of 150 to 200 feet, and to connect with the inside works. Until now the mine has been worked through a tunnel on the ledge in which, at about 50 feet from its mouth, a shaft, first vertical, and then inclined on the dip of the ledge, was sunk, with drifts on each side at two different points. The ledge dipping at an angle of 60° to 70°, though in the same direction as the side of the hill on which it is found, the distance from the surface to the ledge increases in depth. It is from this hill-slope that the present tunnel is run, and it is intended to connect the inside works with it, and thus facilitate extraction and obtain a good ventilation.

*Jersey district.*—In the latter part of May this new mining-district was organized, after the discovery of several ledges. It is situated in the Fish Creek range, close to the old road from Unionville to Austin, and about forty miles southeast of Unionville, being in Humboldt County, about five miles from its eastern boundary. The ores are said to be argentiferous-lead ores. The Lander and the Union Flag are two locations claimed to have yielded ores assaying from $100 to $200 in silver and about 60 per cent. of lead.

In *Gold Run district*, at Greggville, the Gregg or Manati mine, was worked until July, when it was stopped, the resources of the owner, Mr. Gregg, having been exhausted by prospecting for his ledge, which was lost by a break or slip. In the fall a friend helped him, and after another search he succeeded in recovering the vein, as rich, it is said, as ever. A great deal of work has been done in the mine. There is a shaft 260 feet deep, with several levels. About $40,000 have been extracted, all from high-grade ores, shipped for reduction. The low-grade ores are still on the dumps. The pay-streak is small and rich.

Agnew & Sterbey, two miners, have also been working with success a small vein about half a mile from the Manati mine. They obtain a high-grade ore, which pays them well on shipment to reduction-works.

The Picard or Thiers mine, has also been worked all the year. It is in the vicinity of the Manati, and in a granite formation. The vein is small and rich. The owners work it themselves, select the best ore, and ship it to reduction-works. Some 60 tons of the second-class ore of this mine was sold to the Golconda, or Holt's mill. Mr. Holt has built in his mill a reverberatory furnace with two hearths, and works small lots of ore picked up in the country around, *i. e.*, from Gold Run, Sierra, and Paradise districts. The results obtained are said to be as good, if not more regular, than those obtained in drop-furnaces. He uses cheap labor (Chinamen) for firing, and principally sage-brush for fuel. For part of the year he can run the mill by water-power.

The Golconda mine has not been worked this year, except so far as was necessary to maintain title. Application for a patent has been made. The south extension, purchased last year by the Old Guard Mining Company, of San Francisco, was worked in the first part of 1874, and some rock was taken out. Large quantities of water coming in, work was abandoned, at first with the intention of resuming it, but this has not been done.

The bullion shipments from Golconda Station during the year amount to about $5,000.

*Winnemucca district* is seeing better days. Last September work was commenced on the Pride of the Mountain, an old mine, which is now in the hands of the company owning the Humboldt Reduction-Works, at Winnemucca, and has been successfully worked since. In sinking a shaft on the ledge a cross-cut was run at a depth of 75 feet, and another ledge was found 6 feet from the first. A force of about four miners and ten Chinese have been taking out a daily average of three tons of gold-quartz, which yields about $100 per ton.

Encouraged by this success, old prospectors went to work in the neighborhood, and their search resulted in the discovery of a ledge called the Champion. A shaft about 38 feet deep has been sunk, which developed a ledge 2½ to 3 feet wide on the foot-hills of Winnemucca Mountain, about three miles from the town. About two tons, which were tried at the reduction-works, are reported to have assayed about $100 per ton, mostly in gold.

*Crystal district.*—This new mining-district was organized by the proprietors of the Humboldt Reduction-Works, at Winnemucca. It is situated on the Fremont range, northeast of the town. The Louise was located and prospected by a shaft about 30 feet deep. The ore obtained contains about $60 per ton in silver.

*Shipments of Humboldt County bullion at Winnemucca for* 1874.

| Month | Amount | Month | Amount |
|---|---|---|---|
| January | $7,100 | August | $10,900 |
| February | 6,630 | September | 5,950 |
| March | 21,625 | October | 4,850 |
| April | 1,000 | November | 11,710 |
| May | 2,300 | December | 9,000 |
| June | 6,000 | | |
| July | 6,830 | Total | 93,895 |

These shipments were made mostly by the Humboldt Reduction-Works. In connection with this, it ought to be stated that these works were rented by the Rye Patch Mining Company at the beginning of the year, when their mine yielded twice as much rock as could be worked at their own mill. The works were returned to their owners in April, but the latter having claimed damages for deterioration of the machinery, which were not conceded by the Rye Patch Company, a lawsuit was the result. The case was decided during the year in favor of the Humboldt Reduction-Works.

*Paradise district* was organized during the first part of the year or the latter part of 1873. It is situated about twenty-five miles in a northeasterly direction from Winnemucca, in the range of mountains on the east side of Little Humboldt River, and southeast of Paradise Valley. The claims most worked are the Credit Mobilier and the Governor Flanders. The first has a tunnel about 185 feet long on the ledge, and a shaft about 80 feet deep. The ledge is about 2 feet wide, and carries gold and silver. The second has a shaft 50 feet deep and drifts at the bottom running for 25 feet on each side along the ledge. The latter is broken up, and quartz-bowlders are found in it, which are said to be rich, chiefly in silver-sulphurets. The course of the ledges are northeast. They stand nearly vertical, the foot-wall being metamorphosedsand stone, the hanging-wall slate. They are yet in the hands of the discoverers. A good many other claims have been located, but as yet little work has been done on them.

*Columbia district.*—More than a year ago Mr. Vary, the owner of a ranch on Bartlett Creek, in the northern part of the county, discovered a gold-bearing ledge. After prospecting it, he put up an arrastra, and worked several tons of ore, which yielded very well in gold. He named the mine Badger. A mining-district was then organized, which was called Columbia. Last October W. A. Bollinger, of San Francisco, purchased the mine and formed a company, which was incorporated in California. The company proposes putting up a mill next season. This knowledge coming to the outside world, brought into the district many prospectors and all the accessories drawn together by a mining excitement. Many ledges were located, and a town was started, called Varyville. A weekly express runs at present between this new camp and Winnemucca. The ledges discovered are said to be mostly small, but rich in gold.

*Snow Creek district.*—This new district is about ten miles from Bartlett Creek. Some ledges were discovered in the latter part of the summer, and the district was organized. The most promising claim at present is the Belle of the West, on which a shaft 15 feet deep has been sunk. The ore is said to be argentiferous galena, with other minerals.

In *Sierra district* work was commenced in April, on the Lang Syne mine, in the lower tunnel, where there is a vein about 5 feet wide bearing gold. A force of four men was employed, and considerable ore was extracted. The death of the principal owner of the mine has caused a suspension of operations. Four tons of the ore extracted was sent to the Winnemucca Mills, giving very satisfactory results.

The mill in this district built for the Paul process was bought by the company at sheriff's sale.

The Last Chance mine, located not far from the Tallulah, was steadily worked by the owners until last June, when they made a shipment of 10 tons of the first-class ore, which assayed about $300 per ton. This favorable result caused them to increase since that time the force at work on the mine, and more regular shipments to reduction-works have been made.

The Tallulah mine has not been worked this year. It is yet owned by a San Francisco company.

The Auburn has been worked at intervals by the owners and discoverers. At the Thacker, two and four miners have been employed for some time. A shaft 50 feet deep on the ledge shows the latter from 18 inches to 2 feet wide, containing gold-quartz, with some galena, and giving some very good assays.

In the East range, in Inksip Cañon, which is nearly opposite to Star Cañon in the Humboldt range, a copper-ledge was discovered last May. The discoverers, who were wood-choppers, did but little work on it. They sold their claim, which they called the Ella Bruce, to parties in Unionville, who have now two men at work sinking a shaft. The indications for a large copper-ledge are good. No district has been organized.

*Central district.*—This district was the scene of active operations during the first part of the year, there being fifteen men at work on the Marietta and the Golden Age, two on the Dutchman, and two on the Teamster, to provide ore for the mill in the district. It is said that the company owning the mill and furnace could not reduce the ores successfully. The concern having got in debt, the mill, furnace, the interest in three of the mines, and the personal property were attached. The personal property was sold at sheriff's sale, and the rest will have the same fate in due time. The failure is supposed to have arisen partly from the irregular working of the furnace, which could not be made to roast uniformly, and partly for want of assorting the ores properly. The ledges contain poor and very rich rock, which were mixed together without discrimination. After the above occurrence, all the mines were worked at times by the miners owning them, the best ore being shipped for reduction. The mines so worked are the Golden Age, Golden Chariot, Dutchman, Teamster, Marietta, and Solitary.

*Humboldt district.*—During the summer a patent was applied for the Madra ledge, in this district. There was a tunnel of about 60 feet run on the ledge during the year. About a ton of the ore extracted from the mine was sent to New York to be tried, there being, it is said, negotiations going on for the sale of the mine to New York capitalists.

In *Echo district* a large force was employed on the Rye Patch Company's mine until April. The mineral having given out at that time, a few men were kept at work prospecting, and the mill run on tailings. Last September a good body of ore was again struck, and the force was increased to fifteen miners. The daily extraction since then has been about 10 tons. The mill is now run on ore which is amalgamated raw, close attention being given to this process.

The bullion-shipments from Rye Patch have been as follows:

| Month | Amount | Month | Amount |
|---|---|---|---|
| January | $5,630 | August | $2,700 |
| February | 13,500 | September | 2,700 |
| March | 9,000 | October | 8,050 |
| April | 2,600 | November | 25,800 |
| May | 2,200 | December | 20,100 |
| June | 1,350 | | |
| July | 5,740 | Total | 99,370 |

In *El Dorado district* two miners have been at work on some old locations, which have attracted the attention of moneyed men. The district lies between Echo and Humboldt districts.

In *Star district* the Sheba and the De Sota have been worked most of the year by a few men taking out ore, of which the first class is shipped to reduction-works, and the rest passed through the Krom concentrators, the concentrations being also shipped.

In the Sheba mine a ledge in the eastern part of the mine was followed by a shaft 50 feet deep, where a fine body of ore was found, running northward toward the mountain, and southward toward the creek, and dipping eastward at a small angle from the vertical. For the economical extraction of this ore, a shaft was sunk from the surface above the creek, and the ledge was reached at a depth of about 50 feet. The ledge contains bodies of rich ore, which, however, again pinch out.

The De Sota is worked with a small force, which keeps prospecting, and occasionally strikes bunches of ore.

The American Basin mine, owned by the Krom Concentrator Company, is an old mine, on which considerable work had been done formerly by driving a tunnel through very hard and flinty rock to reach the ledge. It has been prospected by a small force this year. The ledge is small, and the mineral, when found, is usually rich. A few tons were concentrated at the mill of the company.

The Krom Concentrator Mill was worked in a small way during most of the summer, operations being limited for want of ore. It was run, when there was enough ore on hand, during the day only, and shut down at night. When the water gave out in the latter part of summer, the reservoir was allowed to fill, and, when opened, the mill could be run for about two hours at a time. The company is said to contemplate the removal of the mill to some mining-camp where a larger supply of ore can be obtained for concentration. The value of ores obtained from the mines may be put down as follows:

| | |
|---|---|
| From the Sheba mine: | |
| 10 tons first-class, at about $330 per ton<br>199 tons second and third class | $17,630 |
| From the De Soto mine: | |
| 10 tons first-class, at about $550 per ton<br>199 tons second and third class | 18,860 |
| Total | 36,490 |

A foundery has been commenced on a small scale at Mill City. It is expected to have a furnace for melting all old cast iron gathered from old machinery and cast-off shoes and dies at mills in the country along the railroad, and of this it is intended to cast new shoes and dies to supply the local demand.

## ELKO COUNTY.

The only mining operations worth recording have been carried on in *Railroad district*. The principal work in the district has been done by the Empire City Mining Company, which has continued developing its mines, with, however, only indifferent results. Mr. O. H. Hahn of Eureka, who smelted in a short run, from October 19 to November 18, all the ore which the company produced during the whole year, has furnished me with the results, which are embodied in the metallurgical part of this report.

In regard to the mines of the district, Mr. Hahn says:

The principal mines are located in a crescent line around the miners' camp, called Highland, two miles to the southwest of Bullion City. They may be classified as lead and copper mines. The latter contain principally silicate of copper, comparatively speaking a poor copper-ore, yet desirable for fluxing-purposes, with subordinate quantities of the different carbonates and oxides of copper, and hardly any precious metals. The ores occur in segregated veins or deposits in a dolomitic limestone, and form sometimes net-works of considerable extent, as in the instance of the Ella mine, once the most noted of the district. It seems, however, that the working of the copper-mines has not been attended by very gratifying results, for the shipping of copper-ores has ceased entirely, and but few of the mines are being developed now.

At the present time miners and operators have concentrated their attention upon the development of the lead-mines, and, it is to be hoped, with better success.

The ores of the lead-mines form lenticular masses at or near the contact of the crystalline limestone with a dioritic porphyry, and consist of carbonate of lead and of galena, which occurs either in solid blocks or is disseminated in ribbons through the country-rock. Their other mineral associates are silicate of copper, red-copper ore, and brown and calcareous spar, which give the ore a refractory character. Of the precious metals only silver is found in the ore, in greatly varying quantities, ranging from 15 to 150 ounces per ton. The principal mines worked are the Last Chance and Elko tunnel, belonging to the Empire City Company, of New York, and the Webfoot and Tripoli, owned by A. J. Ralston & Co., of San Francisco. Both companies own smelting-works in the town of Bullion.

As before stated, the ores of Railroad district have a refractory character, on account of the copper and the carbonate of lime and magnesia they contain.

The Webfoot Company made but one small run of eight or ten days, during which 120 tons of ore were reduced, yielding 34½ tons of bullion, of an alleged silver-value of $220 per ton. It was the intention of this company, in December, to make another run as soon as iron-pyrites could be procured to reduce the skimmings accumulated. I am not informed whether this second campaign has been made.

*Spruce Mountain district* has been lying idle throughout the year. A re-organization of the Ingot Mining Company was intended, but I am not informed whether it has been carried out.

*Cornucopia district* has continued to attract local attention by the richness of some of its ores; but developments have not progressed sufficiently to permit shipments of importance.

## WHITE PINE COUNTY.

Notes on this county have been furnished to me by Mr. A. J. Brown, Mr. Alexander Trippel, M. E., and various other gentlemen, whose kindness has been acknowledged in the proper places in this report.

The mining-industry of this county is slowly but surely recovering from the depression characteristic of the last four years. No discoveries have been made of new mining-districts within the limits of the county, and no very important developments have been made in any of the mines worked in the old districts; yet progress has been steady and encouraging, and promises well for the future. The winter of 1873 was one of the severest ever known since the settlement of the county. The roads were blocked with immense snow-banks, so as to be totally impassable during the first three months of the year. No ore was delivered at the mills, and the bullion-product of the quarter was from ore on hand at the commencement of the year, and therefore amounts to the meager sum of only $49,688, produced from 1,309 tons of ore. The product of the second quarter was 3,720 tons of ore, yielding $124,827. The third quarter produced 4,158 tons of ore, worth $203,013, and the returns of the last quarter foot up 4,527 tons, yielding $222,002. The bullion-product of the year shows a considerable increase over that of 1872, being $599,530, from 13,714 tons of ore and tailings. The increase is mainly due to the Cherry Creek mines. Renewed confidence in the permanence of the mineral-deposits in depth is being gradually established, particularly in the limestone-districts of White Pine and Robinson.

Three new shaft smelting-furnaces have been built in the county during the year, one in Robinson, one in Diamond, and one in Newark. The first-named has made two successful runs of about fifty days each. The last two, owing to the highly quartzose character of the ores, have not been altogether successful. Both districts yield ores of a good grade in silver and a fair percentage of lead, and with a correct system of dressing, the majority of these ores could be made available for reduction in the blast-furnace.

In *White Pine district* one of the old Mattison furnaces was temporarily repaired and a short run made, for the purpose of testing the practicability of smelting the base ores of the district, the idea having taken root somehow that they were especially refractory. This test was made with the view of erecting improved furnaces in a more eligible locality during the ensuing summer, if the campaign proved successful. None but rich carbonate ores, free from copper, were selected for the trial, and though the operation was carried on by men wholly unacquainted with the practical as well as the scientific details of smelting, yet they claim a success, both metallurgically and financially. No trouble was experienced from wall accretions or salamanders, and the lining of "pancake sandstone" was, apparently, wholly uninjured by the run. Three hundred and fifty tons of ore were run through in twelve days, the average assay-value of which was; lead, 25 per cent.; silver, $70 per ton. The product was 55 tons of metallic lead, worth from $250 to $300 per ton in silver, and about 10 tons of matte, skimmings, and speiss, worth, by assay, $160 per ton. This last product still remains on hand. The loss was excessive, being about 33 per cent., a portion of which will, however, be recovered from the material on hand. Still the result, though partially negative, has had the beneficial effect of reviving confidence in the future of the Base range; and mines in that part of the district that have been lying idle for the last four years have been sought out, and work on them has been renewed in earnest, in many cases with highly gratifying results.

No new mills have been built in the county during the year, and no old ones have been dismantled. There are eighteen in running order, containing 264 stamps. None of them, however, have run on full time, and some have not been in operation at all. They are distributed as follows:

White Pine district: Stamford, 30 stamps, idle; International, 30 stamps, running since April; Smoky, 20 stamps, idle; White Pine, 10 stamps, idle; Metropolitan, 5 stamps, ran May and June; Swansea, 10 stamps, idle; Dayton, 20 stamps, idle; Manhattan, 24 stamps, idle; Monte Cristo, 20 stamps, idle.

Newark district: Newark, 20 stamps, ran May and June.

Robinson district: Watson, 10 stamps, running part of the time on ore from Ward and Nevada districts.

Egan district: One mill, 20 stamps, running part of the time on ore from the Social and Steptoe Company's mines and on Cherry Creek ore.

Cherry Creek district: Exchequer, 5 stamps, idle, and one small concern of 2 stamps, run on ore from the Chance mine.

Schell Creek district has three 5-stamp mills, only one of which has been run during the year.

*Newark district.*—This is an old district, which was discovered and organized in 1866. Several of the most prominent ledges were purchased by the Centenary Company, then doing business in Santa Fé district, Lander County, and its 20-stamp mill, formerly erected in the last-named district, was removed to Newark and erected there. The ores were rich, and the district promised well for a time; but the veins were found to be "bunchy," and some of the lodes were too base for milling. The supply of milling-ore was therefore precarious, and the district fell into disrepute, and was abandoned during the White Pine excitement in 1868. Work was again resumed last spring, and carried on for a time with considerable vigor, but without any marked success. Several private companies are mining in a small way and shipping their ore to Eureka for reduction. Most of them are making money.

In regard to coal on Pancake Mountain, I insert here the following paper, which was sent by Mr. A. J. Brown, of Treasure City, to a late meeting of the Institute of Mining Engineers:

I herewith send to the institute a sample of Pancake coal. It is rather early yet to make any estimate of the future value of the discovery; but it is certainly the most promising vein of coal yet discovered in the State of Nevada, and, I believe, the first true coal found west of the Rocky Mountains, or perhaps west of the Missouri River, unless some of the Utah coals belong to the coal-measures of Carboniferous age. The Pancake vein certainly belongs to the Carboniferous, as rocks of a later age are wholly wanting in this locality.

The Pancake Mountain is a low range of hills situated in the valley about midway between the White Pine and Diamond ranges, and occupies throughout its length of thirty miles or more the basin or trough of a synclinal fold. The higher mountain ranges, both east and west, follow mainly the axes of mountain anticlinals, and exhibit only Devonian, Silurian, and Azoic rocks. About midway between White Pine and Pancake, two or three mounds which are identical, both lithologically and paleontologically, with the limestone of Treasure Hill, crop through the quaternary formation of the valley, and still further west are found dark bituminous shales identical with those found along the east slope of Treasure Hill and under the towns of Hamilton and Eberhardt. Some four miles still further west, and belonging to a much higher geological horizon, we find the coal-formation.

The formation immediately inclosing the veins (of which there are said to be two) is mainly slate, though in places a light-colored rock, which the miners call soapstone, comes in contact with the vein. Impressions of leaves and plants are found in the slates, and a few specimens of *Sigillaria* have been found on the surface in the vicinity. The sandstone-formation, from which is quarried the fire-proof lining used in the Eureka furnaces, overlies the coal-formation; and I think that another bed of sandstone occurs some distance under it, but of this I am not positive.

The vein, where worked, is situated near the north end of Pancake Mountain, about fourteen miles west of Hamilton and twenty miles east of Eureka. The vein strikes north and south and dips quite steeply (40°) to the west. In thickness it varies from 5 to 6 feet, but it is much broken and displaced; and in some places the coal appears to have been destroyed, and a kind of ash fills its place in the vein. Several experiments at coking on a small scale have been tried, and have resulted satisfactorily. The citizens of Eureka are discussing the feasibility of lighting their town with gas made from this coal.

Steam hoisting and pumping machinery has been erected on the mine during the past winter, and the company evidently is in earnest in its determination to find coal in paying quantities, if it exists.

The mine is opened by two inclines on the vein, the deepest of which is down 240 feet.

A party of prospectors has been engaged during the winter in sinking a shaft in the black shale on the Momomoke Mountain in search of coal. This shale is highly bituminous where protected from the air, and consequently burns for a while with a bright flame, but does not burn to an ash—a fact which has lured numbers of men on to spend several thousand dollars at different times in search for coal where none can be found. Query: Do not these two beds of bituminous shales correspond, in appearance, characteristics, and geological place, with the Marcellus and Genesee shales of New York?

The above description of the appearance of the coal-vein, thickness, works, &c., was derived from the owners. I was there early last winter, but the dump was on fire and the works were half full of water, so that I was obliged to return without accomplishing my purpose of seeing for myself. If it is half as good as claimed, it is one of the most important discoveries ever made in the State of Nevada.

The mine has been worked steadily during the year, and has attained a depth of 480 feet, measured on the incline. The vein is about 2 feet thick, of good marketable coal. The present product of about 100 tons per month sells at from $12 to $20 per ton on the dump. The Eureka Consolidated Company is the purchaser.

*White Pine district.*—The North Aurora, of the Eberhardt and Aurora Consolidated Company, is the principal mine worked in this district during the year. The works have been extended north through ore, making connection with the Central shaft and the works of the South Ward Beecher, thereby establishing a continuous ore-connection from the Ward Beecher Consolidated to the O'Neil grade, a total distance of 2,450 feet. The ore-channel has followed the great spar-vein throughout its course, sometimes pinching to a mere seam for a few feet, and then extending to a thickness of from 30 to 90 feet. Three of these large bodies have been found and worked since last April, furnishing about 50 tons of $60 ore daily. There is a full year's supply still in sight. The Central shaft has been sunk another hundred feet, and is now 300 feet deep vertically, with good ore in the bottom. The hoisting-engine has been removed from the Beecher shaft to the Central, and it is the intention to make the latter the main working-shaft. This mine has produced during the year 9,741 tons of ore, which has yielded $420,394. The total yield of this mineral-belt to date has been 70,257 tons of ore, giving a gross yield of $2,836,783, or about $40 per ton. The greatest depth from which ore has been extracted is 230 feet from the surface.

The Mammoth mine has completed a tunnel 700 feet in length, running the whole distance through an immense body of low-grade quartz. At the extreme end, and at a vertical depth of 300 feet from the surface, a considerable body of excellent milling-ore has been found, the assays of which range from $30 to $500 per ton of 2,000 pounds.

The Edgar belonging to the Ward Beecher Consolidated Company, has kept a small prospecting-force employed during the seasons. Drifts have been run in different directions, principally south and east. Some good ore has been found, but only in small bunches. A large body of fair ore is still standing in the old works, which will probably be extracted and worked during the coming summer.

In the Original Hidden Treasure, drifts have been run south from the end of the main tunnel, and an incline has been sunk from one of them 200 feet below the tunnel-level, or 700 feet vertical depth from the surface. Some bunches of ore were found, but they proved unimportant. A stratum of graphitic slate was passed through at a depth of 600 feet from the surface. It was probably a part of the slate-formation overlying the mineral-bearing limestone on the east slope of Treasure Hill. If this surmise is correct, then the evidence would be conclusive that the main tunnel has stopped at least 200 feet short of the main ore-channel.

The Indianapolis, situated on the ridge 700 feet south of the Eberhardt mine, has been worked part of the year. A drift has been run north from the main incline 100 feet, and a very large body of $30 ore has been exposed. The general character of the quartz very much resembles that formerly found in the Eberhardt mine, but it lacks the richness of the chloride of that mine, though the silver exists here also in the form of scales of horn-silver.

On the west side of the White Pine Mountain, the Caroline and French have been worked as usual, each yielding a small quantity of rich ore, enough to make the working of the mines remunerative to the owners.

The Chester, a copper-lode, develops well, so far as worked. Sulphurets of iron and copper exist in the greatest abundance. The vein is about 20 feet thick.

At the north end of the mountain a company has been engaged in running a tunnel to tap the Jennie A. series of ledges at a depth of from 800 to 1,000 feet below the surface.

The Champion lode, on the east slope of the mountain, has been worked since August. A large body of cerussite and galenite was found near the surface. From this three miners extract about 5 tons per day in sinking a small shaft. The ore contains 50 per cent. of lead and $55 per ton in silver.

The C. T. Fay has produced 126 tons of carbonate ore, which has been smelted in the Mattison furnace. The yield by assay was about $90 per ton in silver and 36 per cent. of lead. A small quantity of antimony is present in all the ore from this mine.

The Charter Oak Company has just completed a tunnel 250 feet in length, tapping the vein 150 feet from the surface. The vein is 2½ feet thick where opened by the tunnel, and is very solid and regular. The ore ranges from $40 to $100 per ton.

The Mobile Consolidated Company has run a prospecting-tunnel 400 feet along its vein. The greatest depth at the face of the tunnel is 250 feet vertically. The vein is from 3 to 5 feet thick and well defined. One hundred tons of the ore, smelted at the Mattison furnace, yielded $55 per ton in silver. The dump contains about 300 tons of the same class of ore.

The Oro mine is situated low down near the old town of Swansea, and has been opened to the depth of 125 feet from the surface. The vein is from 3 to 5 feet thick at the lowest point reached, and about 2 feet where first found, 40 feet from the surface. About 200 tons of ore have been extracted since the 1st of September. No ore is taken out except what is actually necessary in sinking the shafts and driving drifts. Two men extract from 3 to 6 tons per day. The average yield of the ore, as it comes from the mine, is from $80 to $100 per ton.

The Imperial, one of the oldest and best known mines in the Base range, has been worked during the fine weather. A tunnel, started at the lowest point attainable, has been run 100 feet for the purpose of opening the vein 80 feet below the old works. The vein will be tapped at 120 feet. This mine in former days yielded about 1,500 tons of valuable ore from the old shallow workings, and they are not yet exhausted. The ore averages about $80 in silver per ton and 10 or 12 per cent. of copper carbonates.

A tunnel has been run along the Onetho lode for 150 feet, disclosing a large amount of low-grade ore. The vein varies from 5 to 8 feet in thickness; but the ore contains only $35 in silver per ton and 22 per cent. of lead, so that it cannot be sold at a profit to custom-furnaces. The ledge runs eastward into Treasure Hill, and it is the intention to follow it with a tunnel as far as it goes. The croppings have been traced on the surface for 3,000 feet, or nearly to the "free-metal" belt.

The Silver Plate, one mile north of Hamilton, has been exploited during the year, and a considerable quantity of excellent milling-ore has been extracted.

I introduce here the notes of Mr. A. Trippel in regard to the Eberhardt and Stanford Mills, which are of technical interest:

The Eberhardt and Aurora Mill, at Eberhardtown, was rebuilt in 1874. It has 30 stamps of 750 pounds each, making 94 drops of 8 inches per minute. Battery-screens, 40 meshes per linear inch. The battery discharges the dry pulp directly into cars, by which it is carried into 16 pans of 5 feet diameter, the mullers making 58 revolutions per minute. There are 8 settlers, making 9 revolutions, and 4 agitators making 16 revolutions. Capacity of mill, 40 tons. Yield, 75 per cent. of assay. Ores, $54. There are 3 tubular boilers, of 50 inches diameter and 16 feet long, and 2 reserve-boilers. Fuel, wood, (soft,) of which 9 to 10 cords are used daily. In this mill the stamps, pans, settlers, and agitators are driven by belts. The mullers last from 3 to 6 weeks. Cost of working a ton of ore, $12. The pans are charged twice a day with 2,500 pounds of ore at a time, and the charge is ground in the pans for three hours.

The following shows the operations of the mill during 1873, and is interesting on account of the relative cost of materials per ton: Treated during the year, 6,133 tons of Eberhardt and Aurora ore, all "chlorides," in dark-colored limestone; average assay-value per ton, $56.78; total value, as per assay, $343,357.27; bullion produced, value, $298,892.82; yield, 87 per cent.; running-days, 162; tons worked daily, 37.8; total cost of reduction, $72,471.21; cost per ton of ore, $11.88; loss of quicksilver per ton, 1.4 pounds.

The following are the items of cost:

| Items. | Total cost. | Per ton of ore. | Percentage of cost per ton. |
|---|---|---|---|
| Wood, 2,084 cords | $13,424 75 | 0.34 cord | 18.52 |
| Quicksilver, 10,708 pounds | 9,583 18 | 1.4 pounds | 13.22 |
| Castings, 34,081 pounds | 4,131 05 | 5.55 pounds | 5.70 |
| Cyanide potassium, 2,608 pounds | 2,137 55 | 0.44 pound | 2.95 |
| Salt, 56,270 pounds | 1,700 34 | 9.17 pounds | 2.35 |
| Labor | 25,447 75 | $4 15 | 35.11 |
| Salaries | 6,150 00 | 1 00 | 8.49 |
| Water-rent | 6,000 00 | 97 | 8.28 |
| Lumber, 1,537 feet | 223 00 | | |
| Kerosene, 200 gallons | 180 00 | | |
| Lard-oil, 185 gallons | 301 00 | | |
| Tallow, 2,034 pounds | 244 02 | | |
| Candles, 73 boxes | 365 00 | | |
| Shovels, 28 | 41 25 | 63 | 5.38 |
| Cam-grease, 28 pounds | 12 50 | | |
| Charcoal, 2,489 bushels | 774 14 | | |
| Castor-oil, 119 gallons | 242 50 | | |
| Brooms, 36 | 30 00 | | |
| Sundries | 1,483 17 | | |
| Total | 72,471 91 | | 100.00 |

NOTE.—Quicksilver was bought by good fortune at 95 cents, the price being at the time $1.05 per pound.

The Stanford Mill, at Eberhardtown, belongs to an English company, and was idle in 1874. The ores worked heretofore in this mill were those from Treasure Hill, chlorides, averaging $38 per ton on about 13,000 tons. The mill has 30 stamps of 750 pounds each, making 96 to 98 drops of 8 to 9 inches per minute. Capacity, 40 to 45 tons per day. Dry-crushing. Three tubular boilers, 50 inches diameter, 16 feet length; engine 4 feet stroke, 20 inches diameter; fuel, soft wood; consumption, 10 cords daily. Average yield, from 82 to 87 per cent. of assay-value. The mill contains 16 Wheeler & Hepburn pans, of 5 feet diameter, 8 settlers, and 4 agitators, all arranged in two rooms at right angles to the battery, a passage being left in the center for cars. Quicksilver consumed per ton of ore, 1½ pounds; salt consumed per ton of ore, 6 pounds; cyanide of potassium consumed per ton, ½ pound. The ore is ground in the pans for 3 to 4 hours.

*Lake district* is so named from two small ponds, a few yards in extent, found within its limits. It is situated in the Robinson range of mountains, about twenty-two miles south of Robinson district, and was discovered and organized in August, 1872. The formation is mainly slate, occasionally capped with fragments of a formerly continuous limestone formation. So far but one lode has been discovered. Its course is north and south, and its length, as far as traced by crossings, is from six to seven miles. It occupies apparently a vast fissure, following the axis of a great anticlinal fold. The vein itself forms a longitudinal valley or depression from 150 to 300 feet wide along the summit of the mountain range. The ore found is principally stromeyerite, and ranges by assay from $30 to $600 per ton in silver. Wood and water are convenient to the mines, and unusually abundant. A great number of locations have been made, and many of them promise well; but the developments are meager.

*Ward district* is situated on the eastern slope of the Robinson or Egan range of mountains, twenty miles south of Robinson district. It was organized in 1872. The mines in this district are owned mostly by the Martin White Silver-Mining Company, of San Francisco. The principal mines are the Paymaster, Mountain Pride, Young America, Mammoth, Caroline, Defiance, Wisconsin, Grampus, Governor, Syndicate, Ben Franklin, Cyclops, Marlborough, Joe Daviess, Ben Vorlick, Ben Lomond, Midlothian, Nelson, Endicott, and Waverly. The company owns the first-named seven mines by purchase, the remainder by location. The development of these mines was commenced on the 15th of October, 1874. A shaft has been sunk on the Paymaster mine to the depth of 190 feet, 48 feet of which was in ore. At the depth of 90 feet the ore is just disappearing from the bottom of the shaft as the vein dips to the eastward. The company has run no levels as yet on any of the mines. One hundred and fifty tons of second-grade ore, that the company had worked at Robinson, averaged $120 per ton. This ore was obtained in sinking the shaft; also 15 tons of ore, which are sacked, and not yet disposed of, which are expected to average $500 per ton. This vein is 36 feet wide. There is also a shaft on the Young America mine 60 feet deep, the last 10 feet being in ore. Quality and richness of ore are the same as in the Paymaster, but the vein is not so wide. Shafts are sunk on the Grampus, Mountain Pride, Defiance, Caroline, Mammoth, and Syndicate mines, to depths of from 40 to 70 feet. All the mines show considerable quantities of ore, principally sulphurets, assaying from $40 to $300 per ton. Work is progressing on the other mines as fast as the weather and circumstances will permit. Next spring and summer the company proposes to erect reduction-works. Plenty of wood and water are in the immediate vicinity of the mines. Two good roads have been built into the district, and boarding and lodging houses, &c., have been erected.

The Watson Company, of Robinson, has also purchased some of the best claims in this district, and developments have been begun.

*Robinson district.*—This district, of which Mineral City is the camp, is located forty-four miles northeast of Hamilton and fifty-six miles southwest of Cherry Creek. The altitude of Mineral City is 6,500 feet, the town being situated in an east and west cañon of a southern spur of the Egan range, and within four miles of Steptoe Valley. I am indebted for notes on this district to A. Trippel, M. E., of Austin. The general formation here is quartzite, alternating with compact and shaly limestone. Several extensive porphyritic dikes traverse the stratified rocks in an east and west direction. The stratified rocks are in many places seen much contorted, especially near Murray Creek, to the eastward of Mineral City. Outcrops of ferruginous gossan are observed for over two miles in length near the town.

Some of the principal mines are:

The Hayes: strike, east and west; contact-vein, between quartzite and porphyry; dip, south; from 1 to 10 feet wide; has ore in quantity, averaging $50 in silver.

The Altman: strike, east and west; dip, south; between porphyry and limestone; has a large outcrop; ore, $30 in silver. This mine has been opened now to a depth of 250 feet from the surface. The explorations show a mass of iron-ore 800 feet in length by 300 feet in thickness, traversed by seams and masses of argentiferous and auriferous lead-ore, varying in thickness from 1 to 30 feet.

A recent location, the Alcyon, has opened into a vein of galena 7 feet in thickness, that prospects well in silver. It is situated only about 200 feet from the New Canton furnace.

The Ward Ellis, similar to the last, but with underlying granite.

The Cash, six miles from town, has ore with more galena, and hence suitable for smelting.

About two miles from town a mineral-belt, carrying chiefly copper-pyrites, can be observed. The lodes are nearly north and south, with easterly dip. They are not developed.

*Nevada district.*—This is a new district, about twelve miles east of Mineral City, and in a similar formation. It contains several promising mines, among which are the Sunrise and Sumner. Both of these show some fine ore, containing chlorides and ruby-silver. One of these mines had just been bonded for $35,000, and the ore was to be worked in the Watson Mill, in Robinson, when Mr. Trippel visited the district. These districts offer fair inducements for mining. It is safe to say that sufficient supplies of smelting-ore, with 25 per cent. of lead and $30 per ton silver, could be furnished for a large smelting-works, at a very low cost. Murray Creek offers an ample water-power. There is plenty of wood near, and the cord can be bought at $5; charcoal at 20 cents per bushel. Barley and hay are cheap, and the freight of bullion to Toano, on the Central Pacific Railroad, as "return freight" from Pioche, is $10 per ton. In May, 1874, there was a 10-stamp wet-crushing mill, the Watson Mill, at Mineral City, and the ruins of a small smelting-works, owned by a party from Ohio, which failed to be successful. Recently the Canton Company has erected a shaft-furnace of 25 tons daily capacity, most of the ore reduced coming from the Altman mine, above described.

*Cherry Creek district.*—The following remarks on this district are partly based on notes from Mr. A. Trippel, M. E., and partly on reports from superintendents of mines.

At the foot of the eastern slope of the Egan range of mountains, and about four miles north of Egan Cañon, through which the old overland route passes, is situated Cherry Creek Camp, a town of some 400 or 500 inhabitants. It has been built since the fall of 1873. Altitude is 6,200 feet. East of the town is Steptoe Valley, at that point some seventeen or eighteen miles wide, and bordered on the east by the Schell Creek range. The geological features in the Egan range north of the cañon consist mainly in a series of argillaceous black slates, clay shales of various colors, quartzites, and limestone, of which a portion has a crystalline texture. These strata begin at the eastern slope with massive quartzites, followed alternately by slates and quartzites, and then by siliceous limestones, on which rests a light-colored shale, extending to the top of the range. They have a nearly uniform dip to the west at an angle of 45° to 50°, and a course nearly due north and south.

Between the strata formed by quartzites and slates on the east and limestone on the west we observe a porphyritic mass, which crosses apparently at an acute angle the lines of stratification in the general direction of northeast and southwest, dipping northwest. In Steptoe Valley itself, and in the opposite Schell Creek range, granite is seen at several places protruding to the height of the limestone. The latter has invariably in these cases a crystalline appearance. On the whole the stratification of the rocks in this part of the Egan range shows more than usual regularity in both course and dip.

There are two principal vein-systems in this part of the range. Both of them have the same course and dip as the stratified rocks. The lower or eastern zone is between quartzite and porphyry, the upper or western one between limestone and shales. Both can be traced northward regularly for several miles, but toward the south, or Egan Cañon, they seem to be thrown out of their course by the branching of the porphyritic dike, of which one branch takes a westerly turn. Several veins opened to the south of Cherry Creek Camp are reported to have a decidedly more easterly and westerly course. That disturbances have taken place may also be seen in the southern part of the Tickup mine, where the vein in its upper parts certainly has been thrown eastward; but it has regained its natural direction near the Midas tunnel, still farther south.

Among the mines belonging to the lower or eastern belt from north to south are the following, several of which are being worked: Victoria, Exchequer, Grand Turk, Geneva, and Ida. The ore in these mines carries iron and copper pyrites and galena, besides true silver-ores—in general, ore which would be designated as "base," the galena yielding a great portion of the silver. The ores also carry gold. They are decidedly such as may be concentrated with advantage. The vein-matter is chiefly quartz, toward the hanging-wall more ferruginous than in other parts of the lode.

The principal mines in the upper or western belt from north to south are the following: Wonder, Baltic, Cherry Creek, Silver Glance, Red Jacket, Mark Twain, Wanderer, Chance, Tickup, Midas, and (over the ridge on the west side,) the Keystone and Pine-Nut. These mines show no very distinct line between vein and foot-wall, the ore penetrating often to some depth in the latter; but the line between the overlying shale and the vein is better marked by smooth clay faces and by a layer of very hard ferruginous quartz. The vein itself shows in various locations lenticular masses of a black limestone, and the so-called "pay-streak" is generally bordered by blackened streaks of quartz.

The general character of the ore from the mines in the upper belt is different from that in the lower. So far the quartz has been more or less impregnated with what is supposed to be silver-copper-glance, or perhaps stetefeldite, *i. e.*, small specks of a black mineral surrounded by a greenish tint of carbonate of copper. In the upper levels the mineral is silver-chloride and carbonate of copper. The northern part of the Baltic shows also some carbonate of lead.

The Baltic is situated in Silver Cañon, four miles north of the town, at an altitude of 8,200 feet. It is opened by a short tunnel and by some open cuts. The vein appears to be from 5 to 10 feet wide, runs north 10° east, and has a variable dip, generally to the west as far as now developed. The overlying shales are covered about 50 feet higher up with limestone. The ore on the dump is of acceptable quality, and may average some $50 silver per ton.

The Chance mine, situated in a cañon about a mile north of the town, and at an altitude of 8,000 feet, had been developed by two inclined shafts, one of which was, at the time of Mr. Trippel's visit, 64 feet in depth.

The vein at the surface was but 6 inches, but in 60 feet has widened to 8 feet, of which 3 feet had rich ore. Out of this shaft considerable ore has been taken, which was milled in a 2-stamp mill, leased by the owners for the purpose of testing their mine. The selected ore from this mine is exceedingly rich, assaying far above $1,000 per ton; all the rest is milled, and yields by pulp-assays from $110 to $128 per ton, of which nearly 90 per cent. is extracted, at a cost for milling of $12 per ton. The chimney of the richer ore in this shaft seems to dip from south to north. In the other shaft the vein appears faulted; the work had not been sufficiently pushed to show the real character of the break. Considering the short time since the mine was opened, and the results thus far attained, it is certainly a very promising one. It is said to have paid all expenses from the beginning. The fine-grained reddish porphyry, of which mention was made above, crosses the vein just at the southern end and divides the Chance mine from the Tickup.

The Tickup had, in May, 1874, three inclined shafts, the northern one being 112 feet down on the vein. There were some drifts from it along the hanging-wall 50 feet in length. It was considered then that the drifts were too much on the hanging-wall side, and cross-cuts through the vein had been commenced. The next shaft south was 55 feet deep, and had a drift northward. From this shaft a quantity of rich ore had been taken, and two car-loads, or about 18 tons, had been shipped to San Francisco, with a return of over $1,300 per ton. The third incline shaft was sunk on the southern end of the claim, and was about 22 feet deep, with very rich ore in the bottom. It is in this part where a dislocation in the upper levels has been observed, and the probability is that the vein below is to the westward.

Next south is a tunnel-claim not now worked, and followed southward by the Midas and Steptoe ledge and tunnel claim. Here a tunnel had been driven about 100 feet in a southwesterly direction. The vein, however, traverses this tunnel in a more southerly direction, and leaves the tunnel-head to the west. A shaft has also been commenced, and if prosecuted, must shortly meet the vein. Good ore can be seen as "float" all along the course of the lode, which seems to be from 5 to 7 feet wide.

Passing the Midas location we cross the ridge, and find on the west side the Pine-Nut, with considerable outcropping ore of a fair quality.

At the end of the year Mr. J. R. Stanford, superintendent of the Geneva Consolidated Silver-Mining Company, reports in regard to the Geneva and Tickup mines:

The Geneva mine shows considerable surface-work, but as it was worked more with a view to taking out ore than for legitimate developments, there is really but little to be said about it. One hundred tons of ore taken out were milled at Egan Cañon, and the pulp-assay was $98 per ton. Ten tons sent to San Francisco realized $245 per ton. A shaft is now being sunk by contract.

The Tickup mine has been worked with a view to development, and now shows a shaft 239 feet deep and 448 feet of levels, there being four in number, three running south and one north. The vein varies in width from 8 inches to 3 feet, and has been as wide as 10 feet. At the bottom of the shaft it now shows 2¼ feet of ore, the average grade of which is $234 per ton, as per pulp-assay of 25 tons worked.

Considerable rich ore has been shipped to San Francisco. The first shipment was 6 tons, which assayed $1,280 per ton. The second shipment was 2¼ tons, assaying $1,320 per ton. A total of $8,900 has been realized from ores shipped away. In bullion, $43,000, the result of 305 tons of ore worked, have been shipped. At present 4 tons of ore per day are being shipped to the mill, the pulp assaying from $345 to $496 per ton. The ores are worked up to 94 and even 97 per cent.

The mine at present looks exceedingly well, and the prospects are very good. A chimney which carries the rich ore is being followed. Ore which assays as high as $3,780 per ton is found in it.

Mr. G. F. Williams, superintendent of the Cherry Creek Mining and Milling Company, sends the following:

This company's Exchequer mine is worked through a tunnel, which has been run in porphyry lying between two veins of quartz. The course of the veins is northeast and southwest. The tunnel is in 270 feet. At this point cross-cuts to both ledges have been made. On the 12th December, 1874, the west ledge was struck. The ore-body is here 9¼ feet wide, about 6 feet of which is good ore, assaying about $90 per ton. Drifting on the ledge to the extent of about 60 feet has been done. Both faces of the drifts are in good ore. The depth of the point where the cross-cuts were made is 160 feet below the surface. Here a winze has also been sunk on the ledge, which is down about 20 feet in good ore. The highest grade of ore assays $400 per ton. No ore has as yet been milled. About 75 tons ore on the dump.

On the 18th of December the east ledge was struck. The average assays from this ledge show a value of $55 per ton; highest assay, $219. Mr. Williams cut into this ledge about 6 feet, and as the face showed very low-grade ore, work on it was discontinued. No drifting on this ledge has been done, as it is impossible to work many men, on account of bad air. The tunnel has no ventilation at present, but an air-shaft to connect with it is being sunk. This will be completed in about forty days.

In regard to the Chance mine, Mr. Frank Hallowell, the superintendent, says:

It was located on what is known as the upper belt, situated about one-half mile north of the Tickup mine, and on the same vein. It crops boldly the whole length of the location, (1,500 feet.) Three openings have been made during the last twelve months, viz, near the south end an open cut and incline 100 feet deep; 400 feet north an incline 120 feet deep, and near the north boundary an open cut, all showing a large amount of ore that will mill from $40 to $60 per ton. The vein runs north 14° east, and dips 45° to the west, the hanging-wall being slate and the foot-wall limestone. There have been 184 tons milled, yelding $15,760, or $85.50 per ton. The highest grade of pulp was $249, the lowest $54 per ton. There are on dump about 100 tons of $50 ore, and 600 to 800 tons which will be worth from $25 to $40 per ton. The latter it will not pay to move until a road is built. A large portion of the ore milled came from the central incline. No levels have been run. The ore is from 1 to 10 feet wide in the different openings. A cross-cut at the depth of 80 feet on the incline (45°) showed quartz 16 feet wide, of which 10 feet was mineral-bearing.

The Star ledge, situated on the same ridge as the Exchequer, and perhaps one-fourth of a mile distant, is now looked upon as the principal mine of the district, and, as far as the shipment of bullion goes, it is certainly entitled to that name. The shipments from this mine for the month of December are reported to amount to about $60,000. The company owning this mine and the one next mentioned below is now taking out from 10 to 15 tons of ore per day, that yields by pulp-assay from $100 to $200 per ton. The shaft is down about 150 feet, with drifts running each way from the shaft following the vein, which ranges from 4 to 6 feet in width.

Adjoining this mine on the west is the Gray Eagle, with a shaft 100 feet deep, showing a beautiful vein from 3 to 5 feet thick, and having all the characteristics of the Star.

The Star Company intends to erect its own mill shortly.

At Egan Cañon, four miles south of Cherry Creek, the San José Company resumed work on its mine, the Gilligan, in December, 1874. This mine was discovered and explored to a considerable extent as early as 1864. It is opened to a greater depth and more extensively along the ledge than any other mine in this part of Nevada, but has been lying idle, with the company's 20-stamp mill, for a number of years. The latter has been running during a part of the past year on a few hundred tons of hard ore and tailings belonging to the company, which produced $8,974 in bullion, and on custom-ore, mostly from the Star mine, at Cherry Creek, which produced $24,427.

*Schell Creek district* was visited in May, 1874, by Mr. A. Trippel, M. E., who has furnished me the notes for the following:

The Schell Creek range runs some eighteen miles east of the Egan range, and, like that, north and south. The mining-town of Schellburn is situated opposite Cherry Creek, somewhat hidden behind a line of foot-hills. North of Schellburn, the mountain range extends for a few miles, having the high Jupiter Peak, 9,700 feet, at its northern extremity. Toward the south the range extends for a long distance, and contains several mining-districts.

The geological features in this range seem to be complicated. The ore-bearing lodes, especially to the southward, are rather irregular and broken up by disturbance in the position of the country-rock. Limestone, siliceous and argillaceous shales, and porphyry are the principal rocks, underlaid (as is proved in several locations on the surface and in tunnels) by granite or rhyolite. The foot-hills to the west of Schell Creek range are entirely composed of a reddish porphyry, the *débris* from which partly fills the space between them and the main range. The latter shows at its base granite, overlain by crystalline limestsone and shales, dipping but slightly toward the east. South of Schellburn, throughout Queen's Spring district, these porphyritic masses seem to branch off in different directions, throwing the country-rock from its previous position.

The metal-bearing lodes in this part of Schell Creek range have a course similar to those in Cherry Creek, *i. e.*, north and south, but the dip is easterly. They are contact-veins, or deposits between limestone and shales, between quartzites and shales, and some between porphyry and one of the above rocks. The ore, as far as it has been extracted, contains but little base metal, and is similar to that in the Chance and Tickup mines, in Cherry Creek, while the vein-matrix is a hard, sometimes ferruginous, quartz.

Generally speaking, but little has been done to develop the district. The only shafts sunk so far are in the McMahon mine, while on three locations tunnels have been driven several hundred feet in length. In one of these, driven about 550 feet below the Woodburn mine, north of the town, the granite is seen rising from near the bottom of the tunnel at its entrance at an angle of 17° toward the summit of the mountains in the east. It is overlain by crystalline limestone. The tunnel driven in the Schell Creek mine, east of the town, passes through black slate and limestone about 340 feet. The La Brosse tunnel, in Queen's Spring district, a few miles south, is driven in 300 feet. It is an exploring-tunnel, and is supposed to cross a number of veins passing through quartzites, limestone, and porphyry.

The mining-claims north of Schellburn are all located in McMahon Cañon, which divides the foot-hills from the main range in a north or south direction. Commencing on the south, the following are some of the claims: Spanish, Independent, Mount Diablo, Woodburn, Mayflower, Crown Point, Sheridan & Muncey, McMahon, and Summit. Nearly all of these claims are high up on the mountain, the Woodburn being located at an altitude of 8,300 feet. The Summit is 8,900 feet high.

The Woodburn is not now worked, but has been opened somewhat on the surface, and seems to run northwest and southeast, with a slight dip eastward. The hanging-wall is shale and the foot-wall compact limestone. The vein itself is several feet thick, and composed of quartz impregnated with fahlore, or perhaps stetefeldtite. Metallic silver can be observed at various spots in thin coatings on the faces of the quartz; also filiform, sometimes together with solid horn-silver. The average value of ore from these surface-workings was about $80 per ton.

The shaft in the McMahon mine is 130 feet deep, and followed the vein for 70 feet, when the latter widened out to a chamber filled with *débris*. A tunnel was driven to intersect the shaft at 150 feet in depth. The vein here is from 6 to 8 feet wide, but has furnished rather lean ores, yielding by pulp-assay from $18 to $30. The mine is not worked now.

At the Summit mine, only surface or prospecting work has been done, and the character of the vein cannot be well seen. The ore from the suface, however, has been of good quality, as shown by pulp-samples.

*Queen's Spring district.*—The geological features in this district, located three miles south of Schellburn, are somewhat similar to those in McMahon Cañon, but quartzite is more predominant as country-rock. The principal mining-claims are the following: El Capitan, International, London, Sweepstakes, Citizen, Nutmeg, Mountain Boy, San Francisco, Savage. At several of these claims prospecting-work has been done, and the ore sent to be milled, but so far all these workings have not been able to supply permanently the 5-stamp mill in Schellburn, belonging to the McMahon Mining Company. The district needs development badly. Some years ago the Tehama Mill, with 20 stamps, was constructed, upon the presumption that it would be fully supplied by a few mines, but it never went into operation.

The McMahon Mill, with 5 stamps, works on custom-ore, and produces, by the Washoe process, from 70 to 75 per cent. of assay-value. The charges are $25 per ton, and the actual cost is said to be about $12.

The *Ruby Hill district*, further south, shows some rich ruby-silver ore, impregnated in crystalline limestone. The mines there are reported to be in litigation and idle for the present.

*Assessor's return of ore worked in White Pine County for the quarter ending March 31, 1874.*

| Name of mine. | Tons. | Pounds. | Gross yield. | Remarks. |
|---|---|---|---|---|
| Baltic | 7 | ........ | $420 00 | |
| Curtiss & Keller | 1 | 400 | 121 00 | |
| Eberhardt and Aurora | 1,161 | 577 | 32,726 07 | |
| Geneva Consolidated | 103 | 900 | 11,500 00 | Cherry Creek. |
| Mountain Chief | 10 | 1,000 | 884 00 | Newark. |
| Rescue | 15 | 300 | 3,048 00 | |
| Silverado | 5 | 1,500 | 877 00 | |
| Wilson | ........ | ........ | 112 70 | |
| Total | 1,309 | 677 | 49,688 77 | |

*Assessor's return of ore worked in White Pine County for the quarter ending June 30, 1874.*

| Name of mine. | Tons. | Pounds. | Gross yield. | Remarks. |
|---|---|---|---|---|
| Battery | 100 | 1,500 | $11,177 00 | |
| Bobtail | 1 | 1,230 | 254 85 | |
| Chance | 94 | 800 | 8,570 00 | Cherry Creek. |
| Chihuahua | 480 | ........ | 15,840 00 | Newark. |
| Dictator | 14 | 600 | 2,215 00 | |
| Eberhardt and Aurora | 2,718 | 1,500 | 73,852 00 | |
| International Consolidated | 10 | 1,000 | 783 00 | |
| Metropolitan Mill | 200 | ........ | 1,627 00 | |
| Mountain Chief | 38 | 100 | 1,784 00 | |
| Page & Whimple | 5 | 1,000 | 527 00 | |
| Rescue | 28 | 300 | 5,278 00 | |
| Sanchez | 10 | 200 | 1,008 00 | |
| Silver Charnott | 10 | 1,000 | 831 00 | Schell Creek. |
| Woodburn | 8 | 300 | 440 00 | Do. |
| Total | 3,720 | 800 | 124,827 10 | |

*Assessor's return of ore worked in White Pine County for the quarter ending September* 30, 1874.

| Name of mine. | Tons. | Pounds. | Gross yield. | Remarks. |
|---|---|---|---|---|
| Battery | 54 | 500 | $3,873 23 | |
| Bartlett | ........ | 760 | 133 60 | |
| Caroline | 11 | 480 | 520 50 | |
| Chance | 108 | 1,900 | 6,180 00 | Cherry Creek. |
| Cherry Creek Mill and Mining Company. | 105 | ........ | 4,744 00 | Do. |
| Dictator | 1 | 1,630 | 266 00 | |
| Eberhardt and Aurora | 2,862 | 150 | 153,816 60 | |
| El Capitan | 5 | 1,100 | 691 00 | Schell Creek. |
| Indian Jim | 2 | 1,669 | 239 50 | |
| Mariposa | 7 | 671 | 887 50 | |
| Maryland | 1 | 1,469 | 560 00 | Pinto. |
| Newark | 36 | 1,500 | 3,352 13 | Newark. |
| Pocotilio | 6 | 1,915 | 560 00 | |
| Rescue | 43 | 821 | 7,257 23 | |
| San José | 750 | ........ | 6,550 00 | Egan. |
| Star | 11 | 800 | 545 00 | Cherry Creek. |
| Tickup | 83 | 100 | 12,400 00 | Do. |
| French | 4 | 1,943 | 834 50 | |
| Total | 4,158 | 1,058 | 203,013 79 | |

A large number of mines have had ores worked, no returns for which have been yet given. The Fay, for instance, had 126 tons worked in the third quarter, the value of which was $90 per ton.

*Assessor's report of ore worked in White Pine County for the quarter ending December* 31, 1874.

| Name of district. | Name of mine. | Tons. | Pounds. | Amount. | Remarks. |
|---|---|---|---|---|---|
| Pinto | Autumn | 2 | 1,733 | $713 16 | Roasting. |
| Do | Alturas | 2 | 1,294 | 791 89 | Do. |
| Newark | Battery | 25 | 180 | 2,736 19 | Smelting. |
| Do | Bennett | 1 | 1,330 | 114 56 | Do. |
| White Pine | Caroline | 7 | 800 | 1,344 00 | Roasting. |
| Cherry Creek | Chance | 6 | 1,241 | 474 02 | Free-milling. |
| Newark | Chihuahua | 70 | ........ | 2,800 00 | Roasting. |
| Do | Tailings | 250 | ........ | 330 00 | |
| Schell Creek | El Capitan | 11 | 500 | 831 00 | Free-milling. |
| Robinson | Elijah | 400 | ........ | 17,500 00 | Smelting. |
| White Pine | Eberhardt and Aurora | 2,759 | 410 | 105,542 40 | Free. |
| Cherry Creek | Grey Eagle | 3 | 1,500 | 485 00 | Free-milling. |
| Pinto | Rescue | 48 | 539 | 9,017 92 | Roasting. |
| Do | Silver Stone | 4 | 654 | 217 42 | Free-milling. |
| Ruby Hill | Silver Wreath and Lookout. | 110 | ........ | 7,500 00 | Do. |
| Egan Cañon | San José | 53 | ........ | 1,441 60 | Do. |
| Cherry Creek | Star | 527 | ........ | 43,800 00 | Do. |
| Do | Tickup | 219 | ........ | 21,000 00 | Do. |
| White Pine | French | 15 | 955 | 4,232 36 | |
| | Outside lots | 9 | 1,302 | 1,130 45 | |
| Total | | 4,527 | 438 | 222,001 97 | |

## NYE COUNTY.

*Philadelphia district.*—Here little has been done but developing the mines. In this direction, however, a great deal has been accomplished. The Belmont property, which comprises 6,000 feet in length of vein, has been opened nearly throughout the length of the claim, and new and substantial hoisting-works have been erected over two new vertical shafts situated near the extremities of the property. Although no large ore-bodies have been discovered, it is claimed that the mine is now in condition to furnish great quantities of rich ore during the next year.

The Monitor-Belmont and El Dorado South are reported to have yielded fair quantities of very rich ores; but the principal work done has here also been prospecting and development.

*Tybo district.*—This new district, which has attracted most attention in Nye County during 1874, was mentioned in my last report, having come into notice about the end of 1873. It has been impossible for me or my assistant, Mr. Eilers, to visit this camp personally during 1874, and I have therefore to be content to give the best information on the subject which I have been able to gather. For most of it I am indebted to a series of articles by J. D. Powers, published toward the end of the year, in the San Francisco Mining and Scientific Press. The district is located about one hundred miles southerly from Eureka. The country-rock consists of limestone and slate traversed by porphyry and of granite. The principal mines are owned by the Tybo Consolidated Company, (limited,) a London organization, formed by Mr. J. B. McGee, formerly superintendent of the Richmond Consolidated Company of Eureka.

The Two G claim consists of 1,200 feet. The vein runs northwest and southeast, and is located on the rocky ridge of a spur of the main mountain. This spur rises abruptly south of the cañon for about 1,000 feet. The Two G had been, at the end of the year, opened by three tunnels, to a depth of 245 feet below the highest outcrop of the vein. The average depth, however, to the floor of tunnel No. 1 is not over 188 feet. This tunnel was commenced in May, and has now reached a length of 463 feet along the vein. About 200 feet of this carries galena and some oxidized ore, the rest a dark, decomposed, quartzose ore, which can be milled to advantage without roasting, and contains an average of perhaps $40 silver per ton. Tunnel No. 2 is located at the original discovery, 40 feet above No. 1, and has been driven over 500 feet along the vein, striking one or two chimneys of 7 to 8 feet in width, containing the same ore as above mentioned. Nos. 1 and 2 are connected by a winze. No. 3 is still higher up, and about 100 feet long. In this part is the richest smelting-ore, said to contain over $100 per ton. From the level of tunnel No. 1 two winzes, having an inclination of about 75°, have been sunk on the vein. They are 163 feet apart, the outer one being down 110 feet. At a depth of 75 feet below the tunnel-floor the two have been connected by a drift, which is in 4 feet of ore. The foot-wall in these works is clearly to be recognized as porphyry; the hanging-wall is siliceous limestone. It is somewhat remarkable that milling-ores occur all along the foot-wall, while baser ores, carrying more galena and other sulphurets, follow the limestone or hanging-wall.

The Casket and Crosby are extensions of the Two G, while the Lafayette appears to be a spur of the same vein. The latter is opened by a shaft 70 feet deep, and a drift running southeast 75 feet. It carries the same ore as the Two G. The total linear extent of these four claims is about 3,000 feet.

For the reduction of the ore, Mr. McGee has built a 30-ton blast-furnace, which is 14 feet high above the tuyeres. The ore is brought in cars from the mouth of the tunnel, immediately behind the furnace, to the charging-floor. Blast is supplied by means of a No. 6 Root blower and a 30-horse-power steam-engine. The blower and engine have a sufficient capacity to supply another and larger furnace with the necessary amount of blast. From the 24th of August, when the furnace was first put in blast, to the 1st of December, 1,260 tons of ore were smelted, which yielded 153 tons of work-lead, containing an average of about 250 ounces silver and 1½ ounces gold per ton. Mr. Daniel Meyer, of San Francisco, through whom the Tybo Company's base bullion was shipped, writes, at the end of December, that he expected the shipments of that month to be 160 additional tons.

During the spring of 1875, Mr. McGee intends to erect a 20-stamp wet-crushing mill, with an engine of 150 horse-power, the whole to cost $60,000. The nature of a large part of the ore requires treatment of this kind. Besides this, the erection of a large shaft-furnace (50 to 60 tons per day) and of a reverberatory for refining the lead is contemplated. The cast-iron fan for the latter is already on the ground.

Three small towns, Upper, Middle, and Lower Tybo, have been built in the same cañon. The last-named is located around the company's works, and will probably be the only permanent one.

In *Morey district*, sixty-five miles south of Eureka and thirty-five miles north of Tybo, the New York Company, which has a monopoly of the district, has run its 20-stamp mill part of the time, with good results. Much annoyance was, however, experienced at various times from the working of the machinery, which is mostly old, and, having been bought up from different mills and fitted into one, does not give the desired satisfaction.

In *Jefferson Cañon*, forty-five miles from Tybo, and nearly south of Austin, the North and South Prussian mines, which contain free milling-ore of fair quality, have been worked to a considerable extent.

The Prussian South is owned by the Jefferson Silver-Mining Company, and comprised 1,200 feet on the southern end of the Prussian lode. The company has sunk an incline 340 feet deep on the ledge, which dips at an angle of 65°. At this depth the first water in the mine was struck. Four levels have been driven from the incline at, respectively, 65 feet, 140 feet, 240 feet, and 340 feet from the surface. The vein throughout these works varies in thickness from 3 to 5 feet, the upper portion carrying reddish decomposed ores, while below antimonial sulphurets have made their appearance. All the levels have been driven considerable distances north and south from the shaft; but little stoping has so far been done. The company has erected a 10-stamp wet-crushing mill, and has worked the decomposed ores of the upper levels, which yielded an average of about $75 per ton in silver. The sulphuret-ores below will of course have to be roasted previous to amalgamation. The yield of the mine during the latter part of the year is said to have been $40,000 per month.

The Prussian North is owned by the Prussian Gold and Silver Mining Company, the claim comprising 1,000 feet. An incline sunk on this claim is 250 feet deep, and at this depth so large an amount of water was met that no further sinking can be carried on until heavy pumping-machinery has been procured. Three levels have been run from the incline, at 65 feet, 100 feet, and 200 feet, respectively, from the surface. The second and third are the longest, being over 250 feet in length. The vein in these works is very much larger than in the claim of the Jefferson Company, varying from 3½ to 8 feet in width, but the ore is not as rich. This company has also a 10-stamp wet-crushing mill, which works 20 tons of decomposed ore per twenty-four hours.

*Lyda Valley** (or Alida Springs) is reached from Austin, by way of Smoky Valley, on the east side of the Toyabe range, to its southern end at San Antonio, across the sandy desert to Silver Peak, and through Clayton Valley across a mountain-pass (8,200) into the settle-

* Notes kindly furnished by Mr. A. Trippel, who visited both Lyda Valley and Gold Mountain in the interest of the Manhattan Company, Austin, and conducted his examination thoroughly, the object being the acquisition of really good property, if such existed there.—R. W. R.

ment, Montezuma Mountain lying to the northeast and Mount Magruder southward. The total distance is one hundred and seventy miles. The settlement, consisting of a few dozen shanties and a half-wrecked 5-stamp mill, lies about one hundred miles northeast of Fort Independence, California. The prevailing rocks in the district are the same (limestones, slates, and quartzites) as are found in so many districts in Central and Eastern Nevada. The general course is northwest and southeast. Alida Valley proper has a length of some six miles east and west, is well wooded, (for Nevada,) and has several fine and abundant springs, furnishing pure water. The slopes of the surrounding hills are covered with sufficient bunch-grass and white sage for pasturage. The numerous mineral veins, of which the Chloride and Brown's mines are best known, produce rich silver-ores, partly mixed with copper, but mostly with lead-ores. From the records of the mill, the pulp-assays were found to be over $200 per ton on an average. Specimens of large size containing nearly pure horn-silver are frequently found. Still the development of the mines has not made any progress. The reasons of this are the great distance to a shipping-point, the comparative isolation of the place, and the fact that the present *modus operandi* in treating these ores is evidently unsatisfactory, certainly so to the miner who supplies the "custom-mill." Mr. Trippel says he is satisfied that the bulk of the ores from this district are better suited for smelting, especially when certain portions of them have been concentrated.

With the material on hand, it would not be difficult to establish and conduct profitably at Alida Valley a concentration and smelting establishment of fair proportions which could afford to pay living rates to the miners for their ore.

*Gold Mountain district* is situated in an isolated group of mountains, not over twenty miles in length, and lying about ten miles east of the California line, southeast of the Alida Valley. The road from the latter passes over Mount Magruder, and thence across a desert of twelve miles, after which it ascends the northwestern part of the Gold Mountain group, which is separated from the southern portion by another sandy desert. The camp, on Gold Mountain proper, is located on the north slope of the southern part, which consists almost entirely of syenitic rock. In that part are found a great number of veins carrying auriferous quartz, which frequently has the appearance of jasper, and is exceedingly hard. The locations made here during 1872, 1873, and 1874 are very numerous, certainly over fifty; but so far no developments of any importance have been made. Nearly all the work done was prospecting-work An arrastra of the most primitive construction has been built. The only shaft sunk to any depth is on the Oriental, which shows a vein of 5 feet, carrying fine gold-quartz, together with the ores of base metals, such as galena, or the products of its decomposition, and some malachite. It is, however, characteristic that the southern part of Gold Mountain carries mostly veins of auriferous quartz, among which are the Kohinoor, Boomerang, Oriental, Nova Zembla, and Borneo, while the northern part of the group, where the prevailing rocks are slates, quartzites, and limestone, carries mostly argentiferous ores, connected with the base metals, especially lead, copper, and antimony. Some of the principal veins in the latter part are the Independent, Blue Wing, Austin, and Good Templar. There are a great number of others. None of these have been opened beyond the ordinary prospect-work.

There is little room for doubt that Gold Mountain district is exceedingly rich in auriferous ores, and that at some future time it will be made productive of the precious metals. The chief difficulty in working the mines there is the great scarcity of water. There are but one or two springs of good water known, and whether such could be found at a reasonable depth in sinking wells has yet to be tested. The quantity of wood is also limited, and its growth is confined to a few locations. Pasturage for live stock is scanty.

Assays of fair samples taken from over fifty claims were made, averaging in gold and silver, together, about $150. The following are the average assay-results of ore from some of the representative mines, of which each one is a strong ledge, with good walls, and with an east and west trend, while the dip seems to be from 60° to 80° south:

*Statement of averages, from assays made.*

| Mine. | Silver. | Gold. | Total. |
|---|---|---|---|
| Oriental, average of 9 samples | $169 19 | $13 47 | $182 66 |
| Borneo, average of 5 samples | 15 07 | 67 80 | 82 87 |
| Good Templar, average of 4 samples | 221 07 | 12 48 | 233 55 |
| Blue Wing, average of 4 samples | 58 57 | 6 27 | 64 84 |
| Nova Zembla, 1 sample | 26 70 | 276 17 | 302 87 |

Grand average of all, $173.36 per ton.

The ore of which assays were made was by no means selected, but apparently of ordinary quality. Selected specimens of Oriental gave $1,370.79. In order to see the effect of concentration on some of these ores, they were washed by hand, with the following results:

| | Assay of concentration. | Obtained per cent. |
|---|---|---|
| Oriental, (A) | $682 49 | 80½ |
| (B) | 446 20 | 81½ |
| (C) | 269 00 | 64 |
| (D) | 701 39 | 72½ |
| No. 3, (lead) | 377 00 | 73 |
| No. 4, (lead) | 86 66 | 57 |
| Good Templar, No. 2 | 1,255 64 | 69 |
| No. 3 | 271 50 | 77 |
| Blue Wing, No. 2, (copper) | 41 47 | 52 |
| Borneo, No. 3, (gold) | 648 17 | 87 |

Several of the samples taken contained the mineral in a partly decomposed state. Considering the primitive method used and the character of the mineral, the results leave no doubt about the feasibility of dressing the ores.

### ESMERALDA COUNTY.

From Esmeralda County I have not had any report, except the general information from *Columbus district* that the Northern Belle, a very rich mine, has been worked throughout almost the whole year, with very gratifying results. The Northern Belle Mill, which has been running since March 1, contains one of the Stetefeldt roasting-furnaces, and is reported to give great satisfaction. The roasting-furnace has been tried up to a capacity of 60 tons per twenty-four hours, and is proved to not only chloridize to a far higher percentage than could be done in the reverberatories formerly used, but also to produce a bullion of much greater fineness. The amount shipped from the works during the year I have been unable to ascertain.

## LINCOLN COUNTY.

*Ely district.*—Nothing of importance has been done in this county outside of Ely district. This district, even, has furnished far less bullion than in former years, both the principal producers, the Raymond & Ely and the Meadow Valley Companies, having done a very large amount of prospecting-work in comparison with the work of exploitation. This became necessary in the case of the former by the exhaustion of the large ore-body, which for several years had furnished such large amounts of silver.

The total product of the district during 1874 was as follows:

Shipped by Wells, Fargo & Co.:

| | |
|---|---|
| January | $241,526 00 |
| February | 170,229 00 |
| March | 152,945 00 |
| April | 198,000 00 |
| May | 217,936 00 |
| June | 100,441 00 |
| July | 118,998 00 |
| August | 75,985 00 |
| September | 104,235 00 |
| October | 98,296 00 |
| November | 88,727 00 |
| December | 77,894 00 |
| Total | 1,645,252 00 |
| Shipped by Pritchard's fast freight, (base bullion) | 236,299 78 |
| Ore, (estimated by H. H. Day, superintendent of Raymond & Ely Company) | 19,653 21 |
| Total | 1,901,204 99 |

Of this the Raymond & Ely Company produced $673,051.27, the remainder is from the Meadow Valley Company's mines and tailings, and from a large number of smaller mines, in regard to which I have no separate data.

The following is the report of Mr. D. M. Tyrrell, general superintendent of the Meadow Valley Company, for the year ending July 31, 1874:

During the year just closed there have been extracted from the mine $5,224\frac{1}{2}\frac{454}{000}$ tons of ore, obtained from the following sections:

| | Tons. |
|---|---|
| No. 3 shaft, 200-foot level | 250 |
| No. 3 shaft, 520-foot level | 250 |
| No. 3 shaft, 630-foot level | $3,214\frac{1}{2}\frac{454}{000}$ |
| No. 3 shaft, 750-foot level | 550 |
| No. 3 shaft, 875-foot level | 500 |
| No. 3 shaft, 875-foot level, south ledge | 460 |
| Total | $5,224\frac{1}{2}\frac{454}{000}$ |

The average yield of the ore has been remarkably uniform, considering the separate localities from which it has been extracted. The company's mine is explored by three shafts, called, respectively, No. 3, No. 5, and No. 7. The most productive portion of the mine has been between the Black shaft (which is located about 300 feet east of No. 3 shaft) and the west line of the company's claim, and has, therefore, been more extensively worked. The West, or No. 3 shaft, has reached the greatest depth of any of the works upon the mine. This shaft is composed of two compartments, $2\frac{8}{12}$ by 5 feet each. From the surface to the 520-foot level it has been carried down on the vein through its various changes of dip, making the shaft very irregular in grade for a distance of 520 feet; but from this level to the bottom it has been carried down on an angle of

65°. At the commencement of the year just closed this shaft had attained a depth of 931 feet, and from the 1st of October (the time sinking was again resumed) up to the 8th day of May, 383 feet had been added to its depth, making its total depth 1,314 feet on the incline, at which point the water-level was reached. With our present appliances on hand, the water could not be raised, and sinking was in consequence discontinued.

Three new levels have been opened from this shaft during the year, one at the depth of 1,000 feet vertically from the surface, another at the depth of 1,100 feet vertically, and another at the depth of 1,235 feet, corresponding with the 1,200-foot level of the Raymond & Ely mine. On the 1,000-foot level, in the latter part of October, a station was opened and a cross-cut started north for the vein, which was carried ahead a distance of 42 feet, at which point the vein was cut; but the line of demarkation was so irregular and imperfect that it was difficult to determine whether we had reached the fissure or not. However, a drift was started west at this point, and as the work progressed regularity rapidly became apparent; the walls also became regular and gradually diverged until the face of the drift showed a well-defined fissure with 2 feet of vein-matter, which retained its width for a distance of 265 feet, at which point the drift cut into a broken cross-course, dipping to the east on an angle of 80°, when work was discontinued in that direction and a winze started down on the vein 200 feet west of the shaft. This winze was carried a distance of 160 feet, at which point a connection was made with the west drift on the 1,100-foot level. This winze showed a regular, well-defined fissure, with 2 feet of vein-matter, but comparatively barren.

On the 1,100-foot level, on the 1st of March, a station was opened and a cross-cut was started north, which was carried ahead a distance of 50 feet, at which point the vein was cut, showing regular, well-defined walls, with 7 feet of vein-matter, giving low assays. A drift was started to the west, following the vein, which was carried ahead a distance of 293 feet. Of this distance 240 feet showed a well-defined fissure with 2 feet of vein-matter, but no ore, and the balance of the distance run was through that broken cross-course which appeared on the level above.

From the end of this drift a cross-cut was run to the north for 35 feet, and another to the southwest 23 feet, which prospecting failed to discover the continuation of the vein, and exploration ceased in that direction.

A winze was started from this drift about 150 feet west of the shaft and carried down on the vein a distance of 158 feet, at which point a connection was made with the west drift on the 1,200-foot level, furnishing a good circulation of air. This winze showed a strong, well-defined vein for the entire distance, but no ore.

A drift was also run to the east on the 1,100-foot level, which has now a length of 290 feet. Eighty feet of this distance was run through a broken cross-course, to the east of which the walls begin to show themselves, and the vein becomes regular and partakes of a more uniform character; and at present the face shows 2 feet of very promising vein-matter, with some small streaks of galena-ore, but of low grade. So far the prospecting done on this level has not resulted in any valuable discovery, but the indications certainly afford an encouraging feature for the future of the mine, and give the greatest reason to argue that the vein will continue persistently as the work progresses downward; but what the results will be, future developments alone can determine.

On the 1st of May, a station was opened on the 1,200-foot level and a cross-cut started north for the vein, which was run a distance of 36 feet, at which point it reached the vein, when a drift was started west on the vein and carried to the west line of the company's claim, a distance of 302 feet. Here the fissure is regular and well defined, showing 2 feet of very promising vein-matter, giving low assays throughout the lowest level as far as explored. The vein maintains itself perfectly, except where these breaks occur, being of full average width, and its walls are regular and well defined, giving the strongest and most reliable evidence of permanency and persistency in depth.

A drift is also being run to the east on the 1,200-foot level, which has now a length of 117 feet, the entire distance run being through a broken cross-course; but the formation is getting more regular and showing evident indications of a near approach to the vein.

On the 12th of May, a cross-cut was started south for the south vein on the 1,200-foot level, which has now a length of 250 feet. This vein runs parallel with and lies about 300 feet south of the main Meadow Valley vein. The ground to the south of the old workings in the upper levels has not been sufficiently prospected to prove whether this is a spur from the main vein or a separate and distinct one. The intervening ground between these veins is quartzite, and has not been sufficiently cut up to demonstrate whether it is an isolated fragment or a continuous part of the country-rock.

Prospecting operations in the upper levels have been vigorously prosecuted during the year. The undeveloped portions of the mine to the east and west of No. 3 shaft have been pretty thoroughly explored, but so far we have not been fortunate enough to uncover any paying bodies of ore. The east drift on ninth level has been carried ahead a distance of 1,144 feet. Three hundred feet of this distance were run through a broken cross-course, and the balance of the distance showed a regular, well-defined vein, with an average width of 3 feet. The filling between the walls showed some flattering indications in places, but at present the formation does not look very promising for ore.

A winze is being sunk from this drift 400 feet east of No. 3 shaft, which has now a depth of 172 feet; the entire distance sunk shows a continuous fissure varying in width from 10 to 30 inches, occasionally showing some very promising vein-matter, but very little ore.

A drift was started to the west on the ninth level, and carried ahead on the vein to the west line of the company's claim, a distance of 291 feet. The entire distance run showed a strong, well-defined vein, with some small strata of base ore, but of low grade. A raise was started from this drift about 50 feet east of the west line of the company's claim, which was carried up a distance of 65 feet, at which point a very good streak of ore 2 feet in width was developed. In raising on it the walls of the fissure diverged until the vein showed a width of 8 feet, when the ore became more widely diffused throughout the general filling of the vein, and as the work progressed the ore gradually died out in a large body of barren vein-matter.

A winze was started from this drift 210 feet west of the shaft, and carried down on the vein a distance of 110 feet, where a connection was made with the west drift on the 1,000-foot level; here the vein is strong and well-defined, but comparatively barren.

The cross-cut which was being run for the south vein on the ninth level, which had attained a length of 50 feet at the time of the last annual report, was carried ahead 285 feet, making its total length 335 feet, at which point it reached the fissure, disclosing a vein of quartz 8 feet in width, with a stratum of good milling-ore 12 inches in width lying on the foot-wall. A drift was started west on the old streak and carried ahead a distance of 50 feet, where the ore gradually died out in barren vein-matter. The drift was continued a distance of 114 feet farther, making its total length 164 feet, but failed to disclose anything of value, when work was discontinued.

A cross-cut was started from the end of this drift and run to the north a distance of 35 feet, and, not meeting with any favorable change, work was suspended at this point, and a raise started on the ore, which was carried up 65 feet with the ore-streak, frequently expanding or contracting to greater or less dimensions as the work progressed, until the ore finally disappeared in hard, barren vein-matter.

A drift was also started and carried to the east on the vein a distance of 210 feet, showing a continuous fissure varying in width from 1 to 6 inches; but the formation being unfavorable for ore, work was in consequence discontinued in that direction.

A drift was also started 50 feet above this level and run to the east a distance of 50 feet, which work failed to disclose anything of value, when operations ceased in that direction.

A winze was started on the south ledge and carried down on the vein a distance of 105 feet. This winze showed a large and very promising vein of quartz for a distance of 85 feet, where the vein-matter gradually disappeared, leaving the bottom of the winze in hard quartzite, when work was discontinued.

The drift which was being run to the east on the eighth level was carried ahead a distance of 162 feet, making its total length 473 feet. The formation in the face not looking very encouraging, work was discontinued.

A winze was sunk from this drift about 300 feet east of No. 3 shaft, and carried down on the vein a distance of 108 feet, connecting with the east drift on the ninth level, furnishing a good circulation of air.

A winze was also sunk from the west drift on the eighth level about 210 feet west of No. 3 shaft, and carried down a distance of 108 feet, where a connection was made with the west drift on the ninth level. Sixty feet of this distance was carried down on a small seam of ore, and the balance of the distance through hard quartzite.

The east drift on the 630-foot level, which showed a length of 502 feet at the time of the last annual report, was carried ahead 241 feet, making its total length 743 feet, at which point a connection was made with the south cross-cut from the summit-shaft, furnishing a good circulation of air. The entire distance run showed a continuous fissure varying in width from 1 up to 18 inches, but no ore to the east of the stope being worked up from this level.

A winze was sunk from the 630-foot level 300 feet east of No. 3 shaft, and carried down to the eighth level a distance of 112 feet. The first 50 feet showed a stratum of very good ore, and the balance of the distance was sunk through quartzite.

A cross-cut was started south on the 630-foot level and carried ahead a distance of 210 feet, which work failed to develop anything of special value, when work was discontinued in that direction, and a drift started west, (on a crossing which showed in the drift,) which was carried ahead 44 feet; and not meeting with anything encouraging, work was discontinued.

On the 520-foot level the east drift was carried ahead 370 feet during the year, making its total length 520 feet. This drift developed a small stratum of very good milling-ore, which proved to be a continuation of the ore-chute worked up from the 630-foot level. Where cut by the drift it showed a length of 80 feet, varying in width from 1 to 8 inches, narrowing down at both ends, and at a distance of 40 feet above the level it finally disappeared in barren vein-matter.

On the 440-foot level a drift was run to the west a distance of 55 feet, which failed to disclose anything of value, when work was discontinued in that direction, and a raise started and carried up on the vein a distance of 54 feet, when a drift was started west from top of raise and run a distance of 71 feet. This prospecting failed to reveal ore in paying quantities, and work was discontinued.

The summit or No. 5 shaft is composed of three compartments. The two hoisting-compartments are $3\frac{1}{4}$ by $4\frac{8}{12}$ feet in the clear, and the pump compartment is $4\frac{8}{12}$ by 5 feet. This shaft is carried down in the north wall of the vein on an angle of 62°, and although the work at this point has been prosecuted with unremitting vigor, owing to the extreme hard rock encountered, but 368 feet have been added to its depth during the year, making its total depth 1,141 feet on the incline. At our present rate of progress it will probably take about seven months to reach the present level of the bottom of No. 3 shaft.

A drift was started from this shaft about 500 feet from the surface and run to the west, following the vein a distance of 188 feet, showing a strong, well-defined fissure, with 2 feet of vein-matter, but no ore.

Two cross-cuts were run south from this shaft to the vein for ventilation, one at the fifth level corresponding with the 630-foot level in No. 3 shaft, a distance of 45 feet, and another at the sixth level corresponding with the ninth level of No. 3 shaft, a distance of 35 feet through quartzite.

The sinking of No. 7 shaft was discontinued last August, not deeming it prudent to continue it downward, as the eastern portion of the company's ground could be prospected at a much less expense from No. 5 shaft. Some little prospecting has been done from No. 7 shaft, but so far our operations have been unsuccessful.

The west drift which was started from third station, No. 7 shaft, was carried ahead 22 feet, making its total length 112 feet. The formation not looking promising, work was suspended and a cross-cut run to the north from the end of this drift, a distance of 47 feet; and not meeting with any favorable indications, work was discontinued.

A prospect-drift was started from fourth station, No. 7 shaft, and run to the east a distance of 187 feet through a large block of unexplored ground; but failing to meet with anything encouraging, work was suspended.

The drift which was being run to the west on eighth level, No. 7 shaft, was extended a distance of 478 feet, making its total length 713 feet. The entire distance run showed a regular, well-defined fissure, with an average width of 18 inches; but during its progress nothing of a particularly encouraging nature was brought to view, and work was discontinued in that direction.

At the present time our resources for ore are limited to the 520-foot level, and the success of the present year's operations will depend upon developments in the unexplored ground to the east of No. 3 shaft; and I am not without hope that further explorations in and about the apparently exhausted sections of the mine will, as has happened in similar cases in times past, open up some additions to what is now in sight. But our chief reliance is upon what is anticipated will be brought to light at a not very remote time in the lowest levels of the mine not yet prospected. The distribution of the ore through the vein is variable, sometimes occurring in seams a few inches in width, with intervening strata of poor rock; sometimes it is found distributed through the entire vein, and at other times it dwindles down to a mere trace, but is generally found in bunches or irregularly-shaped deposits, which are usually found on one wall or the other, but do not seem to follow either uniformly.

The upper levels, although to a great extent worked over several times, have, nevertheless, furnished employment to a considerable number of men, and considerable revenue has been derived from contracts let to individual prospectors, by which they were allowed to extract ore from the abandoned levels of the mine at their own expense, paying the company 40 per cent. of all ore extracted, and the other 60 per cent. to be worked at the company's mill.

The machinery on the different shafts at the mine is all in good running order, but entirely inadequate for the work required of it. I would therefore recommend that the necessary steps be taken to erect substantial machinery for permanent operations.

A summary of the labor performed in the mine during the year shows the extracting of $5{,}224\frac{1451}{2000}$ tons of ore, the running of 6,037 linear feet of drifts, the sinking of 989 feet of winzes, the running of 184 feet of a raise, the sinking and timbering of 751 feet of shafts, and other general repairs in and about the mine.

The mill is at present in good running order, with the exception of the mortars, which are getting pretty well worn. There has been no expense incurred within the last year, except what has been necessary to keep the mill in repair. There have been four new pans and two new settlers put in in place of the old ones, which were entirely worn out and no longer fit for use. In the future I do not anticipate any expense on the mill other than repairs incident to running.

Of the tailings remaining on hand and unworked, there are 11,000 tons of tailings of ore, which give an average assay of $18 per ton, and of tailings of tailings there remain about 20,000 tons, which show an average assay of $12 per ton.

To this I add the following table from the superintendent's report:

*Condensed statement of cost of production, &c., for company year, from August* 1, 1873, *to July* 31, 1874.

ORE STATEMENT.

| | Tons. | Pounds. |
|---|---|---|
| Ore extracted during the year | 5,224 | 1,454 |
| Slime and tailings | 17,239 | 659 |
| Custom-ore | 672 | 1,234 |
| Total | 23,136 | 1,347 |
| Ore worked by company during year, plus moisture, tailings, and custom-ore | 23,136 | 1,347 |

COST PER TON.

| | |
|---|---|
| Extracting | $22 19.3 |
| Prospecting, improvements, and sundries | 48 87.3 |
| Reduction | 13 74.8 |
| Total cost of ore per ton | 84 81.4 |
| Average yield of ore per ton | 97 02.0 |
| Average yield of ore and tailings per ton | 30 42.0 |

WORK OF ASSAY-OFFICE.

| | |
|---|---|
| Number of troy ounces of bullion before melting, fine and base | 1,553,985.00 |
| Number of troy ounces of bullion after melting, fine and base | 1,514,881.50 |
| Average loss in melting, per cent., fine and base | 2.58 |
| Number of ore-assays made | 4,254.00 |
| Number of bars bullion made, fine and base | 790.00 |

The secretary's report contains the following interesting summary of the operations for the year:

SECRETARY'S FINANCIAL REPORTS.

*A statement of the gross proceeds of bullion from the mines of the Meadow Valley Mining Company, and cost of production and reduction of the ores yielding the bullion, for the fiscal year ending July 31, 1874.*

DR.

| | | |
|---|---|---|
| TO MINING DEPARTMENT, VIZ: | | |
| Labor in extracting ores | $92, 719 53 | |
| Mining-supplies | 67, 114 81 | |
| Freight from San Francisco on supplies | 1, 077 80 | |
| Contingent mine-expenses | 3, 513 07 | |
| Mine-salaries | 13, 760 00 | |
| Total expenditures in extraction | 178, 185 31 | |
| Deduct inventory of supplies on hand at date | 5, 602 35 | |
| | | $172, 582 96 |
| TO MILLING DEPARTMENT, VIZ: | | |
| Ore transportation from mine to mill, (ten miles) | 22, 560 49 | |
| Chemicals, quicksilver, and other supplies | 216, 085 66 | |
| Freight from San Francisco on mill-supplies | 16, 563 21 | |
| Labor in reduction of ores | 66, 906 16 | |
| Contingent milling-expenses | 1, 461 00 | |
| Assay-office | 7, 166 01 | |
| Foundery-supplies | 6, 543 86 | |
| Total expenditures in milling department | 337, 286 39 | |
| Deduct inventory of supplies on hand at date | 60, 947 55 | |
| | | 276, 338 84 |
| TO MISCELLANEOUS ACCOUNTS, VIZ: | | |
| Freight on base bullion to San Francisco, and discount on bullion for current year | 58, 690 23 | |
| State of Nevada taxes on bullion | 9, 696 19 | |
| States of Nevada and California property-taxes | 3, 840 90 | |
| Exchange | 195 15 | |
| Insurance premiums on mill-property | 2, 800 00 | |
| Telegrams | 1, 135 46 | |
| General expenses | 18, 043 63 | |
| | | 94, 401 56 |
| Total cost of production and reduction | | 543, 323 36 |
| Gross profit on ore-yield for current year | 166, 389 37 | |
| Less extraordinary expenses in explorations and dead-work | | 158, 633 49 |
| Net profit over all expenditures for current year | | 7, 355 88 |
| | | 709, 312 73 |

CR.

| | | |
|---|---|---|
| BULLION. | | |
| By proceeds of Company's reduction-works at Lyonsville, as per tabular statements of general superintendent and office records for the fiscal year ending this date | | $676, 271 84 |
| MISCELLANEOUS RETURNS. | | |
| By rentals | $350 00 | |
| By sales of materials: | | |
| Mill-supplies | 15, 518 17 | |
| Mine-supplies | 766 86 | |
| By custom-milling: | | |
| Receipts from ore-reduction | 16, 405 86 | |
| | | 33, 040 89 |
| | | 709, 312 73 |

## SUMMARY.

DR. *The Meadow Valley Mining Company in account with its estate for the full corporate term ending July* 31, 1874. CR.

| | | | |
|---|---|---|---|
| PERMANENT INVESTMENTS. | | | |
| In mine-properties: | | | |
| For the fiscal term 1869–'73 | $335,893 12 | | |
| For the fiscal year 1873–'74 | 6,014 46 | | |
| | 341,907 58 | | |
| Less received for adverse claims fiscal year 1871–'72 | 53,811 24 | | |
| | | $288,096 34 | |
| In construction of mining-works: | | | |
| For the fiscal term 1869–'73 | 95,320 27 | | |
| For the fiscal year 1873–'74, (repairs) | 5,741 75 | | |
| | | 101,062 02 | |
| In construction of reduction-works: | | | |
| For the fiscal term 1869–'73 | 253,363 24 | | |
| For the fiscal year 1873–'74 | 3,076 30 | | |
| | | 256,439 54 | |
| Total permanent investments | | | $645,597 90 |
| DIVIDENDS. | | | |
| Paid stockholders for the fiscal term 1869–'73 | | | 1,260,000 00 |
| Total | | | 1,905,597 90 |
| CURRENT RESOURCES, (EXHIBIT NO. 1.) | | | |
| Inventory of supplies at mine | | 5,609 35 | |
| Inventory of supplies at mill | | 60,947 55 | |
| D. M. Tyrrell, general superintendent | | 372 86 | |
| Bullion *in transitu* | | 36,550 02 | |
| Cash | | 22,942 56 | |
| | | | 126,415 34 |
| | | | 2,032,013 24 |

| | | | | |
|---|---|---|---|---|
| By assessments: | | | | |
| For the fiscal term 1869–'71 | | | | $210,000 00 |
| By bullion-product: | | | | |
| For the fiscal term 1869–'73 | | $4,233,296 58 | | |
| For the fiscal year 1873–'74 | | 676,271 84 | | |
| Total gross yield | | | $4,909,568 42 | |
| BULLION COST SHEETS DEDUCTED. | | | | |
| For the fiscal term 1869–'73 | | $2,513,102 43 | | |
| For the fiscal year 1873–'74: | | | | |
| Production and reduction | $543,323 36 | | | |
| Dead-work | 158,633 49 | | | |
| | | 701,956 85 | | |
| Total of cost sheets | | 3,215,059 28 | | |
| Less re-imbursed by sale of mill-supplies, custom-milling, &c.: | | | | |
| For the fiscal term 1869–'73 | $52,457 04 | | | |
| For the fiscal year 1873–'74 | 33,040 89 | | | |
| | | 85,497 93 | | |
| | | | 3,129,561 35 | |
| Total net profits on bullion-yield | | | | 1,780,007 07 |
| Total | | | | 1,990,007 07 |
| CURRENT LIABILITIES, (EXHIBIT NO. 1.) | | | | |
| Superintendent's drafts unpresented | | | | 42,006 17 |
| | | | | 2,032,013 24 |

At the Raymond & Ely mine the number of tons of ore extracted during the year was far less than in any previous year during the existence of the company. The disbursements greatly exceeded the receipts, and assessments became necessary. A large amount of tailings was worked during the year, leaving at the end of the year a small quantity on hand. The original ground of the company has been explored to a depth of 1,200 feet, where a strong body of water was encountered, necessitating the employment of pumping-machinery. This machinery was on the ground at the end of the year, but not yet fully in position.

The following is taken from the report of Mr. H. H. Day, the superintendent, made to the president and directors on December 31, 1874:

I herewith respectfully submit for your consideration a report of the condition of the company's property, and a detailed statement of the work at the mine and mills of the company for the year 1874:

| | |
|---|---|
| Amount of ore extracted, tons | $3,255\frac{886}{2000}$ |
| Amount sent to mills, tons | $3,085\frac{986}{2000}$ |
| Amount reduced at mills, tons | $3,266\frac{546}{2000}$ |
| Average assay-value per ton | $77.29 |
| Average percentage obtained | 62.3 |
| Bullion produced from ore | $162,460.90 |
| Amount of tailings reduced, tons | $49,454\frac{1340}{2000}$ |
| Average assay-value per ton | $31.50 |
| Average percentage obtained | 32.7 |
| Bullion produced from tailings | $510,590.37 |
| Total amount of bullion from all sources | $673,051.27 |

COST OF MINING ORE PER TON.

| | | |
|---|---|---|
| Extracting | $23 06 | |
| Prospecting and dead-work | 50 06 | |
| Improvements and repairs | 36 | |
| Sundries | 5 77 | |
| | | $79 25 |
| Cost of milling ore and tailings per ton | | 4 98 |
| Cost of ore-transportation per ton | | 4 00 |

*Work at the mine.*—The Lightner shaft has attained a total depth of 1,214 feet. At this point the flow of water into the shaft became so great that all further attempts at sinking were abandoned until suitable machinery could be erected to free the shaft from water. Extensive works of exploration have been made on the seventh, eighth, ninth, tenth, and eleventh levels without producing important results. The west ground, that has recently come into the possession of the company, gives fair promise of yielding ore in paying quantities above the water-level. At the water-level the vein has been drifted upon a distance of about 700 feet. Throughout this entire distance the fissure is strong and well defined, maintaining a uniformity of strike and dip, and showing other general characteristics which are highly encouraging for deeper explorations.

The improvements at the mine consist principally in the erection of powerful pumping-machinery, with machine-shop, foundery, and other needed accessories, the whole being very complete.

At the mills there have been no improvements made further than making necessary repairs. Vigorous efforts are being made to obtain Government patent to the mill-sites and water-rights of the company.

In conclusion, I will say that, although the year just closed has not been productive of satisfactory results, there is in the present situation much to inspire hope for the future, and I confidently believe that ere the close of another year the Raymond & Ely Company will be in a highly prosperous condition.

The following table is worthy of careful study, as showing clearly the nature and distribution of the expense of mining and reducing ores and of dead-work in this region:

## RAYMOND & ELY MINING COMPANY.

*General statement showing net cost of mining, prospecting, improvements, milling, legal expense, taxes, &c., and returns from bullion, &c., during fiscal year ending December 31, 1874.*

| Months. | Mine-cost. | | | | | | | Pump-improvement. | | | |
|---|---|---|---|---|---|---|---|---|---|---|---|
| | Tons of ore extracted. | Labor. | Supplies. | Freight. | Improvements and repairs. | General. | Total. | Labor. | Materials and supplies. | Freight. | Total. |
| 1874. | | | | | | | | | | | |
| On hand January 11 | 180.1560 | | $11,576 70 | | | | $11,576 70 | | | | |
| January | 1,119.1990 | $32,896 00 | 8,328 49 | $241 23 | $127 50 | $1,819 06 | 43,412 28 | | | | |
| February | 455.90 | 23,222 75 | 6,996 83 | 939 60 | | 2,075 44 | 33,234 62 | | | | |
| March | 304.30 | 19,411 75 | 3,140 39 | 48 00 | | 1,720 19 | 24,320 33 | | | | |
| April | 57.30 | 11,944 60 | 3,442 57 | 302 49 | 326 82 | 1,671 50 | 17,687 98 | | | | |
| May | 335. | 11,286 00 | 2,268 52 | 153 87 | | 1,419 34 | 15,127 73 | | | | |
| June | 4.1730 | 8,895 00 | 3,044 52 | 32 25 | | 745 58 | 12,717 35 | $200 00 | | | $200 00 |
| July | | 7,828 00 | 622 68 | 110 81 | | 2,734 26 | 11,295 75 | 3,148 00 | $2,575 17 | | 5,723 17 |
| August | | 10,295 25 | 1,458 49 | 249 28 | 1,046 50 | 1,818 08 | 14,867 60 | 9,166 50 | 15,786 95 | | 24,953 45 |
| September | 208.486 | 13,528 00 | 5,182 42 | 284 15 | | 1,998 83 | 20,993 40 | 11,140 50 | 19,434 18 | $7,771 26 | 38,345 94 |
| October | 231.100 | 12,752 00 | 1,928 94 | 114 15 | | 1,613 56 | 16,408 65 | 14,824 62 | 20,419 51 | 2,057 15 | 37,301 28 |
| November | 289.1450 | 17,250 00 | 3,798 08 | 25 25 | | 1,912 23 | 22,995 56 | 10,971 25 | 9,021 44 | 5,373 80 | 25,366 49 |
| December | 250.1160 | 14,794 75 | 5,123 88 | | | 1,672 80 | 21,791 43 | 9,805 25 | 20,032 07 | 18,201 24 | 48,038 56 |
| Balance due Nevada Central Railroad | | | | | | | | | | | |
| Total gross expenses | 3,436.546 | 184,104 10 | 57,537 51 | 2,511 08 | 1,500 82 | 21,400 87 | 266,429 38 | 59,256 11 | 87,269 32 | 33,403 45 | 179,928 89 |
| Deduct cash received | | 611 08 | 1,360 00 | | | | | | | | |
| Deduct stores on hand | | | 5,665 25 | | | | | | | | |
| Total deductions | | 611 08 | 7,025 25 | | | | | | | | |
| Add cash received | | | | | | | 7,636 33 | | | | |
| Total net expenses | | 183,493 02 | 50,512 26 | 2,511 08 | 1,500 82 | 21,400 87 | 258,793 05 | 59,256 11 | 87,269 32 | 33,403 45 | 179,928 89 |

*General statement showing net cost of mining, prospecting, improvements, milling, legal expense, taxes, &c.*—Continued.

| Months. | Mill cost. | | | | | | | | | | General expense. |
|---|---|---|---|---|---|---|---|---|---|---|---|
| | Tons of ore and tailings reduced. | Labor. | Supplies. | Freight. | Quicksilver. | Improvements and repairs. | Ore-hauling. | Salaries. | General. | Total. | Legal. |
| 1874. | | | | | | | | | | | |
| On hand January 11 | | | $59,931 67 | | $34,335 65 | | | | | $94,206 72 | |
| January | 4,304.740 | $9,349 25 | 14,333 05 | $2,239 76 | | $695 50 | $4,064 00 | $750 00 | $112 50 | 31,544 06 | $6,518 50 |
| February | 3,697.1600 | 8,572 48 | 8,457 36 | 17 78 | | 2,435 65 | 23,666 97 | 750 00 | 75 75 | 43,975 99 | 2,383 33 |
| March | 4,934.1600 | 8,964 00 | 8,646 21 | 12 15 | | 1,191 98 | 1,334 98 | 750 00 | 63 75 | 20,963 07 | 2,233 33 |
| April | 4,832.1600 | 8,331 25 | 8,759 78 | 1,547 04 | | | 1,429 33 | 650 00 | 112 75 | 20,830 15 | 2,433 34 |
| May | 4,531.1300 | 8,437 48 | 9,635 78 | 60 49 | 4,752 27 | | 1,340 00 | 650 00 | 100 84 | 24,976 86 | 2,383 34 |
| June | 4,682.800 | 6,476 50 | 9,927 62 | 130 93 | | 3,082 15 | 19 46 | 650 00 | 36 00 | 20,322 66 | 2,483 33 |
| July | 4,920.840 | 7,009 00 | 10,659 60 | 119 19 | | 1,353 47 | | 650 00 | 201 50 | 19,992 76 | 2,383 33 |
| August | 5,763.1200 | 7,023 17 | 4,423 54 | 160 92 | 4,829 60 | 38 00 | | 650 00 | 271 01 | 17,396 24 | 3,733 34 |
| September | 4,521.400 | 6,496 00 | 5,283 95 | | | | 909 78 | 650 00 | 790 65 | 14,130 38 | 2,683 34 |
| October | 4,749.1200 | 6,938 50 | 2,046 51 | 271 43 | | | 871 98 | 650 00 | 29 00 | 10,807 42 | 7,833 33 |
| November | 4,314.400 | 6,875 25 | 5,785 95 | | 5,548 75 | | | 650 00 | 175 75 | 19,035 70 | 3,183 33 |
| December | 4,468.206 | 6,732 75 | 3,570 47 | 1,581 78 | 5,740 00 | | | 650 00 | 27 25 | 18,302 25 | 900 00 |
| Balance due Nevada Central Railroad | | | | | | | 1,455 36 | | | 1,455 36 | |
| Total gross expenses | 52,720.1886 | 91,905 63 | 151,461 49 | 6,141 47 | 55,905 67 | 8,796 75 | 35,091 86 | 8,100 00 | 1,996 75 | 357,999 62 | 39,151 84 |
| Deduct cash received | | | 483 21 | | | 287 98 | | | | | 200 00 |
| Deduct stores on hand | | | 43,302 82 | | 15,229 60 | | | | | | |
| Total deductions | | | 43,785 03 | | 15,229 60 | 287 98 | 22,722 47 | | | 82,025 08 | 200 00 |
| Add cash received | | | | | | | | | | | |
| Total net expenses | | 91,905 63 | 107,676 46 | 6,141 47 | 38,976 07 | 8,508 77 | 12,369 39 | 8,100 00 | 1,996 75 | 275,974 54 | 38,951 84 |

*General statement showing net cost of mining, prospecting, improvements, milling, legal expense, taxes, &c.*—Continued.

| Months. | General expense. | | | | | | | Grand total. | Returns. | | |
|---|---|---|---|---|---|---|---|---|---|---|---|
| | Real estate and title account. | Taxes and insurance. | Bullion discount. | Bullion reclamation. | Interest and discount. | Miscellaneous. | Total. | | Bullion. | Amounts received from sundries. | Total. |
| **1874.** | | | | | | | | | | | |
| On hand January 11 | | | | | | | | $105,843 42 | | | |
| January | $22,163 00 | $2,600 00 | $2,365 00 | | $585 00 | $1,419 25 | $35,650 75 | 110,607 09 | $121,448 05 | | $121,448 95 |
| February | | | 3,910 69 | | 644 18 | 993 78 | 7,931 89 | 85,142 50 | 75,118 65 | $444 60 | 75,563 25 |
| March | 350 00 | 5,017 14 | 5,277 25 | | 642 20 | 901 50 | 14,421 42 | 59,704 82 | 84,572 51 | 155 00 | 84,727 51 |
| April | | | 2,923 91 | $224 31 | 679 36 | 859 90 | 7,120 82 | 45,638 95 | 90,322 80 | 432 23 | 90,755 03 |
| May | 1,135 00 | | 6,310 10 | 2,383 90 | 307 30 | 630 85 | 13,150 49 | 53,255 08 | 93,852 61 | 1,892 50 | 95,745 11 |
| June | | 1,741 83 | 1,898 07 | | 616 24 | 1,011 50 | 7,750 97 | 40,990 98 | 25,519 02 | 160 38 | 25,679 40 |
| July | | 1,607 25 | 637 80 | | | 797 50 | 4,825 88 | 41,837 56 | 23,213 16 | 210 46 | 23,423 62 |
| August | | | 7,016 67 | 2,221 98 | | 803 25 | 13,775 24 | 70,992 53 | 20,339 87 | 894 00 | 21,163 87 |
| September | 510 00 | 1,381 08 | 2,000 50 | | 209 46 | 816 60 | 7,540 93 | 81,010 65 | 26,426 74 | | 26,426 74 |
| October | 7,501 00 | | 1,717 73 | 860 24 | 307 36 | 1,165 58 | 19,385 94 | 83,902 59 | 30,598 54 | 762 11 | 31,360 65 |
| November | 100,150 00 | 5,999 69 | 3,400 13 | 408 46 | 305 47 | 981 65 | 114,428 73 | 181,826 48 | 50,715 87 | 21,113 07 | 71,828 94 |
| December | 12 75 | 944 92 | 3,296 51 | | 620 36 | 3,836 15 | 8,910 69 | 97,042 93 | 30,895 45 | 1,025 87 | 31,921 32 |
| Balance due Nevada Central Railroad | | | | | | | | 1,455 36 | | | |
| Total gross expenses | 131,821 75 | 17,931, 86 | 40,754 27 | 6,098 89 | 4,916 93 | 14,217 51 | 254,893 05 | 1,059,250 94 | 673,016 27 | 27,020 22 | 700,036 49 |
| Deduct cash received | 960 00 | | | | 20,578 07 | 963 16 | 22,701 23 | | | | |
| Deduct stores on hand | | | | | | | | | | | |
| Total deductions | 960 00 | | | | 20,578 07 | 963 16 | 22,701 23 | 112,362 64 | | | |
| Add cash received | | | | | | | | | *740 84 | | |
| Total net expenses | 130,861 75 | 17,931 86 | 40,754 27 | 6,098 89 | | 13,254 35 | 232,191 82 | 946,888 30 | 673,757 11 | | |

*One small bar sold, $35; ore sold, $705.84; total, $740.84.

*Statement of bullion produced for account of the company, for year ending December 31, 1874.*

| Months. | Number of bars. | | Weight, troy ounces. | Fineness. | | Value per oz. | Value. | | |
|---|---|---|---|---|---|---|---|---|---|
| | From— | To— | | Gold, 1,000. | Silver, 1,000. | | Gold. | Silver. | Total. |
| 1874. | | | | | | | | | |
| January | 4,259 | 4,337 | 105,548.50 | 001.33 | 869 | $1 15.06 | $2,905 81 | $118,542 24 | $121,483 05 |
| February | 4,333 | 4,304 | 76,078.50 | 001.51 | 740 | 98.74 | 2,381 21 | 72,737 44 | 75,118 65 |
| March | 4,365 | 4,453 | 80,486.00 | 000.88 | 798 | 1 05.08 | 1,467 31 | 83,105 20 | 84,572 51 |
| April | 4,454 | 4,509 | 74,616.50 | 001. | 920 | 1 24.05 | 1,537 77 | 88,785 03 | 90,322 80 |
| May | 4,510 | 4,573 | 92,474.00 | 002.01 | 753 | 1 01.49 | 3,838 11 | 90,014 50 | 93,852 61 |
| June | 4,574 | 4,587 | 22,164.00 | 001.7 | 859 | 1 14.65 | 783 02 | 24,729 00 | 25,512 02 |
| July | 4,588 | 4,650 | 105,510.00 | 000.3 | 164 | 21.88 | 785 57 | 22,427 59 | 23,213 16 |
| August | 4,651 | 4,665 | 24,035.00 | 001. | 637 | 84.42 | 549 05 | 19,790 82 | 20,339 87 |
| September | 4,666 | 4,681 | 25,938.00 | 000.7 | 773 | 1 01.62 | 421 13 | 26,005 61 | 26,426 74 |
| October | 4,682 | 4,707 | 40,476.00 | 000.6 | 574 | 79.59 | 533 10 | 30,065 44 | 30,598 54 |
| November | 4,708 | 4,739 | 51,469.50 | 001.6 | 735 | 98.53 | 1,757 74 | 48,958 13 | 50,715 87 |
| December | 4,740 | 4,759 | 34,694.00 | 001.6 | 662 | 89.05 | 1,203 81 | 29,691 64 | 30,895 45 |
| | | | 733,490.00 | 001.2 | 709 | 94.22 | 18,163 63 | 654,852 64 | 673,051 27 |

| Months. | Ore extracted. | | Cost. | | | | | | | Ore-summary. | | | | | |
|---|---|---|---|---|---|---|---|---|---|---|---|---|---|---|---|
| | | | Extraction. | | Prospecting and dead-work. | | Repairs and improvements. | General. | Total. | Ore shipped. | | From what source. | | | |
| | | | | | | | | | | | | Panaca mine. | | Burke and Creole mines. | |
| | Tons. | Lbs. | Labor. | Materials. | Labor. | Materials. | | | | Tons. | Lbs. | Tons. | Lbs. | Tons. | Lbs. |
| 1874. | | | | | | | | | | | | | | | |
| January | 1,119 | 1,990 | $18,475 50 | $4,170 37 | $14,385 50 | $3,899 25 | $127 50 | $1,786 16 | $42,844 28 | 1,119 | 1,990 | 1,119 | 1,900 | ........ | ........ |
| February | 455 | 90 | 10,071 00 | 3,630 00 | 13,051 75 | 3,105 60 | ........ | 1,879 89 | 31,738 24 | 455 | 90 | 455 | 90 | ........ | ........ |
| March | 304 | 30 | 5,960 00 | 1,763 00 | 13,451 75 | 2,210 39 | ........ | 1,668 11 | 25,053 25 | 304 | 30 | 304 | 30 | ........ | ........ |
| April | 57 | 30 | 4,206 00 | 1,875 00 | 7,738 60 | 3,171 88 | ........ | 1,288 95 | 18,280 43 | 57 | 30 | 57 | 30 | ........ | ........ |
| May | 335 | ........ | 2,475 00 | 850 00 | 8,811 00 | 1,496 39 | ........ | 1,371 19 | 15,003 58 | 335 | ........ | 327 | 1,315 | 7 | 685 |
| June | 4 | 1,730 | ........ | ........ | 8,895 00 | 3,937 77 | ........ | 693 53 | 13,526 30 | 4 | 1,730 | 4 | 1,730 | ........ | ........ |
| July | ........ | ........ | ........ | ........ | 7,828 00 | 2,560 29 | ........ | 1,191 16 | 11,579 45 | ........ | ........ | ........ | ........ | ........ | ........ |
| August | ........ | ........ | ........ | ........ | 10,295 25 | 2,661 89 | 1,046 50 | 1,525 33 | 15,528 97 | ........ | ........ | ........ | ........ | ........ | ........ |
| September | 208 | 406 | 3,815 00 | 1,800 00 | 9,713 00 | 5,405 70 | ........ | 1,997 33 | 22,731 03 | 208 | 406 | 30 | 120 | 178 | 286 |
| October | 231 | 100 | 3,700 00 | 1,250 00 | 9,052 00 | 1,314 09 | ........ | 1,613 56 | 16,929 65 | 231 | 100 | ........ | ........ | 231 | 100 |
| November | 289 | 1,450 | 4,000 00 | 1,500 00 | 13,250 00 | 4,211 08 | ........ | 1,912 23 | 24,873 31 | 289 | 1,450 | 170 | 1,250 | 119 | 100 |
| December | 250 | 1,160 | 4,000 00 | 1,600 00 | 10,794 75 | 3,523 88 | ........ | 1,872 80 | 21,791 43 | 80 | 1,160 | 47 | 1,130 | 33 | 30 |
| | 3,255 | 986 | 56,702 50 | 18,438 37 | 127,266 60 | 37,498 21 | 1,174 00 | 18,800 24 | 259,879 92 | 3,085 | 986 | 2,516 | 1,785 | 568 | 1,201 |
| Average per ton ore extracted | | | 17 40 | 5 66 | 39 09 | 11 51 | 36 | 5 77 | 79 79 | ........ | ........ | ........ | ........ | ........ | ........ |

NOTE.—170 tons ore in dump at mine.

*Statement showing the work of the company's mills at Bullionville for the year ending December 31, 1874.*

| Months. | Mill. | Ore. | | | | Tailings reduced. | | Time. | | | | | | | | | Total ore and tailings reduced. | | Average pulp-assay per ton. | | Gross assay-value. | | |
|---|---|---|---|---|---|---|---|---|---|---|---|---|---|---|---|---|---|---|---|---|---|---|---|
| | | Received at mills. | | Reduced. | | | | Hours charges were worked. | Pans run. | | Pans stopped. | | Stamps run. | | Stamps stopped. | | | | | | | | |
| | | Tons. | Lbs. | Tons. | Lbs. | Tons. | Lbs. | | Days. | Hours. | Days. | Hours. | Days. | Hours. | Days. | Hours. | Tons. | Lbs. | Ore. | Tailings. | Ore. | Tailings. | Total. |
| 1874. On hand | | 180 | 1,560 | | | | | | | | | | | | | | | | | | | | |
| January | 1 | | | | | 1,153 | 1,200 | 6 | 30 | 5 | | 19 | | | 31 | | 1,153 | 1,200 | | $65 49 | | $75,532 27 | $75,532 27 |
| | 2 | 1,119 | 1,990 | 1,076 | 1,540 | 1,074 | | 6 | 29 | 6½ | 1 | 17½ | 16 | 21 | 14 | 3 | 2,150 | 1,540 | $56 45 | 65 41 | $60,784 22 | 70,249 38 | 131,033 60 |
| February | 1 | | | | | 1,300 | 1,600 | 5 | 26 | 20 | 1 | 4 | | | 28 | | 1,300 | 1,600 | | 48 99 | | 63,726 19 | 63,726 19 |
| | 2 | 455 | 90 | 499 | | 1,898 | | 5 | 26 | 19½ | 1 | 4½ | 9 | 10 | 18 | 14 | 2,397 | | 48 75 | 45 25 | 24,310 25 | 85,834 50 | 110,150 75 |
| March | 1 | | | | | 1,724 | 1,600 | 4 | 30 | | 1 | | | | 31 | | 1,724 | 1,600 | | 39 08 | | 67,405 18 | 67,405 18 |
| | 2 | 304 | 30 | | | 3,210 | | 4 | 29 | 18 | 1 | 6 | | | 31 | | 3,210 | | | 39 08 | | 125,446 80 | 125,446 80 |
| April | 1 | | | | | 1,672 | 1,600 | 4 | 29 | 1½ | | 22½ | | | 30 | | 1,672 | 1,600 | | 43 07 | | 72,949 49 | 72,949 49 |
| | 2 | 57 | 30 | | | 3,160 | | 4 | 29 | 6 | | 18 | | | 30 | | 3,160 | | | 40 03 | | 126,488 19 | 126,488 91 |
| May | 1 | | | | | 1,716 | 1,300 | 4 | 29 | 17½ | 1 | 6¼ | | | 31 | | 1,716 | 1,300 | | 21 81 | | 51,178 63 | 51,178 63 |
| | 2 | 335 | | 894 | | 1,921 | | 5 | 30 | 5½ | | 18¾ | 11 | | 20 | | 2,815 | | 108 11 | 29 46 | 96,650 34 | 56,509 23 | 153,159 57 |
| June | 1 | | | | | 1,658 | 800 | 4 | 28 | 19 | 1 | 5 | | | 30 | | 1,658 | 800 | | 25 29 | | 41,941 92 | 41,941 92 |
| | 2 | 4 | 1,730 | | | 3,024 | | 4 | 28 | 4 | 1 | 20 | | | 30 | | 3,024 | | | 25 33 | | 76,610 45 | 76,610 45 |
| July | 1 | | | | | 1,349 | 840 | 6 | 29 | 17 | 1 | 7 | | | 31 | | 1,349 | 840 | | 27 38 | | 36,958 26 | 36,958 26 |
| | 2 | | | | | 2,571 | | 6 | 29 | 12 | 1 | 12 | | | 31 | | 2,571 | | | 24 13 | | 62,038 69 | 62,038 60 |
| August | 1 | | | | | 1,669 | 1,200 | 4 | 29 | 22 | 1 | 2 | | | 31 | | 1,669 | 1,200 | | 25 82 | | 43,122 41 | 43,122 41 |
| | 2 | | | | | 3,094 | | 4 | 29 | 17 | 1 | 7 | | | 31 | | 3,094 | | | 22 02 | | 68,221 80 | 68,221 80 |
| September | 1 | | | | | 1,571 | 400 | 4 | 27 | 7½ | 2 | 16½ | | | 30 | | 1,571 | 400 | | 28 76 | | 45,192 13 | 45,192 13 |
| | 2 | 208 | 406 | | | 2,950 | | 4 | 27 | 8 | 2 | 16 | | | 31 | | 2,950 | | | 24 38 | | 71,923 08 | 71,923 08 |
| October | 1 | | | | | 1,489 | 1,200 | 4 | 24 | 21 | 6 | 3 | | | 31 | | 1,489 | 1,200 | | 31 02 | | 46,219 60 | 46,219 60 |
| | 2 | 231 | 100 | | | 3,260 | | 4 | 30 | 5 | | 19 | | | 31 | | 3,260 | | | 25 76 | | 83,907 72 | 83,907 72 |
| November | 1 | | | | | 1,707 | 400 | 4 | 29 | 16 | | 8 | | | 30 | | 1,707 | 400 | | 25 61 | | 43,728 79 | 43,728 79 |
| | 2 | 289 | 1,450 | 637 | | 1,970 | | 5 | 27 | 12 | 2 | 12 | 9 | 18 | 20 | 6 | 2,607 | | 90 68 | 23 20 | 57,763 25 | 45,717 76 | 103,481 01 |
| December | 1 | | | | | 1,673 | 1,200 | 4 | 29 | 0½ | 1 | 23½ | | | 31 | | 1,673 | 1,200 | | 21 51 | | 36,004 21 | 36,004 21 |
| | 2 | 80 | 1,160 | 159 | 1,006 | 2,365 | | 4 | 26 | 17 | 4 | 7 | 1 | 16 | 29 | 8 | 2,794 | 1,006 | 79 24 | 22 37 | 12,630 11 | 61,046 45 | 73,685 56 |
| | | 3,966 | 546 | 3,966 | 546 | 49,454 | 1,340 | .... | 689 | 13¼ | 40 | 10½ | 47 | 17 | 682 | 7 | 52,720 | 1,886 | 77 29 | 31 50 | 252,153 17 | 1,557,953 85 | 1,810,107 02 |

*Statement showing the work of the company's mills at Bullionville, &c.*—Continued.

| Months. | Mill. | Bullion extracted. | | | Average percentage extracted. | | | Disbursements for account of mills. | | | | | | Yield per ton. | | |
|---|---|---|---|---|---|---|---|---|---|---|---|---|---|---|---|---|
| | | Ore. | Tailings. | Total. | Ore. | Tailings. | Total. | Hauling. | Labor. | Materials. | Imps. and repairs. | Sundries. | Total. | Ore. | Tailings. | Total. |
| 1874. On hand | | | | | | | | | | $94,266 72 | | | $94,266 72 | | | |
| January | 1<br>2 | $36,151 44 | $85,296 61 | $121,448 05 | 59.5 | 57.9 | 58.7 | $4,064 00 | $9,349 25 | 16,572 81 | $695 50 | $789 35 | 31,470 91 | $33 57 | $37 20 | $36 75 |
| February | 1<br>2 | 13,757 28 | 61,361 37 | 75,118 65 | 56.6 | 41.1 | 43.1 | 23,666 97 | 8,572 48 | 8,475 14 | 2,435 65 | 1,237 02 | 44,387 26 | 27 57 | 19 18 | 20 31 |
| March | 1<br>2 | | 84,572 51 | 84,572 51 | | 43.9 | 43.9 | 1,334 98 | 8,964 00 | 8,658 36 | 1,191 98 | 271 00 | 20,420 32 | | 17 14 | 17 14 |
| April | 1<br>2 | | 90,382 80 | 90,382 80 | | 45.3 | 45.3 | 1,429 33 | 8,331 25 | 10,306 82 | | 684 00 | 20,751 40 | | 18 69 | 18 69 |
| May | 1<br>2 | 60,037 86 | 33,814 75 | 93,852 61 | 62.1 | 31.4 | 45.9 | 1,340 00 | 8,437 48 | 14,448 54 | | 699 84 | 24,925 86 | 67 16 | 9 27 | 20 71 |
| June | 1<br>2 | | 25,512 02 | 25,512 02 | | 21.5 | 21.5 | 19 46 | 6,476 50 | 10,058 55 | 3,062 15 | 677 00 | 20,313 66 | | 5 45 | 5 45 |
| July | 1<br>2 | | 23,913 16 | 23,913 16 | | 23.5 | 23.5 | | 7,009 00 | 10,778 79 | 1,353 47 | 836 50 | 19,977 76 | | 5 92 | 5 92 |
| August | 1<br>2 | | 20,339 87 | 20,339 87 | | 18.3 | 18.3 | | 7,023 17 | 9,414 06 | 38 00 | 701 88 | 17,177 11 | | 4 28 | 4 28 |
| September | 1<br>2 | | 26,426 74 | 26,426 74 | | 22.6 | 22.6 | 909 78 | 6,496 00 | 5,283 95 | | 992 50 | 13,682 23 | | 5 85 | 5 85 |
| October | 1<br>2 | | 30,598 54 | 30,598 54 | | 23.5 | 23.5 | 871 98 | 6,938 50 | 2,317 94 | | 689 00 | 10,817 42 | | 6 44 | 6 44 |
| November | 1<br>2 | 37,884 37 | 12,831 50 | 50,715 87 | 65.6 | 14.4 | 34.4 | | 6,875 25 | 11,334 70 | | 825 75 | 19,035 70 | 59 47 | 3 49 | 13 25 |
| December | 1<br>2 | 9,273 17 | 16,300 50 | 25,573 67 | 73.4 | 16.7 | 23.3 | | 6,732 75 | 10,892 25 | | 677 25 | 18,302 25 | 58 13 | 3 76 | 5 72 |
| | | 157,104 12 | 510,590 37 | 667,694 49 | 62.3 | 32.7 | 36.8 | 33,636 50 | 91,205 63 | 212,808 63 | 8,796 75 | 9,081 09 | 355,528 60 | | | |
| Add bullion from slag and one small bar sold. | | 5,356 78 | | 5,356 78 | | | | | | | | | | | | |
| Add amount due N. C. R. R., ore-hauling, Nov. and Dec., 1874. | | | | | | | | 1,455 36 | | | | | 1,455 36 | | | |
| | | 162,460 90 | 510,590 37 | 673,051 27 | | | | 35,091 86 | 91,205 63 | 212,808 63 | 8,796 75 | 9,081 09 | 356,983 96 | | | |
| Less amount collected and stock on hand. | | | | | | | | | | 59,015 63 | 287 98 | | 59,303 61 | | | |
| Less amount paid N. C. R. R. Co. on account ore-hauling, 1873. | | | | | | | | 22,722 47 | | | | | 22,722 47 | | | |
| Net amounts for the year 1874 | | 162,460 90 | 510,590 37 | 673,051 27 | | | | 12,369 39 | 91,205 63 | 153,793 00 | 8,508 71 | 9,081 09 | 2[illegible]4,947 08 | | | |
| Avg. cost per ton of ore-hauling and ore and tailings worked. | | | | | | | | 4 00 | 1 73 | 2 93 | 15 | 17 | 8 98 | Tailings $4.28, average cost working. | | |

I add the following from the secretary's report for the year ending December 31, 1874:

RECEIPTS.

*From balances of former fiscal years, (liquidated to new current balances per contra:)*

| | | | |
|---|---|---|---|
| Nevada Central Railroad Company— | | | |
| Fiscal year ending December 31, 1873 | $100,000 00 | | |
| Fiscal year ending January 15, 1874 | 139,522 00 | | |
| | | $239,522 00 | |
| Magnet Mining Company | | 140,439 67 | |
| Hermes Mining Company | | 3,064 80 | |
| Pioche Phœnix Mining Company | | 3,180 00 | |
| H. H. Day, general superintendent | | 9 00 | |
| | | | $386,215,47 |
| *From bullion-yield of company's mine:* | | | |
| Product of ores and tailings for current year, as per Exhibit No. 2 | | | 673,051 27 |
| From ore sales | | | 705 84 |
| From water-rents | | | 600 00 |
| From house-rents | | | 360 00 |
| From interest | | | 12,264 13 |
| From sales of mine supplies | | | 1,360 00 |
| From sales of mill supplies | | | 483 21 |
| *From assessments:* | | | |
| Assessment No. 1, of $3 per share | | $90,000 00 | |
| Assessment No. 2, of $3 per share, (paid in at date) | | 83,116 00 | |
| | | | 173,116 00 |
| Total receipts for the term | | | 1,252,155 92 |

UNLIQUIDATED BALANCES.

| | | |
|---|---|---|
| *Superintendent's drafts:* | | |
| On San Francisco office, unpresented | $56,153 51 | |
| *Union Iron Works:* | | |
| Balance on foundery-work | 19,902 81 | |
| *Bank of California:* | | |
| For advances | 49,226 96 | |
| | | 125,283 28 |
| | | 1,377,439 20 |

DISBURSEMENTS.

| | | |
|---|---|---|
| *For balance of last fiscal report, since liquidated:* | | |
| Superintendent's drafts | $13,225 25 | |
| Overdraft on Bank of California | 89,277 84 | |
| | | $102,503 09 |

| | | |
|---|---|---|
| *For mine-properties:* | | |
| Purchase of Magnet and other mining-ground | $131,111 59 | |
| Law expenses, contesting adverse claims | 41,451 84 | |
| | | $172,563 43 |
| *For construction and improvements at mills:* | | |
| Cost of works at Bullionville for current year | | 4,785 25 |
| *For mining:* | | |
| Opening mines, explorations, and extracting ores, as per Exhibit No. 2 | | 253,817 21 |
| *For milling:* | | |
| Reduction of ores at company's mills, as per Exhibit No. 2 | | 256,205 78 |
| *For miscellaneous expenditures:* | | |
| Discount on bullion for current year | | 47,483 87 |
| State of Nevada taxes on bullion | | 8,324 92 |
| State of Nevada and California taxes on property | | 5,006 94 |
| Insurance-premium on mill-property | | 4,600 00 |
| Telegrams to and from company's works | | 420 64 |
| Freight to San Francisco on base-bullion yield | | 1,000 78 |
| General office-expenses | | 10,190 35 |
| *For mine-pump construction:* | | |
| Pumping-machinery, freight to and local works at Pioche | | 199,831 70 |
| Total disbursements for the term | | 1,066,733 96 |

UNLIQUIDATED BALANCES.

| | | |
|---|---|---|
| Nevada Central Railroad Company | $220,863 53 | |
| Magnet Mining Company | 61,079 37 | |
| Hermes Mining Company | 3,064 80 | |
| Pioche Phœnix Mining Company | 3,501 52 | |
| Bullion *in transitu* | 22,196 02 | |
| | | 310,705 24 |
| | | 1,377,439 02 |

***Statement of the gross proceeds of bullion from the mines of the Raymond & Ely Mining Company, and cost of production and reduction of ores yielding the bullion, for the fiscal year ending December* 31, 1874.**

| DR. | | | |
|---|---|---|---|
| **TO MINING DEPARTMENT, VIZ:** | | | |
| Labor in extracting ores | | $183, 493 02 | |
| Mining-supplies | | 45, 335 81 | |
| Freight from San Francisco on supplies | | 2, 511 08 | |
| Surveying | | 810 00 | |
| Ore-assays | | 386 00 | |
| Repairs | | 1, 500 82 | |
| Expenses: | | | |
| General | $4, 640 59 | | |
| Contingent | 1, 664 93 | | |
| Charity | 350 00 | | |
| | | 6, 655 52 | |
| Mine-salaries | | 13, 194 96 | |
| Total expenditures in extraction | | 253, 817 21 | |
| Deduct inventory of supplies on hand at date | | 5, 665 25 | |
| | | | $248, 151 96 |
| **TO MILLING DEPARTMENT, VIZ:** | | | |
| Ore-transportation from mine to mills, (11 miles) | | 33, 636 50 | |
| Chemicals, tools, and other supplies | | 97, 269 82 | |
| Quicksilver | | 15, 130 62 | |
| Freight from San Francisco on supplies | | 6, 141 47 | |
| Milling labor | | 91, 205 63 | |
| Repairs | | 3, 723 52 | |
| General ($670.97) and contingent ($327.25) expenses | | 998 22 | |
| Mill-salaries | | 8, 100 00 | |
| Total expenditures in reduction | | 256, 205 78 | |
| Deduct inventory of supplies on hand at date | | 58, 532 42 | |
| | | | 197, 673 36 |
| **TO MISCELLANEOUS ACCOUNTS, VIZ:** | | | |
| Discount on bullion for current year | | 47, 483 87 | |
| State of Nevada taxes on bullion | | 8, 324 92 | |
| States of Nevada and California property-taxes | | 5, 006 94 | |
| Insurance-premiums on mill-property | | 4, 600 00 | |
| Telegrams to and from company's works | | 420 64 | |
| Freight to San Francisco on base-bullion yield | | 1, 000 78 | |
| General office-expenses | | 10, 190 35 | |
| | | | 77, 027 50 |
| Total cost of production and reduction, including extraordinary expenses in explorations and dead-work | | | 522, 852 82 |
| Net profit over all expenditures for current year | | | 169, 971 63 |
| | | | 692, 824 45 |

| CR. | | |
|---|---|---|
| **BULLION.** | | |
| Proceeds of company's reduction-works at Bullionville, as per tabular statements of general superintendent and office-records for the fiscal year ending this date | | $673, 051 27 |
| **MISCELLANEOUS RETURNS.** | | |
| Ore-sales | $705 84 | |
| Water-rents | 600 00 | |
| House-rents | 360 00 | |
| Interest | 16, 264 13 | |
| Sales of mine-supplies | 1, 360 00 | |
| Sales of mill-supplies | 483 21 | |
| | | 19, 773 18 |
| | | 692, 824 45 |

## SUMMARY.

*The Raymond & Ely Mining Company in account with its estate, for the full corporate term ending December 31, 1874.*

**Dr.**

| | | | |
|---|---|---|---|
| PERMANENT INVESTMENTS. | | | |
| *In mine-properties—* | | | |
| For the fiscal term 1871-'73 | $798, 985 44 | | |
| For the fiscal year, 1874 | 172, 563 43 | | |
| | | $971, 548 87 | |
| *In construction of mining-works—* | | | |
| For the fiscal term 1871-'73 | | 102, 715 12 | |
| *In construction of reduction-works—* | | | |
| For the fiscal term 1871-'73 | $165, 341 62 | | |
| For the fiscal year 1874 | 4, 785 25 | | |
| | | 170, 126 87 | |
| *In mine-pump construction—* | | | |
| For the fiscal year 1874 | | 199, 831 70 | |
| Total permanent investments | | | $1, 444, 222 56 |
| DIVIDENDS. | | | |
| Paid stockholders, the fiscal year 1871 | | 615, 000 00 | |
| Paid stockholders, the fiscal year 1872 | | 2, 070, 000 00 | |
| Paid stockholders, the fiscal year 1873 | | 390, 000 00 | |
| | | | 3, 075, 000 00 |
| Total | | | 4, 519, 222 56 |
| CURRENT RESOURCES. (EXHIBIT NO. 1.) | | | |
| Nevada Central Railroad Company | | 220, 863 53 | |
| Magnet Mining Company | | 61, 079 37 | |
| Hermes Mining Company | | 3, 064 80 | |
| Pioche Phœnix Mining Company | | 3, 501 52 | |
| Inventory of supplies at mine | | 5, 665 25 | |
| Inventory of supplies at mill | | 58, 532 42 | |
| Bullion in transitu | | 22, 196 02 | |
| | | | 374, 902 91 |
| | | | 4, 894, 125 47 |

**Cr.**

| | | | |
|---|---|---|---|
| *By bullion-product—* | | | |
| For the fiscal year 1871 | $1, 361, 626 78 | | |
| For the fiscal year 1872 | 3, 693, 936 33 | | |
| For the fiscal year 1873 | 2, 339, 364 40 | | |
| For the fiscal year 1874 | 673, 051 27 | | |
| | | $8, 067, 980 78 | |
| BULLION-COST SHEETS DEDUCTED. | | | |
| For the fiscal term, 1871-'73— | | | |
| Production and reduction | $2, 969, 702 20 | | |
| For the fiscal year 1874— | | | |
| Production and reduction | 522, 852 82 | | |
| Total of cost-sheets | 3, 492, 555 02 | | |
| Less re-imbursed by sales of ores, mill and mine supplies— | | | |
| For the fiscal term 1871-'73, (ores) $527 25 | | | |
| For the fiscal year 1874, (sundr's) 19,773 18 | | | |
| | 20, 300 43 | | |
| | | 3, 472, 254 59 | |
| Total net profits on bullion-yield | | | $4, 595, 726 19 |
| *By assessments—* | | | |
| No. 1, of $3 per share | | 90, 000 00 | |
| Account of No. 2, of $3 per share, (paid to date) | | 83, 116 00 | |
| | | | 173, 116 00 |
| Total | | | 4, 768, 842 19 |
| CURRENT LIABILITIES. (EXHIBIT NO. 1.) | | | |
| Superintendent's drafts unpresented | | 56, 153 51 | |
| Union Iron-Works | | 19, 902 81 | |
| Bank of California | | 49, 226 96 | |
| | | | 125, 283 28 |
| | | | 4, 894, 125 47 |

ASSETS.

Mines:

| | |
|---|---|
| *Panaca mine.* | |
| 2 engines and 4 boilers, complete in place | $30,000 00 |
| Buildings, including machine-shop, foundery, blacksmith-shop, carpenter-shop, ore-house, coal-house, office at mine, powder-magazine, water-tanks, &c | 34,632 00 |
| Pump-improvement | 169,296 89 |
| Sundries in use | 9,850 00 |
| 2 horses, buggy, and harness | 1,200 00 |
| *Burke mine.* | |
| Whim-house and appurtenances | 500 00 |
| Office and furniture | 3,500 00 |
| *Creole mine.* | |
| Whim and whim-house | 500 00 |
| *Panaca Flat.* | |
| 2 dwelling-houses | 750 00 |
| *Pioche City.* | |
| 1 dwelling-house, lot, and furniture | 750 00 |
| | 250,978 89 |
| Supplies in store-house | 5,665 25 |
| | 256,644 14 |

Mills and miscellaneous property:

| | |
|---|---|
| One 30-stamp mill | $75,000 00 |
| One 20-stamp mill | 40,000 00 |
| Water-ditch, pipe, and privilege | 10,000 00 |
| Office and furniture at mill No. 1 | 1,500 00 |
| Office and store-house at mill No. 2 | 600 00 |
| Assay-office and fixtures, complete | 1,500 00 |
| Blacksmith-shop and tools | 2,000 00 |
| Carpenter-shop and tools | 1,000 00 |
| One dwelling-house | 250 00 |
| One steam-pump house, pipes, &c | 1,500 00 |
| One stable, two horses, and one wagon | 900 00 |
| One Fairbanks large ore-scales | 1,000 00 |
| | 135,250 00 |
| Supplies in store-house | 58,532 42 |
| | 193,782 42 |

# CHAPTER III.

## IDAHO.

Notes on the mining-industry of this Territory have again been collected, at my request, by Mr. A. Wolters, superintendent of the United States assay-office at Boise City.

Owing to the absence of the usual heavy fall rains during the previous year, the supply of water was very limited in nearly all the placer-mining camps during the spring and summer of 1874, and the annual yield of the placers shows a corresponding decrease as compared with that of last year. On the other hand, quartz-mining has been carried on with greater vigor and better success than formerly, and the bullion-product from this source shows a decided increase, though not sufficient to counterbalance the falling-off in the yield of the placer-mines.

The whole production of precious metals in the Territory for the current year, according to the most reliable statistics I have been able to obtain, is $2,000,000; of which amount Owyhee County furnished $900,000; Boise County, $700,000; Alturas County, $150,000; North Idaho and Lemhi County, $250,000.

Several new mining-camps have been established during this and the previous year; but, possessing only veins of galena and copper ore, with no great amount of gold and silver, and lacking facilities for transportation, they will probably not become of great importance until railroads traverse the country and enhance the value of base metals.

## OWYHEE COUNTY.

*Owyhee district.*—For the first time in seven years, I have unfortunately failed to obtain a detailed statement in regard to the annual progress of the mining-industry of this district. All I can say concerning it is, that the various silver-mining companies on War Eagle Mountain have not been very successful, as is shown by the largely-diminished product. But in most of the mines heretofore worked with various degrees of success, energetic prospecting operations have been carried on, with rather favorable results at the end of the year. In addition to these mines, some of those which have been lying idle for years, as for instance the Poor Man, have been opened again, and it is expected that some of them will add considerably to the product of next year. On the whole, the outlook for 1875 appears encouraging.

*South Mountain district.*—This promising camp, which was described in detail in my report for 1872, had been lying idle for nearly two years, when, in the fall of 1874, a San Francisco company succeeded in buying up the only important silver-lead-mines then known, and thus virtually brought the whole camp under its control. This company, which is called the South Mountain Consolidated Mining Company, has since continued the shafts and levels previously constructed, and has succeeded in opening up several very considerable bodies of ore containing a fair percentage of lead and being comparatively very rich in silver. The mines which have been most extensively worked are the Golconda, Bay State, Yreka, Independent, and Numkeg. But some work has also been done on the Grant, the Mount Hood, and the Larry. From the main working-tunnel of the Golconda mine a cross-cut was intended to be run in the fall toward the large ledge known as the Original, in order to explore this mass of iron-stained vein-matter at a depth of some 200 feet from the surface. In the Golconda itself a large body of somewhat siliceous and antimonial lead-ore is in sight. The Bay State is, however, as it was when the shaft was only 40 feet deep, the richest mine of the whole district. This shaft is now 135 feet deep, and levels have been run along the ledge on each side of the shaft for over 200 feet. Rich, though not very large, bodies of carbonate and galena, the latter assaying from $300 to $400 per ton, have been struck in various parts of the mine, the largest, 8 feet wide, having been found lately. The Yreka contains more ore than the Bay State. It is, like that of the Bay State, an excellent smelting-ore, but contains less silver.

The company has also built a large shaft-furnace, making one or two campaigns during the fall; but as an insufficient amount of charcoal had been provided to run through the winter, smelting had to be stopped when the heavy snows set in. The company expects to put up a second and larger furnace in the spring, and also dressing-works, to make available a very large portion of ore which cannot now be smelted profitably, on account of the abundance of gangue-matter which it contains.

## ADA COUNTY.

Some lodes were discovered this season in the vicinity of the Upper Weiser, which carry strong bodies of fine-grained galena, containing from 60 to 75 per cent. of lead and about $20 in gold and silver. A strong outcrop of very fine copper-ore has also been discovered in that neighborhood. Pieces of this ore assayed in the United States assay-office contained 55.2 per cent. of copper and about $28 in gold and silver. The ore is an intimate mixture of chalcosine and bornite, partly converted into carbonates by exposure to the atmosphere and moisture, and a small percentage of pistazite and quartz.

A very fine variety of lignite has been found in different places near Boise City as "float," but so far no regular deposit has been discovered, though there cannot be any doubt as to its existence. The coal, in its external appearance, much resembles cannel coal. It is hard, carries no iron-pyrites, leaves no stain when rubbed, burns with a long, yellowish-white flame, and leaves only from 3 to 4 per cent. of ash. The only drawback to its general use lies in the fact that it crumbles into very small pieces upon long exposure to atmospheric influences. As wood must be hauled a distance of eight to twelve miles over a high mountain-range to Boise City, and accordingly commands high prices, the discoverers of a good-sized vein of this coal would probably find it valuable property, even if it could be used only for domestic purposes.

The only placer-mining done in the county amounting to anything is on the Payette River. Some of the claims, especially those of B. L. Warriner, produce very high-grade gold, the best having assayed as high as .942 fine. The yield is between $5,000 and $10,000. A very small amount of gold, probably less than $1,000, is obtained every spring in Fort Gulch, immediately back of Fort Boise.

## ALTURAS COUNTY.

The principal placer-mines of the county are those owned by Pfeiffer & Co., in Hardscrabble district, and those on the Middle Boise River. The former comprise a large area of placer-ground in the different gulches between Bear and Elk Creeks, with a monopoly of the water from both. The owners have built, at a very small cost, five miles of ditches, capable of carrying between 500 and 600 inches of water, the ground being very favorable for such work. They have been working the placers now for quite a number of years, and always with good financial results, the lowest annual yield ever obtained being $8,000, at an expense of $4,000, while this year they cleaned up over $20,000, at an expense of about $5,000. In working their claims the owners have been so fortunate as to uncover three very promising quartz-lodes, the General Grant, General Sherman, and Poor Man, all of which appear to be valuable veins, well defined, and carrying a strong body of rich ore as far as developed. The surrounding country probably received the greater portion of its placer-gold from these veins.

The placers of the Middle Boise River suffered somewhat from the general scarcity of water and shortness of the season.

In the quartz-mines a good deal of activity prevailed during the entire year, and several of them proved to be first-class veins.

On the Ada Ellmore, work has been prosecuted steadily all the year, with most flattering results. The shaft at the time of Mr. Wolters's visit was down 150 feet, and it was the intention to start another pair of levels. The ground above the 100-foot level was mostly worked out, the east level being still in good ore, while in the west level the ore had pinched out, the indications, however, being favorable to its soon coming in again. The shaft being located right in the gulch in which the lode outcrops, following the gulch in strike, there is of course a very large amount of water to be hoisted, and the former engine being inadequate to perform all the work required, a larger one had to be put up. This was not done as well as it ought to have been, the foundation consisting of only four pieces of timber embedded in loose rock.

The Pittsburgh Company, which runs this claim, is not doing as well as the excellence of the mine would permit, if capital were judiciously expended in thorough opening and substantial plant.

To facilitate and cheapen the transportation of the ore from the mine to the mill, a covered tram-way 300 yards in length has been constructed, connecting the two, enabling the bucket-tender to deliver the ore right at the stamps. The car used was constructed by A. Cramer, foreman of the Valley Gold-Mining Company, and is a very excellent one, as far as easy dumping and running are concerned, a child being perfectly able to handle it.

That portion of the lode lying east of the Pittsburgh Company's ground, owned by the Wahl Brothers, has not been worked until it was leased last spring, by Jacob Reeser, for a term of two years. At the time of Mr. Wolters's visit he had sunk a shaft 50 feet deep, and made preparations to put up an engine for driving the hoisting-works and pump, which were to be erected and put in at the same time. Mr. Reeser intends to sink the shaft during the winter to the depth of 150 feet below the surface, and to start a level there. As the rich body of ore in the Pittsburgh ground has been followed up in the Reeser ground, and pitches toward it, Mr. Reeser will, in all probability, have a valuable mine. Next spring he will also get possession of the 10-stamp mill belonging to the property, which is now being used by the Pittsburgh Company.

The Valley Gold-Mining Company has been steadily at work on the Golden Star lode, treating about 8 tons of ore per day in the mill, which, by means of a chute and tram-way, receives the ore directly from the mine. It is first broken in a Dodge crusher and then worked in two large arrastras. The company has a large overshot water-wheel, which drives all the machinery, and also a saw-mill attached on the outside.

Above the Golden Star are the Comstock and Confederate Star, both lodes which formerly yielded large amounts of rich ore, but are not being worked now.

Below the Golden Star is the Bed-Rock, owned by S. B. Dilley. The tunnel on the lode has been extended 40 feet and a shaft has been sunk below the stope, where the best ore is found. This shaft was, at the time of Mr. Wolters's visit, 5 feet deep, and showed a finely-developed crevice 3 feet wide, with a vein of ore 6 inches on top and about 18 in the bottom.

All these lodes are on the right side of Bear Gulch. On the left we find the Sierra lode, 1,200 feet of which are owned by A. Cramer. The lode runs northeast and southwest, dips 33° north, and averages from 2 to 3 feet in width, with a pay-streak of from 10 to 15 inches. At a convenient height above the creek, where the owner has a mill-site 200 by 300 feet, a cross-cut tunnel has been run north, which intersects the lode at a distance of 345 feet from the mouth, gaining a depth of 100 feet on the incline of the vein and 91 feet perpendicularly. The tunnel was driven in 95 feet beyond the Sierra lode, for the purpose of striking also the Chancey lode, which is owned by the same party.

About 110 feet above this tunnel another cross-cut tunnel has been driven a little north of east, striking the lode 91 feet from the mouth. Levels have been run either way about 100 feet, developing a fine body of ore. About 400 tons were extracted, which yielded in the mill at the rate of $20 to $25 per ton. A distance of 90 feet east of the point where the tunnel strikes the lode, a shaft has been sunk to the depth of 36 feet, showing an 18-inch vein of ore, which furnished 26 tons of ore, milling $17 per ton.

There are now from 30 to 35 tons on hand on the dump, which were taken from a stope below the east level, and are expected to mill about $25 per ton.

In the Vishnu lode, ground has been stoped out 40 feet long and 80 feet high, the ore-vein averaging from 3 to 4 feet in width. In the eastern part of the vein the ore is of low grade. Work will go on in this mine during the winter, as also on the Wizzard King, and Alturas.

Nothing has been done since my last report on the Idaho lode, once the leading one of the camp, but the owners expected to do some work during the winter to hold the property.

On several other veins of minor importance work enough is being done to hold them under the act of Congress of May 10, 1872.

In *Red Warrior district*, work has been going on steadily on the Wide West mine, the Avalanche, and the Bear-Skin. The Wide West mine and mill were last spring sold by the owners to Warren Hussey, of Salt Lake City, for $22,000, and work has been going on ever since. The ore-reserves above the lower tunnel had been nearly exhausted by the former owners, and a shaft had been started near the heading of the lower level and sunk to a depth of 30 feet. This shaft has been continued down by Mr. Baxter, the manager of the concern, to a depth of 80 feet, and levels have been run both ways. A fine body of ore was found, which would have paid handsomely if the owners had furnished Mr. Baxter at the right time the means necessary for the erection of substantial hoisting-works. This, however, has so far not been done, and the shaft making a very large amount of water, which, together with the ore, has to be hoisted with a common windlass, the mine can hardly be expected to pay dividends. Some time ago the superintendent went himself to Salt Lake City to consult and remonstrate with the owners, and I am informed that they acceded to his demands, and gave him sufficient means to put the mine in proper shape for cheap working. If this proves true, Mr. Baxter, who has gathered much experience as a mining and mill man in South America and Nevada, will have no difficulty in placing the Wide West mine on a paying basis.

C. Jacobs, who owns the Victor Mill and the Victor, New York, and Golden Eagle lodes, has done little during the season. From the Golden Eagle 8 tons of ore were taken, which yielded in the Wide West Mill at the rate of $42 per ton.

The Bear-Skin has been worked by Joe Kessler, E. Clough, Joseph Temple, and Dan. Fitchwater, who own 400 feet in the lode. They have run a tunnel 80 feet long, have sunk a shaft 80 feet deep, and have stoped out a piece of ground 35 feet high and 20 feet long, which yielded about $1,000.

The Richmond, being 70 feet below the last-named lode, has also been worked, and has furnished some high-grade ore, milling $75 per ton. During the winter the owners of the two lodes intend to run a tunnel 400 feet long, which will first pass through the Richmond and then strike the Bear-Skin at a depth of 150 feet.

In the Avalanche, work has been prosecuted vigorously most of the time. A cross-cut tunnel has been run, striking the vein in 122 feet, and then connected by shaft with the surface. This shaft has been continued below the tunnel to a depth of 40 feet, and shows a body of ore 1 to 2 feet wide, which, with proper selection, yields in the mill from $30 to $100 per ton.

A new company, styled the Buffalo and Idaho Gold and Silver Mining Company, has been formed by Mr. John Tillman on the Itasca lode, at Rocky Bar, and the Silver Tide lode, at Atlanta. The superintendent of the company, Mr. Cavanagh, appears to be the right man for Rocky Bar, where living from hand to mouth has so long been the general rule in mining operations, as will be seen from the following correspondence of the Boise City Statesman:

"Mr. Cavanagh worked gold-mines in North Carolina at a profit when the ore paid only $2.60 per ton. He did this by working on an extensive scale, always having large reserves of ore on hand. He proposes the same programme here, and tells his company they must not expect a dividend for at least two years, and must during that time expend a large amount of money in opening their mines in depth, so that when dividends commence they will continue, and the working of their mines will not be temporary and spasmodic. He is now at work on the Itasca, situated just below Rocky Bar, and will open it by a series of tunnels and shafts. The lowest tunnel, starting a little above the water-level of Bear Creek, will open the vein 1,000 feet deep, and a shaft will also be sunk near the mouth of the tunnel 500 feet below the water-level, which will make a reserve of ore sufficient to run a 50-stamp mill for years. The Itasca is a large and well-defined fissure-vein, and if worked in this manner will pay handsome dividends. If Mr. Cavanagh's company have the nerve to back him, and we trust they have, a new era of mining prosperity will be commenced among us. The fundamental maxim of all successful mining is to keep works of exploration well in advance of works of extraction. This maxim has been violated in almost every instance in working our mines. They have all been worked in a very improvident manner, taking the ore in sight as far and as fast as it could be found, never having reserves, and anticipating the evil day of sinking shafts and running levels."

In *Hardscrabble district*, Messrs. Pfeiffer & Co. commenced, late in the fall, work on the Poor Man lode, which they discovered this summer in sluicing Poor Man gulch. It is their intention to sink a shaft 100 feet deep during the winter, run levels, and erect a 10-stamp mill to treat the ores from this and the General Grant and General Sherman lodes. On the surface the Poor Man shows a well-defined crevice about 2 feet wide, with from 6 to 8 inches of very rich quartz, full of free gold.

On the General Sherman lode, a shaft had been sunk, some time ago, 36 feet deep, but in sluicing the gulch it has been filled up with gravel.

On the General Grant lode, the tunnel, which left the crevice for a considerable distance, has been somewhat extended toward the lode, without, however, so far striking the latter.

The Dividend lode, in the immediate vicinity of the last-named, has not been worked, but will undoubtedly be soon, as it appears to be a very fine lode. It can easily be traced for nearly 2,000 feet, is opened in two different places, and in both shows a finely-developed vein 3 feet wide, with 7 to 8 inches of ore, which assays very well, both in gold and silver. From a large number of assays, running up as high as $1,200, it would be just to expect that the ore in bulk will yield about $25 per ton by stamp-mill process.

About 150 feet east, and parallel to the Dividend, is the Mariposa lode. It is about 4 feet wide, has a shaft 80 feet deep, carries 8 to 16 inches of quartz, but is not being worked.

No work has been going on on the Ophir lode, though it appears to be the largest vein of that district. It can be traced for about a mile, shows a very strong outcrop, and is opened superficially in many places.

At Bonaparte Hill everything has been very quiet since the Bonaparte Company stopped operations, and there appears to be no prospect for a speedy resumption of work.

*Atlanta district.*—The Atlanta lode has been worked under lease by two different parties, Lantis & Co. working the Monarch Company's ground, and Doctor Marshall and S. S. Mattingly the old Miller ground. The former struck a very rich vein of ore right at the surface, and two men have been kept steadily at work there. The ore-vein consists of a 2½-inch streak of very rich ruby-silver-bearing ore, and another streak of ferruginous quartz, impregnated with native silver, about the same width as the other. At the present time they are reported to have over 10 tons of this ore on hand, which will no doubt yield from $1,500 to $2,000 per ton by proper treatment. The intention of the lessees was to ship this ore late in the fall to San Francisco for reduction, but winter setting in nearly a month earlier than usual, and preventing transportation, this could not be carried out, and had to be deferred until spring. For the same reason they, as well as others, were caught by the first heavy snow-storm without a sufficient supply of mining-supplies to last them during the winter, and being unable to procure them afterward, they had to stop work on the deep tunnel, which they were running to strike the lode 100 feet below the present lowest levels.

No work has been done in the level on the north wall, which is still in barren rock, but that on the south wall has been extended 50 feet, and a small streak of fair milling-ore has been developed. It is to be regretted that difficulties existing among the owners prevent a vigorous and legitimate working of the property, as the lode is most certainly a remarkable one as far as size, extent, and richness of ore are concerned. Under sound and economical management, backed by sufficient capital to open the mine properly, and to erect works suitable for the beneficiation of the ore, the reputation of the Atlanta lode as being inferior to only very few lodes in the world would be speedily and easily established.

The claim adjoining the Monarch Company's ground, owned by Miller, comprises 480 feet, and is also worked under a lease, but the claim being entirely unopened, and the lessees lacking the necessary funds to properly open the vein, they have confined themselves entirely to working at or near the surface. In a depression of the mountain they have run an open cut across the whole vein, showing it to be 60 feet wide. They then ran a level on a streak of ore close to the north wall of the lode, 44 feet long, which shows on an average 6 inches of good ore and a 2-foot streak of bluish quartz, carrying some gold and silver, but not enough to pay for working it. A quantity of the first-class ore taken out in running this level has been shipped to San Francisco, but I have not yet learned the returns, which, however, are expected to be not less than $500 per ton. This level has been run northeast from the cross-cut. On the northwest side there is a vein of ore 16 inches in width, which averages $100 per ton, but is not worked at present. A cross-cut tunnel has been run by the original owners 135 feet, which, when completed to a length of 300 feet, will strike the vein a little over 100 feet below the present workings; but lack of means prevents the lessees from availing themselves of the facilities thus given them to open the mine properly.

West of the Miller ground the Atlanta is claimed for 920 feet more in different small claims; then the outcrop disappears for perhaps 800 or 1,000 feet, but re-appears on the south side of Atlanta Mountain, which slopes toward Yuba River. The lode is taken up here as the William Tell lode, and claimed for a distance of 2,600 feet. The immense outcrop, averaging from 50 to 60 feet in width and sticking clear out of the ground, sometimes 8 to 10 feet high, can be seen from the top of the mountain quite a distance downward, and though no work amounting to anything has been expended on this lode to show and establish its value, there is every probability that proper development will prove this part of the Atlanta as good as the rest.

A short distance above the Monarch Company's claim is the Last Chance lode, owned and worked by E. Heath. He has run a level on the vein 175 feet long, the latter portion of which is, however, partly caved in for a distance of about 40 feet. The vein runs northeast and southwest, pitches south, and is about 4 feet wide. The pay-streak is about 6 inches wide, and will mill $50 per ton. A little stoping has been done, and there are a few tons of good milling-ore on the dumps. By many persons the Last Chance lode is considered a spur or branch of the Atlanta, but, in my opinion, this belief is without foundation, as the former dips south and is a gold-bearing vein, while the Atlanta dips north and carries principally silver-ores proper. There is a difference of about 20° in the strike of the two lodes.

About 50 feet north of the Atlanta is the Silver Tide lode, formerly owned by John L. Tillman, but now by the Buffalo and Idaho Gold and Silver Mining Company. In the summer a shaft was sunk on the lode 22 feet deep. This, however, went outside of the south wall. Owing to this fact, and the large amount of water in it, it was abandoned, and a cross-cut tunnel was run 110 feet long, which the company intends to extend this winter 100 feet farther, in order to strike the vein. The lode shows a large and bold outcrop, and, in an open cut some distance above the mouth of the shaft, a streak of very good-looking ore. The best ore assays about $200 per ton.

The Eclipse lode is a little north of the Silver Tide, shows very little development, and no work will be done beyond representing it.

On the Leonora lode, the cross-cut tunnel has struck the vein, and a level has been run for 50 feet along the vein, developing a pay-streak of 2 to 3 feet of average milling-ore. Besides this no work has been done on the lode the whole year.

On the Tahoma lode, the shaft, 6 by 12 feet, has been sunk down 96 feet. The vein was left at a depth of 63 feet, in order to get a perpendicular hoisting-shaft. Down to a depth of 20 feet the whole vein was taken out, and yielded at the rate of $25 per ton. The south wall is well developed. The north wall has never been reached, though the shaft was started 8 feet wide. A tunnel has been run which comes in right under the shaft, being 103 feet below its mouth. For about 30 feet it runs on the vein, and some very good ore was taken out while running this portion. Afterward it leaves the vein and runs the whole remaining distance in a zigzag course, the owners having apparently changed their minds many times in regard to the proper course to be pursued, and furnishing a fine example of how tunnels ought not to be run.

In the Stanley lode, Ostrom & Marshall had for about two months four men at work at the surface, sinking the shaft down to a depth of 30 feet. A small portion of the ore obtained was worked, but the yield was not ascertained by Mr. Wolters. From the fact, however, that the rest of the ore still lies in the mill unworked, it may be inferred that the first lot did not pay for milling.

This lode and the Tahoma are on the same vein. The first comprises the original location of 1,000 feet. The first extension, also 1,000 feet, is owned by Davis & Weed, the former having 800, the latter 200 feet; and the second extension, again, 1,000 feet, now known as the Tahoma, is owned by Davis, Casey, Dolan & Newton. On the first extension some sluicing has been done, yielding the owners of the Tahoma enough to make a living while they carry on the necessary dead-work in their lode, though they have to collect water all day in a small reservoir, in order to get a supply for sluicing from one to two hours. In working this claim they have uncovered another lode, which they call the Potosi, but nothing has so far been done with it.

In the new mining-camp on Wood River very little has been done this season. There are undoubtedly some lodes which carry a large amount of pure and high-grade galena, rich in silver, and if the much-talked-of construction of the Portland, Dalles and Salt Lake Railroad should ever become an accomplished fact, Wood River is certain to become a flourishing camp.

## BOISE COUNTY.

The mining-interest of this county, depending more than any other on the yield of placer-claims, has of course suffered most by the scarcity of water and shortness of the season, and the production from that source will this year probably fall considerably short of $500,000.

In quartz-mining there has been about the usual amount of activity. At Quartzburg the Gold Hill Company has been mining steadily. The level started on the southwest side of the gulch has been extended 1,000 feet during the year, and is in now altogether 1,800 feet. After having passed through under the next gulch, where the tunnel was only 30 feet from the surface, a body of good ore was struck, which, however, gradually became poorer, and about October 1 just paid expenses. Stoping was then abandoned on this side and operations confined to the further extension of the level. Good ore is expected a couple of hundred feet ahead. The mill in the mean time is kept running on ore from the old Gold Hill, northeast side of gulch, where the company is working in good ground.

An attempt made this summer by a San Francisco party to beneficiate the concentrated tailings proved both a metallurgical and financial failure, but another party had 1,000 pounds shipped to Winnemucca, Nev., and if successful in their test-run, may work all the company have, which is now estimated at about 2,500 tons.

The lawsuit of several years' standing about the Iowa lode, claimed as an independent lode by James Hawley & Co., and as identical with the Lone Star lode by Thomas Mootry, jr., has finally been adjudicated in the Supreme Court, the court holding that plaintiff did not prove the identity of the two lodes, and therefore confirming the judgment of the court below.

The owners of the lode this summer consolidated with Eissler & Co., who own the Growling Go lode, which is the eastern extension of the Gold Hill lode. The new company erected a good, substantial 10-stamp mill, which was completed in September, and contains sufficient room for an increase of the capacity. About 500 tons of ore from the Iowa, which had originally been extracted with a view of sluicing it, and was therefore not at all assorted, were run through the mill and yielded $7,159.89, or at the rate of $14.32 per ton.

The tunnel on the Iowa is now in over 200 feet, and a fine body of ore is exposed all along. Nothing has been done in the Growling Go.

In December all work was discontinued, and parties who pretend to know say that the object in view is to freeze out some of the partners, there being too many of them to work harmoniously together.

*Summit Flat district.*—This district is situated ten miles east of Pioneer City and sixteen miles northeast of Idaho City. The three principal streams of the basin head here, and afford good water-power. Timber is not very abundant, but there is enough red and white fir and small mountain-pine to last a number of years for mining-purposes. The lodes all run northeast and southwest, dip to the north, and are generally not very wide.

The first claims were located in this camp in 1863, when three gold-bearing veins were discovered. The largest and most promising of them was the Mammoth lode. Its outcrop was very large and well defined, and the rock showed free gold to the naked eye. A cross-cut tunnel was run on it intersecting the lode at a depth of 30 feet, and developing a vein 3 feet wide, dipping at an angle of about 30°. A 10-stamp mill, driven by water-power, was erected in 1864. Water, however, being very scarce that year and the mining season unusually late, the owners crushed but very little ore, and being unable to meet their obligations, they were forced to sell out to Jackson & Co. These new owners worked steadily during the summer of 1865, and kept the mill running on ore taken from the Mammoth, the best of which yielded over $75 per ton. Further development of the lode showed that the body of ore was only about 60 feet in length, extending downward about the same distance. Here evidently occurs a fault in the lode, occasioned by the crossing of a vein of trap-rock. After the exhaustion of this ore-body, work was here abandoned, and there was never an attempt made to find the vein again by following the vein of trap with a cross-cut, but two years later a shaft was sunk fifty feet deeper, which did not strike ore; and there was a new cross-cut tunnel pushed ahead within 75 feet of the line of the lode, when lack of funds forced the owners to quit working. They obtained between 200 and 300 tons of ore by sluicing the hill-side below the ledge, which paid well in the mill. Since then no more work has been done on the discovery-claims, but the outcrop appearing again on the first extension west, a distance of about 500 feet from the discovery-shaft, a cross-cut tunnel has been run here, cutting the lode 70 feet from the surface. It proved to be 8 feet wide, but the ore was too poor to pay expenses. A level was then run toward the discovery-claim, and when the line of the latter was nearly reached, a body of good ore 30 feet long and extending nearly to the surface was met with. After exhausting this pocket, the level was extended 200 feet into the discovery-claim, but no more ore was struck. In 1872 a shaft, 22 feet deep, was sunk in the level where the pocket of good ore had been found, and some ore was stoped out which paid well in the mill; but the shaft making water at the rate of a gallon per minute, work was abandoned entirely, and so far has not been resumed.

The entire yield of the lode, including the float-rock obtained by sluicing, is about $50,000. The present owners of the discovery-claims, first extension west, and mill, are Messrs. Clarkson & Brown, both farmers, who have too little time to pay any attention to mining.

The King lode, situated three miles south of the Mammoth, is a well-defined vein, carrying from 2 to 8 inches of ore which mills at the rate of $40 per ton, small lots having averaged as high as $100 per ton. A tunnel 200 feet long cuts the vein at a depth of 60 feet, and follows it some 400 feet into the hill, gaining in all 150 feet in depth from the surface. The ore has been stoped out the whole length of the level, and nearly up to the surface. The rock in the levels and stopes was pretty hard, and in running the last fifty feet every inch had to be blasted, which made mining slow work and rather expensive. There are almost no sulphurets in the ores of this lode, as well as in most of the others. The gold amalgamates readily, and is rather coarse.

This mine, like the Mammoth, has changed owners several times to no advantage, and has been worked only spasmodically since it was opened in 1864. An 8-stamp mill, driven by steam, was erected in 1865, which has been employed this season in crushing rock from other lodes in the vicinity opened lately.

The entire yield of the King lode has been about $40,000.

The Specimen lode is situated about 50 rods south of the King. A few tons of excellent rock were taken out at the surface, but at the depth of 25 feet it gave out, and no more has been found, though four different shafts were sunk, the deepest of which is 70 feet.

The Newfoundland lode, discovered in 1864, is 100 feet south of the King. The ore-vein is about 3 inches wide, and is very irregular. Some ore was taken out, paying very well, but being only in small pockets, mining was found to be too expensive, especially as the rock is very hard and flinty.

The Golden Era lode, about one and three-fourths miles in a southerly direction from the King, was discovered in 1872. It is opened by two tunnels, one cutting the lode quite near the surface, the other about 60 feet deeper. The foot-wall is well defined, but no regular hanging-wall has yet been found. The lode is a good deal broken up and disturbed, apparently owing to the crossing of a vein of trap-rock. The ore is found in sheets or flakes, running diagonally across the opening. Its average value will probably be about $15 per ton, while some of it has paid over $100 in the mill. In all about $4,000 have so far been taken out, and there is a large amount of fair milling-ore in sight. The ore is packed to the mill by mules, at a cost of $5 per ton, as so far no wagon-road has been constructed. In running the lower tunnel, another lode was crossed, which carries some very good rock, and development may prove it valuable.

The Hog 'em lode, situated three-fourths of a mile south of the King Mill, was located in 1873. It can be traced on the surface for 800 feet, and is very favorably located for working. It carries from 8 to 12 inches of good ore. Nine tons of this ore were recently crushed, and yielded $50 per ton. The lode is within 300 yards of the summit of the mountain, and is believed to be the highest in that part of the country.

The Packer lode was discovered in August, 1874. It shows some free gold of the coarse kind, but contains also a large amount of iron pyrites. Seven tons crushed yielded $7 per ton, the sulphurets not being saved.

The Justitia lode, owned by Goodwin, Andrews & Co., is situated about four miles from Pioneer, in Charlotte gulch, being six miles from Summit Flat. While all the aforementioned are gold-bearing lodes, yielding bullion from .670 to .730 fine, this one is a base-metal ledge, carrying a mixture, consisting principally of galena, iron pyrites, and zinc-blende. The lode is opened by a tunnel on the vein, about 300 feet long, which shows it to be from 2 to 6 feet wide. The ore can, of course, only be treated by smelting, but, though there are good local facilities in the shape of water and timber, lack of capital has prevented the owners from erecting such works. A few hundred pounds of the ore were sent to Salt Lake City for reduction, and gave good results.

No paying placer-mines have been found in the Summit Flat country until this season, when a claim was opened and worked on Grimes Creek, about four miles from its source, at a widening of the creek-bottom called the "Meadows." The claim paid well, two men taking out $208 during the last week they worked. In consequence, quite a number of claims have been located, which will be opened next season. A bed-rock flume is also talked of.

In *Banner district* there has not been much accomplished during the year. The mill in process of construction has not yet been finished, the owner, G. W. Crafts, having suffered, like all other business-men in the basin, from the effects of the bad mining season, which prevented those making a living by mining operations, either directly or indirectly, from paying their bills. Work on some of the principal lodes has been kept up pretty steadily, and as they are, in spite of this, still very little developed, Mr. Crafts may yet have good reason to bless the delay in the completion of his mill. This gives him time to develop first his mines to such an extent as to make him but little dependent upon outsiders for supplying his mill with ore.

### IDAHO COUNTY.

After Mr. Leland had obtained possession of the Rescue lode, in 1873, through the courts, he formed a joint-stock company, consisting of citizens of Idaho and Washington Territories. Work was commenced by extending the shaft below the lowest level; but soon the water came in at such a rate that the erection of good substantial hoisting-works became a matter of urgent necessity. An order for the same was forwarded to San Francisco with a view of having the machinery in Warren's early in the fall, but very unfortunately there was so much delay that by the time of its arrival in Lewiston snow had already accumulated on the mountain-ranges to such a depth as to make them impassable for pack-trains. Thus the company were forced to wait until spring, and as the heavy expense of sinking the shaft further without hoisting-machinery excluded the possibility of even making expenses, the board of directors stopped all work in the fall of 1873. In June, 1874, the machinery arrived in Warren's, but, as is nearly always the case when work is suspended for a long time in mining enterprises, a majority of the stockholders lost a good deal of their original interest and confidence; by far the greater portion of the assessments were not paid in, and there was not money enough to pay for the machinery; neither had the company funds to carry on their work. Since then they have been trying to raise money to put the hoisting-works in shape, pay their debts, and develop the mine by further sinking of the shaft and running levels, preparatory to stoping.

Nothing of importance has been done on any other lodes beyond the amount of work necessary to hold them.

### LEMHI COUNTY.

Some work has been going on on several lodes in the vicinity of Salmon City, but I have not been able to get any special information about them.

ONEIDA COUNTY.

A new mining-district has been formed about thirty miles north of Kelton, on the southern slope of the Goose Creek range. Lodes carrying mostly galena with little silver were discovered several years ago, but nothing has been done so far beyond representing them according to law.

---

# CHAPTER IV.

## OREGON.

Operations are still continued, as they have been at intervals for more than twenty years, on the gold-beaches of Northern California and Oregon. The formation at Gold Bluff, in Klamath County, California, is described in a paper by Mr. A. W. Chase, of the United States Coast Survey, published in my last report. The operations in California are distributed along the coast from about fifteen miles below the town of Trinidad to the vicinity of Crescent City, where numerous small parties are at work, earning good wages by rude projects. The principal parties operating in the original Gold Bluff, so called, are Greenbaum & Co., proprietors of the Lower or South claim, and the Upper Bluff Company, each of which owns four miles of the coast. Greenbaum & Co. have been at work here for seventeen years, and are reported to have been steadily successful, taking out sometimes as much as $25,000 in a single year. The gold in both these claims is coarse, and more easily caught than the average generally met with along the coast.

At Rogue River, in Oregon, some thirty miles north of Crescent City, a number of small companies are profitably at work on a small scale, and without the aid of improved mechanical appliances. It is here that the Perseverance Black Sand Company, organized in 1874, in San Francisco, is at work.

I quote from the Mining and Scientific Press the following article, headed "Beach-Mining along our Gold-Coast:"

Oregon beach-mining, after an active period, occurring some twenty years ago, when the cream of these deposits was rudely skimmed off by those in the field, has been sensibly revived within the past year, owing to the favorable results obtained through the use of improved processes and machinery. It was generally supposed until within a few years ago that the greater portion of the gold contained in these beach-sands was saved by the sluices, copper plates, &c., then in use. But thousands of assays and other tests have since proved that in most cases fully four-fifths of this gold is invisible to the naked eye, and was lost in all former workings. When this fact became established, interested parties set to work to solve the problem how to separate these infinitesimal particles of gold from the heavy black sand with which they are associated. And it proved to be one of the most difficult things to accomplish that has yet puzzled the brains of scientific men or honest miners. The extreme lightness of the minute particles of gold and the great weight of the grains of black sand rendered any process of concentration or amalgamation, based upon their relative specific gravity, impracticable; and another serious obstacle was encountered in the existence of a coating or oxidation upon the surface of these tiny, microscopic specks of gold, which prevented amalgamation. Many thousands of dollars were spent in making tests and experiments to obtain an economical method of working. Chlorination and other chemical modes of treatment were found to be too expensive, but they confirmed the assays, showing the extraordinary richness of the sands, and stimulated those interested to further efforts to overcome all obstacles. A vast amount of this sand was proven to contain from $10 to $30 per ton, and results were frequently obtained showing several times these values. The most extensive of these deposits yet developed are located at Randolph, Coos County, Oregon, where the claims of Messrs. Lockhart & Lane are known to contain several millions of tons of black sand worth from $5 to $100 per ton, or on an average, say, of $15 to $20 per ton, leaving enormous profits for working by a process not costing more than $3 per ton. The cost of mining the sand is merely nominal in most cases. Plenty of similar deposits, though of less extent, exist at various localities along the coast from Humboldt Bay to the Columbia River, and in fact extending north to Alaska. The most extensive deposits are those of the ancient ocean-beaches, formed before the late upheavals of the coast-line, and lying back of and above the present ocean-beach. But there are immense quantities of rich sand upon the present beach also, that will be worked at great profit, as they are more accessible and easier mined. Indeed, the quantity of black sand in these beach-deposits, ancient and modern, carrying from $5 to $10 per ton, may be considered almost inexhaustible. But these placers have lain for many years almost unnoticed and unworked for want of a process and machinery that would save this fine, flour gold. So many have had their wits at work upon the problem within the last two years that perseverance has at length been rewarded by success, and a new branch of mining-industry developed which is of far greater magnitude and importance than the public or even the mining portion of the community are aware of. This microscopic or "flour" gold exists in large quantities in all our gravel and placer diggings, as well as in the tailings of our quartz-mills, and is lost to a great extent as mining-operations are now conducted. The successful treatment of the black sand of the ocean-beach is much more difficult than of any other class of auriferous deposits, and hence the process which overcomes such obstacles will easily conquer those involved in ordinary placer-mining, and millions of dollars, now lost annually, will hereafter be saved by the new treatment. The great difficulty experienced in saving these infinitesimal particles of gold arises from the fact that they are so light that the current of water employed in ordinary mining-operations carries them off, their gravity not being sufficient to overcome the force of the current. This has been obviated by effecting amalgamation before subjecting the sands or pulp to a stream of water, which result is secured by evaporating the quicksilver under water. Heating the sand or pulp, containing a proper quantity of quicksilver with plenty of water, up to the boiling-point, causes the mercury to expand and diffuse itself all through the mass in minute globules, which unite with the fine particles of gold, provided the latter are in a proper condition to amalgamate. The heating aids to remove the coating upon the gold; but to effect a thorough preparation for amalgamation, the sand is first subjected for some hours to a chemical preparation, which is not expensive, though very powerful. The next and final difficulty to be overcome is to collect the fine particles of amalgam (some of which are still too fine to be visible to the naked eye) together, after the boiling, without too great a loss. The loss of quicksilver by the mills in Washoe averages about one and a half pounds per ton of ore, while a loss of as many ounces in working black sand would leave little or no profit to the miner; because, being amalgam, the loss *in gold* would be too great.

The Perseverance Black Sand Mining Company, recently incorporated in this city, has succeeded, after spending more than a year's time and several thousand dollars, in perfecting a machine for separating these fine particles of quicksilver or amalgam from the heavy black sand without material loss, and at a trifling expense. This company is about to commence operations on a large scale with the new process and machinery upon their claims near the mouth of Rogue River, where they have secured extensive deposits of these sands as rich as any on the coast. Through the aid of this improvement, it is thought a new era will be introduced into mining, by saving at least 90 per cent. of the microscopic gold hitherto lost through the imperfect processes and machinery in use. The ordinary black sands, as stated above, contain from $5 to $15 per ton; and many deposits, containing thousands of tons, run up to twice and even three times these figures. The cost of working by the new method will be from $2 to $3 per ton, including mining, which in most cases is merely nominal.

Operations similar to those of the Perseverance Company are soon to be started on the Lockhart claim, near Randolph, by Mr. John Bray, who has been experimenting on the sand with flattering success. The owner of that claim is now in this city making preparations for the necessary machinery, which is to be forwarded to the mine and set to work as soon as possible.

As a field for mining, this northern gold-coast presents some peculiar advantages for men of small means. In the first place, it is easily and cheaply accessible; then, no great amount of capital is required for outfitting purposes; while, again, there is not much trouble in securing claims for working, the chances of taking up ground or buying from former locators being good at most points along the coast.

At one time there was an effort made to take up the strip of ocean-beach between the line of high and low water, under the law of Congress granting to the State all swamp and overflowed lands between its borders. But the surveyor-general, acting under advice from the Commissioner of the General Land-Office, decided that ocean-beach did not come within the purview of that law, which was designed to cover only the overflowed grounds along our bays and other inland waters. This places these modern sea-beaches, wherever found to be auriferous, among other mineral lands belonging to the public domain. Hence, they are open to location by the miners, who may occupy and hold them under such local regulations as they may themselves see fit to adopt, so long as these conform to the statutes enacted by the General and State governments.

The Perseverance Company alluded to in the foregoing paragraphs employs a process which was for some time invested with more or less mystery. The following account of it is believed to be full and explicit. It was furnished by the president of the company:

The sand is first screened so as to materially reduce its bulk, and then subjected for about twenty-four hours to a solution composed of caustic potash and common salt in proper quantities. This is for the purpose of removing any coating or oxide that may be upon the gold, and destroying sulphur and other base substances which would be absorbed by the quicksilver to its injury. The pulp is then heated in a pan by a jet of steam, (being constantly stirred,) which takes but a few minutes, as it should not be too hot, and then the quicksilver is poured in and the steam and agitation continued from fifteen to thirty minutes, when the gold becomes thoroughly amalgamated, and the pulp is discharged into a vat to cool before putting it through the separating-sluice. The heat has the effect of expanding or partially vaporizing the quicksilver, and, aided by the agitation, distributes it all through the sand-pulp in very fine particles, like flour, where it meets and amalgamates with the equally fine and universally diffused particles of gold.

Care must be exercised in the heating, as too much heat flours the quicksilver more than is necessary to secure perfect amalgamation, and thereby increases the difficulty of collecting it together again without serious loss. When we consider the fact that the amount of heat required to raise the temperature of one pound of water from 32° to 212° will raise that of about 30 pounds of quicksilver through the same range, the reason for this caution is obvious. There is reason to believe that most of our mill-men have fallen into the common error of using too much heat in amalgamating the precious metals.

It is proper to state here that great care is taken to purify the quicksilver, by proper means, before it is put into the pulp, upon every occasion of using it, as success depends much upon having the quicksilver in good condition. Retorting alone is not always sufficient, as some of the base metals volatilize and pass over with the mercurial vapor.

The last and most difficult operation is that of separating the fine particles of amalgam and quicksilver from the heavy black sand and collecting them together again without too great loss; for it must be remembered that black sand cannot be treated like quartz-pulp, on account of its greater specific gravity. The fine particles of gold and quicksilver cannot be wholly precipitated by agitation or a concentrating motion, because they are so minute and light in comparison to the grains of sand, that they will not settle by their own gravity through the heavy sand as through quartz-pulp.

This difficulty is overcome and separation secured without material loss, by this company, by means of a system of galvanized-copper rollers, grooved spirally, and placed side by side and in layers one above the other, so as to "break joints" and not quite touch together, and extending across the sluice, which is three feet or more in width. A screen is placed over them to distribute the sand and water as they fall upon them, and a galvanized-copper plate beneath to catch the quicksilver as it drips from them. Two or three layers of these rollers are thus arranged at two or more places, a few feet apart, in the sluice, and drop-riffles or wells are sunk across the bottom a little below the copper plates, to receive and retain the quicksilver as it runs off from them. The rollers are one foot in length and one and a half inches in diameter, and hollow, six or eight of which are laid side by side and end to end, extending across the width of the sluice. These, as also the copper plates, are kept in a highly-sensitized condition and free from verdigris, and as the pulp passes down, over, and between their multiplied surfaces, it necessarily brings the fine particles of quicksilver into contact with some one of them to which they will adhere, before passing through the whole of them as arranged in the sluice. As the quicksilver accumulates upon the rollers it drops from the under side, and the amalgam is cleaned from them in the same manner as from the plates.

These amalgamating-rollers and their application to both the cradle and the sluice are the invention of William Sublett, of this city, who has applied for a patent therefor, and have been freely tested by the Perseverance Company.

North of Rogue River the beach is worked to a very limited extent at Port Oxford and Cape Blanco. Randolph, situated a short distance south of Coos Bay, has been for a long time an active mining-center for this class of mining, but at present there is little doing in the immediate vicinity of the town, the principal work being performed upon the Lane and Lockhardt claims, which are located on the ancient beach, now two miles inland. These claims were located seven years ago, and have since been opened by means of tunnels, the old beach being here buried under from 20 to 60 feet of sand. This deposit is said to have an average breadth of about 350 feet and a depth of about 6 feet. It is shaped like a wedge, with its broader edge toward the sea, and tapering in the opposite direction to a point. The claim has the reputation of having produced a great deal of gold, though, as in all other cases in this kind of mining, the owners are convinced that but a small percentage of what the sand contained has really been saved. This ancient, or back beach, as it is called, though not traced for more than three or four miles in the neighborhood of Randolph, has been found at several points up and down the coast; and since it promises to be more profitable to the miners than the front or present beach, particularly considering that the latter has been frequently reworked, and thus relieved of a good deal of its gold, while the former remains still nearly virgin ground and but little explored, it is likely that operations on the back beach will be more extended during the next few years.

*Eastern Oregon.*—For reports of progress and prospects in Eastern Oregon, I am indebted to Messrs. Reynolds and Packwood, of Baker City, and to Mr. George Gillespie. In combining and condensing the valuable material furnished by these gentlemen, I shall not attempt to give separately the source of each statement, since, in numerous instances, the same ground is covered and the same facts are communicated by more than one of them.

Auburn is the oldest placer-mining camp in Eastern Oregon, and has passed through the usual vicissitudes attendant upon the prosecution of surface-mining with limited water-facilities. The resources of the district are, however, by no means exhausted. On the contrary, it promises, with the aid of a new enterprise started last year, to become again profitably productive. A company of gentlemen from Marysville, Cal., purchased last year, for a nominal sum, the water rights and ditches of the Auburn Canal Company, and now have in construction, in Blue Cañon, a large flume, which will open up an extensive area of rich placer-ground, which never has been available heretofore, since it could not be worked without such a flume. This new territory, together with the accumulated tailings of twelve years' placer-workings, will, it is believed, abundantly repay the projectors and furnish employment for a large number of workmen. There have been otherwise no new discoveries in this camp. The past year of 1874 afforded fair wages to a small number of men, the supply of water being about as usual.

At Fort Sumter, in Gimletville, the placers are principally worked by Chinamen, with a few white miners, and are said to have paid well during the past season. The quartz-lodes in this vicinity are idle for want of capital and enterprise on the part of the owners. All along Burnt River the small placer-camps, such as Bull Run, Winter's Diggings, &c., have been worked as usual on a limited scale, and with fair results. They seem almost inexhaustible under this treatment.

At Clark's Creek the force of white men and Chinamen was not as great as last season, but the yield of gold-dust seems to have been about the same. Mormon, or Humboldt, Basin maintains its former position as a steadily-productive camp, though the supply of water during the last season was not abundant.

The mining-interests of El Dorado district have been for the two years preceding 1874 materially discouraged by a litigation concerning the Burnt River ditch property, which was settled by compromise in the spring of 1874, Messrs. Packwood & Carter, the original proprietors, resuming possession and prosecuting with energy the incomplete work. Since July they have built and put in operation twelve additional miles, there being an extension in the upper end connecting with the main ditch; and at the beginning of this year they had under contract another extension of twelve miles on the extreme lower end to convey water to Amelia district. The ditch was opened a little before the 1st of July, and water was run until about the 20th of November. At least two months and a half of the mining season was lost, but it is estimated that the business of the season amounted to over $20,000 at the rates charged; and it is expected that, with the present improvements and some additional reservoirs, this ditch will furnish and find purchasers for $50,000 to $75,000 worth of water yearly for many years to come, the amount of placer-grounds tributary to it being very large. The ditch is about 8 feet wide at the top, 6 feet at the bottom, and 2 feet deep, with an average grade of about 6.4 feet to the mile. It will measure, including the portion now in construction and to be finished May 1, 1875, not far from one hundred and thirty-five miles in length, being one of the longest and best constructed works of the kind on the Pacific coast. It was first projected and commenced more than ten years ago, by Mr. Packwood and others, and has cost, up to the present time, not far from $500,000 in coin. Water is sold at 20 cents per inch for ten hours, and is used for the hydraulic process with a pressure of from 100 feet to 125 feet, and in heads of 100 inches and upward to the pipe. The mining-ground in El Dorado district is from 10 to 40 feet deep. It is expected that this year, with the increased supply of water furnished by the completion of the ditch, both Shasta and Rich Creeks will be successfully exploited, which will double the yield of gold in the district.

Messrs. Lynn & Co., who own the old Reeves ditch, have cleaned and enlarged it, and will have about two months' water on the Rattle snake and Rich Creek ground from this source.

Messrs. Campbell & Co., of Quartz gulch, have been using 125 inches of water during the past season, and intend to employ 200 inches next year. Iron gulch has been mined for years by Chinamen on the bars; next season the main gulch will be opened and flumed and piped. In Amelia the placers have paid fairly, but the natural supply of water is very limited, lasting only a few days in the spring. Messrs. Barnhart, Quinn, & Gaffney, have very rich ground on Discovery gulch. Their claim is said to have paid $50 per day to the hand by piping during the short season. The deep-gravel or cement beds at Amelia have been mined in a peculiar way, determined by the brevity of the water-season. During the past winter the cement has been excavated and thrown up in piles, where it slakes or disintegrates by alternate freezing and thawing, and will be run, in the spring, through sluices, which are found to give very profitable results. Hereafter water-power will be available from the Packwood & Carter ditch extension to Amelia; and it is probable that this will lead to a new era of mining in Eastern Oregon, namely, the extensive development of the cement-gravel beds, and the working of the gravel in mills or pans. These cement-beds extend for ten or twelve miles from Rich Creek on the west to Fourth of July on the east, varying in thickness from 10 inches to 10 feet and upward. They lie near the surface, and every gulch and creek cutting through them has been found to contain in places very rich deposits of gold, and almost universally sufficient gold to pay for working. The bed-rock immediately under the cement has frequently yielded from $1 to $5 per pan. The cement itself has been tested only in the manner already alluded to, by stacking it and allowing it to disintegrate during the winter-season. I am not certain that this method will not prove, after all, the best; though, with the continuous supply of water throughout the larger portion of the year, it will not be necessary to confine the period of extraction to the winter only, or the period of working to a few weeks in the spring. Probably the disintegration of the cement could be effected at any season by allowing it to lie a certain time and occasionally wetting it down. In many places locations comprising from 50 to 1,000 acres in a body could be made, but the work of mining in this cement will probably have to be carried on with at least occasional bank-blasting, owing to the extreme hardness of the material. If it should prove advantageous to work it in mills, it is said that water-power for driving them can be abundantly supplied by the Packwood & Carter ditch. The resources of the El Dorado and Amelia districts will now be more thoroughly prospected than ever before, and this impulse given to mining by the completion of the ditch, which is expected to furnish from 1,500 to 2,500 inches of water daily, from April to December, will make itself felt in the largely-increased product of gold.

Messrs. William H. Packwood and Alexander Stewart, whose names are indissolubly associated with the development of the mineral-resources of Baker and Union Counties, were the projectors and builders of the ditch from Eagle Creek, in the latter county, to Sparta, a distance of forty-five miles, through the most difficult ground for such an enterprise that could have been found in the State. The entire line was completed between the 11th of May and the 16th of October, 1871, on which latter date water was brought through and sold. It has proved of great value to the region traversed and highly remunerative to its owners.

Twelve miles from Sparta, on the same spur of the Eagle Creek Mountain, is the quartz-mining camp of Hog 'em. Gold was found here as early as 1862, and in the following year the limited alluvial deposits surrounding the high peak known as Hog 'em Hill began to be actually worked. It is said that the surface-ground yielded about half a million dollars. It is now practically exhausted, and attention has been for several years concentrated upon the quartz-veins which originally furnished the gold. Many veins were found upon the mountain, among which may be named the Knight, Abernethy, North Star, &c. A 5-stamp mill was erected, and, while developments were carried on in a cheap and desultory manner, it is believed that considerable profit was realized, without, however, leading to extensive and systematic work. The Summit lode, high up on the hill, and its south extension, known as the Big Giant, now constitutes the basis of the principal mining-operation which has survived to the present time, and promises to be permanently profitable. This vein was discovered in 1870, and found, though small on the surface, to increase in width and richness with greater depth. Messrs. Packwood & Stewart having acquired a title, have consolidated with it so much of the surrounding claims as to prevent danger of litigation, and have prosecuted the development of the mine with energy, erecting offices and other necessary buildings, with a complete 10-stamp mill, sinking the shaft to a depth of 350 feet, exploring the vein thoroughly with levels and winzes, and constructing a right of road through twelve miles of Goose Creek Cañon, which enables them to obtain supplies without difficulty throughout the winter. Up to February, 1874, the work of stoping above the 200-foot level had produced, from an average vein of 20 inches thickness, between 3,000 and 4,000 tons of ore, yielding over $16 per ton in the mill, or something over $60,000 in the aggregate. The vein below the 200-foot level increases in width, the average, from the depth of

250 feet to that of 300 feet, being fully 4 feet. In the bottom of the shaft, at 350 feet, the lode was reported to be 5 feet wide, showing free gold and sulphurets, and carrying a zone of 18 inches, which is believed to be worth about $300 per ton, while the average of the whole ledge is $30. These figures are, of course, deduced from assays only. It is expected that the operations of 1875, including the extension of the shaft to 400 feet in depth and the exploitation by levels and stoping of the lower 200 feet, will furnish more than double the quantity of ore per running foot which was obtained in the upper levels, and that the ore will be also richer, judging by the indications in the shaft. The lode is well defined throughout, and carries a clay selvage or gouge 6 inches and upward in thickness. In this clay occurs considerable pyrites, whether auriferous or not has not been determined. Mr. John Griffin, formerly superintendent of the Virtue mine, is conducting operations. He reported, at a depth of 320 feet, that in sinking the last 100 feet he had had to extract no waste-rock, and had been obliged to use but little powder.

The Virtue mine, near Baker City, and about twenty miles south of Hog 'em Hill, I believe is now owned by a San Francisco company. Early last season operations were suspended for six months, for the purpose of increasing the plant, with a view to working on a much larger scale. Late in the summer the company started again, with ten additional stamps, making twenty in all. I have the report of one month's yield, amounting to $16,000, a sum which, from all I can learn, may be relied upon, in the absence of any unforeseen hinderances, as a fair monthly average for the present year. The gold-bullion of the Virtue mine is very fine, being worth $17 per ounce. Mr. M. Hyde, formerly of Owyhee, has succeeded Mr. Jackson as superintendent.

The James Gordon lode, eight miles south of the Virtue, has been opened to the depth of 120 feet in the discovery-shaft. Some ore milled in this mine yielded $30 per ton, but active operations have been suspended, owing, it is said, to dissensions among the owners. Reports from various quarters concur in pronouncing the mine a very valuable one.

Several new lodes in the vicinity of the Virtue have given excellent prospects—the Ironstone, Jim Jam, Lady Bertha, Senator Jones, Manning & Ellis, and others.

Rye Valley, which has been alluded to repeatedly in former reports, is one of the few districts in Eastern Oregon which contain silver-ores as well as gold. The bullion from the placer-mines is largely alloyed with silver, being worth only from $9 to $11 per ounce. The alluvial mining, though hindered by the severity of the seasons, is regularly maintained, with a notably uniform, though moderate, profit. A new enterprise, in the form of a bed-rock flume three-quarters of a mile long, 28 inches wide, and 32 inches deep, is completed in this district, the owners having 160 acres of mining-ground and a twelve years' deposit of tailings as basis for their undertaking. Several new discoveries of quartz-lodes have been made in the Granite Mountain, among which may be named the Wild Raccoon and the Comet. On the Lafayette, Messrs. Davidson & Co. have done considerable work, and are now running a small 5-stamp steam-mill, with good results, on ore from the Lafayette, Washington, and Mountain. The process saves only free gold, which is, however, somewhat alloyed with silver, the bullion being worth $12 per ounce. The Green Discovery and Monumental lodes, operated by Messrs. Green, Macdonald & Co., who are, if I am not mistaken, representatives of a San Francisco company, and owners also of the

Macedonia and other lodes, are the leading mines of Rye Valley, and promise to be, during the present year, extremely productive. On the Green Discovery a tunnel 500 feet long opens the vein to a depth of 300 feet below the surface. The force at work in the mine at the beginning of this year was only nine men. It was intended, however, to push with vigor, night and day, the extraction of ore, so as to supply the new reduction-works. The mill erected last fall, under the superintendence of Mr. W. L. Burnham, has five stamps, two Horn pans 7 feet in diameter, and one large settler 9 feet in diameter, all new. The capacity is 12 tons in twenty-four hours. The mill is located about 150 feet from the mouth of the Green Discovery tunnel, so that the mine-cars can run directly to the building. It commenced running about the 20th of December, and up to the first part of January, the date of my latest advices, there had been no clean-up, but pulp-assays ran from $300 to $500 per ton. This is not surprising, considering the nature of the ore, in which wire-silver, silver-glance, and chloride of silver are frequently to be seen. The bullion is also auriferous, being worth from $2.50 to $4 per ounce. There were on the 1st of January at the mills about 300 tons of ore from the Green Discovery, upon which mine the principal work is done, though the Monumental has furnished also very rich specimens. The steam-power in the mill is adequate for a plant of double the present capacity, and an enlargement will probably be made as a result of the present operations. There has been considerable excitement over the brilliant prospects of this enterprise, and many prospectors have been at work in the neighborhood. A mountain or large mass of gypsum has been discovered, which is interesting for the present, mainly as a mineralogical curiosity. I do not know the conditions of its occurrence; but it may possibly be worth while to suggest a careful examination in the neighborhood for rocks containing salt, which will be metallurgically a valuable addition to the resources of the district.

At Conner Creek the prospects of the mines are very promising. Messrs. E. M. White and Joseph Myrick have moved to this locality their 10-stamp steam-mill, formerly at Gem, and have become part owners in the Conner Creek mine on the main lode. The new company (known, I believe, as Wood, White & Co.) has now 15 stamps, 5 run by water and 10 by steam, crushing 20 tons of ore per twenty-four hours, and producing $15 per ton. The bullion is .827 fine. The workings on the mine are now 100 feet deep, and the ledge is reported to be 6 feet wide. Messrs. Hover & Williams, on the extension of the same lode, have a 5-stamp water-mill, crushing 6 tons per twenty-four hours, and producing $15 per ton. The lode is reported to be 6 feet wide at the depth of 80 feet. The fineness of the bullion is .834. I am indebted to Mr. T. F. Hover for information relative to this and other claims.

The bars on Snake River in this vicinity are worked when the condition of the stream permits, and continue to be profitable, though the industry is necessarily a precarious and intermittent one. The gold is .845 fine. Several parties are building ditches—one already completed, being three miles long—to bring water to these bars. Across Snake River, near the old Brownlee Valley, and politically, therefore, in the Territory of Idaho, though practically belonging to this region, a lode was discovered last summer 6 feet wide, and carrying galena of $42 gold and $12 silver per ton.

At Gem and Sparta, in Union County, the Eagle Creek Canal, above alluded to, has done a fair business, though these camps, once populous, are now occupied by a comparatively small population.

Along the western side of Powder River Valley, the Pocahontas, Rock Creek, Little and Big Muddy, Wolf Creek, and North Powder, there have been during the season the usual supply of water and the usual amount of scattered placer-mining carried on by individuals or small partnerships, and producing, in the aggregate, considerable sums, though very difficult to trace in detail.

From Grant County I have no news of importance.

The great necessity of Eastern Oregon is improved communications with the outside world. The present general depression of business-interests throughout the country will probably delay the execution of the scheme for a railroad connecting Dalles, in Oregon, with Salt Lake. But sooner or later this project will doubtless be realized, to the great advantage of the agricultural and mining industries of this region.

---

# CHAPTER V.

## MONTANA.

The collection of mining-statistics for this Territory has again been intrusted to Mr. William F. Wheeler, of Helena, who has furnished me with the following carefully-prepared statement of the product of the precious metals and the accompanying synopsis of the mining-operations in the Territory during the year:

| | |
|---|---|
| Gold-shipments by express | $2,511,276 |
| Silver-shipments by express, (refined bars) | 16,766 |
| | 2,528,042 |
| Add one-third for undervaluation, for amounts taken out of the country in private hands, and values still in the hands of miners | 842,680 |
| From the various freight companies it has been ascertained that 1,580 tons of selected silver-ores in sacks have been shipped, (more than twice the quantity of the previous year,) about two-thirds of which went down the Missouri River and the rest by Franklin, the terminus of the Utah Northern Railroad, and by Corinne, on the Central Pacific Railroad. The average value of this selected ore is estimated at $300 per ton | 474,000 |
| Total gold and silver | 3,844,722 |

Forty tons of copper-ore were shipped, which assayed 40 per cent. of copper, and, calculated at $150 per ton, were worth $6,000. At least 1,000 tons of silver-ore were prepared for shipment, but, owing to lack of transportation, this ore is now stored and will be shipped in the spring. I do not add this to the product of the year, because nothing has been realized from it which would add to the prosperity of the country.

The means of transportation in Montana to navigation or railroad are only mule and ox teams. The distance to the nearest water-transportation last year was two hundred and twenty-five miles, to Carroll, on the Missouri, and those to Franklin and Corinne, on the railroads, were respectively four hundred and four hundred and fifty miles. From these figures the pressing necessity of rail communication can be easily seen.

For ten years past no winter has furnished so much snow as the present, and it is certain that this will have a beneficial influence on next summer's placer-mining. A line of boats, partly, if not wholly, owned in the Territory, will run up the Missouri River to Fort Benton during the next summer, and from the heavy winter snows it is to be presumed that navigation will be good until fall. The distance from Helena, whence most of the silver-ores are shipped, to Fort Benton is but one hundred and fifty miles, and it is believed that next summer every pound of rich ore raised can be shipped in good time to reach an eastern market.

The ores from Madison, Beaver Head, and the southern part of Deer Lodge Counties, about 500 tons, were shipped in 1874 to Franklin and Corinne, and it is probable that a much larger amount will be sent that way this year.

The scarcity of transportation has led to the construction of several small silver-mills, where comparatively low-grade ores are now being successfully worked.

Two or three new districts have been discovered in the past year, which promise well.

The following description of the Clark's Fork district has been furnished by the discoverers. The country is at present inaccessible to wagons.

*Clark's Fork mines, on Crow reservation.*—The existence of silver-lead lodes on Clark's Fork, a tributary of the Yellowstone, has been alluded to in Hayden's Report for 1872, (published in 1873,) page 47.

From Mr. John Barnett, an old prospector in this region, Mr. Wheeler procured some fine specimens of ore, containing from 70 to 80 per cent. of lead, which he had found at the headwaters of Soda Butte Creek, a tributary of Clark's Fork of the Yellowstone, near the east boundary of the National Park.

*Blackmore* and *New World* mining-districts have been located here, taking in twelve miles from east to west and six from north to south. They contain a large number of well-defined leads. The mountains, as Mr. Barnett describes them, are covered with washed bowlders of galena, and the veins, so far as tested, are from 6 to 25 feet wide.

A tunnel across the Mammoth vein, in the New World district, shows 25 feet of solid ore, and the width of the vein is not yet known.

The New Caledonia shows a well-defined 6-foot vein. The Great Republic, Greeley, Iron-Clad, Houston, Woody, Silver Zone, Silver Gift, Blackfoot, Shoo Fly, Alta California, Alta Montana, and a large number of other veins have been located, but not developed. There is an abundance of wood and water for all purposes. A few miles of road-building would make the mines accessible for wagons. The ores could then be hauled in vast quantities to or near the Crow agency, on the Yellowstone, and then shipped by regular trains to Carroll, on the Missouri, from which point regular lines of boats would take them to the East. From crude assays these ores will yield from 50 to 200 ounces of silver per ton, and from 70 to 80 per cent. of lead. The mines are about one hundred and ten miles southeast of Bozeman.

On Sixteen-Mile Creek, a tributary of the Missouri River, in Gallatin County, a very fine galena-lode, carrying a fair proportion of silver, has been prospected by a shaft sufficiently to show that it is valuable, by reason of the width of the vein and the quantity of ore it contains.

A new silver-district, called the *Silver Lake* district, in Deer Lodge County, in the mountains southwest of Deer Lodge City, promises well;

but the work done during the past year has not been sufficient to test its value.

*Trapper district*, in Beaverhead County, discovered in 1873, promsies to be one of the best silver-districts in Montana. The ore is of high grade, and several hundred tons of it were shipped last year. Several hundred men have been steadily employed in raising ore during the present winter, and a large amount will be ready for shipment in the spring. A good deal of work is also being done in Beaverhead County, in the *Blue Wing*, *Bannack*, and *Argenta* districts.

Messrs. Dahler & Armstrong erected sampling-works in the Trapper district in 1874. They have been run successfully and profitably. This winter the same parties are constructing smelting-works, to be started in the spring. They are also preparing to work the gold-veins at the Upper Silver Star district, in Madison County, and for that purpose are refitting the old Trivitt Mill.

At Rochester, in Madison County, where gold-quartz mining was formerly carried on with some success, while for the past few years it has been a failure, Dr. Getchell, Milo Courtwright, and Mr. Tennant have again revived the business, and by a more thorough knowledge and superior skill they have achieved success, and the camp has resumed its former prosperity.

In Deer Lodge County, regular silver-quartz mining is being carried on in the Vipond, Butte City, Silver Bow, Moose Creek, and Philipsburgh districts. A large amount of ore will be raised the coming season.

At Philipsburgh, the Cole Saunders Mill is at work again, after a year's interruption, turning out successfully refined silver bars. The process is that of dry-crushing, chloridizing, and amalgamating. The mill has a capacity of five stamps, and crushes five tons a day, but five more stamps are now under erection. The ore is from the famous Speckled Trout and other mines in the vicinity. A second mill will be put in operation there in the spring. The mines were described at some length in my report for 1872. The district is looked upon as one of the best in the Territory.

In Lewis and Clarke County, the *Ten-Mile*, *Red Mountain*, and *Vaughn* districts are the most noted. Nearly one-third of the ore shipped from the Territory last year was from these districts. The ore from Vaughn's mine, named the Little Jennie, over 150 tons of which were shipped last year, assayed, on an average, $600 per ton. Mr. Vaughn has two other mines, which appear to be of nearly equal value, but they have not been as thoroughly tested.

At the eastern end of Red Mountain, adjoining the Vaughn district, are the mines of Bismarck Hill and Providence districts, partly in Jefferson and partly in Lewis and Clarke County. The silver-veins in these districts are numerous and valuable, and were described in my last report.

In Jefferson County, immediately east of Helena, are the oldest and most numerous silver-veins, and considerable work has been done on them during the year. All were described in my reports for 1872 and 1873. Concentrating-works have been erected on the Argentum and Legal-Tender mines in this county, and the results have been very satisfactory.

The following is a description of the operations at the Legal-Tender mine for 1874:

Messrs. Lewis, Bull & Co., the owners, have, during the year, erected steam hoisting-works of 16-horse-power, at a cost of $9,000. They have also added to their improvements works for wet-concentration, consisting of a 10-horse-power boiler and engine, a Blake ore-crusher, and a set of five Cornish jigs, at a cost of $4,000. With the new hoisting-works they have increased the depth of their main shaft from their 160-foot level to the 366-foot level, and are still pushing the work of sinking. The amount of ore raised and sold during the year is as follows:

| | |
|---|---|
| 257 tons shipped to Freiberg, Germany, assay-value, $284.69 per ton | $73,165 77 |
| 169¾ tons, second class, sold to home reduction-works, assay value | 13,837 92 |
| Total, 426¾ tons | 87,003 69 |

The works for concentrating have proved a success, enabling the owners to make first-class shipping-ore from low-grade second-class, by separating the galena from the zinc-blende and gangue.

The low-grade ores of the Legal-Tender mine have been successfully and profitably worked by the Kemp Bros. during the year 1874. They claim to have obtained from 80 to 90 per cent. of the assay-value of the silver in the low-grade ores of this mine, and during the present winter they have improved their machinery for working these and similar ores. They claim to have discovered a new and cheaper process than heretofore known of working low-grade ores at a profit.

Very few new mills have been erected in the Territory during the last year for working either gold or silver ores, but additions have been made to old mills, and improved machinery has been introduced. Some of the old gold-mills and several new arrastras have been successfully at work on quartz from the older well-developed mines of the Territory.

Among these mills are those at the Red Bluff, Mother Hendricks, and Lost mines; Cissler & Zin's new 20-stamp mill for crushing gold-quartz, in Madison County; the mills at the Cable mine, in Deer Lodge County; those at Unionville, in Lewis and Clarke County; and the mill of Blacker & Keating, at Radersburgh, in Jefferson County.

At the Rumley lead, which has been fully described in my last report, and is situated on the Boulder River, in Jefferson County, work has been actively prosecuted during the past summer. In running across the vein but one wall-rock has been found in a distance of 60 feet. Over a thousand tons of ore now lie on the dump. This ore carries about 50 per cent of galena and 50 to 100 ounces of silver per ton. The proprietors intend to erect smelting-works in the spring to reduce their ores.

Near Jefferson City, William Nowlan, one of the men who by their persistent endeavors have proved their faith in the mines of the country, has put in operation a smelter during this winter, which furnishes now considerable "base bullion."

A water-power has been completed at Jefferson City, which is to be used in connection with a Stetefeldt furnace. The works are intended to be in operation in the summer.

The yield from gold-quartz, by mills and arrastras, I believe to exceed that of former years. It is steadily increasing, while it is evident that gold-placer mining is on the decrease.

Messrs. Blacker & Keating, in Jefferson County, and the National Mining and Exploring Company at Unionville, near Helena, in Lewis and Clarke County, have uniformly, for half a dozen years past, been the most successful gold-quartz miners in the Territory. The first party has cleared from forty to fifty thousand dollars, the latter about $100,000, per year for several years past.

During last fall the National Mining and Exploring Company of Unionville has procured new machinery of every description for developing its mine. A new engine and iron track, with iron cars, for hoisting and dumping its ores, a new pulsometer for pumping, and two new Burleigh drills, for working underground, are among the improvements made for the year. It is believed that thereby the profits of the company will be largely increased.

The decrease of the gold-yield of Montana is entirely due to placer-mining. The time has passed when short ditches could be built by two or three men to bring water into narrow and rich gulches and to bars. These have been practically worked out. As in California, a new era of placer-mining must now begin. Rich companies, with abundant capital, are now constructing long and large ditches from the rivers and the larger streams to the gold-bars which abound in the numerous valleys of Montana, and which it does not pay to work by individual effort. Many of the Montana miners have visited California, and there they have seen the superiority of large streams of water run through "Little Giants" under high pressure. These men are all anxious to effect consolidations of the mining-ground heretofore owned in small parcels by many individuals, and thus to secure the benefits of the new and improved system of working hydraulic mines. It is therefore to be expected that hydraulic mining will improve, and that the annual yield will hereafter be increased, provided labor does not remain too high. During the last three or four years many laborers have left the country in consequence of the rich gulches being worked out, and there is for this reason now a scarcity of miners, which keeps up the price of labor. It is doubtful whether the old force of laboring-men can be secured again before a railroad reaches the country. Meanwhile it is fortunate that water is usually sufficiently abundant in the Territory to permit the introduction of the California methods of hydraulic mining.

About two years ago Mr. Wheeler made some experiments in coking Montana coal. Mr. Thomas H. Clark, a mechanic in Kemp Bros.' blacksmith-shop, heated some coal which they were using for blacksmith-purposes in a partly-closed retort for several hours, and produced a fair coke. Then Mr. Hobart, who was furnishing stone-coal from his mine at Mullen's Pass, on the summit of the Rocky Mountains, sixteen miles from Helena, was induced to try coking. He did so, and produced a tolerable article, which was tried in the foundery of Hon. A. J. Davis. This gentleman declared that he could use it, but he suggested that the coal should be burned longer. Fortunately a man was found who had formerly been employed in burning coke at Johnstown, in Pennsylvania. He went to Mr. Hobart's mine, and engaged in the business. His coke is good, and has been used ever since in the iron-founderies with complete success. Since that time the importation of anthracite coal, which was formerly in exclusive use here for foundery-purposes, has entirely ceased.

Thus it appears that the Montana coal, which is abundant in all parts of the Territory, can be used, as it now is, for blacksmithing, for making steam, for locomotive and steamboat purposes, as well as for making gas and smelting iron, galena, and, in short, for every purpose for which soft Pennsylvania coal can be used. According to the reports of the Secretary of the Interior, about seventy thousand square miles, or half of Montana, is underlaid with coal-beds, but this I regard as an overestimate.

Iron-ore is equally abundant, and in time Montana can alone supply the needs of the western half of the continent with iron.

# CHAPTER VI.

## UTAH.

The production of gold and silver during 1874 has been more than twice that of the previous year. This result is due to the development in Bingham of extraordinarily large bodies of low-grade lead-ores, which can be mined very cheaply, and to the energetic prosecution of mining in innumerable small mines, carrying ores of good grade, and belonging to individual miners or partnerships. Of the large companies, the mines of which carry high-grade ores, only very few, like the Flagstaff, Prince of Wales, and Winnamuck, have furnished large amounts of ore, while some others, like the Emma, Davenport, Miller, and many in the more southern districts, have taken out little or none.

A few of the smelting-works in the Territory have been kept regularly at work, producing the large amount of "base bullion," *i. e.*, argentiferous lead, given below; but the majority have only worked periodically, and many have not been started at all.

The following smelting-works may be said to have run regularly:

1. The Winnamuck Works, in Bingham Cañon. The operations at these works are given in detail in the metallurgical portion of this report.

2. The Sheridan Hill Works, at West Jordan, have four rectangular furnaces, with water-jackets and six tuyeres each. Those in Nos. 1, 2, and 3 have 2 inches diameter; those in No. 4, 3½ inches. The tuyeres are cast in with the jackets, there being two in each of the sides and one in the back plate. The furnaces are 9 feet high from the slag-top to the charging-door. Their size is 30 by 30 inches in the hearth; above they are widened by means of a flat bosh to 4 by 4 feet. Their capacity is 15 to 18 tons of ore per day. The ore-charge consists generally of 50 per cent. of Neptune and Kempton ore, containing from 40 to 50 per cent. of lead and from 16 to 22 ounces of silver per ton. The other 50 per cent. consist of the following ores, or whatever may be on hand of them:

| | |
|---|---|
| Dressed Emma ore, carrying 40 per cent. lead...... | 80 to 90 oz. silver. |
| Richmond, (Little Cottonwood,) carrying 24 per cent. lead.......................................... | 50 oz. silver. |
| McKay, carrying 24 per cent. lead ................ | 47 to 50 oz. silver. |
| Davenport, carrying 8 to 15 to 24 per cent. lead .... | 43 to 60 oz. silver. |
| Toledo, carrying no lead.... ...................... | 90 oz. silver. |
| Nez Percés, carrying 32 per cent. lead ............ | 16 oz. silver. |
| City Rock, carrying 10 per cent. lead............... | 29 oz. silver. |

The mixture of these ores is, of course, always varying, and has to be made according to the quality and quantity on hand. This ore-charge is then mixed with 15 to 20 per cent. of iron-ore, 10 per cent. of lime, and 15 per cent. of slag, from the same operation. The fuel used is coke, 7 per cent.; charcoal, 4 per cent.; and 3.75 per cent. of stone-coal are consumed under the boilers of the 80-horse-power engine at the works. This engine drives, besides a No. 5 and a No. 4 Root blower, a large Woodward pump, supplying a tank, and a Knowles pump, which supplies the boilers. There is a small Woodward pump in reserve at the works.

3. The Galena Smelting-Works, belonging to a Boston company and managed by Messrs. Carson & Buzzo, contain six furnaces, three of

which are similar to those at the Sheridan Hill Works, having, however, round water-jackets. The other three are constructed according to a patent of Mr. McKenzie, and are not at all suitable for lead-smelting. The original design of these furnaces was intended for use in the re-melting of cast-iron, for which it is well adapted.

The works are smelting ores from the Jordan and Galena mines, which are very much poorer in silver than those from the Neptune and Kempton. Together with these ores the works are smelting richer silver-ores brought from the Cottonwood mines. Up to the time of Mr. Eilers' visit to the Territory, these works used a steam-engine as a motive-power, but it was the intention to use water-power as soon as a large ditch, bringing water from Little Cottonwood Creek, should be completed. This ditch was reported to cost $100,000, which, to say the least, is certainly a large outlay for power in a mining-country.

4. The Last Chance Works, treating principally ores from the Flagstaff mine, in Little Cottonwood, and also some purchased ore, have been running nearly throughout 1874, but were stopped toward its close. They have furnished a larger amount of base bullion than any other works in the Territory; but, like most of the other works, they seem to have run without any considerable profit. The furnaces at these works also are provided with water-jackets, and have on top of the stacks a curious contrivance intended to catch the dust, an object which it fails to accomplish with any degree of perfection.

5. The Mountain Chief, Saturn, Wahsatch, and a number of smaller works in the vicinity of Salt Lake, have been stopped almost throughout the year.

The Waterman and the Chicago Works, on Rush Lake, have been running, the first for a short time, the last almost throughout the year. They are mentioned in another part of this report.

6. The Germania Refining Works have worked with only half capacity. Most of their base bullion was obtained from Nevada, the whole quantity bought by the works during 1874 amounting to 3,725 tons, of which 2,005 tons were from Nevada and 1,720 tons from Utah. During the fall this company commenced to smelt ore in addition to carrying on their business of refining. For this purpose they built a large circular blast-furnace with water-jackets, a description and drawing of which is given in another part of this report. They also introduced Faber Du Faur's retorts for the purpose of treating by distillation, henceforward, the zinc-silver alloy, which has heretofore been run through small shaft-furnaces, necessitating, in this manner, the total loss of the zinc employed. These works are treated in a separate article in the metallurgical portion of this report.

From the best authorities at my command, and by comparing and rectifying various statements heretofore made public, I estimate the production of Utah during 1874 as follows, (I should add that I am obliged to Mr. George J. Johnson, of Salt Lake City, for the statement of railroad shipments appended below:)

*Production of gold and silver in Utah during* 1874.

| | |
|---|---|
| Gold-dust by express, (and 10 per cent. for undervaluation) | $92,093 |
| Silver-bullion by express, 642,754* ounces, at $1.25, (partly gold) | 803,442 |

*Of which 449,912 ounces silver and 9,460 ounces gold were shipped by the Germania Works.

| | |
|---|---|
| Base bullion shipped out of the Territory, 14,056 tons, at $200* | $2,811,200 |
| Ore shipped, 8,262 tons, at $80† | 660,960 |
| Gold and silver in copper-matte, shipped by the Germania Refining-Works, 4,235 ounces silver and 18 ounces gold | 5,664 |
| Total | 4,373,359 |
| Deduct silver-bullion shipped by the Germania and included above, but extracted from Nevada lead | 461,758 |
| Total production of gold and silver | 3,911,601 |

*Production of lead in Utah during* 1874.

| | |
|---|---|
| Base bullion as above, 14,058 tons, at $80 | $1,124,640 |
| Refined and hard lead from Germania Works, exclusive of that produced from Nevada bullion, say, 1,418 tons, at $128† | 181,504 |
| Lead in ore shipped, 30 per cent. of 8,262 tons, at $50 | 123,900 |
| Total | 1,430,044 |

*Production of copper in Utah during* 1874.

| | |
|---|---|
| Copper-matte shipped by Germania Works, 60 tons, at $90 | $5,400 |
| Copper-ore, (as per Salt Lake Tribune,) 460 tons, at $51.50 | 23,690 |
| Total | 29,090 |

The amounts of ore, bullion, and lead shipped from the Territory from January 1, 1874, to January 1, 1875, are as follows:

| Month. | Ore. | Bullion. | Lead. |
|---|---|---|---|
| | *Pounds.* | *Pounds.* | *Pounds.* |
| January | 991,000 | 1,301,457 | 700,000 |
| February | 840,000 | 1,660,000 | 640,000 |
| March | 1,191,820 | 1,600,000 | 320,000 |
| April | 1,203,340 | 1,874,638 | 952,000 |
| May | 877,886 | 1,857,275 | 240,000 |
| June | 777,650 | 2,650,784 | 560,221 |
| July | 2,029,560 | 2,487,989 | 480,069 |
| August | 2,307,745 | 2,842,547 | 829,498 |
| September | 3,522,319 | 3,449,402 | 1,069,756 |
| October | 1,180,000 | 2,544,529 | 803,889 |
| November | 988,972 | 2,140,000 | 916,525 |
| December | 613,132 | 3,703,473 | 262,024 |
| | 16,523,424 | 28,112,094 | ‡7,773,982 |

*Little and Big Cottonwood districts.*—The production of these districts has been less than in previous years, principally on account of the exhaustion of the large ore-body in the Emma and of the stoppage of

* Estimated.

† Estimated at the same rate as the average of all classes of lead from all sources shipped by the works during the year.

‡ Of this, 1,418 tons were produced from Utah, the rest from imported Nevada "base bullion."

several other heretofore producing mines. I am indebted for notes on the district to Mr. J. H. Morton, a resident of Alta City, who has had unusual facilities for acquainting himself with the facts set forth.

The topographical features of the Little Cottonwood have been so often described that little need be said on that subject. The cañon has a length of about twelve miles from the head above Grizzly Flat to the mouth. High, barren mountains wall it in on either side from Granite to Central City. Where the mountains form a semicircle near its head the town of Alta is located. Here the mountains are very precipitous, and from every side in winter converge the avalanches that have been such a terror to the people. The winter of 1874-'75 is one that will long be remembered, as the snow-slides have been more frequent and more destructive than ever before, penetrating even into the center of the town. They have left a feeling of insecurity of life that will probably be a serious draw-back to the growth of the place.

The necessity of some means of cheap and constant communication with Salt Lake Valley for the transportation of ores and mining-supplies is yearly becoming more pressing, and as the natural supply of timber becomes exhausted it must be met in one form or another. The Wahsatch and Jordan Valley Railroad, commenced two years ago, has been built to a short distance above Granite, where, either owing to a lack of funds or indisposition of the company to push the work farther, it is allowed to end. The most important advantage of the road is thus lost, as the old expensive plan of sacking the ore has still to be adhered to. This of itself is, according to Mr. Morton, an expense sufficient to almost finish the road in two years. He says the cost of sacking the ore is, on an average, at a low estimate, about $3 per ton, and adds that this would sum up for the mines in the course of a year as follows:

| | |
|---|---|
| Flagstaff, 14,000 tons, at $3 | $42,000 |
| South Star and Titus, 3,000 tons, at $3 | 9,000 |
| Toledo, 2,500 tons, at $3 | 7,500 |
| Vallejo, 5,000 tons, at $3 | 15,000 |
| Highland Chief, 2,500 tons, at $3 | 7,500 |
| Prince of Wales, 2,000 tons, at $3 | 6,000 |
| All other mines, 10,000 tons, at $3 | 30,000 |
| Total expense for sacks and sacking ore | 103,500 |

The cost of building, shedding, and equipping narrow-gauge railroad, Mr. Morton estimates as follows: Three miles, at $15,000 per mile, 45,000; and five miles, at $50,000 per mile, 250,000; total cost to Grizzly Flat, $295,000. This would place the railroad within easy reach of every mine in the district. The ores could then be mostly shipped in bulk, and ores that are now of too low grade to pay for handling could be made profitable. The price now paid for shipping ore from Alta to the end of the railroad is about $5 per ton; the railroad could take it at, say, $2. This would effect a saving of $3 on sacks and $3 on freight, which would enable the miners to utilize much of the low-grade ore of the district.

This, however, would not be the only saving which a railroad would effect. Timbers now cost from 18 to 30 cents per linear foot; with railroad-communication those prices should be cut down to about 10 cents per linear foot. Lumber of all kinds costs $45 per thousand; with a railroad it would not cost over $27 to $30. All this, coupled with the disadvantages of snow-blockades, points to the necessity of a well-constructed and completely-shedded railroad from Alta to Sandy, and indicates that it would be a profitable investment, if the mines continued to be actively worked.

The rocks at the base of the Wahsatch range appear to be uniform wherever they appear, and consist chiefly of granite, often quite massive, and again appearing to possess very regular bedded lines. At the mouth of the cañon the granites are flanked by enormous masses of conglomerate, which extend west, and finally dip beneath the drift of the valley. Upon the eastern slope of the granite mass the texture varies from a fine-grained gneiss to an extremely coarse porphyritic granite. These rocks are overlain by deep beds of hornblendic schists and by quartzites, traversed from east to west by veins of ferruginous quartz and dykes of porphyry. This is again overlain by a bed of crystalline limestone, and above this appears a thin bed of slate, followed by a heavy bed of quartzite. This is followed by a third small bed of slate, above which and dipping to the northeast is the great bed of ore-bearing limestone, known on the north side of the cañon as Emma Mountain and upon the south as Emerald Hill. In this formation, though not exclusively, are found the ores of the district, and though much to the contrary has been said and written in regard to the nature of these deposits, I still regard them as irregular chimneys, and newer developments support this view. The irregularity of the ore-bodies, which often appear to give out in every direction, and often as suddenly open out again, show that a series of pre-existing caves has been afterward filled with the ore-matter. This is particularly conspicuous in the mines of Emma Hill along the lower ore-zone. Those of the upper group present similar features, and, however vein-like they appear near the surface, they uniformly develop this character as depth is gained upon them.

In the quartzite formation the Toledo and Emily Mines present the characteristics of true veins, and the Crown Prince and Frederick those of contact-veins, having quartzite hanging and limestone foot walls.

The Lexington, Sedon, Skipper, and City Rocks mines appear to be veins, occupying a portion of the fissures along porphyry-dikes. In the three first named the ores are disseminated through a soft, decomposed porphyritic gangue, and in the City Rock the ore occurs in a similar manner, but more generally in a clay, evidently formed by the decomposition of the feldspar. Though the masses of unchanged porphyry are less frequent and less extensive in the City Rock mine than in the others, still there is every evidence that it is a dike-fissure, from which part of the intrusive rock has been removed by decomposition and the action of mineral solutions, while the ores have been deposited in its stead. The connection visible between the ore-formations of the district and the intrusive dikes and the manner in which the ores occur, associated with the porphyry, suggest the idea that they are secondary effects of the porphyritic eruption, and were deposited shortly after the eruption ceased. That the mode of their formation was by infiltration from below does not, I think, admit of a doubt. Mr. Morton, who also holds this view, infers that the irregular deposits found in the limestone formations may yet be traced down to true veins, which is not impossible, though I cannot venture to predict it upon the strength of present indications.

In the granite mountain, known as Virginia Mountain, the ore-deposits exhibit the feature of true veins. This vein of granite extends from the upper part of Alta City, south and east, covering an area of six by two miles, and as far as it traverses the Little Cottonwood district it is cut at short intervals by veins carrying ore so different from that found in the limestone, that a piece of it can be at once recognized as coming from this locality. Following the strike of these veins into the limestone formations of Emerald Hill, we find their counterparts existing in two cases, at least, just where the veins should outcrop, were they continuous fissures, passing from the granite into the limestone. However, they do not appear to be true veins outside of the granite formation; but in the limestone appear as stock-works or chimneys, the silica being mostly present in pulverulent form. In the ores found in the granite the quartz is solid. If this should be received as an indication of the condition of all the ore-deposits of the district, it might be predicted that, as depth is gained, the character of the deposits will gradually change from that of irregular deposits and chimneys to a more vein-like structure near and in the granite, which is believed to underlie the entire district. Such, so far as I am acquainted with the facts, is the whole argument which can fairly be urged at present.

The Crown Prince and Frederick mines have lain idle during 1874. The Frederick is developed to the depth of 200 feet, and produced in 1872 about 700 tons of ore. They are both supposed to be on the same vein, which is a contact-vein, between quartzite and limestone, the limestone forming the foot-wall. It is said that the owners will soon commence operations again.

The Enterprise, below the Crown Prince, was worked during the summer of 1874. The ore is of good grade, but the shipments have, so far, been light.

The Toledo mine is remarkable as being the only mine in quartzite in the district that has been worked to any extent. The works consist of an inclined shaft, 280 feet in depth, and numerous drifts and winzes along the vein. It appears to be a true vein, and is traced on the surface for a distance of 800 feet. The ore is chiefly a mixed oxide of iron and quartz, carrying about 100 ounces of silver and a very small percentage of lead. It varies from a hard brown oxide of iron to a pulverulent siliceous substance, carrying from 100 to 300 ounces of silver to the ton. The average daily product is about 10 tons. There are several buildings at the mine, and a tram-way has been built for delivering the ore at the bottom of the hill.

The Emily belongs to the Emily Silver-Mining Company of Pittsburgh, Pa. The lode shows some characteristics of a true vein. The works consist of two tunnels run on the vein, and connected by a winze. The ore contains, besides silver and lead, copper, antimony, iron-pyrites, and zinc-blende. The shipments show an average value of about 90 ounces of silver to the ton and 38 per cent. of lead. Two houses have been built upon the mine, both of which have been destroyed by the snow.

The Flagstaff mine is probably the most extensive deposit of rich ore yet found in the Territory. The works have now attained a considerable depth, and are pushed along the strike of the vein for a great distance. The ore-body varies from a couple of inches to 18 feet in width, showing an average width through the lower works of about 7 feet. The ore is chiefly a carbonate of lead, with considerable galena and oxide of iron. The Company is preparing to erect hoisting-works and to sink a new shaft on the mine, which, when completed, will be a decided improvement over the present under-ground whim. The future of this mine looks bright, indeed, as all the lower levels are said to be in ore and the mine in splendid condition. The official statement in regard to ore produced from the mine and argentiferous lead obtained at the smelting-works (Last Chance) is as follows: Ore produced, 14,767 tons; argentiferous lead, 3,521 tons; average value of lead, $207.25 per ton.

The South Star and Titus mines are situated upon the same line of deposits as the Flagstaff. The character of the ore is similar, though of lower grade, owing to the presence of an excess of iron-ore. The deposit is very irregular, and is faulted at a depth of about 30 feet from the surface, the foot-wall evidently having been raised about 45 feet. The fault, being caused by a lifting force from below, has left the entire upper portion of the vein at about the height mentioned above the lower part. The mine appears to be extensive and to improve with depth, but has been worked in an unskillful manner. Two shafts have been sunk from the surface. The first having been sunk as an incline upon the vein, the second was commenced likewise upon the vein, which it followed to the line of the fault, and then continued down in the hanging-wall to within about 10 feet of the bottom of the old discovery-shaft. A tunnel was run to cut the vein at a depth of about 200 feet from the surface, but was stopped within about 30 feet of the ore, and an incline was sunk about 30 feet from the face of the work to a depth of 135 feet. Its dip was out of the hill and away from the mine. What was the object of sinking this incline it is difficult to say. The ore consists chiefly of carbonate of lead, galena, and plumbic-ocher. There are also large quantities of wulfenite (molybdate of lead) disseminated through the ocherous ores, with occasional specimens of sulphate of lead. The average value of the ore may be placed at about $25 in silver and 30 per cent. of lead. The developments consist of the two shafts before mentioned, a tunnel 390 feet long, and one level run 79 feet along the vein. The rest of the work, being in the ore, is irregular, and has been done without any regard to the future development of the mine. A force is now working upon the tunnel, and the company intends to put the works in better shape during the coming summer.

The Vallejo mine is the next along the ore-belt eastward. It was originally a tunnel-location, and obtained by purchase 390 feet of the South Star and Titus ground. The condition of the deposit and character of the ore are both similar to the South Star mine, though the ore shipped is of somewhat better grade, probably from more careful assorting. The company has made extensive improvements, and everything appears to be in good condition. There is one of Halliday's wire-rope tram-ways, capable of delivering 50 tons per day, at this mine. The average yield of this mine is about 10 tons per day with the present force of eight men, but there is no reason to doubt that with a more vigorous development the quantity could be doubled.

The Caledonia lies next toward the east on the same ore-zone. This mine is developed to the depth of 400 feet, the hoisting being done by a whim. The deposit is similar to both the last-described mines, but is rather more irregular. The shipments have been continued steadily up to the time of the heavy snow-storms, when the mine was shut down for the winter.

The Monitor and Magnet, after two years of inactivity, have again resumed work, with very flattering prospects. The ore is of high grade, averaging 90 ounces of silver per ton and 40 per cent. of lead. The mine is the property of the Monitor and Magnet Company, of Salt Lake City.

The North Star or Bruner mine has lately changed hands, and is now being worked partly on shares. The ore is chiefly plumbic ocher and hydrated oxide of iron, and contains from 10 to 60 ounces of silver per ton and about 30 per cent. of lead. The mine is developed to a depth of 390 feet by an inclined shaft, and over 4,000 feet of drifts and winzes. It is also connected with the Illinois tunnel by a side drift about 200 feet in length. The ore-deposit is evidently badly faulted, as in the South Star and Vallejo mines. In this connection fuller notes will be found in the description of the Emma mine.

The Cincinnati is located about 400 feet east of the Bruner, and has been lately started again. The developments consist of an incline and about 150 feet of drifts. The ore is chiefly galena, and carries about 70 ounces of silver to the ton and about 45 per cent. of lead.

The Highland Chief is situated about 1,200 feet farther up the mountain than the Vallejo mine, and is the first mine on the upper ore-belt. It is developed to a depth of 500 feet by an inclined shaft. The vein is a contact one, between the white and blue limestone, the ore expanding and becoming richer as depth is gained, though this is only the case in the portions of the deposit situated in the blue limestone. Where it is surrounded by white limestone the ores are poor, and the vein is comparatively small. The average daily yield is about eight tons, of an assay-value of 80 ounces silver and 40 per cent. of lead. The mine employs about 20 men.

The Winsor Utah group, consisting of the Savage, Hiawatha, Last Chance, and Montezuma mines, has produced only a limited quantity of ore, the assay-value being about 60 ounces of silver per ton. The works consist of inclined shafts on each of the mines and a tunnel connecting the Savage and Montezuma. About fifteen men are employed here.

The McKay and Revolution, situated about 1,000 feet east of the Winsor mines, are simply two chimneys of ore, connected by stringers. The product during the time they have been worked was about 1,000 tons, the ore assaying about 100 ounces of silver and 30 per cent. of lead. They are now idle.

The Stoker mine is still farther east, on the same belt, and is developed by an incline 460 feet in depth. The ore is milling-ore, and carries about 40 ounces of silver to the ton.

The Davenport belongs to an English company, and has been worked only part of the time during the past year. The product was about 180 tons of ore, assaying 140 ounces of silver and 30 per cent. of lead.

The City Rocks, also owned by English capitalists, has been worked during the greater part of the year. The product was about 700 tons, assaying from 30 ounces to 300 ounces of silver per ton, and about 10 per cent. of lead.

The Grizzly mine has an immense body of ore, but for some cause has lain idle during the entire year. It is owned by Warren Hussey and others.

The Prince of Wales, though not properly one of the Little Cottonwood mines, is still so close to the borders of the district as to be classed as belonging to the same ore-belt that the Winsor, Utah, and Davenport mines are found in. It has produced during the year 2,000 tons of ore, assaying 135 ounces of silver to the ton and 35 per cent. lead. It is developed to a depth of about 600 feet. Walker Bros., of Salt Lake City, are the owners.

The Emma mine has been worked during the greater part of the year. The product has been about 6,000 tons of second-class ore, which was concentrated, and about 300 tons of first-class ore. Work is now entirely suspended, and several attachments are placed upon the mine.

The developments consist of several drifts, run along the deposit as far as ore could be found, one shaft, and the incline known as the Attwood winze. The shaft has a depth of about 280 feet from the tunnel-floor, and is in hard white limestone near the bottom. The fault-line is very distinct, and the groovings on the walls of the fault indicate an upward throw of what is now the upper portion of the vein, together with the country-rock, in which it is inclosed. By this it is evident that the vein was, so to speak, cut in two, the upper portion being carried still higher and the lower parts being only slightly, if at all, displaced. This would naturally leave stringers of ore scattered along the line of break and mix the vein-matter and broken fragments of limestone together in a brecciated mass, such as we now find. This feature also occurs in the South Star and Titus mines, which are also faulted probably by the same upheaval that has faulted the Emma. The South Star mine having passed the fault, which was here very slight, furnishes us with an example of the probable condition of the Emma mine. It is to be regretted that the mine has not been in the hands of energetic men, who would have pierced the fault at any cost while the ore lasted above. The company has, however, but 150 feet of actual sinking to show for three years' work upon a mine which, when it first came into its possession, promised to be one of the leading silver-mines of the world. That the ore-deposit will yet be found below the present workings, I have reason to believe, and that the ores are as valuable below as above, we may infer from the richness of the stringers and branches found near the line of the fault. This fault is plainly traceable through the North Star mine and Illinois tunnel.*

The Wellington mine, situated about a mile south of the town of Alta, has produced during the year about 70 tons of ore, assaying about 65 ounces of silver and 50 per cent. lead. The mine has been idle with the exception of a few weeks during the summer. The ore-body is irregular, and hidden in white limestone.

The Skipper mine is developed to the depth of about 400 feet, and has a continuous vein, evidently once filled with porphyry. The ore assays from 35 to 150 ounces of silver and 10 per cent. lead. The mine is situated upon the northern end of Peruvian Hill, about three-quarters of a mile southwest of Alta City.

In connection with what has been said of the Emma mine before, I insert here an article on the same subject, published lately by Mr. J. H. Morton, in the Engineering and Mining Journal:

There often seem to be vicissitudes in the working-life of a mine, like that in the life of an individual who, after a long struggle, becomes suddenly rich, and again, after a short period of flashy display and recklessness, becomes suddenly poor again, to the astonishment of himself and those around him. This has been the experience of the Emma Company, starting out, in 1871, with a mine whose deposit was then described in the reports as a lake of mineral, whose value was considered as scarcely overestimated in the price paid for it, ($5,000,000;) a mine so vast that one of the directors declared that 100 or even 1,000 tons extracted scarcely made an impression upon it. It seems strange that this magnificent property should have been exhausted in the space of two years, and that now the only prospect seems to be a winding up of the concern and an abandonment of the property they paid such an enormous price to obtain. That the estimates placed upon this property were too high in the first place, is evident; for, although it is said that about $2,000,000 have been already produced, still this amount could not be said to have been in sight when the mine was purchased, and, as any mining-engineer knows, this amount should have been in sight to justify the purchase of the mine at the price paid for it, as it is always a correct rule in limestone

* The notes upon which this opinion is based are those of Mr. Morton, whose intelligence and opportunity of personal examination give me reason to put confidence in his views. As neither Mr. Eilers nor I have examined the Emma since there was any evidence of a supposed "fault" in the deposit, my judgment must be kept somewhat in suspense. But I heartily adhere to what I have said in condemnation of the lack of thorough explorations during the past three years.—R. W. R.

formations to base the value of the mine upon the ore, and at the utmost to allow not more than two and one-half times the value of the ore in sight as the value of the mine. If this $2,000,000 had been in sight in the mine at the time of its purchase, and all questions as to title definitely settled, then the purchase-money was still excessive, as the developments were insufficiently forwarded to tell the actual condition of the mine on either its dip or strike. To say that it was part of the same vein that comprised the Flagstaff, South Star, Vallejo, Caledonia, and North Star mines, was to start from a hypothesis not yet proved to be more than a strong conjecture, and we have seen how often the strongest conjectures in mining have failed. It only illustrates the great risk of expending large sums of money where the prospect of remuneration is even in part based upon hypothetical conclusions.

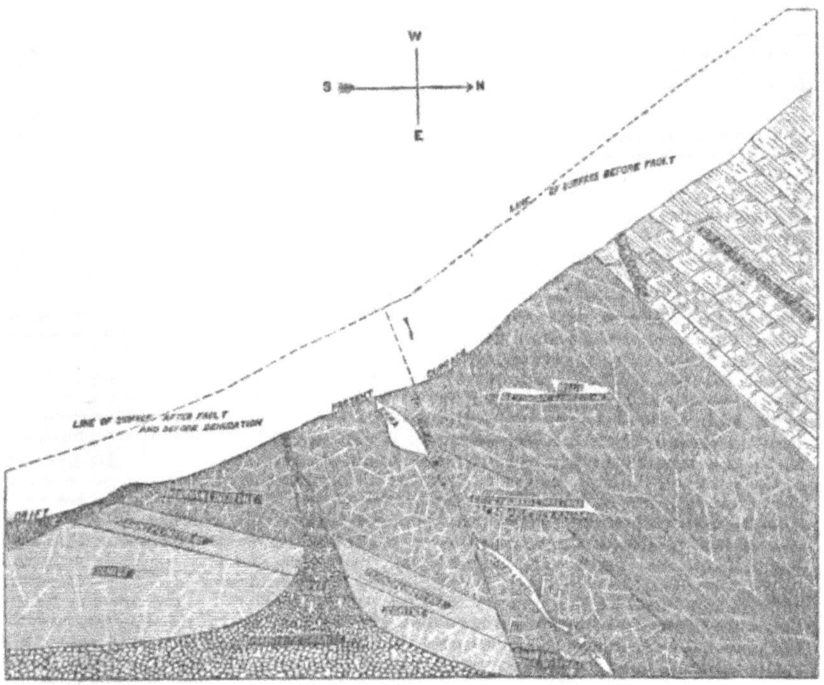


CONJECTURAL GEOLOGICAL SECTION, INCLUDING THE EMMA MINE.

The present condition of the mine may be taken from the reports of different persons made to the company.

Mr. George Attwood, the general manager, reports: "The property is in a most unfortunate condition, all accessible pay-ore having been extracted."

Mr. E. S. Blackwell reports: "The future of the mine depends entirely upon virgin ground."

Mr. Clarence King reports: "The great Emma 'bonanza,' the object of such wide celebrity, the basis of such extravagant promises, is, with insignificant exceptions, worked out."

Mr. A. Murray, F. R. S., reports: "In my opinion, the famous Emma mine is exhausted, and nothing more is to be expected from it but the leavings of the old workings, the scrapings of the walls, the ore which may have been entombed by the cave, the old fillings, and the second-class ore on the dump."

These were the reports made in 1873, and if the condition could be even more discouraging, it certainly was so at the time of my visit in December, 1874. The engine-shaft had been then sunk to a depth of about 82 feet below where the last trace of ore had been found, and was in hard crystalline limestone; from that point down in the old workings, the creeping of the ground was slowly crushing the strongest timbers; the walls, wherever accessible, were picked clean, and all the drifts ended in solid limestone, without a trace of ore in sight. In the old workings a mass of broken timbers, limestone, and mine-fillings, principally the refuse breccia that had been once rejected in sorting the ore, was all that remained of the famous lake of mineral upon which the small matter of extracting a thousand tons or so was to make no impression!

But is the Emma mine really exhausted? So far as the works have anything to show, it is; but unless geological evidences are utterly at fault, it is not; and here a few words in regard to the ore-formations of the district and their mode of occurrence may serve to explain a feature that is not generally understood. About a year ago I published a geological map of the district, and explained it by a set of sections and two articles published in the Utah Mining Gazette. In the first article I showed that the Granite Mountain, to the east of the Emma, was undoubtedly an intrusive one, and subsequent observations have not only confirmed the opinion I then expressed, but have also fully satisfied me of the correctness of an opinion I then held, viz, that the numerous dikes of porphyry which traverse the sedimentary rocks of the district are all connected with the mass of this mountain, and are, in fact, spurs of the granite mass itself. These, in two instances, are first seen at the base of the mountain, putting out into the sedimentary rocks as apparently true granites, and gradually changing in appearance until the true porphyritic structure and appearance entirely prevail. They continue, and finally give out in the limestone or quartzites, and in no instance have I been able to trace any connection between the porphyries and the Western Granite Mountains. Another phase is, that the ore-deposits all seem to be of more recent origin than the porphyry dikes, and to be in almost all cases where the deposits are important in more or less intimate connection with, or in close proximity to, the dikes, in several marked instances occupying the fissures of the dikes from which a part of the porphyry has been removed by pseudo-morphic action. It would seem as if the formation of the ore-deposits was a secondary effect of the porphyritic eruption. The strike of the veins is also conformable to that of the dikes, except in one or two unimportant instances.

It will be seen, by reference to the geological section accompanying this paper, that the Emma mine is also connected with a dike of porphyry, and, as far as the dike can be traced upon the surface, or in the workings of the adjacent tunnels, its direction of strike is in conformity with that of the Emma ore-vein. That it has been instrumental in the formation of the vein is a highly probable conjecture, as it has been traced to near the mouth of the Vallejo tunnel, and I have also found in the ores of the mine slight traces of matter similar to that of which the dike is composed. Passing to the north of the ore-deposit is a deeply-marked line of fault. This fault was, I believe, first found in the main shaft of the mine and afterward in the Illinois tunnel, where it is exposed by a level run by the North Star Company for a distance of about 300 feet, and also by a shaft sunk from the tunnel-level to a depth of about 90 feet. The grooving of the walls of the fault and the condition of the faulted rocks go to show that the throw was from above, or rather a sinking of what might be called the hanging-wall of the break. This, as will be observed, crosses the line of dip of the vein, and has, without doubt, carried that portion of the vein contained in the sunken rocks to a considerable depth below its original position, leaving a dirty trail of ore for some distance below the upper portion of the vein, and also crushing into the lower part of the ore-chamber a sufficient quantity of limestone to fill that portion and give it the appearance of a limestone floor. So compact and hard has this become, that only a close examination can detect the difference between it and the crystalline lime of the wall-rocks. This compact nature has been caused probably by the heat generated by the friction of the walls or surfaces of the fault, and the same cause has produced another effect upon the ores contained in the lower part of the mine—that is, changing the ores from sulphides into oxides. To no other cause can we attribute the anomalous condition of the ores below the water-level of the mine, for long and almost universal experience has shown that in all ordinary cases the change of ore at the water-level is from oxides to sulphides; but here is a case so unusual that, so far as I know, no explanation has ever been offered before, and in examining the break, as exposed in the Illinois tunnel, we find the friction-surfaces of the rocks to have been changed into a compact marble, on each side of the break, of the most beautiful description, owing to crystallizing together of fragments of the different-colored rocks. This feature does not appear, to any extent at least, in the mine, owing to the more homogeneous condition of the rocks lying near the granites, which were probably metamorphosed long before the formation of the ore-deposit.

It will be observed that in the geological section I have carried out a continuation of the vein below the break, marked "supposed lower ore-body." While this may be considered pure theory, there are good reasons for supposing the ore to continue below the fault. It happened that at the time of my visit to the Emma mine I was engaged at the South Star and Titus mines, and found so many points of resemblance between the break in the two mines that I at once set to work to trace out, if possible, the connection of the fault in the South Star with that in the Emma. I have succeeded thus far in tracing it from the Flagstaff, on the west, to the Vallejo, and from the Emma, on the east to the Caledonia, through the North Star, Monitor, and Magnet mines. It is also said to exist in the Caledonia, though I have not seen it myself. If this should prove to be correct, then we have a distinct chain of evidence through all the mines situated upon the Emma vein of the existence of this remarkable displacement, slight in the Flagstaff, being a throw of less than 20 feet, increasing in the South Star to about 45 feet, and of unknown depth in all the mines to the east of the Vallejo. It has been proved in the South Star, and the ore found to continue uninterruptedly downward, thus proving that the rocks have in former times contained a continuous vein of ore, which was subsequently broken and the lower portion carried downward, as shown in the geological section. That the entire chain of mines belong to the vein appears to me very probable, as in all cases, even in the western end of the Emma, there is clear evidence of a difference between the foot and hanging walls, forming what may be termed a strata-vein, or a vein lying between the beds of limestone. The hanging-wall is undoubtedly of Devonian age, though highly metamorphosed on the eastern end; while the foot-wall may probably be termed Silurian, though I am inclined to refer it also to the Devonian era.

THEORETICAL CONCLUSIONS.

We have shown that the connection of the line of fault is continued throughout the entire length of the vein, as far as known; that the fault has been passed in the South Star mine, and, so far as developed, the ore continues downward; that the vein is probably a contact one between different beds of limestone, and we might even extend these points of resemblance by showing a similarity in the ores of the different mines; but the evidence given may be supposed to convey all that is required to give strong grounds for believing that the Emma mine is not exhausted, and that there is good reason for a further extension of explorations downward.

*American Fork district.*—Here the Miller, Wyoming, Pittsburgh, and Wild Dutchman have been worked during the year, with, however, but indifferent success. Late in the year a new discovery, apparently of considerable importance, was made. This is the Queen of the West, located on a spur of Miller Hill, and not far from the Miller mine. Mr. Hillegeist, the superintendent of the Sultana Smelting-Works, reports to me that in the first few days after the discovery 4 feet of solid and comparatively high-grade galena were exposed for a length of 40 feet along the surface. The mine was bought shortly after the discovery by Salt Lake parties. The Miller was explored during the whole year with a small force of miners, and ore was principally taken from the new tunnel near the crest of the hill, which was mentioned in my last report. The Wyoming is now also owned by the Miller company, and is, in fact, mostly on the original location of the Miller mine. It has also furnished some ore. The Pittsburgh was in litigation at the end of the year. None of the ore of this mine, which is of low grade, and present in large quantities, has, to my knowledge, been reduced. The Wild Dutchman has furnished several small lots of rich ore to the Sultana Smelting-Works.

The Miller Company's Sultana Smelting-Works have made only one run during the year, which lasted from July 8 to September 10. Mr. W. Hillegeist, the superintendent of the works, has kindly furnished me the following report:

Material smelted: Ore, sluiced matte and slag from dump, &c., 1,697,902 pounds, containing, lead, 546,925 pounds; silver, 22,301 ounces.

| Product of— | Silver. | Lead. |
|---|---|---|
| | *Ounces.* | *Pounds.* |
| Work-lead, 431,652 pounds, containing | 18,277 | 431,652 |
| Matte, 77,175 pounds, containing | 836 | 12,446 |
| Furnace residues, &c., 43,200 pounds, containing | 544 | 16,160 |
| Ore-dust, 8,170 pounds, containing | 80 | 3,200 |
| | 19,737 | 463,458 |

The loss was therefore 11.5 per cent. of the silver and 15 per cent. of lead.

The total dust collected was 83,360 pounds, which, with the exception of the small quantity above given, was all thrown back at once into the furnaces. The quantities and cost of materials consumed, and the expense for labor, &c., were as follows:

| | |
|---|---|
| 161 cords wood | $643 88 |
| 56,071 bushels charcoal | 7,289 25 |
| Coke | (?) |
| 1,734 pieces fire-brick | 173 40 |
| 26 tons clay | 78 00 |
| Pay-roll, including superintendent's salary | 6,418 52 |
| | 14,603 05 |

The work-lead produced was sold in Salt Lake City for $39,112.38.

*Bingham Cañon*, or *West Mountain district*, has furnished most of the lead-ore smelted in the vicinity of Salt Lake. In this respect the district is fast becoming the most important in the Territory. It is true, the lead-ores of Bingham Cañon are, with the exception of those from one or two mines only, very poor in silver, but their average contents of lead are high, and they carry frequently considerable quantities of iron in some form. Since some of the mines, notably the Winnamuck, Neptune, Kempton, Spanish, and Utah, have passed through the oxidized ores and entered sulphurets. The prevalence of iron as sulphuret in the ore has become so pronounced as to render large quantities of the ores from these mines worthless as lead-ores. As the pyritous ore, however, in nearly all cases contains enough silver to pay for mining, transportation to furnaces, and heap-roasting, it will always be valuable as a flux for the shaft-furnace process, especially as it occurs in great part mixed with little gangue and in large bodies. At the same time, very large and solid bodies of galena have been found in the same mines. In some of them this galena is considerably mixed with zinc-blende, and in that case dressing-works will, of course, have to be called into requisition to remove the objectionable mineral. On the whole, it may be said that at least one-half of the ore in the Bingham Cañon mines ought to be dressed before smelting, and a commencement in that direction has already been made. Besides the large dressing-works of the Utah Company, (English,) two very small (and imperfect) ones have been built for experimental purposes in the vicinity of the Spanish mine. The Utah works were running for a while in the summer on ore from the mine of the same company, and did good work. But this ore changed very suddenly in the stopes to almost solid zinc-blende, containing very little galena, and this so intimately mixed with the blende, that it was no longer profitable to treat it, especially as the silver contents of this ore were no higher than of the ore formerly treated. Later, after the Utah Company had stopped operations in its own mines, the dressing-works did custom-work for a short time.

None of the dressing-works in the district have been running regularly during the year, but sufficient has been done to show to those interested the desirability of a general use of dressing-machinery in the camp.

No new mines of great importance have been developed in the district during the year, but in some of the older ones quite important discoveries have been made by explorations in depth. As most of the Bingham Cañon mines have been described in former reports, I shall here only speak of those which, either by their large yield or by extraordinary discoveries, have maintained or acquired prominence.

The Winnamuck mine, which, on account of the abundance and richness of its ores, has been for several years the most noted in the district, had at the end of the year no considerable amount of oxidized ores in sight, and had entered sulphurets in the lower levels. This change coming upon a somewhat improvident management rather suddenly, the company found itself in a position where new devices had to be decided upon to render its business profitable. For this purpose, Mr. C. A. Stetefeldt, M. E., of San Francisco, was employed, who made a report upon the condition of the mine and the situation of the business. Mr. Stetefeldt has kindly furnished me his report, the substance of which is herewith given.

The developments on the Winnamuck mine are confined to the eastern slope of the cañon, and only to that part which lies above the water-level, as was easy of access by tunnels driven on the ledge, the lowest of which is 47 feet above the creek in the cañon. In this tunnel, about 250 feet from its mouth, an incline has been sunk on the foot-wall of the ore-chimney to a depth of 130 feet. Here water was struck, which made further sinking inconvenient and impracticable, as all the hoisting had to be done by hand, and a large amount of water was to be expected, with increased depth. About 100 feet below the lower tunnel drifts have been started from the incline to the northwest and southeast beyond the limits of the ore-body. This is the whole extent of explorations in depth, and must be the guide in forming a judgment as to future prospects. At present only one ore-chimney is positively known. It was struck in the lower tunnel, about 210 feet from its mouth, and extended upward about 500 feet to the surface. Its horizontal course is northwest to southeast. The quartzite foot-wall dips 45° northeast above the lower tunnel, but below this point the dip becomes much steeper. The ore-channel itself has an inclination of 60° to the northwest. The greatest extent of the ore-chimney, horizontally, has been 240 feet in one of the upper levels, and the maximum width of the ore was 20 feet. In the drift below the lower tunnel the ore-body measures, horizontally, about 170 feet, with a maximum width of 6 to 10 feet.

Southeast of the ore-chimney the formation has been greatly disturbed by a series of faults, running nearly north and south, which have thrown up the strata lying east of the faults. Consequently, in following these faults, the contact-line between quartzite and slate, where alone ore may be expected to be found, was not recovered. In order to do this, a raise should be made from the lower tunnel. Mr. Clayton, who resides in Salt Lake City, and who has studied this fault closely, professes himself willing, at any time, to designate the proper point from which explorations for this purpose should be started. But it is by no means sure that any ore will be found in this direction. One important fact is, however, fully established, namely, that these faults do not disturb the great ore-chimney, described above, in the least, and that there is no reason to doubt its continuance in depth.

Unfortunately, the mineralogical character of the ore has changed already in the Winnamuck mine, before reaching the water-level. In the upper tunnels the ore was oxidized, or so-called carbonate-ore, consisting mainly of carbonate and some sulphate of lead, with silver and a small amount of gold—the silver being present as chloride—mixed with 25 to 60 per cent. of quartz and clay. In the vicinity of the lower tunnel, however, the carbonate-ore gives out entirely, and in this tunnel, and below it, is found a mass of sulphurets, composed of pyrites of iron, galena, zinc-blende, and pyrites of copper, which occur in part massive and in part inclosed in a very hard quartz. All these sulphurets carry silver and some gold. The relative proportion of these sulphurets below the lower tunnel is estimated by Mr. Stetefeldt to be about as follows: 72 per cent. pyrites of iron, 22 per cent. galena, 4½ per cent. zinc-blende, 1½ per cent. pyrites of copper. The proportion of gold to silver is about 1 ounce gold for every 320 ounces silver. A peculiar feature is the distribution of the precious metals in the ore-chimney. As stated above, all of the sulphurets carry silver, but the amount of silver varies between wide limits at different points in the ore-body, without any perceptible change in the appearance of the ore. So sudden are these changes, that if ore at one point contains, for instance, 100 ounces silver per ton, it often falls to 10 ounces at a distance of only a few feet from the first point. Hence, assorting the ore according to its physical characteristics is impracticable, and only by frequent assays can the richer ore be distinguished from that of low-grade.

As a natural sequence of disintegration, the carbonate-ores are higher in lead and silver than the sulphuret-ores. This fact does not appear very plainly from the average assays given below, but it must be borne in mind that, for certain reasons, only high-grade sulphuret-ore was taken out, while it was profitable to utilize carbonate-ore of much lower grade. Below the lower tunnel the value of the sulphurets was found as follows, viz: Average sample from incline contained 19.3 per cent. lead and 20.4 ounces silver per ton; average sample from drift 100 feet below lower tunnel contained 15 per cent. lead and 21.8 ounces silver. Sample from the same drift, southeast of incline, rich in galena, contained 38 per cent. lead and 37.9 ounces silver. Sample from cross-cut southwest of incline, mostly pyrites of iron, contained 4 per cent. lead and 58.3 ounces silver. These latter assays indicate the existence of accumulations of high-grade ore, and must be looked upon as very favorable.

Since January 1, 1872, there were extracted from the mine—

1872. 3,900 tons carbonate-ore, with 34.98 per cent. lead, 51.46 ounces silver.

1873. 4,400½ tons carbonate-ore, with 24.70 per cent. lead, 65.54 ounces silver.

488 tons first-class sulphuret, with 14.42 per cent. lead, 46.90 ounces silver.

205 tons second-class sulphuret, with 6.61 per cent. lead, 15.88 ounces silver.

267 tons quartz-ore, with 29.82 ounces silver.

230 tons assorted, from old dump, 20 per cent. lead, 38 ounces silver.

1874. 2,540 tons carbonate-ore, 23.60 per cent. lead, 54.41 ounces silver.

1,018 tons sulphuret-ore, 12.30 per cent. lead, 55.97 ounces silver.

Total, 13,048½ tons.

The expense of extracting the ore, exclusive of prospecting, was, in 1872, $5.92 per ton; 1873, $6.52 per ton; 1874, $8.88 per ton. The cost of extraction was greater in 1874, because much of the carbonate ore was mined from old stopes, which caused more labor and necessitated retimbering.

The present condition of the mine may be stated in a few words. The carbonate-ore above the lower tunnel has been nearly exhausted. There were, on the 15th of January, 1875, only about 150 to 200 tons of this ore left in the mine. The sulphuret-ore above the lower tunnel was only in part extracted, and what remains may be estimated at 1,000 tons, or perhaps more. But this by itself is not available with the present means of reduction. Between the lower tunnel and the lowest drift, Mr. Stetefeldt estimates an amount of sulphuret-ore of from 4,000 to 5,000 tons. This ore is also at present not available either for extraction or for smelting. From these facts it becomes evident that the Winnamuck mine must turn a new page in its history; it must cease for the present to be productive, and considerable money and time must be expended before returns can again be expected.

The change in character of the ore necessitates also a great change in the method of reduction hitherto employed. It is, therefore, to no purpose to describe and criticise at length the past smelting operations. The carbonate-ores, although refractory on account of a large percentage of quartz, offered no especial difficulties in smelting if mixed with the proper fluxes. It was also profitable to smelt a certain amount of sulphuret-ore, after previous roasting in heaps, or a partial roasting in the reverberatory furnace, together with carbonate-ore, as thereby a saving in iron-ore could be effected. But when we have to treat the sulphuret-ore alone, we find heap-roasting altogether insufficient, and a complete roasting in the reverberatory furnace too expensive, and, besides, the percentage of lead in the sulphuret-ore is too low for smelting.

The cost of smelting in 1873 was $27.48 per ton. During 1874 the yearly statement does not clearly show the real cost of smelting, as heavy expenses for permanent improvements have been put to the account of the smelter, which should have been specified separately.

The losses of metal in smelting have been enormous. The accounts of 1873 show a loss of nearly 20 per cent. in silver and over 25 per cent. in lead. During 1874 sufficient data could not be found in the company's works to calculate the loss, but as much sulphuret-ore has been worked, the losses may be still greater. These losses are caused, first, by dust escaping from the blast-furnaces, in consequence of insufficient dust-chamber capacity; second, by slags high in silver and lead; third, by formation of a large amount of matte, in consequence of imperfect roasting of sulphuret-ore; fourth, by the formation of incrustations in the furnace caused by volatilization of zinc.

*Yearly statements of* 1873 *and* 1874.

The following summary of the yearly statements shows the operations of the company as given:

In 1873 were smelted—

5,119.56 tons Winnamuck ore.
153.16 tons ore bought.
683.64 tons ore from Spanish and Dixon mines.

5,956.36 tons.

Cost of production:

| | |
|---|---|
| Assaying | $1,536 64 |
| Bullion-shipment | 4,608 88 |
| Repairs | 1,517 71 |
| Smelting | 150,754 34 |
| Mine | 70,011 56 |
| General expenses | 21,490 15 |
| Salary-account | 4,900 00 |

| | |
|---|---|
| Loss of material in smelting | $12,975 69 |
| Unspecified items | 327 80 |
| | 268,122 08 |
| Profit | 169,899 33 |
| Production | 438,022 01 |

In 1874 were smelted—

2,098.158 tons bought ore.
752.162 tons Wahsatch ore.
2,978.814 tons Winnamuck ore.

5,829.134 tons.

Besides about 400 tons of roasted matte.

Cost of production:

| | |
|---|---|
| Bullion-shipment | $2,064 79 |
| Traveling | 360 00 |
| Salary | 8,167 36 |
| Assaying | 659 48 |
| General expenses | 5,486 81 |
| Smelting | 213,844 97 |
| Interest | 2,958 41 |
| Wahsatch mine | 7,843 43 |
| Legal expenses | 675 00 |
| Loss of material in mine | 31,602 23 |
| Loss in smelting | 13,879 94 |
| Profit | 3,164 24 |
| Production | 290,706 66 |

This statement is not very satisfactory, and a great many expenses have been charged to the smelting which should be specified, as, for instance, repairs and permanent improvements.

As far as can be learned, there were expended in 1874—

| | |
|---|---|
| For water-pipes | $1,400 00 |
| Reverberatory furnaces | 7,000 00 |
| Building | 1,500 00 |
| Roasting-stalls | 500 00 |
| Dust-chamber | 1,200 00 |
| New engine, blower, including building, &c | 12,000 00 |
| | 23,600 00 |

Besides the ore in the mine, there were on hand the 15th of January, 1875, (estimated:)

| | Per cent. lead. | Ounces silver. |
|---|---|---|
| 300 tons Winnamuck carbonates, with | 15 | 70 |
| 100 tons Spanish ore | 28 | 11 |
| 250 tons roasted sulphurets | 16 | 55 |
| 100 tons matte | 10 | 80 |
| 50 tons sulphuret ore | | 35 |
| 40 tons Titus ore | 34 | 31 |

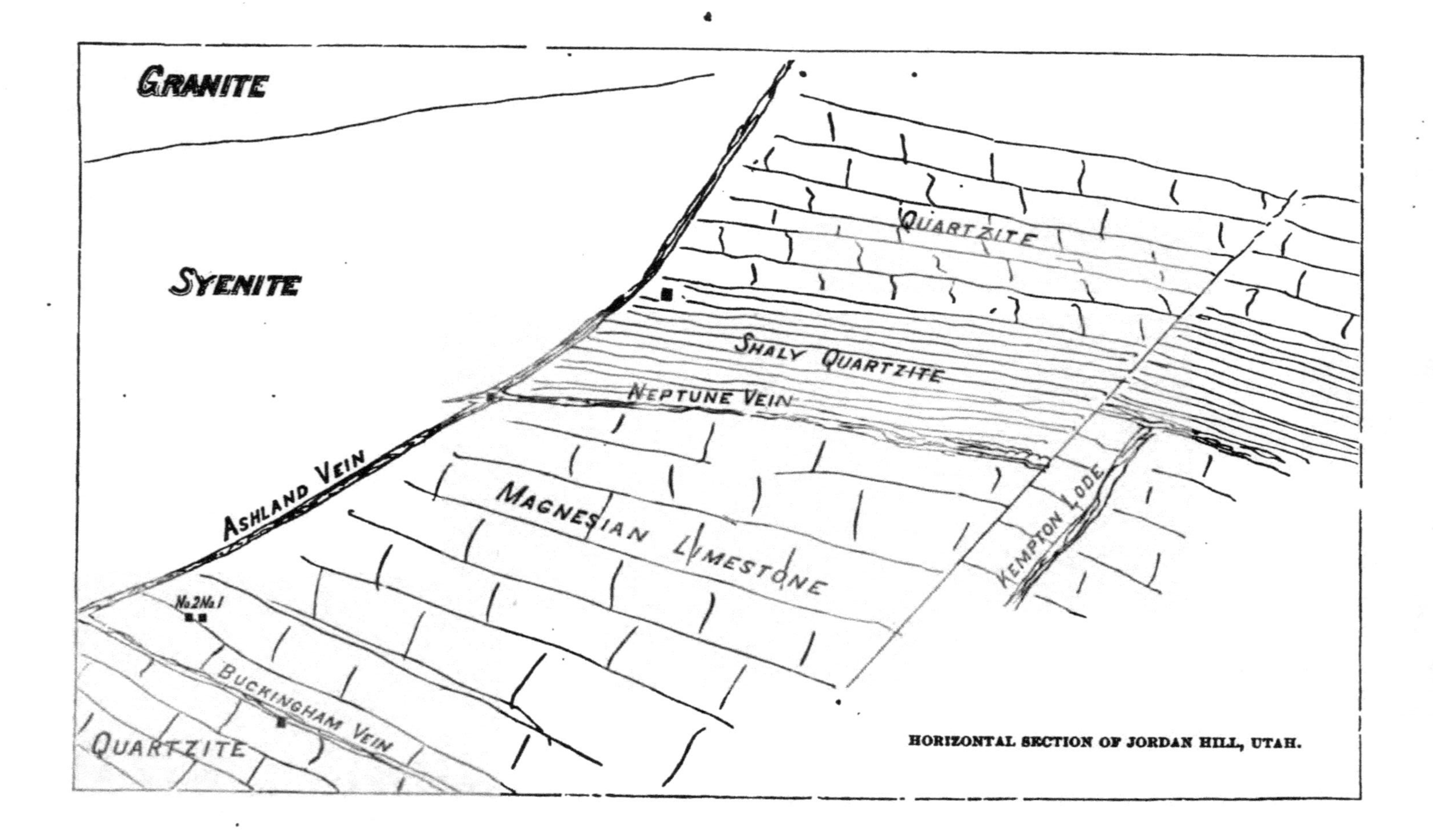


HORIZONTAL SECTION OF JORDAN HILL, UTAH.

Mr. Stetefeldt, in the report before me, states his views as to the best method of working the Winnamuck ores in the future. He recommends a dressing of the ore, thereby separating galena, blende, and iron-pyrites. The galena and blende are then, according to his plan, to be sold to other works, (as the quantities would be too small to supply the Winnamuck Smelting-Works,) and the argentiferous and auriferous pyrites of iron are to be subjected to a chloridizing roasting in a Stetefeldt furnace, and either amalgamated or subjected to leaching with hyposulphite of soda.

After Mr. Stetefeldt's visit to the works, smelting operations were stopped, the ore on hand was sold, and only mining operations were continued, 10 to 15 tons of ore being raised per day and sold to other works. Mr. Stetefeldt's plan of beneficiating the ores in the future is held under consideration by the directors of the company.

Next in importance to the Winnamuck mine are the Neptune and Kempton. They lie on the south side of Jordan Hill, at the head of the main cañon. The geological structure of this part of Jordan Hill is shown in the annexed rough sketch, for which I am indebted to Dr. W. Bredemeyer, the superintendent of the mine.

The Neptune runs northeast and southwest, and dips about 39° to the northwest. It is, however, crossed by a fault, the general line of which is nearly north and south. By this fault the vein has been dislocated considerably, the northeastern portion lying about 120 to 150 feet farther to the north than the southwestern part. The exact distance is not known, the fault not appearing on the surface, and communication with the northeastern portion in depth having only been made through the Kempton. The foot-wall of the Neptune is dolomitic limestone, the hanging-wall a slaty quartzite, which farther to the northwest becomes very solid. The vein dips and strikes with these strata in the main, but frequently bulges out into either. Large dikes of porphyry, which are seen in various localities in this part of the West Mountain district, as well as the syenitic porphyry cutting off the sedimentary strata to the northwest, are apparently connected with the upheaval of these strata, and the faults occurring in them.

The work on the Neptune has been mostly done for purposes of exploration and development. It consists of five inclines, sunk at short distances from each other on the dip of the vein to depths varying from 110 to 120 feet. From these inclines numerous drifts have been run, aggregating more than 1,000 feet in length. An amazing quantity of ore has been developed by this work, the vein showing a width of from 5 to 40 feet, in which the solid ore is often from 3 to 20 feet thick. In the upper levels the ore is carbonate and sulphate of lead, but below the second level it consists of galena, iron-pyrites, and copper-pyrites, mostly containing little gangue. So far not much zinc-blende is visible. Although no stoping of any moment was being done at the time of Mr. Eilers's visit to the mine, it produced, together with the Kempton, 40 tons of first-class and 20 tons of second-class ore per day. The first-class ore from the Neptune assays, according to Dr. Bredemeyer, from 53 to 65 per cent. of lead and 27 to 31 ounces of silver and 0.15 to 0.30 ounce of gold per ton, and the second class from 30 to 35 per cent. of lead and 12 to 14 ounces of silver. In all this ore, but especially in the second class, is a large amount of iron-pyrites, sufficient in all cases to permit of roasting it in open heaps.

The Kempton, which, like the Neptune, is owned by the Schoenberg Brothers, of New York, lies wholly in the magnesian limestone on the east side of the fault spoken of above. It runs about north 15° east and dips 80° to the northwest, and joins the northeast portion of the Neptune, as indicated in the sketch. It has been explored by one main incline 210 feet deep, from which drifts 600 feet long have been run along the vein. It is from 3 to 10 feet wide, the largest expansion having taken place in the lowest level, where 10 feet of solid galena are in sight. The ore from the Kempton is also separated into two classes, the assay of the first class being given by Dr. Bredemeyer as 53 to 74 per cent. of lead and 35 ounces of silver per ton, and of the second class as 32 to 45 per cent. of lead and 11 to 13 ounces of silver per ton.

Dr. Bredemeyer calculates that in the fall of 1874 about 40,000 tons of ore were in sight in these two mines, and although Mr. Eilers professes himself not prepared to indorse these figures from any personal measurement, he reports the quantities of ore in sight as extraordinarily large.

A tunnel is being run from the mountain-side below for the Neptune. This tunnel starts 227 feet vertically below the level of the present shaft-house, and will bring in 377 feet of "backs" on the vein, measured on the incline. It will be 615 feet long to the foot-wall of the vein, and through it all the ore will be brought out after its completion. At present the ore is hoisted from the Neptune through the main incline by means of a 15-horse-power engine, and from the Kempton by means of a horse-whim. The timbering in both mines is done in a substantial, workmanlike manner, and Dr. Bredemeyer is bringing them both up rapidly to the standard of well-planned and regularly-conducted mines.

The smelting-works which have treated most of the ore from the Neptune and Kempton are the Sheridan Hill Works, located on the Jordan River, at the crossing of the Bingham Cañon Railroad. They belong in part to the parties owning the mines, and have treated, besides the Neptune and Kempton ore, other ores, richer in silver, in order to increase the value of the work-lead or "base-bullion" produced. The works contain four small shaft-furnaces, of from 20 to 25 tons capacity per day, with water-jackets and splendid machinery for crushing, pumping, and blowing. But the tuyeres are cast in the same piece with the jackets, and frequent trouble is the result. The crucibles are too shallow for the use of the automatic tap, and the latter is, therefore, also a source of annoyance. These works have, like the Winnamuck, begun to treat the resulting matte, but have no dust-chambers, though the location of the furnaces presents excellent facilities for their construction. The following statement of monthly production throughout the year has been kindly furnished by the manager, Mr. E. Schoenberg:

*Statement of "base-bullion" production of the Sheridan Hill Mining and Smelting Company for the year ending December 31, 1874, (values at Salt Lake City prices.)*

| Months. | Car loads shipped. | Weight of bullion. | Value, (currency.) | | | Total value. |
|---|---|---|---|---|---|---|
| | | | Gold. | Silver. | Lead. | |
| 1874. | | *Pounds.* | | | | |
| January | 12 | 253, 650 | $557 30 | $23, 678 29 | $10, 100 00 | $34, 335 59 |
| February | 6 | 124, 556 | 679 32 | 5, 002 56 | 4, 969 76 | 10, 651 64 |
| March | 6 | 120, 352 | 449 16 | 5, 298 86 | 4, 814 00 | 10, 562 02 |
| April | 11 | 224, 282 | 764 26 | 16, 087 66 | 8, 938 84 | 25, 790 76 |
| May | 16 | 342, 238 | 1, 317 60 | 15, 416 10 | 13, 650 32 | 30, 384 02 |
| June | 22 | 456, 887 | 1, 407 20 | 27, 365 14 | 18, 219 48 | 46, 991 82 |
| July | 27 | 556, 275 | 1, 468 54 | 37, 632. 00 | 22, 179 00 | 61, 279 54 |
| August | 34 | 706, 115 | 1, 864 12 | 49, 679 42 | 28, 143 80 | 79, 687 34 |
| September | 32 | 667, 125 | 2, 201 49 | 54, 120 51 | 26, 581 00 | 82, 903 00 |
| October | 28 | 565, 406 | 3, 109 72 | 33, 864 99 | 22, 560 00 | 59, 534 71 |
| November | 13 | 264, 433 | 1, 250 80 | 8, 217 24 | 10, 539 43 | 20, 027 47 |
| December | 17 | 341, 735 | 1, 541 23 | 20, 923 04 | 13, 625 80 | 36, 090 07 |
| Total | 224 | 4, 623, 054 | 16, 610 74 | 297, 285 81 | 184, 341 43 | 498, 237 99 |

In the same vicinity as the Neptune and Kempton, but more to the west, and along the contact of the syenitic porphyry on the west and magnesian limestone on the east, occurs the Ashland. It runs at the lower part of Jordan Hill, for the first 200 feet north 36° east, and for the next 1,200 feet north 28° east. It dips from 70° to 80° to the northwest, and is generally from 3 to 4 feet wide; but there is at least one ore-chamber exposed in it, at the crossing of the Ashland tunnel No. 1 with the shaft, where the vein for 70 feet in length is from 30 to 40 feet wide. The best ore assays here, according to Dr. Bredemeyer, from 50 to 60 per cent. of lead and 36 ounces of silver per ton. The minerals in the vein are carbonate of lead and galena, with ocherous iron-ore and iron-pyrites. The ore is divided into three classes: No 1 assaying from 40 to 60 per cent. of lead and 18 to 36 ounces silver, and up to $5 in gold per ton; No 2 assaying from 30 to 50 per cent. of lead, 16 to 21 ounces of silver per ton, and containing 5 to 12 per cent. of iron-pyrites; No. 3 assaying from 30 to 40 per cent. of lead, from 12 to 18 ounces of silver per ton, and containing no pyrites. The vein has been exploited by a main incline sunk to a depth of 200 feet vertically below the surface. Fifty-two feet vertically below the mouth of this incline, and 85 feet distant on a horizontal line, the discovery-tunnel has been run in on the vein 161 feet, crossing the incline. Thirty-four feet vertically below this tunnel No. 3 has been run in 140 feet. It is connected with the tunnel above by two winzes. Forty-three feet vertically below tunnel No. 3 No. 1 is driven in over 420 feet, also crossing the main incline. It is connected with the discovery-tunnel by the discovery-shaft. On the Ashland No. 2 tunnel No. 2 is driven in 120 feet, and about 300 feet of drifting has been done besides. Dr. Bredemeyer reported this property in condition to furnish 20 tons per day in the fall. At the entrance of the discovery-tunnel the Ashland crosses the Neptune at an angle of 40°, the Neptune being cut off by it toward the southwest. A porphyry-dike, running from southwest to northeast, crosses the Ashland tunnel No. 1 at a distance of about 100 feet south of the main shaft. The same dike crosses the discovery-tunnel about 120 feet from its mouth; then it runs across the main Ashland shaft, and is seen in all the levels of the Neptune. In the Ashland tunnel No. 1, after crossing this dike, a large ore-chamber was found immediately on its upper side.

From the foregoing it will be seen that this part of Bingham Cañon is geologically most interesting, and future developments will no doubt bring out additional important features.

The Galena and Jordan, which were fully described in my last report, have continued to furnish throughout the year large amounts of ore, high in lead and low in silver. The sulphurets have also been reached in these mines.

The Galena Works, which treat the ore from these mines, have run with very few interruptions throughout the year, and have furnished more lead than any other works in the Territory. The precise amount has, however, not been reported to me. Financially, the works have not been successful, and are not likely to be so as long as the dust is permitted to fly from the chimney in thick clouds and the matte remains without proper treatment. Besides, the construction of the furnaces is susceptible of much improvement, and the frequent long stoppages, generally recurring once or twice a day, for purposes of cleaning out accretions formed just below and above the tuyeres, ought to be avoided by the better composition of charges and a more careful appliance of blast.

The Spanish mine has also been worked during most of the year, and has reached sulphurets in depth. Much ore has been shipped from this mine, and very large quantities are reported in sight.

Besides the above, a very large number of mines have been worked in the cañon, most of which have been described in former reports. With very few exceptions, the ores from all these are tolerably rich in lead, but carry only small quantities of silver. Since in so many mines a mixture of sulphuret of lead, iron, zinc, and copper has been reached, dressing becomes of the utmost importance, and the steps so far taken in this direction by a few will no doubt lead in the immediate future to a general introduction of this branch of metallurgy. The only draw-back to the introduction of dressing-works in the district is the scarcity of water, the water-courses in the upper portion of the cañon being sufficient for the supply of small works only, while the ore-supply would require several very large establishments. These will eventually, no doubt, be located on the Jordan River, in the Salt Lake Valley, the Bingham Cañon Narrow-Gauge Railroad furnishing cheap transportation.

Some placer-mining for gold has been done, as in former years, in various parts of the district, but the result was only the small amount of gold mentioned in another part of this report.

*Ophir district.*—Here considerable activity has prevailed during most of the year. This district embraces Dry Cañon and East Cañon. In the former, the mines furnishing the greatest quantity of ore have been the Hidden Treasure, Chicago, Flavilla, and Queen of the Hills. The Mono, which furnished considerable quantities of extraordinarily high-grade ores during the previous year, was, in 1874, sold to a company. Shortly after the event the mine began furnishing ores of very much lower grade only, and even these were reported scarce in the last quarter of the year. At the end of the year the mine was shut down and in litigation.

The Hidden Treasure is located at the head of the cañon, as described in former reports. The road from the mouth of the cañon up to the mine is very steep and rugged, especially the part below Jacob City, which in many places follows the dry bed of the creek, passing several narrow and steep gates in the limestone layers, where in the wet season the stream forms cascades. Ore-transportation in the ordinary way is therefore difficult, and the owners of the Chicago mine, located immediately below the Hidden Treasure, did, therefore, a wise thing in erecting their wire tram-way for that purpose.

The Hidden Treasure is called a contact-vein, between shale on the hanging and limestone on the foot wall. Properly it is no vein at all, but an irregular deposit between the strata of the two rocks mentioned. It does not even always follow the contact between two specific strata, but leaves the line, which it may have occupied for some distance, quite suddenly, taking a course at right angles to the main one in some rent of the foot-wall, which it follows then to the next line of bedding below, to resume here a course parallel to the former one. Besides this, it is by no means continuous, the ferruginous vein-matter disappearing sometimes altogether. In other places it opens up into considerable chambers, which may or may not be in the general line of stratification. They are caverns formed by the dissolution of the lime, which have been afterward filled with the mineral, and not fissures formed by volcanic agency and afterward filled.

During the year the tunnel driven in across the limestone for 420 feet has been connected with the irregular works above, and the extraction of ore is thereby very much facilitated. Mr. George P. Lockwood, who was superintendent of the company's smelting-works during the latter part of the year, sums up the work done in the mine during 1874 as follows:

An average of 40 men have been employed during the most of the time. In the principal incline, the Summit, a depth of over 500 feet has been reached, effecting connection with the tunnel. About 200 feet of this sinking was done during 1874. The Lawrence shaft has reached a depth of over 300 feet, 250 feet of which was done in the past year. In the Magazine 140 feet have been added, making it now 180 feet deep. In addition, drifts have been run connecting all the shafts, and opening above 600 feet of ground along the vein east and west. A tunnel was run in from the western face of the hill 420 feet, which enters the ore-channel through the hanging-wall. From the point of intersection drifts have been run east and west on the course of the deposit, the west one intersecting the Summit shaft at the bottom, affording ventilation and cheapening the mining. For some months previous to the connection with the tunnel work was suspended in the bottom of the mine on account of foul air and the expense of raising the ore. The dip of the vein was so variable, that three whims were necessary for hoisting to the surface. This limited the ore-supply, which can be readily quadrupled in the coming year. The ores are carbonate, sulphate, and sulphuret of lead, and the various oxidized ores of copper, such as malachite, azurite, and red oxide. Copper and iron sulphurets occur sparingly. The gangue is hydrated oxide of iron and calcspar, both carrying a little silver, and forming, often in small threads, the only connection between the ore-bodies.

The Waterman Smelting-Works, which belong to the same parties as the mine, and treat its ores, are located nine miles distant, at the northern end of Rush or Stockton Lake. They contain one shaft-furnace, connected with a very efficient condensation-chamber, described in the metallurgical portion of this report. The furnace is a round one, having at the tuyeres a diameter of 3 feet 4 inches. Height from bottom of hearth to slag-spout, 22 inches; to center of tuyeres, 33 inches; from tuyeres to charge-door, 11 feet. There are 4 water tuyeres, with 3-inch nozzles. A Baker rotary-blower, running at 96 to 102 revolutions per minute, supplies the blast, at a pressure of $\frac{3}{8}$ to $\frac{1}{2}$ inch mercury. During the year the furnace was in blast 170 days. There were consumed—

| | | | | | | |
|---|---|---|---|---|---|---|
| Hidden Treasure ore. | 3,239$\frac{435}{2000}$ | tons, assaying | 26 | ounces silver and | 38 | per ct. lead. |
| Purchased ore ...... | 100 | tons, assaying | 25 | ounces silver and | 15 | per ct. lead. |
| Dust from chamber . | 296$\frac{1338}{2000}$ | tons, assaying | 28 | ounces silver and | 25 | per ct. lead. |
| Total ore..... | 3,635$\frac{1773}{2000}$ | tons, averaging | 26.1 | ounces silver and | 36.4 | per ct. lead. |

Fuel: Wood for engine, 425 cords, at $5; charcoal, 98,811 bushels, at 18 to 22.5 cents, (exclusive of waste, which was not less than 10 per cent.)

There were produced: Work-lead, 17,539 bars, weighing 749$\frac{123}{2000}$ tons, assaying 70 ounces silver per ton; matte, 285 tons, assaying 35 ounces silver, 35 per cent. lead, and 15 to 35 per cent. copper; speiss, about 15 tons, contents not given. Neither matte nor speiss has been reworked. Over 750,000 pounds (nearly 11 per cent.) of dust and fumes have been gathered. The average moisture in the ore, as brought to the works, was over 20 per cent. in the winter and 12 per cent. in the summer. The average moisture of the ore, as it went into the furnace after having been partially dried, was, according to Mr. Lockwood, about 15 per cent. The freight for ore from the mine to the furnace was from $4.50 to $8. per ton, according to the condition of the roads; that for the work-lead to Salt Lake City was $7 per ton. The latter has been much reduced since the Utah Western Narrow-Gauge Railroad has been finished to a point half-way to the works. The Chicago mine is reported to have now reached a depth of 1,100 feet. It has produced a large amount of ore, which was shipped over its tram-way down the steep part of the cañon, and thence by teams to the Chicago Smelting-Works, on Rush Lake. These works have smelted, besides the ore from the Chicago mine, an equal if not greater amount of purchased ore from all parts of the Territory, and some even from Montana. As the location of the work is not at all favorable for the working of custom-ores, except those coming from the immediately-surrounding districts, it is, in my opinion, doubtful whether the treatment of bought ore could have been very profitable, especially as ore-prices were high in Salt Lake during nearly the whole year, and a great part of the ore had to be brought to the works from the city by teams. The ore smelted and the product of the works during the year 1874 was, according to Mr. E. I. Dowlen, the accountant of the Chicago Company, as follows: Ore smelted, 7,641 tons; lead produced, 188 car-loads, (11 tons each,) worth $405,190.75. Whether this value includes only the silver, or both silver and lead, is not specified in Mr. Dowlen's report. I suppose the latter.

Only about 450 tons of flue-dust have been saved in the very imperfect condensing-apparatus at the works, and 240 tons of copper-matte, carrying 60 ounces of silver per ton, have been produced and not yet reworked.

The Flavilla and Queen of the Hills have been consolidated during the year. This property is now opened to a depth of 500 feet, and has yielded considerable good ore during 1874. A tram-way was built late in the year from the mines down to Gisborn's toll-road, and a new engine has been put over the shaft for hoisting-purposes.

Besides the above, the Sacramento, the Mount Savage Company's mine, the Jefferson, and Brooklyn have yielded fair quantities of ore, and a large number of mines have been worked more or less, without having realized as yet the high hopes of their owners, though many are in a fair way to success.

In East Cañon, a large number of the mines, fully described in my two preceding reports, have been worked during the greater part of the year. Prominent in the production of ore have been the Zella, Lion, Blue-Wing, Tiger, and Sunnyside, all of which furnish milling-ores. The Miner's Delight, which carries smelting-ores of low grade, has also produced considerable ore, but of a character which renders dressing before smelting a necessity. Captain Longmaid, formerly of the Utah Silver-Mining Company's Works at Bingham, took charge of the Miner's Delight in the fall of 1874, and it was reported at that time that the erection of concentration-works would at once be commenced. But this has not been the case, and the execution of the plan is said to be deferred till spring. The Pioneer Mill has been kept running nearly throughout the year, but was stopped in December, on account of the scarcity of water. The Baltic Mill, which has also been in operation during a considerable part of the year, shared the same fate.

*Columbia district*, which was organized in the latter part of 1871, in the mountains south of Rush Lake, has attracted some attention during the present year, especially since the Ohio Silver-Mining and Smelting Company has bought a number of mines here, comprising the Chanticleer, Buckskin, Idaho, Washington, and Chimney Corner. Other mines of some note are the Champion, Montana, Augusta, Mammoth, Western World. Altogether, over 300 locations have been made. The country-rock of the district is granite and slate, and the ores are low-grade argentiferous lead-ores, which can, no doubt, be profitably reduced as soon as the Utah Western Railroad, now in course of construction, reaches the district.

*Clifton mining-district* is located one hundred and fifty miles southwest of Salt Lake City, in the Goshoot Mountains, and covers an area of twelve square miles. Nearly 600 locations have been made here. The country-rock is granite, quartzite, and limestone, and the ores occurring in these rocks are principally carbonates of lead and galena, though some milling-ores occur. Wood and water are reported abundant.

In *Tintic district*, seventy miles south of Salt Lake City, mining has not been prosecuted with the energy of former years, principally, it seems, on account of bad management of the more important mining properties, and also in some measure by reason of the considerable distance of the district from the railroad. For a brief description of the district, I quote from a late publication by Mr. J. C. Cameron, a mining-engineer, of Salt Lake City:

*Tintic mining-district* is situated in Juab County, Utah, and distant from Salt Lake City about seventy miles, in a southerly direction. It is approached therefrom by the Utah Southern Railroad, via Provo, thence by stage or other conveyance.

This mining-district is located in and embraces a portion of the Oquirrh range of mountains, which also contains on the north the Ophir, Dry Cañon, Stockton, Tooele, and West Mountain mining-districts.

The most productive and valuable silver-bearing mines which have been opened and worked within Tintic district are the Mammoth Copperopolis, Sunbeam, Gold Hill, Julian Lane, Norwegian, and Eureka Hill. The greatest bulk of the ores extracted belong to what are known as the free-milling class, and are found in veins of 1 to 10 feet in width and over, coursing north a few degrees east, dipping nearly vertically, and following the stratification of the belt or zone in which they are contained. This is composed chiefly of a fine-grained grayish granite. The matrix of the veins consists chiefly of a granulated quartz, carrying in the best mines $40 to $80 in silver, some gold, and occasionally a large percentage of copper carbonates, sulphurets, oxide of iron, zinc-blende, bismuth, with comparatively small quantities of lead, either carbonate or galena.

The granite of this district forms a junction, and is capped in some instances upon its eastern limit by a belt of limestone, of the Devonian and Carboniferous periods, interfoliated by numerous strata of quartzite, (evidently an altered sandstone.) These belts do not conformably follow each other, and are all more or less metamorphosed. The limestone belt contains the Mammoth Copperopolis, Eureka Hill, Mineral Hill, and others.

The Mammoth Copperopolis, owned by a London company of the same name, is a very large vein of ore, and opened by tunnels, under the superintendence of Thomas Couch.

The Eureka Hill mines were purchased by Capt. E. B. Ward, of Detroit, Mich., and are opened to an extent of 2,000 linear feet.

The Mineral Hill mines are worked by shafts and tunnels to an extent of 1,300 linear feet, and are being prepared for extensive operations during the winter. Hoisting-works are being erected, and additional improvements valuable to the mine are in progress. They are owned by Leetham Brothers Mining and Smelting Company, and under the management of John Leetham, esq., a member of the firm.

The Crismon Mammoth is evidently a northerly extension of the Mammoth Copperopolis; is opened by shafts and tunnels, exhibiting at one place a vein of 43 feet in width. The ore extracted is rich in copper carbonates, gold and silver ore, selected specimens of which gave $43,000 in gold. It is owned by Crismon Brothers *et al.*, who have lately leased and put in operation the Germania Copper-Smelting Company's furnace, for the reduction of these ores.

There are in this district a number of very promising mines, which will undoubtedly prove valuable properties to their owners. The facilities possessed by Tintic for the conversion of the crude ores into commercial values consists of four first-class mills and five substantial blast-furnaces. The greater proportion of the ores exposed belongs to the milling class, and the large quantities attainable at a very moderate expense in regard to mining and transportation of product evince conclusively an inefficient milling capacity. Three of the furnaces are erected for the reduction of copper-ores, which are abundant, and will form a valuable auxiliary to the interest of the district.

The Wyoming Company own at Hommansville a 10-stamp mill with Stetefeldt furnace attached, and are working ores chiefly from their own mines, and small lots by purchase.

The Shoebridge Company's mill is situated about six miles southwest from Diamond City, (the headquarters of the southern portion of the district,) and consists of a 15-stamp mill and Aiken furnace. It is working ore from the company's own mines, as well as purchased lots.

The Tintic Mining and Milling Company, provincially known as Miller's mill, are working 10-stamps by the wet process, and use ores from the Mayflower, Gold Hill, and Tesora mines.

The Copperopolis mill consists of ten stamps, erected for the purpose of working the ores from the mine of the same name. This company also owns two copper-furnaces. Owing to some difficulty with the London management, no mining is carried on at present. This company's property is generally considered a valuable one, and it is currently reported that operations will be resumed shortly.

In *Camp Floyd district*, fifty-five miles southwest of Salt Lake City, on the southern slope of the Oquirrh Mountains, mining operations have been unimportant throughout the year, though the 20-stamp mill of the Camp Floyd Silver-Mining Company of London has been running for a short time. Only the smallest part of the ore worked, however, came from the mines of the immediate vicinity.

*Nebo district* is eighty-five miles south of Salt Lake City, and within a mile and a half of the proposed extension of the Utah Southern Railroad. It is about nine miles long and five miles wide, and contains principally lead-ores poor in silver. The distance from Tintic is between fifteen and twenty miles, two valleys, the Goshorn and Juab, intervening between the two districts. The mines of Nebo are found in a compact gray limestone, the strike of the ore-deposits following the stratification of the country-rock, which is southwest and northeast. The dip is northeast, the angles varying from 20° to 17°. Most of the mines so far opened are in the vicinity of Bear Cañon. Few of them have been worked to any extent, the most promising enterprise in the district being the Mount Nebo tunnel, by which it is intended to intersect the Mountain Queen, Olive Branch, and Hagar ore-deposits. All these deposits contain near their outcrops carbonates of lead ranging from 1 to 6 feet in width, and assaying high in lead but low in silver. The advent of a railroad in this vicinity will no doubt have to be awaited before mining can be rendered profitable.

*Star district*, about two hundred and twenty miles a little west of south of Salt Lake City, contains a large number of veins carrying smelting-ores, which are richer in silver than most of the Utah lead-ores. None of the mines are developed to any great depth. Considerable work has been done on those belonging to Campbell & Co.—the Harrisburgh, Knarah, Little Bilk, Central, and Mountaineer. Two of the shafts on these mines are down 100 feet, and considerable bodies of rich galena have been found. On another mine of the same company, the Big Mormon, the shaft is down 150 feet. On the Burning Moscow, belonging to O'Niel & Co., the shaft is 170 feet deep, and contains a 4-foot vein of carbonate of lead. The Elephant mine has been sunk upon to a depth of 40 feet, and is reported to show a 12-foot vein of $100 ore. From the Mammoth, situated in Middle Camp, 200 tons of high-grade ore are reported to have been shipped in December. The Savage contains, at a depth of 170 feet, a 4-foot vein of galena and carbonate, and more ore is reported to have been shipped from this mine than from any other in the camp. The Baltimore mine, with a shaft of 70 feet in depth, has made several shipments, reported to have assayed $180 per ton. There are a number of other veins which have shipped ore of good grade, all of which has been smelted at the shaft-furnace located at Shauntie

Springs. Very little outside capital seems so far to have come into the district, though it is certainly a promising one.

In *Parley's Park* considerable work has been done during the year. The McHenry mine, which raised such an excitement in Salt Lake some years ago, has been opened by two tunnels—an upper one, which is started in west of the large outcrop, and a lower one, which is located immediately under the bald rock. Ore has been encountered in both. In the latter part of the year the ore was quarried from the surface, many deep holes being bored at a time and exploded together by means of electricity. A large quantity of ore was thus taken out, which was assorted, so as to make three classes, said to assay respectively $400, $100, and $20 per ton. To treat this ore an excellent 20-stamp mill, with the necessary pans and settlers, has been built, which was expected to commence running on September 1. But for some reason it had not done any work at the end of the year, so far as I have learned.

On the Flagstaff, high up on the divide, there is an incline 260 feet deep, which is in ore all the way. The latter is on an average 4 feet thick. There are also 10 or 12 prospecting-shafts along the vein, and 1,000 to 1,200 tons of ore, reported to assay from $30 to $60 per ton, have been taken out. The company was building a 20-stamp mill during the summer, which is situated in Pike City, about six miles from Kimball's Station, and near the McHenry Mill. The distance from the mine to the mill is about three miles. This mill has been running during the latter part of the year, and seems to have been well supplied with ore.

The Ontario, which also contains free-milling ore, is opened by a number of tunnels and several shafts. The mine has been worked during almost the entire year, and a considerable amount of ore has been taken out, the first quality of which assays over $100 per ton.

The Pioneer mine contains smelting-ores which assay from 15 to 30 ounces in silver and 20 to 30 per cent. in lead. It has been worked steadily.

The Walker & Webster is another mine carrying smelting-ores, and opened by several tunnels. The ore assays from 20 to 60 ounces of silver and about 40 per cent. of lead. According to Mr. J. E. Clayton, there are thousands of tons of ore in sight in this mine. It was worked during the fore part of the year, but was stopped in the summer, when Mr. Eilers visited the vicinity.

The Piñon is opened by four tunnels and two shafts. It contains smelting-ore which, when assorted, contains 45 per cent. of lead and 42 ounces of silver. About 450 tons have been shipped, and 6,000 tons were reported in sight in August. The mine was not worked at that time. Besides the above, there are at least 50 other mines, mostly developed to an inconsiderable extent. A few of them, however, have small quantities of ore on the dumps.

## THE CASTLE VALLEY AND SAN PETE COAL-FIELDS.

The Castle Valley and San Pete coal-fields have attracted considerable attention during the past year, and are, no doubt, of the utmost importance to the lead and silver industries of the Territory; but besides this, they are of very great value in connection with the future iron-industry of Utah, of the development of which there can be no doubt, in view of the immense iron-deposits in a large part of the southern half of the country. In these fields occur coals which permit of coking. Considering that, besides the beds of Trinidad, in Colorado, and one at Mullen's Pass, Montana, these veins are the only ones known in the far

West where large quantities of good coke can be made, their importance for industrial purposes is at once apparent.

Mr. J. E. Clayton, of Salt Lake City, has kindly furnished me with notes of a late visit to the Castle Valley field, and also with the accompanying ideal section.

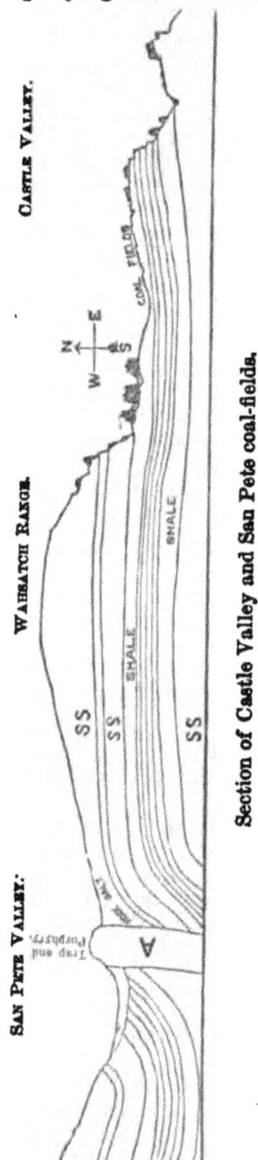

Section of Castle Valley and San Pete coal-fields.

The Castle Valley coal-field is situated at the base of the Wahsatch range, and fifty-five miles, by the Gunnison trail, east of Salina, on the Sevier River. The pass through the mountains, by the way of Salt Creek and one of its southern branches, is one of the best for a wagon or railroad in any portion of the Wahsatch range.

The formation is sandstone, with two very thick beds of shale or indurated mud. There is but little trace of stratification in these shale-beds, but they are cracked into all sorts of irregular shapes. There are occasionally calcareous nodules embedded in them, but very few fossils have been found. Mr. Clayton found a few *Ostreæ* at the bottom of the upper shale-bed, next to the zone of sandstone that contains the principal coal-measures, and also found a pavement-scale of a shark or ganoid, as well as a petrified trunk of a palm-tree, one foot in diameter and 8 to 10 feet long, the latter embedded in the sandstone below the upper shale-zone. Both the sandstone and shale contain very little lime. The formation is divided into three zones of sandstone and two of shale, as shown in the section. The lower sandstone was not examined.

The lower shale-bed is at least 600 feet thick, and is darker than the upper one, its color along the exposures being dark-bluish gray. The upper shale-bed is lighter colored on the surface, and dark gray below. The middle sandstone, lying between these two shale-beds, is about 800 feet thick, and in this are the great coal-seams of Castle Valley. There are from three to five seams, the two lower being the most important, on account of their size, uniformity, and superior quality.

No. 1, the lower bed of coal, is very hard and compact, but little disposed to slack, and in some places along the face of the bluff, where the sandstone stands in perpendicular cliffs, the coal-seam shows faces almost flush with the overlying cliffs. The floor of the seam is a hard grayish-white sandstone, made up of clay and sand, for from 2 to 6 feet below the coal, at which point the clay disappears and the underlying sandstone is yellowish gray, part of it being quite hard and semi-vitreous. The roof is composed of the ordinary gray sandstone of the middle zone. The thickness of seam No. 1 varies from 3 to 8 feet, and will average probably 5 feet of good coal. There is a small seam of black shale near the roof from 1 to 4 inches in thickness. There is no other waste in the bed. A firm sandstone roof rests immediately on the coal. This coal burns freely, with very little smoke, leaves 3½ per cent. of white ash, and is remarkably free from sulphur or iron. For the generation of steam, and for domestic use, it has no superior in this country.

No. 2 lies about 40 feet above the one just described, and contains 6 to 20 feet of clean, solid coal, without a seam of shale in it; but there is quite a bed of shale overlying the coal. This overlying shale contains a number of small seams of coal, from 4 inches to 4 feet thick, and is from 20 to 40 feet thick at the point where a cut was made into it. At one point measured, this No. 2 coal-bed was found 21 feet 6 inches thick, and without a trace of shale in it. The floor is like that of No. 1 bed, a white argillaceous sandstone. The quality of the coal is not as good as that of the lower bed. It is softer, more inclined to slack, and has more ash, (6 to 10 per cent.,) but is also free from sulphur or iron.

Nos. 3 and 4 lie in the upper beds of the middle zone of sandstone. They are not so regular or uniform in size as those described before, but the coal is of the same general quality. These beds vary from 1 to 4 feet in thickness, judging by the exposures examined, except at one or two places, where the outcrop is 6 to 8 feet thick.

The principal reason that the upper beds could not be examined as well as the two lower ones is, that they have been burned out over areas of several thousand acres, and several places were found where they were still burning. The burnt districts could be plainly traced by the color of the sandstone, which is brick-red in these localities.

These overlying coal-seams must have been very thick in places, judging from the effects of the fire on the sandstone. There are places where the latter is not only burned red, but is actually glazed and vitrified by the heat. Masses of glazed fragments, adhering together in large bowlders, can be seen in the side ravines of the main cañons.

The position of these beds is horizontal on a line northeast and southwest, but in the opposite direction, northwest and southeast, they lie in gentle waves, making very flat anticlinal and synclinal axes, as shown in the section.

The available area of the coal-fields is about sixty square miles. The extent, so far as at present known, is twelve miles long, northeast to southwest, by four to eight miles wide from northwest to southeast.

Nos. 5 and 6 are two well-marked coal-seams, about 1,200 to 1,500 feet above the middle zone. These beds lie nearly horizontal, showing a slight rise toward the west, but, after passing the center of the mountain, the beds all dip in a curved line toward the Sevier River and the San Pete Valley, which is a long, narrow valley inclosed between two branches of the Wahsatch range. Along the eastern side of this valley there is a great fault or break in the formation, and extensive outflows of porphyry, trap, and occasionally true lava, have taken place. The position of the sedimentary beds east of the break is well exposed in Salina Cañon, but their positions west of the eruptive rocks can only be inferred by the exposures of the beds west of the valley. The San Pete coal-measures are, in Mr. Clayton's opinion, the same as those shown in the mountain section as Nos. 5 and 6. In the prolongation of the Castle Valley section to the west, across Sevier River Valley, the position of large salt-deposits, just east of the eruptive rocks, will be remarked. It is readily seen why this salt was so deposited. The dip of the beds west of the summit of the east range, underlaid by shale, brought the drainage down against the hot eruptive rocks; the water was vaporized and escaped upward through the crushed and broken shale, and the salt crystallized in the interstices of the broken mass.

Large deposits of gypsum also occur along the line of fracture, near the salt.

To render the Castle Valley coal-fields available, a railroad about eighty-five miles long will be required to connect with the Utah Southern Road at Sevier River Bridge. If the Central Pacific Railroad Company should build a branch starting from Wells Station, or from Tecoma, they would have no obstacles until they reached Salina Cañon, a series of low gaps through the Nevada ranges, making the road virtually a valley line from Tecoma to Gunnison's, or Salina Pass, in the East Wahsatch range. From Castle Valley to Denver, to connect with the Kansas Pacific, is the only difficult portion. This line would, however, be two hundred miles shorter than the present Pacific lines.

In regard to the San Pete coal-field, I am indebted to Mr. George P. Lockwood, M. E., of Salt Lake City, for the following further information:

The San Pete coal-field is located in San Pete Valley, ninety miles south and ten miles east of Salt Lake City. As far as known, it comprises 4,080 acres on the west side of the valley. The sandstone in which the coal is found is estimated to be here 1,500 to 2,000 feet thick. In these rocks, five distinct coal-veins have been found, which vary in thickness from 8 inches to 4 feet. The two principal veins are respectively 4 feet and 26 inches thick, and are separated by 8 inches of limestone. The strike of the veins is about due north and south, and they dip at a slight angle to the west. The coal-measures are cut by four cañons at right angles to the strike of the veins. Most of the developments are at those points where the veins are exposed on the sides of the cañon.

On the north, in New Cañon, the vein has been opened by three working-levels—50, 100, and 400 feet long, respectively. In Old Cañon, one mile south, there are three levels opened, of 50, 200, and 800 feet. In Big Cañon, about two miles farther south, there are three openings of a limited depth. The 4-foot vein has increased at this point to 5 feet. In Axe-Handle Cañon, about three miles farther south, the vein has been opened and found to be 6 feet thick, the interstratum of limestone having here disappeared. This shows an increasing strength in the vein in going from north to south. From two to three miles north of New Cañon the vein is found near the top of the mountains, while below Axe-Handle Cañon it appears in the foot-hills. The 800-foot level at Old Cañon runs north, about parallel with the strike of the vein.

At the heading of this level, coal was mined for 20 feet on either side. The roof in this chamber and in the level is so firm as to need but little support, showing that a minimum amount of coal need be left in the workings as pillars, and wide chambers can be laid out. The coal separates clean from the walls, and the floor is firm and regular.

In regard to the estimated amount of coal in these fields, I gather the following from the report of Mr. Isaac Stone, M. E., F. G. S., of North Wales: "The length from north to south is about eight and one-half miles, and the width from five to six miles. The outcroppings of the coal, at right angles to its strike, looking from the east, may be distinctly traced. From what I saw of the strata underlying the most southern portion of the property, I have no doubt that the same coal will be found there also, making a total length of eight and one-half miles. * * * The strata, too, parallel with the dip-line to the west, can be clearly and distinctly seen, and a definite conclusion may be arrived at as to their continuity westward." Altogether, Mr. Stone estimates the area of the coal-fields to be at least 7,760 acres, with an average thickness of fine-working coal over the whole area of 3 feet 6 inches. Future developments may increase the area as well as strength of vein. It is generally believed that, by sinking, new veins will be found, which would greatly increase the resources of the fields.

The coal is pronounced by H. S. Poole, F. R. S., of London, Prof. F. A. Genth, of the University of Pennsylvania, and Mr. Stone, to be bituminous. It is hard, dense, and of uniform structure. Color, black; streak, black; luster, medium to bright. Specific gravity, undetermined, but low. Cleavage, uniformly parallel to the faces of the cube.

Chemical examinations have shown the following results:

By H. S. Poole, F. R. S.:

| | | | |
|---|---|---|---|
| Top coal: | Volatile matter | 35.50 | (includes HO.) |
| | Coke | 64.50 | (includes ash.) |
| Bottom coal: | Volatile matter | 33.70 | (includes HO.) |
| | Fixed carbon | 54.29 | |
| | Ash | 12.01 | |
| Middle coal: | Volatile matter | 32.00 | (includes HO.) |
| | Fixed carbon | 56.80 | |
| | Ash | 11.20 | |

Average percentage of coke, 66.13 per cent.

By Prof. F. A. Genth:

| | | | |
|---|---|---|---|
| Moisture | 1.16 per cent. | Ash | 11.18 per cent. |
| Volatile matter | 32.91 per cent. | Coal contains 2.88 per cent. of sulphur. | |
| Fixed carbon | 54.75 per cent. | | |

By Fred. Claudet, of London:

| | | | |
|---|---|---|---|
| Ash | 10.54 per cent. | Sulphur | 1.80 per cent. |
| Coke | 62.10 per cent. | | |

By Dr. John Percy, London:

| | | | |
|---|---|---|---|
| Coke | 62.00 per cent. | Ash | 10.00 per cent. |
| Volatile matter | 28.00 per cent. | | |

Dr. Genth says that "the coking qualities of this coal are not inferior to those of the best Pittsburgh coal, and that the coke made of the same is apparently of excellent quality, and sufficiently dense to bear the burden of a blast-furnace." Some of the coke, manufactured in a crude way, was used in the blast-furnaces of the Germania Works, six miles south of Salt Lake City. When mixed with one-half of Saint Louis coke it worked well. Alone, it matted, forming a cake through which the blast could not penetrate. This indicates plainly that the coal had not been exposed to a sufficient heat, and there is no doubt that upon a better coking it would be found suitable, at least, for the use of the lead blast-furnaces, when the charge-column is not more than 12 to 20 feet high. Its principal demerit now seems to be lack of tenacity. As a steam-generator it is claimed that two tons of San Pete coal equal three tons of Wyoming lignite.

The sandstones occurring in the coal-measures are of a light-gray color. During 1875 it is intended to connect these fields by a narrow-gauge road with the Utah Southern Railroad, at Nephi, and thus to bring the coal into the market.

# CHAPTER VII.

## COLORADO.

The collection of statistics for this Territory I have intrusted to Mr. Theodore F. Van Wagenen, M. E., who, in connection with his duties as editor of the Mining Review, has had unequaled facilities for forming an opinion as to the actual state of the mining industry at the end of the year.

The yield of the mines of Colorado (gold, silver, lead, and copper) is summed up in the following table:

| | |
|---|---|
| Gold-bullion from smelting and amalgamating works | $422,563 |
| Silver-bullion from smelting and amalgamating works | 1,983,207 |
| Gold-bullion from stamp-mills | 1,297,425 |
| Gold-bullion from placer-workings | 382,500 |
| Ore and matte shipped out of the Territory | 1,102,815 |
| Pig-lead | 73,676 |
| Copper | 100,197 |
| Total value, (coin) | 5,362,383 |
| To this may be added coal, about | 1,600,000 |
| Total | 6,962,383 |

The yield for 1874, classed under the head of the four metals, makes the following showing:

| | | |
|---|---|---|
| Gold | $2,102,487 | |
| Silver, (with some gold, amount unknown, in matte) | 3,086,023 | |
| Total for precious metals | | $5,188,510 |
| Copper | $100,197 | |
| Lead | 73,676 | |
| | | 173,873 |
| Total | | 5,362,383 |

It is difficult to divide this amount among the several counties. The difficulty arises from the fact that a very large quantity of ore mined in Boulder, Clear Creek, and Park Counties is treated at works located outside of those counties, and these works are not in every case able or willing to give the exact quantities bought from each locality. The following table has been compiled with extreme care, and elsewhere is given the authority for all the figures used. Most of them have been furnished officially. Those not obtained in that manner are marked as "estimated."

| | |
|---|---|
| Clear Creek County | $2,203,947 |
| Gilpin County | 1,631,863 |
| Park County | 596,392 |
| Boulder County | 539,870 |
| Lake County | 223,503 |
| Summit County | 126,188 |
| Southern counties | 40,620 |
| | 5,362,383 |

This is the first year in which the production of silver has been greater than that of gold. In view of the facts that Gilpin County is now the only prominent gold-district in the Territory; that the placer-yield shows but a small gain on the figures of the last two years; that the silver-mines around Georgetown are proving more and more valuable and extensive every month, and that of the new districts opened since 1871 nearly all are argentiferous, it does not seem probable that the gold-yield of the future for this Territory will experience more than the natural increase caused by deeper and more extensive working of old mines, while there is every probability that the silver-yield will increase yearly from the production of old mines and developments in those of later location.

The item of lead is almost wholly a new one. In the Mining Review of January, 1874, Mr. Van Wagenen placed the value of this metal raised during 1873 at $703. There is good reason now to believe that that figure was too small by $5,000 at least. The amount given for 1874 is warranted by the largely-increased percentage of lead-ores mined and by the successful operation of the Lincoln City Smelting-Works, in Summit County. It is expected that furnaces for smelting low-grade galena-ores will be erected during the current year in the Snake River district, and the yield of this metal will in that case increase largely.

The copper-production is derived entirely from Gilpin and Park Counties. I have no positive authority for the figures given, but have calculated them from the known quantities of gold-ores treated in the former and from the matte produced at the Alma Works, in the latter.

The quantity of ores and matte shipped out of the Territory for treatment shows a large decrease on the figure for last year. The falling-off is, however, entirely in the item of matte; but, at the same time, the ratio of argentiferous ores shipped to the total yield of argentiferous ores is less than it was during 1873. This fact may be taken as an indication of a healthier condition of the home reduction-works, and a growing inability of outside smelting-works to offer such prices for ores as will draw them away from Colorado reducing-establishments. Ores are now shipped to Chicago, Ill.; Wyandotte, Mich.; Pittsburgh, Ohio; Swansea, in Wales, and Germany. At the first point the failure of Mr. S. P. Lunt indicates that the prices paid for ores were too high. At the second, Colorado ores are only bought to work with the ores of Lake Superior, which are poor in lead, and yield no profit (or but a small one) to the buyer. At the third, it is probable that some profit is obtained; while in Germany labor is so cheap and the working is technically so perfect, that ores can be treated from all parts of the world. I am inclined to believe that in a few years but little ore will go out of Colorado, except to Germany or to points in the East where fuel and labor are extraordinarily cheap.

The ruling prices for gold and silver ores at the close of the year are given in the following tables. They are the highest prices paid at that time at Central and Georgetown.

| SILVER-ORES. | Per ton. |
|---|---|
| 100 ounces, at $0.67 per ounce | $67 |
| 150 ounces, at $0.82 per ounce | 123 |
| 200 ounces, at $0.91 per ounce | 182 |
| 300 ounces, at $0.98 per ounce | 294 |
| 400 ounces, at $1.02 per ounce | 408 |
| 700 ounces, at $1.07 per ounce | 749 |
| 1,000 ounces, at $1.10 per ounce | 1,100 |

## GILPIN COUNTY.

The production of this county has been as follows:

| | |
|---|---|
| Gold | $1,525,447 |
| Silver | 50,965 |
| Copper | 55,451 |
| | 1,631,863 |

Late in the year a contract was made with the Pittsburgh Lead Company, by the Clifton Mining Company, by which the latter was to furnish to the former 10 tons of galena per day. I cannot learn whether these shipments commenced before the close of the year. If they did the amount sent was small, and could hardly have added more than $2,500 to the figures given above.

Mining has been very active in the county during the year, and the old and standard mines have done exceptionally well. In quite a number work is going on at a depth of 500 feet and over, without diminution in the size of the ore-veins or any indications of exhaustion. On the contrary, the deepest workings in the Territory, viz, those on Quartz Hill, show as strong and well-defined crevices as could be desired.

The quantity of milling-ore broken has been about 115,000 tons, and of smelting-ore about 3,500 tons. All this, excepting about 125 tons, has been treated in the county, the former in stamp-mills and the latter at the Boston and Colorado Smelting-Works. Comparing this figures with those of former years, it is evident that the ratio of milling to smelting ores is gradually changing, the quantity of the former increasing, while that of the latter shows but little gain. This is by no means an evidence of a growing poverty in the ores, but is a direct result of great improvements in the amalgamation-mills, and of the fact that several large companies are working their own mills on a very large scale, and are therefore able to treat rock that formerly was found profitless, and was therefore thrown away. Careful investigation into the mill-returns of a number of mines shows the average value of mill-ore has not decreased from what it was near the surface, excepting, of course, the decomposed gossan found directly on top of the lodes, which often was worth hundreds of dollars per ton.

Gilpin County has been so thoroughly prospected that almost nothing in the way of new discoveries of note has taken place during 1874. At the close of the year, under the operation of the new mining law,* a large number of old claims that had been untouched for a long time were re-opened, and as a result good ore in paying quantities was found in many. In consequence the list of working mines is now larger than it has been for many years, and the bullion-product is increasing rapidly, as shown by the following bank-shipments:

| | | | |
|---|---|---|---|
| January | $59,940 | August | $86,710 |
| February | 81,445 | September | 93,855 |
| March | 97,290 | October | 113,170 |
| April | 91,075 | November | 86,370 |
| May | 119,135 | December | 125,740 |
| June | 118,925 | | |
| July | 93,270 | Total, (currency) | 1,169,925 |

* The final extension of the time for doing work on old claims, to maintain possessory title, expired January 1, 1875.—R. W. R.

The water-supply for milling-purposes has been of late the cause of considerable discussion in Central, and steps are now being taken to provide for an increased quantity by enlarging the consolidated ditch, and thereby bringing in a full head from Upper Fall River. This will permit the continuous working of a number of stamp-mills, which can now run during a portion of the year only.

The business of the Boston and Colorado Smelting-Works, at Black Hawk, is summed up as follows:

| | |
|---|---|
| Silver | $1,103,487 |
| Gold and copper | 535,390 |
| Total, (currency) | 1,638,877 |

The total number of tons of ore treated has been about 6,000. Of the matte produced, about 1,500 tons have been refined to copper bottoms, which are made nearly free of silver, and consist almost entirely of gold and copper. The value of this product is the second figure given above.

The ores worked have been mined in the four counties of Gilpin, Boulder, Clear Creek, and Park, in about the following proportions:

| | |
|---|---|
| Gilpin | $542,000 |
| Clear Creek | 422,000 |
| Park | 452,000 |
| Boulder | 222,000 |
| | 1,638,000 |

The supply from Gilpin County consists of gold-ore carrying between fifty and sixty thousand dollars' worth of silver. The Clear Creek quota consisted of $375,000 worth of silver-ores, and nearly $50,000 in matte, produced at the Swansea and Whale Mills. That from Park County was all matte, produced at the company's works, at Alma; while the Cariboo and Gold Hill mines furnished the amount from Boulder County.

Favorably situated as they are, these works have the command of a large area of mining-country, and can draw supplies from all sides. With the increased facilities of direct railroad communication with Georgetown, the probability is that during the current year they will treat $1,500,000 worth of ores mined outside of Gilpin County. The present capacity is 40 tons per day, from which are produced six to seven tons of matte, worth, on an average, $1,200 per ton. This matte formerly contained a high percentage of copper; but owing to the fact that during the last year the smelting-ore yielded by the Gilpin County mines has not been found so rich in copper as before, that metal has been replaced to some extent by iron.

The separation-works in connection with this establishment are now in complete running order, and are proving very successful. The original plan of refining both gold and silver in Black Hawk has been abandoned, and metallic silver alone is produced. The system adopted is that of Ziervogel, to which Augustin's process is added, in order to extract the last percentage of silver from the gold and copper. This latter, almost in the pureness of an alloy, is shipped to Boston, and there refined. The former, from the condition of cement-silver, is melted and cast into bars, which, as a rule, are over .990 fine, and are worth from $1,200 to $1,500 each.

Of the many mines worked in Gilpin County, I will mention here the following:

The Ophir claim, on the Burroughs lode, has been sunk upon to the depth of 750 feet, and is the deepest mine in Colorado. Since Messrs. Roberts & Co., the lessees, took hold of the property, in May, 1873, it has yielded $160,000 worth of gold. The exposure of ore in the lowest workings is as good as any ever found in the mine, the mill-ore carrying at the rate of about 13 ounces per cord, equal to $30 per ton. The percentage of smelting-ore raised is small, but of good average grade, carrying $100 to $140 per ton.

The First National claim, on the Kansas lode, has a shaft of 530 feet, at the bottom of which the ore is 3 feet wide, solid, and assays about $80 per ton for the milling-rock. The same ore-body, exposed in the 480-foot level, shows there a width of 4 to 5 feet of ore, averaging 12 to 14 ounces per cord.

The Waterman claim, on the Kansas lode, has a shaft 560 feet deep. The length of the claim is 300 feet. Reports from this mine for the first nine months of 1874 give the production and expenses as follows:

Production:

| | Average yield. | Total yield. | Value. |
|---|---|---|---|
| Stamp-ore, 5,400 tons | $13 20 | 4,296 ounces. | $71,468 38 |
| Smelting-ore | | | 4,146 48 |
| Produced by sublessees | | | 2,891 61 |
| | | | 78,506 47 |

Expenses:

| | | |
|---|---|---|
| Wages | $3,779 40 | |
| Contracts, &c. | 14,188 15 | |
| Hauling and milling | 15,786 02 | |
| Repairs and supplies | 3,617 02 | |
| Wood | 1,182 35 | |
| General expenses | 2,818 81 | |
| | | 41,371 75 |
| Leaving a profit on the nine months' work of | | 37,134 72 |

Four claims on the Gunnell lode have been worked during the year to a greater or less extent, viz, the Coleman, Gunnell Central, University, and Gunnell Gold Company claims. The result of the year's operations has been satisfactory to all, though the actual product of the mine has not been large, as several months were required to place the property in order and clear it of water. The yield may be stated for the year at about $40,000, with every prospect of a heavy increase for 1875. The main shaft was at the close of the year nearly 650 feet deep, and was sinking on a good body of ore of average grade.

The Fisk mine is cut by the Bobtail tunnel 275 feet deep, and about 700 feet from the mouth of the tunnel. It is under lease to Mr. George W. Mabee, who has been working it on a small scale since May, 1873. During these twenty months the mine has yielded 3,200 tons of milling and 50 tons smelting ore, from which $67,000 worth of bullion has been produced. The gross yield of the property during 1874 was $44,941.99. The average value of the mill-ore has been $21.38, and of the smelting-ore $124.63 per ton. The expenses per ton have been, for mining, $8; hauling and milling, $4.86.

The improvements on the Buell property, which is perhaps better known as the Leavitt, have amounted to over $100,000 during the year. They comprise an elegantly fitted up stamp-mill (12 batteries) and new hoisting and pumping machinery. The main shaft has been sunk to the depth of 450 feet, and the lode opened very extensively on either side. The vein continues as large as ever, and has yielded an average of 90 to 100 tons of ore daily throughout the year, representing an output of bullion of not less than $200,000. Under the name of the Buell mine are comprehended, besides the Leavitt, a number of other properties, mostly those owned by the original Kip & Buell Company. These are now consolidated under one name. The mine is justly considered one of the finest in the Territory.

The following is an extract from the report of Mr. Gray, territorial assayer at Central, on this property:

The property now known as the Buell mine consists of about 3,000 consecutive feet along the Leavitt, Kip, Vasa, and U. P. R. veins. Immediately adjoining are 67,500 feet of patented mill-site property, upon which are a one-story frame assay-office and a story and a half mill-building, 50 by 150 feet, whose massive stone walls and tin-covered roof render it nearly, if not quite, fire-proof from external exposures. In the west end of this building is the main working-shaft, through which all ore mined in the lodes is hoisted. This and the hoisting-apparatus proper take up about one-fourth of the building, the engines, boilers, and stamp-mill occupying the remainder.

The main shaft, which has now reached a depth of about 430 feet, is divided by means of heavy plank partitions into three compartments, through two of which the ore is hoisted, the third being used for a ladder-way. Through the two larger compartments, which are each 3 by 5 feet in size in the clear, are run iron safety-cages, similar to those used in California and Nevada. These cages carry an iron car capable of holding 2,000 pounds of ore, each bar being 4 feet long, 2 feet wide, and 2 feet high. The cages run alternately—that is, one ascends as the other descends. The cars are filled in any level required, run to and on the cages, raised to the surface and dumped behind the stamps in the mill. The hoister is a 70-horse-power vertical double-acting engine, with link-motion attachments, with which is connected—by a 30-inch rubber belt running over a 4-foot pulley—a 9-foot pulley attached to the shaft of the hoisting-apparatus. On either side of this latter pulley, and running independently on the same shaft, are two large drums or cylinders, over which run hoisting-cables of English manufacture, consisting of steel wire twisted into a flat rope of 3 inches width by half an inch thickness, and having a breaking strain of fifty tons. The action of the hoisting drums is regulated by means of the usual friction-bands. This engine and hoister are capable of raising 40,000 pounds at a time if necessary, and of delivering at the mouth of the shaft 100 tons ore in every twelve hours.

In the mine are three direct-acting steam-pumps, being located, one at the 130, one at the 300, and the third at the 400 foot level. The steam supplying these pumps is conveyed through 2-inch pipes run from the main boiler at the surface, thence down the shaft through the ladder-way compartment. Below each pump is a cistern receiving the water from its adjacent level, and also from the pump next below, whence it is carried through 4-inch cast-iron pipes to the surface, where it is used in the mill, &c., as will be described further along.

At a depth of about 30 feet, a level has been run easterly from the shaft a distance of 600 feet, which level is used as a water-course or stollen, to catch all surface-water that may find its way into the mine.

In the mill-building proper are three tubular and two flue boilers, having an aggregate of 350-horse-power capacity. Four of these boilers are incased in brick in pairs; the fifth is also inclosed in brick-work, and is kept as a reserve in case of accident to either of the others. These three sets of boilers are so arranged that they may be run together or independently of each other, as circumstances may require.

The stamp-mill is run by an 80-horse-power horizontal expansion-engine, having a 36-inch stroke. It is connected with the line-shaft by a 30-inch leather belt running from a 5-foot to a 13-foot pulley. The mill contains 60 stamps in all, of which one-half are 500 and the remainder 650 pounds weight each. They have a fall of about 18 inches, the heavier ones 25 and the lighter ones 30 drops per minute. They are divided into four sections of 15 stamps, each section containing three 5-stamp batteries. The mortar-beds or foundations consist of 2-inch plank set on end, and resting on a 2-foot-square timber. The stamps are raised by the ordinary cams, keyed to 4-inch wrought-iron shafts, each shaft extending the length of the section. On each end of these shafts are keyed 5-foot cast-iron pulleys, over which run 15-inch leather belts to 4-foot pulleys on the line-shaft. These pulleys are provided with friction-clutches, (similar to the Mason clutch,) which can be thrown on or off at will, thus permitting the several sections of stamps to be operated independently of each other. On the ends of the cam-shafts and behind the pulleys are cast-iron ratchet-wheels, each provided with a tongue so arranged that at the first backward movement of the cam-shaft (caused by the weight of the stamp) it falls, holding the stamps in suspension. This arrangement obviates the frequent breakage of cams by the falling stamps. Good California blankets are used to collect the tailings which have passed over the amalgamating-tables. The blankets are washed out about every half hour. In the northeast corner of the mill-building are located three Freiberg pans of 5 feet diameter and 2 feet depth each. These pans are run by means of an 8-inch belt, direct from the line-shaft, and are used for working over the blanket-tailings. In these pans the tailings are reduced to an impalpable pulp by a grinding-process obtained by means of two heavy stone drags in each pan kept rotating over the iron bottom of the pan. When the tailings have thus been reduced to a sufficient degree of fineness, quicksilver is introduced and amalgamation obtained.

Behind the stamps, and about 10 feet above the feeding-floor, is a tramway running from the quartz-floor at the mouth of the shaft through the entire length of the mill. Over this the ore is conveyed from the shaft to the stamps. Two Dodge crushers are used for breaking the large lumps of ore.

When the mill had been about completed, the question of an adequate water-supply was found to be a very serious one. The quantity required for all purposes is about 20,000 gallons per diem, and the greatest that could ever be obtained was about 22,000 gallons. It was, therefore, necessary to use the same water many times, and in order that this might be done, a series of settling-tanks, having an aggregate capacity, it is estimated, of about 500,000 gallons, was constructed just outside of the building. These tanks are sluiced out every twenty-four hours. The total distance traveled by the water passing through these tanks, from its leaving the stamps to its return to them, is about 400 feet, and it re-enters the mill, after the settling-process, as clear as the purest spring-water. These tanks or reservoirs are built of 2-inch plank, having double sides, which are filled in with manure, rendering them both water-tight and frost-proof.

The entire cost of the works above described (exclusive of any mining-expenses) has been about $200,000. This property has been worked uninterruptedly and profitably for several years past, and the mill has been kept constantly at work since its completion, under the immediate supervision of its sole owner, Hon. Bela S. Buell.

The Bobtail has been steadily worked throughout the year, with success. In addition to the work being done by the Bobtail Company proper, several parties of leasers are working the lode from the surface and are finding good ore. The stopes above and below the tunnel-level in the main mine are looking well and yielding the usual quantity of excellent ore. I am unable to obtain exact figures of the yield, but estimate it at between $75,000 and $100,000 for the year. The Bobtail ore is, as a rule, of high grade, and the proportion of smelting to milling ore is larger than that of any other mine in the district.

The great Gregory vein has, during the last year, been drained and placed in working order. Of the various claims upon it, those of the Narragansett, Consolidated Gregory, Briggs & Smith, and Parmelee are most worthy of mention. The vein is now yielding about 150 tons daily of excellent ore, as good as was ever produced, nearly one-half of which is coming from the Briggs. The latter has been sunk upon nearly 500 feet, and is furnishing some very rich ore. About ten months ago a pocket of rich ore, showing a great abundance of free gold, was struck, which yielded considerable ore before it was exhausted.

There are a large number of other mines in operation in this county. The most noted are the Prize, American Flag, Pewabic, Kent County, Bates, Gardiner, Cooley, Clayton, Register, Smith, Emmett, and Casto. Of other mining enterprises in operation may be mentioned the German, Quartz Hill, and La Crosse tunnels. The first is an old enterprise, which has just been revived. The tunnel is to run under Central City and Lake Hill. The Quartz Hill tunnel was, at the close of the year, 830 feet in length, and has been driven steadily. Its objective point is the Gardiner and Roderick Dhu belt of veins, which, it is estimated, will be cut 1,800 feet from the mouth, and at a depth of about 600 feet. The La Crosse tunnel enters Quartz Hill on its northerly face, about

1,200 feet farther up the gulch than the Quartz Hill tunnel, and 300 feet higher. It is now something over 1,000 feet in length, and has cut the Kansas at 125 feet, the Monroe at 200 feet, the Burrough at 450 feet, and the Old Missouri at 550 feet. Since cutting the Missouri no other veins have been intersected up to date, but a deposit of auriferous quartz and gangue-rock without defined walls has been met, which has not yet been passed through. The whole of this material, for over 100 feet in width, yields from 1 to 2 ounces gold per cord.

### CLEAR CREEK COUNTY.

The production of the mines of this county has been as follows:

| | |
|---|---|
| Silver-bullion | $555,268 49 |
| Ores treated in Colorado out of the county | 520,968 60 |
| Ores shipped out of the Territory | 1,085,210 88 |
| Placer and stamp gold | 42,500 00 |
| Total, (coin) | 2,203,947 97 |

The gain for the year has been principally in the first two items, which indicates a growing activity of the home reduction-works. Clear Creek County is now the banner mining-county of Colorado, and will probably hold that position until surpassed by one of the larger southern or western counties. In no district in the Territory is there a greater activity in the mines or a larger percentage of successful operations. Its milling facilities have also increased greatly during the past year, though as yet they are not equal to the treatment of the ore produced, as will be seen by the large item of shipments, both to works in other counties and out of the Territory.

The total number of tons mined and reduced or sold has been 9,490, worth, on an average, $232 per ton. This includes 200 tons of gold-ore from Empire of very low grade. Of this amount 3,275 tons, averaging in value $186 per ton, has been treated at works in the county; 2,024 tons, worth about $230 per ton, was reduced in works in other counties, and the remainder, 4,191 tons, worth about $260 per ton, was shipped to Chicago, Detroit, Pittsburgh, Swansea, and Germany.

The amount of lead contained in the ores shipped is valued at $48,239, being equal to an average yield of 20 per cent. for the ores producing it. This item includes all the ores shipped to the Golden City Smelting-Works. The estimate may be $10,000 less then the actual amount. About $300,000 worth of ore has been shipped away from Georgetown during the year, of which there is no accurate register of value in lead to be had.

The ore-shipments from Georgetown amount to more than 10,000,000 pounds. Of this amount, 4,128,775 pounds was sent out of the Territory, and the remainder has been taken to Black Hawk, Golden City, and Denver for reduction. Of the ore sent East, by far the largest quantity has gone to Chicago; but since the failure of these works more has gone to Germany. The Pittsburgh Lead Company is now making vigorous efforts, by means of both traveling and resident agents, to obtain galena-ores in large quantities from this and other counties.

*Mines and mining-works.*—The yield of the Terrible mine for 1874, is shown in the following table, which has been furnished by the kindness of Mr. George Teal, the superintendent:

| | Pounds. | Price received per ton. | Value, (currency.) |
|---|---|---|---|
| First class | 193,520 | $550 | $53,218 |
| Second class | 768,570 | 150 | 57,642 |
| Third class | 5,102,000 | 12 | 30,612 |
| Fourth class | 2,982,000 | 11 | 14,960 |
| | 9,046,090 | | 156,432 |
| Costs, &c | | | 68,000 |
| Profit | | | 88,432 |

The actual coin-value of this yield is about $203,000, of which amount $6,000 represents the value of the lead, the remainder being that of the silver.

The concentration-works in connection with this mine went into operation about July 1, and closed the season's work late in October, having proved an undoubted success. The material treated was the third and fourth classes of the above table, which have heretofore been thrown away as worthless or stacked on the dump or along the course of the suspended tram-way leading from the mouth of the tunnel to the gulch. At the commencement of the year the stock of this ore on hand was estimated at 3,500 tons, of an average value of $12 per ton. This stock was increased by the addition during the year of the 4,541 tons given in the above table, third and fourth, making, in all, about 8,000 tons. During the season about 2,500 tons were passed through the works, producing the following classes of marketable material:

| Class. | Pounds. | Assay. | Value received. |
|---|---|---|---|
| | | *Ounces per ton.* | |
| A | 47,253 | 325 | $7,488 25 |
| B | 97,141 | 279 | 2,602 50 |
| C | 9,488 | 500 | 2,466 00 |
| D | 190,420 | 199 | 15,614 50 |
| E | 12,619 | 408 | 2,625 50 |
| F | 133,135 | 142 | 7,130 50 |
| G | 15,002 | 211 | 1,140 00 |
| H | 380,000 | 150 | 9,500 00 |
| | 885,058=442½ tons | | 58,827 25 |
| Expenses | | | 7,328 61 |
| | | | 51,498 64 |

It will be seen that about 6 tons of the crude material has been concentrated into 1 ton of salable product. The composition of the eight classes is as follows:

A.—Impure galena, 60 per cent. lead.

B.—Second class, from Cornish jigs, 20 per cent. lead and 30 per cent. zinc.

C.—First-class pickings, 45 per cent. lead, 10 per cent. zinc.

D.—Second-class pickings, 10 per cent. lead, 40 per cent. zinc.

E.—Iron and copper pyrites from the enriching jigs, carrying the brittle silver.

F.—Zinc-blende from enriching and automatic jigs, (nearly pure.)

G.—Best work from slime-tables.

H.—Savings from picking-table, (to be recrushed and re-treated.)

The Polar Star mine has produced during the year ore to the coin-value of $25,766.22, which sold in Georgetown for $19,820.17. The previous yield of the mine brings this sum to about $40,000 for the two years it has been worked. It was discovered in 1872, but "pay" was not reached in considerable quantity till late in 1873. It is opened by four shafts, and an adit-level from the west, which, following the vein for 500 feet, strikes the deepest shaft 180 feet from the surface. A shaft is now sinking from this level, which at the close of the year was 40 feet deep, and carried a good vein of the same class of ore as found above. A second adit has been commenced 150 feet below the first, commencing 400 feet farther west on the vein. The lode has been found to be a very wide one, carrying a "gouge" 3 to 7 feet broad, and so easily worked that but little blasting is required. The owner, Mr. C. S. Stowell, believing that the property is one of value, proposes to develop it steadily and extensively. It is justly considered as one of the best mines in the county.

From the Pelican and the Dives it has been impracticable to obtain returns, owing to the litigation now in progress between them. In spite, however, of the many difficulties under which they have been worked during the year, consisting of injunctions on the richest parts of the properties, their combined yield has been greater than that of any other two mines in Colorado. I estimate the value of the ore taken out and sold at not less than $650,000—a sum which might easily have been doubled had there been no restrictions on the workings. The quantity of ore in sight is enormous; but as neither party is allowed to attack some of the best bodies, it is not for the present available. Still, in both the claims those portions not under injunctions are yielding handsomely, providing sufficient means to pay the host of lawyers who are eating up the profits. Probably this fact will indicate more clearly than any other the true resources of the lodes. Late in the year tunnel No. 3 struck the Pelican, and disclosed on the north wall several seams of mineral, the total width of which was over 12 feet. The main workings on this mine are now between 400 and 500 feet deep, and the property has been well explored to that depth and for fully 300 feet on either side of the shaft. The total length of drifts and levels is almost 3,500 feet.

The Silver Plume has yielded during the year 250 tons of ore, which sold for an average of $240 per ton, netting to the owners about $60,000 in cash.

On the Coldstream mine 300 feet of levels and 150 feet of shafting were opened during 1874. The main shaft is now 150 feet deep. At this depth the crevice is more regular and larger than above, being from 3 to 6 feet wide, with both walls well defined. Near the surface, there was very little gangue-rock, and often little or no indication of a foot-wall; now, however, good gangue is appearing, which is quite similar to that of the Pelican. Although the mineral is not so rich as it was above, it occurs more regularly. At the depth of 100 feet a zone of lean ores was passed through, since which a considerable body of galena has been developed in the third, or lower, level. The drift east of the discovery-shaft, only recently opened, shows quite as well as the drifts west, and there is every reason to believe that the eastern part of the lode will prove the richer. The vein improves with depth in all parts of its workings, and gives indications of large bodies of mineral below. No fair estimate can be made of the yield of the mine, on account of the litigations with adverse claimants. Large quantities of ore have been taken from the lode of which no account can be had. I think it sufficiently within bounds, however, to say that under unrestricted working the mine would have paid for its own development and yielded since its discovery, three years ago, $2,500 per month.

Of the large number of tunnels in this district, three are especially worthy of mention, the Burleigh, Diamond, and Ocean Wave. Of the rest, not much that is new can be said. The Baltimore has not progressed any during the year, but some of the lodes it has cut have been worked more or less, and good ore has been taken from them. The Lebanon has been driven ahead steadily, until within a few months ago, and is disregarding all of the main lodes cut, making its objective point the Hise, now over 100 feet in advance of the headings. The Marshall remains at the length given for last year, 1,300 feet, not having made any progress into the mountain. Work has been done, however, on several of the lodes it has crossed, and with varying success. The Eclipse has been idle for over a year. The Douglas has not been driven ahead, but the lode cut about a year ago in the headings, and supposed to be the New Philadelphia, is energetically worked, and carries in places a good body of ore, not very rich in silver, but auriferous to a small extent.

The Burleigh is now in about 1,800 feet. About 1,750 feet from the mouth a large vein was cut, which at the point of intersection was 16 feet wide, and carried a seam of galena and zinc-blende, not very rich in silver. This vein is supposed to be the Cashier. Though the strike has not been productive of any ore, it is an encouraging evidence of the continuance of the ore-bodies in depth in the veins on Sherman Mountain. When the district is supplied with the concentrating facilities it has so long been in need of, this tunnel will afford a most economical means of ore-extraction for this and the Bush lode, (intersected 900 feet from its mouth,) both of which are capable of producing large quantities of low-grade ore carrying a high percentage of galena and zinc-blende.

The Diamond tunnel, located in Cherokee gulch, and directly below the Pelican and Dives, is now in a little over 800 feet, and is progressing at the rate of about 40 feet per month. It is estimated by the company owning it that the lodes just mentioned are between 200 and 300 feet farther on, which distance will be passed before the close of 1875. The tunnel has already cut five well-defined veins, all showing well, and carrying more or less ore at the point of intersection. As soon as spring comes, and the requisite ventilation can be gained from the water-power passing directly in front of the tunnel, active work will be commenced on several of them.

The Ocean Wave tunnel, driven on the vein of the same name, was, at the close of the year, 840 feet long, and was calculated to be not over 50 feet from the south wall of the Equator. Numerous bodies of mineral have been met with in driving this adit, which have aided to a considerable extent in defraying expenses. The Equator will be cut about 550 deep, and the many well-known veins beyond it at a still greater depth.

*Mills and milling.*—The Stewart Silver-Reduction Company has treated during the year 2,647 tons of ore, of an average value of $180 per ton, producing silver-bullion to the amount of $436,181.43, (coin,) as follows:

| Month | Amount | Month | Amount |
|---|---|---|---|
| January | $7,283 54 | July | $48,799 20 |
| February | 3,260 29 | August | 51,081 31 |
| March | 40,037 26 | September | 57,729 24 |
| April | 42,862 69 | October | 36,265 71 |
| May | 36,862 68 | November | 43,490 53 |
| June | 48,399 12 | December | 20,109 86 |
| | | | 436,181 43 |

The works have been extensively increased and altered during the year, and now contain three single reverberatories and one which is a combination of three low-arch reverberatories, patented by Mr. Stewart in the spring of 1874. A complete set of vats has also been put in for the Hunt & Douglass process, and the works are now buying gold-ores from Empire and working them with the silver-ores of Georgetown. The establishment has a capacity for light* ores of from 15 to 20 tons per day.

In May, 1874, Mr. Crosby began the remodeling of his mill, at the head of Georgetown, and in September began buying ores. The mill is fitted with 15 stamps, one Stewart furnace, and four combination pans, and has a capacity of from 10 to 12 tons daily. Its arrangement is excellent, and perhaps no mill in Colorado is more conveniently constructed. From September 7 till the close of the year, 427 tons of ore were worked, producing 46,744 ounces of silver, worth, in coin, about $56,514.

Early in the year the Pelican Mining Company took possession of the old mill of the What Cheer Company, and, after working on it for most of the remainder of the year, has now nearly finished its renovation. It contains five Brückner cylinders, eight barrels, and two Ball pulverizers, and has an estimated capacity of 60 tons per week. For several months during the year it has been run on one or two cylinders, and has turned out about $75,000 worth of bullion. The ores worked were from the Terrible mine, and have consisted largely of very heavy material, carrying from 40 to 70 per cent. of zinc and lead. It is the intention of the company to use this mill solely for the treatment of the Pelican second and third class ores, which are comparatively light, and easy of reduction.

### BOULDER COUNTY.

This county embraces the Cariboo, or Grand Island, Gold Hill, Sunshine, and Ward districts, and a small area of country in which occur good placer-diggings. Its production has been as follows:

| | |
|---|---|
| Silver-bullion | $216,000 |
| Gold-bullion | 145,000 |
| Ores shipped | 225,522 |
| | 586,522 |

Cariboo, the most prosperous mining-camp in Northern Colorado, is a well-established and promising town. Almost without exception, exploitation has shown that the veins are of a good class, and worthy of deep and extensive mining. In spite of the great altitude of the mines, and the consequent difficulties attending their working, the results have been encouraging enough to induce the investment of considerable outside capital.

The "tellurium excitement" on Gold Hill has almost died out, and unless the new reduction-works create a revolution in the treatment of its ores, the district is likely to relapse into the condition it was in previously to the discovery of the famous Red Cloud. The search for these ores has been persistently and energetically kept up, but nothing like the mine just mentioned (except the Cold Spring) has been the result. Undoubted tellurides of gold and silver have been found in other veins, but they have been in very small quantities, and mingled so

* This is the term applied to the ores containing small percentages of lead and zinc. —R. W. R.

abundantly with galena, zinc-blende, and iron and copper pyrites, that the ores have been of too low grade to permit profitable treatment.

The discovery of the Sunshine mines did much to draw away the floating population from Gold Hill. In this new district the discoveries have been very numerous, and the surface-ores disclosed are remarkably rich; but as yet the developments do not warrant anything more than a favorable anticipation. In their comparatively low position and easy accessibility lies the greatest advantage of the Sunshine mines.

*Quartz-mines.*—The Sherman vein is parallel with the Cariboo, and about 100 feet east of it. The main working-shaft is down 160 feet, from which levels north and south have been run to a total length of 310 feet. Its product for 1874 has been 220 tons of first-class ore, estimated roughly at $40,000. Besides this, which represents the only class of ore that can profitably be shipped to Black Hawk for treatment, there is considerable lower-grade ore in and around the mine, awaiting the erection of a mill by the company owning it. Early in the year this vein was tapped by the Cariboo tunnel, and found to be as productive below as on the surface. The Sherman and No Name were during the year sold to an eastern company, but they are worked under separate management. The price reported at the time of the sale was $80,000. The Sherman is a very strong vein, showing as far as developed a crevice of 5 to 8 feet in width, easily worked, and very regular in its ore-bearing quality.

The No Name belongs to the cross-course series of veins on Cariboo Mountain, having approximately a northeast and southwest trend and intersecting the Cariboo and Sherman diagonally. Its crossing with the first-named has been extensively explored, and along the line the Cariboo is considerably faulted and thrown down the hill, thus showing it (the Cariboo) to be the older vein. The No Name has been developed to a greater extent than any other lode in this district, the Cariboo alone excepted, and has uniformly paid well for the labor expended. The main shaft is 320 feet deep, and the total length of shafts and drifts is nearly 800 feet. The product for the year is placed at 4,000 tons, one-half of which is low-grade milling-rock, and not yet available. The remainder is ore that has been shipped to Black Hawk or to the Nederland Mill. The total value of ore mined is estimated by the owner at $400,000. The first-class ore from this mine has been sold for over $1,300 per ton; second-class, $250 to $400, and third-class $30 to $75.

The main shaft on the Cariboo mine had been sunk at the close of the year to a depth of 420 feet, and explorations had been extended east and west for a total distance of about 700 feet. The mine has during the year given steady employment to from 60 to 70 men, and as nearly as can be learned it has produced about 1,800 tons of ore, which, treated at the mill, has produced about $130,000 worth of silver-bullion. Of course a very large quantity of lower-grade material has been broken and raised, but the greater part of this has been stacked for future handling in concentration-works.

In view of the enormous price paid for the mine, the operations of the year can hardly be called encouraging, and it is generally believed that during 1875 some new arrangements will be made, or that the mine will pass into other hands. The Cariboo vein, as might have been expected, has not proved so productive of rich ores, when depth was gained, as it was on and near the surface. The ore now taken out will average by mill-samples between 60 and 90 ounces per ton. Occasional pockets of richer material are found; but they do not exist in quantities sufficient to raise the general average to 100 ounces. At the same time the stretches of barren ground are long and extensive. On the other hand, the ore is a true amalgamating product, containing perhaps a little more copper than is desirable, but easily worked up to 85 and 90 per cent. of the assay, if ordinary care is taken in the cylinders and pans. Mr. N. H. Cone, formerly of Georgetown, is now in charge of the mill, and under his care the bullion-yield has increased somewhat, and, at the same time, the bars have improved in fineness.

The Red Cloud mine, which, on account of the remarkable deposit of tellurium minerals found in it, was, two years ago, one of the most famous in the country, was shut down in September last, and will probably not be re-opened until a market for poor gold-ores is opened on or near Gold Hill. At that time the main shaft was 430 feet deep, and fully 1,000 feet of levels and winzes had been driven and sunk. Ore was found almost continuously in all parts of the mine; but the wonderfully rich pocket of tellurides near the surface has never been duplicated. Interspersed and scattered throughout 3 and 4 feet of gangue-rock, auriferous pyrites everywhere took the place of the richer ore; and it was probably because of inability to make it pay that the mine was closed. As the vein does not appear to have been lost, work may be begun again; but it seems quite probable that the owners will postpone further operations until some other enterprise shows that the tellurides may be found in depth.

On the western face of the porphyry-dike that forms the south wall of the Red Cloud is the Cold Spring, similar to the Red Cloud in almost every characteristic, and apparently quite as likely to come to the same end. I am not positively informed as to the exact condition of this mine. At the surface the display of tellurides equaled, if it did not surpass, that in the Red Cloud, but during the last six months but little has been taken from the mine.

These two mines have produced, as nearly as can be ascertained, a total of about $600,000. This was milled from about 400 tons of ore, which gives an average of $1,500 per ton. Some ore sold was paid for at the rate of from $5,000 to $20,000 per ton, but this was, of course, quite exceptional. The occurrence of two such rich pockets, so close together and having so many points of similarity, is, to say the least, very interesting. It is hardly supposable that these are the only two on the hill; and the day may not be far off when others will be discovered, and the tellurium excitement on Gold Hill revived. Meanwhile it is removed to the new camp of Sunshine, close by.

*Sunshine district.*—In March, 1874, the first discovery was made in this prominent little camp, by D. C. Patterson, an old Boulder County prospector. Sunshine is on the eastern terminus of a strong spur from the range, which a couple of miles westward is called Gold Hill. The same characteristics of vein-formation and ore that are found in the old camp are met with on the new, which is evidently but a continuation of the Gold Hill belt. Geological details will therefore be unnecessary, as the peculiarities of the tellurium-mines have been described in former reports.

The first discovery by Patterson did not create much excitement, beyond a surprise that gold and silver lodes should be found so close to the plains. Later in the season, however, the American was located; and as this mine showed considerable quantities of free gold in its outcroppings, the news spread rapidly, and prospectors hurried in from every point of the compass. In a short time numerous new and promising veins were uncovered, all of which were prolific in large assays, free gold, tellurium-ores, &c., and the usual concomitants of a mining excitement took possession of the sunny hill-side which had for so many years been passed over scornfully by the miner and prospector on their way to the older discoveries beyond and above.

The prominent mines, and their condition at the close of the year, are briefly summarized as follows:

The American has a shaft 100 feet deep. At 50 feet a level 160 feet long shows a strong vein of scattered ore, consisting of free gold and auriferous iron and copper pyrites. The latter are abundant, forming the bulk of the ore, and containing in scattered bunches the silver and gold tellurides and whatever free gold is found at that depth. At the depth of 100 feet a second level was started, which is now about 30 feet in length, and displays the same characteristics as that above. Some months ago the mine was sold, for $17,000, to Mr. Hiram Hitchcock, of New York, and is now under the charge of Mr. J. Alden Smith.

The Grand View has a shaft 40 feet deep. The vein is well defined, carrying free gold at the surface, and bunches or specks of tellurides below. It is reported to have been sold for $50,000 to Cincinnati parties.

The Sunshine has a shaft 30 feet deep, and is opened also by a short adit-level running on the vein.

The Osceola has a shaft 60 feet deep. The vein is well defined, carrying both gold and silver.

The Young America has a shaft 30 feet deep. Free gold occurs on and near the surface, changing into pyritous ores below, in which enough tellurides are found to furnish the basis for large assays and an excellent reputation. A second shaft is down about 40 feet, and displays a vein of similar character.

Among other prominent claims may be mentioned the Idaho, Katie King, Silver Dale, Warsaw, Saint Vrain, Paymaster, Glendale, Hawk-Eye, Denver, Baxter, White Crow, and Dead Medicine. Altogether not less than three hundred claims have been located, among which, as usual, are very many absolutely worthless, and a few that, under proper circumstances, may be made profitable. A rough estimate of the actual amount of precious metals taken out of ores from this district places the figure between $20,000 and $30,000, the largest part of which has been produced from surface-ores carrying free gold, and worked at the small works of Mr. Smith, at Sunshine.

No bodies of tellurium-ores of the size and richness found in the Red Cloud and Cold Spring mines, on Gold Hill, have been found in any mines in Sunshine. So long, however, as any of that delusive mineral is met with, each miner fully believes he is to be the fortunate finder of one, and is tempted to continue hunting after high-grade "bonanzas," instead of taking advantage of the abundance of lower grades, until money and credit are both gone.

It seems probable that the Sunshine mines will prove in course of development to be producers of a grade of ore not very difficult to handle, and not averaging over $50 per ton. The proximity of the mines to the plains, and the consequent comparative cheapness of supplies of all kinds, ought to make the camp a successful one.

*Mills and metallurgical works.*—The facilities for the treatment of ores in Boulder County remain, as they have been, exceedingly limited; and to this fact may be attributed the very slow growth of the mining-industry. There is, however, a better outlook for the future, as shown by what has been done during the past year, though, as in olden times, the county is still regarded as an excellent field of operations for "patent processes." At Cariboo work has already commenced on a new mill for the treatment of No Name and Sherman ores, under the direction of Mr. Cash, formerly of Central. As that gentleman has strong proclivities toward chlorination, and made an undoubted success with that system on Central City gold-ores four years ago, it may be inferred that the new works will be built more or less for that style of treatment. The Cariboo ores are, as a rule, exceedingly easy to treat, and should, on general principles, be susceptible of reduction by chlorination as well as by amalgamation, though it does not appear why expensive chlorination by means of chlorine gas should be preferred to chloridizing roasting. In Four-Mile Creek are located the chlorination-works of the Ohio and Colorado Company. The establishment is built with a special view to treat Gold Hill ores, which are supposed to be more complicated than any others in Colorado. It would be premature to express an opinion as to the probabilities of success or failure, as there has hardly been a sufficient period since the time of their firing up to furnish data for an intelligent opinion. Their calculated capacity is from 15 to 20 tons per day.

Boyd's Smelting-Works, at Boulder City, though finished early in the year, are not in operation. The system employed is based on a patent process held by Mr. Boyd, by which ore, after being melted, is made to pass through a bath of molten lead, which latter is supposed to extract the silver and gold. After being in the market for a short time as a purchaser, the works closed, and it was asserted that the miners would not sell for the prices offered. Whether this was the whole reason cannot be determined, but I am inclined to believe otherwise. The very peculiar system employed and the curious chemical reactions it requires, would seem amply sufficient to close up any smelting-establishment in the country.

PARK, SUMMIT, AND LAKE COUNTIES.

The mineral-resources of these three counties are far less developed than those of the counties above mentioned, and may conveniently be considered under one head. Though divided by the main range of the Rocky Mountains, the lodes are all on the same belt, and the placer-ground is formed by the disintegration of one continuous course of veins. These are found on both sides of the main ridge, and on its outlying spurs for from five to twelve miles on either side. The district embraces about four hundred square miles of granite and gneissic formation, cut by gold and silver veins at nearly every point. On these veins there are over nine thousand locations. There are also forty-eight square miles of limestone and sandstone formation, containing segregated deposits of the precious metals, and 250,000 acres of placer-ground, fully two-thirds of which are under exploitation. The geological center is at Mount Lincoln, on the divide, from the slopes of which four large rivers take their rise, viz, the Platte, Arkansas, Blue, and Eagle. The first three are rich in placer-gold; the latter is as yet almost unexplored.

In order to cover more completely the mining-industry of these three counties, and to show the progress made during 1874, I will place the subject under the two heads of placer and lode workings, which may again be subdivided as follows:

Placer-workings:

1. The Arkansas Valley.
2. The Platte Valley.
3. The Blue Valley.

Lode-workings:

1. Snake River district, (Summit County.)
2. Breckenridge district, (Summit County.)
3. Hall and Geneva district, (Park County.)
4. Mosquito Range district, (Park County.)
5. Upper Arkansas district, (Lake County.)

The production of the mines and placers in these counties during 1874 was as follows:

| | |
|---|---|
| Gold, (gulch and bar) | $230,000 |
| Gold, (stamp and arrastra) | 130,000 |
| Silver | 565,860 |
| Lead | 35,653 |
| Copper | 49,720 |
| Total, (currency) | 1,011,233 |

All the silver-ore mined has been reduced in works outside of the district, and the same may be said of the copper.

*Placer-workings.*—1. *Arkansas.*—The operations in this river and its branches have been confined to sluicing and hydraulic mining in California, Colorado, Cash, Chalk, Iowa, and Clear Creeks. The bed of the river is also being worked to a moderate extent. The result of the season's work has been about $80,000 worth of gold, although the supply of water was small and the summer shorter than usual. Most of the old ground is now considered as worked out, and, as a consequence, attention is given to the upper levels of the creeks mentioned, and to the bars of the main stream. New ditches are being dug, to carry water to the heads of the richest gulches, and large areas of ground in the Arkansas Valley proper are being taken up and prepared for work next season. The success which has attended the introduction of the automatic boom (described in my last year's report) in the Blue Valley has induced several parties to try this method on the poor grounds in the Arkansas, and it is confidently expected that the result will be a large increase in the gold washed during 1875. The most important new enterprise is that of the Oro Ditch Flume and Mining Company, which has already built a ditch nine miles long, tapping the main stream near its head, and which will carry sufficient water to wash the upper ground on most of the eastern tributaries of the Arkansas.

2. *Platte.*—The production of gold from this stream during the year has amounted to $70,000. Almost the entire length of Montgomery gulch from Hoosier Pass down to and even below Fair Play, a distance of over twelve miles, is occupied by working-claims, some of which are operated extensively. Tarryall Creek also has during the last year been the scene of renewed activity, and many fine claims have been in operation, from its head down to Hamilton. The fortunes of the latter town, which has lain dead for many years, are once more on the increase, and as but few auriferous gulches in Colorado have yielded more freely in their time than Tarryall, the owners of ground there are esteemed among the fortunate ones.

Numerous minor gulches and ravines among the tributaries of the Platte were also worked last year with success. Labor is now 100 per cent. cheaper than in the early days, and the appliances for placer-working are so greatly improved in every way that a very large extent of auriferous ground may now be washed that formerly would not afford a profit.

Most of the parties operating are still using the hydraulic in preference to the boom method, as the ground is generally too level for the latter. In Snowstorm gulch, however, large reservoirs have been built, and booming is carried on with great success.

The ground of Messrs. Mills & Hodges, between Alma and Dudley, proved the most productive last year. They are operating against the left bank of the stream, and have a beast from 30 to 100 feet high for many hundred feet in length. The ground is comparatively free from bowlders, and is broken by the hydraulic stream with great ease. The resulting gravel is washed into narrow flumes, which empty into the main creek.

The Fair Play Gold-Mining Company owns the largest bar in Colorado, and will operate on a very extensive scale. The association owns 1,100 acres of land opposite and below the town of Fair Play, in a claim about five miles in length and 2,000 feet in width. It is supplied with two flumes, the lower of which is 4,000 feet in length, 6 feet wide, and 7 feet high; but the ground is so level that bed-rock has not yet been reached, and it is estimated that over 5,000 feet more will have to be driven before it will be gained. At present, only surface-washings are made, and the yield is, of course, not what it will be when the bottom of the bar is reached. Water is brought from the stream above through four large ditches, and gains at the upper workings a head of 140 feet, while at the lower the head will be over 220 feet. Two miles of conducting-pipe are now laid for present workings. The pipe is made of sheet-iron, in sections 20 feet long, which slip into each other, and taper from the reservoir to the present workings from 22 to 8 inches. This claim will be reopened early in May, and will employ 150 men.

In Beaver Creek, Messrs. Freeman & Pease commenced to place in order their 600-acre claim last year, and expect to be working on a large scale next season. The ground in this gulch is very deep, but rich, and an immense quantity of material can be washed down. It was in the early days a noted creek, but, owing to the high cost of working it, has lain idle for many years. It will now be opened systematically and with capital, and may be expected to produce heavily.

3. *Blue.*—Nothing new of note is to be reported from the placer-workings of this valley. The production of 1874 amounted to $70,000, and was taken from the old and standard claims in French, Illinois, Iowa, Lomox, Georgia, Swan, and Indiana gulches. Early in the year a project was set on foot to carry a ditch from the Upper Swan River into the head of Georgia gulch, which is the richest gulch in this valley. It has not yet been completed, nor am I able to ascertain whether it was commenced. The new ground to be won would amply warrant a large expenditure of money to bring water into it. In the operations on the Blue and its tributaries the system of booming, which was introduced some two years ago, has almost completely superseded all other methods of washing where there is sufficient inclination of the ground.

*Lode-workings.*—1. *Snake River district.*—This district, one of the earliest discovered in Colorado, has developed more slowly than any other, principally by reason of its great inaccessibility and the heavy and unpromising character of the ore. In fact, it was in the Coaley lode, on Glacier Mountain, that the first discovery of silver in the Territory was made; but the quantity was small, and for many years it has only been known in the dreams of prospectors. The district includes the camps of Montezuma and Peru. The ores are heavy, carrying both zinc and lead, and as a rule are not rich in silver. There is no market as yet nearer than Georgetown, and transportation across the Snowy range is only possible in summer, and then at a cost of $18 to $20 per ton. Early in 1874 an excitement sprang up concerning the mines, and the district was invaded by about two hundred prospectors from Georgetown, South Park, and Denver, who continued there until driven out by the deepening snows on the range. The result of the excitement was the discovery of a number of new lodes, the re-opening of many of the old ones, and a lively interest among smelters and millers. There is now a company formed, with the purpose, as soon as the weather will permit, of erecting smelting-works at Peru to handle the heavy galena-ores of that and neighboring camps. Last year these Peru lodes were considerably developed. They are undoubtedly of great strength and value, carrying very heavy bodies of galena, that will assay from 20 to 60 ounces of silver per ton.

The Comstock property is the only claim in this district upon which continuous work has been done for the past two years. It is owned by the Boston Silver-Mining Company, which has a large and fine mill, and is now engaged in driving a tunnel for the vein, which will cut it 600 feet deep. It is now within 300 feet of the lode. It is reported that this company will open its works for custom-ore next season.

2. *Breckenridge district.*—The ores of this district are of low grade, (in silver,) but very pure galenas. The veins are numerous, but only a few have been worked, as until lately no market has been found for the ores. In the summer of 1873, Messrs. Spears & Conant put up a small reverberatory furnace at Lincoln City, (French gulch,) which, however, did not prove successful. In 1874 the mode of treatment was altered by the introduction of a blast-furnace, which, not proving satisfactory alone, was supplemented by the addition of two Drummond lead-furnaces. These last did well, and by September they were fired up in earnest, and continued running steadily and successfully till the close of the year. The product of the four months' run was 216 tons of work-lead, carrying from 60 to 80 ounces of silver per ton. The ore is derived mainly from the Cincinnati or Robley lode, which has been developed to a considerable extent, and is capable of producing 10 tons daily of almost pure galena. The works having proved successful, owners of other similar veins are preparing to open them next summer, and the company will probably enlarge its facilities in order to work the increased supply. The work-lead is shipped by train across Hamilton Pass to the South Park, and from there it is carried to Denver, where it takes rail to Chicago, bringing 5 cents to 6 cents per pound, after refining, exclusive of the value of silver contained in it. Breckenridge district contains also a number of gold-veins, but none are being worked.

3. *Hall and Geneva gulches* are located at the extreme northwest corner of Park County and on the eastern flank of the range. The ores are galena, gray copper, and zinc-blende, in a gangue largely composed of sulphate of baryta. Bismuth has been met with in a number of the veins, (in a mineral supposed to be bismuth-silver,) and iron, generally in an oxidized instead of a sulphureted condition. The latter occurs both in regular veins of micaceous iron and in deposits of ocher and bog iron. Zinc is not a prominent mineral, nor met with in quantities sufficient to cause much trouble in smelting. But, on the other hand, baryta forms so large a portion of the gangue of the ores as to make them very intractable. To add further to the difficulties in treating the product of this district, the gray copper, which carries the bulk of the silver, occurs so intimately mingled and interspersed with the heavy-spar that hand-dressing cannot be resorted to. At the same time water-separation is rendered impossible by the fact that the two minerals have

nearly the same specific gravity. It is not surprising, therefore, that as yet no successful method of smelting has been reached.

The Hall Valley Smelting and Mining Company has control of most of the rich and valuable mines in Hall's gulch, and also owns the best of those outside. It has spent money freely in developing its property, and has a considerable quantity of ore in sight, besides several hundred tons at the dressing and smelting floors of the works. The latter are very extensive in plan, having a calculated capacity of 40 tons per day. They are fitted up without regard to expense. Extensive separating and sizing apparatus (modeled on German systems) receives the ore after it passes through the crusher and rolls. From these the mineral is carried on trucks to the furnace-room. Here the trouble commences. Last summer three large and handsome cupolas were erected, but on trial did not prove successful. Later in the year the erection of reverberatories was begun, and at its close they had not yet been put in operation. I am of the opinion that, with the complex ores to be handled, no simple smelting can be carried on until a large percentage of the heavy-spar is separated. This will doubtless prove a difficult task. The prominent mines of the district are the Whale, Cold Spring, Liftwick, Revenue, Congress, and Treasure Vault.

The Whale is opened very extensively, and is said to show a million dollars' worth of ore in the various levels. It crosses the top ridge of the main divide diagonally, and is attacked very advantageously by adit-levels on the course of the vein.

A full description of this remarkable lode, by Mr. J. L. Jernegan, M. E., in charge of the mine, is herewith given:

The Whale lode occurs in the main range of the Rocky Mountains, Park County, Colorado Territory, at the Hall Valley. It has been opened up and worked to some extent by the Whale mine, situated some 11,300 feet above the level of the sea. The mine, with 700 feet of the lode on either side of the discovery-shaft, is the property of the Hall Valley Silver-Lead Mining and Smelting Company.

The adjacent country-rock is a compact, fine-grained gneiss; its general strike is about north and south, dipping steeply to the west. In the immediate vicinity of the mine the strata are very nearly vertical. The gneiss incloses numberless veins of a reddish granite, in which the feldspar largely predominates over the quartz and mica, the latter oftentimes being scarcely perceptible. As a general rule, these granitic veins seem to more frequently follow the lines of stratification of the gneiss than otherwise. They are of varying width, some large, some small; the larger are from 5 to 6 feet wide. Gash veins of feldspar and quartz are also of very frequent occurrence.

The general trend of the lode is about northeast and southwest, intersecting the country-rock at about an angle of 45° to its stratification. It dips to the northwest at an angle of 65°. I am unable to accurately state how far this lode has been traced upon the surface, but it is at least well known to extend entirely through the mountain upon which the mine is situated, and is traceable over the ridge by a hollow extending for some distance along the surface, in which the lode crops out. This hollow in the surface-rock has probably been formed through decomposition of the vein-matter by atmospherical influences. The width of the crevice is variable, generally between 5 and 10 feet; the pay-vein varies from an inch, or less, to 36 inches. The vein is generally accompanied, both on foot and hanging walls, by a whitish, semi-decomposed rock, composed of quartz and decomposed feldspar, and in some places—where the process of decomposition has not reached as advanced a stage—

a light-greenish or black mica. This rock, therefore, is in places composed of all the constituents of gneiss, and is undoubtedly nothing more nor less than the same in a decomposed condition. It is often very strongly impregnated with iron pyrites, generally crystallized in minute cubes, and it is probably owing to the presence of this mineral that the rock is in such a state of decomposition. The vein is frequently separated from this decomposed country-rock by a "gouge" of clayey substance.

The accompanying sketch will serve to give the reader a more lucid idea of the general structure of the lode.

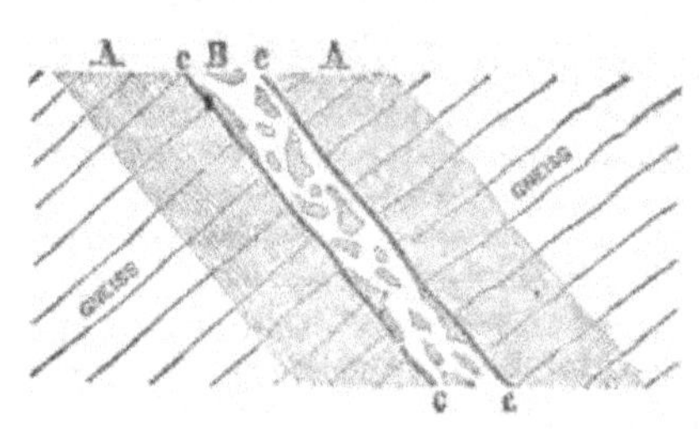


CROSS-SECTION OF WHALE LODE.—A. Decomposed gneiss impregnated with iron pyrites. B. Vein of barite inclosing irregular bunches of galena and gray copper. c c. Clay "gouge."

The occurring minerals, including also those constituting the decomposed walls of the vein, are quartz, decomposed feldspar, mica, clay, barite, dolomite, (brown spar,) iron pyrites, galena, gray copper, (tetrahedrite,) copper pyrites, malachite, azurite, and sulphate of copper.

The feldspar and mica are confined to the decomposed wall-rocks. Quartz occurs both in the wall-rocks and the ore-vein. Iron pyrites seem to be almost entirely confined to the decomposed wall-rocks, mere traces, if any, occurring in the ore-vein. Barite, brown spar, galena, and gray-copper pyrites are present only in the ore-vein. The clay constitutes the "gouge" on either side of the ore-vein. The minerals of secondary formation, malachite, azurite, and sulphate of copper, are also confined to the ore-vein, and occur principally in portions of the lode not far from the surface.

Of the metalliferous minerals, galena and gray copper are of the most frequent occurrence. Barite is the principal and almost only gangue where the ore-vein is of any considerable width; where pinched, however, quartz and brown spar are associated with it, the barite sometimes disappearing, especially if the vein carries no ore.

The entire, or almost entire, absence of iron pyrites in the ore-vein is rather singular, especially as the decomposed gneiss on either side is frequently strongly impregnated with this mineral. This circumstance might be explained by impregnation from the vein outward into the wall-rocks, accompanied, perhaps, by deposition of other mineral in the space formerly occupied by the iron pyrites.

Crystallized occur: barite, quartz, brown spar, galena, tetrahedrite, (seldom,) and iron pyrites. The barite often in large tabular crystals; brown spar in small rhombohedrons; quartz in six-sided prisms, terminated with six-sided pyramids; galena in cubes modified by the octahedron; tetrahedrite in modified tetrahedrons, and iron pyrites in minute cubes in the decomposed wall-rocks only. The crystals of quartz are oftentimes colored a reddish-brown or yellow, through the agency of oxide of iron, and the crystals of barite a pinkish color. The presence of oxide of iron is very general in the upper portions of the vein, near the surface, imparting to the crevice-matter and wall-rocks a more or less reddish-brown tint, thus forming a very distinct gossan, or "iron hat."

Of pseudomorphism, one instance only has come within my observation; it was copper pyrites after tetrahedrite.

The general structure of the vein is porphyritic; snowy-white barite inclosing irregular lumps of galena and gray copper. The banded structure, however, is not uncommon for short distances, especially where the vein is pinched. The presence of the banded structure in this lode speaks in favor of its belonging to the class of true fissure-veins, on which point, I think, there can be no doubt.

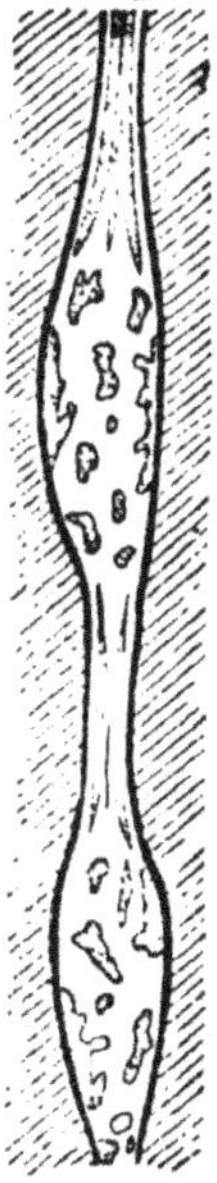

It has previously been intimated that the vein varies in width, being much wider in some places than in others, as represented herewith. Its porphyritic structure is shown in the wider parts, and the banded where it is pinched. This frequently-recurring inequality in the width of the vein may be owing to a lateral movement of the fissure-walls subsequent to the formation of the crevice. Smooth and polished slickensides are often observable, and their occurrence speaks somewhat in confirmation of the above view. Striated slickensides, however, from which the direction of the movement might possibly be deduced, have never come within my observation.

Pinch.

Heaves and slides are, up to the present time, unknown to the lode. The vein sometimes splits and incloses a fragment of the country-rock, called by the miners a "horse." Spurs and stringers also occur. In drift No. 1 of the mine, 73 feet from the mouth and in the hanging-wall, a spur, measuring about 24 inches in width, comes in at about an angle of 45° and joins the main vein. The course of its strike is about north and south, consequently parallel to the stratification of the country-rock; its dip also coincides with the same, namely, to the west. Its mineral character agrees with that of the main vein, the gangue consisting of barite; the metalliferous minerals are galena and gray copper. Thus far, I have not been able to trace this branch of the vein into the lower workings of the mine.



Horse.

Some 235 feet from mouth of same drift, the vein throws off another spur at about an angle of 30° into the foot-wall, dipping north 75° west, therefore opposite to dip of main vein. In drift No. 2, 50 feet vertical distance in a straight line beneath the same point in drift No. 1, the same spur can be perceived coming again into the main vein on the foot-wall, it having a dip of about 25°, which would bring them together somewhere between the second and third levels. This spur or branch-vein must consequently describe an arc of a circle, as represented in Fig. 3, which is an ideal cross-section: *a*, main vein; *b*, branch vein. Whether, in driving the first and second levels further to the south-west on the line of the main vein, this branch vein will again come in and join the vein, remains to be seen.

As regards the distribution of metalliferous mineral within the vein, it seems to be more or less influenced by the nature of the country-rock, gangue, and width of crevice. Where the crevice is wide, the gangue consists almost entirely of crystalline barite; the adjoining country-rock is soft and decomposed, and at the same time strongly impregnated with iron pyrites, and the metalliferous minerals seem to be present in much greater proportion. Where the vein is pinched, the wall-rocks are unconformable, hard, and of a less decomposed nature, a light-greenish or black mica forming one of its principal constituents, deficient in iron pyrites, and at the same time contain numerous little stringers of a grayish quartz. The gray copper oftentimes seems to be present in proportionately larger quantities in the narrower parts of the vein; or, more strictly speaking, the relative proportion of gray copper to galena is greater than in the wider portions of the vein. Where barite occurs in large crystals, the metalliferous minerals are more scattered and less frequent. Where the vein throws off spurs or divides, in taking in a "horse," it is almost always poorer. It is probable that, in further exploitation, the ore will be found to be chiefly confined to zones, or chimneys, as already the first, second, and third levels have been drifted through considerable ground on the vein that has proved quite barren of ore, though the crevice has always remained well defined.

The most striking characteristic of this lode is a crevice almost entirely filled with the mineral barite, carrying argentiferous galena and gray copper. The preponderance of barite over all the other minerals of the vein is of great significance in regard to the dressing and subsequent metallurgical treatment of the ore.

The Whale lode might perhaps be well compared to the so-called barytic-lead formation of Freiberg, in Saxony, of which the principal minerals are barite, fluorite, quartz, galena, and zinc-blende. Common to both are barite, quartz, and galena.

The geological history of this lode seems to have been simple, and somewhat as follows: 1. Formation of the fissure in the country-rock by means of plutonic agency, and, perhaps, accompanied at the same time by a lateral movement of one or both of the fissure-walls. 2. Deposition from solution of minerals now composing the vein-material on the sides of the fissure. A third period might, perhaps, be added, viz, the impregnation of the country-rock, from the vein outward, with iron pyrites.

The above might, of course, be extended into many more periods, were the effects of chemical changes to be taken into consideration. I do not wish to be understood, by deposition of minerals now composing the vein, as assuming that the minerals were actually deposited from their solutions in the very form and state of chemical composition in which they are at present found; it is by far more probable that many and various chemical changes have taken place since the time of their first crystallization upon the sides of the crevice.

There are numerous other veins in the vicinity of the Whale lode which have been more or less developed, and almost all of them are similar to the Whale in many of their characteristics, especially in the frequent occurrence of barite, which is, to say the least, worthy of notice.

Both the Cold Spring and Leftwick are well opened, and in the former a very large body of almost pure galena was found.

The Revenue is at the head of Geneva gulch, at an altitude of over 12,000 feet above the sea, and its shaft and ore-house are directly on top of the range. It has been opened for about 400 feet and to a depth of about 65 feet, and carries a very regular seam of galena and gray copper, averaging 4 to 6 inches, and worth in silver about $120 per ton.

The Congress is a parallel lode, not over 30 feet to the southeast of the Revenue, and of the same general character.

The Treasure Vault contains bismuth-silver-ores, and at the time of its discovery created considerable excitement. Several other veins of the same character have been found, and it is now known that there is a well-defined belt of lodes carrying ores of this metal at the head of Geneva gulch. They have a uniform course nearly north and south, crossing the others, and give every indication of more recent formation. As yet they have produced but little beyond specimens, but their location and the poverty of their owners will easily account for that. During last summer about 1,500 pounds of "slide" found on the slope below the Hidden Treasure, and supposed naturally to have come from it, were taken to Georgetown and sold at the rate of $400 per ton.

The ore has been examined, and approaches in its analysis very nearly that of the mineral schapbachite.

4. *Mosquito district.*—It is now over two years since the limestone formations of the Mosquito range were found to contain deposits of silver and gold ores. At first it was expected that the surface-pockets discovered would disappear in depth, or that it would be impossible to trace them, but time and development have given a very different aspect to the mines of this district. While there are still no greater assurances of continuance in the segregations, nor any evidence that they lead to permanent fissures in the underlying strata of crystalline rocks, the deposits have proved so extensive and so easily worked that they rank, deservedly, among the best mines in the Territory.

The Mosquito range is a spur of the great continental divide, breaking off from the latter about twelve miles north of Fair Play, and coursing nearly south for forty or fifty miles. From the head of this spur (Mount Lincoln) the ridge sinks slowly southward until it assumes the character of a low divide at the southwestern corner of the South Park. There it bends to the east, and, gradually rising again, terminates in Pike's Peak, one of the three great eminences of the Rocky Mountains. Along its entire course it forms the divide between the South Platte and the Arkansas, and the western and southern boundary of the South Park. The floor of the latter is of sedimentary origin, consisting of sandstones and limestones, and these, abutting against the Mosquito range, are tilted up and form a portion of its eastern slope. The formation is shown very clearly on the sides of Mounts Lincoln and Bross, and is as follows, commencing with the lowest:

First. Mica schist, from 600 to 700 feet thick, and carrying minute quantities of gold and silver ores, but too finely disseminated to be of value. This stratum is supposed to be underlaid by granite. It is the lowest known rock of the Mosquito range.

Secondly. Quartzite, about 500 feet thick, carrying auriferous iron and copper pyrites in apparently true fissure-veins.

Thirdly. White limestone, 300 feet thick, slightly metalliferous.

Fourthly. Blue limestone, 300 feet thick. This stratum carries all the rich silver-mines, the best being found near its upper face.

Fifthly. Sandstone, partially decomposed; from 40 to 50 feet wide; carrying throughout grains of iron and copper pyrites, which, however, are extremely poor in the precious metals.

Sixthly. Porphyry. This rock forms the caps of the two above-mentioned peaks, being cut away in the depression between them. It is completely disintegrated on and near the surface, and marked by the action of glaciers. When polished it is found seamed with slate and quartz dikes, and the whole mass is sensibly impregnated with silver and gold. On Mount Lincoln this porphyry cap is about 200 feet thick, while on

Mount Bross it is about 400 feet, thus showing an inclination of the strata to the south of about 15°.

All the great silver-mines are found, as has been stated, in the blue-limestone belt. Outside of that no deposits have been discovered of any value. The ore occurs in beds and chambers parallel with the dip of the strata, in apparently true fissures striking across the strike of the limestone, and in irregular deposits or pockets, having no general course or pitch, and often unconnected with each other in the same mine by any seam or fissure by which to follow from one to the other. Sometimes these pockets are very large, extending for long distances into and along the hill, and again they are but little vuggs in the rock, entirely alone, and separated by hundreds of feet from any other. Very few, if any, of the mines show any regularity, either in the course or pitch of their ore-bodies, and often the mineral is so disseminated throughout the rock that what might be termed the vein-material is many feet in width and entirely without any well-defined boundary. In fact, so irregular are the ore-channels, that it is often an impossible matter to tell where the pay-material ceases, without assays or mill-tests.

The ores furnished by the Lincoln and Bross mines are generally sulphides. Copper, lead, iron, and antimony in a sulphureted or oxidized condition, form the mass of the material, and in these the silver is distributed as glance, native metal, and perhaps a little chloride, though the occurrence of the latter compound is very doubtful. Galena exists in very small quantities, copper to a higher percentage, while probably the largest proportion of base metals is in the various minerals of iron. All these deposits are accompanied with the gangue of heavy-spar, which often furnishes a clew by which to trace out hidden bodies of mineral, and is confidently followed by the miner as a sure guide.

The following is a short description of the most prominent mines:

The Moose deposit crops out on the northeast face of Mount Bross, and has been extensively developed for a distance of 500 feet along its course. The vein of ore lies nearly horizontal, and is very regular in size and character. But little ore has been produced during the year, but the seam has been developed steadily. Over a million dollars' worth of ore is in sight, with no diminution in the size of the vein. This is undoubtedly the most extensive deposit of the district, and has proved to be more regular and even in its course than any other.

The Hiawatha is a segregation of pockets, some of which have yielded thousands of dollars. It is worked after no regular system, but followed wherever the leaders and stringers indicate a continuance of the deposit. As nearly as I can learn, its yield has amounted to considerably over $200,000 since first opened.

The Dolly Varden, Russia, Lincoln, Montezuma, and Elephant are all of the same class. Detailed descriptions of their workings would sound like accounts of indiscriminate quarrying. Such is the nature of the deposits that no system can be followed out and no calculations made upon their future. It will not be above the mark to state that the mines just mentioned have yielded during the year 1874 about $300,000 worth of ore.

Lower down on the mountain, and in the stratum of quartzite, are located the gold-veins the disintegration of which has undoubtedly furnished the float-gold of the Platte placers. They are mostly old discoveries, and in the early days yielded richly from the decomposed surface-quartz. When unchanged iron and copper ores were struck, they were deserted, and supposed to be valueless. When, however, the Alma and Dudley Smelting-Works were built, a demand arose for pyrites for fluxing, and the lodes were re-opened to furnish these minerals. It was then discovered that the ore was still auriferous; and since that time they have been continuously and steadily producing.

A list of this class comprises the Phillips, Orphan Boy, War Eagle, and a number of others of minor importance. The ore is mostly iron pyrites, carrying from ½ to 2 ounces gold per ton, and a small percentage of copper.

The mines of Mosquito district have produced during the year about 3,000 tons of silver-ore, of an average value of $140 per ton, and $50,000 worth of auriferous pyrites. The production has been somewhat in excess of the demand; and there can be no doubt that if smelting-facilities were greater, the mines could with great ease be made to double their yield. It is estimated that at the close of the year nearly $2,000,000 worth of ore was in sight in the various mines or stacked on the dumps.

As just intimated, the reducing-facilities of the South Park are hardly adequate to the supply of ore. The only works in operation during the year were the Alma Works, which produced $452,000 worth of matte. This was shipped to Black Hawk for separation. The process is an exact copy of that in use by the Boston and Colorado Smelting Company, viz, roasting and matting, and has been found to be completely successful. The Mount Lincoln Smelting-Works, at Dudley, for some reason, have lain idle during the entire year, while the Holland Works did not prove a success, owing to the lack of a sufficient quantity of galena.

The general condition of the district, however, is satisfactory and encouraging. In spite of the lack of capital to develop the mines and to carry on ample reducing-works, it has sprung in two years to the position of the third camp in the Territory, and, owing to the ease with which the rock may be worked, affords more than usual facilities for self-development. Prospecting all along the eastern flank of the Mosquito range for twelve to fifteen miles from its head shows that it is metalliferous to a high degree, and capable of yielding immense quantities of ore for a long time. The question as to the permanence of the deposits is one not likely now to trouble the miner, as there are evidences everywhere that they are extensive and numerous.

5. *Upper Arkansas.*—This district embraces about a dozen gulches among the headwaters of the Arkansas, carrying both gold and silver veins. Its production last year was small, not over $145,000, which was derived mainly from the gold-mines. There is as yet no market for silver-ore, and most of that which has been produced has been shipped to Golden City for treatment. The belt at the upper end of the valley is undoubtedly a re appearance of the same belt that courses through Montezuma and Breckenridge, and is almost exclusively argentiferous. Lower down gold-veins appear, which are most strongly developed in California and Colorado gulches. The occurrence of ores of nickel and cobalt in the ores of the upper valley is perhaps the only matter of mineralogical interest that has been shown by the year's operations. These metals are found with argentiferous galenas, but in small quantities. From a number of tons of Home Stake ore, treated by Mr. West, at Golden City, about 500 pounds of nickel speiss was run out, carrying from 2 to 12 per cent. of that metal. The great distance of the mines from railroads prevents this metal from affording any profit to the miner.

### THE SAN JUAN COUNTRY.

The first excitement over the San Juan mines has died out, but some encouragement is given for future *bona-fide* operations. The entire district of Southwestern Colorado is now known to be very prolific of gold and silver veins, and there can hardly be a doubt that in a few years it will be proved to be a very promising mining-field. So far the majority of the discoveries are of silver, and the mines bear more resemblance to those of Georgetown than to those of any other district in Colorado. But there are many varieties of formations and classes of veins, some of which are entirely distinct from any yet found in the Territory. As a whole, the country may be described as exceedingly rugged and mountainous, and comparatively bare of vegetation and timber. The rocks are principally the granites, gneisses, and quartzites found elsewhere in the Sierra Madre range, with here and there croppings of trap and other volcanic rocks. The mineral-veins do not show much regularity in position or course, and if surface-indications prove reliable, will be found faulted and irregular below. As a rule they are of great strength, and are frequently found standing above the surface for long distances. The gold and silver lodes are not separated into belts or districts, but the former, except in the Summit district, appear to be accidents, and will perhaps be found to change to silver at 100 or 200 feet in depth, as has been the case in so many veins in lower Clear Creek County.

The locations so far made number nearly two thousand, four-fifths of which are silver-lodes. A number of these are, of course, double—that is, two or more on the same vein—so that the actual number of distinct veins discovered has probably not been over twelve hundred. The country is now divided into the following districts: Summit, Decatur, Alamosa, Telluric, Sangre de Christo, Lake, Uncompahgre, Humboldt, Adams, El Dorado, Mosca, La Plata, Mancos, Animas, and Eureka.

Geographically, it is divided into two almost equal divisions by the main range of the Rocky Mountains, on the western slopes of which are the mines around and in Baker Park, and those on the tributaries of the Uncompahgre and Gunnison. East of the range are the mines clustered around the upper waters of the Alamosa and Rio Grande del Norte.

Concerning the first the following extract from a letter received by the Georgetown Mining Review will give all that I am able to obtain of a reliable and satisfactory nature:

The mining-belt of San Juan is twenty-five miles broad, commencing at Mineral Creek. The country-rock is principally composed of granite. There is also a rock, locally called hornstone, which is black in color, and very hard and tough. This is the general rock of the country, although there are a number of places where syenite distinctly occurs. The vein-material is generally massive white quartz, filled with minerals of all kinds, iron and copper pyrites, zinc-blende, galena, gray copper, antimony, and, in fact, I doubt if there is a known mineral that cannot be found in greater or less quantities. Calc-spar also occurs in some of the veins, in a thin streak by itself, dividing the seams of ore. At Mineral Creek there are a number of veins opened, showing micaceous iron and a little galena. These are the Baker Park mines, and the belt extends north for a distance of about three miles to Hazleton Mountain, which carries the first rich belt, and although the veins are narrow, (not averaging more than 3 to 5 inches of ore,) a number of them are working to a good profit. The best of these are the Grey Eagle, Susquehanna, and Aspen. The gray copper in these veins assays from 500 to 2,300 ounces per ton; the galena runs lightly, 50 to 80 ounces. A small quantity of native silver has been taken out of the Aspen. In this belt there is a streak of zinc-blende, almost pure, and about 4 feet in width. It assays too light to be of any value with no market for zinc.

From here to Minnie Creek the lodes are of huge dimensions, carrying galena poor in silver; but this statement has some notable exceptions. The Pride of the West, owned by two Georgetown miners, is perhaps the best lode in San Juan, the ore struck being 7 feet wide, of solid galena, interspersed with gray copper, and running well, (at least in assays.) The Green Mountain lode is a similar vein, but has not the same true vein appearance, nor does it carry as large a body of ore as the Pride of the West. However, its gray copper assays from 800 to 1,500 ounces per ton, and the galena about 100 ounces. These lodes are located a little north of the head of Cunningham Gulch, and from here to Minnie Creek the veins are frequent, and many of them are of immense size, but assay low, (from 6 to 40 ounces per ton.) Almost any piece of rock you may pick up on the hill-side will assay about 6 ounces, and more than half the slide is made up of rock filled with ore. I have seen from 100 to 180 ounces taken from such pieces.

The mines on Minnie Creek are of large size, carrying considerable rich ore, and crossing each other at every conceivable angle. Consequently there are lots of disputes, and fine chances for law-suits. I did not take anything here, but went on to Eureka district, which, I think, will prove to be the best camp in San Juan. Here the veins have regular courses, (northeast to southwest.) The change is sudden and remarkable. Just south of Eureka Creek they assay light, run in all directions, and are almost numberless; while on the north side of Eureka and Niagara Creeks they have a true course, are large, and about 300 feet apart. The Emma Dean is the first of this peculiar class; the law makes the locations 1,500 by 300 feet, but 4,000 feet of the Emma Dean have been sold up to date, and war to the knife has been declared by the rival owners, all of whom are too poor to go into court. On this vein are situated the Pavilion, Lowlander, and two others, whose names I have forgotten. Like all others in the belt, its outcropping can be plainly seen from a distance of from five to ten miles, according to the location of the observer.

The next is the La Plata Grande, partially owned by your correspondent, which has the same characteristics as the Emma Dean. The ore is composed of iron and copper pyrites, gray copper, galena, and zinc-blende, set in pure white quartz. The Emma Dean has two seams of ore, one on the north and one on the south wall. The former has not yet been opened sufficiently to determine its character, but the latter has a 5-foot vein of iron and copper, filled with gray copper, which in the center is solid, 12 inches in width, and ranging from this to 6 inches, but 18 inches of the vein is fully one-half gray copper, and assays from 95 to 2,500 ounces per ton, according to its freedom from pyrites. The La Plata Grande carries a 6-foot streak of white quartz, filled with galena, gray copper, and zinc-blende.

The Great Western is the next lode to the north, and is just 300 feet from the La Plata, and parallel, carrying the same kind of mineral, but more solid. We took out some chunks that gave us a job to turn over with levers. The Cañon claim is upon the same vein, one on one side and one on the other, of the Animas River. These three discoveries are on ground where a wagon can be taken. There is just room between the drifts and river for a dump. As they cross the stream at nearly a right angle, they can be worked by drifts, and being only 300 feet apart, they hold 600 feet by 1,500, and on this tract is situated the only timber in the neighborhood, which is of considerable importance.

Just to the north is the Crispin. Besides other streaks it shows 2 feet 4 inches of white quartz, so filled with gray copper that you can't set two fingers endwise on the face without touching that mineral. It assays from 700 to 1,700 ounces per ton. On the same vein are situated the Silver and Garibaldi claims.

Again to the north is the Hartman lode, a good vein, with 5 feet at least of ore, and 8 inches of gray copper. From here up to where Van Giesen is working, the veins are narrower, but where Van is operating a new belt sets in, but the veins are not so rich or wide, nor do they hold so true a course as those on Eureka.

Across the Saguacherange, on the Uncompahgre, is another good belt, on which is located the Poughkeepsie. Assays have been had out of this vein running up to 10,000 ounces per ton. It carries silver-glance in considerable quantities. Thirty thousand dollars have been offered for it and refused; $15,000 have been offered for the Hartman and refused; $40,000 asked by the owners of the Crispin, and $20,000 offered by Pennsylvania parties.

Of the mines on the eastern slope, those in Summit and adjacent districts are the most promising. What is known as the South Mountain appears to be the center of a broad belt of mines, mainly auriferous, of which the Little Annie may be taken as a type. This location is on a broad and well-defined vein of quartz, carrying free gold and pyrites, and has already, despite the small amount of work done on it, proved of great value. The neighboring veins appear to be in almost all respects similar, and the district on the whole seems of unusual promise.

Being located so many hundred miles away from the rest of the world, and lying under the many disadvantages caused by hard winters and almost impassable mountain-barriers, the San Juan mines have as yet no market for ores, and, until this necessity is supplied, can hardly be expected to add much to the bullion-product of the Territory. During last year about 25 tons of silver-ore that would reach $300 to $400 per ton were shipped to various points in the East for reduction, and several minor lots have been smelted at Golden, Denver, and Black Hawk. In Baker Park two smelting-furnaces were built in 1874, but they have not proved successful. In Summit district a stamp-mill for gold-quartz has been erected, and several other similar mills are contemplated.

The silver-ores appear to be rather complex in composition, and of that nature which the smelter denominates as "heavy." They will therefore give plenty of trouble to inexperienced metallurgists, and be the cause, undoubtedly, of many failures. The abundance of copper, lead, and zinc in nearly all the mines ought, however, fully to compensate for the extra expense involved in a complete process, so soon as connection with the rest of the world makes it possible to send them to a market.

The latest interesting development in these mines is the discovery in the Hotchkiss lode of tellurides of gold and silver. It has been confirmed by Mr. Schirmer, of the Denver mint, and therefore may be considered as an undoubted fact. Judging by the present condition of the other tellurium-districts of the Territory, I cannot regard the discovery as likely to prove commercially important.

### FREMONT COUNTY.

For the following statistics in regard to the mineral-production of this county I am indebted to Mr. R. N. Clark, the superintendent of the Cañon City coal-mines:

| | Tons. |
|---|---|
| The Cañon City coal-field yielded in 1874 | 19,385 |
| The Cañon City coal-field yielded in 1873 | 12,909 |
| Increase | 6,476 |

Which was all shipped over the Denver and Rio Grande Railroad. In addition to this, the road shipped 129 tons of silver-ore, from the Hardscrabble district, which yielded by mill-returns $21,986, and 8½ tons of copper-ore, yielding about 30 per cent. of copper. The above is the total yield of silver from this county.

### GENERAL NOTES ON THE TREATMENT OF COLORADO ORES.*

Whether Colorado ores are really as rebellious as in past times they have been supposed to be, or whether this Territory is supposed to be a legitimate and proper field for the trial of every new and outrageous idea that is hatched by would-be metallurgists, it is certainly true that the metallurgical industry here grows slowly, and does not show anything like the advance that is necessary to keep pace with the increasing ore-product. This is evidenced by the quantity of high-grade ores that are still shipped to the Eastern States and to Germany for treatment.

Still native metallurgical science advances, though very irregularly, and in most cases the advance is gained by bitter experience. Perhaps the most discouraging feature of the subject is the fact that in the new San Juan districts some of the same errors that were made in the early days of Clear Creek and Gilpin Counties are being repeated, and under circumstances which offer no excuse. Even in the older towns and under the shadow of former failures can be found evidences of the same disregard of science and experience which nearly wrecked the young community in 1862-'66.

* These are the suggestions and opinions of Mr. Van Wagenen. I can only say that they seem to me, in the main, very reasonable. But I do not feel sure that the future of Colorado metallurgy will be in all respects what he prophesies.—R. W. R.

From the work of the year I gather the following results:

The stack-furnace (Stetefeldt or Airey) for roasting purposes seems to have proved unequal to the treatment of the blendic ores of Colorado. At last its most ardent admirers have given it up, and have gone back to the old reverberatory or modifications of it. As a system, roasting and amalgamation is slowly losing favor, and bids fair in a few years to give way before methods of treatment that will save not only the precious metals, but also lead, zinc, and copper. The introduction in Stewart's Works of the Hunt & Douglass process is an effort to compromise on this point by saving copper. As lead, the most abundant base metal in the district in which these works are located, is unprovided for, I am inclined to think that the addition will only effect a postponement in the general change. As a system, however, roasting and amalgamation has been greatly improved during the last two years. The contest between the friends of the reverberatory and of the Brückner cylinder has resulted in a close study and improvements of each, so that a saving of 88 to 90 per cent. is now expected, where before mill-men thought they were doing well to get 80 per cent. The combination of three reverberatories, known as the Stewart furnace, works well, and though presenting no new ideas, may be considered as an advance on the plain single-hearth furnace.

The matte-producing works show no particular improvement, and have made none. They have met with the usual amount of success, and are well proved to be the correct system for all classes of ores not too heavy in zinc or lead. In the treatment of the zinc-blende so abundant in many parts of the Territory, Mr. West, of the Golden Smelting Company, has furnished to the Georgetown Mining Review some interesting results of his experience at Golden, which, as a corroboration of well-known metallurgical experience at other points, appear to be worth reproducing here. Speaking of his intentions when the Golden Works were being built, Mr. West says:

The general character of the ore in the several districts was taken into account when planning the works in Golden, with the view of treating such as would afford the largest supply. Subsequent experience taught me that I had overestimated the supply of ore carrying a high percentage of lead. It was supposed that ore could be obtained that would afford a mixture of 25 per cent. galena sufficient to collect the precious metals and yield a profitable return from the lead. Had such been possible, no difficulty could have arisen, but when it was proved that less than 15 per cent. of galena was all that we could procure, and often this would carry an equal percentage of zinc, the loss in silver was too high to allow of profitable operation.

On a run of 200 tons of ore that averaged 25 per cent. galena and 7 per cent. blende, the yield was equal to 90 per cent. of the lead, 97 of the silver, and the whole of the gold and copper. Another run of 90 tons of ore, in which the galena averaged 15 per cent. of the mass and the blende 14 per cent., the yield was only 47 per cent. of the lead and 87 of the silver. Such a loss would be insupportable, and if a better class of ore could not be obtained, or some improvement in reduction, the business would not be profitable.

Care in assaying the fumes and slag, for the purpose of checking the loss, showed the slag to be so rich as to account for all or nearly all of the loss. The slag, according to the condition of the furnaces, assayed from 6 to 18 ounces of silver per ton, while the fumes from the chambers and flues assayed only from 7 to 9 ounces. The proportion of fumes to the mass is too small to render it of account. Most of the time the furnace worked well, and the slag flowed quite freely, and did not, to all appearances, contain any matte.

A careful examination under the best glass at command did not discover any grains of matte, but upon borrowing a microscope, and further following up the examination, crystals of blende could be detected in almost every piece. Here was apparent the true cause of loss. The crystals of blende had passed through the furnace without change or decomposition, and the law of specific gravity had prevented them from sinking into the matte.

Evidently a more pefect decomposition of the blende before passing to the smelting-furnace was necessary, or a better result could not be obtained. Finer grinding and repeated roasting did not act as well as was anticipated, and some agent was necessary to assist the decomposition. Knowing the action of silica at high temperatures to be quite effective in decomposing many sulphates, I determined to use a larger proportion in the roasting-charge and increase the heat to semi-fusion. The result more than justified my conclusions. The slags were free from crystals of blende, and the assay fell from 7 ounces to 3 ounces of silver per ton. I continued this course, carefully noting results, until called upon to take charge of the Denver Works.

Here I found 70 tons of siliceous ore which had been ground quite fine, and as the previous superintendent had used up all ore that could be easily smelted, I had all the blende and quartz to work. Taking the mass, and mixing in proportions with iron-oxides so that the quartz would form 40 per cent. of the charge passed to the roasting-furnace, I mixed the blende in quantity up to 17 per cent., the galena averging about the same. This mixture was carefully roasted and fused, and the whole fritted together to prevent loss in the blast-furnace from the blowing-out of fine ore. The fritted lumps examined showed that the blende was decomposed, and a vitreous mass of silicates of iron, lead, and zinc formed. This mass, when passed to the smelting-furnace, was easily reduced, and a clean slag was formed, carrying less than 1 ounce of silver to the ton.

The silica, then, had proved a friend. The high temperature necessary for fritting had decomposed the blende and formed a union of the zinc-oxide and silica, or, in other words, the affinity of silica for the zinc-oxide was stronger than that of the sulphuric acid. The latter had been expelled and a new combination had been formed. No doubt the lead present claimed the silver in this new relationship, and in the reactions in the smelting-furnace the lead and silver were reduced to the metallic state. In this case the perfect result is undoubtedly due to the finely-ground silica, and in my future operations I increased the quantity as well as grinding finer, and have had no further trouble from loss.

The agents at Georgetown know how freely I bought of blendic ores the past summer, often taking ore containing 30 per cent. zinc, and making no reduction on that account, and, after an imperfect clean-up, I have only a loss in the season's operations of less than 5 per cent. of the silver purchased, after going through the operations of smelting, separating, and refining.

The following table, prepared by Mr. Van Wagenen, shows the present extent and nature of metallurgical works—other than stamp-mills—in Colorado:

*Table showing the name, character, and location of the various reduction-works in Colorado.*

| Name of works. | System employed. | Capacity. | Plant. | Location. |
|---|---|---|---|---|
| | | *Tons.* | | |
| Boston and Colorado Smelting Company. | Roasting and copper matte. | 40 | 2 calcining and 2 matting furnaces; separating-works. | Black Hawk. |
| Alma Smelting Company | ......do | 15 | 1 calcining and 1 matting furnace. | Alma. |
| * United States General Smelting and Mining Company. | ......do | 12 | ......do | Spanish Bar. |
| * Swansea (Colorado) Smelting-Works. | ......do | 12 | ......do | Empire. |
| Stewart's Silver-Reducing Company. | Roasting and amalgamation. | 15 | 4 calcining-furnaces and 8 pans, Hunt & Douglass. | Georgetown. |
| Judd & Crosby | ......do | 12 | 1 calcining-furnace and 3 pans. | Do. |
| Pelican | ......do | 15 | 5 cylinders and 5 barrels. | Do. |
| Nederland | ......do | 15 | 4 cylinders and 8 pans | Nederland. |
| * Hall Valley Works | Lead-reduction | 40 | 3 cupolas | Hall Gulch. |
| * Golden Smelting-Works | ......do | 12 | 1 slag-furnace and 1 stack (blast) furnace. | Golden. |
| * Denver Smelting-Works | ......do | 20 | 2 roasting-furnaces, 1 stack, and Balbach refining-plant. | Denver. |
| * Lincoln City Smelting-Works. | ......do | 10 | 2 Drummond furnaces | French Gold. |
| * Boyd's Smelting-Works | Peculiar | 15 | Reverberatory, stack, and fusion furnaces. | Boulder. |
| * Ohio and Colorado Smelting Company. | Chlorination | 20 | ...... | Salina. |
| * Denver Smelting and Concentration Company. | Concentration | 20 | 1 lead-furnace and concentration-machinery. | Denver. |
| * Collom's Works | ......do | 15 | ...... | Idaho. |
| * Collom's Works | ......do | 10 | ...... | Black Hawk. |

* These works were not in operation during the entire year.

# CHAPTER VIII.

## ARIZONA.

From the southern portion of this Territory have come frequent reports during the year just past of the revival of a once flourishing mining-industry, which had, however, for years been actually wiped out of existence by the Apache Indians. These have now been quiet, thanks to the skill and energy of General Crook, for more than a year, and miners and prospectors have in large numbers flocked into the country, re-occupying old districts and locating new ones.

Considering the isolated position of the country, it must be acknowledged that mining affairs have changed for the better with wonderful rapidity; and though great results in the way of bullion-shipments have, of course, not been reached in so short a time, it is confidently expected that another year will permanently incorporate Southern Arizona in the list of bullion-producing districts.

Of the many discoveries made, I mention the following briefly, hoping that next year I shall be able to record the favorable results of mining and metallurgical operations which are now anticipated.

The Ostrich gold-vein, located in a range of mountains about eighty miles southwest of Tucson, is no doubt at present the most important discovery in Southern Arizona. The vein has been found along a distance of about 6,000 feet, and a great many locations have been made. It is from 6 to 12 feet wide, and the ore is reported to assay from $40 to $100 per ton in gold, and to contain, besides, considerable silver. A number of shafts, from 20 to 50 feet in depth, have been sunk on various claims, that of Mr. D. C. Thompson being the best developed. This is the original Ostrich, which has been worked by Mr. Thompson for several months during the latter part of the year. The same gentleman has now 8 arrastras on ore from this mine, and this number was to be considerably increased. Messrs. Carr & Hopkins, who have lately bought one of the claims on the Ostrich, intend to put up a stamp-mill immediately.

The old Trench mine in the Patagonia Mountains, which was described in a former report, has been leased by a Mexican, Señor Padres, who has had a large number of workmen employed in mining and smelting. The latter operation has as yet been conducted on a limited scale, the furnaces being only small; but larger smelting-works are soon to be erected. Meanwhile several wagon-loads of argentiferous lead have been shipped from the mine. The main shaft on it is 240 feet deep, where the vein is 6 feet wide, carrying ore assaying $70 silver per ton. On the extension of the original Trench, Señor Padres is sinking a shaft. Wood and water are comparatively plenty in this vicinity.

The old Mowry mine, which is now in possession of the Patagonia Company, lies five miles south of the Trench, on the southeast spurs of the Patagonia Mountains. The country-rock here is granite and limestone, and the ore is argentiferous galena, carbonate of lead, and iron-sulphurets. It is low-grade in silver, but generally rich in lead.

In the same mountains, Messrs. D. A. Bennett and E. N. Fish have commenced mining operations on the San José and the Santa Maria, the shafts on which, at the end of the year, were respectively 100 feet and 250 feet deep. The ore is argentiferous gelana, some of which has been smelted by Señor Padres. The owners of these mines expect to have smelting-works of their own in the near future.

In the Santa Rita Mountains considerable prospecting and mining have been done. One of the prominent mines here located is the Lost Ledge. It is situated on the south slope of the mountains, and about five miles north of Sonoita Creek. It was discovered a number of years ago, but after a shaft 12 feet deep had been sunk, mining had to be abandoned on account of the danger from Apaches. The vein was less than 1 foot thick at the surface, but in the bottom of the shaft it had increased to 3 feet. Lately it has been taken up by several residents of Tucson, among whom are Messrs. John H. Archibald and E. N. Fish, who have concluded to begin at once the sinking of the shaft to a depth of at least 100 feet.

Considerable prospecting has been done on the east side of the Santa Rita Mountains, not far from Crittenden, by D. B. Rea, A. Smith, and P. J. Hand. This district is wholly new; that is, there is no evidence in it of any work in former times, though the distance southeast from Tucson is only about thirty-five miles. The prospectors named have taken up nine claims, some of which contain gold and others silver. Along the cañon, on the sides of which the quartz-claims are located, there is reported to be a considerable extent of good placer-ground. The veins east of the cañon are described as gold, those west as silver veins. There is plenty of grass, oak and pine timber, as well as water, in the country. Besides the parties named, two others have lately prospected the country and located a large number of additional gold and silver veins.

On the Salero Hill, a south spur of the Santa Rita Mountains, a new mining-camp (Boyleston) was established during the latter part of the year. The district is called Truman district. A large number of lodes have here been opened, which show argentiferous galena of excellent quality. The country-rock is granite, with porphyry. The Salero lode has a northeast and southwest course, while some twenty others run nearly due east and west. Some of these veins carry very quartzose amalgamating ores, but most of the others furnish ores which will have to be treated by smelting. In at least one of the veins, the Serena, gold occurs in addition to the silver. There is abundant evidence in this vicinity that the mines have been formerly worked, most probably by the early Jesuits who visited this country.

The Arivaypa Cañon, east of San Pedro, has also been prospected by Mr. Atkinson and others, and a number of copper and silver veins have here been located and slightly worked. In short, most of the older and once famous silver-districts to the south, southeast, and southwest of Tucson have been invaded during the year by prospectors and miners. Old claims have been relocated under new names, and energetic work has been done in many instances. Beside this some new fields have been opened, and it is only fair to expect that the mining-industry of this region will henceforth, if undisturbed by savage foes, pursue a course of progress which adverse circumstances have heretofore rendered impracticable.

In the immediate vicinity of Tucson and all around it a number of copper-veins have been discovered and worked. To reduce these ores Messrs. Tully, Ochoa & Co., of Tucson, have erected small smelting-works in the city, and a few short campaigns have been made.

The most astonishing discoveries have, however, been made during the last year just west of the line between New Mexico and Arizona, in the mountains north of the Gila River, and in the vicinity of its tributaries, the Rio Francisco, Prieto, and Bonito.

The following interesting description of the recent developments in this wonderful copper-region in Arizona is contributed by Mr. A. Harnickell, whose connection with the trade as one of our largest dealers in copper, and particularly his connection with these mines, afford him unusual opportunity for knowing that whereof he speaks. The abundance and richness of these ores bid fair to make this district famous throughout the world.

Before the rebellion already the Santa Rita and Hanover mines in New Mexico were largely exploited. An account of them appears in one of the former reports on mining-statistics. The war broke up this industry, but it has now been resumed in New Mexico and the adjacent portion of Arizona, and bids fair to assume very large proportions; indeed, when transportation-facilities are improved, a business is likely to grow up in that region hardly inferior to that of Chili, in copper-produce, and of greater magnitude, and more profitable, than that of Lake Superior.

The mountains in which the veins and deposits of copper occur, lie north of the Gila River, and between its tributary streams, the Rio Francisco, Prieto, and Bonito, in Arizona, extending to near Silver City, in New Mexico, and, although at a considerable altitude, they are easily accessible, well watered, timbered, and even fertile. The copper-ores, at any depth thus far reached by the miner, are all of the rich decomposed varieties. The nature of the copper-ore in the veins appears the same on top of the mountain as in the gorge 1,000 feet below; the same a few feet below the outcrop as at the bottom of a shaft; richer, by far, and in greater volume than in the famous mine of Urmeneta in Chili. Solid masses of red oxide, copper-glance, and true carbonate are the regular ores of the veins, as distinctly separate from the varied gangue-rocks of clay, limestone, &c., as the most economical miner could wish, and lavished upon the mountains in truly gigantic proportions. Yellow pyrites are not found as yet, but in several places so much oxidized iron occurs with the glance as to indicate there a transformation from pyritous minerals. The average yield of the ore dressed by hand is 35, 50, and 70 per cent. of copper; while, unlike sulphuret-ores, these oxidized ores can be smelted almost as readily and cheaply as the concentrated native copper-mineral of Lake Superior, which, in fact, does not average much higher in percentage of copper.

It is obvious that this great wealth of copper, the richest formation thus far discovered on this continent, must attract attention. But, owing to the distance from railroads, and the greater difficulty than with precious metals of marketing the products, no great influx of mining-adventurers has taken place; but, better than this, commercial enterprise has taken hold of some of the mining-claims, working them with capital, skilled labor, and good management.

Work has been resumed in the Santa Rita and enlargement is contemplated. At other places in New Mexico copper is now being mined and smelted, the San José and Chino mines yielding wonderfully rich ores, while prospectors have discovered other promising croppings and veins in the Burro Mountains, and, farther east, in the Organ Mountains.

The great mines, however, are over the border in Arizona, within the net of the Gila streams, south of the Sierra Blanca and east of the Cordilleras de Gila, being situate, politically, in the White Mountain Indian reservation. Croppings and deposits of carbonate in various places and directions invite and amply merit thorough geological prospecting. Thus far, however, only the oro-hydrography of the region has been ascertained and reduced to accurate maps for the use of the Government, and not yet published. This labor, as well as many other difficult tasks, was performed by that splendid corps of explorers, Lieutenant Wheeler's expedition. Two mines, or veins, have been sufficiently prospected and explored; and these alone demonstrate that we have here the wealth of the Chilian mines concentrated in a few miles.

The Longfellow mine, situate some ten miles west of the post-office town of Clifton, is a curiosity in its way, and unlike anything thus far found in copper formations. The length of cropping stripped thus far, simply because it is all that appears on the surface, and satisfied all curiosity, is only 250 feet. The length of copper-bearing outcrops, in extension of this, however, is admitted to show thousands of feet, giving the idea of a great vein having given rise to them. The ore cropped out along the slope of a mountain and followed the turn of the mountain. The miners have labored hard to find the direction of their vein proper, if it be a vein, but without success. Wherever they sunk or tunneled on the slope of the hill, 60, 80, 100 feet, and more, below the outcrop, and without any dead-work, they broke out ore; penetrating 70 feet into the mountain, at a short distance below the outcrop, nothing but ore was found, and the place has thus necessarily been turned into an open quarry, and engineering operations adjourned to ten years hence. The thing resembles a large iron-ore bank, and, indeed, iron and clay occur with the copper-ore. Some 75 tons of it, undressed, were shipped to Baltimore and yielded 35 per cent. of copper. Since then, most of the ore with gangue is thrown aside and only the copper-glance and red oxide transported to the smelting-works at Clifton, where the Mexican blast-furnaces at first used—worked by hand-bellows—have given way to reverberating furnaces run by Welsh smelters from Baltimore, who have built a stack 120 feet high and make their own brick. A good water-power, furnished by the Rio Fresco, drives, crushes, &c., and may finally be used for pressure-blast engines should half-high furnaces be erected for quick work. Wood being plenty, of great pyrometric value, (mesquit,) and only a limited business contemplated at present, the reverberatory furnaces are now most convenient. The stock of ore in dumps ready for smelting, or in course of transportation by huge wagons from the mine to Clifton, is 1,600 tons, which, it is calculated, will produce 1,500,000 pounds of pig-copper. The mining, or rather quarrying, goes so much ahead of the capacity of smelting and transportation that a pause had to be made, and now it is likely that the miners will have a mind and leisure to push investigation into the lay and dip and bearing of their ore-deposit and to prospect the continuations of it.

The crude pig-copper produced was shipped to Baltimore—some 200,000 pounds—and being refined proved soft and good in quality, as did also that from the New Mexico mines. This is due to the fact that neither antimony, arsenic, nickel, nor tin occurs with the ores of the region.

While this mountain of ore should prepare us for surprises in that locality, it is totally eclipsed by the Caronado mines, some three miles west of the Longfellow, and discovered by the party working the latter. The discovery had been kept secret until the land had been cut off from the Indian reservation by the President of the United States and restored to the public domain. This fact being advised by telegraph and swift expresses, a relocation was made by the discoverer, thus securing a virgin title that can never be disturbed.

Here we have a true vein, in a limestone and granite formation, cutting mountains and gorges 9,000 feet long as the crow flies, and probably much longer, as a mountain of green carbonates, some miles beyond, seems to lie in the same direction. Gay-colored croppings of carbonate plainly define and picture out the course of the vein. Six different names had to be given to the successive locations, viz, Boulder, Horseshoe, Coronado, Copper Crown, Crown Reef, Matilda. The width of croppings varies, averaging 30 feet, widest 135 feet, and narrowest 2 feet at the commencement, which is in Twin Cañon. The vein runs along both sides of the cañon, plainly visible here, of solid red oxide, then ascends the mountain on both sides, one of them rising 1,000 feet perpendicular, trial-pits showing copper-glance in limestone and other ores of copper along the whole course of the vein for over 13,000 feet superficial. The main work has been done on the Horseshoe, where the croppings are wide, specimens from the whole width of which, carbonates, assayed over 50 per cent. An adit was here cut 15 feet below outcrop, the bottom of which was found to be solid copper-glance for 20 feet into the vein, being as far as the work was carried up to the time of my envoy's departure. Curiosity prompted him to turn the adit into a wide open cut, and he found that the smaller veins cropping out had at the depth of 15 feet already run together into one vein, and to all appearances this may continue for the whole width of 135 feet. This show is enormous, almost incredible, but there it now lies bare, ready for anybody's inspection. Enough has been done to show a gigantic ore-course, bared in the cañon at 1,000 feet below the highest point, and the same ore shows everywhere. The general course of the vein is northerly, but it varies much from a straight line, and at one point is covered for 600 feet by a land-slide.

It is intended to proceed at once with mining and road-making, the natural outlet being at the Gila, below the mouth of the Fresco, where smelting-furnaces and water-works can be built, wood and clay being plenty.

The only great drawback of the mines, at present, is the distance of the location from railroad transportation. The projected line of the Southern Pacific Railroad runs within a few miles of the mines. That railroad built, ore could simply be shipped to a Texas port, and thence to Baltimore and Europe.

Thus far the cost of mining and smelting has been 5 cents per pound of copper, and the transportation to Baltimore 6 cents per pound. The distances are: From Clifton to Silver City, one hundred and twenty miles; Silver City to Las Cruces, one hundred and fifteen miles; from there to terminus of railroad in Colorado, six hundred and fifty miles. This distance will be shortened as the railroad progresses toward Santa Fé. All these are mail-routes, but the merchandise is transported during eight months of the year by ox and mule transportation, which take copper as return freight at four to five cents, and extra at six cents per pound. The Coronado Company, however, contemplate running a train sufficient to carry 2,000,000 pounds of copper to market.

With regard to the mines in Yuma County, I am again indebted for information to Mr. F. H. Goodwin.

The *Castle Dome district* appears to maintain its predominance as the most important one in the county. It is situated only forty-five miles northeast of Yuma, on the western slope of the Castle Dome range. Between twenty and thirty mines have been worked here during the year. Most of these are located on six different parallel veins. All of them contain argentiferous galena. At the mine of Messrs. Miller & Hopkins shafts have been sunk to a depth of 250 feet, and stopes have been opened and partly worked out on the 60, 100, 160, and 220 foot levels. The ore has all been shipped for reduction to San Francisco, where, in the early part of 1875, the owners of the mine erected a smelting-furnace, and reduced about 200 tons of their ore. After this they erected furnaces at Castle Dome Landing, on the Colorado, about twenty miles from the mines. At the end of the year these had not commenced smelting, but were expected to do so very soon.

Besides this mine, the Flora Temple, the mines of the California Mining Company, those of the Chelson Bros., the McKenzie Mining Company, W. T. Miller, Bettis & Goodwin, Roberts & Helm, and several others have been worked. The monthly shipment of ore to San Francisco during the year has amounted to about 150 tons. Most of the mines are easily worked, at small expense.

At the end of the year, Mr. Goodwin reports, there were on the various dumps of Castle Dome district about 500 tons ready for shipment or reduction, and as much more at Castle Dome Landing. The same gentleman reports the shipments from the county during the year as follows:

| | |
|---|---|
| Argentiferous lead, 20 tons, estimated at $75 per ton........ | $1,500 |
| Gold-dust .................................................. | 50,000 |
| Argentiferous galena, 1,800 tons, estimated silver-value $40 per ton................................................ | 72,000 |
| Silver and copper ores, 200 tons, estimated silver and gold value, $70 per ton.................................................. | 14,000 |
| Copper, 5 tons, estimated value, $200 per ton.............. | 1,000 |
| | 138,500 |

To which ought to be added the amount of gold carried off by Indians and Mexicans, which is estimated to be at least $50,000.

About ten miles northeast of Castle Dome is the *Eureka district*, which contains the same character of ore as the foregoing. A few mines have been worked here.

South of Castle Dome a new district, the *Montezuma*, has been discovered and organized during the year. The ores occurring here are silver-bearing copper-ores, containing some gold. Mr. W. P. Miller, who discovered the district, has a shaft down on one of his claims about 90 feet deep. The width of the ore averages about 4 feet.

Messrs. Baker & Goodwin have done a large amount of work on two mines which they own here. Their ore has been shipped to San Francisco. At the end of the year they were erecting small Mexican smelting-furnaces to reduce their ore at the mines.

In the *San Domingo district* but little has been done during the year, on account of its great distance from a market. The veins are here generally large, but the ore does not assay sufficiently high to cover the immense cost of transportation.

The placer-mines of the county, which, as explained in former reports, are mostly worked by "dry-washing," have continued to yield profitable returns to the miners, who are mostly Mexicans and Indians. The yield during the year brought into Yuma was about $50,000; but, as formerly, the larger part of the gold is reported to be carried by the miners themselves into Mexico.

From Yavapai County, Mr. Henry A. Bigelow has kindly furnished me with notes on the mining-industry. This, together with information from various other trustworthy sources, is incorporated in the following:

The cessation of the Indian hostilities has had a most beneficial influence on the mining-industry of the county. This has not been felt so much through the investment of outside capital as through the resumption of many small mining enterprises, which are conducted by the inhabitants of the country. This is true of placer mining as well as of lode-mining, many new arrastras having been employed to reduce the gold-ores on the spot.

The Vulture Mining Company, near Wickenburgh, has not resumed operations, and has even, it is claimed, neglected to do the work required by Congress, so that on the 1st of January the property was open for relocation. Advantage has been taken of this state of affairs by Dr. Jones and others, who have relocated the claim, and expect to utilize the immense masses of low-grade ores in sight in the mine.

The 10-stamp mill of Mr. Smith, which works gold-ores taken from the extension of the original Vulture, has been running steadily throughout the year, with profitable results. This is the only mill in the county which has run during the past year, but at the end of the year two more 10-stamp mills, both within ten miles of Prescott, were expected to begin the reduction of ore at once. At one of these mills, that of Mr. Frederick, on Hassayampa Creek, about 600 tons of $35 rock were piled up at the time.

At the Weaver mines, those curious placer-deposits of which I have spoken in former reports, a number of miners have worked during the year, making small wages. Many promising quartz-lodes have of late been discovered in this vicinity, but nothing of note has been done on them.

Little has been accomplished in the *Bradshaw district* beyond the running of some arrastras. The great silver-mine of the district, the Tiger, has about 700 tons of ore on the dump, which is not reduced for want of capital. Some developments in other silver-mines have been made in the county during the past year, and it is now certain that the Sierra Prieta is rather an argentiferous than an auriferous range, as the early settlers supposed it to be. The fact that most of the mines, the surface-ores of which yielded gold worth $17 per ounce, are running, at a depth of from 50 to 60 feet, into silver-ores, and yield bullion worth only $12 to $13 per ounce, together with the fact that so many silver-mines carrying almost no gold, or very small quantities of it, have been found, points in this direction. Mr. Bigelow reports the product of Yavapai County, which he can authenticate, as $227,000 in gold dust and bullion.

In Mojave County, the mining-districts in the Wallapai or Cerbat range have been somewhat neglected during the year, on account of some new and excellent discoveries further to the south. In Mineral Park, however, the Keystone Company is building a 5-stamp mill. It has sunk a shaft 180 feet deep on its mine, taking out ore of good quality. In regard to the smelting-works in the Cerbat range I have not had any information, but as I could not trace any shipments of lead from there, I presume that the work done by them has been inconsiderable. Colonel Buell's mill seems to have been imperfectly supplied with ore during the year, and has made only occasional runs. The fact is that too little capital has come into this part of the country to render lasting developments of the mines possible. At the same time, it is certain that the mines of these districts are as good as any in Arizona, and ought to yield largely.

Five miles east of Mineral Park the Hackberry mine has come into notice. It is located on a remarkably regular vein, which carries ore said to assay from $200 up into the thousands per ton. A large body of such ore is reported in sight.

The new discoveries mentioned above have been made in and about the *Hope district*, located about fifty miles south of the Cerbat range. Here a number of rich lodes have been discovered, which, so far as developed, promise to become paying mines. Thirty miles to the south the most important discovery of the year has been made, that of the McCracken mine. The distance from the mine to the Colorado River is thirty miles, and here a new settlement, Aubury City, has been made. The works to treat the ores from the McCracken and other mines of the vicinity are to be erected here. The McCracken mine is located on an immense vein, and a shaft sunk to the depth of 100 feet has developed an abundance of rich ores. The vein can be traced for a long distance, and a large number of locations have been made along it. The original discovery is reported to have been "bonded" by a California company for $150,000.

Greenwood City is located about twelve miles from the McCracken mine, on the Big Sandy. The 10-stamp mill formerly belonging to the old Moss Mining Company, and located at Hardyville, has been erected here to crush the ore from the Greenwood gold-mine. This ore was expected to yield about $60 per ton, but the first mill-run made convinced the owners that the value of the ore was much less, and that it would be unprofitable to treat it without closely assorting it.

The yield of Arizona Territory in gold and silver during 1874, I estimate to be at least $487,000. Besides this, considerable values of lead have been shipped, of which I am, however, unable to form an estimate. The copper-shipments of the Territory amount to not less than $90,000, and there is every reason to expect a large increase during the coming year.

# CHAPTER IX.

## PROGRESS OF THE METALLURGY OF THE WEST DURING 1874.*

The year 1874 marks a decided advance in the metallurgy of the West, in two directions. On the one hand, the technical management has been very materially improved, and on the other, the production has been largely increased over that of the previous year.

The technical improvements of greatest importance are the introduction of condensation chambers or flues at several works and the further treatment of the lead-matte produced in the ore-smelting. These two are of general importance, and although by no means introduced at the majority of works, their adoption at the few where their great importance has begun to be understood will no doubt bring about their universal introduction. To these may be added a third step forward, which for some localities is vital enough, though for others it must still remain doubtful whether it is to be regarded as an improvement or not. This is the introduction, at several works in Utah and elsewhere, of water-jackets, in addition to the water-tuyeres used heretofore.

Finally, the year 1874 has witnessed the successful introduction, in one Territory, of the Ziervogel process, by means of which argentiferous and auriferous copper mattes, heretofore sent to England and Germany for further treatment, are now deprived of their silver in the most satisfactory manner *in loco*.

Of contrivances for the purpose of collecting the dust and condensing the fumes, two deserve special attention. These are the long flues built by the Richmond and Eureka Companies, at Eureka, Nev., and the peculiar chamber constructed by Mr. Ayres, of the Waterman Smelting-Works, at Stockton, Utah.

The flue at the Richmond works is 800 feet long, and ends in a wooden stack, 40 feet high, which stands on the side of the mountain behind the works, and the top of which is about 200 feet higher than the charge-doors of the furnaces. It receives the fumes from three large furnaces, which smelt 150 tons of ore (mostly fine) per day. The 250-feet flues along the back of the furnaces and thence to the hill-side are constructed of strong sheet-iron, plates of which are riveted together in the shape of a pentagonal prism, the two upper corners being slightly rounded off. By means of iron rods this part of the flue is suspended horizontally from wooden trestles, with the sharp angle downward. About 4 feet below the lower extremity of the flue, a car-track runs along the entire length. At intervals of a few feet, small sliding doors are inserted along one side and at its lower edge, so that by opening the same the dust can be drawn into the car below. The size of the flue in this part is amply sufficient to receive all the fumes. The accompanying sketch, Fig. 1, shows the size as well as mode of construction.

Further on, where the canal enters under ground, it is still larger, being 9 feet wide on top and 8 feet deep; but from this point on, all the

* This chapter, prepared by Mr. A. Eilers, and presented also at the February meeting of the American Institute of Mining Engineers, needs no words from me to secure for it the attention of metallurgists at home and abroad.

way up the hill to the wooden chimney, it is simply a culvert in the ground, without lining of any kind, and closed on top by a sheet-iron cover.

In 1873, the total loss at the Richmond works was 20 per cent. of the assay-value of the ore; in 1874, after the flue had been put up, the loss was reduced to 12 per cent., of which a large proportion is accounted for in the speiss. Only the first 250 feet are cleaned frequently; the portion under ground requires this only at long intervals. Yet in the sheet-iron portion alone there are saved from 9 to 10 tons of dust per day, when the three furnaces are running. The deposit here obtained assays invariably higher in the precious metals than the ore smelted, and if we assume its value as only $55 per ton, it is easily seen what an immense saving is brought about by this simple contrivance.

The Eureka flue is also 800 feet long, and constructed of galvanized sheet-iron throughout.

The condensation-chamber at the Waterman Works, near Stockton, Utah, is substantially represented by the accompanying cuts, (Figs. 2 and 3.) At the time of my visit to the works the furnace was in blast, and it was, therefore, impracticable to obtain dimensions inside of the chamber. Some of these were, however, furnished to me afterward by Mr. George P. Lockwood, who superintended the smelting-works at the time. The accompanying cuts show approximately the construction of the chamber, which is partly novel, at least as far as the use of the cylinder with archimedean screw is concerned.

In starting the furnace *F*, the chamber is cold, and there is an insufficient draught through the chimney *D*, as well as danger of explosions from the flaming furnace. Therefore, the top of the furnace is at first kept partly open. In from eight to twelve hours the chamber is sufficiently warmed, and the flaming of the furnace has ceased so much that the top of the furnace can be closed and the gases directed through the chamber. Passing through *A* and *B*, there is no escape, except through the revolving cylinder *E*, in which, as two-thirds of its size are immersed in water, the flames are thoroughly cooled, and the dust precipitated either in the water or on the walls of *C*.

The walls of the two compartments of the chamber are cooled and kept wet by the jets *a a*, which throw water against them in a continuous shower. The uncondensed gases then pass off into the chimney *D*.

The cylinder in this chamber revolves sixty-five times per minute, when in action. The velocity imparted to it must, of course, be commensurate with the quantity of gases evolved from the furnace.

The bottom of the chamber slopes from all sides to the discharge-valve *H*, through which the accumulated dust is drawn off every twenty-four hours into settling-tanks outside of the furnace-building. From these the clear water is drawn off every twelve hours. The bottom of the condensing-chamber is again filled, by means of a hose, after every cleaning.

To keep the water at the level *o*, an escape is provided to run off the water brought in continually by the jets *a a*. The roof of the chamber is constructed of slightly-arched ⅜-inch boiler-iron plates, which are laid on loosely.

There are several improvements which suggest themselves for this chamber, as, for instance, some way of preventing the settling of dust in *A*, which can be effected by bringing the top of the arch under *A* to an angle of about 45°, or by inclining the whole canal steeply upward or downward. The size of the chambers is also smaller than it ought to be. In fact, I do not doubt that with two chambers at least 16 by 16 feet and 25 feet high, to be cooled by sprinklers from above, the no doubt troublesome cylinder and screw might be dispensed with altogether.

As the chamber is now, however, it does very good work, saving, on the whole, about 11 per cent. of the ore smelted as dust. There are no smelting-works in Utah of which as favorable a record in this respect can be given.

In the further treatment of the first or lead matte, a beginning has been made by several works in Utah, where an increased quantity of sulphurets has been encountered as the mines advanced in depth. Lead-matte has been made for years in the West, and the total loss incurred by disregarding it has been large enough, as I have shown in a former paper; but the amounts of silver and lead thus squandered at individual works were not sufficiently important in the eyes of the inexperienced to induce them to treat a product of which they knew nothing, and which for years was designated as "iron." Since, however, more abundant sulphurets have begun to occur in the ore, a very largely increased production of matte has taken place, which in some cases has been equal by weight to the product of lead.

Mr. Wartenweiler, of the Winnamuck, first began, in Utah, to roast his matte and to use it in the ordinary way in subsequent ore-smeltings as a very welcome flux. Others, and notably the Sheridan Hill Works, have since followed his example. All have found that besides the advantage of extracting the larger part of the lead and silver from the matte, there are other very material gains in doing so. Mr. Wartenweiler says that by using the matte he did not only lessen the quantity of costly iron-ore flux from 20 per cent. to 3.5 per cent. of the charge, but that also the quantity of fuel used per ton of smelting-mixture was very largely reduced. His data permit the calculation of the exact saving in the consumption of fuel, which is 28 per cent. of the quantity formerly used. In other words, while 409 pounds of coke per ton, or 20.4 per cent. of the charge, were used before roasted matte formed a part of the mixture, only 293 pounds per ton, or 14.6 per cent. of the charge, were necessary after the change was made. The total additional cost in rehandling and roasting the matte (portions of it three and four times) is certainly not more than $4 per ton. Eventually, of course, a small fraction of the original bulk remains as argentiferous and often auriferous copper matte, which is not further treated and sold in that shape.

The introduction of water-jackets instead of fire-proof material used in the smelting-zone of the shaft-furnace is another step forward, which, at least for Utah and Cerro Gordo, where fire-brick or other refractory material is excessively high, cannot be regarded as of doubtful merit. These jackets are simply hollow iron castings, occupying about 3 feet 6 inches in height on the outside of the hottest part of the furnaces. The water-space is from 2½ to 4 inches. The tuyeres were first cast in one piece with the jacket-sections where the furnaces were round, or the side and back castings where they were rectangular. But it was soon found far better to leave openings and insert wrought-iron tuyeres. For when a cast tuyere cracked, which was frequently the case, the whole section or side of the jacket, as the case might be, had to be removed, which is no easy matter with a furnace in blast. Otherwise the appliance proved very satisfactory, the increase in the consumption of fuel being insignificant; and there is no doubt that it will be largely used in the future. Where the water used holds appreciable quantities of mineral in solution, a form of the jacket preferable to the present one will undoubtedly be simple cast plates provided with a water-trough at the lower extremity to carry off the water, which is continually forced in many small streams against the top.

The introduction of the Ziervogel process, for the extraction of silver from copper mattes, at the works of the Boston and Colorado Company, at Black Hawk, Colorado, has been a perfect success, both technically and financially.

The increase in the production of argentiferous and auriferous lead in the far West over that of the previous year has been large. This was not so much due to the erection of additional works as to better management in keeping a number of those in existence regularly at work. The following is a comparative statement of the production during the years 1873 and 1874:

*Production of work-lead.*

| Where produced. | 1873. | | 1874. | |
|---|---|---|---|---|
| | Number of tons. | Gold, silver, and lead—value. | Number of tons. | Gold, silver, and lead—value. |
| Nevada | 12,812 | $5,063,235 | 11,516 | $3,865,419 |
| Utah | 9,566 | 2,901,191 | 15,474 | 4,332,720 |
| California | 4,000 | 920,000 | 5,095 | 1,680,000 |
| Montana, Colorado, and other sources, (estimated) | 300 | 144,000 | 375 | 180,000 |
| Total | 26,678 | 9,028,426 | 32,460 | 10,058,139 |

# CHAPTER X.

## THE DISTILLATION OF ZINC-SILVER ALLOY.

The subject of this chapter* may be more precisely stated as the American method of treating by distillation the zinc-silver-lead alloy obtained in the desilverization of lead.

Although this process has been in successful operation for nearly five years, superseding, gradually, all other processes having in view the same end in almost every lead-refinery in the country, no description of it, giving economical data, has ever appeared in print, nor has the process been known in its details to any considerable number of engineers beside those conducting it at their works.

A discussion of this subject at the present time becomes especially interesting in view of a late publication in the Prussian ministerial organ for mining and metallurgy, by Dr. Wedding, in which the method of treating the zinc-silver-lead alloy by distillation, as finally settled upon at Tarnowitz, is very minutely and fully described. It is the more interesting, because in that article reference is made to the American method in an unfavorable light, while, at the same time, it is clear from the article itself that the most important details of the process, aside from the fact that it is conducted in black-lead retorts, were not at all understood. In publishing herewith the details of our American method,

* It was prepared chiefly by Mr. A. Eilers, and has also been read as a paper before the American Institute of Mining Engineers.

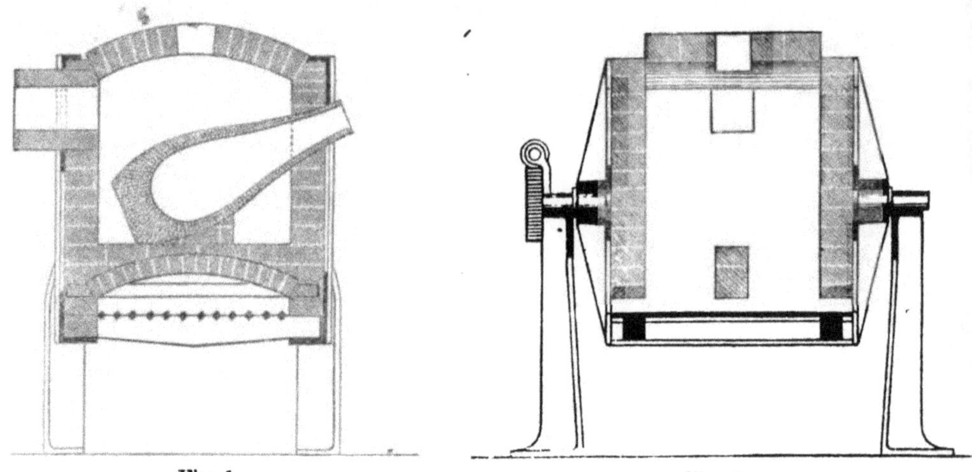

Fig. 1.　　　　　　Fig. 2.

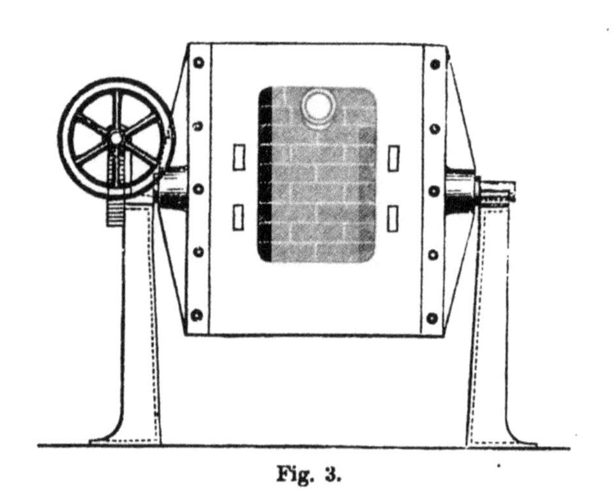

Fig. 3.

Feet.

12 6 0 1 2 3 4 5 6 7 8 9 10

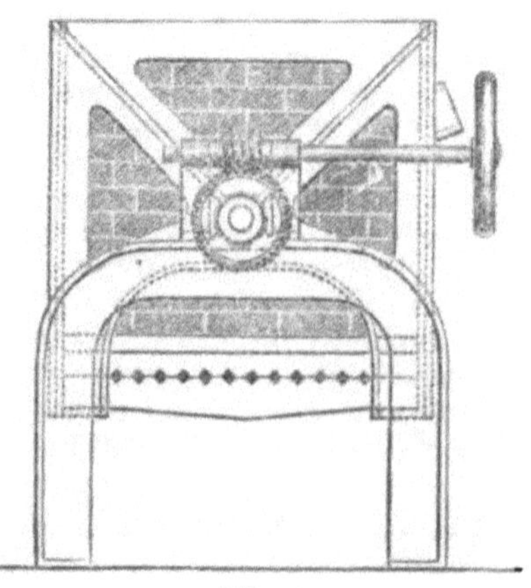

Fig. 4.

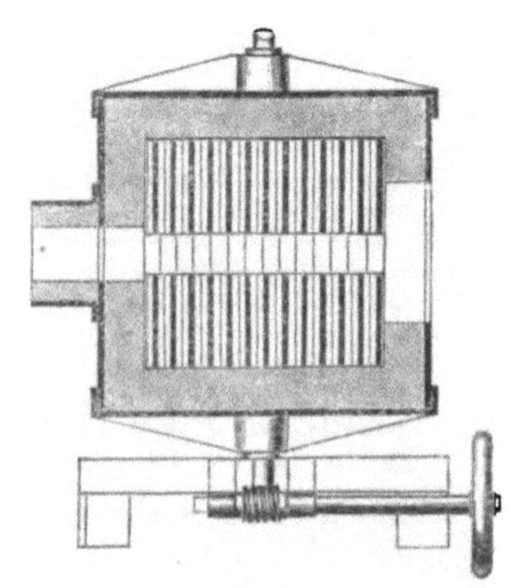

Fig. 5.

FABER DU FAUR'S TILTING DISTILLATION COKE-FURNACE.

one of my objects is to invite comparisons between our simple and direct process and the elaborate one at Tarnowitz.

In extracting the silver from work-lead by means of zinc, it is always the intention, in this country, to obtain a zinc-crust so rich in silver that the rich lead resulting from subsequent distillation contains from 8 to 10 per cent. of silver, or even more. Where the ordinary work-lead of the western smelting-works, containing from 100 to 300 ounces of silver per ton, is treated, from 1.4 to 3 per cent. of zinc is used; and it is only necessary to use the second and third additions of zinc again in a subsequent operation as addition No. 1, in order to bring them up to the required standard, the No. 1 of the first operation being already rich enough for the purpose contemplated. When poorer work-lead is treated, (which, by the way, does not often occur,) the same object is reached by the repeated use of the No. 1 zinc addition as above mentioned in regard to additions 2 and 3 of the ordinary process. One object in making the zinc-crust so rich in silver is to render it less liable to oxidation in the following liquation; another, to shorten and therefore cheapen the distillation itself.

The rich zinc-crust is liquated at some works in reverberatories, at others in kettles standing for that purpose near the large desilverization-kettles. It is, however, always the aim not to produce any oxides, and for that reason the temperature is kept exceedingly low, and access of air is limited as much as possible. In this fact lies the fundamental difference between our American distillation and that at Tarnowitz. In carrying liquation not nearly so far, and therefore not producing any oxides, we get rid at once of an immense amount of work, which the oxides formed at Tarnowitz occasion; and our immediate product of silver reaches, in consequence, a far higher percentage.

The liquated zinc-crust was subjected to distillation before 1870, by Mr. Balbach, of Newark, but the process was at that time very expensive, in consequence of the frequent breaking of retorts, which cost from $14 to $16 apiece. The retorts then used were made of the same material and of the same shape as those used to-day. They are made of New Jersey clay and chamotte, and contain about 25 per cent. of plumbago. But at that time the neck of the retort had to be freed from the surrounding brick and lowered every time at the end of a distillation, and this handling of the retort while white-hot caused frequent breakage. This most serious objection to the process was, however, removed in 1870, by the invention of Mr. A. Faber du Faur's tilting-retort furnace, which has since been introduced in the majority of works in this country. The accompanying drawings, Figures 1, 2, 3, 4, 5, represent this furnace and require no further explanation.

The process of distillation now is as follows: The retort-furnace is heated gradually by means of coke* until the retort has become dark-red. Then it is charged by means of a small copper shovel with liquated zinc-crust, which has previously been subjected to what is incorrectly called granulation, *i. e.*, after having been taken from the liquation-furnace it is spread, still soft, on a clean iron plate in front of it, and here cut up, by means of a shovel, into pieces of about 1 to 1½ cubic inches. According to the richness of the alloy and the size of the retort, a charge filling the retort to the neck consists of from 250 to 400 pounds of alloy, with which from 3 to 5 pounds of small charcoal, of bean to nut size, have been mixed. Next the condenser is put on. This may

* At one works in the West crude petroleum is said to have been used instead of coke. This is probably so far advantageous, as thereby the formation of slag-accretions on the outside of the retort is prevented.

either be made for the purpose, being in that case simply a truncated cone of fire-clay, about 2 feet long and of an inside diameter at the base a little larger than the outside diameter of the retort, or it may be an old retort, which it is unsafe to expose to the white-heat required for distillation, and which is thus made to do duty for a while longer. The temperature is then at once raised to white-heat, and kept so until the distillation is complete. The operation lasts from eight to ten hours, according to the percentage of zinc in the alloy. During all this time it is only necessary to keep the retort uniformly at a white-heat. If this is neglected, a crust of chilled alloy is apt to form on top of the metal-bath, which, upon a renewed raising of the temperature, would cause an explosion in consequence of zinc-fumes suddenly developed under the crust. An occasional introduction of a small iron rod into the retort through the condenser serves to show the workman whether he has kept the temperature high enough. Experienced men never make a mistake in this respect. The metallic zinc, collecting in the condenser, and retained there by a rim of blue powder and oxide of zinc, forming around the mouth, is from time to time tapped, and the blue powder and oxide are quickly scraped into iron vessels, from which the air can be excluded, the object being to prevent the oxidation of the blue powder. When sufficient metallic zinc has thus been collected, it is remelted in a kettle under a coal covering, the oxide and impurities are taken off, and the metal is cast into plates, which are again used for desilverization. From 40 to 50 per cent., and sometimes more, of the zinc originally added to the work-lead is thus regained in the form of plates, which contain only a trace of silver. The blue powder and oxide, containing no more silver than the metal, and comprising about 10 to 20 per cent. of the original zinc, are sold to zinc-works. Thus, about 50 to 70 per cent. of the original zinc is obtained again, the remainder having been partly retained by the desilverized lead, the contents of which in all cases amount to the somewhat constant figure of 0.7 to 0.8 per cent. of zinc to the whole mass of lead, and partly lost as oxide escaping from the mouth of the condenser.

When, in spite of a continued white-heat, the zinc-vapors are developed only very sparingly, the process is carried as far as policy permits, the rich lead containing then still a trace of zinc. At the same time it is desired that the zinc-contents of the rich lead should not be more than a trace, in order that serious losses from this cause may be avoided in the subsequent cupellation. The condenser is then taken off, so that the zinc-fumes still in the retort may more readily escape, and the furnace is left to itself for a few minutes. Meanwhile a small two-wheeled wagon, carrying a cast-iron pot lined with molder's sand of the iron-casting houses, is brought in front of the retort, and by tilting the whole furnace the rich lead is transferred in a stream to the kettle. After having here cooled a while, the metal is poured into lead-molds, previously washed inside with lime-milk, and well warmed, by tilting wagon and basin together. These molds are only half filled, in order to produce thin bars, which are handier afterward for gradual addition on the English cupelling-hearth. The residue remaining in the retort after the discharge of the rich lead, and consisting of a little charcoal and slag, is scraped out with an iron hook, while the retort is yet tilted. The larger pieces of coal go back into the retort in the next distillation. The smaller stuff and slag is kept separate, and is afterward added in the smelting, during which the rich litharge is reduced, or sometimes it is immersed in poor lead. The entire quantity produced during any one distillation should, after sifting out the large coal, not weigh over a pound or two. If no dust or dirt has been allowed to get into the alloy before distillation, and if the temperature has been kept high enough during that process, including the discharge of the retort, the remaining scraps will always be found insignificant.

A handful of fine charcoal-dust is now thrown into the discharged retort, the object being to prevent the oxidation of small lead globules, because litharge once formed would soon destroy the retort. Next, the furnace is turned back to its original position, the grate is cleaned, accretions of melted ashes which may have formed on the sides are broken off, there is new coke added, and the retort is at once filled with a new charge. A retort outlasts now from fifteen to thirty, or an average of about twenty, distillations, the retorts becoming unserviceable principally on account of accretions on the outside, which are melted coke-ashes. To obviate this, firing with crude petroleum and flame-fire from gas-generators have been proposed. Both ways are, no doubt, practicable, and the latter especially will result in a large saving in fuel.

The following are the results of two campaigns of the Pennsylvania Lead Company, at Pittsburgh, for which I am indebted to Mr. E. F. Eurich, the metallurgist and superintendent of the works. In one of these, unrefined work-lead, as it comes from the shaft-furnace of the company, was treated; in the other, work-lead refined before melting it down in the desilverization-kettle.

I.—DESILVERIZATION OF WORK-LEAD DIRECT FROM THE SHAFT-FURNACE.

| | | |
|---|---|---|
| To the kettle: Impure work-lead | 87,294 lbs | |
| Taken off: "Schlicker," (cupreous oxide) | 3,497 lbs | |
| | | Ag. oz. |
| Remains: Pure work-lead | 83,797 lbs | with 6,305.6 |
| To this was added: Zinc | 1,760 lbs | =2.1 per cent. |
| The zinc crust after liquation was | 9,525 lbs | |
| "Abstrich" from dezincation of poor lead | 7,810 lbs | |
| Oxides and metallic lead from market-kettle | 1,000 lbs | |
| Lead from liquation of zinc-crust | 808 lbs | |
| Market-lead | 67,104 lbs | |

DISTILLATION OF LIQUATED ZINC-CRUST.

The liquated zinc-crust was subjected to distillation in twenty-seven charges. Average charge, 353 pounds of alloy, with $\frac{3}{4}$ pound of small charcoal. In twenty-four hours two distillations were effected in each retort.

| | | |
|---|---|---|
| Charged: | Liquated zinc-crust | 9,525 pounds. |
| | Charcoal | 108 pounds. |
| Result: | Rich lead | 7,609 pounds. |
| | Metallic scraps | 390 pounds. |
| | Charcoal, with little metal | not weighed. |
| | Metallic zinc | 770 pounds. |
| | Blue powder and oxide | not weighed. |

Coke used, 410.4 bushels, at 40 pounds = 1.7 pounds per pound of zinc-crust.

The metallic scraps were immersed in poor lead on the cupelling test, and then cupelling was continued in the usual way, by adding rich-lead bars from time to time. By the immersion of the scraps in poor lead, 230 ounces of silver were extracted from them.

Results:

1. *Control of silver.*

| | | |
|---|---|---|
| In refined work-lead | | 6,305.6 oz. |
| Obtained and proved: | | |
| Silver tapped from test, 6,098.75 ounces, at .980 fine | 6,031.66 oz. | |
| Small pieces of silver from test, 150 ounces, at .970 fine | 146.50 oz. | |
| Directly-obtained silver | 6,178.16 oz. | |
| Silver in market-lead, 0.33 ounce per ton in 67,104 pounds | 11.18 oz. | |
| | | 6,189.34 oz. |
| This leaves in litharge, hearth, retort-scraps, oxides, and scum from immersion, and liquation-lead | | 116.26 oz. |
| | | 6,305.6 oz. |

And the direct product of silver is 98.1 per cent.

2. *Control of lead.*

| | | |
|---|---|---|
| Unrefined work-lead | | 87,294 lbs. |
| Obtained and accounted for: | | |
| "Schlicker," 3,497 pounds, at 80 per cent | 2,797 lbs. | |
| Lead in zinc-crust | 7,765 lbs. | |
| Soft market-lead | 67,104 lbs. | |
| Oxide and scum from market-kettle 1,000 pounds, at 95 per cent | 950 lbs. | |
| Liquation-lead | 808 lbs. | |
| Oxides from dezincation, 7,810 pounds, at 80 per cent | 6,248 lbs. | |
| | | 85,672 lbs. |
| Loss, about 1.9 per cent | | 1,248 lbs. |

II.—DESILVERIZATION OF REFINED WORK-LEAD.

| | |
|---|---|
| To the kettle: Lead | 62,895 pounds, with silver 6,165.9 ounces. |
| Added: Zinc | 1,260 pounds. |
| Produced: Liquated zinc-crust | 6,362 pounds. |
| "Abstrich" from dezincation of poor lead | 3,500 pounds. |
| Oxides and metallic lead from market-kettle | 700 pounds. |
| Market-lead | 53,420 pounds. |

DISTILLATION OF LIQUATED ZINC-CRUST.

The liquated zinc-crust was subjected to distillation in twenty charges. Charges and time required were the same as in the first campaign.

| | | |
|---|---|---|
| Charged: | Liquated zinc-crust | 6,362 pounds. |
| | Charcoal | 80 pounds. |
| Result: | Rich lead | 5,221 pounds. |
| | Metallic scraps, charcoal, zinc, and oxides | not weighed. |

Coke used, 276 bushels, at 40 pounds = 1.73 pounds per pound of cru-t.

The residue in the retort after the discharge of the rich lead, *i. e.*, metallic scraps and charcoal impregnated with metal, was not divided into two classes, as in the former case, but was all kept together, to be added in the reduction of the rich litharge at some future time. There was, therefore, no immersion in poor lead in this case, and, consequently, a smaller direct product of silver than in the previous campaign.

Results:

1. *Control of silver.*

| | | |
|---|---|---|
| In refined work-lead | | 6, 165. 9 oz. |
| Obtained and proved: | | |
| Silver tapped from test 5,714.5 oz., at .989 fine | 5, 645. 9 oz. | |
| Small silver pieces from test, 115 oz., at .970 fine | 111. 5 oz. | |
| Directly obtained | 5, 757. 4 oz. | |
| In litharge, 5,209 pounds, at 30 ounces per ton | 78. 0 oz. | |
| In market-lead, 53,420 pounds, at 0.33 ounce per ton | 8. 9 oz. | |
| | | 5, 844. 3 oz. |
| This leaves in hearth and retort-scraps | | 321. 6 oz. |
| | | 6, 165. 9 oz. |

And the direct product of silver is 93.3 per cent.

2. *Control of lead.*

| | | |
|---|---|---|
| Refined work-lead | | 62, 895 lbs. |
| Obtained and accounted for: | | |
| Market-lead | 53, 420 lbs. | |
| "Abstrich" from dezincation, 3,500 pounds, at 80 per cent | 2, 800 lbs. | |
| Oxide and metallic scum from market-kettle, 700 pounds, at 95 per cent | 665 lbs. | |
| Lead in zinc-crust | 5, 002 lbs. | |
| | | 61, 887 lbs. |
| Loss, 1.7 per cent | | 1, 008 lbs. |
| | | 62, 895 lbs. |

In both campaigns above cited the loss of lead, which will take place upon reduction and further treatment of oxides and other intermediate products, is not taken into account, as it could not be directly ascertained. From former experience, however, the total loss of lead in refining, *i. e.*, adding to what is given here, the loss in the further treatment of all middle products, is from 3 to 4 per cent. of the original weight of the unrefined work-lead.

It is to be regretted that the direct proof could not be furnished that the silver, *not directly* produced, is really all in the various intermediate products given above. But as long as the works are not so situated that these by-products can be worked by themselves, this cannot be done, and we must be satisfied to find at the end of each year, when the balance is struck, that the supposition has been correct.

Although not strictly within the scope of this article, I will add here, for the satisfaction of those who think that no market-lead pure enough for the manufacture of white lead is made in this country, that the Pittsburgh works, as well as two others, do regularly produce an article of unsurpassed purity.

Since new and large desilverization-kettles have been introduced and so set as to prevent an inconvenient cooling of the upper part during skimming, the silver-contents of the refined lead have been brought down to the low figures of from 4 to 8 grams in a ton. These limits can now be maintained regularly. The following is a late analysis by Dr. O. Wuth, of market-lead produced from Utah and Colorado ores, and subjected to desilverization by zinc without a preparatory refining. The sample was taken from one out of ten charges, all which were made up of lead obtained from the same ores. This lead is used by the Pittsburgh White-Lead Works, which, I believe, are substantially under the same control as the smelting-works:

| | |
|---|---|
| Ag | 0.00042 |
| Sb | 0.00051 |
| Cu | 0.00007 |
| Zn | 0.00038 |
| Fe | trace. |
| S | 0.00018 |
| Pb | 99.99844 |
| | 100.00000 |

By comparison with the analysis of the great majority of the foreign brands, used for the same purpose, it will be observed that the Pittsburgh lead is superior, and ranks with the best made in any part of the world.

The only unsatisfactory feature of this American method of distillation is the large consumption of fuel which is connected with it in the present apparatus. This has led to the construction of other furnaces, which are heated either by flame-fire or gas, and though at present the general results obtained are still doubtful, the saving of fuel intended is absolutely proved.

Of these furnaces I will here mention the two following:

1. The furnace constructed lately by Mr. Faber du Faur, which is represented in Figs. 6, 7, and 8.

The furnace is heavily armed with cast-iron plates and cross-pieces.

The same apparatus can, by slight alterations, be adapted to gas-fire. In this case the grate is lowered about a foot, and provided with under and over blast from hot-blast pipes arranged in the flue *d*. The top wind, to ignite the gases, would have to enter in jets arranged across the inner edge of the fire-bridge. The better way, however, proposed by Mr. Faber du Faur, is to have a separate and stationary gas-apparatus, so arranged that when the retort-furnace is tilted the fire-bridge will be lifted off from the combustion-apparatus.

2. The furnace constructed by Mr. W. M. Brodie, first at the Montgomery works, near Bloomfield, N. J., and lately at the works of the Messrs. Tatham, near Philadelphia. It is represented in Figs. 9, 10, and 11.

This furnace contains six graphite retorts of the common pattern used in this country heretofore for the purpose of treating zinc-crust. They are arranged in two rows, the upper ones lying over the spaces between the lower ones, in the position shown in the drawing. Fire-brick arches keep the flame from striking directly against the retorts. These are

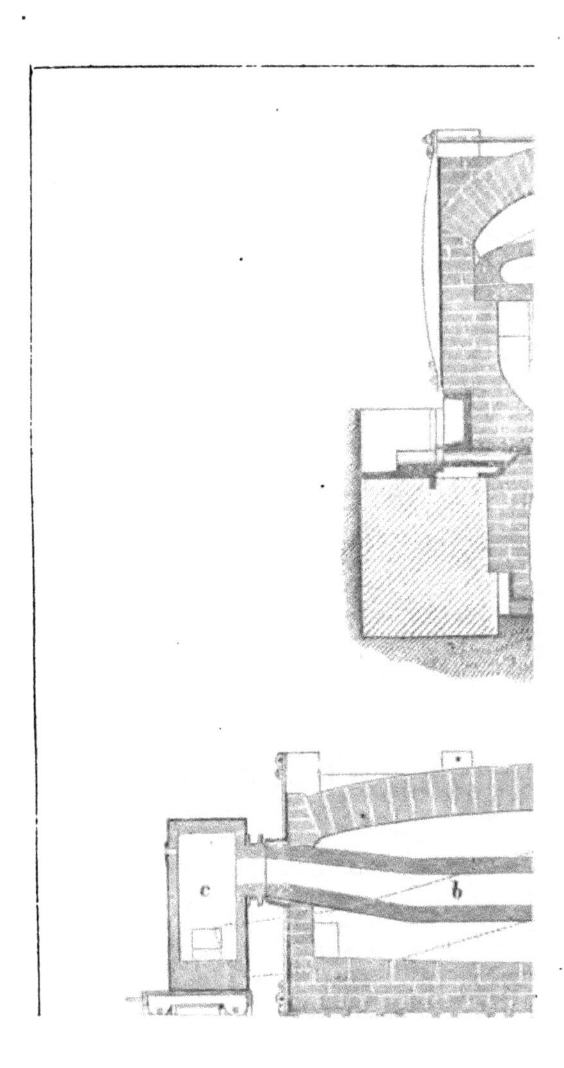
c
b

tapped, when ready for discharge, by means of ¼-inch holes bored in the lower edge of the bottom, the rich lead being conducted into molds through black-lead gutters. The fire-grate is provided with under-blast, which is heated in pipes lying in the main flue. The six retorts hold 2,600 to 3,000 pounds of zinc-crust, according to the degree of liquation that it has been subjected to. The time required for one operation in this furnace is from twelve to twenty hours, according to the material operated upon. Clean scum, resulting from refined work-lead, requires about twelve hours; impure crust, resulting from badly-refined, especially from cupriferous, work-lead, requires naturally a much longer time. The former leaves, after tapping, scarcely any residue but charcoal in the retorts; the latter leaves a considerable quantity of dross. It is, therefore, policy to refine the lead as completely as possible before adding the zinc, even if there is so little dross in the work-lead, as received from the shaft-furnaces, that it would apparently not pay to subject it to a refining before the addition of zinc. The quantity of fuel used in this furnace is 2,240 pounds of soft coal per ton of zinc-crust. One man attends with ease to the six retorts. The product of silver and zinc is not different from that obtained in the Faber du Faur furnace; but the time required to finish an operation is far longer in this case. But perhaps this may be remedied by a larger grate-surface, and by having the retorts made in the future with a prolongation in the bottom, the object being to render possible such a position of the retort in the furnace that the flame can play around every part of it which contains metal.

---

# CHAPTER XI.

## SILVER-LEAD SMELTING AT THE WINNAMUCK SMELTING-WORKS.

In a previous report these works, and the process in operation at the same in 1874, have been described in detail. Since then, however, the changed character of the ores has rendered necessary material additions to the works, and alterations in the process heretofore followed.

For the following details of the work done during 1874, and the present aspect of affairs, I am indebted to Mr. Alfred Wartenweiler, the metallurgist in charge. This gentleman was the first in Utah to introduce matte-roasting processes, and to subject the lead-copper matte formed in the ore-smelting to intelligent further treatment. Since then, several other works have imitated him in this respect.

The ores smelted during 1874 came partly from the Winnamuck and the Wahsatch, and partly from various other mines, the latter having been bought, principally, toward the end of the year.

The Winnamuck ore, mined during the year, may be divided into two classes, that from the old upper workings of the mine being mostly oxidized carbonate and sulphate of lead; the other, from the lower portion of the mine near water-level, being a mixture of galena, iron and copper pyrites, and zinc-blende. These two classes were carefully kept apart at the mine, in order to facilitate as good and perfect a working of the last-mentioned class as could be done under existing circumstances, *i. e.*, with imperfect appliances, procured at the spur of the moment, and under the pressure of supplying the furnaces with ore as soon as possible. The average assay of sulphureted ore during the year was, in silver, 55.5 ounces per ton, and 16 per cent. of lead; of oxidized ore, in silver, 52.48 ounces, and 23.13 per cent. of lead. The records show, however, great differences in values between the beginning and the end of the year. While in January and February, ore with 30 to 35 per cent. of lead and 30 to 50 ounces of silver per ton was smelted, November and December furnished ore, the average contents of which were 15 to 20 per cent. of lead, and 50 to 75 ounces of silver. In fact, it has been considered a rule at this mine that the higher the ore assays in silver, the lower is the percentage of lead, and the reverse.

Silica and alumina represent the gangue. The average of six different tests made shows:

| | | | |
|---|---|---|---|
| $SiO_3$ | 41.16 | $Al_2O_3$ | 8.31 |
| $Fe_2O_3$ | 3.99 | CaO | 2.10 |

The Winnamuck ore showing a gradual decline in its contents of lead, it became necessary to either purchase another mine, containing ore with a high percentage of lead, or to purchase such ores. In January, 1874, the company bought the Wahsatch mine, situated on the eastern slope of the range. For a few months the ore from this mine was of great assistance to the smelting, as it contained 48 to 52 per cent. of lead, and 9 to 12 ounces of silver per ton. But 500 tons of such ore, and about 240 tons of second-class, containing 35 per cent. of lead, exhausted the mine.

Spanish ore, of which there had been about 400 tons lying at the works for over two years, was next used to make up the deficiency in lead. But although it fulfilled the requirements in this respect, it certainly did not improve the smelting otherwise, as it contained 32.5 per cent. $SiO_3$. A few other small lots of ore were bought, such as Nez Percé and Utah, but these, too, were of such a composition as to add only more difficulties to the smelting operations. The amount of flux used during the first eight months of 1874 was enormous, amounting often to 33 per cent. of the ore-charge, and consisting of hematite, at $18 per ton, and limestone, at $4 per ton. Later in the year, after roasting-processes had once been thoroughly established, the amount of fluxing-material was reduced 50 per cent. and more.

The iron-ore received from Wyoming during the year was of a very inferior character. Up to June, it averaged 65.6 per cent. of metallic iron; from June to November, only 54.9 per cent. Of late, large cobbles of the hardest flint are intermixed with the hematite. After November, the use of iron-ore in the furnaces was stopped entirely. In its stead "South Star" and "Titus" ores were bought. They are known as probably the most basic ores of the Cottonwoods. Large quantities of roasted ore and matte were used at the same time.

A trial of some Tintic iron-ore was made earlier in the fall, with such limited quantities as to leave the *result* uncertain. An analysis of this ore showed 69 per cent. of sesquioxide of iron and 4.2 per cent. of lime.

The principal fuel used was coke, the supply of charcoal being limited and very uncertain. It has, however, been found very advantageous to use some charcoal mixed with the coke. This has, of course, the effect of lightening the column of charges, and of leaving the smelting-mixture less packed. It therefore permits a more perfect play of the blast than the use of coke alone, which renders the columns in the furnace so dense that the blast cannot sufficiently penetrate.

The waste and shortage on both coke and charcoal has been tremendous. The charcoal delivered at the works was mostly soft and spongy, and its calorific effect was, therefore, very unsatisfactory. One-half of it was reloaded from broad-gauge railroad-cars into narrow-gauge cars, and a large amount of fine coal was thus made. Not less than 24 per cent. of waste was measured out during the year.

Two different kinds of coke have been delivered at the works—Saint Louis and Connellsville. The supply of the former was stopped early in spring, as it was found unfit for smelting-purposes. Pieces of slate, from the size of hazel-nuts to ten and fifteen pounds weight, were picked out from this fuel.

The coke lately delivered by Morgan & Co., of Pittsburgh, was a very fine article, but when finally laid down at the Winnamuck works, hard and dense as it was, a waste of from 10 to 12 per cent. would be found. Loading, two or more transfers, unloading, &c., converted a large amount into powder, and tons upon tons have been carted away. Such a loss is hard to bear, where the price of the article in question runs up to $35 per ton. To remedy it in some measure, perhaps the only way would be to mix the dust with tar, and to press it into small bricks.

The rapid burning out of the fire material has been encountered by the use of cast-iron water-jackets of 3 feet 6 inches in height, the inside plates being ½ inch and the outside one ¾ inch thick. The water-space is 3 inches. These jackets are made in six sections, answering to the periphery of the furnace. They are held in position by a stout iron band. The first, constructed in 1873, has been replaced by another pattern, principally because the former ones were cast with tuyere and jacket all in one piece. The tuyere projected about 14 inches into the furnace, and as it was the most exposed part, it very soon cracked and was found leaky. This necessarily required the instant removal of the jacket, which, with a furnace in full blast, was an exceedingly ugly job, causing much loss of time. The jackets were, therefore, cast with simply an opening for a tuyere in the center of each section. By this means two great advantages were gained. First, wrought-iron tuyeres could be used independent of the jacket; and second, by selecting tuyeres of not less than 20 inches in length, the section of the furnace at the tuyere-line could be changed *ad libitum*, at the option of the metallurgist, at any time during a campaign. During a period of seven months, after the introduction of this improvement, not one tuyere or water-jacket has been found in bad condition. The furnaces could make campaigns of from sixty to ninety days. Many objections, on theoretical grounds, have been raised to the jackets. But, considering the expenses incurred every month in rebuilding and repairing a brick furnace, the economical advantage in works situated like the Winnamuck is decidedly on the side of the water-jackets, even if we admit an increased consumption of fuel. Comparing the records of the previous smelting in fire-brick furnaces with those obtained by smelting with the help of jackets, the increase in fuel used figures up to about $\frac{1}{18}$, but by starting a furnace carefully, the difference is only nominal.* After two or three days run of a furnace, the inside of the jacket is coated with what might be termed artificial galena, from 4 to 6 inches thick, preventing the coroding of the iron perfectly, and being a good non-conductor of heat.

Up to the spring of 1874, no attempt was ever made to make use of the large amounts of sulphureted ore from the lower portion of the mine.

* Mr. Wartenweiler seems to overlook the fact that his comparison with former campaigns is not a fair one, because in the late campaigns he had the advantage of using partially-roasted ore and matte in his charges, which tends greatly to diminish the consumption of fuel.—R. W. R.

In May, a small reverberatory furnace was constructed, at great expense, caused mainly by costly and difficult excavations. This furnace, roasting only very limited quantities and losing a large portion of the heat in the flues, proved naturally a very expensive experiment. The same difficulties experienced in building the first presenting themselves for any lateral extension of a second furnace, a double-hearth reverberatory was constructed; the lower one being at a height of 3 feet 4 inches from the ground, and having an area of 9 by 17 feet; the upper one is worked from a platform, and has a roasting-surface of 9 by 15 feet. The fire-box, built outside and independent of the furnace, delivers its flame over a perforated fire-bridge, whence it passes over the ore on the lower hearth, ascends through a narrow flue, 6 feet by 6 inches in section, and 8 feet 6 inches high, into the upper hearth, here over another layer of ore, and then turns over the arch of the upper chamber into the flues.

A great saving of labor and fuel has been effected in this manner, the ore coming into the finishing-hearth already partially roasted. Whenever a roasted charge is drawn, the whole column of ore passes along the entire length of the two hearths, and another charge is fed in from the top. The two chambers contain always four charges, of one ton each. The entire cost, stack excepted, of this furnace amounted to $1,650. The fuel required for twenty-four hours consisted of 1,400 pounds of coal, (lignite,) and there were four laborers required to roast 8,000 pounds of ore. The old reverberatory required the same amount of fuel and three laborers, who could roast only 5,000 to 6,000 pounds of the same ore.

All the ore coming from the lower levels of the mine was run along a track directly over the track of the double hearth, and dumped over an iron screen, allowing half-inch pieces to pass. The fine ore falling on the top of the furnace near the feed-hole, rendered the filling in of a charge a job of only a few minutes' duration. The coarse pieces passed over the screen into a chute, which delivered the ore to a flat below the works, to be here roasted in stalls or pits.

Owing to the character of the ore, the roasting of a furnace-charge to a point where it is called "dead" was impossible. There might have been a more satisfactory result had all the particles of ore been of nearly the same size. As it came to hand, the size varied from dust to half-inch. It was a combination of 25 per cent. galena, 45 per cent. iron and copper pyrites, and 5 to 10 per cent. zinc-blende; and whoever has tried to roast such a mixture will understand the difficulty in the way of a quick and, at the same time, good roasting. The consequence was, of course, the formation of a large amount of accretions gathering in the blast-furnace, near the feed-hole, in such quantities as to interrupt the campaign. After four or five weeks this mass generally had to be removed by blowing the furnace down partially and breaking the lumps off. The total cost of this imperfect roasting in reverberatories was $5.10 per ton of ore, coal costing $12 per ton and labor $3.50 per day.

The pits, or roasting-heaps, were put up in the usual manner, on a double layer of wood. On eight cords of this there were generally piled about 250 tons of coarse ore; four to six chimneys, consisting of small sticks of wood, having been constructed in the heaps to facilitate the lighting of the heap at the start, and insure a good draught. Such heaps usually burn from two to three months, and were put up at a cost of 75 cents per ton. At the time of writing this article, 400 tons of this roasted ore are being shipped, having been sold to other smelting-works, at a net profit to the Winnamuck Company of $40 per ton.

The first roasting-stalls put up were built of pieces of slag, cemented by a little mortar, and consisted simply of two side walls, 5 feet 6 inches high and 18 inches thick, with an inclined floor between. They were used principally for the roasting of matte, receiving at the starting of a fire 1½ cords of wood and 60 to 70 tons of matte, which burned through in about four weeks. The cost of roasting in these stalls was $1 per ton.

Two covered or arched stalls, having a stack in the center to carry off the fumes, gave especially satisfactory results. The average cost of construction of the stalls was $58 apiece. An experiment made with a shaft roasting-furnace, with interior fire-place, to be charged at the top and drawn at the bottom, was not very successful. As long as ore was roasted in it, carrying a considerable portion of galena, it did very well; but, with a large amount of pyrites intermixed, the heat was so raised as to bake and sinter the whole contents, which had then to be removed with bars. This proved too expensive, although the furnace did not consume any fuel after being started. When the draught was shut off below, and a number of openings in the side walls were relied upon to furnish a limited amount of air, the latter could not penetrate to the center, while the ore remained unaltered.

To obtain a tolerably good product in only one fire, all the roasted ore or matte from heaps, stalls, or kilns had to be picked over, and all the unroasted and partially-roasted stuff had to be put back into a second fire.

The average assay of lead-matte (first matte) produced during the year was 12 per cent. of lead and 56 ounces of silver per ton.

By the concentration of this first matte, produced during a period of three months, about 30 tons of copper-matte, containing 12 per cent. copper and 130 ounces silver per ton, were obtained. This product was sold at about $115 per ton.

Slag-assays show during the year an average of 3.68 per cent. lead and 2.41 ounces silver per ton.

A dust-chamber was added to the works in January, 1874, but being of an inferior construction, and, first of all, too small, it never did the services required of such an apparatus, and the saving of flue-dust did not amount to even 50 per cent. of the whole. About 110 tons were collected during the year, assaying in silver 24.1 ounces, in lead 26.5 per cent. It seems a very curious phenomenon that the contents in lead of the flue-dust should be higher than those of the ore, and this can only be explained by the fact that the ore was mostly in the shape of fine carbonate and sulphate, and that, as the condensing capacity of the chamber was small, the lighter gangue was blown beyond the chamber, while the heavier lead-ore particles remained.

An entirely different slag from that produced when iron-ore and limestone were the only fluxing-materials was, of course, obtained after roasting processes had been thoroughly established.

An analysis made in June, 1874, shows 38 per cent. silica; another, commenced in November, but never finished, is as follows:

| | |
|---|---|
| Silica | 25.31 |
| Protoxide of iron | 41.97 |
| Alumina | not determined. |
| Sulphide of iron | 6.25 |
| Sulphide of lead | not determined, but present. |
| Lime | 21.06 |
| Zinc | 0.56 |

This seems to show that the roasting was entirely insufficient.

The following is an average of ore, fluxes, and fuel since the introduction of roasting:

| | Pounds. |
|---|---|
| Winnamuck oxidized ore | 110 |
| Winnamuck roasted ore | 110 |
| Outside ores | 180 |
| Flue-dust | 20 |
| Iron-ore | 20 |
| Lime | 40 |
| Slag | 80 |
| | 560 |
| Coke | 70 |
| Charcoal | 1 1-5 bushel. |

Charges before roasted ore was used, early in 1874, were as follows:

| | Pounds. |
|---|---|
| Winnamuck oxidized ore | 200 |
| Wahsatch ore | 100 |
| Iron-ore | 90 |
| Lime | 40 |
| Slag | 20 |
| | 450 |
| Coke | 80 |
| Charcoal | 1 1-5 bushel. |

It is plainly seen that a great advantage was gained by the introduction of roasting over the former way of smelting. It not only reduced the consumption of costly hematite to a minimum, but there was a decided gain in the use of fuel. The roasted ore and matte imparted such a heat to the entire smelting-charge, that the weight of charges could be increased by about one-third, and at the same time a reduction of the quantity of fuel was rendered possible.

The average number of charges put through in twenty-four hours was one hundred and five.

The quantity of matte produced with such charges was about equal to that of the work-lead. This was certainly a very great, but only a temporary, loss. About 400 tons of matte were worked over during the year.

The following ores and matte were smelted during 1874:

| | Tons net. |
|---|---|
| Winnamuck oxidized ore | 2,124 76 |
| Winnamuck roasted ore | 708.65 |
| Winnamuck roasted matte | 400.00 |
| Wahsatch ore | 747.48 |
| Ore from different mines | 1,699.65 |
| Total amount smelted | 5,680.54 |

The consumption of smelting-materials is recorded as follows:

| | Tons. |
|---|---|
| Ore | 5,680.54 |
| Iron-ore | 1,550.59 |
| Limestone | 909.45 |
| Slag | 921.78 |
| Total material smelted | 9,062.36 |

Of fuel there was used, including shortage and waste—

| | |
|---|---|
| Coke | 1,723.23 tons. |
| Charcoal | 107,521 bushels. |

The losses of lead and silver can be calculated only approximately, there being still about 130 tons of lead-matte on hand, containing 70 to 80 ounces of silver and 12 per cent. of lead. Leaving this matte out of the question altogether, the losses of silver amount to 14.6 per cent. and of lead to 21.1 per cent. It is clear that, after thoroughly sampling the matte, and giving it its actual value in dollars and cents, these figures will be materially modified.

The total silver-lead production during 1874 amounts to 893.601 tons, representing a money-value of $290,706.66.

The machinery at the Winnamuck Works has been increased in August by a new 60-horse-power engine and boiler, and a new No. 5 Root blower. The old 20-horse-power engine, after over three years' continual use, was beginning to give out, being too weak for the severe work required of it.

The total improvements at the works and additions thereto, made during the year, are enumerated as follows:

One 60-horse-power steam-engine, boiler, and buildings.
One No. 5 Root blower, counter-shafts, &c.
Two reverberatory furnaces, flues, stack, and buildings.
Eight roasting-stalls, and roasting-kiln.
One dust-chamber, (sheet-iron.)

These, together with an entirely new water-course, represent an expenditure of nearly $40,000. The old No. 4 and No. 5 Root blowers are still at the works, and intended for auxiliary blast-engines.

In conclusion, it should be said that the whole future of the Winnamuck Works rests with a complete system of appliances for separating the sulphureted ores. The time is very near when no other ores but these can be extracted from the mine.

It is not concentration, as commonly understood, which is needed, but a separation of the different useful minerals. The zinc-blende, even, is valuable in this case, and must not be thrown away.

A separation of the minerals in the Winnamuck ore, effected in a gold-pan, some time ago, illustrated this fact plainly enough.

Three different samples were obtained by the washing, the assays of which showed as follows:

| | | |
|---|---|---|
| Galena | 68 per cent. lead, 62 | ounces silver. |
| Pyrites | 59.5 | ounces silver. |
| Zinc-blende | 46.1 | ounces silver. |

This, of course, does not represent an average of the ore in the mine. The data are simply given to show the relation of the different minerals as to their contents in silver. If the zinc-blende can be sold at reasonable prices to amalgamating-mills provided with a Stetefeldt or any other good roasting-furnace, there will be no difficulty in treating the two remaining portions profitably by smelting, and making a complete success of the whole enterprise. This way of working and preparing ores in a cheap manner previous to subjecting them to metallurgical treatment is, however, advisable, not only for the Winnamuck Works, but really for the majority of the Utah mines. Many of the Bingham Cañon mines have already encountered the above mixture of sulphurets in their lower levels, and all the rest are almost certain to find it, or perhaps something worse, in the near future.

# CHAPTER XII.

## THE GERMANIA REFINING AND DESILVERIZATION WORKS, UTAH.*

These works are situated about six miles south of Salt Lake City, at Flach's Station, on the Utah Southern Railroad. They were established only a few years ago, and, on account of the enormous competition in the business of desilverization and refining of lead, they have, like most works of the kind in the country, been unable to run continuously. Furthermore, the question of fuel is a very serious one for the works, and on this account principally they are now stopped, waiting for the completion of the Utah Southern Railroad down to the San Pete coal-fields, in Southern Utah. After this road is finished, it is expected that coke, which now costs over $30 a ton, will be obtained for from $10 to $12, and at this price the works will be able to run profitably.

Until December of last year, the works had been running most of the time at about half capacity, receiving for treatment principally argentiferous lead from the Utah Smelting-Works and from those of Eureka, Nevada. In the latter part of 1874, they bought for treatment from the Richmond Company, of Eureka, a considerable quantity of lead which had been already refined to a certain degree previous to leaving Eureka.

The works were originally projected to desilverize and refine only according to the so-called Flach system; but at the end of 1874 a blast-furnace for smelting the lead-ores of the vicinity was erected, and several campaigns were made, and the last remnant of the Flach system, namely, the treatment of the liquated zinc-silver alloy, together with very basic slags, in low blast-furnaces, was abandoned. In its stead the distilling-furnace of Faber du Faur was introduced.

The treatment of the argentiferous lead has hitherto been as follows: Twenty tons of bars are melted down slowly in a large reverberatory, locally called the A furnace. According to the amount of impurities in the lead, the molten metal remains here, exposed to a low heat and plenty of air, for a shorter or longer time, the average being about eighteen hours. After the first six or seven hours the covering of oxides, containing most of the copper and antimony and much lead, is taken off. It amounts to from 1.5 to 2 per cent. of the original weight put into the furnace. The oxides forming after this are not taken off, but left in the furnace, after tapping the lead into the desilverization-kettles. The object of this is to utilize in the refining of the second charge the tendency of litharge to give off its oxygen to substances more easily oxidized. The drawing off from the furnace into the desilverization-kettles is effected by means of a partly-covered spout which, on account of the unfortunate location of the A furnace, is about 40 feet long, and occasions, therefore, the formation of a great deal more oxides and scraps than are desirable.

There are five desilverization-kettles. The two large ones, the so-

* This article was prepared for my report by Mr. A. Eilers, who visited the works frequently during last summer, and wishes to acknowledge the kindness of Mr. A. von Weise, the superintendent, and Mr. G. Billing, the manager of the works, in furnishing detailed information.

GERMANIA WORKS—DESILVERIZATION KETTLES.

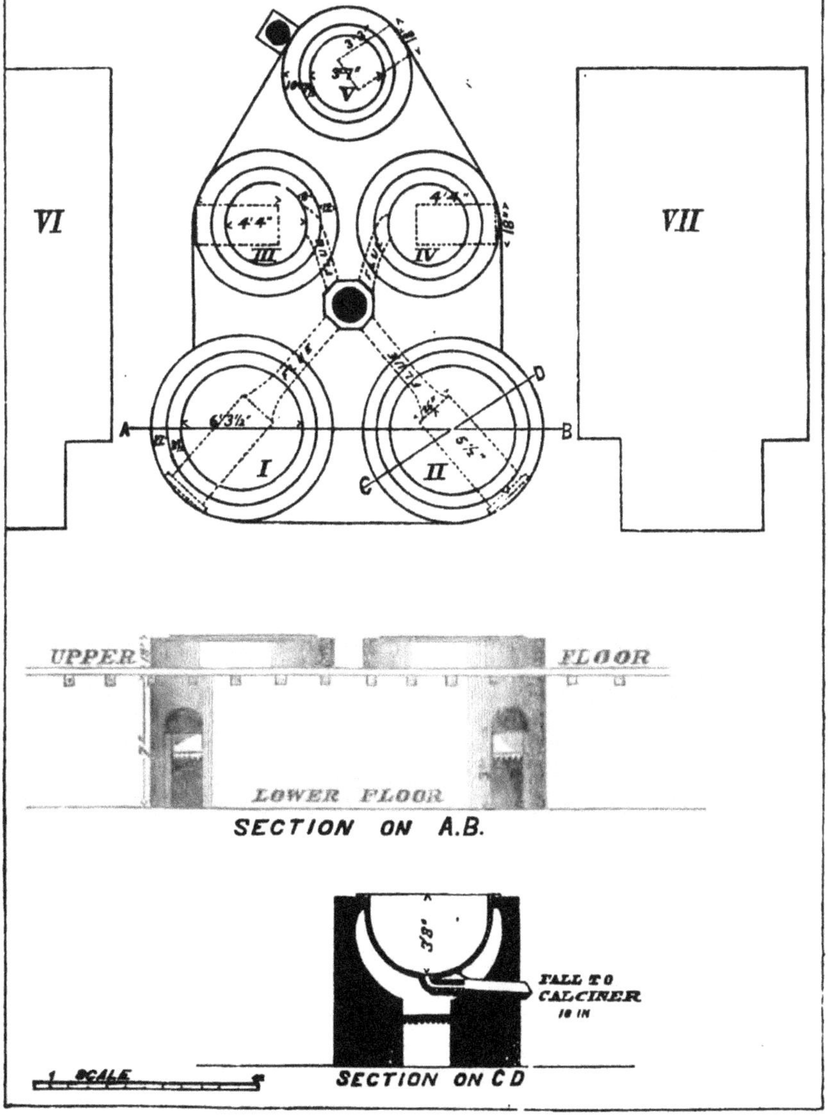

VI

VII

V

III

IV

I

II

UPPER FLOOR

LOWER FLOOR

SECTION ON A.B.

FALL TO CALCINER 10 IN

SCALE

SECTION ON C D

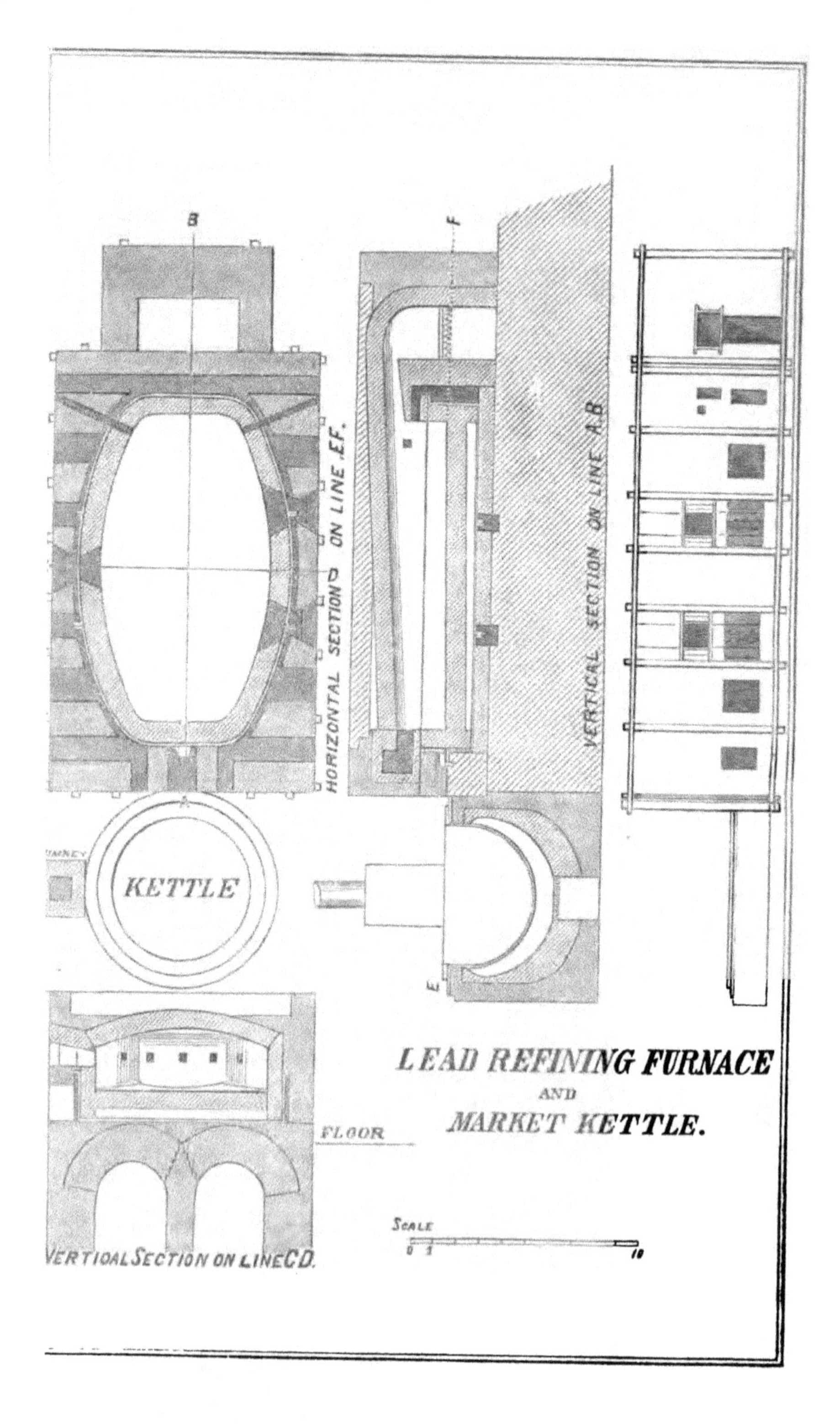

LEAD REFINING FURNACE AND MARKET KETTLE.

called zinc-pots, hold 42,000 pounds of metal each. They are walled in, side by side, and immediately in front of them are the smaller liquation-kettles, Nos. 3 and 4, which hold 7 tons each. In front of these is the smallest kettle, holding 4 tons, which is used for liquation to dryness. The whole arrangement is in the shape of a triangle, Nos. 1 and 2 forming the base, and No. 5 the apex, as shown in the accompanying sketch.

The lead treated here contains from $150 to $350 per ton in gold and silver. According to the richness of the material to be treated, there is from 1.8 to 2.6 per cent. of zinc added in either No. 1 or No. 2 kettle. The zinc is added in three, sometimes four, portions. Instead of using pure zinc for all these additions, the practice of using the second and third scum of a previous operation for the first, and sometimes also for the second, addition in a subsequent operation has, during the last year, been introduced. By this means a very large saving of zinc has been effected. Each addition is mixed with the lead for from one half to three-quarters of an hour, the temperature being maintained above the melting-point of zinc. Then the fire is drawn and the kettles are allowed to stand long enough to permit the charge to cool, and the zinc-silver alloy to rise to the surface, the lead below remaining, of course, liquid. This requires, in the summer, not less than four hours, and in the winter only about two. The scum is now taken off with perforated ladles and transferred to pot 3 or 4. If it is the scum resulting from the first zinc addition, it is, after a partial liquation, immediately transferred to pot No. 5, where the liquation is finished. The lead resulting from liquation is transferred from No. 3 or 4 back into No. 1 or 2 before the second addition of the zinc is introduced. The scum resulting from the second, third, or fourth addition is not liquated, but used again in a subsequent operation, in place of a first zinc addition, as mentioned above. The dry zinc scum from No. 5 has heretofore been run through a small shaft-furnace, together with very basic slag, an exceedingly low pressure of blast being used. The result of this smelting was rich lead and slag and a very impure oxide of zinc, which was gathered in the condensation-chamber; but as this process, compared with the distillation of the zinc scum in retorts, occasions a loss of not less than from $15,000 to $25,000 a year, (most of the zinc being virtually lost forever,) it has now been abandoned.

The now desilverized lead remaining in kettle 1 or 2 is tapped into one of the two refining reverberatories, (calcining-furnaces,) which are located on each side and below the kettles. One of these, together with the market-kettle, is shown in the sketch appended. They hold from 18 to 19 tons of lead. The hearth consists, as does also that of the A furnace above mentioned, of an iron pan cast in three pieces, (as shown in the sketch,) in which the hearth proper is prepared by first putting into the bottom of the pan a layer of coke, next one of brasque, and, finally, one of fire-brick, put on end in the shape of an inverted arch. Much trouble has been experienced in these furnaces, as well as in the A furnace, by the rising of the hearth through lead which penetrated below it, and on which it would float. This has now been obviated by boring holes into the angles formed by the sides and bottom of the pan, so that when lead penetrates to the pan it cannot collect in it, but runs out of the holes, and is seen in time to prevent further mischief.

In these furnaces the lead remains from eighteen to twenty-two hours, a low temperature being maintained, and a strong draught of air being directed on the surface by opening air-holes left in the walls just beyond the fire-bridge. Abstrich is taken only once after the first ten or twelve hours. It amounts to from 2,200 to 3,400 pounds, and contains about 50 per cent. of lead, and the greater proportion of the zinc and antimony. Sometimes this abstrich is liquated in the double-hearth liquation furnace, a drawing of which is given, and the remainder is reduced for hard lead; but usually it is reduced as it comes from the reverberatory. After this dross is taken off, the metal bath is rabbled and the samples are taken. The dross forming after this is allowed to remain in the furnace after the lead is tapped into the poling or market kettle, for the same purpose as mentioned above in regard to the dross in the first calcining or A furnace, namely, to utilize the oxygen of the litharge for the next charge. In the market kettle the lead is now poled by means of short green sticks of wood, which are held down to the bottom of the kettle with an apparatus constructed for that purpose. Each pole requires about an hour to burn out, and from three to four are required to give to the lead the desired purity. When this is reached the lead is ladled into molds, the bars are dressed, and are ready for market.

Most of the lead made at these works has been sold to white-lead manufacturers in this country, and the establishment has acquired an enviable reputation for the purity of its product.

The dross formed during the poling in the market-kettle amounts to from 1,100 to 2,000 pounds. It is directly reduced in a reverberatory or shaft furnace, the resulting lead being fit for making soft lead.

The whole of the first dross taken from the first calcining or A furnace is smelted, together with iron pyrites, for lead and copper matte, the latter being concentrated from three to four times and sold. Various lots sold have contained about 40 per cent. of copper, 1.18 ounces, 1.02 ounces, 1.02 ounces of gold, respectively, and 113.54 ounces, 88 ounces, 94.66 ounces of silver per ton, respectively. The rich lead resulting from the shaft-furnaces which treat the zinc-silver alloy is cupelled on English cupels, of which two work at a time. They have the usual construction shown in the annexed sketch, the necessary air being thrown on the metal bath by means of a steam-jet. They are usually in blast five days, receiving in that time 180 bars of 65 pounds each of rich lead. One of the cupels is then bored, the silver run into molds, and the bars are fed into the other cupel, where the refining of the whole is finished, producing a button of silver of from 8,000 to 10,000 ounces, and .992 to .995 fine in silver and gold. But the proportion of silver and gold varies, of course, very considerably, .981 in silver having been the highest yet obtained.

To illustrate portions of the process further, I add the following figures, for which I am indebted to Mr. A. v. Weise:

Melted down in A furnace, 41,614 pounds of "base bullion," containing, silver, 5,700 grams; gold, 110 grams, per 1,000 kilograms.

After *abzug* was taken in the A furnace, and the remaining metal tapped into pot No. 1, three additions of zinciferous scum and zinc were made, with the following results:

1. Added: Zinc-scum of a previous operation, (second and third additions,) not liquated, 4,000 pounds. Remained in lead after first zinc-crust was taken off: silver, 1,360 grams per 1,000 kilograms.

2. Zinc, 600 pounds. Remained in lead after taking off second zinc-crust: silver, 20 grams per 1,000 kilograms.

3. Zinc, 80 pounds. Remained in lead, a trace of silver.

In another lot of 40,120 pounds of work-lead, containing 1,980 grams of silver and 10 grams of gold per 1,000 kilograms, the desilverization took place as follows:

1. Added: Zinc-scum from previous operation, 3,000 pounds. Left in lead after skimming: silver, 1,160 grams per 1,000 kilograms.

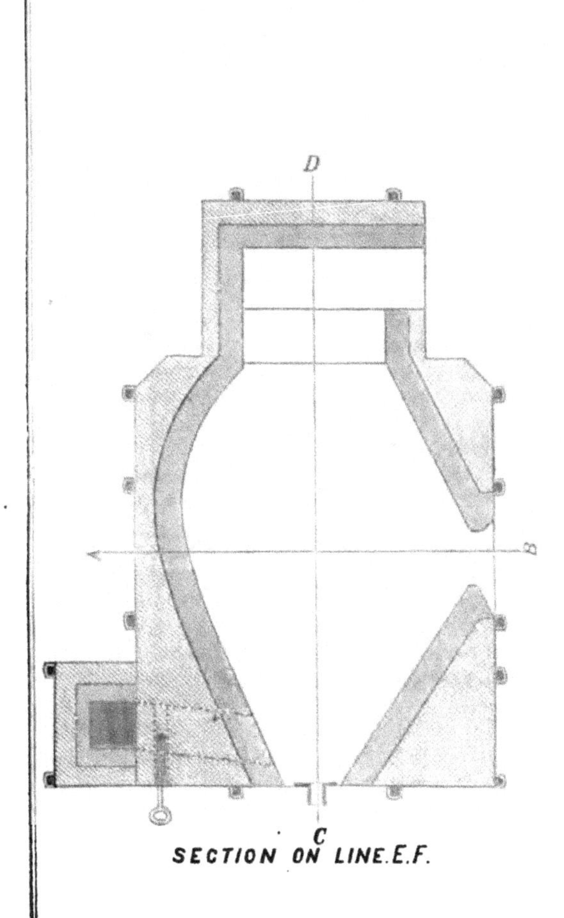


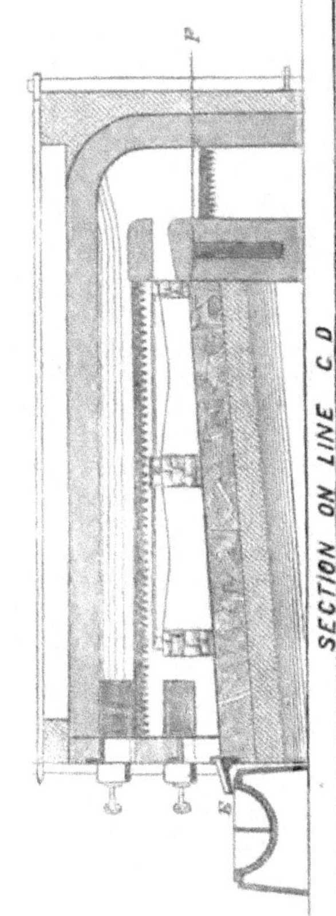


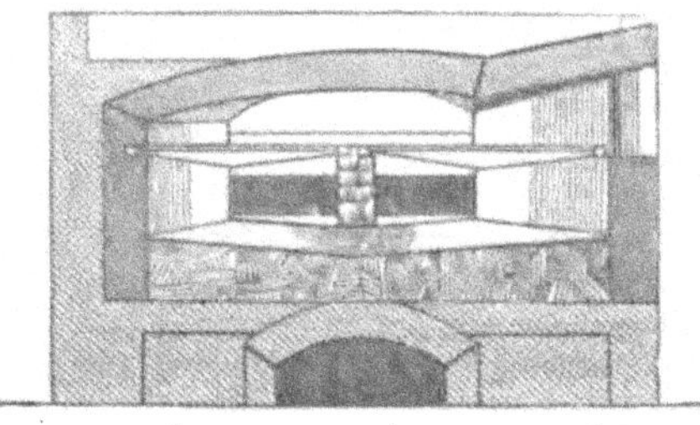


**DOUBLE HEARTH LIQUATION FURNACE.**

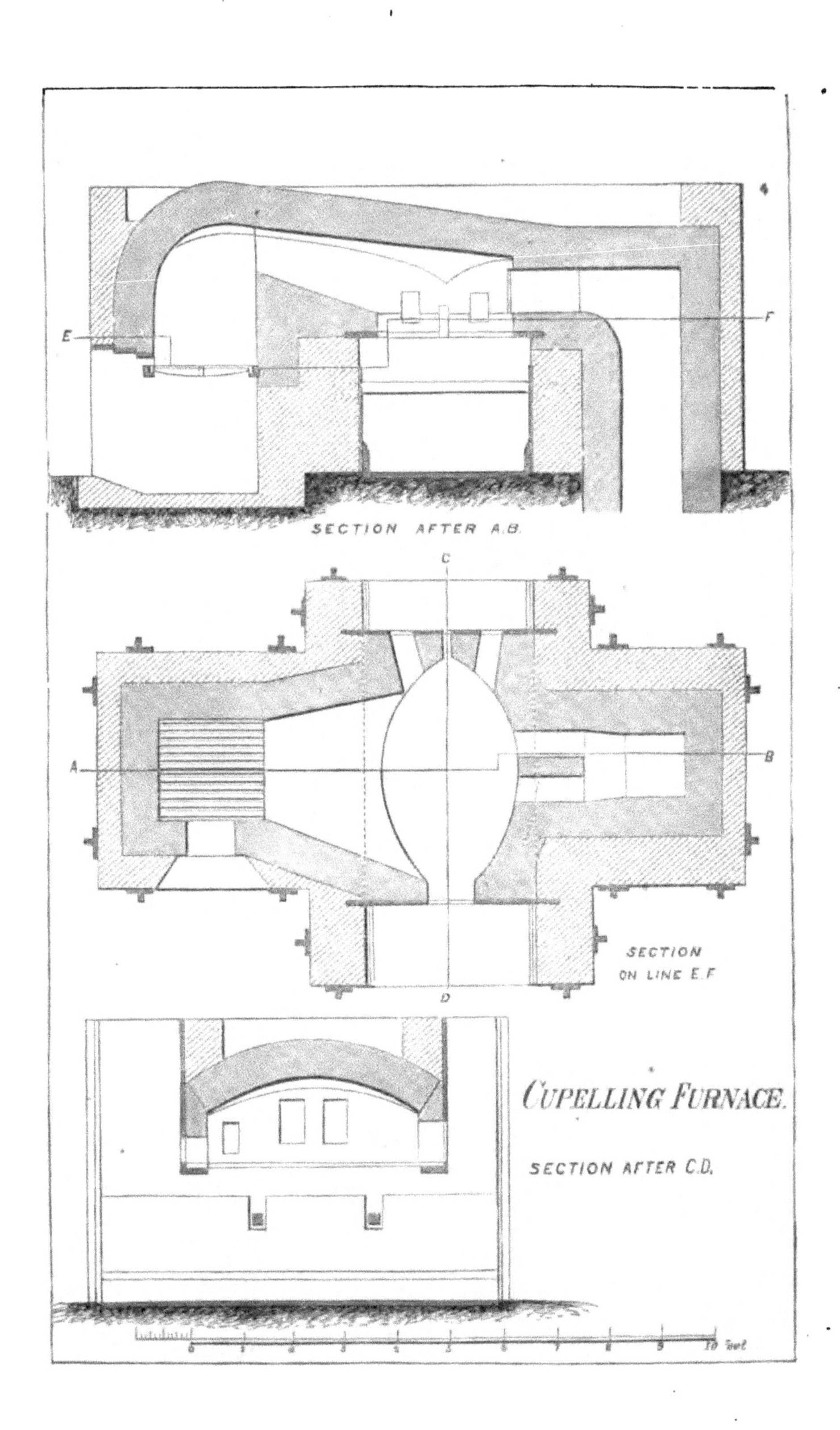
4
E
F
SECTION AFTER A.B.
C
A
B
SECTION
ON LINE E.F
D
CUPELLING FURNACE.
SECTION AFTER C.D.
0
1
2
3
4
5
6
7
8
9
10 Feet

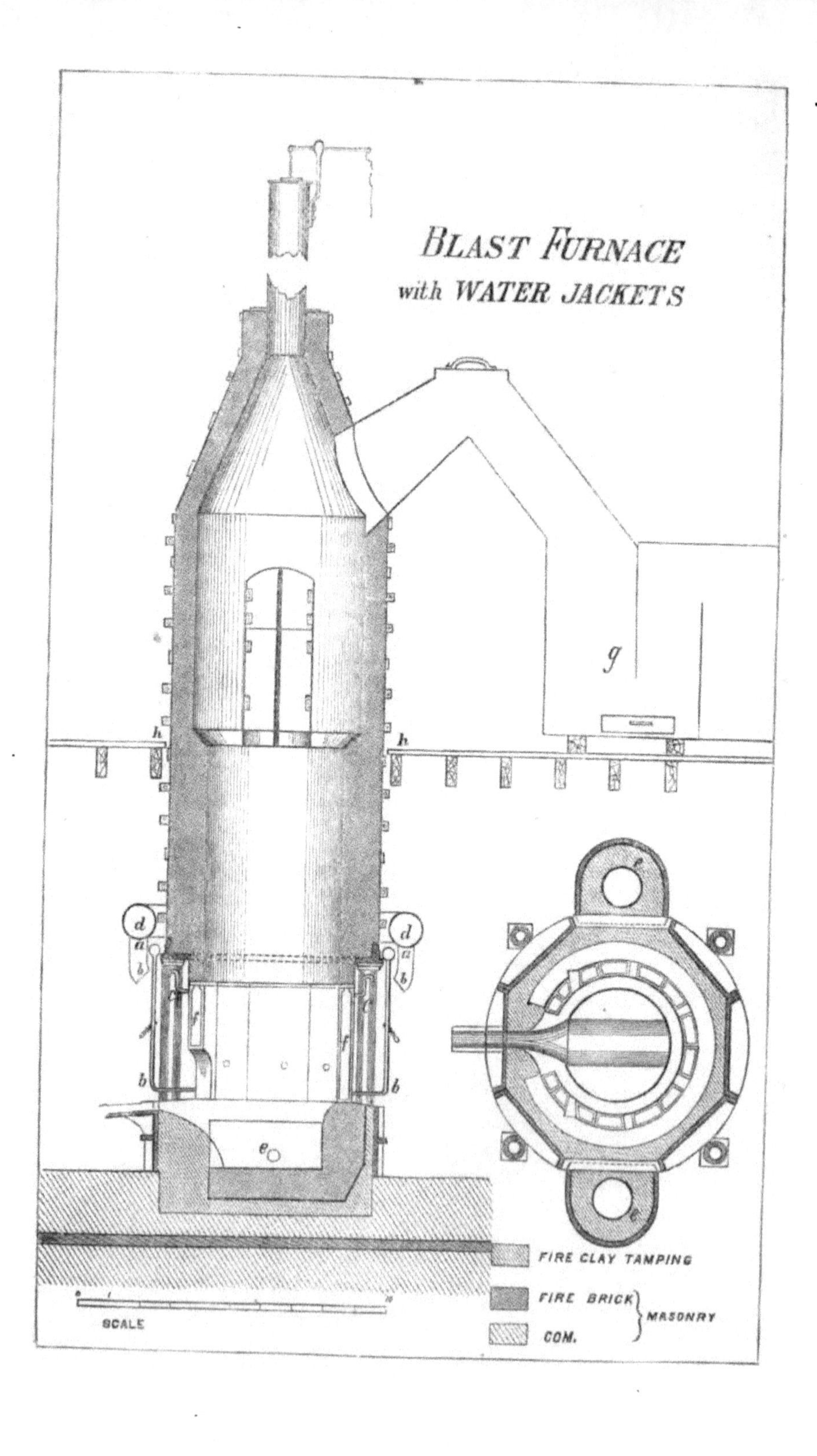
BLAST FURNACE
with WATER JACKETS
g
h
h
d
a
b
d
a
b
c
c
f
f
b
b
e
e
e
FIRE CLAY TAMPING
FIRE BRICK
MASONRY
COM.
SCALE

2. Zinc, 600 pounds. Left in lead after skimming: silver, 30 grams per 1,000 kilograms.

3. Zinc, 125 pounds. Left in lead after skimming: silver, 6 grams per 1,000 kilograms.

In no instance are there more than 8 grams of silver per 1,000 kilograms left in the lead. Of work-lead of the above quality, there were desilverized in successive campaigns 287,383 pounds, for which purpose 1.8 per cent. of zinc was used.

For another lot of 402,442 pounds of Richmond work-lead, containing 4,300 grams silver and 125 grams gold per 1,000 kilograms, 2.3 per cent. of zinc was used.

Still another lot of 402,224 of partly-refined Richmond lead, containing 4,256.7 grams silver, and 127.45 grams gold per 1,000 kilograms, required 2.6 per cent. zinc for complete desilverization.

In the refining-furnace there were taken off after the desilverization, 31,700 pounds of dross, and from the market-kettle 20,353 pounds, and 76.25 per cent. of the original charge was produced as market-lead, containing 6 grams of silver per 1,000 kilograms.

In a single charge of this lot, which was exceptionally poled four times in the market-kettle, the different polings produced the following amounts of dross:

| | |
|---|---|
| First poling | 1,301 pounds. |
| Second poling | 881 pounds. |
| Third poling | 671 pounds. |
| Fourth poling | 290 pounds. |
| Total | 3,143 pounds. |

The quantity of zinc-silver alloy taken from the whole lot of Richmond lead last under consideration amounted to 28,571 pounds, and contained 8.5 per cent. of silver. This was run through the small shaft-furnaces above mentioned.

The new shaft-furnace built at the works for the purpose of ore-smelting is furnished with all the new improvements, including water-jackets and siphon-tap. Its construction is shown in the annexed sketch, and does not require any special description.

The Germania Works bought for treatment during the year 1874, the following amounts of

CRUDE LEAD.

| Locality. | Number of tons. | Coin-value of gold and silver. | Coin-value of lead. | Total coin-value. |
|---|---|---|---|---|
| From Nevada | 2,005 | $461,758 55 | $150,375 | $612,133 55 |
| From Utah | 1,725 | 208,554 57 | 107,500 | *316,054 57 |
| Total | 3,725 | 670,313 12 | 257,875 | 928,188 12 |

The works shipped during the same time—

SILVER AND GOLD.

| Quantity. | Silver. | Gold. |
|---|---|---|
| | *Ounces.* | *Ounces.* |
| 6 car-loads of copper-matte, containing | 4,235.38 | 18.05 |
| 28 car-loads "base bullion," containing | 39,504.23 | 363.60 |
| 369 bars silver and gold, containing | 449,911.87 | 9,460.05 |
| Total | 493,651.48 | 9,841.70 |
| Coin-value | $622,009 86 | $201,754 95 |

LEAD, MATTE, AND "BASE BULLION."

| Quantity. | Number of tons. | Estimated value of lead and copper. |
|---|---|---|
| 173 car-loads common lead | 1,895 | $232,137 50 |
| 154 car-loads selected lead | 1,690 | 228,150 00 |
| 26 car-loads hard lead | 285 | 32,775 00 |
| 6 car-loads copper-matte | 60 | † 4,320 00 |
| 28 car-loads "base bullion" | 302 | ‡ 24,160 00 |
| Total | 4,232 | 521,542 50 |

RECAPITULATION.

| | |
|---|---|
| Value of gold shipped | $201,754 95 |
| Value of silver shipped | 622,000 86 |
| Valve of lead and copper | 521,542 50 |
| Total | 1,345,298 31 |

* The lead from Utah was bought on a currency basis. I have given the respective amounts in gold at 1.12 currency.

† Value of copper. ‡ Value of lead.

# CHAPTER XIII.

## A CAMPAIGN IN RAILROAD DISTRICT, NEVADA.

This chapter was written at my request by Mr. O. H. Hahn, of Eureka, and read at the February meeting of the American Institute of Mining Engineers. The smelting-campaign described, if not highly satisfactory, was at least very interesting in a technical point of view, and is therefore worthy of record and study.

The lead-ores in this district occur in lenticular masses along and near the contact of crystalline limestone with a dioritic porphyry, and consist of agentiferous carbonate of lead and galena, the latter occurring in solid blocks, or in ribbons running through the limestone. Associated with these are silicate, carbonates and suboxide of copper, and brown and calc spar. Adding to this that the proportion of galena in the ore is small, and that no iron pyrites exist in the district, it is at once evident that the ore-mixture is difficult to treat. For a successful reduction by smelting, it would be necessary to introduce into the charge, first, the proper quantities of sulphur-bearing ingredients to effect a complete separation of copper and lead; and, secondly, sufficient silica and iron-oxide to form with the lime and magnesia of the ore a slag of the required fluidity. The first condition could be fulfilled only very imperfectly, there being no iron pyrites in Railroad nor in the surrounding districts. The nearest occurrence is in Battle Mountain, whence a supply could not be procured in time. The second condition, *i. e.*, the addition of iron-ore and quartz, was easily complied with, there being an abundance of these materials in the vicinity.

The furnace of the Empire City Company is a small shaft-furnace, with an area of hearth of 6 square feet, and is supplied with blast by a No. 4 Root blower. In spite of this small size, the furnace smelted during this campaign, which lasted thirty-one days, an average of 28 tons of ore and 13 tons of fluxes—together, 41 tons of charge per twenty-four hours.

The following tables give the complete data in regard to the smelting-process and the results obtained:

*I.—Exhibit of the daily consumption and production at the Empire City Company's furnace.*

| Date. | Bushels of charcoal consumed. | Coke, pounds. | Elko-tunnel ore. | | | Custom-ores. | | Fluxes. | | | No. of bars produced. | Weight of bars in pounds. | No. of car-load. | No. of bars in car-load. | Weight of car-load in pounds. | Silver contents of car-load in ounces. |
|---|---|---|---|---|---|---|---|---|---|---|---|---|---|---|---|---|
| | | | Galena. | Coarse. | Screenings. | Lyon. | Black Warrior. | Quartz. | Iron-ore. | Slag. | | | | | | |
| October 19 | 560 | | 2, 192 | | 10, 542 | | 1, 321 | | 1, 786 | 14, 336 | 7 | 682 | | | | |
| 20 | 637 | | 5, 348 | 10, 200 | 27, 945 | | 5, 561 | | 12, 593 | 16, 320 | 22 | 2, 143 | | | | |
| 21 | 341 | 3, 887 | | 23, 000 | 27, 847 | | 5, 571 | | 17, 548 | 14, 336 | 37 | 3, 604 | | | | |
| 22 | 279 | 5, 359 | | 35, 000 | | | 2, 048 | | 3, 899 | 14, 912 | 45 | 4, 383 | | | | |
| 23 | 290 | 5, 543 | | 21, 000 | 23, 772 | | 1, 229 | | 5, 749 | 15, 424 | 57 | 5, 620 | | | | |
| 24 | 268 | 5, 129 | | 15, 000 | 38, 488 | | | | 6, 824 | 14, 272 | 96 | 9, 465 | 1 | 205 | 20, 079 | 1, 036. 58 |
| 25 | 296 | 5, 681 | | 16, 000 | 45, 280 | | | | 7, 798 | 15, 808 | 66 | 6, 499 | | | | |
| 26 | 319 | 6, 118 | | 12, 000 | 33, 960 | | 491 | | 6, 824 | 17, 024 | 105 | 10, 340 | 2 | 205 | 20, 103 | 1, 062. 94 |
| 27 | 291 | 5, 589 | | 12, 000 | 33, 960 | | 2, 458 | | 7, 311 | 15, 552 | 58 | 5, 712 | | | | |
| 28 | 320 | 6, 118 | | 12, 000 | 33, 960 | | 2, 458 | | 7, 311 | 17, 024 | 85 | 8, 371 | | | | |
| 29 | 3[illegible]8 | 6, 670 | | 10, 000 | 28, 300 | | 2, 048 | | 6, 093 | 18, 560 | 82 | 8, 075 | 3 | 202 | 20, 093 | 1, 059. 90 |
| 30 | 362 | 5, 704 | | 6, 800 | 45, 280 | | 2, 867 | | 8, 104 | 17, 024 | 70 | 6, 893 | | | | |
| 31 | 975 | | | 7, 000 | 45, 280 | | 2, 704 | | 8, 164 | 24, 960 | 61 | 6, 025 | | | | |
| November 1 | 967 | | 4, 680 | 7, 600 | 50, 940 | | 3, 072 | | 10, 967 | 24, 768 | 95 | 9, 385 | 4 | 204 | 20, 028 | 1, 031. 44 |
| 2 | 955 | | 780 | 12, 000 | 41, 884 | | 2, 867 | | 8, 530 | 24, 448 | 80 | 7, 753 | | | | |
| 3 | 953 | | 1, 300 | 14, 000 | 39, 620 | | 2, 867 | | 6, 824 | 23, 348 | 91 | 8, 772 | 5 | 205 | 20, 059 | 997. 93 |
| 4 | 937 | | 1, 820 | 16, 000 | 45, 280 | | 3, 277 | | 7, 798 | 12, 000 | 105 | 10, 122 | | | | |
| 5 | 947 | 759 | 2, 600 | 19, 000 | 53, 883 | | 3, 564 | | 7, 555 | 11, 616 | 113 | 10, 893 | 6 | 208 | 20, 088 | 1, 021. 97 |
| 6 | 768 | | 5, 400 | 15, 000 | 54, 336 | | 497 | 3, 600 | 8, 225 | 10, 880 | 146 | 14, 074 | | | | |
| 7 | 737 | | 3, 180 | 12, 000 | 58, 864 | | | 3, 400 | 8, 286 | 16, 080 | 115 | 11, 086 | 7 | 208 | 20, 039 | 961. 87 |
| 8 | 790 | | 4, 160 | 9, 000 | 61, 128 | | | 4, 000 | 8, 286 | 15, 168 | 135 | 13, 014 | 8 | 208 | 20, 025 | 936. 17 |
| 9 | 723 | | 1, 8[illegible]0 | 7, 000 | 56, 600 | 3, 200 | | 2, 000 | 6, 093 | 15, 792 | 119 | 11, 669 | | | | |
| 10 | 743 | | 1, 040 | 7, 000 | 56, 600 | 2, 400 | 2, 400 | 800 | 6, 336 | 15, 504 | 104 | 10, 117 | 9 | 207 | 20, 092 | 931. 76 |
| 11 | 806 | | 1, 560 | 9, 000 | 61, 128 | | 3, 600 | | 7, 555 | 13, 440 | 96 | 9, 369 | | | | |
| 12 | 766 | | 2, 080 | 9, 000 | 56, 600 | | 3, 600 | | 8, 530 | 12, 768 | 112 | 10, 931 | 10 | 204 | 20, 079 | 1, 029. 04 |
| 13 | 746 | | 2, 340 | 10, 000 | 52, 072 | | 3, 200 | | 10, 723 | 12, 432 | 75 | 7, 638 | | | | |
| 14 | 720 | | 3, 640 | 10, 000 | 44, 246 | | 2, 400 | | 9, 017 | 12, 000 | 71 | 6, 946 | | | | |
| 15 | 8[illegible]5 | | | 14, 000 | 39, 718 | | 2, 400 | | 10, 967 | 10, 560 | 64 | 6, 261 | 11 | 205 | 20, 056 | 987. 75 |
| 16 | 900 | | | 14, 000 | 39, 718 | | 2, 400 | | 10, 602 | 12, 800 | 77 | 7, 677 | | | | |
| 17 | 871 | | | 16, 000 | 53, 302 | | 1, 880 | 1, 200 | 11, 698 | 11, 616 | 66 | 6, 580 | 12 | 201 | 20, 089 | 1, 039. 60 |
| 18 | 728 | | 3, 120 | 18, 600 | 25, 227 | | | 2, 000 | 8, 164 | 9, 824 | 156 | 15, 556 | 13 | 149 | 14, 825 | 807. 96 |
| Total | 20, 228 | 56, 557 | 47, 060 | 399, 200 | 1, 285, 760 | 5, 600 | 66, 780 | 17, 000 | 256, 220 | 481, 596 | 2, 611 | 255, 655 | .... | 2, 611 | 255, 655 | 12, 904. 91 |
| Add 15 per cent. waste | 3, 034 | | | | In addition there were produced: 54,000 pounds skimmings, estimated to contain | | | | | | | | | | 32, 400 | 2, 323. 08 |
| | | | | | In addition there were produced: 32,703 pounds matte, known to contain | | | | | | | | | | 8, 023 | 953. 78 |
| | 23, 262 | | | | Sum total | | | | | | | | | | 296, 078 | 16, 181. 77 |

REMARKS.—The Lyon ore was used on account of its ferruginous character. The Black Warrior is a quartzose carbonate-of-lead ore. What I designate as quartz is really a trachytic rock, which was used whenever the Black Warrior ore gave out. The iron-ore is a hematite. Only the impure slag, that is, the slag which ran along with the matte into the pot, was used over again.

II.—*Exhibit of the working-results at the Empire City Company's furnace, Railroad district, Nevada, during the run from October 19 till November 18, 1874.*

| Name of ore. | Percentage of moisture. | Percentage of lead in ore. | Ounces of silver per ton of ore. | Gross weight of ore, pounds. | Net weight of ore, pounds. | Lead contents of ore in pounds. | Silver contents of ore in pounds. | Yield of ore. | |
|---|---|---|---|---|---|---|---|---|---|
| | | | | | | | | Pounds of lead. | Ounces of silver. |
| Galena | | 69.4 | 41.19 | 47,060 | 47,060 | 32,650 | 969.20 | | |
| Coarse | 2 | 30.2 | 24.79 | 399,200 | 391,216 | 118,147 | 4,849.12 | Bullion ... 255,655 | 12,904.91 |
| Screenings | 11 | 26.8 | 18.95 | 1,285,760 | 1,144,326 | 306,679 | 10,842.58 | Skimmings 32,400 | 2,323.08 |
| Black Warrior | 7 | 27.5 | 16.71 | 66,780 | 62,105 | 17,104 | 518.92 | Matte .... 8,023 | 953.78 |
| Lyon | | 9.7 | 6.75 | 5,600 | 5,600 | 543 | 18.90 | | |
| | | | | 1,804,400 | 1,650,307 | 475,132 | 17,198.72 | 296,078 | 16,181.77 |
| Yield | | | | | | 296,078 | 16,181.77 | | |
| Loss | | | | | | 179,054 | 1,016.95 | | |
| Loss in about 2,000,000 pounds of slag, as authenticated by assay | | 5.5 | 1.12 | | | 110,000 | 1,120.00 | | |
| Unexplained loss | | | | | | 69,054 | | | |

It will first be noticed, from Exhibit I, that the total amount of sulphur-bearing material in the ore was only 2.6 per cent. The consequence was the formation of a very large amount (54,000 pounds) of skimmings, which are an alloy of metallic lead and copper, mixed with a little matte. This disagreeable product being lighter than metallic lead, floats on the top of the lead-basin (in this case an automatic tap) and threatens continually to clog up the tap-hole, as it requires a higher heat than lead to remain liquid. Without a chance to melt down the whole product at a high heat in a separate apparatus, it is impossible to take a correct sample of it. From various assays made of samples taken at random, I estimate the contents of lead at 60 per cent., and of silver at 86.04 ounces per ton.

The average of all assays made of samples from 180 sacks of matte, produced in this campaign, indicates:

| | |
|---|---|
| Cu | 48.0 per cent. |
| Pb | 24.5 per cent. |
| Ag | 58.33 ounces per ton. |

Of flue-dust there were only inconsiderable quantities formed, but the following interesting investigations were made with a sample taken from the roof of the furnace-building:

The sample was divided into a fine and a coarser portion, by sifting through a No. 80 and a No. 30 sieve. The former yielded by assay: Pb., 24.7 per cent., Ag., 15.3 ounces, per ton; the latter, Pb., 10.5 per cent., Ag., 7.29 ounces, per ton.

The slag from the coarse sample indicated by its color a larger percentage of copper than that from the fine portion. In looking over the data shown by the exhibits, there are several points which would deserve investigation by such members of the profession as are not continually beset by the duties of active practice. It appears, for instance, that although there was not enough sulphur in the charge to cover all the copper and carry it into a matte, not alone a portion of lead, but also of iron, reduced from the iron-ore, entered the matte, while a portion of lead-oxide entered, instead of the protoxide of iron, into the slag. From this fact, it would almost appear that Fournet's law is not quite as reliable under all conditions as we have been taught. From the large proportion of skimmings (metallic lead and copper) formed, and the presence of so much PbS in the matte, it is also clear that the precipitating action of metallic copper on sulphide of lead is extremely imperfect.

To remedy the unwelcome occurrences which I have mentioned, it would have been necessary, in my opinion, to add to the same charge, as above given, a sufficient quantity of only partially-roasted iron pyrites to furnish enough sulphur for the purpose of carrying all the copper into the matte, and at the same time enough of an easily-reducible oxide of iron to prevent the entrance of lead and the accompanying silver in the slag. No doubt there would have been much more lead carried into the matte in such a case; but from this it can be easily extracted after roasting by means of a second smelting, while from the slag it rarely pays in the West to extract the lead once taken up in it.

Another interesting feature in the above exhibits is the really very large loss of lead, which can only partly be accounted for, while that of silver does not at all correspond. The latter is, in fact, very small and all authenticated by assay.

For the sake of completeness, I will add some items of expense:

| | |
|---|---|
| Price of charcoal | 29 cents per bushel. |
| Price of coke | $40.00 per ton. |
| Price of iron-ore, delivered | 4.50 per ton. |
| Cost of hauling ore | 2.25 per ton. |

The daily expenses during the run were, at the furnace:

| | |
|---|---|
| Labor | $103 50 |
| Charcoal and coke | 254 10 |
| Wood for engine | 16 00 |
| Fluxes, and other materials, repairs, &c | 30 00 |
| Total | 403 60 |

This amount, divided by 27.93, the number of tons of ore smelted daily, gives:

| | |
|---|---|
| Per ton | $14 45 |
| General expenses | 1 50 |
| Mining expenses, at $5,000 per month | 16 66 |
| Hauling | 2 25 |
| Total expense per ton of ore | 34 86 |

---

# CHAPTER XIV.

## THE CONSTRUCTION AND OPERATION OF A SLAG-HEARTH.

This chapter, written by Mr. J. Y. Bergen, jr., is here introduced on account of its eminently practical character, which will render it extremely valuable, I think, to many lead-metallurgists; although, in the far West proper, there are as yet few localities where the process here described can be employed with advantage.

The utilization of lead-slags produced in smelting the purer galenas is a problem of peculiar interest to the American metallurgist, as he is often called upon to deal with slags from Scotch-American hearths and reverberatory furnaces smelting for the production of a corroding lead direct from the ore.

The purity of the galenas obtained in the Illinois and Wisconsin mines, more notable yet in those from the "soft Missouri lead" region, leaving cupola-smelting entirely out of the question, accumulates large quantities of slag valuable only for the production of a hard or "scoria" lead.

For the treatment of these slags, furnaces of cupola-form are generally inadmissible, as being too expensive to build, and unsuited to the smelting of slags containing, as is often the case, fragments of fire-brick and other rubbish in such quantity as to quickly foul the furnace and prevent long campaigns. The old English slag-hearth, however, is so slow in its operation and entails so much expense to the amount of lead produced, as to have become almost obsolete where rapidity of production is an important consideration. But the low first cost and convenient operation of this form of furnace have been happily combined with increased production in the construction adopted by the best smelters in the Wisconsin mines, and lately introduced into the lead-mines of Missouri; a construction which is, I think, the best in use for smelting-works of moderate size.

The slags with which the furnace has to deal, whether from Scotch hearth or "air-furnace," are closely similar in character. The former occur in coarse lumps, composed of varying admixtures of sulphide, sulphate, oxide, and silicate of lead, agglomerated with more or less completely scorified particles of gangue, and containing numerous shots of metallic lead. Air-furnace slag is rather more finely divided, contains fewer lead-shots, but a higher percentage of lead, ranging from 33 to 46 per cent. metallic, while Scotch-hearth slags yield 30 to 37 per cent. Hearth-ends, or "bottoms," from the air-furnace, are somewhat more infusible than ordinary slags, and contain 26 to 30 per cent. of lead in metallic shots and strings, and in the form of red lead; also scorified with silica. (See tables of assays and analyses.)

The charge for the furnace will be composed of the hearth-ends and slags above mentioned and a few others, with fuel, (coke;) the use of fluxes being, in general, unnecessary.

The dimensions of the furnace are:

| | Inches. | | Inches. |
|---|---|---|---|
| Width of hearth | 26 | Slope of bed-plate to 1 foot. | 1½ |
| Depth, front to back | 36 | Size of bed-plate | 40×48 |
| Height | 46 | Size of front-plate | 30×36 |
| Diameter of tuyere | 3 | Long diameter of lead-pot. | 36 |
| Height of tuyere above bed-plate | 10 | Front to back of lead-pot.. | 12 |
| Height of port | 9 | Depth of lead-pot | 12 |
| Width of port | 12 | Thickness of lining, back and sides | 9 |
| Prolonged axis of tuyere strikes front lining of furnace above port | 2 | Thickness of lining, front.. | 5 |

The cast-iron bed-plate, which forms the sole, or hearth proper, rests by its whole under surface on tightly-rammed clay, or stone laid in clay, except in front, where it overlaps the brim of the cast-iron lead-pot. The fire-brick lining rests directly on the bed-plate, and should be laid of brick of the ordinary form, with all the side and back courses headers.

On the bed plate, over the whole bottom of the furnace, is tamped a brasque or a moistened mixture of equal parts, by bulk, of clay and coke rubbish, to the thickness, at the back of the hearth, of five inches, thinning at the port to one inch. This brasque is not disturbed or renewed excepting when it becomes necessary to reline the furnace. The lining is repaired by patching with fire-clay and bats of fire-brick, from time to time, (as will be hereafter mentioned,) the original rectangular form and vertical sides of the furnace being as nearly as possible preserved. I have found the smelting of the furnace less satisfactory where the form of the horizontal section approximated to the oval than where the corners were, as usual, left unfilled. A stack, in the form of a hood of suitable size, serves to convey the fume away from the furnace-top—preferably into a condensing-flue, but ordinarily into the open air. To protect the workmen from escaping fumes, the hood should be as low as it can conveniently be allowed in front; and, to give facilities for the use of bars at either side, in front, it should be supported by a very flat arch or a cross-beam of proper dimensions.

Preparation for smelting is begun by ramming the port full of stiff clay, through which, at the bottom, a wooden tap-plug is introduced and driven well up into the center of the hearth. Sticks of dry kindling-wood are then stood in an inverted-V shape, thus, Λ, from front to back of the furnace, so as to form a temporary tunnel for the passage of the blast from tuyere to port. A shovelful of live coals is then thrown in at the back over the mouth of the tuyere, and the furnace filled to the top with clean charcoal; half-blast is then put on till the charcoal appears thoroughly ignited and burns over the top with a blue flame. Coke should in the mean time be spread to the depth of six inches over the charcoal, a little more thickly behind, and the full smelting-blast admitted.

Charging slag evenly over the top of the furnace, and especially behind, may at once be begun; and in the beginning, for forming a nose, the cleanest obtainable slag, of moderate fineness, should be employed. Coarse brouss charged at the outset, as described by Percy, in his account of smelting on the English hearth, has not, so far as my observation goes, produced so good a nose, nor formed it so quickly, as good slag will do. Brouss from the slag-hearth is mostly in rather larger lumps, and these lumps charged at the outset will not infrequently settle in front of the nozzle and choke the blast.

The normal charge, of about four shovels of slag to one of coke, may be given from this time, care being taken to dispose the fuel for the most part in the front and along the center, the slag around the back and sides of the top, and to keep the furnace even full. Blow-holes should not be allowed to form, but the blast kept even over the top, only strongest in front. At the outset the top may be left rather bright, but when a full smelting-heat has been established, the top should be kept dull by sufficient charging to prevent undue volatilization and unnecessary consumption of fuel. When the material charged has first to be sluiced or jigged, it will be found advantageous to charge it wet.

The tap plug may be drawn half or three-quarters of an hour after the admission of the full blast; and, if necessary, a steel bar should be driven through the opening left, well up into the furnace, and quickly withdrawn. The lead-pot should, in the mean time, have been rammed full of finely-broken charcoal, over which the stream of melted "black slag" will pass into the water-tank, through which flows a slow current of water.

The lead reduced settles at once through the charcoal filter, and passes through an opening in the partition into the molding-end of the lead-pot. The stream of the black slag should be kept running freely by bridging over with a billet of wood or charcoal brand, and by "snouting" with a chisel-pointed steel bar, when it passes over the front of the pot. The operation of the furnace may be judged by the appearance of the stream at this point. The slag should run in a thin stream of a bright straw-yellow color. Fuming of the slag indicates the passage of lead-matte, to be remedied by the addition of desulphurizing fluxes, or the charging of more highly oxidized slags.

A sluggish, dull-red flow of slag causes mechanical loss of lead in shots carried over, and indicates too low a heat, caused in most instances by insufficient blast. This difficulty can often be remedied by breaking off the point of the nose with a long bar introduced into the furnace through the tuyere from behind. The pressure of blast should be as great as can be employed without blowing particles of coke and slag out of the furnace-top. Nine ounces to the square inch will smelt more rapidly, and with less fuel, than a less pressure, but more than this amount would not generally be advantageous.

When a slag continues to run sluggishly, and the operation with proper charging proceeds too slowly, the trouble is usually to be found in the shape of dirt, ashes, and other rubbish in the slag. It will then be found necessary to sluice at least the finer portions of it, but indiscriminate sluicing of slag is to be avoided, since air-furnace slags, as usually sold, contain a large amount of lead-material, as fine powder in the fume, or "arsenic," from the stacks, with a value of 36 to 38 per cent. lead. Slime-tanks should be connected with the sluice-discharge, and the material collected in them concentrated on suitable jigs. Slime collected from washing ordinary air-furnace slags will give from 4 or 5 up to 18 per cent. lead. Scotch hearth slags never require sluicing, and where these or other clean slags constitute the bulk of the charge, dirtier slags, "chats," clayey carbonate ores of lead, and other impure lead-containing ores and furnace-products, may be worked to advantage on the slag-hearth. It will not in general be found possible to maintain an invariable composition for the black slag; the diminished rate of production, cost of fluxes and additional fuel required, will more than counterbalance the slight saving in percentage of lead and protection of the furnace-lining from attack of basic slag.

It will, however, be found advisable to mix siliceous and aluminous bottoms with lime-containing slags, and to maintain as nearly as possible a constant percentage of lead in the charge. This percentage will vary in practice between the limits of 35 and 40 per cent., yielding in furnace 20 to 25 per cent., or a 75-pound pig of lead to each 300 to 375 pounds of slag charged. Of the 15 per cent. loss, 3 to 6 per cent. is due to scorified lead, carried off in black slag, the remainder to volatilization and mechanical loss. Hearth-ends and brouss from the slag-furnace should be recharged during the campaign, and in small quantities at a time. The lumps of brouss will be found serviceable, charged in the back corners to keep them open.

The rate of production of the furnace will increase, so that while the first eight hours running will produce 40 pigs of 75 pounds each, the next eight hours, with same slag, will produce 60 pigs. Lean slag, low blast, or poor fuel will decrease the production.

Care should be taken, after the furnace has reached its maximum speed, to keep the lead-pot full of clean charcoal, otherwise much dross will be produced. It will be found advantageous to clean out a pot entirely and refill with fresh charcoal once in two or three hours, and clean lead can then be molded throughout the campaign. The duration of the campaign may be prolonged to some days, but by the end of the sixteen hours already described the bottom of the furnace will usually have become more or less fouled with pieces of brick and other rubbish, and the lining in places eaten into holes. For these reasons, it is not usually found advisable to prolong the campaign beyond this point, and the smelter will begin the operation of burning down without further charging, which is continued till black slag ceases to run, when the port is opened, and the loose contents of the furnace drawn out through the opening, after first cleaning out and refilling it with charcoal rubbish.

Three or four pailfuls of water are then thrown in at the top and the furnace allowed to cool till morning, when it is stubbed out with the sledge and steel bars, the holes in the lining patched, and the port filled, when firing is begun as before.

The work of the furnace and smelting-expense may be summarized thus:

Thirty thousand to thirty-seven thousand pounds of slag, of assay-value of 35 to 40 per cent., can be smelted in sixteen hours, producing 7,500 pounds hard lead, at a cost of—

| | |
|---|---|
| Wages of 2 chargers, at $3 | $6 00 |
| Wages of 4 helpers, at $1.75 | 7 00 |
| Wages of one yard-hand, at $1.50 | 1 50 |
| 100 bushels coke, at 25 cents per bushel | 25 00 |
| 22 bushels charcoal, at 12½ cents per bushel | 2 75 |
| Expense for 10-horse-power engine | 8 00 |
| Oil, tools, and sharpening | 30 |
| Total | 50 55 |
| Smelting-expenses to 1,000 pounds lead | 6 73 |

Lead, lead-dross, (85 per cent. metallic,) and black slag, besides the brouss obtained in stubbing out, are the furnace-products obtained. The lead cannot, at present difference of price from corroding lead, (⅜ cent to ½ cent,) be profitably refined, and is sold as hard lead. Unsalable drossy pigs and dross are not to be thrown back in the furnace-top, but should be reserved for a separate reduction, which is to be performed with charcoal by aid of a light blast. Of lead, 600 to 800 pounds per hour can thus be reduced from dross, with charcoal rubbish, useless for other purposes, with very little loss of lead. Loss of lead in the black slag cannot be entirely prevented, but care should be taken to prevent lead from being mechanically carried over, and black slag containing many shots should be recharged at once, or after concentration, by jigging. Condensing-apparatus attached to the stacks would (where the extent of the operation would warrant) quickly pay for the cost of erection.

The percentage composition of the various slags charged in, and obtained as products of, the slag-furnace, I am not able to give in full, not having had recent access to facilities for complete analyses.

Four analyses of galenas and three of lead slags are appended, with the addition of some results of many crucible-assays performed with soda, argols, borax, and metallic iron, as recommended by Mitchell:

*Galena and slag analyses, from report of State geological survey of Missouri, 1873-'74.*

| Ingredients. | A | B | C | D | 1 | 2 | 3 |
|---|---|---|---|---|---|---|---|
| Siliceous | 0.61 | 0.71 | 0.15 | 0.21 | 23.98 | 14.15 | 12.32 |
| Lead, (metallic) | 84.06 | ...... | ...... | ...... | 38.09 | 55.45 | 53.58 |
| Iron, (metallic) | 0.16 | 1.95 | 0.24 | 0.48 | 3.09 | 2.08 | 1.80 |
| Zinc, (metallic) | 0.94 | 0.73 | Trace. | Trace. | 8.13 | 5.43 | 9.23 |
| Lime | 0.42 | 0.86 | ...... | ...... | 15.72 | 11.25 | 10.37 |
| Silver | ½ oz. to ton. | ½ oz. to ton. | None. | Trace. | ...... | ...... | ...... |

A.—Galena as delivered to furnace, Holman diggings, Granby.
B.—Galena as delivered to furnace, Swindle diggings, Joplin.
C.—Galena, Walker diggings, Miller County, Missouri.
D.—Galena, New Granby diggings, Morgan County, Missouri.
1.—Scotch-American hearth slag, "Riggins & Chapman," Joplin.
2. 3.—Air-furnace slag, "Jasper Land & Mining Co.," Joplin.

Galenas A and B, from the southwestern mines, occurring in chert and cherty limestone of Lower Carboniferous age, show comparatively high siliceous contents, while C and D, which are typical galenas of the Central Missouri lead-region, show a much lower percentage of silica, being mined from magnesian limestones of Lower Silurian or Calciferous age, and occurring commonly with barytiferous gangue. Zinc is rare, and, when present, mostly in small quantities in the central region; hence its galenas are free from this metal, and it is not generally to be found in the slags of the region.

Iron as pyrites, or ferruginous "joint-clay," is a nearly universal accompaniment of galena.

Of the slags, No. 1 may be observed to differ from 2 and 3 in being poorer in lead, with a consequent higher percentage of foreign matter, lime being an especially abundant constituent, as a result of the practice of dusting over the charge for the Scotch-American hearth with slacked lime, to "set up" the slag, and, perhaps, as a desulphurizer. All three of the slags given were produced from the smelting of a galena substantially like B, and consist, essentially, of the impurities of that galena concentrated by themselves, and more or less scorified with lead. No considerable amount of unaltered galena will be found in well-smelted slag, and few metallic shots in air-furnace slag. Iron will be observed to have been much less concentrated in the slag than the other impurities, an appearance of things which may be partly explained by the fact that the galena B is somewhat more pyritiferous than other Joplin galenas; but part of the explanation must lie in the absorption of more or less iron alloyed with the lead produced in the first smelting. It may be remarked, in passing, that 2 and 3 are too rich for average air-furnace slags by about 8 per cent. of lead.

Clean galena at Joplin, smelted in air-furnaces, yields to a 1,500-pound charge of galena about 85 pounds slag, or 5⅔ per cent., but much earthy or siliceous matter in the charge may leave as much as 10 per cent. slag.

Crucible-assays of various slags and furnace-products, conducted by the method of Mitchell above mentioned, (and mostly in duplicate or triplicate,) have given me results from which the following are selected:

| | Metallic lead. |
|---|---|
| Fume from air-furnace stack, Joplin | 0.385 |
| Fume from air-furnace stack, Morgan County | 0.372 |
| Average of several tons air-furnace slag, Morgan County | 0.44 |
| Same from smelting of coarser or "chunk" mineral | 0.412 |
| Average of 125,000 pounds Scotch-American hearth slag, Joplin | 0.35 |
| Tailings from machine-jigging of above | 0.20 |
| Scotch-hearth slag, Morgan County | 0.37 |
| Skimming from hand-jigging of same | 0.306 |
| Same air-furnace slag, Cole County | 0.463 |
| Same from another furnace | 0.45 |
| Slime from sluicing air-furnace slag | 0.18 |
| Lean brouss from slag-hearth | 0.127 |
| Black slag, free from metallic shots | 0.027 |
| Same with few shots, average of 200,000 pounds | 0.037 |
| Air-furnace bottoms, Morgan County | 0.262 |
| Dross from lead pot of slag-furnace | 0.852 |

Besides the ingredients already given in the complete analyses, other elements occur in notable quantity, and I have found the black slag obtained from smelting a mixture of slags from nearly all the furnaces of the Central Missouri region, to yield decisive reactions for silica, lead, iron, lime, magnesia, baryta, alumina, manganese, and sulphur.

The utilization of this black slag, except for mechanical purposes, such as road-making, is not often possible; in some cases, small quantities may be used as a diluting flux, a purpose for which it is fitted by its ready fusibility, as well as by being already saturated with lead.

The question of fuel has not, so far, been discussed in this paper, the use of coke being assumed, except where charcoal is used as a kindling-material at the outset. Charcoal may, however, be exclusively employed in cases where coke is not readily obtainable. For this purpose, the furnace should be at least a foot higher, to produce the proper heat, and prevent mechanical loss at the top. In any event, the use of charcoal will, while producing a rather better quality of lead, cause greater expense for fuel, of which a larger percentage to slag than that (10 to 12½) of coke, by weight, must be employed. Working the furnace will also become more difficult, and the lessened heat and more difficult fluxing of black slag will cause diminished production.

# CHAPTER XV.

## ROCKY MOUNTAIN COAL AND COKE.

The fuel question being a very important one in connection with the metallurgy of the West, I print here the following extracts from a paper recently read before the American Philosophical Society, by Mr. Blodget Britton, of the Iron-Masters' Laboratory, Philadelphia, and a report of some analyses of Trinidad coal, made at the Denver (Colorado) School of Mines.

The four coals from east of the Rocky Mountains, and on the line of the Union Pacific Railroad, exhibited at the meeting of the society October 6, 1874, I have since analyzed for metallurgical purposes, with the following results:

I.—*Carbon coal from the mine at Carbon.*

The sample consisted of several pieces, and weighed 12 pounds.

| | |
|---|---|
| Water | 12.50 |
| Volatile combustible matter | 35.47 |
| Fixed carbon | 44.96 |
| Ash | 7.07 |
| | 100.00 |

100 parts of the raw coal gave of coke ........ 52.03

| | | | | |
|---|---|---|---|---|
| The coke was composed of, carbon | 86.42 | | | |
| ash | 13.58 | | | |
| | 100.00 | including | Sulphur | 1.03 |
| | | | Phosphorus | Trace. |

II.—*Coal from Alury mine.*

The sample consisted of several pieces of fine stuff, and weighed 21¼ pounds.

| | |
|---|---|
| Water | 12.95 |
| Volatile combustible matter | 32.54 |
| Fixed carbon | 44.56 |
| Ash | 9.95 |
| | 100.00 |

100 parts of the raw coal gave of coke ........ 54.51

| | | | | |
|---|---|---|---|---|
| The coke was composed of, carbon | 81.75 | | | |
| ash | 18.25 | | | |
| | 100.00 | including | Sulphur | 0.29 |
| | | | Phosphorus | 0.04 |

III.—*Coal No. 3, from mine at Rock Spring.*

The sample consisted of a single piece, and weighed 18¼ pounds.

| | |
|---|---|
| Water | 13.40 |
| Volatile combustible matter | 35.25 |
| Fixed carbon | 49.81 |
| Ash | 1.54 |
| | 100.00 |
| 100 parts of the raw coal gave of coke | 51.35 |

The coke was composed of, carbon ...... 97.01
ash ......... 2.99

100.00 including { Sulphur ............ 0.63; Phosphorus ........ 0.02 }

IV.—*Coal from Excelsior mine at Rock Spring.*

The sample consisted of several pieces and fine stuff, and weighed 16¼ pounds.

| | |
|---|---|
| Water | 10.10 |
| Volatile combustible matter | 36.76 |
| Fixed carbon | 51.03 |
| Ash | 2.11 |
| | 100.00 |
| 100 parts of the raw coal gave of coke | 53.14 |

The coke was composed of, carbon ...... 96.03
ash ......... 3.97

100.00 including { Sulphur ............ 0.92; Phosphorus.........Trace. }

NOTES.—The coals swelled very little during the coking. When powdered they each agglutinated and formed into a strongly-adhering mass. The coke resembled in appearance the kind produced from the average bituminous coals of Western Pennsylvania.

A portion of the sample from the Carbon mine was subjected for an hour and a half to a temperature of 178° Fahr., and lost in weight 5.72 per cent.; subjected one hour more to a temperature of 280° Fahr., the loss was increased to 7.31; and again for two hours more to the same temperature, the whole loss was found to be 7.55. Another portion of the same sample was then subjected for three hours to a temperature of 500° Fahr., and the loss was 9.55. The watery vapor was all condensed in a cold glass tube, the tube was carefully weighed and then the water was evaporated from it, and when cold was weighed again; from the loss the weight of the water was ascertained. The coal was then weighed, and its loss was found to correspond very nearly with the weight of the water, thus showing that at a temperature of 500° Fahr. little less than water was expelled. A portion of the same coal was immediately put into another tube and subjected for a moment to a low red heat, when more water passed off and condensed in the cold part of the tube; subjected for another moment to a little higher temperature, a dark-brown oil passed off and condensed partly with the water and partly in the space of the tube between the water and the coal. The oil emitted a strong odor, the same as the oil produced by distillation from the brown friable lignites of Southern Arkansas and Texas, and, also, the Breckinridge coal of Kentucky, and some of the cannel coals of West Virginia. The other three coals produced water and oil in like manner, at a low red heat. These coals are certainly not lignites. No mineral resin or amber-like substance could be detected in them; they were black, apparently not friable, and had all the appearance of belonging to the true bituminous series. If dried at a temperature of about 500° Fahr., or something above, to expel the excess of water, they, beyond doubt, would answer for puddling iron and the purposes of the blacksmith; and their cokes, it is equally certain, would answer for producing pig-iron in the blast-furnace.

In order to ascertain the value of these coals for steam and illuminating purposes, samples were sent to Dr. Charles M. Cresson for examination, and subjoined is a copy of his report:

"OFFICE AND LABORATORY,
"*No. 417 Walnut Street, Philadelphia, December 5, 1874.*

"Coals marked 'Carbon mine,' 'Excelsior,' 'Mine No. 3,' and 'Alury' have been examined as to their fitness for the production of steam and for their suitability for producing illuminating-gas, Pittsburgh (Pennsylvania) gas-coal being used as the standard of comparison.

"The following results have been obtained:

| Coal. | Water evaporated by one pound of coal. | Gas to the pound of coal when all of the gas is worked off. | Value of 5 cubic feet of gas. | Value of 5 cubic feet when the amount is limited to 4.4 cubic feet per pound of coals. |
|---|---|---|---|---|
| | *Pounds.* | *Cu. feet.* | *Candles* | *Candles.* |
| Carbon mine | 13.42 | 5.71 | 9.52 | 10.30 |
| Excelsior | 13.53 | 5.54 | 11.80 | 12. |
| Mine No. 3 | 12.65 | 6.06 | 7.80 | 12.30 |
| Alury | 12.77 | 5.77 | 6.90 | 9. |
| Pennsylvania gas-coal | 14.67 | 5.2 | 12. | 14. |

"The heating-power of these coals compares favorably with that had from the majority of semi-bituminous and many bituminous coals. They should be burned in boilers adapted for use with bituminous coals.

"As gas-coals, Excelsior and Mine No. 3 possess fair qualities. They yield a very large amount of gas, and with a little enrichment (either by the admixture of cannel or a small amount of oils) will prove serviceable to the gas-maker.

"If these samples are from outcrop or from near the surface, it will most likely be found that the quality of the coal will improve as it is obtained from a greater depth; so that, without any limitation in the quantity of product, they will compare more favorably with the eastern bituminous coals for gas-purposes.

"CHARLES M. CRESSON, M. D."

The above speaks for itself as to the value of these coals, and I have no remarks to make on Mr. Britton's examination, except as to the occurrence of mineral resin or an amber-like substance in the coal, which Mr. B. has not found. Mr. A. Eilers, my deputy, however, reports the occurrence of mineral resin as quite frequent in the Carbon as well as in the Rock Springs coal, and other observers have reported the same. As to coking these coals, experiments made on a considerable scale at Rock Springs have so far not yielded a coke fit for blast-furnace use.

In the same connection, the following analyses of Trinidad coal and coke, made at the request of William Lawson, proprietor of the Swansea Smelting-Works, at Denver, are of great interest. The coke very much resembled that made from the bituminous coals of Pennsylvania. Pieces of the coke appeared to resist the blows of a hammer to about the same extent as those cokes. It was very slaty.

*Proximate analysis.*

| | Specific gravity. | Fixed carbon. | Volatile matter. | Ash. | Per cent. coke. | Moisture. | Weight of one cubic foot. |
|---|---|---|---|---|---|---|---|
| | | | | | | | *Pounds.* |
| Coal | 1.286 | 54.10 | 26.98 | 18.98 | 73.08 | .84 | 79.80 |
| Coke | 1.398 | 78.97 | 2.03 | 19.00 | ........ | ........ | 87.37 |

Color of ash, light buff.

*Ultimate analysis.*

| | Carbon. | Hydrogen. | Nitrogen. | Oxygen. | Sulphur. | Combined water. | Free hydrogen. |
|---|---|---|---|---|---|---|---|
| Coal | 57.42 | 4.64 | 1.50 | 16.20 | 1.26 | 17.38 | 2.615 |
| Coal * | 70.87 | 5.72 | 1.85 | 20.01 | 1.55 | 22.51 | 3.22 |

Ratio of oxygen to hydrogen, as 3.4 is to 1.

* Calculated free from ash.

*Calorific power.*

| | Carbon. | Hydrogen. | Coal. | Pyrometric power. |
|---|---|---|---|---|
| Caloric unit | 8080 | 34,462 | 5550 | 2652° C. |

The amount of combined water was determined by adding to the total amount of oxygen one-eighth of its weight of hydrogen, the surplus hydrogen being considered as available for heating-purposes. The heating-power of the coal was ascertained by multiplying the respective caloric units of carbon and hydrogen by the amount of each of these constituents proved by the analysis. The pyrometric power, or actual temperature, which could be attained during the combustion of the coal, was calculated by multiplying the amounts of the various gases formed and required during the combustion by their respective specific heats, and dividing their sum by the actual caloric unit of one unit of the fuel. The temperature of the surplus or available heat remaining after all the gases resulting from the combustion had absorbed their due amount of heat, thus lowering the actual temperature, was, in this way, found to be 2652° C.

It is noticed that both the coal and coke contain a very large amount of ash. In consideration of there having been found large pieces of slaty mineral in the coke, and of there being very much of it disseminated throughout, it is suggested that the coal should be washed, or that "slate-pickers" be employed. How much the operation will tend to diminish the amount of ash, or what percentage of the coal will be lost in washing, cannot be predicted. Experience teaches, however, that many slaty coals, without washing, are entirely unfit for the production of coke. We may hence infer that a prior washing will greatly diminish the amount of ash. As a large percentage of the coke appeared to be in every respect equal to Connellsville coke, capable of supporting burden in the furnace, being made from an excellent *smithy* coal, it is probable that a superior article of coke can be made from the coal in question with properly-constructed pressure-ovens. Taking the coke with its present ash-content, the comparative value of its carbon is readily calculated.

$$\left.\frac{87.45}{78.97}\right\} \text{per cent.} \left\{\begin{matrix}\text{carbon in Connellsville coke} \\ \text{carbon in the Trinidad coke}\end{matrix}\right\} = 1.10.$$

Twenty dollars per ton for Trinidad coke is equivalent to $22 per ton for Connellsville coke.

In order to ascertain whether this large amount of ash would merely amount to a weight loss, or would, when acted upon by heat, influence the successful operation of the blast-furnace, an analysis of it will be made.

For reverberatory furnaces the coal is a most excellent fuel, giving a long, highly-carbureted flame. The coke obtained from the destructive distillation of a weighed amount of the coal by a *gradual* elevation of temperature, as is done in coke-ovens, had less of its carbon eliminated in combination with hydrogen than would have been the case had the experiment been tried with a view to ascertaining the gas-producing qualities of the coal.

In determining the sulphur by the fusion-process, the crucible was heated in a muffle instead of over a Bunsen burner. This effectually prevents any sulphur which might be in the gas from combining with any part of the alkaline mixture. It is probable that even the small percentage of sulphur which was found, exists mainly as a sulphate in the ash, and not as a sulphide.

# CHAPTER XVI.

## SEPARATION OF GRAY-COPPER ORE FROM BARYTES.

The above subject is so very important to many miners of the West that I have thought it advisable to publish here the following article, prepared by Mr. Frederick Sturm for a European technical paper. If his plan of removing heavy-spar from silver-ores should succeed, it would confer a great benefit on many miners, particularly in Colorado.

The gray-copper ore (fahlore) obtained at the Kogler mines is found in the limestone formation, and is, for the greater part, accompanied by barytes, (heavy-spar,) which occurs in pockets and nodules, but mostly in widely-diverging threads. The slight difference in specific gravity of these two minerals, fahlore being 4.5 to 4.85, barytes 4.3 to 4.7, renders their separation impossible by the wet way on sieves and buddles. Heretofore the numerous experiments conducted for years, and at much expense, to effect a mechanical separation of the ores have failed.

When different kinds of barytes, especially the crystallized and crystalline varieties, are heated before the blow-pipe, they decrepitate and crumble up into a powder, which consists of small rhombic plates. This led me to think that this property of barytes could be employed for its separation from the fahlore, which is able to resist a moderate heat.

The first experiments, undertaken with small quantities of ore and incomplete apparatus, led me to hope for favorable results from this method of separation; hence I undertook the following more accurate experiments with larger quantities of the ore. The ore experimented upon contained from 75 to 80 per cent. barytes, and was sorted according to the size of the lumps, from 3 to 5 millimeters, (say $\frac{1}{10}$ to $\frac{3}{16}$ inch,) and placed in covered cast-iron pans, where it was calcined with occasional stirring until the decrepitation had nearly ceased. It was then passed through a sieve, along with pieces of fahlore and some pieces of lime, which had not been entirely removed in its preparation for the experiment. Some of the finer pieces of fahlore had broken off and gone through the sieve with the barytes, and hence the latter was passed through two more sieves, one with meshes 1 millimeter wide and the other with half-millimeter meshes. Fahlore sand constitutes the residue. Thus three end-products were obtained, (lumps of fahlore, fahlore sand, and barytes powder.) The table shows the results of my experiments. The 10 per cent. of loss is attributable to loss of moisture, spattering of the barytes, and what goes off as dust. The time required depends on the firmness and on the size of the lumps. In the experiment above described a hundred-weight required 40 minutes, so that 18 charges could be worked in twelve hours.

| | Quantity. | Metal in hundred-weight. | |
|---|---|---|---|
| | | Silver. | Copper. |
| | *Pounds.* | *Mint-pounds.* | *Pounds.* |
| Ore taken for experiment | 100 | 0.045 to 0.048 | 4 to 5¼ |
| Obtained from this: | | | |
| Lumps of fahlore | 20 to 22 | 0.248 to 0.254 | 20 to 20¼ |
| Fahlore sand | 4 to 8 | 0.018 to 0.080 | 2 to 7 |
| Barytes powder | 61 to 64 | 0.010 | ........ |
| Total average | 90 | ........ | ........ |

The results would be still more favorable if large quantities were operated upon in a flame-furnace, arranged to be worked continuously, and with improved contrivances for separating and sifting. The advantages of the method are briefly these: 1. Simplicity and rapidity of the process, combined with almost perfect separation. 2. Small expense. 3. Large amount of metal obtained. 4. Poorer ores can be profitably employed.

The author will continue his experiments on concentrating the fahlore in the fine, sandy state.

# CHAPTER XVII.

## THE PATCHEN PROCESS.

The increased richness of the ores obtained in the bodies on the Comstock lode has given additional importance to the large percentage of loss incurred by the Washoe process, as ordinarily employed. A new process, patented by Mr. Abel Patchen, has given, according to the published reports, very favorable experimental results, and is to be put in operation, I understand, on a large scale. The working-tests, as published in the Mining and Scientific Press of San Francisco, are as follows:

*Results from eight tons of Belcher ore, worked by Patchen process, in charges of 2,000 pounds each, occupying five hours to each charge.*

| Quantity. | Pulp-assays. | | | Tailings. | | | Per cent. of bullion extracted. | | |
|---|---|---|---|---|---|---|---|---|---|
| | Gold. | Silver. | Total. | Gold. | Silver. | Total. | Gold. | Silver. | Total. |
| One ton | $28 94 | $15 50 | $44 44 | $1 55 | $0 93 | $2 48 | 94.7 | 94 | 94.4 |
| One ton | 28 94 | 15 50 | 44 44 | 1 55 | 93 | 2 48 | 94.7 | 94 | 94.4 |
| One ton | 25 80 | 15 50 | 41 30 | 82 | 1 37 | 2 19 | 96.8 | 91 | 94.4 |
| One ton | 28 94 | 19 91 | 48 85 | 21 | 1 03 | 1 24 | 99.3 | 94.8 | 97.4 |
| One ton | 26 87 | 15 64 | 42 51 | 76 | 64 | 1 40 | 97.2 | 95.9 | 96.7 |
| One ton | 22 74 | 13 45 | 36 19 | 82 | 1 31 | 2 13 | 96.3 | 90 | 94 |
| One ton | 24 80 | 17 58 | 42 38 | 82 | 1 22 | 2 04 | 96.7 | 93 | 95.2 |
| One ton | 22 74 | 15 64 | 38 38 | 82 | 1 24 | 2 06 | 96.3 | 92 | 94.6 |
| Total | 209 77 | 128 72 | 338 49 | 7 35 | 8 67 | 16 02 | ....... | ....... | ....... |
| Average per ton | 26 22 | 16 09 | 42 31 | 92 | 1 08 | 2 00 | 96.5 | 93.2 | 95.26 |

*Working-results of Hale & Norcross and Savage mines.*

| Company. | Time, years. | Tons. | Value. | Pulp-assay, per ton. | | | Bullion produced, per ton. | | |
|---|---|---|---|---|---|---|---|---|---|
| | | | | Gold. | Silver. | Total. | Gold. | Silver. | Total. |
| Hale & Norcross | 8 | 299,929 | $11,479,372 | $12 11 | $26 16 | $38 27 | $9 24 | $15 92 | $25 16 |
| Savage | 5 | 211,941 | 8,060,734 | 10 94 | 27 09 | 38 03 | 8 35 | 16 83 | 25·18 |

| Company. | Time, years. | Tons. | Per cent. produced. | | | Value of tailings. | | |
|---|---|---|---|---|---|---|---|---|
| | | | Gold. | Silver. | Total. | Gold. | Silver. | Total. |
| Hale & Norcross | 8 | 299,929 | 76 | 61 | 65.2 | $2 87 | $10 24 | $13 11 |
| Savage | 5 | 211,941 | 76.3 | 62 | 66.2 | 2 59 | 10 26 | 12 85 |

*Statement showing saving which, it is calculated, might be effected by Patchen process.*

| Company. | Tons. | Value per ton. | Bullion produced. | Pulp-assay. | Increase, at 90 per cent. | Saving of quicksilver, at $2.50 per ton. | Total saving. |
|---|---|---|---|---|---|---|---|
| Belcher | 166, 857 | $54 84 | $9, 150, 372 | $13, 071, 960 | $2, 614, 392 | $417, 142 | $3, 031, 534 |
| Crown Point | 169, 824 | 39 74 | 6, 752, 735 | 9, 646, 764 | 1, 929, 353 | 424, 560 | 2, 353, 913 |
| Consolidated Virginia | 89, 783 | 55 48 | 4, 981, 090 | 7, 115, 843 | 1, 423, 169 | 224, 467 | 1, 647, 636 |
| Other companies | 101, 159 | 15 00 | 1, 516, 586 | 2, 166, 551 | 433, 110 | 252, 890 | 686, 000 |
| Total | 527, 623 | ........ | 22, 400, 783 | 32, 001, 120 | 6, 400, 024 | 1, 319, 059 | 7, 719, 083 |

Mr. Louis A. Garnett and Prof. Thomas Price superintended the experiments on the Belcher ore above tabulated. Mr. Garnett says:

For the purpose of comparison, I have examined into the milling-results obtained by the process heretofore and at present employed on the Comstock ores, and while I find that certain mills occasionally obtain 75 to 80 per cent., the average is between 65 and 70 per cent. only. For November, the average result of the Belcher, from 12,334 tons, assaying $55 per ton, was 66⅔ per cent., and for December, from 12,200 tons, assaying $45 per ton, 67 per cent., being, respectively, 28⅓ and 26½ per cent. below the results given above. Only two companies seem, however, to have furnished in their annual reports the gold and silver in the pulp-assay separately, which is necessary to determine accurately the percentage of the results obtained. These are the Hale & Norcross and the Savage, and they confirm each other to a remarkable degree.

It is added that further improvements have been made, calculated to increase the efficiency and economy of the process.

From all that I can learn concerning the process, it involves no new metallurgical principle or reaction. Mr. Patchen simply treats the ore with chemicals in a separate pan from that in which the amalgamation takes place. He uses two pans and a settler. The ore is transferred from the battery to a pan lined with wood or copper and provided with a heating-chamber. Here it is treated with chemicals and subsequently run into an ordinary iron pan, where quicksilver is added and an amalgamation effected, after which the pulp is run into the settler in the usual way. Patchen's patent covers the combination of the chloridizing-pan lined with wood or copper, with a heater, and the ordinary amalgamating-pan and settler. No particular chemicals are claimed. The patentee appears to prefer the treatment invented by Henry Janin, involving the use of dichloride of copper; and the company owning the Patchen process has already paid a considerable sum for the use of the Janin patent. It has also been obliged to compromise with Mr. C. H. Aaron, the patentee of a process similar to Patchen's, and it is said that Mr. Morris, the patentee of a wooden-bottom pan, has also presented his claims.

No full description of the Patchen process had been published up to the end of February, 1875; but the patent itself limits the claim to the features I have mentioned. It would not surprise me if the use of the dichloride of copper, the right to which has been purchased by the company from Mr. Janin, should after all be discarded in practice. Theoretically it is a beautiful thing; practically, trouble may come from the readiness with which the dichloride is converted into an oxychloride, unfavorably affecting the quicksilver, but remaining inert toward the silver-sulphurets. This suggestion does not originate with me; the difficulty referred to was actually encountered in testing the Janin process. I do not see what there is in Mr. Patchen's method to prevent its arising to annoy him as it annoyed his predecessors; and if he relies upon this peculiarity of his proposed method, he may experience a decisive failure. Mr. Patchen, however, claims to save quicksilver by not grinding his ore at all in either pan. This necessitates the employment of a very fine screen in the battery, a measure which will seriously affect the crushing-capacity of the mills, and therefore may prove in practice to be sometimes poor economy. The present Comstock Mills use coarse screens and do considerable grinding in the pans. It must also be borne in mind that in working large quantities of ore, less care is employed than in making small experimental tests, and that the latter therefore give, as a rule, results more favorable than are obtained in general practice. It has been shown in my former reports, and particularly in the chapter in my last year's report by Mr. J. M. Adams, that a very high efficiency (over 90 per cent.) can be attained by the Washoe process, when the skill of the manager and the faithfulness of the workmen co-operate with a real desire to extract as much of the precious metals as possible in the first operation. This has not been the condition hitherto in the Comstock Mills. It has been the interest of the mill-proprietors to leave a considerable percentage in the tailings, and to favor the impression that this loss is due to the inevitable limitations of the process itself, which is not true. The Patchen process will undoubtedly effect a saving both in gold, silver, and quicksilver; so would the ordinary Washoe process, if managed closely, with finer crushing, slower working, and a judicious use of chemicals in the pans. Whether the separation of the chloridizing reactions from amalgamation by the employment of separate pans will effect an extra yield sufficient to compensate for extra plant and time, does not appear from any comparison of test-experiments with the unfavorable results of intentionally careless working in the present Comstock Mills. I think it not improbable, however, that such will be the case, and that particularly for the treatment of ores of such high grades as are now produced in the Comstock mines, this, or some similar modification of the ordinary method, will be found profitable to the mine-owners, and will be, in the course of time, forced upon the mill-owners. But it is worth inquiring, whether, after all, the Stetefeldt furnace and chloridizing roasting are not what the high-grade Comstock ores require.

# PART III.

## MISCELLANEOUS.

# CHAPTER XVIII.

## GEOLOGY OF THE SIERRA NEVADA IN ITS RELATION TO VEIN-MINING.

BY AMOS BOWMAN.*

My notes of a reconnaissance of the Georgetown Divide, El Dorado County, California, falling under this general head, together with about 500 specimens, collected and finally placed in the State mining and geological collection at Sacramento, embrace considerable material toward a geological section of the Sierra Nevada, at about the middle of its extent. After locating the mines of the divide on a large scale, as on the map accompanying my report to the California Water Company, the first business was to locate and map in a similar manner, but on a smaller scale, such other mines as these were directly related to, and the position of both classes in the parent slate formation.

The subject is considered under the following heads: Vein-systems and their origin; the country-rock or matrix; mineral contents of the veins; surface geology.

As all these veins are acknowledged results of dynamical causes, it was necessary to consider specifically their dynamical history, with that of the slates, involving their relations as observed to the surrounding older and newer formations. The structure of the range and the geographical outlines of the gold-bearing slates on its western flank, &c., are, therefore, concisely stated. What is said under historical geology is as specific and brief as it could be to lay the foundation of the conclusions, which, it will be admitted, are as practical as they are important to the pursuit of mining in this region.

The miner, it is true, deals principally in hard knocks or physics, but not entirely in physical problems. The moment he begins to read his book in its most practical part, to question concerning the contents of veins:—"Where is the gold?" "Whither do its flakes, sheets, and chimneys extend?" "Where did it come from?" and "What minerals is it associated with?"—he leaves physics and goes into the chemistry of nature. As it would be impossible to comprehend the phenomena of the seam-diggings without some idea of vein-geology in its purely chemical phase, I have applied to the subject the best lights extant, derived from concurrent observations of others, with myself, in this field of practical science.

### VEIN-SYSTEMS, THEIR ORIGIN AND RELATIONS.

*Structure of the Sierra Nevada as related to mineral-belts—leading features.*—There are three summits instead of two in the region of the head-waters of the Middle and South Forks of the American.

The great snowy belt, so prominently visible from the capital and from the valley opposite Georgetown divide, as the dominant snow-field

* This chapter remains in the words of Mr. Bowman, and rests upon his authority. It is worthy of attention as an intelligent and skillful attempt at a much-needed comprehensive generalization.—R. W. R.

of the Sierra Nevada in this latitude, and of which Tell's Mountain is the most prominent culmination, lies in the third or western summit. It is in this and in the adjacent central summit that the abundant water-supplies are stored until late summer in the form of perpetual snow, which are utilized by the California Water Company.

The third or western summit starts out, like the eastern summit, of which Job's Peak is the culminating point, in the form of a spur; yet opposite a portion of Lake Tahoe it is the highest of the three ranges. It is visible from the north end of Lake Tahoe on its eastern slope as the highest summit range, carrying the largest amount of snow anywhere visible from the lake. It is clothed in white the summer through, and therein constitutes one of the principal charms of this, the noblest of mountain lakes on the continent of North America.

The two eastern summits are granitic. The western, or Tell's Mountain range, sometimes called the Conness range, contains beds of gneiss, formed of the component materials of granite in a rudely-stratified form.

The Sierra Nevada range continues to the north, as already remarked, in two different directions—northward in the Warner range and the Cascade or the Blue (?) Mountains of Oregon; northwestwardly in the direction of the Lassen and Shasta Buttes. The latter are close on the eastern flank of the great area of crystalline rocks of the Trinity and Scott Mountains.

This continuation of the gold-bearing slate formation is seen to extend to the coast, not far from the Oregon boundary, in one grand, swelling dome, gentler than that of the Sierra Nevada, terminating toward the north at Bald Mountain, 2,800 feet high, in the rear of Port Orford.

Both of these last-named ranges, as well as the Sierra Nevada, united at Lake Tahoe, contain cores of granite, viz: the Warner range, just west of Alkali Lake, in Surprise Valley, (visited by me in 1863,) and the Middle Age or Secondary coast-range of the Trinity, referred to above, in the divide between Cottonwood Creek, Shasta, and Trinity River, (visited in 1871,) and further northward into Oregon, in spots.*

It is to the southward of Georgetown divide, and far south of the middle of their geographical extent, that the Sierra Nevada culminates. At a point near Owen's Lake, opposite Tulare Lake, the northerly-trending mountains of the plateau of Nevada culminate in the White Mountains, and the northwesterly-trending Sierra Nevada range culminates in Fisherman's Peak.†

It will be observed that the axes, or efforts, of uplift developed the forms represented in the diagram.

Simply two different axes of uplift, at an angle of 40° to each other, and each having its parallels to the dominant wave or crest; into this, it becomes evident, from a glance at the diagram, all the mountain systems of the Pacific slope of the North American Corderillera resolve themselves. In examining the geology of any portion, therefore, it will be

* So reported to me at Port Orford, in 1873, by an intelligent prospector and gravel-miner, who had explored the entire region of the auriferous coast-gravels of the Coast range of Oregon, (Mr. Potts.)

† So named, by the party that first ascended it, in honor of a fisherman, who fished for fame by naming the biggest mountain in America after himself, and locating it at Mount Whitney, six miles from Fisherman's Peak. *Vide* account of King's ascent of Mount Whitney, supposing it to be Fisherman's Peak, by J. D. Hague, in the Overland Monthly of October, 1873; and of an ascent of the same mountain by W. A. Goodyear, the accidental discoverer of this error, before the California Academy of Sciences in 1873, copied in the American Journal of Sciences, October, 1873. See, also, King's explanation, attributing the error in the "Central Map" to local attraction, in Mountaineering, 1873.

only necessary for us first to identify clearly the older axes of the two. The mapping and sectionizing of the flanking sedimentary deposits, entered upon in the light of the testimony of the Sierra Nevada and its related groups, may then become a comparatively easy task.

The northerly trends of the plateau of Nevada had the effect to interrupt the course and deflect the uplift of the Sierra Nevada in a remarkable manner. At the point of interruption and deflection, (G,) the effort disclosed in the uplift of the Sierra Nevada appears to have spent its forces in attaining the highest culmination of the range.

A valley was formed on the California side of the range, opening northward, like those of Eel and Salinas Rivers—as if it were intended to connect with the Klamath. (Note the striking identity of the axis of the Klamath and San Joaquin Rivers, on the map.) But the Siskiyou Mountains, near Weaverville, developed a spur (A B) to seaward, in sympathy with the Monterey and San Francisco arm of granite bending around from the Tejon. Thus was laid the foundation for the future very peculiar topography of the great valley of California. The date of their uplift was Tertiary.

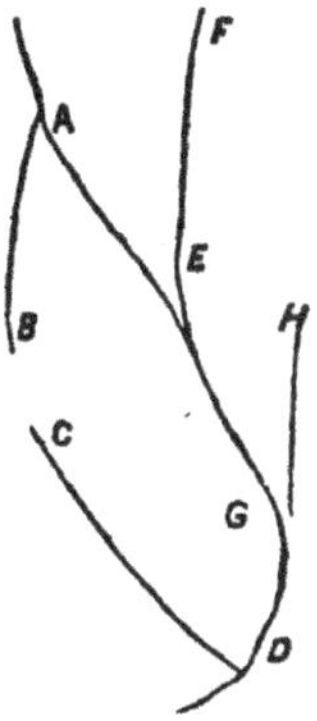

Axes of uplift in California.

The extensive volcanic plateau, east of Shasta and Lessen Peaks, extending to Goose Lake—the Modoc country—scarcely explored before the late Modoc war, represented by the letters A, E, F, owes its origin to the manner in which the two separately-occurring forces of uplift affected that inter-montane area. The older or northwesterly axis was that of the Sierra, the date Cretaceous; the newer or northerly, that of the Cascades, the date Tertiary in part, being also the date of origin of the Modoc plateau.

The Bernardino Sierra, continuing westerly from D to the Pacific coast, north of Los Angeles, is of granite. (See Blake's reconnaissance, containing a geological map of the Tejon region.) It is flanked by newer sedimentary deposits, forming the coast mountains from Los Angeles to San Francisco. In a similar manner the Trinity granite mountains have their continuations to the southward in the newer sedimentary rocks forming the coast mountains extending from Trinity River to San Francisco.

The northerly trending mountains of Nevada were, according to the testimony of their fossils, uplifted in the earlier Mesozoic, (or Secondary Time,) the Sierra Nevada later in the Cretaceous, and the coast ranges still later in the Tertiary and Post-Tertiary.

From this order of succession we receive the suggestion that the two different axes had certain of their parallels of a widely-different age, pointing to identical causes that operated intermittently throughout many geological periods. These older and newer parallels or groups can be very easily distinguished:

1. By their geographical position in relation to the continental core.
2. By their lithological composition.
3. Still more certainly and easily by their fossiliferous contents.

*Vein-systems of the Sierra Nevada—formation and strike.*—The geography of the auriferous slates of the Sierra Nevada and of the associated granites, with the relative position of the axes of elevation and depression bounding them on the east and west, and the actual position of the veins in this remarkable golden basin of the Mesozoic Sea, which is now the western slope of the Sierra Nevada, instrumentically located, plotted, and presented to a vision undimmed by any fogs of hypothesis, as in the accompanying map, gives us the first definite idea of the geological relations of our famous gold-mines.

The slate and granite areas are given as approximately located by the State geological survey. The veins are given as located by surveys of the United States mineral deputies.

A zone of complete flexure and breakage of the auriferous slates, displayed in the mother lode and its continuations, is shown by the map to be as near as can be midway between the axis of greatest uplift of the main or central summit and the axis of greatest depression, which is evinced in the drainage-bed of Sacramento Valley—Sacramento River itself.

On plotting the veins in their true geographical position to the mother lode south, and to the Nevada County and Sierra County quartz-lodes north of the Amador mine, we see, besides the general connection and relationship of worked veins, a forking of the vein-system, which is displayed in the direction of the veins as well as in the geographical position of the vein-regions where the veins have proved rich enough to be worked.

The figures on the map designate mines, which are similarly numbered in the printed reference list.

As the Sierra Nevada range itself continues to the northward in two different directions, (northward in the Warner range and Cascade Mountains, of Oregon, and northwestward in the Lassen and Shasta Butte culminations,) the forking of the vein-belt and predominating directions of the veins themselves, as here developed, are not otherwise than might be expected from a study of the axes of uplift that caused the fissures in the auriferous slate formation.

The diagram in the upper right hand corner was constructed by laying some tracing-paper over the map, and drawing on it outlines, which included all of the veins that are plotted. By cutting this out and laying it over that, it will be found to cover nearly every vein I have located, including every vein-mine that could be crowded into the space, for which a United States mineral survey and patent were, up to the date of plotting, recorded in the office of the United States surveyor-general at San Francisco.

The strike of the slates is, in the main, represented by the general direction of the two formations printed black or shaded, viz: north 40° west. Both in the map and in the small diagram in the right, the preponderance of veins keeping company with the "mother lode" in the same general direction is made apparent. Several of the characteristic variations in direction are repeated in the little diagram, by way of suggestion as to the probable dynamical cause or causes. The strike of the slates of course implies merely the bending and corrugation of the sediments at right angles to the lateral pressure that caused the principal uplift, the axis of which is seen to be parallel. That fissures were caused in the last stages of this bending is positively established. The particular process by which the metamorphic belts, seam-belts, quartz-ribbons, and the great quartz-veins, like the mother lode, in the strike of the slates, were first formed and then filled, solidified, or decomposed as we see them, may be regarded partly as the mechanical result of different degrees of continuousness of fracture, depending perhaps upon the thickness of the slate formation or their position relative to the bottom or top of the plication in which they occur.

The age or origin of all these different systems of fissures cannot have been precisely the same. The dynamical cause of one series was not

Map s

nornia, by Amos Bow

the same as that of another series. In general those parallel to each other were formed at the same time, and the age was that of the related axes of uplift. So that even modern fissures may be shown to have occurred in ancient rocks.

The filling of the fissures with a variety of minerals, such as we find in the veins, was a different matter entirely, belonging distinctly to another, a succeeding, period and in different systems varies accordingly. Probably the time will come when the observations of miners touching the different contents of cross-veins, and of geologists touching the order of uplift—the longitudinal fractures, the twists and the transverse fractures of the slates, as they were experienced in different portions of this Mesozoic basin since it became dry land—may make out clearly the priority and exact relations of the several systems.

The longitudinal fractures represented by the mother lode were probably the oldest, telling the story of the uplift of the Sierra. The twist and transverse fractures which evince a relationship either to the northerly or northwesterly trends of the Sierra will have to be studied probably in connection with the history of the volcanic peaks, the uplift of the Cascade Mountains, and, in the case of the eastern slope base-metal ranges, in connection with that of the plateau mountains of Nevada.

The totals below show about the proportion of veins occurring in each system described, except, perhaps, in regard to system No. 5. That might be subdivided, as it represents double the sweep of horizon of either of the other systems, including the veins intermediate between the others in course, both the northeasterly and the northwesterly twist fractures.

1. Northwesterly fractures or veins within 15° of the general trend of the Sierra, (north 15° west to north 45° west; average, north 30° west,) embracing a sweep of 30°, (mother lode system:) 2,* Potts; 5, Kelsey; 7, Arbena; 8, Gray Eagle; 15, Northern Light; 21, Salathiel; 35, Galena; 36, Sebastopol; 47, Penon Blanco; 49, Greenwood; 52, Oneida; 62, Keystone; 65, Spring Hill and Geneva; 87, Newtown; 88, S. Bright; 89, Saint Lawrence; 90, Cederberg; 114, Rocky Bend; 126, Gover; 130, Tecumseh; 136, Wisconsin; 172, Union; 194, Eureka; 201, Copp; 202, Secret Cañon; 212, El Dorado; 214, Coyote Hill; 215, Talsig; 217, Greenwood; 233, Cosumnes; 243, Last Chance; 260, Marietta; 289, Empire; 292, Sulphuret; 294, Mahony—total, 35.

2. Northwesterly transverse fractures or veins within 15° of a right angle to the trend of the Sierra, (north 45° east to north 75° east; average, north 60° east:) 14, Kelly; 22, Rising Sun; 50, Spring Hill; 97, Norambaqua; 193, Eclipse; 199, Keystone; 209, Chauleur; 315, Old Pioneer—total, 8.

3. Northerly fractures or veins within 15° of a northerly trend, (north 15° west to north 16° east,) embracing a sweep of 30°, (Grass Valley or Virginia system:) 22, Norridgewock; 30, Spring Valley; 31, Venus; 32, Stanton & Allison; 34, Auroral Star; 44, Rough and Ready; 65, Original Amador; 73, Dry Company; 74, Stanislaus; 80, Hancock & Tibbetts; 112, Nisbet; 115, Lone Jack; 120, Sliger; 134, Everlasting; 173, Poorman; 182, Carson; 184, Wolverine; 195, Confidence; 198, Yellow Jacket; 203, N. Confidente; 211, Shores; 233, Cosumnes; 246, Plymouth Rock; 277, Fort Yuma; 305, R. R. Hill; 319, Uncle Sam—total, 26.

4. Northerly transverse fractures or veins within 15° of a right angle to the northerly trending axes of uplift, (north 75° east to south 75°

* The numbers prefaced to these titles are the numbers of the United States patents taken or applied for, after due survey, by which, among other things, the course of the lode is defined.

east,) embracing a sweep of 30°: 3 and 4, Oaks, Reese & Jones; 39, Epperson; 54, State Ledge; 81, Union; 132, Tyson; 191, Keystone; 226, Wet Gulch; 238, Ophir; 298, Hancock & Watson—total, 10.

5. Twist fractures or veins running from north 15° east to north 45° east (30°) and north 45° west to north 75° west, (30°,) making, in all, a horizon of 60°, or as much as Nos. 1 and 3 together, the veins of which are neither parallel nor transverse to either of the principal axes of uplift: 9, Schofield; 45, Eureka; 84, Enterprise; 98, New York Hill; 115, Moorehouse; 124, Bangbart; 128, Calaveras; 131, Bobby Burns; 138, Independent; 165, Lucan; 175, Stickle; 189, St. John; 204, Butcher Boy; 208, C. Baker; 210, Waters; 219, Green, Walter; 233, Cosumnes; 249, Boree; 251, Crœsus; 303, Mammoth; 320, Doctor Hill—total, 22.

## VEIN-SYSTEMS, THEIR CONTENTS AND PARALLELISMS.

Having shown the geographical position and relations and the geological origin of these vein-fissures in the auriferous slates, it will be in order now briefly to consider how far these, and other circumstances connected with their origin, may have had something to do with the quality or quantity of their contents.

There are, at least, four well-defined parallel mineral-belts, running in a northerly and southerly direction, represented in the veins of the Sierra Nevada, yielding, respectively, copper, gold, base metals, and silver.

Beginning at the sea-coast and including, with those of the Sierra Nevada the entire series, the order is as follows:

*Cordilleran mineral-belts.*

| System. | Minerals. | Period of fracture and deposit. |
|---|---|---|
| 1. Coast range | Quicksilver, tin, chromic iron | Tertiary. |
| 2. Sac and San Joaquin Valley | Brown coal | Tertiary and Post-Tertiary. |
| 3. Sierra Nevada: | | |
| (a) Foot-hills | Copper | Oldest Cretaceous. |
| (b) Mid slope | Gold mineralized by sulphur and iron, the system extending through to Western Mexico. | Cretaceous. |
| (c) Eastern slope | Copper and lead | Cretaceous in part. |
| (d) Eastern slope | Silver, with very little base metal, frequently wholly or partly inclosed in volcanic rock, cutting through volcanic dikes. | Tertiary. |
| 4. Basin of Mexico, Arizona, Eastern Nevada, Idaho. | Silver, associated with base metals | Devonian rocks. |
| 5. New Mexico, Utah, Western Montana. | Rocky Mountain argentiferous galena | Mesozoic and Paleozoic rocks. |
| 6. New Mexico, Colorado, Wyoming, Montana. | Rock Mountain gold, with base metals | Do. |

The parallelism of these belts is clearly referable to the structural features of the country. It is related not merely to the mountain ranges, but to the succeeding formations of different ages, which in those ranges were uplifted.

From the Silurian to the Post-Tertiary, the gradual land-making to the westward, and the insular spots so far as recognized, all reproduce the same remarkable unity of parallelisms in which we find our gold and silver veins are concerned. In crossing the country from west to east, we traverse a whole series of formations, while by following roads parallel to the mountain ranges, we may travel continuously upon the outcrops of the same age for a thousand miles or more.

The grand fact that the axis of the Cordillera of North and South America is continued into Asia, where it had undoubtedly a great deal to do—in its direct continuation as well as in parallel uplifts—in shaping that continent, bearing upon its flanks everywhere formations producing gold and silver, and, furthermore, that this is the identical axis which, passing around the world, divides it into one hemisphere nearly all land and another nearly all water, leads the miner of the Sierra Nevada at once into the study of the profoundest cosmical problems. And if he could read unerringly, from the contents of the veins, the simple original cause of this line of fractures, he might contribute as effectively toward a solution of one of the unsolved problems of space as did the miners of the Erzgebirge, who founded the science of geology, toward a solution of the problem of time.

*Periods of deposit.*—Two periods of vein-deposit—in general accompanied by the ejection of igneous rocks, affording the conditions of solfataric action favorable to metalliferous deposit—are promulgated by Clarence King, to wit, the "Late Jurassic" and Tertiary.

The Cretaceous age of the principal gold-veins of the Sierra Nevada, it will be seen, is definitely limited and fixed. The rocks being Jurassic, (next preceding in age,) the Tertiary (the next following formation) furnishes us with concentrated auriferous gravels that originated from the denudation of the veins in question, the intervening period being the Cretaceous.

The age of the fissures in question is always the same as the axis of uplift or fracture. See, beside, structure of the Sierra, under *Vein-systems;* also, *Stratigraphy,* further on.

The accompanying *Tabular exhibit of veins and mining* shows that one series, or half of the mother lode system of veins, is older than the other. (See Pine Tree.) And under *Mineral contents of veins*, further on, will be found reasons for the conclusion that the companion talcose veins and seam-belts belonging to the same general system are more recent than the mother lode.

The copper-zone veins of the foot-hills, if not older than the great central vein-system of the western slope, occur at least in a zone of older (Triassic) rocks.

The Tertiary age of the Comstock system of veins, on the eastern slope of the Sierra, is based on the assumption that the accompanying eruptive rocks are Tertiary. It is not distinctly stated whether their synchronous occurrence with, or relations to, the Pliocene lavas of the Sierra, or any other determined formation, have ever been made out. Richthofen's determinations of the age of the "propyllites" and "rhyolites" have been shown by Dr. Blake to be as hypothetical as the new names he gives these rocks are unnecessary and confusing.

These veins are evidently not newer than the Tertiary; nor can they be older than the Cretaceous, to which period those fractures on the eastern slope that are related to the main or northwesterly axis of the Sierra must be referred. This embraces apparently several of the base-metal mines, the Santa Maria and Exchequer, in the *Tabular exhibit.*

The northerly trending axes and vein-fissures of the Sierra Nevada generally, whether on the eastern or western slope, would appear to be newer than the northwesterly trends, for the following reasons: 1. From the fact that the northerly trending uplifts, of which the Cascade range in Oregon is the continuation, as has been shown by Prof. Joseph Le Conte, belong to the Tertiary; the Cascades at the Dalles being Miocene. 2. That they intersect volcanic dikes which are Tertiary. 3. The northwesterly trends are undoubtedly Cretaceous.

The age of the transverse fissures of axial system No. 2 above may be the same as that of No. 1, "Cretaceous," for the reason that the uplift of the Sierra was uneven, being 15,000 at Fisherman's Peak, and only 6,000 at Beckworth's Pass. Transverse or twist fractures must necessarily occur under such conditions in a longitudinal dome not evenly raised at its extremities by the lateral pressure. And a similar state of things might be referred to in connection with the northerly transverse fractures, No. 4. But better evidence is dersirable in regard to the age of both these and the twist fractures, No. 5. The mineral contents and and intersections with other systems or with dikes, may hereafter furnish better means of determining them.

*Fissure-systems, as typified by districts or by general cause of origin.*—I need not repeat here the geological characteristics of mines presented in the tabular exhibit; but an arrangement according to the contents and yield of mines typified by well-known districts will help out systematization by comparison. We may designate the vein-systems according to districts; or, as related to strike, regardless of age or contents, pointing to identical continued, or rather repeated, dynamical causes of origin, thus:

1. Northwesterly strike: (1.) Mother lode system—(*a*) in the southern counties; (*b*) at Grass Valley; (*c*) accompanying talcose series. (2) Base-metal system of the foot-hills.

2. Western slope, cross-fractures generally: Eureka and Sierra Buttes systems.

3. Northerly strike: (*a*) Comstock lode system; (*b*) on western slope; (*c*) eastern slope, base metal—classifying the subordinate transverse and related twist fractures as belonging to the main system in connection with which underground testimony may prove that they arose.

*Points applying to Georgetown divide.*—While nearly everything that has been said or tabulated relating to vein-systems or mining not upon Georgetown divide thus far finds its application on Georgetown divide, (as represented in the detailed description of mines in the Georgetown report,) there are several points to which local attention ought to be specially directed:

1. The descriptions of the Quartz Hill, Saint Lawrence, Taylor, and Sliger veins, compared with those of the mother lode and Grass Valley veins of a northerly or northwesterly strike, show that the great fissure-zone of the Sierra is continuously represented in all its geological characteristics, accompanied by wealth in gold.

2. The related talcose-vein series on Georgetown divide accompanies the mother lode through the southern mines. The phenomena of the seam-diggings are associated, however, not with the mother lode alone, but also other veins of the mother lode system. Belonging, as remarked, in part at least, to a later period than the mother lode system, they are due to the chemical conditions described under *Mineral contents of veins*, further on.

3. Lenticular masses, or chimneys of quartz, often having a feather-edge, are characteristic of the best mines worked on either slope of the Sierra. In fissure-veins the fissure always continues, while the pay and even the quartz may pinch out or develope fantastic forms and strange affinities especially in the hanging-wall of the country-rock. Both the quartz and the pay are as likely to be repeated in adjacent chimneys in the same fissure prolifically as to yield, within certain limits, as though there were no ore-bodies nor chimneys of quartz, and the pay were only found in sheets resembling veins of coal. These things are in the nature of fissure-vein deposits.

While spaces of 1,000 to 2,000 feet, however, upon the mother lode, or any other vein, may be found rich, or in the form of chimneys or lenticular masses, yielding rich pay, the extensions beyond are generally more likely than not to be poor for a short distance.

4. The tabular exhibit affords proof of something like a general rule, that the quartz widens and also increases in richness in depth. There are exceptions; but the results of experiences, cited from so many localities, cannot be resisted. In a majority of instances noted the pay-chimneys dip north.

5. Parallel veins adjacent to rich deposits are usually comparatively barren, even if uniting into the same vein in depth. Instances where veins of the character of the seam-diggings unite with some main quartz-vein adjacent in depth are not infrequent in the tabular exhibit.

The importance of any given belt as an ore-channel in depth can only be inferred from these concomitant indications, taken along with the yield of the seams. Never is the insignificant quantity of quartz in place conclusive as against the existence of gold-deposits belonging to what might once have been perhaps a well-defined vein. (See various talcose slate and seam-diggings in the tabular exhibit and under *Contents of veins.*)

### THE COUNTRY-ROCK, OR MATRIX.

*Geographical outlines and relations of the auriferous slate-formation of the western slope of the Sierra Nevada.*—I have now pointed out the position and the relations of all the notable veins and seam-belts, and delineated the character of the deposits developed by mining. The conditions under which the miner operates in seeking gold in veins would be very imperfectly stated without some account of the country-rock, in which these veins are distributed, and in which they apparently originated. The geography and relations to neighboring formations of the world-famed auriferous slates of the Sierra Nevada should have been determined, or become the property of the public, in outline at least, a great many years ago.*

*Historical position of the slates—three ages.*—While the slates in the longitudinally central or vein-bearing portions of the Sierra are accompanied by Jurassic fossils, (as in Mariposa County,) the siliceous strata developed near the summit (as at Redding Soda Springs) have an evident relationship in their strike to the older triassic and permo-carboniferous rocks found fossiliferous and also gold-bearing in Plumas County.

Other fossils of the period next preceding the Jurassic have been found, in the form of impressions on the surface of the slates, at several localities in the region embraced in the map of Georgetown divide.

I refer to the Triassic fossils found at Coloma, on Placerville divide, and contributed by John Conness to Doctor Trask, (goniatites;) also fossils found at a point two miles west of Spanish Flat, on Georgetown divide, by Gorham Blake, (viz, of a cephalopod which could not be distinguished from belemnite found on the Mariposa estate, and a gonia-

* PUBLISHED SOURCES.—P. T. Tyson, in a report to the Secretary of War, in 1849, described the principal lithological characteristics and the physical relief of the Sierra Nevada. William P. Blake and others, in the Pacific Railroad reports dating down to 1862, and the members of the geological survey of California since 1860, besides various travelers and contributors to the United States records of mining statistics, and to the journals of the day, have contributed to the general knowledge on the subject. The State of California has appropriated over $250,000 for the publication of such information lying at the foundation of her peculiar industries. After twelve years of patience with the administration of this work, no geological map nor systematic memoir of the geology of any portion of the State having appeared in that time, the legislature of 1873-'74 discontinued the appropriation.

tite, found also in the Humboldt Mountains of Nevada.) Whitney, in volume 1, Geology of California, sets this formation down as corresponding to the Upper Trias beds of Hallstädt and St. Cassian in the Alps.

Near the western base of the Sierra, in Butte County, there are still older rocks than the Triassic. I refer to those containing the Carboniferous fossils found at Pence's ranch, near the foot of Table Mountain. Not far from the summit of the Sierra, in Plumas County, there are found fossiliferous rocks of the same remote age.

If the strike of the slates, then, associated with localities definitely located as to their geological position, by fossils, may be taken as an index of the age of the rocks at the base and summit of the range, respectively, on Georgetown divide, it would appear that there are *older rocks than the central zone of gold-bearing slates, both at the base and the summit*—the eastern and western margins of the ancient basin in which were formed the auriferous slates of the Sierra Nevada.

The older rocks of the Redding Soda Springs, at the head of Forest Hill divide, are strikingly different in appearance from the slates occupying the greater portion of the slope of the Sierra westward of them. They are white, highly siliceous, and crystalline in their character. These characteristics are observable in a less marked degree on Georgetown divide, where thinly-bedded series of the same character of rocks are found in the Tell's Mountain range, interstratified with gneissoid rocks and mica slates, showing a close relationship in their lithological character to the granites themselves. J. E. Clayton confirms me in having observed a similar series of (probably ancient) crystalline rocks, near the summit of the Sierra, on its western slope; and he further states that their lithological character suggests to him their possible identity with a series of rocks found by him across the mountains, on the Nevada side, at Silver Peak, containing an abundance of fossils of Silurian age.

Pilot Hill, situated near the edge of the granites to the west of the auriferous-slate formation, is composed of a peculiar crystalline rock, differing entirely from anything found anywhere near the center of the geological basin of slates.

Concerning the geological age of these rocks of the western slope, we have, then, two points established:

1. That they are situated in a granitic basin, the rim of which, on the east and west sides, is as old as the Carboniferous period; its lithological relations to the Devonian and Silurian rocks of Nevada never having been followed out. The rocks at both the western and eastern rims are crystalline, and different in character from those in the center.

2. That in it were deposited sediments which formed slate rocks of the *three grand ancient subdivisions* of geology, viz, the Carboniferous, the Triassic, and the Jurassic, distributed as follows:

(*a*) Of the Carboniferous period, the outcrops of which have been identified both near the eastern and western rims of the basin.

(*b*) Of the main body of auriferous slates, composed of two periods, viz, the Triassic and Jurassic, the geographical distribution of which, from the evidence of the fossils found, is mixed; so that we must look for the present to lithology as our only guide to their historical position.

(*c*) The localities, so far as determined, belonging to the newest period, (the Jurassic,) are situated principally near the center of the basin; and the oldest of the two named formations, as developed by means of fossiliferous evidence found on Georgetown divide and in Plumas County, at points well toward the eastern rim, and also well toward the western rim, (at Coloma,) underlies the main body of the Jurassic slates.

Under *Physical geography*, in the report on Georgetown divide, I have described the surface undulations that are everywhere associated with higher or lower degrees of metamorphism; under *Stratigraphy* will be found the key to the foldings of the slates which took place in the center and throughout the whole width of the basin here referred to; and under *Lithology* will be found facts furnishing our only remaining clew, in the absence of thorough paleontological research, to further the details concerning the age of the auriferous rocks.

Descriptions of the fossils referred to will be found in *Paleontology*, volumes 1 and 2 of the Geological Survey of California, and in connection with Blake's and Newberry's geological reports in the United States Pacific Railroad Explorations.

*Stratigraphy of the Sierra.*—That the slate-forming muds were in general deposited in pretty deep water throughout a long period of time we have good reason to believe, from their consistency and from their probable thickness. No careful geological measurement of the thickness in feet has ever been attempted. Yet it would seem that in the extreme simplicity and regularity of the bedding of the slates as indicated at the surface, along the entire western slope of the range, we should find it not difficult to hit upon an explanation of the precise method of granitic uplift whereby the off-shore strata of mud became shaped into a dome of such unparalleled sublimity as that witnessed in a profile section across the Sierra Nevada. This dome presents to our view an unbroken and almost perfect arc, over a base of seventy miles, having an altitude at the head of Georgetown divide of 8,000 to 10,000 feet.

A stratigraphical section of the western slope of the Sierra, as laid down in profile, shows the position of the gold-bearing slates overlying the "primitive" or Paleozoic gneiss and crystalline schists of the summit-belt in the great dome of the Sierra to be very far from regular, or conformable to that remarkable vertical dip which is the principal scenic and mining feature of the slates, especially toward the foot-hills.

Between Tell's Mountain range and the rich-vein region of the mother lode and its continuation northward, there are, on Placerville divide, many square miles—hundreds, I might have said—of gneissoid micaceous slates, lying nearly horizontal, or even dipping to the westward.

This is a noteworthy exception to what has hitherto been regarded as the rule, that of an unvarying easterly dip at a steep angle.

We find, then, in the stratigraphy of the auriferous slates, not a continuous vast bedding of such astounding thickness as would be implied in a conformable stratification with the dip from base to summit, but every evidence of folding, like that which took place in the Alleghanies, and has been demonstrated by Rogers. While at the eastern end, the geological base of the series, we find the slates very plainly and very naturally superimposed upon siliceous and crystalline schists, the gneissoid strata referred to; and these in turn upon the foundation rock of identical mineralogical constitution, the granitic core of the Sierra Nevada.

Where the dip of the slates in the basin is steep, the angle nevertheless varies greatly. Sometimes it is to the east, and one-quarter of a mile off it is to the west again. The angle of easterly dip varies from 40° to 90°. In all probability longitudinal zones, of low angles and high angles of dip, might be traced for some distance in the line of the strike of the slates.

In making a cross-section, all that can be said is that the prevailing dip is to the east, which is equivalent to saying that the eastern end of the slope is lifted, and, at the same time, the apex of each anticlinal is shoved away, in many instances, from the main axis of the Sierra.

At Greenwood and numerous other localities occur sandstones and conglomerates, in various stages of metamorphism, dipping conformably with the slates. This fact shows conclusively that the lamination of the slate is actual bedding, and not to be ascribed to cleavage from lateral pressure.

It has been remarked that the dip at the bottom of El Dorado Cañon, on Forest Hill divide, 1,200 feet deep, is steeper at the bottom than at the top. Curves of this character are universal in all the cañons in either slope, and are observable in cuts, shafts, and ravines in the mining-region, within distances apart of 10 to 50 feet vertical, and are quite as perceptible as in the cañons 1,000 feet deep. The convexity of the curve in these is frequently toward the west, however, as well as toward the east, as at the Doncaster mine. Such convexity furnishes us with a clew for the study of stratigraphic details, inasmuch as the western convexity must generally belong to the western half of an anticlinal, and the eastern convexity must, in like manner, belong to the eastern half of an anticlinal.

At Grass Valley, Nevada County, Professor Silliman recognized a synclinal or geological valley, in the dip and strike of the veins mentioned, occurring entirely conformably to that of the slates in general between the New York Hill vein and the Allison Ranch vein; and a saddle or anticlinal in the valley between the veins of Cincinnati Hill and Massachusetts Hill, which is repeated below and to the westward of these claims by the elevation of the syenitic mass in which the Norambagua occurs—something like the following, in a section east and west:

The course and dip of the Grass Valley veins, he further says, are conformable with the rocks, and "the streams have in general excavated their twenty valleys in a like conformable manner."

At the Princton mine, Mariposa County, situated in the center of Bear Valley, Mariposa County—a valley trending northwest and southeast, parallel to the Sierra—Professor Blake also observed a stratigraphic section of the auriferous slates at right angles to the crest of the Sierra, namely, from the Bear Creek Mountains on the west to the Mount Bullion range on the east. He here recognized a plication, or folding, in the form of a simple anticlinal. At the vein, which is conformable in strike and dip to the slates, the latter are soft and finely laminated, light-colored or drab at the surface, and black in depth, with numerous intercalations of sandy layers, passing into coarse grits, sandstones, or conglomerates.

In both of the bounding ridges there are only heavy metamorphic conglomerates. Magnesian rocks accompany the vein in the region of the soft shales in the valley.

In nearly all the high ridges on Georgetown divide, enumerated as trending parallel to the Sierra, there is observed a high degree of *siliceous* metamorphism, accompanied either by porphyritic or crystalline rocks; while the intermediate spaces consist of soft light or dark shales, showing zones of *basic* metamorphism, and containing hydrous-magnesian minerals.

Between Pilot Hill and the Little South Fork of the Middle Fork there are no outcrops of granite, except possibly some insignificantly small ones. East of that point all is granite; slate is the exception.

The American River, between Cape Horn (above Colfax) and Folsom, runs in general along the strike of the slates. It follows the line of their strike for a mile or two, and then, turning abruptly, crosses the strike at right angles, or in the direction of the slope of the Sierra, for a few hundred yards, only to resume afresh its former course in the strike of slates in the whole of the next long reach of the river. This is characteristic of that portion of the American River for a distance of thirty-five or forty miles by the river. At Wild Goose, the American follows this line of strike near the junction of the slates with the granite, affording excellent opportunity for a stratigraphic section at the base of the Sierra, or the western rim of our auriferous basin. One mile from Auburn, just before coming to Smith's house, on the American River, on the old road to Coloma, occurs a dike of fine-grained syenite. Trappean intrusions join the granite abruptly, and quartz is abundant in this region.

For a similar section of the slates near the eastern rim of the basin, Tell's Mountain range offers a good exposure. Here becomes evident the grand fact that the slates near the summit are thinly bedded, and are in large part carried away by denudation; for we can see to what extent denudation has left the granite exposed. The dip at Tell's Mountain is at a very low angle to the east.

On Placerville divide the granite of the summit reaches far down the slope, and the South Fork of the American appears to have followed for some distance near the northern rim of a promontory of this rock. The dip of the slates along the South Fork of the American, adjacent to the granite outcrop, is, accordingly, changed to all angles, and nearly all points of the compass, from west to south and east. In the Slate Mountain range, just across the river on Georgetown divide, the dip is southerly and westerly. For ten miles on the Placerville road, above Brockliss's Bridge, the dip is at a low angle to the west.

Where the slates are not associated with granitic outcrops there is a remarkable regularity in their strike. Over a section nearly fifty miles east and west in a straight line, there occurs no other geological feature so prominently to the observer. The direction, of course, varies in places, but only slightly. I have found nothing anywhere near the central or western portions of the slate-basin indicative of any cataclysm such as the "immense edgewise longitudinal thrust" which the mass of the Sierra "must have undergone," by which "vast bodies of strata, once continuous for hundreds of miles, have been torn asunder, portions engulfed, and the remainder twisted so as to lie at all angles with regard to the original direction of the mass," but "not so far removed as to leave any doubts of their having once been parts of the same continuous formation," pictured by Professor Whitney, in volume 1, *Geology of California*. (See remarks on limestone, below.)

The most prominent scenic features of the foot-hill region are the long lines of slates, standing on end like gravestones, continuously in the same general line of strike from Calaveras to Butte County. The general parallelism and perfect regularity of strike is, as remarked, the grand feature of the auriferous-slate formation, over the entire western slope. The local variation of strike in the Slate Mountain range, (accounted for,) is the only exception I have seen in about ten thousand square miles of the formation.

William P. Blake, in his visit to the mining-region, in 1864, recognized another of the characteristic physical features of the western slope of the Sierra. He correctly described the enormous erosions in the auriferous slates as having taken place in "one unbroken plateau or slope." Whatever may have been the method of tilting the slates underwent, their plateau-slope character, as shown in one grandly regular, easy line of profile from base to summit, on which the undulations described under *Physical geology* are insignificant, is a feature everywhere noticeable to the geological observer. But it is nowhere else so unmistakably recognized, and so striking to the eye, as from the dome of the State capitol, at Sacramento. The parallel lines of this slope are, from that point, seen to lie level behind one another, like the lines of a level plain.

Professor Blake clearly testifies, further, that the auriferous slates trend in a northwesterly course "with great regularity, and without any abrupt local plications or disturbances of the beds." "The plications, where they exist, are upon a magnificent scale, and very regular."—(*Geological Reconnaissance of California.*)

The same idea of plateau character connected with the slate formation has frequently suggested itself to me from local observation of the bed-rock flats associated with the gravel courses of ancient rivers.

I need not enter here into any particulars concerning the forces which caused the original foldings of the slates, the formation of this plateau or gradual slope, nor the character of the sediments which made folding and plateau formation possible and unavoidable. For the operation of the same, ever-enduring, simple physical forces, see Dana's *Geology.*

The occasional parallel granite belts or dikes, intermediate between the summit and base of the Sierra on other divides, and on this one, especially near the eastern and western rims, show that whatever may have been the origin of the granite, (whether it was Plutonic metamorphism or eruptive ejection,) the slates at mid-slope must have been extensively folded by the operation of the same agencies which enabled the granites to protrude.

If it was eruptive ejection, then the slates were folded and lifted along with them. If it was Plutonic metamorphism of deeper-seated strata, the strata were corrugated, and the lower metamorphosed ones bent upward, and afterward denuded off at the surface, and so exposed.

In neither case is the *rationale* of the slates at mid-slope altered. The thickness of the slates, however, might be differently concluded upon. On the latter hypothesis, the frequent exposure of granites in parallel belts conformable to the slate, as on the Grass Valley divide, would imply thin bedding. In the tabular exhibit of mines, the probable limits of thickness are referred to as from one and a half to three miles, vertical. That it could have been more is not impossible; though the probabilities would be against that supposition until the fact could be demonstrated by stratigraphical evidence. In prosecuting this inquiry, it should not be forgotten that the midslope region was in the Triassic and Jurassic periods, probably in the center of a valley, and the position of an axis of depression like that of Sacramento Valley, having older rocks, as elsewhere demonstrated, on the east and west.

The folding and general stratigraphic character of the slates of the region, according to these facts, is represented by the foregoing section across the Sierra Nevada.

The granites of the summit and of Folsom were the abutments of the arch; and the sediments of the region, which were originally sinking at mid-slope, forming a geological valley, must have begun to arch up and fold the moment the abutments began to be brought nearer together.

The question arises, Where are situated the oldest rocks of the mining region? Are the veins, which most concern us, in the older or newer rocks—at the top or bottom of the Jurassic series?

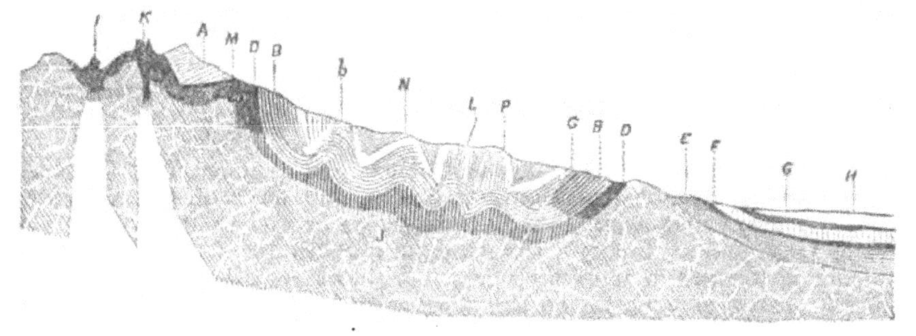


STRATIGRAPHIC SECTION.

A. Tell's Mountain; nearly horizontal slates of Tell's Mountain and Mount Dana; probably as old as the Carboniferous.
B B. Robb's Mountain; Triassic of Colfax and of Plumas County.
*b.* Sand Mountain.
C. Pilot Hill; Triassic of Coloma; copper-veins.
D D. Carboniferous; limestones of Plumas County and Pilot Hill; limestones of Butte County.
E. Cretaceous, marine; Folsom.
F. Tertiary, accompanied by coal-deposits; partially marine; partially like H.
G. Post-Tertiary clays, adobes, loams, and gravel.
H. Peat of the Tule formation, percolated by fresh or brackish water; lacustrine deposits, like G, forming the basin of Sacramento River.
J. Granite, underlying all.
I K. Volcanic craters and overflows at the summit, damming the northern end of Lake Tahoe, and filling up or capping all the gravel-filled cañons of the Tertiary period; being of the same age as the line of demarkation, F G.
L. Jurassic of Spanish Flat, Kelsey, and Spanish Dry Diggings range; position of the Georgetown seam-belts.
M. Gneiss of Tell's Mountain and Brockliss's Bridge; probably older than the Carboniferous rocks.
N. Grey Eagle Hill; horizontal slates flanking the granite and gneiss country on Placerville divide, adjacent to Georgetown divide.
P. Goat Mountain and Greenwood trend.
I. Mount Pluto; volcanic peak at the North end of Lake Tahoe.
K. Twin Peaks, head of Blackwood Valley, Lake Tahoe.

From the above section and from the remarks made under *Historical geology*, it will appear that their position is most probably in the newer rocks, or at the top of the series. The oldest rocks lie near the base, and near the summit of the Sierra; those inclosing the principal gold-veins are near mid-slope.

Neither on the Yuba nor in the basin of the American is there found anything like a continuous limestone formation. Limestone occurs in small lenticular masses of white and crystalline rock, conformably to the dip and strike of the slates. A large cave exists in a body of this character at Alabaster Cave, near Pilot Hill. On the road from Auburn to Georgetown, limestone has been quarried and manufactured into lime extensively for many years. Dolomite also occurs in Amador County, in narrow, snow-white veins, traversing talcose and chloritic rocks, and bearing coarse, free gold.

A zone of limestone-country, several miles wide and many miles in length, occurs further south, in Tuolumne and Stanislaus River Basins; and bedded limestone is again found in Butte and Shasta Counties, where it is fossiliferous and of Carboniferous age.

In the limestone-region of Tuolumne and Calaveras Counties this rock is conformable to and parallel with the slates. Like the slates, it is generally vertical in its planes of structure. These are planes of bedding. At Abbey's Ferry, limestone is found in connection with mica slate and granite. Though to some extent metamorphic, it is more regularly stratified than the limestones of the Tejon, where the metamorphic action was more intense. That of Tuolumne has many blue layers and veins, all trending with the strike of the slates.—(Blake's *Reconnaissance*.)

The limestone-outcrops near the Coloma and Auburn lime-kilns on Georgetown road, at Yankee Jim's, Colfax, and on Wolf Creek below Grass Valley; also near Black's Bridge, six or seven miles above Nevada City, and at Emory's Crossing, on the Middle Yuba, continuing thence northward in spots for thirty miles, and crossing Feather River below Strawberry Valley on the Marysville and La Porte road, (being claimed as a quarry by Butts and Diamond, near Camptonville, &c.,) are examples of geographical distribution.

These deposits are obviously not all of one continuous formation, nor the result of any reasonably imaginable dislocation. The most that can be said of them is that there is a zone of interrupted or insular limestone intercalations continuing toward the north, in a line of island reefs parallel to the northerly trend of the course of the Sierra near the head of the American and Yuba Basins. There is no reason, indeed, why local masses of limestone should not have been forming throughout all of the three periods represented by the auriferous slates.

There is near Ringgold a disconnected mass of limestone about one mile square, the beds striking east of north and dipping east 50°, which is considered by Whitney to be in the same strike of the slates as that at Indian Diggings, eighteen miles southeast of Ringgold, at Cave Valley, at the lime-kilns on Wolf Creek, Nevada County, and that of the limestones of Pence's ranch, Butte County, having accordingly "the position that one connected group should hold." But there is no such connected group that can be reconciled in position with the strike of the slates, so remarkable in its regularity. Another belt, running nearly parallel to the last named, "situated about ten miles farther west," crosses the South Fork of the American at Salmon Falls, connecting with a similar deposit at Clarksville, eight miles southeast of Folsom. Unfortunately none of these intermediate belts stratigraphically connect the southern limestone-region of Tuolumne with the Carboniferous limestones of the north.

The granite-belt of the foot-hills lies west of Auburn and west of Coloma, extending to the edge of the valley as far as any rock is visible.

This belt is, geologically, the most important erupted axis of the entire region. As remarked, it is the first rock met with on leaving the plains. It continues north, as represented in the general vein-map, in the strike of the slates, west of the Grass Valley series.*

At Logtown, seven miles southwest of Placerville, several gold-bearing veins occur in granite. Four of these, the Empire, Pocahontas, Excelsior, and El Dorado, strike northwest and southeast, parallel with the mother-lode series.

Grizzly Flat, sixteen miles east-southeast from Placerville, has veins also in granite, containing sulphurets of lead and zinc.

The strike of the slates at Placerville is north 21° west; elsewhere from north 20° to north 30° west. At Sarahsville, on Forest Hill divide, the strike is north 5° to 10° west.

---

* Compare with north and northwest axes, in the diagram on a previous page.

The strike of the slates further north than Georgetown, as at Grass Valley, becomes more nearly north and south.

*Lithology and distribution of rocks in detail.*—I collected for notice under this head about five hundred specimens of rock and vein-material.

The former are important, as showing the details of a geological section across the entire width of the auriferous slates of the Sierra Nevada. They are all located by approximate measurements connected with fixed points upon the Georgetown map.

The latter represent, beside the vein-material, the peculiarities of the country-rock in which different mines, at nearly all points on the divide, are situated. They consist of ores, minerals, gangues, and adjacent country-rock associated with veins and seams, accompanying the sections and descriptions of mines.*

With the exception of the underlying granites and the occasionally overlying erupted trachytes, all the rocks on the divide fall under the heads of fragmentary and metamorphic.

Clay slates, talcose slates, and sometimes mica slates, alternate in bands, running with the general strike of the slates.

There is a low degree of metamorphism near the center of the slope. The slates are generally light-colored, and thinly laminated. For many miles, where the color is light, the slates have apparently changed but very little from the original sediment. Except in the matter of consolidation, the same holds true over half the country.

The more highly metamorphosed portions are adjacent to the vein or seam systems, or to ridges of greenstone. The latter is itself metamorphic, and sometimes crystallized out into definite diorite; sometimes crypto-crystalline, as aphanite.

These "dikes" are considered by many geologists and others as intrusive, consequently as belonging to a later period of eruption through fissures. I have not hesitated in expressing my opinion that they are metamorphic, as I have seen repeated evidences of their being formed in much the same manner, and by the same causes, as the veins themselves, and the gouge associated with the veins. The traps and diorites are found in a thousand places in all stages of transition, from indisputable slate to indisputable trap or diorite which cannot be distinguished lithologically from the "massive" or "eruptive" rock. The specimens I have collected will themselves testify to this fact.

The "dikes" occur in parallel lenses from 6 to 10 and 20 feet wide. They are repeated irregularly, sometimes every few hundred yards; generally associated with ridges.

Both the slates and greenstones are traversed by quartz-veins which are full of crystallized iron pyrites, associated with gold. In some metamorphic zones the country rock changes into serpentine, or serpentinoid rock, or into "soapstone," a rock of soapy feel, more or less approaching the mineral steatite.

The slates of certain zones, especially where there is a basic metamorphism, decompose very easily at the surface, to a depth of 20 and even 150 feet, so that they can be readily excavated with a pick, or with the hydraulic pipe.

The character of the slates is in all the gradations from roofing-slate to sandstone and conglomerate. The latter is sometimes so highly metamorphosed as to be barely recognizable with certainty. Masses of sandstone occur in various stages of metamorphism, patches retaining almost entirely their original condition.

---

*They have been donated to the State by direction of the president of the California Water Company, and are on exhibition in the mining and geological collection connected with the State library at Sacramento.

On Irish Creek the slates are like roofing-slate. At Kelsey's and at Georgetown they are more talcose and magnesian. Generally around Georgetown they are very slightly metamorphosed, and either light-colored or black and compact. West of Greenwood and east of Forney's there is a great deal of metamorphism. At Sarahsville, on Forest Hill divide, the slates are light-colored and talcose. In the neighborhood of Auburn and northward, there are hard clay slates.

Several of the most noteworthy localities of metamorphism into serpentine are at Bald Hill, on Georgetown divide, and Brimstone Plains, on the road from Sarahsville (on Forest Hill divide) to Independence. This serpentine "forms the largest mass of the kind in the State," being lithologically, according to Professor Blake, identical with that found at Fort Point, San Francisco.

Masses of serpentine also occur in the midst of other metamorphic rocks near the junction of the slates with the granite, north of Auburn, and associated with the seam-diggings at numerous points on the divide.

The granite of the foot-hill belt between Auburn and Sacramento is itself traversed by feldspathic or granitic veins. It weathers in large, round blocks, looking like great bowlders scattered over the surface.

On Placerville divide the granite sets in opposite Coloma, cutting off there, as well as on Georgetown divide, the basin of the slates from the valley.

Toward the eastward the granite again sets in between Sportsman's Hall and Brockliss's Bridge, on Placerville divide, and on the east slope of Mount Robb on Georgetown divide.

In the basin of the Rubicon, at the head of Georgetown divide, the granite is remarkable, on account of showing immense cleavage-lines which have been cut into by streams, and followed for a short distance only before leaving them again to seek their nearest course by gravitation to the river. In overlooking this country from a high mountain, it is almost impossible to recognize, or with the eye to follow, the course of the streams in the valleys and mountain-sides, owing to the abundant repetition of partially-eroded cleavage-courses like little Yosemite Valleys.

### MINERAL CONTENTS OF VEINS.

*First principles concerning ore-channels.*—When the truth first dawned upon the placer-miner that there were ancient rivers in the hills richer than those of "'49," he was not slow to discover or take advantage of the first principles of hydraulic mining. There was surely a channel; that channel had a definite (though winding) course; it had a deep gutter; and it had a rim, or rim-rock, on either side.

The man who now expects to pursue profitably the business of hydraulic mining, without regarding these first principles, would be deemed a strange phenomenon if he did not sooner or later outlive his luck.

We have found in these slates great gold-deposits such as the world never knew before California "came out." We have formed an idea of the outlines, the changes, and of the position (in time) of this formation which has yielded the world $1,000,000,000 in twenty-three years; and we have discovered that the source of all the gold is in certain subterranean *ore-channels* as well defined in many respects as are those of the ancient rivers.

It may appear some day, when we know a little more about it, that the man who would expect to pursue profitably the business of vein-mining, without regarding or understanding the nature of these channels, should also sooner or later expect to outlive his luck.

*Distribution of gold.*—Of the fifty or sixty elementary substances to which chemistry reduces matter, we find gold to be one of the heaviest, scarcest, and most independent of alliance or tendency to mineral intermarriage. Yet, like silver, the next following as a precious metal—and as truly as iron, clay, and quartz—it is universally distributed in the material of the earth's crust.

Silver is found in solution in common sea-water, the world's envelope. Gold, though scarcer, we now begin to realize is found in all countries. It has, furthermore, been found crystallized in veins or precipitations belonging to all geological ages, including even the Post-Tertiary, in California.

It could not be otherwise, for gold and silver are both soluble, along with quartz, in the natural waters of the earth. They are extracted from their ores in California and Nevada largely "in the wet way."

In the following tables are presented, as far as known, the instrumentalities that undoubtedly effect the transformation of gold, &c., underground, from a solid to a fluid state, and *vice versa*, being a list of the reagents producing the solution and precipitation of gold, iron, and quartz in nature.*

*How gold, iron, and silica are made soluble and dissolved in water.*

[Words in roman refer to laboratory processes; those in italic to processes occurring in nature.]

| Reagents which, coming in contact, produce chemical action. | | Result as fluids, or in solution, as— |
|---|---|---|
| As solids. | Solvent. | |
| 1. Iron, metallic | Sulphuric acid and water | Sulphate of iron. |
| 2. Iron, oxide | *Same, from decomposition of pyrites* | Do. |
| 3. Iron, sulphuret | *Alkaline, carbonate, and sulphate waters** | Do. |
| 4. Iron, sulphate | *Water* | Do. |
| 1. Gold, metallic | Nitrohydrochloric acid (evolving chlorine) with water. | Chloride (sesqui) of gold. |
| 2. Gold, metallic | Chlorine gas, (from salt and sulphuric acid.) | Chloride (terchloride) of gold. |
| Gold, metallic | *In nature, same, arising from the decomposition of pyrites and the never-wanting chloride of sodium.* | Do. |
| 3. Gold, metallic | *Sulphate or sulphide of iron, in some way.*† | Disulphuret or sesquisulphuret (?) surrendering sulphur to iron on depositing. |
| 4. Gold, metallic | *Persulphate of iron* ‡ | Do. |
| 5. Gold, chloride, (sesqui) | *Water* | Chloride, (sesqui.) |
| 1. Silica | *Alkaline (or basic) waters* | Alkaline (or basic) waters. |

* Sterry Hunt.
† Hypothesis of John Arthur Phillips.
‡ Similar idea to that of Phillips, definitely confirmed by laboratory process of Wurz.

* See my Report rendered in 1870, Chapter LXI, p. 450, from which I quote the following:

"Professor Bischoff, the eminent chemist and geologist, has found that sulphide of gold is slightly soluble in pure water. It is now also known that chloride of gold will co-exist in very dilute solutions with protosalts of iron, provided there are present an alkaline carbonate and a large excess of carbonic acid. If the sulphide of gold is required in solution, it is only necessary to charge the solution with an excess of sulphureted hydrogen. In the same connection should be mentioned the discovery, from a different quarter, that metallic gold is soluble in solutions of the persalts of iron."

In the same connection I cited the paper read in 1868 by Mr. Cosmo Newberry before the Royal Society of Victoria (Australia) on the formation of gold nuggets in the auriferous drifts, and remarked:

"The arguments and facts contained in that paper strongly support the theory of the growth of nuggets, and furnish a simple analogy in the precipitation of metallic gold from solution by the reducing action of organic matter, in the presence of a nucleus. * * * To some such process of solution and deposition we may ascribe the presence of gold in rocks and veins; its alteration and reduction to metallic form ('free gold') in the upper or exposed parts of such rocks or veins; and, finally, the still greater purity and size of its particles, so frequently noticed in placers."—R. W. R.

***How gold, iron, and silica are precipitated when they have been dissolved.***

[Words in roman refer to laboratory processes; those in italic to processes occurring in nature.]

| The reagents which, coming in contact, produce chemical action. | | Result. |
|---|---|---|
| As fluids, (in solution.) | Precipitant. | As solids. |
| 1. *Iron, sulphate, in water* .. | *Organic matter** ............... | *Pyrites.* |
| 2. *Iron, sulphate and silicate solutions.* | *Organic matter** ............... | *Anhydrous oxide of iron.*‡ |
| 1. *Gold, chloride,* (sesqui) .. | *Proto-salts* (sulphates) *of iron, and heat.* | *Metallic gold.* |
| 2. *Gold, chloride, (terchloride.)* | *Sulphate of iron* ............... | *Metallic gold, a black or brown powder, being the chlorination process.* |
| 3. *Gold, chloride, (sesqui)* .. | *Hydrosulphuric acid, with heat.* | *Disulphuret of gold; a black powder.* |
| 4. *Same* .................. | *Hydrosulphuric acid, without heat.* | *Sesquisulphuret of gold; a dark-brown powder.* |
| 5. *Same* .................. | *Same* ......................... | *A black pulverulent chloride (?), which when heated evolves fumes of hydrosulphuric acid, leaving metallic gold.* |
| 6. *Same* .................. | Hydrosulphate of ammonia... | Same. |
| 7. *Same* .................. | Protochloride of tin.......... | "Purple powder of Cassius," (oxide ?) |
| 8. *Same (terchloride) boiling hot.* | *Sulphydric acid* ............... | *Brown sulphuret of gold, AuS.* |
| 9. *Same, cold and dilute* .. | *Same* ......................... | *Black persulphuret, $AuS_2$.* |
| 10. Disulphide (?) associated with sulphate of iron.† | *Organic matter** ............... | Metallic. |
| 1. Silica, dissolved in alkaline or basic waters. | Acid waters carrying sulphates of iron, &c. | Silica. |

* Sterry Hunt.
† Hypothesis of John Arthur Phillips.
‡ As at Steamboat Springs, according to J. A. Phillips.

*Processes of chemical concentration.*—Going back to the dark ages of geology, and remembering the high specific gravities and first affinities of these metals, we must first conceive all the gold there was within some vertical miles of the surface as pretty evenly disseminated through the semi-molten azoic mud.

Along with the fluvial replacements of material attending the everlasting rising and sinking of lands, chemical affinity or solution has kept this gold and silver moving ever since, wherever water moves, precisely like all the rest of the fifty or sixty elements of matter, and conforming faithfully now and forever to the laws of physics and chemistry that govern all matter.

This is a *résumé* of the best judgment on the subject of noted mining geologists in this and older mining countries, where the geology of silver and gold has been studied for hundreds of years.

Prof. Sterry Hunt, of Montreal, has published an outline of the processes by which the siliceous, calcareous, and argillaceous rocks, that form so large a part of the earth's crust, may have been generated from a primitive fused mass, and therewith has indicated the origin of the salts of the ocean. The first precipitates from the ocean would, according to Hunt, have contained most of the metals. In the subsequent resolution and deposition of these precipitates is to be found an explanation of the origin of metalliferous deposits, and of their distribution in various formations, either as integral parts of the strata or as deposits in veins, the former channels of mineral springs.

"The metals of the Quebec group," he conceives, were originally brought to the surface in watery solution, from which they were "separated by the reducing agency of *organic matter* in the form of *sulphurets*, or in the native state, and mingled with the contemporaneous sediments. During the subsequent metamorphism of the strata, these metallic materials were taken into solution by alkaline carbonates or sulphurets and redeposited in fissures."

*Mechanical concomitants of the process of concentration.*—To what depth the original mechanical concentrations with water in the soft crust of the globe extended is a matter of slight practical importance. By the same laws of physics that now exist, concentrations undoubtedly took place in the Paleozoic slates, that were washed into basins and in many places sorted into layers like the Mansfield Prussian copper-schists.

While this action took place at the surface, precisely as now, under the surface chemical affinity and chemical action must have set in and operated from the earliest dawn of creation precisely as now. And the results of chemical affinity being quite the same, whether in the wet way, or under pressure, or by fire, the methods of the chemical concentration of gold were also in general the same.

In the usual and very natural method of drying, wrinkling, and surface oscillation that has caused the principal mountains and valleys of the globe, these slate-muds of the Sierra might have been sinking for a long while, and *piling up* thicker and thicker, at the same time that the axis of the Sierra Nevada was rising. This we can infer from the fact that the valley of Alta California has itself undergone such a process, being a valley of depression, corresponding to the elevation to the eastward.

The coal-mines of Mount Diablo and Corral Hollow will easily convince any observer of the latter. The coal-veins deposited on the top of the Cretaceous hills are seen to pitch under San Joaquin Valley, where they have been explored, and also worked.

The Jurassic and Triassic muds were no sooner deposited, probably, than they were disturbed. The Sierra Nevada certainly began to rise before the Cretaceous rocks on their flanks (near Shasta, Folsom, &c.) were deposited, since we see the latter lying nearly horizontal on the upturned edges of the slates. Veins are the infiltrated cracks. What stronger evidence is needed, then, of the history of the gold-bearing veins of the Sierra than that of their general parallelism to the axis of uplift and depression?

As soon as these breaks or cracks commenced forming, quartz, iron, and other minerals commenced precipitating in them. As soon as the proper mechanical and chemical conditions were supplied, from that time forward, and so long as the same laws of chemistry and physics remained in force, the same process must have continued, and must ever continue.

The present day not being exceptional in that respect, there is no reason why vein-concentrations should not be going on now.

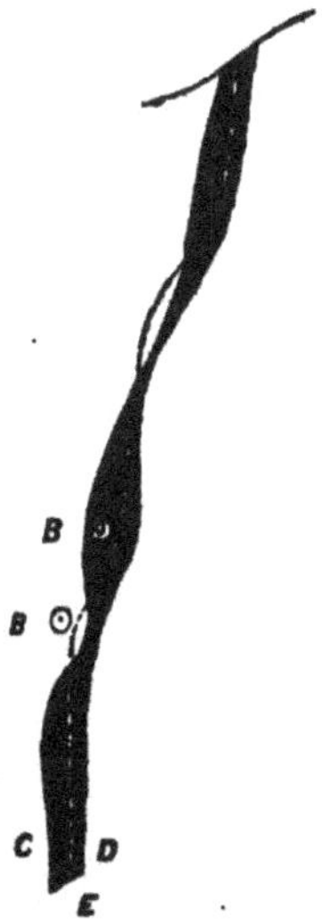

The reason why we find lenticular "ore-bodies" or chimneys that pinch out in most of the great fissure-veins of the tabular exhibit, is made plain by the accompanying diagram. The dotted line E represents the original fissure. A dislocation of the country-rock takes place, whereby the angle at the upper B in the foot-wall slides down, and is now at B in the hanging-wall. Of course, the hanging-wall and the foot-wall—originally identical—after this cease to be parallel. The resulting lense or chimney-shaped spaces were the recipients of mineral-bearing waters. Gouge was added to quartz, by continued rubbing or dislocation of the walls.

Dislocations causing lens and chimney-shaped spaces.

*The mineralizers of gold: silica, iron, alkalies, and sulphur, in argillaceous sediments.*—As the gold in quartz-veins is attributed to chemical concentration in the wet way, we may reasonably regard the argillaceous-slate formation as the matrix of the gold-bearing veins. For if the gold-solutions were derived from the underlying granite, or from traps more modernly erupted through the granite, we should find as many gold-bearing veins in the granite or granite associated with trap independently of the slates as in the slates themselves. Gold-bearing veins are found, it is true, in granite, on Placerville divide at Logtown, as shown in the vein-plottings, and at Meadow Lake, on the head-waters of the South Yuba. But the fact remains that the great paying veins of California, Australia, and the Ural are in slate.

As already remarked, gold is pretty universally distributed the world over, but not in a concentrated state. It is most concentrated where the chemical conditions, accompanied by certain mechanical conditions, were most favorable for concentration and precipitation. Precisely as in the ancient river placers, it was most concentrated where the mechanical conditions alone were most favorable for such concentration or separation from the accompanying rock.

We have here what was originally light mud, impregnated throughout with iron, gold, and silver; besides the alumina, magnesia, silica, and lime, which formed its principal constituents. Even where vein-formation did not occur after the consolidation of the sediment, iron pyrites have very generally crystallized out in the slates. Such is also the fact in a more limited degree in the granites on the western slope of the Sierra. Quartz, iron, gold, and magnesia especially, being easily soluble under the conditions to which they were subjected, formed the concentrations seen in the vein-material and gouges associated with the metamorphosed seam-belts, (the several hundred specimens of which, collected by the writer, will speak for themselves.)

The gold is found precipitated as metallic gold, free from combination with any of the above elements. Even when entirely inclosed in pyrites, it is granular upon disintegration of the latter. It is sometimes visible in undecomposed pyrites, with and without the assistance of a magnifying-glass. (See instances in the tabular exhibit, Tuolumne County.)

Fourcroy's *General System of Chemical Knowledge*, published in 1804, has the following passage: "Bergmann observes that the gold which is extracted from auriferous pyrites by digestion in nitric acid is in small angular grains, which proves that this metal existed in the state of simple mixture, and not of composition in the pyrites."

Leading mineralogists of the present day entertain a similar view. In arriving at nature's method of precipitating the gold, we may set it down as conclusive, then, that the gold precipitates first. But the precipitation of iron-sulphuret is, perhaps, almost simultaneous. It appears to be the conclusion or concomitant of the same chemical reaction that precipitates the gold.

*Philosophy of action—"aggregate results."**—Voltaic electricity may be called the soul of the earth; and it is, like all other things in nature, *dual*, or positive and negative in its manifestations.

The electro-positive and electro-negative principle is not only at the

* I venture to repeat here the remark at the beginning of this chapter, that Mr. Bowman is responsible for the theories it contains, among which, with a vast amount that is both brilliant and useful, I find some things which strike me as over-bold and premature. The attempted analogy with the organic world in the following table is, of these, the most fanciful.—R. W. R.

foundation of all chemical action, but it suggests to us the *modus operandi* of the vein-chemistry of the Sierra in a manner interesting to those engaged in following the deposits of gold underground.

The earth, we find, in the order and plan of nature, presents itself to us in three grand aggregates, the "elements" of Aristotle, viz, *land*, *sea*, and *air*, to which should be added, perhaps, a fourth, the *organic world*. Each of these four great natural aggregates of the chemical elements is found to possess a feature chemically peculiar to it, viz, land has *silica* as constituting 50 per cent. of all the rocks, or 60 per cent., if the limestones be excepted; water has *hydrogen;* the air has *oxygen*, and the organic world has *carbon*. According to their leading ingredients, the earth and air would have to be considered electro-negative, or as most allied to the acid class, (in the following table;) the sea and the organic world, as electro-positive, or as most allied to the basic class.

*The dual principle of electro-magnetism or chemical affinity as operating in fissure-fillings characteristic of the mother lode.*

| The two classes of elements acting upon each other. | Test to recognize. | Function in chemical geology. | If decomposed by voltaic electricity, they— | | Corresponding— | | | |
|---|---|---|---|---|---|---|---|---|
| | | | | | In aggregates of nature to— | | In broader terms. | |
| *Basic*, or alkaline class of elements and compounds in solution: Alumina, potash, soda, lime, magnesia, iron, and the metals generally. | Change red litmus paper to blue. | Dissolve or change silica, &c. .......... | Go to the negative pole. | Called electro-positive by electricians. | The sea . | Fluid . | The active principle—recreative in effect. | Masculine function in the organic world. |
| *Acid* class of elements or compounds in solution: Silica, &c. | Redden blue litmus paper. | Solidify or perpetuate in the form of silica chiefly, being the agency through which new mineral forms are constantly crystallized out, whenever the electro-positive or basic elements are brought in contact by circulating waters. | Go to the positive pole. | Called electro-negative. | The land. | Solid . . | The passive or conservative principle. | Feminine function in the organic world. |

*The results of chemical action, as witnessed in connection with the mother-lode system of veins.*—Those are apparently only "aggregate results" which we encounter in the seam-diggings, or companion talcose veins alongside of the solid quartz-veins of the mother-lode system.

The one set of veins, the earlier, is acid, the other, basic in its principal constituents. Both contain gold; the former is found in sulphurets, the latter, more frequently in the form of free gold, associated with metallic oxides, carbonates, and hydrous magnesian minerals, silicates of the bases mentioned in the preceding table.

What is the history, then, of the two different solutions and precipitations of gold?

Simply that the acid and basic conditions alternated; that while the one endured, both iron and gold were in a fluid condition, and when the other intervened, they were precipitated.

They so remained, notwithstanding the fact that the accompanying quartz and sulphurets of the companion talcose veins and related seam-diggings were subsequently decomposed, partially dissolved out, removed, and replaced by the hydrous magnesian minerals, the silicates of the bases which accomplished the metamorphism.

This order of events did not in any manner interfere with the subsequent infiltration of siliceous waters, and the formation of other or barren seams that are often found in the same neighborhood as the decomposed quartz-seams, and kidneys in which the gold is found in sheets and pockets.

Basic or alkaline waters, in brief, changed, or carried quartz.

Acid waters carried gold, dissolved in sulphate of iron, or in the form of chlorides.

Where the two met, and the acid or electro-negative solutions were strongest, they solidified in the manner, the effect of which is designed to be illustrated by the table, both the quartz and the gold, with iron in the form of sulphurets.

Where, on the other hand, the basic solutions were strongest, there resulted decomposition and recreation generally, solution and change of quartz and the transformation of everything into hydrous silicates of magnesia, accompanied by transformation of the sulphurets of iron and copper into carbonates and soluble sulphates, which were removed by water, leaving only the gold and oxides of iron.

*Mineralogical contents of veins and parageneses.*—The mineralogical contents of veins are attributable to particular chemical combinations, which can be traced. In mineralogy this is called *paragenesis.** It corresponds to the intermarriages of individuals, while the vein-systems are tribes, where king and queen rule alternately; and the grand aggregates of nature are the four nations in which silica, hydrogen, oxygen, and carbon are the absolute monarchs, as already explained.

In many cases, the solutions in circulation could form quartz, &c., only to one side of the ore-channel, on account of the country-rock having been previously impregnated, perhaps, with different precipitating ingredients.

Further details showing the peculiar contents of veins and seams on Georgetown divide, and in their related vein-systems, are given in the tabular exhibit.

*Age of gold-bearing formations, predicated upon the origin of gold.*—We find veins containing gold and silver on the eastern slope of the Sierra Nevada in comparatively modern volcanic rocks.

---

* Very interesting treatises on the subject have been written by Breithaupt and others.

We have found gold not only in the "dark age," Silurian slates of the Ural, and of Australia, as promulgated by Murchison, but in the "middle age," Jurassic and Triassic slates of California, and now even in the placer-pyrites of the Pliocene rivers of the Pacific slope.

The latter occurrence has been repeatedly reported in the pyrites that crystallize in carbonized woods found in placer-mining. We are assured that after a sufficient degree of care has been taken to exclude all possible mechanical admixture, some of the placer-pyrites in Nevada and Sierra Counties, separated by specific gravity in water, are still rich in gold. J. Arthur Phillips indorses this view, upon the strength of facts observed by him while in California.

It is hardly necessary, then, to enter into the question so tenaciously argued by Murchison, and some others since 1849, concerning the predominating ancient or modern age of the principal gold-bearing rocks. Geologists deduce a law where they have found a repetition of "the indications." Such indications had been observed by Murchison, which seemed to place the gold-bearing rocks of all the gold-producing countries known uniformly in the form of slates, into the dark ages of geological history.

In the new edition of his *Siluria*, Murchison modified, however, the views first put forth by him as to the distribution of gold in the earth's crust. His more recent conclusions are:

1. That looking to the world at large, the auriferous vein-stones in the Lower Silurian rocks contain the greatest quantity of gold.

2. That, where certain igneous eruptions penetrated the Secondary deposits, the latter have been rendered auriferous for a limited distance only beyond the junction of the two rocks.

3. That the general axiom before insisted upon remains, that all Secondary and Tertiary deposits (except auriferous detritus in the latter) not so especially affected never contain gold.

4. That no unaltered purely aqueous sediment ever contains gold; or, in other words, that the granites and diorites have been the chief gold-producers, and that auriferous-quartz leads in Paleozoic rocks are the result of heat and chemical agency.

The law of gold distribution and concentration is a broader one. It is as broad as that of the distribution and concentration of any other mineral.

Murchison's reasoning concerning the origin of the gold-deposits of California would be that the greenstone dikes, with their associated metamorphism, constitute the immediate cause of the impregnation of gold.

As I have shown, these "dikes" are really attending phenomena. If metamorphic in origin, (as I believe from my own observation,) the causes of such metamorphism were probably no less deep-seated than if they had been erupted bodily.

The result, then, so far as the concentration of gold in veins in association with greenstone dikes is concerned, would be quite the same; even to the conclusion that we must regard the granitoid rocks as the source of the gold, (as of every other mineral ingredient pertaining to the original surface of the earth,) though not in any concentrated form.

Sedimentary concentration, such as Murchison claims peculiarly for the Silurian age, doubtless took place. But it continued with equal effect, as is conclusively shown by the facts presented in this chapter, down into the Jurassic. It is in the auriferous-slate region that the great gold-veins of California occur, while the granites of California, are, as a rule, comparatively barren.

To sedimentary concentrations, deposited in the Carboniferous, Triassic, and Jurassic slates in the western parallels of the Cordilleran axis, we may with some show of reason ascribe the immediate sources of the enrichment of the gold-veins of California.

## SURFACE GEOLOGY.

*Ancient and modern valleys.*—Topographically, as well as geologically, there is a marked difference between the Georgetown and Placerville divides. Standing upon the summit of Grey Eagle Hill or of Robb's Mountain, a view west and southwest shows the former to be irregular in outline, the latter a sloping plain, with only an occasional knob of metamorphic slate rising above the general level.

Geologically the former is uncovered slates, the "bed-rock" formation of the Sierra Nevada; the latter a continuous gravel and lava deposit, covering a hundred square miles or more, beginning at the summit, and at its western end blending with the flat plain of the undenuded slates of Placerville divide, above Mount Thompson.

While Georgetown divide was a *divide* of the Pliocene period, Placerville divide was a *valley* of the same period. In it were heaped up the gravels of the Pliocene river-filling period, and upon the top of this poured the volcanic outflows of the eruptive period, which closed the Pliocene, and marked the revolution which brought about the new order of things—the present topographical frame-work of existing watershed, valley, and stream.

From the summits of Georgetown divide, looking toward the south, the Placerville gravel-ranges fall prominently into view, and the gravel-pits, exposing the bed-rock of the ancient channel in a straight line, which jumps from hill to hill and from range to range, as if a ruled line had been drawn there in profile, present the most striking and interesting objects of landscape to the eye.

*Molten volcanic matter.*—On visiting Placerville itself, and ascending the high gravel-range referred to, which trends westward just south of the town, you will find all of the characteristics of the ancient river of the Yuba Basin—an immense deposit of quartz, gravel, and rock, covered by a mass of volcanic rock.

So heavy and almost solidly continuous is the trachytic breccia in the vicinity of the reservoir—so entirely free of broken and rounded corners belonging to a transported breccia—that this difference at once appears. The lava of Placerville must have reached its position to a degree in a molten state, part of it entirely unmixed with river-water. This character of the lava-flows of the Sierra is repeated and intensified toward the south in Calaveras and Stanislaus Counties; while toward the north, on the Yuba, the volcanic matter is generally (not always) in the form of volcanic washed bowlders, in an ashy cement.

The volcanic matter of the divide was derived from the summit. In localities in the vicinity of Tell's Mountain the remnants of genuine molten trachytic lava-flows (or dikes?) exist, which were scraped over by the glaciers of the glacial period; but it is almost certain that the rock is not in place.

*Comparison of ancient and modern drainage.*—To the north of Georgetown divide was another valley, now marked by the gravels of Forest Hill and Iowa Hill divide.

A comparison of the drainage of the Pliocene with that of the Post-Pliocene and recent periods can be made from the following diagram, where the ancient stream is marked by a dotted line and the present one by a continuous line. The dark shading represents *cañon;* the light, *plateau.*

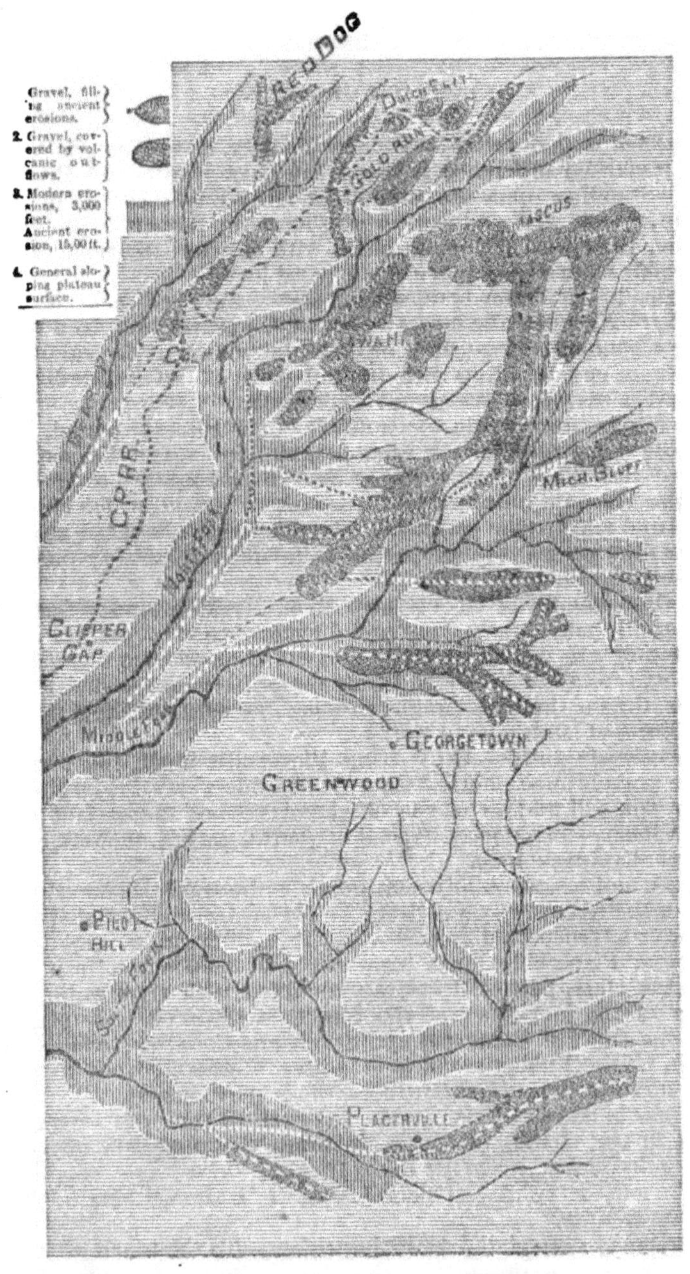


Diagram of the Pliocene North, Middle, and South Forks of the American River, showing the relations to and from the present streams; scale, 6 miles to 1 inch.

The Middle Fork of the American, it will be seen, ran near its present route, but in the main a little further north; the South Fork was located further south than the present South Fork of the American.

The ancient side cañons are seen to exist in a manner corresponding with the cañons of Pilot Creek, Otter Creek, Cañon Creek, &c.

*Constituents of superficial deposits.*—The material of the gravel-deposits consists of:

1. Gravel from the metamorphic slates, chiefly diorites and siliceous schists.

2. Washed bowlders of volcanic rock; trachytic porphyry, or basalt, some of which is known as black lava. In lava-capped hills of the mining-region the colors are generally owing to the various degrees of oxidation of the protoxides and magnetic oxides of iron contained in these lavas. In other cases the volcanic rock is from an ashy or leaden color to iron-gray. The trachytic porphyry, found in the form of volcanic bowlders, is often considerably lighter, and of a faint reddish tint.

3. Ancient-river sandstones, consisting:

(*a*) Of sandstones of granitic origin. At Yankee Jim's, on the authority of Professor Blake, several layers of placer-deposit consist of fine granitic sand. On the authority also of Doctor Willey and Henry G. Hanks, there is in the deep placer-mines at Michigan Bluffs a seam-deposit formed of the component parts of granite. This is stained yellow and red with oxide of iron. In portions of the placer-beds the sand is colored black by infiltrated oxide of manganese, looking at a little distance like black sand or lignite.

(*b*) Sandstone of volcanic origin, known as white lava. The "white lava" is a fine, gritty consolidation of volcanic sand and ash. It makes a tolerable building-stone, being easily worked. It is used for this purpose at Diamond and Shingle Springs; is whiter and less compact than the porphyritic rock quarried at Green Valley, near Bridgeport, Solano County, which is said to harden on weathering. Some of the best buildings at Mokelumne Hill, Calaveras County, are likewise constructed of volcanic material. This rock is frequently found subsequently metamorphosed into complicated forms, showing crystallizations of sanidin and zeolites, so as to resemble the more compact varieties of trachyte.

4. The soil of the country, consisting of decomposed slates to a depth of 5 or 10 feet. It is from cherry-red to brown, yellow, or nearly white, depending upon the amount of iron in different localities.

*Explanation of figures on the map.*

2 Potts, (gold,) N. 43° W.
4 Oaks, Reese & Jones, S. 82° W.
5 Kelsey, S. 12 30° E. and S. 26° E.
7 Arbena, N. 30½° W.
8 Gray Eagle, N. 29½° W.
9 Schofield, N. 56° W.
13 Kate Kearney, N. 59½° W.
14 Kelly, S. 57½° W.
15 Northern Light, S. 29½° E.
21 Salathiel, S. 26° E.
22 Norridgewock, S. 12¼° E.
23 Rising Sun, N. 74° E.
30 Spring Valley, S. 13° E.
31 Venus, N. 12° E.
32 Stanton & Allison, N. 9° E.
34 Auroral Star, N. 3° E.
35 Galena, N. 41½° W.
36 Sebastopol, N. 43½° W.
39 Epperson, E. 8° W.
44 Rough and Ready, S. 10° E.
45 Eureka, S. 40° W. and S. 16° W.
48 Penon Blanco, S. 37½° E.
49 Greenwood, S. 38° E.
50 Spring Hill, N. 77° E. and N. 57° E.
52 Oneida, S. 28½° E. and S. 31° E.
54 Slate Ledge, S. 88° E.
62 Keystone, N. 32° W.
65 Original Amador, N. 3° W.
66 Spring Hill and Geneva, N. 8° W.
73 Dry Company, N. 2° W.
74 Stanislaus, S. 10° E.
80 Hancock & Tibbetts, N. 2½° W.
84 Enterprise, N. 67° 30′ W.
87 Newtown (copper,) N. 27° W.
88 S. Bright, N. 43¾° W.
89 Saint Lawrence, N. 15° W.
90 Cederberg, N. 15½° W.
91 Union, N. 82½° E.
97 Norambagua, N. E.
98 New York Hill, N. 55° W.
112 Nisbet, N. 6° W.
114 Rocky Bend, N. 38° W.
115 Lone Jack, N. 8° E.
116 Moorehouse, N. 18½° E.
120 Sliger, N.
124 Banghart, N. 60° W. and N. 87½° W.
126 Gover, N. 14½½° W.
128 Calveras, N. 48½° W.
130 Tecumseh (silver and copper,) N. 28° W
131 Bobby Burns, N. 52° W.
132 Tyson, (iron and copper,) N. 75¾° W.
134 Everlasting, N. 30¼° W. and N. 11¾° W.
136 Wisconsin, S. 20° E.
138 Independence, S. 20° E.
165 Lucan, N. 16½° E.
172 Union, (copper,) N. 45° W.
173 Poorman, N. 6° E.
175 Stickle, N. 51¾° W.
178 Finnegan, S. 72° E.
182 Carson, N. 8¾° W.
184 Wolverine, N. 4½° W.
189 Saint John, N. 34° W. and N. 61° W.
191 San Bruna, N. 88¼° E.
193 Eclipse, N. 67° E.
194 Eureka, N. 26¼° W.
195 Confidence, N. 14° W.
198 Yellow Jacket, N. 2° W.
199 Keystone, N. 54½° E.
201 Copp, N. 27½° W.
202 Secret Cañon, N. 21° E.
203 North Confidence, N. 8½° W.
204 Butcher Boy, N. 69° W.
208 C. Baker, N. 69° W.
209 Chaleur, N. 51° E.
210 Waters, N. 64½° W.
211 Shores, N. 9° W.
212 El Dorado, N. 27° E.
214 Coyote Hill, N. 19° W.
215 Salsig, N. 39° W.
217 Greenwood, N. 36¼° E.
219 Green Walter, N. 57½° W.
226 Wet Gulch, N. 82° W.
228 Maryland, N. 24° W.
233 Cosumnes (copper,) N. 24° W., N. 68° W., and N. 4½° E.
238 Ophir, N. 86° W., N. 60° W., and N. 76° W.
243 Last Chance (copper,) N. 23¾° W.
246 Plymouth Rock, N.
249 Bovee, N. 48° W.
251 Crœsus, N. 56¾° W.
260 Marietta, (gold and silver,) N. 26° W.
277 Fort Yuma, N. 7½° E.
289 Empire, N. 18¼° W.
292 Sulphuret, N. 15° W.
294 Mahoney, N. 34½° W.
298 Hancock & Watson, N. 88½° W.
303 Mammoth, N. 69½° W.
305 R. R. Hill, N.
315 Old Pioneer, N. 47½° E.
319 Uncle Sam, N. 9½° E.
320 Doctor Hill, N. 38° E.

NING

ous | Gold, plat
and | sulphure
by | dalusite,
| actinolit

ear of them any
resinous luster.

LOPE

CAL FEATURE

TER.

phous mass. Su
nd antimony in
ilver. Vitreous
ruby-silver, horn
changing to car
hs at Gold Hill.

lver, and sulphu

Miargyrite, da
rets, and silver-g
ent., containing
ites, and pink c

ray sulphide of
nd red oxides of

bonates, and gra
masses, 3 to 6 fe
lesite, galena, an

HER L

FO

In chimneys
pay often ri

# CHAPTER XIX.

## THE HISTORY OF THE RELATIVE VALUES OF GOLD AND SILVER.

This chapter constitutes a portion of an address prepared by me for the New Haven meeting (February, 1875) of the American Institute of Mining Engineers. As I attempted, in other parts of the same address, to show, the present position of the mining and metallurgical industries of this country offers in several respects most important indications of radical change. This is an epoch for more than one branch of these industries. We are commencing again to export copper; we have shifted the main production of lead from the Mississippi Valley to the far West, besides developing a new production of that metal in Missouri; we have seen the price of quicksilver go up and the price of iron go down to such an extent as to affect profoundly, on the one hand, the great business of extracting silver by amalgamation through the Washoe process in this country and the patio process in Mexico, and, on the other hand, the various manufactures employing iron as a raw material, above all, the manufacture of steel. But the epoch which has occurred in the relative value of gold and silver, in consequence of causes which I shall attempt to explain, and among which the extraordinary increase in the silver-product of Nevada is not the least, is perhaps the most striking feature in the review of our situation. Since my official duties in connection with the mining-industry of the Pacific slope and the great interior basin have led me to pay particular attention to this subject, I have thought it not inappropriate to compile a somewhat extended account of its history.

It is a topic of peculiar interest, because of the general use of these two metals as standards of value in the exchange of commodities. Neither of them is suitable to be, for the political economist, a real standard of value. That standard is rather to be sought in some more universal product of labor, such as wheat, the relation of which to the rate of wages is said, on high authority,* to have been substantially unaltered through centuries, one bushel of wheat representing one day's labor. But the most convenient measure of exchange has been furnished by the precious metals, and hence these have been employed as standards of value. Unfortunately, the use of both of them as standards has involved a new element of complexity—the fluctuation in their relative value. It is this element which I wish to trace at present, quite independently of the general relation between money and labor and other commodities. Not the purchasing power of the precious metals measured in terms of wages or supplies, but the purchasing power of each of them measured in terms of the other, is the question to be considered, and the use of these metals as currency gives to the inquiry more than a merely speculative interest, for the attempt to maintain a double standard involves the perpetual re-adjustment of the relation by law, and the alteration of coinage accordingly. Hence the civilized nations have gradually come, with few exceptions, to the adoption of gold as the standard, and the employment of silver as the material for subsidiary or token coinage. The illustration of this principle is scarcely required; yet a simple statement may not be out of place. If the law

* I owe the suggestion to a conversation with Mr. Abram S. Hewitt.

says that I may offer in all payments requiring dollars either gold dollars or silver dollars, as I choose, then both coins are legal tender, and there is a double standard. At the same time, the law must have fixed, for the protection of my creditors, the exact amount of fine gold in the gold dollar and of fine silver in the silver dollar—in other words, it must fix, for the purpose of coinage, the relative value of these metals. Now, so long as this is also the market-value, or nearly so, it is a matter of indifference to me whether I give and take silver or gold. But there may be a change in the market-value. For instance, a great deal of silver may be wanted by manufacturers, or for shipment in commerce to countries, like India, where gold is not available for the purpose; and the parties desiring silver may be willing to pay more than a gold dollar for the silver dollar, in order that they may melt or export the latter. In that case, I would prefer, of course, to sell my silver dollar, pay to my creditor the gold dollar, and keep the surplus. Everybody feeling the same impulse, the silver coinage disappears and is melted or exported. If the silver coins smaller than a dollar are also of the same fineness and relative value, this disappearance of silver causes great inconvenience, because people can no longer "make change," and the natural result is the general use of tokens, dinner-tickets, shin-plasters, &c.—a fractional currency, in short, which is not legal tender at all, and which, when it proceeds from individuals or private corporations, has but a local circulation. Recognizing this evil, all governments provide for a regular token-currency for small units of money, which shall be legal tender for small amounts only, but is redeemable by the mints in larger amounts. There is not a cent's worth of nickel or copper in the cent, or a dime's worth of paper in the ten-cent note of postal currency. The same principle is extended to the smaller silver coins of most nations, and constitutes, in fact, a recognition of the law of political economy that there should be but one standard of value in coinage. What is universally done with regard to small coins should be done for all silver coins. They should be subsidiary or token coins, legal tender only in small amounts, and redeemable in gold at the mint.

But whether the token coinage is confined to the smallest denominations or includes all silver coinage, the question remains, How shall it be maintained? How shall its value be fixed? In this country we have seen how the copper cent disappeared, because the price of copper rose until a hundred copper cents could be sold in the market for more than a dollar. The same fate will inevitably overtake any coinage which is undervalued by law. Hence, to prevent the evils resulting from this source, it is necessary to fix upon the subsidiary coinage a value which shall be permanently in excess of its market-value as metal. Yet, on the other hand, it is, for reasons which I will not fully discuss here, not advisable to debase a silver coinage or reduce its weight so far as to ignore entirely its real value. This may be done with the smallest coins; but it should not be done, for instance, with dollars and half-dollars. A very important reason for this is the increased temptation and facility thus afforded to counterfeiting. The imitation of a die is not a difficult matter; and the security of the public against counterfeit coins is not so much in the design upon the coin, the workmanship of which may be, when genuine, much defaced by wear, as in the easily-ascertainable weight and fineness of the metal, and in the fact that it would not pay individuals to manufacture coins of the same weight and composition. For any coin of higher denomination than five cents, a considerable overvaluation would lead to extensive and dangerous counterfeiting by the use of the genuine metallic alloy, or one nearly approaching to it in fineness.

We have, then, our two limits for the legal valuation of coins; and the problem is to get as near as possible to the relative market-value of gold and silver, and yet always keep on the safe side; which is, as we have seen, the side of the overvaluation of the silver or subsidiary coinage. These considerations render the history of the relative values of gold and silver an important study.

I am indebted, in the investigation of this subject, to various financial and economical authorities, including the parliamentary blue books and reports, from the days of Sir Isaac Newton, and congressional documents, from the famous report of John Quincy Adams down to our own day. Mr. E. B. Elliott, the well-known statistician of the Treasury Department, has kindly furnished valuable suggestions and information. But the most recent, compendious, and satisfactory discussion of the special branch of the question to which I shall now devote myself is an article by Professor Soetbeer, of the University of Göttingen, one of the leading European writers on banking, currency, and exchange. This distinguished economist has examined with great care the accessible sources of information, and illuminated by his acute criticism many sources of error. Large portions of the present discussion will be little more than translations, compilations, and re-arrangements of what he has said; but I may indulge in the expression of views, arguments, or prophecies for which neither of the authorities upon whom I have drawn could fairly be held responsible.

With these preliminary remarks I present several tables, and remarks upon them, illustrating the history of the relative value of gold and silver.

TABLE I.—*Ancient period.*

| Date. | Ratio. | Authority. |
|---|---|---|
| B. C. | | |
| 1600 | 13.33 | Inscriptions at Karnak, tribute-lists of Thutmosis, (Brandis.) |
| 708 | 13.33 | Cuneiform inscriptions on plates found in foundation of Khorsabad. |
| | 13.33 | Ancient Persian coins, gold darics at 8.3 grams = 20 silver siglos, at 5.5 grams. |
| 440 | 13.00 | Herodotus's account of Indian tributes, 360 gold talents = 4680 silver. |
| 400 | 13.33 | Standard in Asia, according to Xenophon. |
| 400 | 12.00 | Standard in Greece, according to "Hipparchus," an alleged essay by Plato. |
| 404-336 | 12.00<br>13.00<br>13.33 | Values in Greece, from the Peloponnesian war to the time of Alexander, according to hints in Greek writers. There were variations under special contracts. Unit, the silver drachma. |
| 338-326 | 11.50 | Special contracts in Greece. |
| 323-43 | 12.50 | Standard in Egypt, under the Ptolemies. |
| 218 | 17.14 | Rate at Rome, fixed for coinage of gold scruples. Violent and temporary. |
| 100 | 11.91 | General rate of gold pound to silver sesterces at Rome to date. |
| 100 | 8.00 | About this time sudden influx of gold from Aquileja, temporarily reducing the relative value of the metal. |
| 58-49 | 8.93 | Great sums of gold brought by Cæsar from Gaul. |
| 29 | 12.00 | Normal rate in the last days of the republic. |
| A. D. | | |
| 1-37 | 11.97 | Rate under Augustus and Tiberius. — During the reigns of these emperors, however, the silver coinage was debased; hence the value of the precious metals, pure, was as 1 to 11 and less. |
| 37-41 | 12.17 | Rate under Caligula. |
| 54-68 | 11.80 | Rate under Nero. |
| 69-79 | 11.54 | Rate under Vespasian. |
| 81-96 | 11.30 | Rate under Domitian. |
| 138-161 | 11.98 | Rate under Antoninus. |
| 319 | 14.40 | Rate according to coinage of Constantine and his successors. Temporary. (?) |

NOTE.—In all these tables the figures in the second column show how many times as valuable as a given weight of fine silver would be an equal weight of fine gold.

REMARKS.—It appears from a study of this table that the ancient and long-established relation between silver and gold in the Orient was 1:13.33. The stability of this relation must be referred, probably, to the limited nature of commerce and its control by the sovereigns. Gold, by reason of its specific gravity and non-liability to oxidation, is found, as silver is not, in alluvial deposits, and would naturally be first brought into use among barbarous nations. But the production of silver by rude metallurgical processes (requiring much lower temperature than the metallurgy of iron) would not be a difficult discovery; and we find that at a very early period this metal also was in general use. The greater scarcity of gold and its superior qualities, especially at a time when the chief uses of both metals were for jewelry and ornament, naturally established for it a higher value; and, in the absence of sudden demands or fluctuations in supply, such as active industry and free commerce would produce, the relative value of the two metals might remain, for purposes of trade or tribute, where it was once fixed by arbitrary authority. We see, however, that in Greece, Egypt, and Italy the value referred to did not vary, as a rule, very far from the ancient rule; and this may justify us in supposing that the relation of 1:12 or 1:13 approximately represented the cost of producing silver, as compared with gold. The variations from this ratio, noted in the table, are generally due either to the sudden influx of gold from new quarters or to the arbitrary action of governments for purposes of profit in coinage. In those days the laws of political economy were not so well understood as now; and, indeed, until a very recent period, tampering with coinage has been a favorite method of replenishing the treasury of the sovereign.

TABLE II.—*The middle ages.*

| Date. | Ratio. | Authority. |
|---|---|---|
| A. D. | | |
| 864 | 12 | Probable ratio, as shown by the *Edictum Pistense*, under the Carlovingian dynasty. |
| 1104–1494 | 9–12 | Variable and apparently arbitrary British mint-edicts. |
| 1260 | 10.5 | Average ratio in the commercial cities of Italy. |
| 1351 | 12.3 | Ratio in North Germany, as shown by very accurate rules of the Lübeck mint, corroborated, in the main, by the accounts of the Teutonic Order of Knights, averaged in periods of forty years. |
| 1375 | 12.4 | |
| 1403 | 12.8 | |
| 1411 | 12.0 | |
| 1451 | 11.7 | |
| 1463 | 11.6 | |
| 1455–1494 | 10.5 | Ratio according to the accounts of the Teutonic Order of Knights. |
| 1469–1508 | 9.2 | |
| 1497 | 10.7 | Ratio established by Isabella, in Spain, (edict of Medina.) |
| 1500 | 10.5 | Ratio in Germany, according to Adam Riese's Arithmetic. |

REMARKS.—The data on which this table is constructed are not all of equal authority. Particularly in England there appears to have been a considerable undervaluation of gold for purposes of coinage. This would tend to keep the silver coinage at home, and to repay to the government the cost of obtaining silver for that purpose. But it is not necessary here to discuss the special meaning of the British figures. It is safe to say that, down to the fifteenth century, the general market-ratio of silver to gold was not far from 1:12. The serious variation shown in the thirteenth century, when, according to Italian economists, the average ratio in Milan, Florence, Lucca, Rome, and Naples was as low as 1:10.5, indicates a much greater supply of gold in that region at that time than was available in the more warlike and less commercial parts of Europe. Indeed, at the period referred to, these Italian cities were notably prosperous in trade and manufactures; and we may well believe that gold flowed in abundance to them, as to points where it could be safely and profitably invested.

There is no doubt, however, that during the fifteenth century the relative value of gold declined from about 12 to the neighborhood of 10, probably by reason of the scanty supply of silver, and the growing demand for that metal in manufactures (especially by the silversmiths) and in oriental commerce. The ancient sources of silver in Thessaly and Spain had been to a great extent exhausted; the New World had not begun to yield its argent treasure; and the mines of Saxony and Bohemia had not reached their maximum productiveness. It will be seen in the following period, as at others in this history, that the development of new fields and new methods of industry checked the tendency created by the currents of commerce and restored the balance of value so seriously disturbed.

TABLE III.—*From the discovery of America to the opening of the mines in California and Australia.*

| Date. | Ratio. | Authority. |
|---|---|---|
| A. D. | | |
| 1526 | 11.30 | Apparent relation of market-value, as deduced from the British mint-regulations, some absurd and unsuccessful experiments in coinage being disregarded. |
| 1543 | 11.10 | |
| 1561 | 11.70 | French mint-regulations. |
| 1575 | 11.68 | |
| 1551 | 11.17 | German imperial mint-regulations. |
| 1559 | 11.44 | |
| 1604 | 12.10 | British mint-regulations—experiments disregarded. |
| 1612 | 13.30 | |
| 1619 | 13.35 | |
| 1623 | 11.74 | Upper German regulations. |
| 1640 | 13.51 | French mint-regulations. |
| 1665 | 15.10 | |
| 1667 | 14.15 | Upper German regulations. |
| 1669 | 15.11 | |
| 1670 | 14.50 | British regulations. |
| 1679 | 15.00 | French regulations. |
| 1680 | 15.40 | |
| 1687–1700 | 14.97 | Ratios calculated from the biweekly quotations of the Hamburg prices-current, giving the value of the gold ducats of Holland in silver thalers, down to 1771, and, after that, in fine silver bars. The nominal par of exchange during this period was 1:14.80; and the quotations show the variations of the market rate in percentage above or below this. At par, 6 silver marks-banco were equivalent to one ducat, 68 20-47 ducats containing one mark (weight) of fine gold, and 27¾ silver marks banco containing one mark (weight) of fine silver. Hence, 6 × 68 20-47 ÷ 27¾ = 14.80, the par ratio. |
| 1701–1720 | 15.21 | |
| 1721–1740 | 15.08 | |
| 1741–1790 | 14.74 | |
| 1791–1800 | 15.42 | |
| 1801–1810 | 15.61 | |
| 1811–1820 | 15.51 | |
| 1821–1830 | 15.80 | |
| 1831–1840 | 15.67 | |
| 1841–1850 | 15.83 | |

REMARKS.—The first part of this table, like those which precede it, is compiled from imperfect data, in the absence of such direct evidence of ruling market-values as is furnished from 1687 to 1840 by the Hamburg quotations, the appearance of which is a significant indication of the extension of commerce and manufactures. All conclusions drawn from the mint-regulations of countries maintaining a double standard, are affected with a double uncertainty. In the first place, the frequent practice of enforcing, for reasons of state, arbitrary relations of value in coinage may mask the real market ratio; and, in the second place, the lack of free commerce and swift intercommunication among the nations may retard the evils resulting from such arbitrary action, and postpone the necessity of legislative remedy. Hence the change, as shown by the mint-regulations, may apparently occur some time after the real change in the relative value of the metals. Certain wild experiments with the coinage, particularly in England, have been omitted from the table, as having no real connection with the actual relative value. But if we do not scan the dates too closely, nor lay too much significance upon the maintenance of a certain rate in one country after it had been abandoned in another, the figures above given, down to 1680, may furnish us an instructive picture of the effect of the discovery, conquest, and plunder of South America upon the commercial relations of Europe.

The earliest effect of this discovery was the shipment of gold to Spain during the first quarter of the sixteenth century. At the beginning of that century, as shown in Table II, the ratio of silver to gold was about 1:10.7 or 1:10.5. The average annual influx of gold to Spain, down to 1527, probably did not exceed $400,000. But this supply, though it seems small to us now, would doubtless have depressed still further the value of gold, but for the remarkable productiveness of the Bohemian and Saxon silver-mines. The joint result of both causes was a notable increase of the amount of precious metals in circulation, and hence a rise in general prices, with which we have here nothing to do. The effect on the ratio between gold and silver was a gradual increase in the value of gold.

About 1550, the annual supply of gold may be roughly estimated to have been $400,000 and that of silver $2,600,000. By 1600, the production had become about $1,200,000 gold and $10,000,000 silver; in 1650, about $2,300,000 gold and $15,000,000 silver. This increase in the supply of gold is largely due to the development of the Guinea coast and the productiveness of the Hungarian gold-mines. The increase in the supply of silver is due to the introduction of amalgamation, and its use, principally in the form of the patio process, for the cheap production of silver in Peru and Mexico.

It is scarcely necessary to point out that the relative value of the two metals is not in the simple inverse ratio of their supply. The question of cost of production and of superiority in qualities adapting them for different purposes of currency, exportation, ornament, and manufacture must always play an important part; and even after all these elements have been calculated, it would be difficult to fix or foretell the relative price of the metals, since the local, temporary, and fluctuating demands and so-called "movements of specie" come in as disturbing phenomena of that incalculable element, the human will.

Thus we find that in the latter half of the sixteenth and the first half of the seventeenth century the production of silver was about seven times as great as that of gold. The drain of silver to the East Indies had already begun, and a large part of the product of South America found its way thither, returning in varied forms of wealth and luxury to enrich and enervate the Mediterranean nations. Yet the remainder of the silver-product was great enough to cause an increasing surplus of that metal, and a consequent fall in its relative price. It was, moreover, in many countries a period of war, and the large expenditures incurred led to a general increase in the gold coinage. The sudden and violent changes shown by some of the continental mint-regulations are indicative rather of financial and legislative necessities than of commercial changes. The English regulations seem to have been, on the whole, more stable, and represent probably with greater accuracy the steady enhancement of gold, the value of which, as compared with silver, advanced during the seventeenth century about 30 per cent., and during the sixteenth and seventeenth centuries together, about 45 per cent.

A general tendency in the same direction is visible in the eighteenth century; but here there is much greater stability, and the rise of gold is frequently checked.

Aided by the exact data drawn from the Hamburg quotations, we can trace the causes of fluctuation more accurately than in former years. I repeat that we are not here concerned with the general rise in prices produced by the greater volume of metallic currency and the multiplication of its efficiency through banking and credit. Our object is to study simply the agencies at work upon the relative value of the metals; and we find in this period, as in former ones, counteracting causes with a general resultant effect. The influx of silver from South America to Europe was continued and increased down to the beginning of the nineteenth century; but the export to Asia of piasters coined from American silver assumed colossal proportions, averaging, as Professor Soetbeer estimates, from 1690 to 1800, about $6,000,000 annually. Moreover, there was a considerable importation of gold from Asia, in return, to Spain and England. In China and Japan, according to the celebrated report of Sir Isaac Newton, master of the British mint in 1717, the pound of fine gold was worth but nine to ten pounds of fine silver, while the ratio in Europe was about 1:15. A merchant buying his goods in India with silver, and selling them in Europe for silver or for gold, with which he purchased silver again for a new venture, would gain 50 per cent. in exchange alone, besides the fair profit to the trade; and, of course, it would be worth while for him, after completing in the East his cargo of bulkier products, to lay out the remainder of his cash capital in the purchase of gold, which could be easily stowed and transported. This explains why, in the piracy and naval warfare of those times, the East Indiamen, going or coming, were considered so desirable as prizes. In either case, they were tolerably certain to carry rich booty of silver or of gold.

But besides the export of silver to India, and the imports of gold from India, there was for a short period a notable influx of gold from America—principally from New Granada and Brazil. The Brazilian supply was greatest from 1749 to 1761, and at that period occurred the unusual phenomenon of a rise of more than 4 per cent. in the relative value of silver, due to this cause. From 1701 to 1748 the average ratio was 1:15.19; from 1751 to 1755 it was but 1:14.53. This reaction, on the other hand, was checked by the demand for gold in the United Kingdom for the purpose of coinage and for the payment of heavy war subsidies to the continental nations during the struggle with Napoleon. The Bank of England, it is true, suspended specie payments from 1797 to 1819, and during this period the circulation of gold in Great Britain was reduced; but the subsidies and hoarding kept up its price; and the resumption of specie payments caused a sharp rise, so that the average ratio for 1821 was 1:15.95. The contemporaneous falling off in the supply from America, Africa, and India would doubtless have carried this movement still further, but for the unexpected development of a new source of production in Siberia and the Ural. In the absence of the political and commercial causes already enumerated as operating to enhance gold, this sudden influx of it would have mightily diminished its price. In fact, however, the ratio of value was not essentially changed, though the proportion of production was for the time being revolutionized. At the end of the eighteenth century the whole annual product was estimated at about $15,000,000 gold and $40,000,000 silver. In 1846, on the other hand, by reason of the supply from Russia, the gold-product was estimated at about $43,000,000 gold, while the silver had not increased—had, in fact, slightly fallen off. At the beginning of the nineteenth century, in other words, the production of the precious metals was about 27.8 per cent. gold and 72.2 per cent. silver. By 1846 it had become about 52.3 per cent. of gold to 47.7 of silver. Yet the relations of value had changed. The average of 1846 shows a ratio of 1:15.66.

From 1841 to the present time the relative values are calculated from the London quotations, as explained fully below.

TABLE IV.—*From* 1851 *to* 1874, *inclusive.*

| Date. | Ratio. | Authority. |
|---|---|---|
| 1851 | 15.46 | The London quotations. These give the price of a given weight of standard silver in shillings and pence sterling. Bearing in mind that there is in Great Britain no charge for coinage, and, hence, that the price referred to varies exactly as the market-value of the metals, we can calculate the ratio as follows: The standard gold is $\frac{11}{12}$ fine, and its value is fixed at 77*s.* 10½*d.*, or 934.5 pence per ounce troy. Hence the value of an ounce of fine gold is $\frac{12}{11}$ of this sum, or 1019.45 pence. The standard silver, on the other hand, is $\frac{37}{40}$ fine; hence an ounce of fine silver is worth 1.081 times as much as an ounce of standard silver. If the fixed value of an ounce of fine gold be divided by 1.081 times the quoted price of an ounce of standard silver, the quotient is the ratio desired. Thus, if $x$ be the quoted price per ounce in pence, $\frac{1019.45}{1.081\,x}=\frac{943}{x}$ (very nearly) is the ratio. Briefly, dividing 943 by the price in pence of an ounce of standard silver gives the ratio correctly to the second decimal place. London being the acknowledged center of the commercial world, this ratio determines the relative value of the metals among civilized nations. The table shows annual averages only. The lowest *monthly* value of gold was 15.12 in May, 1859, and the highest 16.35 in October, 1874. The annual average for 1874 here given is calculated upon the prices of eleven months ending November 30. |
| 1852 | 15.57 | |
| 1853 | 15.33 | |
| 1854 | 15.33 | |
| 1855 | 15.36 | |
| 1856 | 15.33 | |
| 1857 | 15.27 | |
| 1858 | 15.36 | |
| 1859 | 15.21 | |
| 1860 | 15.30 | |
| 1861 | 15.47 | |
| 1862 | 15.36 | |
| 1863 | 15.38 | |
| 1864 | 15.40 | |
| 1865 | 15.33 | |
| 1866 | 15.44 | |
| 1867 | 15.57 | |
| 1868 | 15.60 | |
| 1869 | 15.60 | |
| 1870 | 15.60 | |
| 1871 | 15.59 | |
| 1872 | 15.63 | |
| 1873 | 15.90 | |
| 1874 | 16.15 | |

REMARKS.—The discovery of gold in California in 1847 and in Australia in 1851 threw an unexampled quantity of this metal into circulation. While the annual product at the beginning of the century had been only about $15,000,000, and in 1846 had been swelled by the Russian placers to the then astonishing amount of $43,000,000, it had reached in 1853, according to Professor Soetbeer, a value of more than $165,000,000, while the annual silver-product was scarcely more than $40,000,000. This enormous disproportion, coupled with the continuance of vast shipments of silver to India, naturally caused many economists to anticipate a serious and permanent depreciation of gold. It appears from the table that this did not take place in such a degree as expected—and for two reasons. First, the supply of the precious metals, particularly silver, on hand in Europe was so great that even the immense production of the gold-mines in California and Australia could not immediately cause an overwhelming revulsion. General prices, of course, advanced; but that is not the point now in question. Chevalier estimates that the gold-production from 1500 to 1847, inclusive, had been nearly $3,000,000,000, of which (according to other authorities) something less than $2,000,000,000 had come from Spanish America, which had furnished in the same period about $5,000,000,000 in silver. Of the latter, perhaps $1,000,000,000 had been exported to India; but the stock of both metals still remaining in Europe was large; and the introduction of silver-plated wares liberated much silver which had been kept in the form of solid ware. But this would not long have sufficed to retard the fall of gold but for another cause, namely, the enormous coinage of gold in the countries employing the French system. Before discussing this important element, it is well to point out that the gold-production of California and Victoria suffered a diminution, through the exhaustion of the richest superficial placers, say about 1860, which was but partially compensated by the new discoveries in New Zealand, Idaho, Montana, &c. Moreover, after 1866 the shipments of silver to India notably declined, and finally the production of silver from Nevada and other

Territories advanced to colossal proportions. It was in the period preceding these changes that the French coinage-system exercised its restraining influence against the depreciation of gold. By the French law of 1803, the franc was fixed at five grams of standard silver, .900 fine, and it was provided that the kilogram of standard gold, of the same fineness, should be coined into 155 twenty-franc pieces. In other words, the relative value of the metals was established by law as 1 : 15.5. By the payment of a small coinage-charge, any one could obtain at Paris coin of either metal in exchange for fine bars of the same. Now, by reference to Table IV, above, it will be seen that the ratio between gold and silver in the market, which had been up to the beginning of the nineteenth century always less than 15.5, ranged steadily above that point. Hence, there was no inducement to coin gold. The gold coins themselves commanded a premium for melting. From 1825 to 1848, inclusive, the French coinage was only 268,000,000 francs in gold, against 2,380,000,000 francs in silver. The law of 1803 had been passed at an unlucky period, and its effect for nearly fifty years was to drive gold from circulation and flood Western Europe with a redundant silver currency. But it undoubtedly operated during that period to check the depreciation of silver by employing an excess of it as currency, and by diminishing the similar employment of gold, and thus increasing the available supply of that metal for other uses.

The reaction caused by the influx of American and Australian gold was immediate. So soon as fine gold was enough cheaper in the market than 3,444⅘ francs per kilogram to make it profitable to holders of it to take coin for it at that rate, the amount of gold presented for coinage in France was immense. From 1851 to 1867 there were coined in that country more than 5,806,000,000 francs gold, against 383,000,000 francs silver; and the silver coinage rapidly disappeared. Of $800,000,000 silver shipped to India in the period just named, probably more than half was melted down from the coinage of the franc-using countries—France, Italy, Belgium, and Switzerland. Thus the new supplies of gold found a new application, and the eastern demand for silver found a new supply, so that the value of gold was kept at 15.09 to 15.21, in spite of the excessive influx of it, which would otherwise have caused it to sink much lower. The practical abolition of the double standard by the United States, in 1853, and the increased gold coinage of Great Britain and her colonies, were subordinate causes, assisting this tendency. But the French system, serving for a while as a breakwater against the natural effects of great commercial facts, was itself swept away by the tide. The disappearance of a silver coinage is more inconvenient to the masses than the lack of gold. The lack of silver for currency brought about the famous convention of December, 1865, at which France, Italy, Belgium, and Switzerland united in the attempt to regulate again the relative values of the metals in coinage. But for the obstinacy of France, the cumbrous and fluctuating system of a double standard would have been given up. What was done was to diminish the value of the silver coinage, or rather to establish a coinage of pieces from a half franc to two francs, which, being legally as coins worth more than they were actually, could not be remelted or exported with profit. It was a pity that even a measure of this kind was carried out in the most unfortunate way. By changing at that time the weight of the gold instead of the silver coinage, the franc-countries might have obtained a system in which the weights of coins would have been a direct relation to the metric system; and if the double standard had been definitely abolished, and gold adopted as the sole standard, a foundation would have been laid for that international system which still appears an unattainable blessing, a system, namely, by which the different nations, each coining its own money, might maintain a certain uniform fineness in gold coinage, and accept as legal tender all gold coins according to their weight in metric units. The gram of standard gold would then be the international unit of money, with which all coins would gradually be brought into simple, integral relations.

But the convention of 1865 chose a different course, and it had hardly taken the step of debasing the silver coinage, to counteract the effects of the enhancement of silver, when a revolution took place, and a depreciation of silver was inaugurated, which has continued to the present time. The causes of this change, some of which have been already alluded to, deserve closer attention.

The exports of silver to India down to 1856 were to pay for products of the East exported to Europe and America. During our civil war, the extraordinary production of cotton in India increased this balance of trade, and thus augmented the drain of silver in that direction. Moreover, between 1856 and 1866, large sums of silver went to India as loans, to be employed in the construction of an extensive railway-system. These special causes ceased after 1866, and the payment of the interest on the Indian loans is now, so far as it goes, an element on the other side of the balance of exchange. The demand for silver in India is, however, by no means extinct; and one effect of the increased development of internal trade and industry in that region will be to maintain this demand. Professor Soetbeer, in proof of the intimate relation between the rate of Indian exchange and the ratio of silver to gold in London, calculates the following five-year averages from the London prices, and the Calcutta quotations for six-month drafts:

| Average of period. | Drafts on London in Calcutta. | Price of the ounce standard silver in London. | Ratio of gold to silver. |
|---|---|---|---|
| | *Pence per rupee.* | *Pence.* | |
| 1856–'60 | 25¼ | 61⅛ | 15.33 |
| 1861–'65 | 25 | 61¼ | 15.40 |
| 1866–'70 | 23⅛ | 60⅝ | 15.59 |
| 1873 | 23 | 59¼ | 15.91 |

The second great cause of depreciation of silver was the colossal production from the United States. The product of this country, as I have shown in another place, was in 1861 but $2,000,000. From 1862 to 1866, inclusive, it was $45,000,000, or an average of $9,000,000 per annum. In 1867, it was $13,500,000; in 1868, $12,000,000; in 1869, $13,000,000; in 1870, $16,000,000; in 1871, $22,000,000; in 1872, $25,750,000; in 1873, $36,500,000. The declared importations into Great Britain were:

| | |
|---|---|
| 1870 | £3,387,000 |
| 1871 | 5,689,000 |
| 1872 | 4,575,000 |
| 1873 | 5,992,000 |

The great *bonanzas* discovered in the Comstock lode of Nevada have been the chief cause of the latest increase; but the extraordinary and rapid development of the silver-lead-smelting industry in Nevada (Eureka) and Utah has also been an important factor.

A third cause of the depreciation of silver was the abandonment by Germany, after the Franco-German war, of the silver-standard, and the issue of a standard gold coinage for the empire. Professor Soetbeer estimates that in November, 1874, there was still in circulation in the different German states about $230,000,000 in silver coin, less than half of which would probably remain in circulation after recoinage, the remainder being thrown upon the market for exportation. The effect of the change is twofold—the liberation of silver from circulation as coin and the increased demand for gold. The amount of imperial gold coinage resolved upon in 1871 was 400,000,000 thalers, (about $290,000,000,) of which 362,000,000 had been coined up to November, 1874. It was evident, however, that at least 500,000,000 thalers would be required; and this would tend to continue the special demand for gold.

The countries of the franc-system, with double standard, have, however, again furnished a check to the prevailing tendency which has, doubtless, greatly modified the fall in silver. In these countries, the ratio being fixed at 15.50 by law, it has been since 1866 profitable to have silver coined. In 1872 the coinage of silver in the Netherlands was 33,540,945 florins; in 1872 and 1873, all the states of the franc-system coined some 200,000,000 francs in silver. But this could not continue, and in 1874 a limit was provisionally set to the coinage of five-franc pieces. A new convention is said to have been arranged, the results of which will be very important to the question now under consideration. The depreciation of silver may be retarded, if the coinage of it as a legal-tender currency at a rate above its market-value is continued. But this policy, whatever its immediate conveniences, is contrary to the true principles and tendencies of political economy. The demonetization of silver and the universal adoption of gold as a standard currency, with silver as a material of subsidiary coinage, is that to which all nations must come. The ratio of 15.5 is too near the average market ratio for the last century to be safely adopted. But the ratio of 15, as proposed by Mr. Elliott and other American writers on this subject, and partly adopted in our coinage, is reasonably certain to remain permanently below the market ratio, and hence to serve satisfactorily in coinage, while it possesses practical value, as an integer, simply related to the decimal-system.

A review of the whole history shows us that the natural tendency has been, since industry and commerce were organized, toward a depreciation of silver; that this tendency has been periodically counteracted by three great causes—the successive discovery of new gold-fields, (Africa, Brazil, the Ural, California, Australia, New Zealand, Idaho, Montana, &c.,) the exportation of silver to the East, and its elastic employment for coinage. It is highly improbable that new Californias will be opened, though the present production from quartz and hydraulic mines, in this country at least, is likely to be maintained, and even increased; the demand for India will not be augmented, but, by growing reciprocity of commerce, probably decreased, though slowly, and the use of silver in coinage will probably also remain stationary or decline. On the other hand, the production of silver will probably increase, though the enormous re-enforcements from this or that great *bonanza* will be temporary in their effects. Looking at the subject on a larger scale, it is not this or that single mine, but the vast area of silver-bearing territory in this country and Mexico which may be expected to develop and maintain an increased production. What may be the metallic treasures of Asia we scarcely know as yet, but their exploration will, at all events, for many years to come, produce nothing that the vast population of that region will not immediately swallow up.

All general considerations so far seem to point to a permanent depreciation of silver. Not that the price of November, 1874, the lowest ever reached in history by the free action of commercial and industrial causes, will be maintained or further decreased; but that the ratio ruling a dozen years ago is not likely to be regained, and the low price of silver must be accepted by miners and metallurgists as a part of the economical problems submitted to them. When I say that the gross product of a silver mine to-day is worth in gold 9 per cent. less than it would have been in 1859, it will be seen that this is a question intimately connected with profits and dividends. Indeed, the probable suspension of mining and metallurgical operations which cannot afford this loss will, perhaps, be one reaction tending to diminish the production, and thus check the depreciation of silver. In whatever way the general current flows, it is sure to be interrupted by such eddies, the results, indirectly, of its own motion.

---

# CHAPTER XX.

## RECENT IMPROVEMENTS IN MINING AND MILLING MACHINERY IN THE PACIFIC STATES.

By WILLIAM P. BLAKE.

The recent extraordinary developments of large bodies of rich silver-ore at great depths in the Comstock lode, Nevada, have given a great impetus to the construction of mining-machinery on the Pacific coast. The increasing depth and extent of other mines and the discovery of new districts have also had a great effect in stimulating and extending the industry. First street in San Francisco, where most of the founderies are located, is the chief center of production, but there are establishments at Sacramento, Marysville, and other chief cities and mining-centers in California, at Virginia City, Nevada, and at Salt Lake City, Utah. Considerable quantities of machinery for Utah and for Montana and Colorado are manufactured at Omaha, Chicago, Pittsburgh, and other points in the States of the Mississippi Valley. But San Francisco takes the lead, not only in the amount of machinery made, but in new forms and details of construction especially adapted to the wants of the miners and the exigencies of each case.

The great advance of late may be said to be in the construction of pumping and hoisting machinery for operating at great depths, and in the arrangement of large mills, so that the ore may be handled and worked at a minimum cost for hand-labor and in a more systematic and satisfactory manner. The enormous sums of silver and gold which some of the claims are yielding permit of the construction of mills in the very best manner, the question of cost of any desirable arrangement or combination of machinery being insignificant compared with the production and the saving which a proper construction permits. The machines and the mill described in the following pages may be regarded as the best types of their kind now in use.

### PUMPING-MACHINERY.

The largest-sized pump for any mine on the Pacific coast has recently been built by the Union Iron-Works, of San Francisco, for the Raymond & Ely mine, in Pioche district, Nevada. (For descriptions of this mine see the Commissioner's Report of 1873.)

The depth to be unwatered from the surface is 1,400 feet, and the main object in view is to free the mine so as to prospect deeper.

There are six pumps, each 12 inches in diameter and 8 feet stroke, arranged with hoist-gearing upon the driving-frame, so that they can be readily handled in the shaft. The pump-rods of wood are in 20-foot lengths, 8 by 8 inches and 12 by 12 inches, with scarf-joints, strengthened with iron plates about 15 feet long, of 1 by 8 inch iron. These are firmly secured on both sides of the rod by 1-inch bolts at intervals of 10 inches.

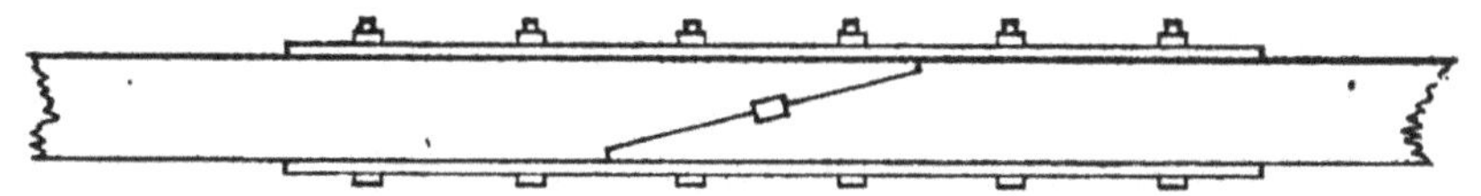

The engine, which deserves special mention, is one of the most approved forms of Booth & Co. It is horizontal, the cylinder 26 inches in diameter, and the stroke 72 inches. It is fitted with balanced poppet-valves, operated automatically by means of cross cut-off and the Scott & Eckart governor. The bed-plate is Scott's improved modification of the Corliss pattern. The crank is forged iron, but the piston-rod, crank-pin, and valve-stems are of steel. The main shaft is 14 inches in diameter, the fly-wheel 24 feet in diameter, and weighs, in its finished state, 60,000 pounds, or 30 tons. As a wheel of this extraordinary size and weight could not be transported in one piece over the mountains, it was made in segments, afterward carefully fitted and bolted together and finished, so that it could be taken apart and set up again with accuracy.

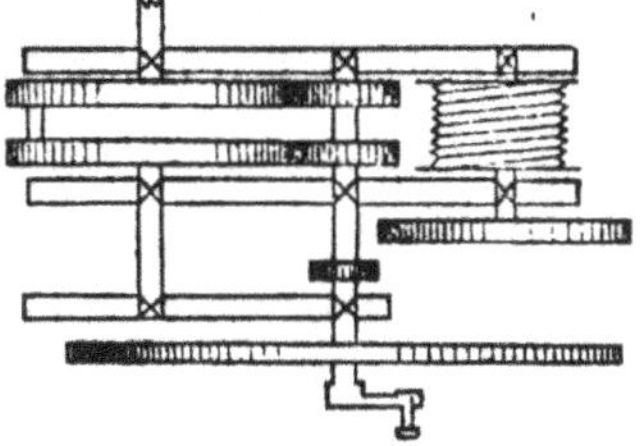

Another pump is building by the same works for the California mine, in which the connection with the engine is made direct, without any gearing. The cylinder is placed horizontally, and the piston connects with the top of the pump-bob. The cylinder is 8 feet long and 40 inches in diameter, and is lined with steel. Pump-bob upright 15 feet long, and horizontal arm 13 feet.

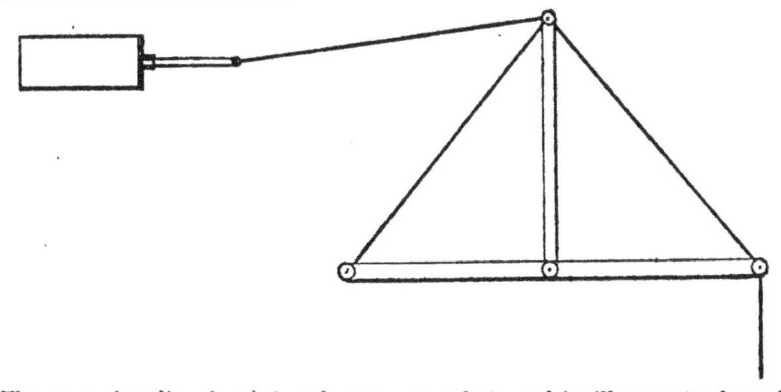

The annexed outline sketch (not drawn accurately to scale) will serve to show the arrangement of the Raymond & Ely engine and pump-gearing, and of the hoisting-arrangement behind, which ordinarily is out of gear, but can be connected with a pinion on the main shaft.

The main pinions, working into the 12-foot pump-wheels, are 36 inches in diameter, 13 inches face, 4 inches pitch, with the teeth accurately cut. These gear into two mortise-wheels 12 feet 3 inches to pitch-line, carrying a crank-pin 7 inches in diameter between them. These wheels are accurately turned all over, and the teeth are formed with great care of selected pieces of second-growth hickory, boiled in oil, and driven in with a 20-pound sledge-hammer. The shafts of these pump-wheels are 12 inches in diameter, and are well supported by double bearings. The foundation-bolts are 23 feet long. The hoist-drum is 6 feet in diameter, and is chased for 1¾-inch round wire rope.

The pump-bob timber is 26 inches by 24 inches and 26 feet long. It is fitted with double stays in front and is heavily braced. The journals are of wrought iron, 13 inches in diameter through the timbers and 12 inches in the boxes. All of the brass-boxes are lined, having bottom as well as side brasses, and every part was designed to have a central strain, the pins, 7-inch, all being supported on both sides.

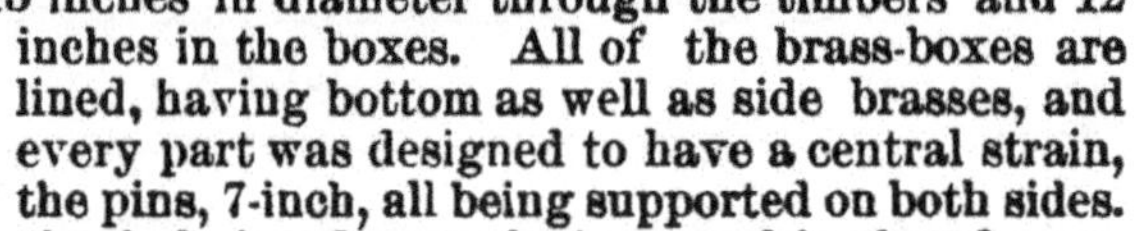

The main bearing of the bob is of wrought iron, and is placed accurately in the intersection of the center or medial lines instead of below them, as formerly.

A duplicate of this pump and gearing, excepting the teeth of the pump-wheels, which are of iron instead of wood, has been made by Booth & Co. for the Baltimore-Consolidated mine, near Gold Hill, Nevada. These are the two largest and most complete pumping-engines of the kind in the Pacific States.

### HOISTING-MACHINERY.

The increasing depth of the principal mines, and the large quantities of ore to be hoisted, necessitate much more powerful engines and heavier gearing than have been in use heretofore. The largest hoisting-gear yet built was completed recently for the Savage mine, Comstock lode, Virginia City, Nevada, by Booth & Co., of the Union Iron-Works. This mine has attained a depth of 2,200 feet, but with the new hoisting-arrangements it is expected to push the incline to a depth of 4,500 feet from the surface.

The engines are horizontal and double cylinders, 24 inches in diameter and 48 inches stroke, with valve-motion, and the frame same as built by the firm for the pumping-machinery of the Richmond Consolidated, except that the cut-off is worked by means of a lever alongside of the reversing-gear. The engine-shaft has a diameter of 11 inches. There are two fly-wheels, each 14 feet in diameter and weighing 15,000 pounds. Their position on the shaft and the arrangement of the other parts of the gear may be seen by inspecting the diagram.

The motion is transmitted to the reel by a single pinion-wheel, 55 inches in diameter, 19 inches face, 6 inches pitch, with the teeth accurately cut. This works into a large gear-wheel on the reel-shaft, 21 feet 8 inches in diameter, with same face and pitch as the pinion. There are 132 teeth, accurately cut, and the wheel is turned all over. It weighs, when finished, 66,641 pounds. The winding-drum is conical and designed for round cable. It is 23 feet in diameter at the larger end and 14 feet at the smaller. Length on the tapering surface, 15 feet. It is cut with a 2¼-inch spiral groove, to suit a 2-inch round wire cable. It has 79 grooves, and it will hold 4,590 feet of cable. The cable-tapers from 2 inches to 1¾ inches in diameter. The reel-shaft is 13

SAVAGE MINE HOISTING-GEAR.

Scale 1/8" to 1'.

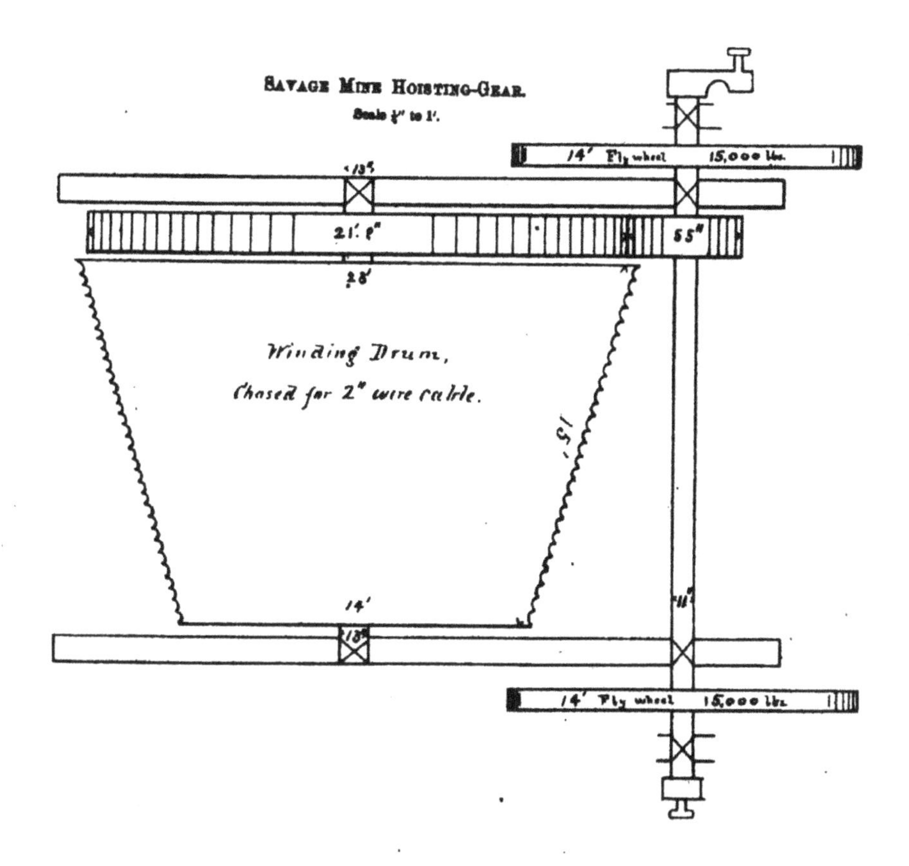

CONSOLIDATED VIRGINIA.

Sketch-plan of mill, showing the arrangement of the machinery, pans, &c. Not drawn to scale.

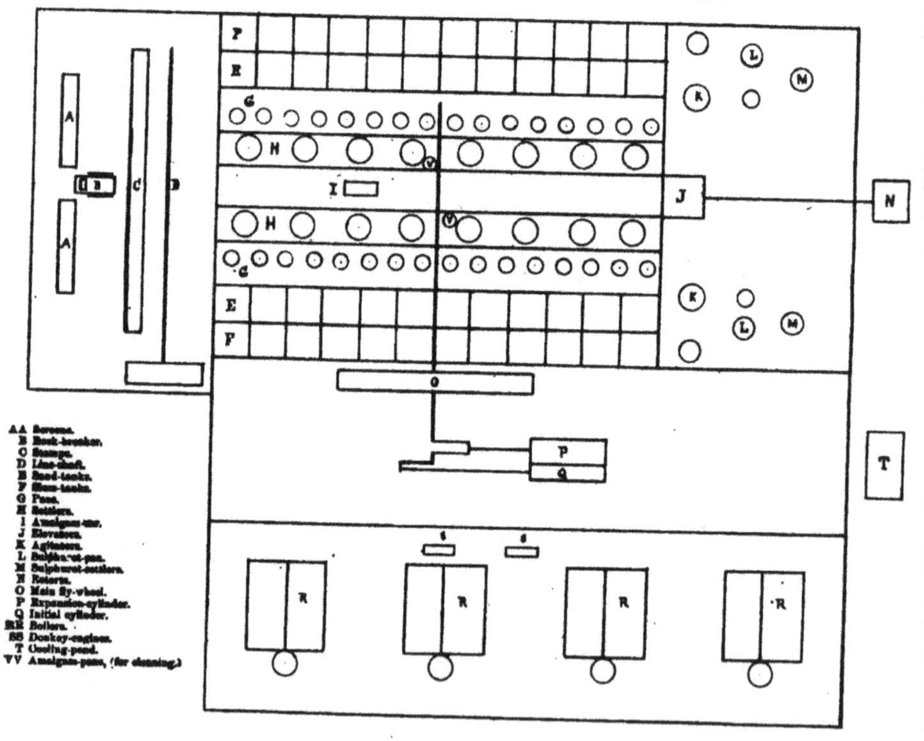

inches in diameter. The tapering reel, of cast iron, is covered with an oak-wood casing 6 inches thick, in which the grooves are cut, as shown in the partial section of the reel, herewith.

This oak casing is seasoned and prepared by Wood's process, to prevent shrinking and swelling.

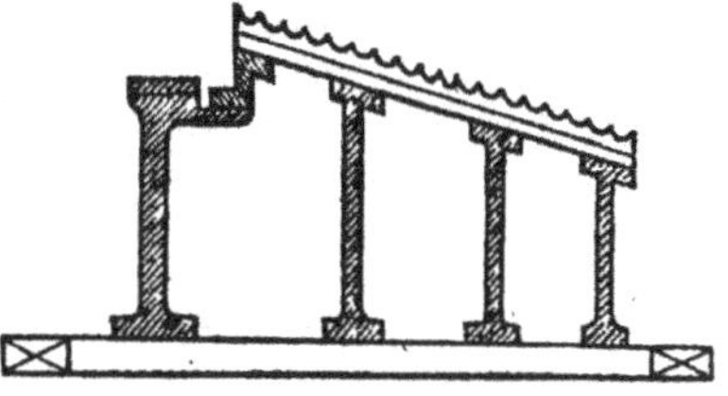

The overhead sheave is 14 feet in diameter, and is accurately turned, as is also that at the head of the incline, which is 12 feet in diameter.

The car and rope are not balanced. The car alone weighs 6,700 pounds, and carries a load of 10 tons. It is arranged so as to dump automatically.

### UNDER-GROUND PORTABLE HOIST.

The same firm have made a small portable hoisting-engine for use under-ground to draw from winzes and sumps. It is run by compressed air, and can be placed in drifts. It weighs about 1,200 pounds, and is compact enough to slide in between the timbering, as it is usually set in the Comstock mines. The compressors are placed at the surface, and the air is conveyed below in iron pipes from 4 to 6 inches in diameter, branching to 1½ inches at the machine. The cylinders are 4 by 10 inches. Cranks on drum-shaft. The drum is 20 inches in diameter and 16 inches long. It is provided with link-motion and a brake. The outside dimensions are 39 inches wide, 30 inches high, and 49 inches long. It will hoist 400 pounds 100 feet, and is very useful for work below the levels at a distance from the shaft.

Another small engine, for use in the drifts to drive ventilating-fans, is run by compressed air. It is vertical in construction; cylinders 4 inches in diameter and 6 inches stroke; fly-wheel 48 inches in diameter. A jack-screw in the cylinder-head serves to hold the engine in place usually between the timbers.

### AIR-COMPRESSORS.

For air-compressors the Union Iron-Works prefer the Rand and Waring combined with their form of engine. They are horizontal and double-acting; steam-cylinder 16 by 30 inches, and air-cylinder 16½ by 30 inches; fly-wheel 12 feet in diameter, weighing 15,000 pounds, on an 8-inch shaft. There are two sizes, the smaller being 12 by 30 inches cylinder, and fly-wheel 10 feet in diameter. Twenty revolutions of the small size give a result equal to a No. 4 Burleigh at thirty-two revolutions. The air is delivered perfectly cool and dry. They are used by the Ophir, Consolidated Virginia, Belcher, Chollar, Imperial, and Sierra Nevada Companies.

The air-pipes are 6 inches in diameter, and are made at the Golden State Foundery. Twenty-six hundred feet were recently made there for the Imperial silver-mine. They are of boiler-iron lap-welded.

### SILVER-REDUCTION MILL.

The most complete silver-mill in the world has just been constructed by H. J. Booth & Co. for the Consolidated Virginia mine, Virginia City, Nev., to work the ores direct from the dumps of the mine, without any hauling by teams or railway-transportation, thus in the outset effecting a most important saving, not only in the cost of moving the ore, but in the waste which inevitably attends it. The cars dump over the "grizzlies" into the ore-bins. The grizzly is simply a coarse screen for separating the coarse ore from that which is already sufficiently small to pass through the self-feeding apparatus. It borrows its name from the grating used by the placer-miners to throw out the bowlders from the sluices. As made for this mill, they are revolving, each about 16 feet long, and constructed of heavy bar-iron 3 inches wide by ⅞ inch thick, placed on edge, 3 inches apart, and held securely by bolts and thimbles. They are about 30 inches in diameter. The ore which drops through these revolving screens goes directly to the ore-bins. The details of the mill are as follows:

The coarse ore separated from the fine by the grizzlies goes to the Blake rock-breaker, which delivers to the ore-bins. From the ore-bins the ore passes to the self-feeding batteries, from thence to the sand-tanks and to the slum-tanks behind them. The total fall of the ore through the mill by these different stages is 52 feet 4 inches.

The rock-breaker is the 15 by 9 inches size; and, inasmuch as all the fine ore is separated and only the large masses reach the breaker, its capacity is fully equal to the demand upon it.

There are 60 self-feeding stamps, 850 pounds weight, 7 inches drop, and 85 drops to a minute. There are 32 improved pans, with 5-foot mullers, making 85 turns.

A charge consists of 1¾ tons of pulp, and is renewed each five hours, making the capacity for each pan 8⅖ tons a day. There are 16 settlers, 9 feet in diameter, 36 inches deep, 14 revolutions; two clean-up pans, each 48 inches in diameter and 12 inches deep, making 16 revolutions; four agitators, 8 feet in diameter, 5 feet deep, 21 turns; four sulphuret-pans, 5 feet in diameter, 24 inches deep, 50 turns; two sulphuret-pan settlers, 8 feet in diameter, 24 inches deep, 14 turns; one hydraulic elevator, for handling amalgam; six retorts, 14 inches by 4 feet long.

All the quicksilver is moved by pumps, which force it into the distributing-tank, from which it is drawn into the charge, then into the pans, the settlers, and the strainers, from which it flows back to the receiving-tank, which supplies the pump. It thus makes the circuit without being handled, or spilled and wasted, leaving nothing behind but the amalgam and such portions as may be carried off in the tailings. This is a great labor-saving arrangement, for the weight of the quantity of quicksilver required to work 240 tons of 140-dollar ore is not less than 32 tons. In the old mills, the quicksilver was carried about in pails and flasks, and there was inevitably a great waste by spilling.

The steam-engine for this mill is a low-pressure compound condensing one, with the same style of frame, valves, and governor before described as built by the firm of Booth & Co. for the large pumping-engine. The initial cylinder is 24 inches in diameter and 48 inches stroke. Speed, 55 revolutions per minute. The air-pump is worked from the cross-head; cranks are set opposite each other. The motion is perfect. In starting the mill with the new and rough shoes, the friction of the shafting-stamps and breaker, 666 horse-power were developed, and after running for a time and getting smoothed down 580 horse-power are required. It required 100 horse-power for the stamps and breaker when new, and 17½ horse-power to each pan, with its proportion of settlers, agitator, sulphuret-pan, and quicksilver-pump.

The boilers are eight in number, each 54 inches in diameter, 16 feet long, 3½-inch tubes. Drum, 36 by 14 feet; mud-drum, 24 by 13½ feet.

CONSOLIDATED MILL.

Sketch-section of one tier of pans and apparatus, showing relative position of the pans, settlers, distributers, and quicksilver-pump. (Not to scale.)

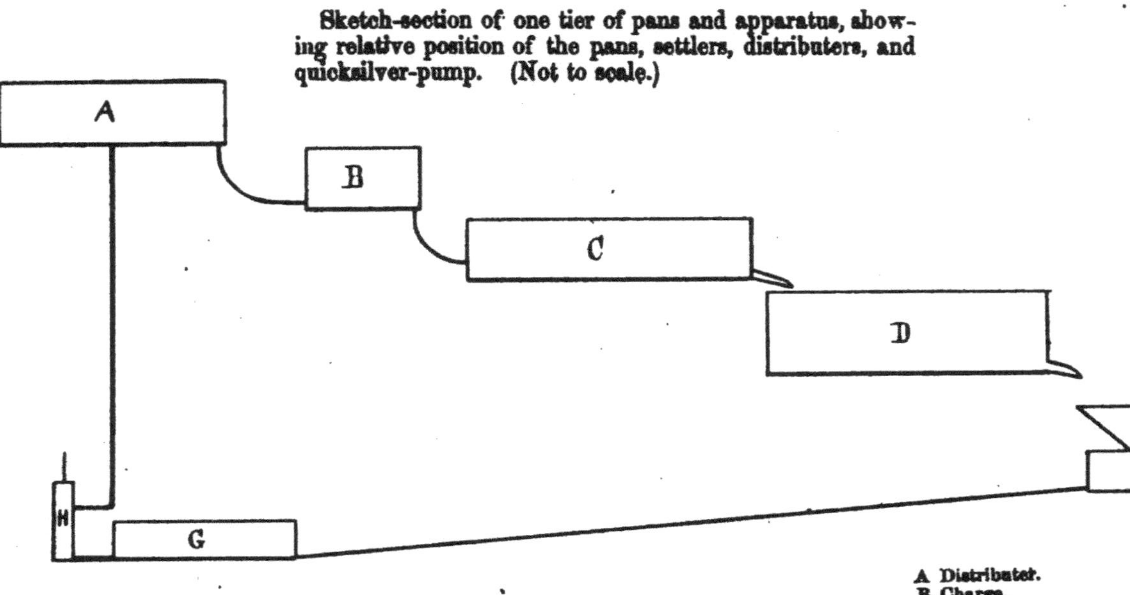


A Distributer.
B Charge.
C Pan.
D Settler.
E Strainer, (quicksilver.)
F Tank.
G Receiving-tank.
H Quicksilver-pump.

They are set in four pairs, each pair being detachable and complete in themselves. Twenty-eight cords of wood are required daily. Twenty-two inches of vacuum were obtained, but it was found more economical to carry 18 inches and feed the water hotter to the boilers. The condensing-water is cooled by a pond open to the air. It enters at one end and flows back to the condenser at the other end perfectly cool. The engine gives 20 horse-power to a cord of wood, while the best old steam-mills give only 12 horse-power and an average duty of 10.

The stamps at this mill crush five tons each of the ore in twenty-four hours, while the duty at the best old steam-mills ranges from 1½ to 2½ tons to a stamp. Each pan works 8¾ tons in twenty-four hours, giving the mill a capacity of 260 to 280 tons a day.

Some of the special and new points in this mill may be further enumerated.

1. The ore loaded in the mine goes to the bins without handling over.

2. The "grizzlies" separate the coarse from the fine ore, and deliver only the coarse ore, requiring breaking, to the breaker.

3. This separation enables one man to handle all the coarse rock that 260 tons of ore yield.

4. The self-feeders, rendered practicable by the uniformity in size of the ore, enable the sixty stamps to be supplied uniformly with the attention of one man only. There is here a great saving of labor: one man at the breaker, three men (eight-hour shifts) at feeders; total, four men in twenty-four hours, (the breaker working but eight hours.) In the old mills: four men at breaker; at feeders, four men to a shift; total, sixteen men.

5. The slums are delivered in tanks behind and on the same level with the sand-tanks; hence sands and slums are worked at the same time, saving the labor of hauling the slums from a lower level to the pans.

6. All the quicksilver is handled by machinery, saving all the manual labor hitherto required for this purpose.

7. The amalgam is transported on a car holding two tons, and is lowered by machinery.

8. The engine is a condensing compound, and the water is cooled in a simple, economical manner.

These improvements combine to produce the extraordinary and very satisfactory results cited, of which the designers and makers may be justly very proud.

The ore so worked ranges about $140 per ton in value. The bullion is turned out .937 fine.

This mill commenced running on the 8th of January, 1875. Its cost was, approximately, $350,000. A larger, and perhaps more complete, mill is now (April, 1875) building by Booth & Co. for the California mine.

---

# CHAPTER XXI.

## MISCELLANEOUS STATISTICS.

### GOLD AND SILVER PRODUCTION FOR 1874.

Mr. J. J. Valentine, general superintendent of Wells, Fargo & Co., published recently the company's annual statement of precious metals produced in the States and Territories west of the Missouri River, including British Columbia, during the year 1874, which shows an aggregate yield of $74,401,055, being an excess of $2,142,362 over 1873. California, Nevada, Utah, Colorado, and British Columbia increased;

Oregon, Washington, Idaho, Montana, Arizona, and Mexico (west coast) decreased. The increase in Nevada and Colorado is merely nominal, but in California and Utah it is $3,100,000, three-fourths of which is to the credit of California.

*Statement of the amount of precious metals produced in the States and Territories west of the Missouri River during the year* 1874.

| States and Territories. | Gold dust and bullion by express. | Gold dust and bullion by other conveyances. | Silver-bullion by express. | Ores and base bullion by freight. | Totals. |
|---|---|---|---|---|---|
| California | $16, 015, 568 | $1, 601, 556 | $967, 857 | $1, 715, 550 | $20, 300, 531 |
| Nevada | 345, 394 | 34, 539 | 30, 954, 602 | 4, 117, 698 | 35, 452, 233 |
| Oregon | 553, 564 | 55, 356 | 150 | ............ | 609, 070 |
| Washington | 141, 396 | 14, 139 | ............ | ............ | 155, 535 |
| Idaho | 1, 207, 667 | 120, 765 | 551, 572 | ............ | 1, 880, 004 |
| Montana | 2, 581, 362 | 258, 136 | ............ | 600, 000 | 3, 439, 498 |
| Utah | 83, 721 | 8, 372 | 746, 565 | 5, 072, 620 | 5, 911, 278 |
| Arizona | 23, 333 | 2, 333 | 400 | ............ | 26, 066 |
| Colorado | 1, 590, 700 | ............ | 1, 745, 705 | 855, 000 | 4, 191, 405 |
| Mexico | 84, 655 | ............ | 714, 223 | ............ | 798, 878 |
| British Columbia | 1, 487, 473 | 148, 747 | 357 | ............ | 1, 636, 557 |
| Grand total | 24, 114, 833 | 2, 243, 943 | 35, 681, 411 | 12, 360, 868 | 74, 401, 055 |

My own estimate, which differs in some respects from that of Mr. Valentine, has been given in the introductory letter at the beginning of this report, as follows:

| | | |
|---|---|---|
| Arizona | | $487, 000 |
| California | | 20, 300, 531 |
| Colorado | | 5, 188, 510 |
| Idaho | | 1, 880, 004 |
| Montana | | 3, 844, 722 |
| Nevada | | 35, 452, 233 |
| New Mexico | | 500, 000 |
| Oregon | $609, 070 | |
| Washington | 154, 535 | |
| | | 763, 605 |
| Utah | | 3, 911, 601 |
| Wyoming and other sources | | 100, 000 |
| | | 72, 428, 206 |

TREASURE EXPORTS AND RECEIPTS AT SAN FRANCISCO.

[From Commercial Herald and Market Review, January 14, 1875.]

*Treasure exports.*

Our treasure exports during the last three years have been as follows, exclusive of shipments through the United States mail:

| | 1872. | 1873. | 1874. |
|---|---|---|---|
| To New York | $4, 055, 565 46 | $14, 597, 895 76 | $20, 689, 627 86 |
| To England | 2, 262, 302 25 | 667, 109 81 | 184, 755 63 |
| To China | 7, 476, 862 72 | 6, 335, 353 50 | 8, 324, 675 54 |
| To Japan | 10, 212, 049 63 | 2, 206, 157 12 | 41, 360 00 |
| To Panama | 56, 679 82 | ............ | ............ |
| To other countries | 5, 266, 075 76 | 908, 609 49 | 940, 213 20 |
| Totals | 29, 330, 435 64 | 24, 715, 125 68 | 30, 180, 632 23 |

The comparative descriptions of our exports of treasure by the above table were as follows:

| | 1872. | 1873. | 1874. |
|---|---|---|---|
| Gold bars | $11, 910, 565 | $2, 828, 682 | $3, 295, 837 |
| Silver bars | 7, 913, 391 | 8, 457, 769 | 9, 492, 719 |
| Gold coin | 7, 888, 620 | 9, 076, 173 | 10, 849, 561 |
| Mexican dollars | 1, 427, 441 | 3, 779, 063 | 2, 253, 341 |
| Gold-dust | 37, 007 | 77, 645 | 82, 212 |
| Silver coin | .......... | 106, 5[illegible]9 | 58, 425 |
| Trade-dollars | 153, 412 | 389, 234 | 4, 018, 517 |
| Currency | .......... | .......... | 130, 000 |
| Totals | 29, 330, 436 | 24, 715, 125 | 30, 180, 632 |

*Combined exports.*

The combined exports, treasure and merchandise, exclusive of overland railroad, during the past twelve months, as compared with the same time in 1872 and 1873, were as follows:

| | 1872. | 1873. | 1874. |
|---|---|---|---|
| Treasure exports | $29, 330, 436 | $24, 715, 126 | $30, 180, 632 |
| Merchandise exports | 23, 793, 530 | 31, 160, 208 | 28, 405, 248 |
| Totals | 53, 123, 966 | 55, 875, 334 | 58, 585, 880 |

*Receipts of treasure.*

The following table comprises the receipts of treasure in this city, through Wells, Fargo & Co.'s Express, during 1874:

*From the northern and southern mines.*

| 1874. | Silver bullion. | Gold bars, &c. | Coin. | Totals. |
|---|---|---|---|---|
| January | $903, 801 | $423, 898 | $626, 052 | $1, 953, 751 |
| February | 1, 696, 461 | 452, 174 | 540, 951 | 2, 689, 586 |
| March | 1, 439, 139 | 548, 592 | 474, 463 | 2, 462, 194 |
| April | 1, 829, 101 | 678, 741 | 578, 409 | 3, 086, 251 |
| May | 1, 920, 180 | 792, 522 | 740, 657 | 3, 453, 359 |
| June | 1, 882, 003 | 703, 973 | 834, 874 | 3, 420, 850 |
| July | 1, 417, 262 | 751, 448 | 773, 273 | 2, 941, 983 |
| August | 1, 495, 898 | 709, 418 | 893, 021 | 3, 098, 337 |
| September | 1, 651, 267 | 654, 398 | 945, 106 | 3, 250, 771 |
| October | 1, 517, 725 | 514, 418 | 930, 474 | 2, 962, 617 |
| November | 1, 4[illegible]1, 210 | 432, 338 | 905, 798 | 2, 789, 396 |
| December | 1, 065, 007 | 430, 954 | 1, 016, 274 | 2, 512, 235 |
| Totals | 18, 260, 054 | 7, 092, 924 | 9, 259, 352 | 34, 621, 330 |
| Totals, 1873 | 11, 749, 320 | 8, 290, 258 | 6, 636, 143 | 26, 675, 721 |
| Totals, 1872 | 6, 386, 794 | 14, 843, 835 | 6, 769, 641 | 28, 000, 270 |
| Totals, 1871 | 14, 609, 809 | 13, 872, 648 | 7, 125, 928 | 35, 608, 385 |
| Totals, 1870 | 14, 152, 984 | 17, 762, 131 | 6, 487, 037 | 38, 402, 152 |

*From the northern coast.*

| 1874. | Silver bullion. | Gold bars, &c. | Coin. | Totals. |
|---|---|---|---|---|
| January | | $135,300 | $52,427 | $187,727 |
| February | | 57,478 | 144,802 | 202,280 |
| March | | 63,798 | 69,383 | 133,181 |
| April | | 97,545 | 28,487 | 126,032 |
| May | | 132,011 | 57,275 | 189,286 |
| June | | 107,820 | 50,493 | 158,313 |
| July | | 137,422 | 49,354 | 186,776 |
| August | | 199,066 | 23,892 | 222,958 |
| September | | 118,520 | 20,619 | 139,139 |
| October | | 230,664 | 41,925 | 272,589 |
| November | $300 | 177,805 | 78,175 | 256,280 |
| December | | 91,001 | 40,650 | 131,651 |
| Totals | 300 | 1,548,430 | 657,482 | 2,206,212 |
| Totals, 1873 | 4,200 | 1,441,438 | 878,377 | 2,324,015 |
| Totals, 1872 | | 2,305,414 | 661,889 | 2,967,303 |
| Totals, 1871 | 9,785 | 2,552,668 | 708,096 | 3,270,549 |
| Totals, 1870 | | 3,380,566 | 532,901 | 3,913,467 |

*From the southern coast.*

| 1874. | Silver bullion. | Gold bars, &c. | Coin. | Totals. |
|---|---|---|---|---|
| January | $363 | $14,292 | $36,383 | $51,038 |
| February | | 20,772 | 29,191 | 49,963 |
| March | | 5,159 | 31,797 | 36,956 |
| April | 1,346 | 24,693 | 54,136 | 80,175 |
| May | 4,508 | 24,606 | 39,395 | 68,509 |
| June | 1,230 | 30,478 | 37,619 | 69,327 |
| July | 1,300 | 14,790 | 35,207 | 51,297 |
| August | | 10,318 | 17,206 | 27,524 |
| September | 800 | 17,360 | 15,189 | 33,349 |
| October | | 20,870 | 37,647 | 58,517 |
| November | | 4,827 | 21,523 | 26,350 |
| December | | 20,280 | 28,172 | 48,452 |
| Totals | 9,547 | 208,445 | 383,465 | 601,457 |
| Totals, 1873 | 3,688 | 180,537 | 570,013 | 754,238 |
| Totals, 1872 | 3,884 | 274,249 | 564,477 | 842,610 |
| Totals, 1871 | 5,750 | 347,627 | 551,413 | 904,790 |
| Totals, 1870 | | 399,888 | 844,548 | 1,244,436 |

## *Treasure product, imports, &c.*

The receipts of treasure from all sources, through Wells, Fargo & Co.'s Express, during the past twelve months, as compared with the same period in 1873, have been as follows:

| | 1873. | 1874. |
|---|---|---|
| From northern and southern mines | $26,675,721 | $34,621,330 |
| Coastwise, north and south | 2,079,958 | 1,475,931 |
| Imports, foreign | 5,539,147 | 5,473,482 |
| Totals | 34,294,826 | 41,570,743 |

*Movement of coin in the interior.*

The following has been the circulation of coin, through Wells, Fargo & Co.'s Express, during 1874:

| | To interior. | From interior and coastwise. |
|---|---|---|
| January | $1, 394, 487 | $714, 862 |
| February | 1, 108, 611 | 714, 944 |
| March | 963, 314 | 575, 643 |
| April | 1, 312, 070 | 661, 032 |
| May | 1, 573, 629 | 837, 327 |
| June | 1, 552, 281 | 922, 986 |
| July | 1, 917, 597 | 857, 834 |
| August | 2, 001, 436 | 934, 119 |
| September | 2, 055, 615 | 980, 914 |
| October | 2, 294, 193 | 1, 010, 046 |
| November | 1, 950, 220 | 1, 005, 496 |
| December | 1, 813, 910 | 1, 085, 096 |
| Totals, 1874 | 19, 937, 363 | 10, 300, 299 |
| Totals, 1873 | 16, 428, 233 | 8, 084, 523 |
| Increase in 1874 | 3, 509, 130 | 2, 215, 776 |

*Currency movement.*

The annexed table exhibits the interior and coastwise receipts, (Wells, Fargo & Co.,) imports, (foreign,) and exports for the years 1872, 1873, and 1874:

| | 1872. | 1873. | 1874. |
|---|---|---|---|
| Interior receipts | $30, 478, 248 | $28, 755, 679 | $36, 097, 261 |
| Imports, (foreign) | 8, 060, 412 | 5, 539, 147 | 5, 473, 482 |
| Total | 38, 528, 660 | 34, 294, 826 | 41, 570, 743 |
| Exports | 29, 330, 436 | 24, 715, 126 | 30, 180, 632 |
| Currency movement | 9, 208, 224 | 9, 529, 700 | 11, 390, 111 |

*Mint statistics.*

The coinage at the branch mint in this city for 1874 compares with that in 1871, 1872, and 1873 as follows:

| | 1871. | 1872. | 1873. | 1874. |
|---|---|---|---|---|
| January | $1, 570, 000 | $840, 750 | $900, 000 | $1, 994, 000 |
| February | 1, 171, 725 | 1, 210, 000 | 1, 219, 400 | 279, 000 |
| March | 965, 000 | 1, 127, 750 | 1, 140, 000 | 3, 958, 000 |
| April | 1, 800, 000 | 1, 420, 000 | 1, 282, 000 | 1, 752, 000 |
| May | 2, 178, 050 | 2, 020, 000 | 2, 772, 000 | 367, 000 |
| June | 881, 000 | 686, 000 | 652, 000 | 2, 393, 000 |
| July | 2, 700, 000 | 2, 245, 000 | 3, 082, 000 | 2, 309, 000 |
| August | 1, 900, 000 | 730, 000 | 2, 131, 000 | 4, 320, 000 |
| September | 2, 210, 000 | 1, 264, 500 | 2, 264, 500 | 2, 570, 000 |
| October | 1, 689, 000 | 1, 895, 000 | 2, 658, 000 | 3, 204, 000 |
| November | 1, 684, 000 | 1, 525, 000 | 254, 500 | 96, 000 |
| December | 1, 218, 000 | 1, 436, 600 | 3, 720, 000 | 4, 087, 000 |
| Totals | 20, 026, 775 | 16, 389, 600 | 22, 075, 400 | 27, 329, 000 |

The description of coinage for 1872, 1873, and 1874 was as follows:

| | 1872. | 1873. | 1874. |
|---|---|---|---|
| Double eagles | $15, 600, 000 | $20, 812, 000 | $24, 375, 000 |
| Eagles | 173, 000 | 120, 000 | 50, 000 |
| Half-eagles | 202, 000 | 155, 000 | 35, 000 |
| Quarter-eagles | 25, 000 | 67, 500 | |
| Half-dollars | 290, 000 | 116, 500 | 197, 000 |
| Quarter-dollars | 26, 250 | 39, 000 | 98, 000 |
| Dimes | 19, 000 | 45, 500 | 24, 000 |
| Half-dimes | 36, 350 | 16, 200 | |
| Silver dollars | 9, 000 | 700 | |
| Trade-dollars | | 703, 000 | 2, 550, 000 |
| Totals | 16, 380, 600 | 22, 075, 400 | 27, 329, 000 |

*Annual receipts of coal at San Francisco.*

| Years. | Mount Diablo. | Coos Bay. | Bellingham Bay. | Vancouver Island. | Chili. | Australia. | English. | Cumberland. | Anthracite. | Queen Charlotte Island. | Sitka. | Seattle. | Rocky Mountain. | Saghalien. | Fuca Straits. | Japan. | Total. |
|---|---|---|---|---|---|---|---|---|---|---|---|---|---|---|---|---|---|
| | *Tons.* | *Tons.* | *Tons.* | *Tons.* | *Tons.* | *Tons.* | *Tons.* | *Tons.* | *Tons.* | *Tons.* | *Tons.* | *Tons.* | *Tons.* | *Tons.* | *Tons.* | *Tons.* | *Tons.* |
| 1860 | | 3, 145 | 5, 490 | 6, 655 | 1, 900 | 7, 850 | 6, 640 | 5, 970 | 39, 985 | | | | | | | | 77, 635 |
| 1861 | 6, 620 | 4, 630 | 10, 055 | 6, 475 | 12, 495 | 23, 370 | 23, 565 | 2, 975 | 26, 060 | | | | | | | | 110, 245 |
| 1862 | 23, 400 | 2, 815 | 10, 050 | 8, 870 | 5, 110 | 12, 590 | 16, 055 | 4, 970 | 36, 685 | | | | | | | | 120, 545 |
| 1863 | 43, 200 | 1, 185 | 7, 750 | 5, 745 | 1, 790 | 16, 890 | 14, 660 | 5, 670 | 38, 660 | | | | | | | | 135, 550 |
| 1864 | 50, 700 | 1, 200 | 11, 845 | 12, 785 | 2, 323 | 21, 160 | 18, 330 | 7, 275 | 41, 680 | | | | | | | | 167, 298 |
| 1865 | 60, 530 | 1, 500 | 14, 446 | 18, 181 | 1, 410 | 17, 610 | 9, 655 | 4, 230 | [illegible], 585 | | | | | | | | 150, 147 |
| 1866 | 84, 020 | 2, 120 | 11, 380 | 10, 852 | 1, 480 | 53, 700 | 7, 400 | 9, 524 | 12, 124 | | | | | | | | 192, 601 |
| 1867 | 109, 490 | 5, 415 | 8, 899 | 14, 829 | 14, 949 | 26, 619 | 7, 302 | 12, 177 | 48, 518 | | | | | 218 | 509 | | 248, 925 |
| 1868 | 132, 537 | 10, 524 | 13, 866 | 23, 348 | 8, 511 | 31, 590 | 29, 561 | 2, 292 | 29, 592 | | | | | 204 | | | 282, 025 |
| 1869 | 148, 722 | 14, 824 | 20, 552 | 14, 880 | 1, 114 | 75, 115 | 17, 386 | 11, 536 | 24, 844 | | | | | | | | 328, 973 |
| 1870 | 129, 761 | 20, 567 | 14, 355 | 12, 640 | 7, 350 | 83, 982 | 31, 196 | 9, 322 | 21, 320 | | | | | | | | 320, 493 |
| 1871 | 133, 485 | 28, 690 | 20, 284 | 15, 621 | 4, 161 | 38, 942 | 54, 191 | 6, 060 | 7, 231 | 565 | 18 | 4, 918 | 1, 025 | | | | 315, 194 |
| 1872 | 177, 232 | 32, 562 | 4, 100 | 26, 008 | 3, 682 | 115, 332 | 29, 190 | 10, 051 | 19, 618 | | | 14, 830 | 1, 862 | | | | 434, 467 |
| 1873 | 171, 741 | 38, 066 | 21, 211 | 31, 435 | 400 | 96, 435 | 52, 616 | 8, 857 | 18, 295 | | | 13, 572 | 1, 904 | | | 50 | 454, 582 |
| 1874 | 206, 255 | 44, 857 | 13, 685 | 51, 017 | | 139, 109 | 37, 826 | 14, 475 | 14, 263 | | | 9, 027 | 433 | | | | 531, 947 |

## *Quicksilver product of California for* 1874.

The following statement was prepared for me by Mr. Charles R. Yale, of the San Francisco Mining and Scientific Press. I regret that a more extended review of the quicksilver industry of the year, from the same skillful and experienced hand, miscarried in the mails, and was not received for use in the earlier chapters of this report:

| Mine. | County. | Flasks. |
|---|---|---|
| New Almaden | Santa Clara | 9, 084 |
| New Idria | Fresno | 7, 000 |
| Cerro Bonito | ...do | 900 |
| California | Napa | 3, 000 |
| Redington | Lake | 7, 200 |
| Manhattan | Napa | 620 |
| Phœnix | ...do | 685 |
| Buckeye | Colusa | 700 |
| Missouri | Sonoma | 200 |
| Oakland | ...do | 307 |
| California Borax | Lake | 570 |
| Great Western | ...do | 1, 464 |
| Saint John | Solano | 1, 900 |
| Washington | Napa | 200 |
| Other mines, producing less than 100 flasks each. | San Luis Obispo, Napa, Sonoma, and Colusa. | 324 |
| Total | | 34, 154 |
| Estimated product of mines refusing to make returns | | 200 |
| Total flasks | | 34, 254 |

*Quotations, dividends, and assessments upon stock dealt in at the San Francisco Stock and Exchange Board.*

[From the San Francisco Stock Report, December 25, 1874.]

| Name of company. | Bid. | Asked. | No. of assessments. | No. of feet in mine. | Total number of shares. | Total amount per share. | Dividends. | Total amount of assessments. | Amount of dividends disbursed. |
|---|---|---|---|---|---|---|---|---|---|
| CALIFORNIA MINES. | | | | | | | | | |
| Alpine | | | 7 | 1, 200 | 12, 000 | $6 50 | | $78, 000 | |
| Consolidated Amador | | | | 1, 850 | 30, 000 | | 9 | | $210, 000 |
| Bellevue | 1¼ | | 10 | 8, 000 | 20, 000 | 5 10 | | 101, 000 | |
| Cederberg | | | 1 | | 24, 000 | 50 | 4 | 12, 000 | 24, 000 |
| Calistoga | | | | 3, 600 | 60, 000 | | | | |
| Chariot Mill | 5 | | | | 30, 000 | | 3 | | 39, 000 |
| Eureka | 9½ | | | 1, 680 | 20, 000 | | 76 | | 2, 094, 000 |
| Genesee Valley | | | | | | | | | |
| Independent Gold-Mining Co | ½ | | 7 | 1, 800 | 25, 000 | 3 32 | | 83, 000 | |
| Keystone, (quartz) | | | | 10, 000 | | | 10 | | 5, 000 |
| Magenta | 2 | | 1 | | | 50 | | 10, 000 | |
| Mansfield | | | | 3, 300 | 33, 000 | | | | |
| Saint Lawrence Gold-Mining Co | | | 7 | 3, 000 | 10, 000 | 5 00 | | 50, 000 | |
| Saint Patrick Gold-Mining Co | 1 | | 9 | 18, 000 | 5, 000 | 7 00 | | 140, 000 | |
| Tecumseh | | | 9 | 3, 000 | 30, 000 | 2 40 | | 72, 000 | |
| Yule Gravel | | | 5 | 400 | 10, 000 | 90 | 9 | 9, 000 | 40, 000 |
| NEVADA MINES. | | | | | | | | | |
| Cope district: | | | | | | | | | |
| Excelsior | | | 1 | | 12, 000 | 2 00 | | 20, 000 | |
| Echo district: | | | | | | | | | |
| Rye Patch Mill and Mining Co | 3¼ | 3½ | 3 | 1, 600 | 20, 000 | 2 25 | 1 | 67, 500 | 15, 000 |
| Ely district. | | | | | | | | | |
| Amador Tunnel | | | 2 | 3, 000 | 30, 000 | 2 00 | | 60, 000 | |
| American Flag | 2 | 2¼ | 6 | | 40, 000 | 6 50 | | 195, 000 | |
| Alps | 40c. | ½ | 7 | 800 | 30, 000 | 3 25 | | 97, 500 | |
| Bowery | 30c. | | 6 | | 30, 000 | 3 15 | | 94, 500 | |

*Quotations, &c., at the San Francisco Stock and Exchange Board*—Continued.

| Name of company. | Bid. | Asked. | No. of assessments. | No. of feet in mine. | Total number of shares. | Total amount per share. | Dividends. | Total amount of assessments. | Amount of dividends disbursed. |
|---|---|---|---|---|---|---|---|---|---|
| Chapman |  |  | 4 |  | 30,000 | $1 25 |  | $37,500 |  |
| Cherry Creek | 1¼ | 1½ | 1 | 1,600 | 30,000 | 15 |  | 4,500 |  |
| Chief, (east extension) |  |  | 3 |  |  | 46 |  | 14,400 |  |
| Chief of the Hill |  | 15c. | 5 |  | 30,000 | 2 25 |  | 67,500 |  |
| Condor |  |  | 3 |  | 25,000 | 1 50 |  | 37,500 |  |
| Caroline | 30c. |  | 5 |  |  | 1 95 |  | 58,500 |  |
| Pioche, (west extension) |  |  | 5 |  | 35,000 | 1 60 |  | 56,000 |  |
| Harper |  |  | 1 |  | 30,000 | 10 |  |  |  |
| Hahn & Hunt | ½ |  | 10 | 3,600 | 30,000 | 7 05 |  | 279,000 |  |
| Ingomar | 35c. | 35c. | 7 | 1,000 | 40,000 | 1 75 |  | 70,000 |  |
| Ivanhoe |  |  | 2 |  | 30,000 | 50 |  | 15,000 |  |
| Jolly Traveler |  |  | 2 |  |  |  |  |  |  |
| Kentucky |  |  | 7 | 1,000 | 30,000 | 4 25 |  | 127,500 |  |
| Lehigh |  |  |  | 1,000 | 30,000 |  |  |  |  |
| Lillian Hall |  |  | 2 | 1,000 | 15,000 | 75 |  |  |  |
| Louise |  |  | 3 | 2,400 | 30,000 | 63 |  | 18,995 |  |
| Meadow Valley | 6 | 6½ | 7 |  | 60,000 | 3 00 | 17 | 18,000 | $1,200,000 |
| Marion |  |  | 1 |  | 30,000 | 25 |  | 18,000 |  |
| Meadow Valley, (east) |  |  | 2 |  |  | 60 |  |  |  |
| Mocking-Bird |  |  |  | 1,200 | 30,000 |  |  |  |  |
| Newark | 1 | 1 | 9 | 800 | 32,000 | 8 05 |  | 256,600 |  |
| Orient |  |  | 2 | 1,000 | 20,000 | 35 |  | 7,000 |  |
| Page & Panaca | 10c. |  | 8 | 2,400 | 40,000 | 4 75 |  | 190,000 |  |
| Peavine |  |  | 5 | 1,000 | 30,000 | 1 25 |  | 37,500 |  |
| Pioche | 4 |  | 8 | 1,000 | 20,000 | 9 00 | 3 | 180,000 | 60,000 |
| Portland | ½ |  | 5 |  |  | 2 36 |  | 71,000 |  |
| Raymond & Ely | 22½ |  | 2 | 5,000 | 30,000 | 6 00 | 23 | 180,000 | 3,075,000 |
| Silver Peak | ½ | 1 | 5 |  | 30,000 | 3 00 |  | 90,000 |  |
| Stanford |  |  | 1 |  |  | 10 |  | 3,000 |  |
| Sterling |  |  | 1 |  | 30,000 | 50 |  | 15,000 |  |
| Spring Mount |  |  | 3 |  | 35,000 | 1 25 |  | 43,750 |  |
| Spring Mountain Tunnel |  |  | 8 |  | 20,000 | 7 25 |  | 33,000 |  |
| Washington and Creole | ⅞ | 1½ | 13 | 1,520 | 30,000 | 8 75 |  | 262,500 |  |
| Watson | ⅜ |  | 1 |  | 30,000 | 1 00 |  | 30,000 |  |
| Esmeralda: |  |  |  |  |  |  |  |  |  |
| Juniata Consolidated |  |  | 2 | 5,000 | 50,000 | 1 65 |  | 82,500 |  |
| Eureka district: |  |  |  |  |  |  |  |  |  |
| Adams Mill |  |  | 5 |  | 50,000 | 1 00 |  | 50,000 |  |
| Eureka Consolidated | 13¼ | 13½ |  |  | 50,000 |  | 14 |  | 650,000 |
| Jackson |  |  | 6 |  | 50,000 | 1 05 |  | 52,500 |  |
| Phœnix | ½ |  | 14 |  | 50,000 | 6 75 |  | 337,500 |  |
| Star Consolidated |  |  | 3 | 18,000 | 50,000 | 60 |  | 30,000 |  |
| Philadelphia district: |  |  |  |  |  |  |  |  |  |
| Belmont | 13½ | 14 | 4 |  | 50,000 | 4 50 |  | 225,000 |  |
| El Dorado, (north) | 1¼ | 1½ | 1 |  | 25,000 | 50 |  | 12,500 |  |
| El Dorado, (south,) Consolidated | 2 | 2¼ | 4 | 2,600 | 40,000 | 5 00 |  | 200,000 |  |
| Josephine |  |  | 1 |  | 25,000 | 15 |  | 3,750 |  |
| Monitor Belmont | 2½ | 2½ | 3 |  |  | 1 50 |  | 75,000 |  |
| North Belmont | 15c. |  | 2 |  | 50,000 | 20 |  | 10,000 |  |
| Prussian | 4 | 4½ | 2 |  |  | 50 |  |  |  |
| Quintero | 25c. | 25c. | 3 | 2,000 | 50,000 | 30 |  | 15,000 |  |
| Washoe: |  |  |  |  |  |  |  |  |  |
| Alamo |  |  | 1 |  |  | 25 |  | 7,500 |  |
| American Flat, (south) |  |  |  |  | 48,000 |  |  |  |  |
| Alpha Consolidated | 25 | 27 | 6 | 300 | 30,000 | 5 00 |  | 150,000 |  |
| Alta | ½ |  | 1 | 3,600 | 36,000 | 10 |  | 3,600 |  |
| American Consolidated |  |  |  |  |  |  |  |  |  |
| American Flat | 7 | 7½ | 4 |  | 30,000 | 2 50 |  | 75,000 |  |
| Andes | 3½ | 4 | 3 | 1,000 | 50,000 | 1 50 |  | 75,005 |  |
| Bacon Mill and Mining Co | 7¾ | 8¾ |  | 65 | 4,000 |  |  |  |  |
| Baltic Consolidated |  |  |  | 3,000 | 37,500 |  |  |  |  |
| Baltimore Consolidated | 7¾ | 8 | 7 |  | 54,000 | 6 50 |  | 351,000 |  |
| Best & Belcher | 64 | 68 | 9 | 545 | 100,800 | 6 08 |  | 136,192 |  |
| Belcher | 50 | 55 | 8 | 1,040 | 104,000 | 6 35 | 33 | 660,400 | 14,135,000 |
| Bowers |  |  |  | 20 | 5,000 |  |  |  |  |
| Bullion | 18 | 18½ | 45 | 2,500 | 100,000 | 18 02 |  | 1,802,000 |  |
| Buckeye |  |  | 12 |  | 16,000 | 11 75 |  | 188,000 |  |
| Caledonia | 24 | 26 | 9 | 5,000 | 20,000 | 25 00 |  | 500,000 |  |
| California | 475 | 490 |  | 600 | 108,000 |  |  |  |  |
| Central Comstock |  |  |  |  |  |  |  |  |  |
| Challenge Consolidated | 12 | 13 |  | 90 | 50,000 |  |  |  |  |
| Chollar Potosi | 80 | 86 | 5 | 2,800 | 28,000 | 26 50 | 39 | 742,000 | 3,080,000 |
| Confidence | 46 | 49 | 10 | 130 | 24,960 | 9 75 | 6 | 243,260 | 78,000 |

*Quotations, &c., at the San Francisco Stock and Exchange Board*—Continued.

| Name of company. | Bid. | Asked. | No. of assessments. | No. of feet in mine. | Total number of shares. | Total amount per share. | Dividends. | Total amount of assessments. | Amount of dividends disbursed. |
|---|---|---|---|---|---|---|---|---|---|
| Consolidated Virginia | 492 | 502½ | 15 | 730 | 108,000 | $17 40 | 8 | $411,200 | $2,592,000 |
| Consolidated Gold Hill, (quartz) | | | 1 | 34½ | 20,000 | 75 | | | |
| Consolidated Washoe | | | | 4,200 | 40,000 | | | | |
| Crown Point Extension | | | | 1,000 | 40,000 | | | | |
| Crown Point | 47 | 49¼ | 21 | 600 | 100,000 | 6 28 | 49 | 623,370 | 11,388,000 |
| Crown Point Ravine | ⅜ | ½ | 1 | | | 50 | | 15,000 | |
| Daney | 2¼ | | 11 | 2,000 | 24,000 | 8 25 | 2 | 198,000 | 56,000 |
| Dardanelles | | 21 | 1 | 200 | 60,000 | 1 00 | | 60,000 | |
| Dayton | 4½ | 4¾ | 1 | 1,600 | 100,000 | 1 00 | | 100,000 | |
| Dexter | | | | 1,200 | 60,000 | | | | |
| Green | | | 1 | 1,200 | 24,000 | 50 | | 10,000 | |
| Eclipse, (Winters & Plato) | 13 | 15¼ | 1 | 70 | 25,000 | 50 | | 12,500 | |
| Empire Mill | 16 | 16½ | 16 | 75 | 50,000 | 7 82 | 21 | 391,400 | 713,500 |
| Europa | | | 2 | 1,000 | 20,000 | 50 | | 10,000 | |
| Exchequer | 240 | 240 | 10 | 400 | 8,000 | 22 50 | | 180,000 | |
| "420" | | | 8 | 420 | 30,000 | | | | |
| Fairmount | | | | | | | | | |
| Florida | | | | | 50,000 | | | | |
| Flowery | | | 1 | 3,600 | 30,000 | | | | |
| Franklin | ⅛ | | | | 30,000 | | | | |
| Genesee | | | | | 30,000 | | | | |
| Globe | | | 8 | | 20,000 | 6 00 | | 120,000 | |
| Globe Consolidated | | | 4 | | 38,000 | 3 00 | | 114,000 | |
| Golden Swan | | | | | 50,000 | | | | |
| Gould & Curry | 32 | 34 | 23 | 879 | 48,000 | 32 20 | 36 | 1,532,000 | 3,826,800 |
| Hale & Norcross | 63 | 66 | 44 | 400 | 16,000 | 51 63 | 36 | 1,450,000 | 1,598,000 |
| Imperial | 17 | 18½ | 20 | 184 | 100,000 | 12 70 | 30 | 1,270,000 | 1,067,500 |
| Independent and Omega | | | | 60,000 | | | | | |
| Indus | 3 | | 2 | | 30,000 | 20 | | 6,000 | |
| Insurance | | | 2 | 950 | 30,000 | 35 | | 10,500 | |
| Jacob Little | | | 2 | 2,000 | 37,200 | 1 25 | | 46,500 | |
| Julia | 10 | 11 | 20 | 1,860 | 30,000 | 14 29 | | 428,700 | |
| Justice | 74 | 80 | 12 | 2,000 | 21,000 | 20 50 | | 431,500 | |
| Kentuck | 20 | 21 | 11 | | 30,000 | 8 00 | 32 | 240,000 | 1,252,000 |
| Keystone | | | | | | | | | |
| Knickerbocker | 5¼ | 6 | 10 | 95 | 24,000 | 10 50 | | 252,000 | |
| Kossuth | 2⅝ | | 2 | 2,700 | 27,000 | 50 | | 54,000 | |
| Lady Bryan | | | 4 | 1,200 | 50,000 | 2 00 | | 100,000 | |
| Lady Washington | 4¼ | 4⅜ | 1 | | 36,000 | 50 | | 18,000 | |
| Leo | 1 | 1¼ | 3 | 1,200 | | 85 | | 27,200 | |
| Lower Comstock | | | | | 86,000 | | | 20,000 | |
| Magenta | | | | | | | | | |
| Mides | | | | | 30,000 | | | | |
| Mint | 35c. | 40c. | 8 | | 50,000 | 80 | | 40,000 | |
| McMeans | | | | 3,600 | 36,000 | | | | |
| Nevada | | | | 3,000 | 40,000 | | | | |
| New York Consolidated | 3¾ | 4 | 11 | 3,600 | 3,600 | 5 75 | | 207,000 | |
| North Carson | | | | 1,500 | 100,000 | | | | |
| Occidental | 3⅛ | 4 | 2 | 800 | 40,000 | 1 25 | | 42,500 | 20,000 |
| Original Gold Hill | | | 2 | | 30,000 | 1 00 | | 30,000 | |
| Ophir | 176 | 182 | 28 | 1,400 | 100,800 | 17 33 | 22 | 1,832,800 | 1,394,400 |
| Overmann | 73 | 76 | 30 | 1,200 | 38,400 | 39 86 | | 1,531,080 | |
| Patten | | | | 1,200 | 50,000 | | | | |
| Phil. Sheridan | 1 | 1¼ | 1 | 1,200 | 24,000 | 50 | | 12,000 | |
| Pictou | ¾ | | 6 | 2,000 | 30,000 | 1 27½ | | 38,200 | |
| Rock Island | 5½ | 5¾ | 5 | | 24,000 | 4 50 | | 108,000 | |
| Savage | 130 | 142 | 16 | 800 | 16,000 | 109 00 | 52 | 1,744,000 | 4,460,000 |
| Segregated Belcher | 125 | 135 | 14 | 180 | 6,400 | 33 25 | | 212,800 | |
| Senator | 1¼ | 1¾ | 10 | | 24,000 | 3 15 | | 75,600 | |
| Sierra Nevada | 24¾ | 26½ | 39 | | 100,000 | 31 00 | 11 | 700,000 | 102,500 |
| Segregated Caledonia | 1½ | | 2 | | 10,000 | | | | |
| Segregated Rock Island | | | | | 60,000 | | | | |
| South Comstock | 2¼ | | 1 | 2,000 | 40,000 | 50 | | 20,000 | |
| South Overmann | | | 1 | | 30,000 | 50 | | 15,000 | |
| South Star | | | | 1,200 | | | | | |
| Sutro | 3¼ | 3½ | 1 | | 24,000 | 50 | | 12,000 | |
| Silver Cloud | | | 1 | | 32,000 | 25 | | 8,000 | |
| Silver Hill | 12½ | 13½ | 4 | | 54,000 | 8 00 | | 432,000 | |
| Succor Mill and Mining Co | 5¾ | 5¾ | 10 | 7,600 | 22,800 | 4 75 | 2 | 273,900 | 22,800 |
| Trench | | | 1 | 20 | 5,000 | 1 00 | | 5,000 | |
| Tyler | 25c. | | 7 | 2,200 | 33,000 | 1 90 | | 62,700 | |
| Union Consolidated | 86 | 92 | 6 | 825 | 20,000 | 3 00 | | 60,000 | |
| Utah | 10½ | 12½ | 7 | | 20,000 | 5 00 | | 100,000 | |

*Quotations, &c., at the San Francisco Stock and Exchange Board*—Continued.

| Name of company. | Bid. | Asked. | No. of assessments. | No. of feet in mine. | Total number of shares. | Total amount per share. | Dividends. | Total amount of assessments. | Amount of dividends disbursed. |
|---|---|---|---|---|---|---|---|---|---|
| Woodville | 3¼ | 3¼ | 8 | 1,400 | 28,000 | $7 75 | .... | $249,000 | .......... |
| Yellow Jacket | 150 | 154 | 19 | 1,200 | 24,000 | 88 25 | 25 | 2,118,000 | $2,184,000 |
| White Pine district: | | | | | | | | | |
| General Lee | ½ | ...... | 6 | 1,000 | 20,000 | 75 | .... | 10,000 | .......... |
| Hayes | 1½ | ...... | 5 | ........ | ........ | 1 25 | .... | 50,000 | .......... |
| Mammoth | 25c. | ...... | 18 | 1,800 | 36,000 | 2 55 | .... | 91,800 | .......... |
| McMahon | ...... | ...... | 7 | ........ | ........ | 1 75 | .... | 52,500 | .......... |
| Noonday | ...... | ...... | 15 | 1,000 | 20,000 | 3 02 | .... | 60,300 | .......... |
| Original Hidden Treasure | 3 | 3¼ | 11 | ........ | 21,333 | 15 50 | 1 | 330,061 | 31,999 |
| Silver Wave | ...... | ...... | 11 | 1,600 | 20,000 | 8 00 | .... | 160,000 | .......... |
| Ward Beecher | ...... | ...... | 3 | ........ | 24,000 | 3 50 | .... | 84,000 | .......... |
| IDAHO MINES. | | | | | | | | | |
| Empire | ½ | ⅝ | 8 | ........ | 25,000 | 10 50 | .... | 250,000 | .......... |
| Golden Chariot | 3½ | 3¾ | 11 | 750 | 10,000 | 28 00 | 12 | 510,000 | 410,000 |
| Ida Elmore | 1⅛ | ...... | 15 | ........ | 30,000 | 14 16 | 6 | 575,000 | 60,000 |
| Mahogany | 4½ | ...... | 14 | 720 | 25,000 | 13 90 | .... | 348,800 | 15,000 |
| North Oro Fino | ...... | ...... | 8 | 600 | 10,000 | 50 | .... | 5,000 | .......... |
| Poorman | ...... | ...... | 1 | ........ | 50,000 | 50 | .... | .......... | .......... |
| Red Jacket | 6 | ...... | 5 | ........ | ........ | 4 50 | .... | 90,000 | .......... |
| Silver Cord | ½ | ...... | 6 | ........ | 24,000 | 4 50 | .... | 78,000 | .......... |
| South Chariot | 1⅛ | ...... | 11 | 660 | 20,000 | 9 00 | .... | 225,000 | .......... |
| War Eagle | 3¼ | 3⅝ | 8 | 1,000 | 10,000 | 10 00 | .... | 100,000 | .......... |
| OREGON MINES. | | | | | | | | | |
| Virtue | ...... | ...... | 4 | 2,600 | 20,000 | 6 00 | ... | 120,000 | .......... |
| UTAH MINES. | | | | | | | | | |
| Deseret Consolidated | ...... | ...... | 1 | 2,400 | 30,000 | 20 | .... | 6,000 | .......... |
| Wellington | ...... | ...... | 4 | ........ | 50,000 | 10 | .... | 55,000 | .......... |

*Highest and lowest prices of mining-stocks for the twelve months ending December* 31, 1874.

[Compiled by J. Henry Applegate, jr., for the Commercial Herald and Market Review.]

| Companies. | January. | | February. | | March. | | April. | |
|---|---|---|---|---|---|---|---|---|
| | Highest price. | Lowest price. | Highest price. | Lowest price. | Highest price. | Lowest price. | Highest price. | Lowest price. |
| Alpha | $100 00 | $50 00 | {$60 00 / 16 00} | {$57 00 / 8 00} | $16 50 | $11 50 | $16 50 | $11 50 |
| Alps | 1 00 | 60 | 62½ | 50 | 75 | 40 | 75 | 40 |
| American Flag | 8 00 | 4 12½ | 5 25 | 4 00 | 5 12½ | 4 12½ | 5 00 | 3 50 |
| Arizona and Utah | 3 25 | 1 75 | 2 00 | 1 25 | ........ | ........ | ........ | ........ |
| Adams Hill | 10 | 10 | ........ | ........ | ........ | ........ | ........ | ........ |
| Alamo | 6 75 | 4 00 | 10 00 | 7 00 | 5 00 | 5 00 | ........ | ........ |
| American Flat | 7 75 | 5 00 | 6 00 | 5 00 | 7 00 | 5 00 | 6 25 | 5 50 |
| American Consolidated | 3 00 | 75 | 1 00 | 1 00 | ........ | ........ | ........ | ........ |
| Andes | 4 25 | 1 25 | 2 00 | 1 25 | 1 75 | 75 | 1 62½ | 1 00 |
| Alta | ........ | ........ | ........ | ........ | ........ | ........ | ........ | ........ |
| American Flat, (south) | ........ | ........ | ........ | ........ | ........ | ........ | ........ | ........ |
| Belcher | 120 00 | 95 50 | 103 25 | 81 00 | 89 50 | 77 00 | 91 25 | 76 00 |
| Baltimore Consolidated | 9 50 | 6 12½ | 8 00 | 5 00 | 8 50 | 6 00 | 7 87½ | 6 25 |
| Belmont | 31 00 | 5 25 | 9 00 | 3 75 | 7 75 | 6 00 | 7 87½ | 5 00 |
| Best and Belcher | 43 00 | 22 00 | 28 00 | 22 00 | 37 00 | 20 00 | 37 00 | 26 50 |
| Buckeye | 2 50 | 1 37½ | 2 50 | 2 00 | 2 00 | 2 00 | 2 00 | 1 00 |
| Bacon | 7 00 | 4 75 | ........ | ........ | 5 50 | 4 00 | ........ | ........ |
| Bowery | 1 00 | 50 | 75 | 35 | 2 50 | 50 | 50 | 20 |
| Bullion | 40 00 | 25 50 | {37 00 / 8 00} | {22 00 / 7 50} | } 9 00 | 6 75 | 9 00 | 8 00 |
| Baltic Consolidated | 2 00 | 2 00 | ........ | ........ | ........ | ........ | ........ | ........ |
| Bellevue | ........ | ........ | ........ | ........ | 2 25 | 2 25 | 2 00 | 1 75 |

*Highest and lowest price of mining-stocks for* 1874—Continued.

| Companies. | January. | | February. | | March. | | April. | |
|---|---|---|---|---|---|---|---|---|
| | Highest price. | Lowest price. | Highest price. | Lowest price. | Highest price. | Lowest price. | Highest price. | Lowest price. |
| Bowers | | | | | | | | |
| Consolidated Virginia | $97 50 | $54 00 | $70 00 | $62 50 | $86 75 | $67 00 | $90 00 | $82 50 |
| Chollar Potosi | 95 00 | 63 00 | 74 00 | 62 00 | 70 00 | 60 00 | 71 00 | 60 50 |
| Crown Point | 135 00 | 87 00 | 105 00 | 82 00 | 95 50 | 86 00 | 97 00 | 80 00 |
| Confidence | 16 50 | 8 50 | 11 00 | 7 50 | 11 25 | 7 75 | 10 25 | 8 12½ |
| Caledonia | 44 00 | 24 00 | 39 50 | 25 00 | 38 00 | 28 00 | 39 00 | 20 50 |
| Cederberg | 4 50 | 3 00 | 3 00 | 3 00 | 3 00 | 2 75 | 3 50 | 3 00 |
| Chief of the Hill | 1 75 | 75 | 1 00 | 10 | 90 | 50 | 62½ | 50 |
| Charter Oak | 1 50 | 50 | 1 00 | 1 00 | | | 1 00 | 1 00 |
| Chief, (east extension) | 20 | 15 | 20 | 20 | 20 | 12½ | | |
| Central | 130 00 | 90 00 | | | | | | |
| Central No. 2 | 46 00 | 36 00 | | | | | | |
| Caroline | 3 50 | 50 | 1 00 | 12 50 | 1 00 | 25 | 35 | 25 |
| Chariot Mill | 7 25 | 4 50 | 6 25 | 5 00 | 5 90 | 5 00 | 5 00 | 5 00 |
| Challenge | 6 25 | 5 00 | | | | | | |
| Consolidated Washoe | 2 50 | 1 25 | | | 6 00 | 5 00 | | |
| California | 48 50 | 30 50 | 39 75 | 34 00 | 50 00 | 33 00 | 46 00 | 40 50 |
| Chapman | 50 | 50 | | | | | | |
| Consolidated Amador | | | | | 35 00 | 35 00 | | |
| Condor | | | | | | | | |
| Cherry Creek | | | | | | | | |
| Crown Point Ravine | | | | | | | | |
| Calaveras | | | | | | | | |
| Calistoga | | | | | | | | |
| Dayton | 10 00 | 5 50 | 8 00 | 7 00 | 9 00 | 6 50 | 7 25 | 5 87½ |
| Dardanelles | 7 00 | 4 00 | 5 00 | 4 50 | 5 00 | 2 12½ | 5 00 | 4 50 |
| Dexter | 2 50 | 2 00 | 2 25 | 2 25 | | | | |
| Daney | 4 50 | 2 00 | 4 50 | 3 12½ | 4 00 | 2 00 | 3 00 | 2 25 |
| Empire Mill | 8 00 | 5 00 | 5 50 | 4 37½ | 5 75 | 4 50 | 5 37½ | 4 25 |
| Eureka Consolidated | 15 00 | 10 00 | 14 25 | 12 75 | 15 00 | 12 50 | 18 00 | 14 25 |
| Exchequer | 53 00 | 25 00 | 27 50 | 12 50 | 44 00 | 22 00 | 44 00 | 32 50 |
| Eclipse | 8 00 | 5 50 | 6 00 | 4 75 | 6 50 | 5 00 | 5 75 | 4 50 |
| Eureka | 18 00 | 9 00 | 11 50 | 9 00 | 10 75 | 10 00 | 9 50 | 9 00 |
| Empire | 9 25 | 4 50 | 8 50 | 7 25 | 9 00 | 5 75 | 8 25 | 6 50 |
| El Dorado (south) Consolidated | 9 50 | 4 87½ | 7 25 | 6 00 | 7 25 | 6 25 | 8 75 | 6 87½ |
| El Dorado (north) Consolidated | 5 50 | 5 00 | 5 00 | 5 00 | | | 2 50 | 2 50 |
| Europa | | | | | 2 75 | 2 00 | | |
| Empress | | | | | | | | |
| Franklin | 3 00 | 1 50 | 2 50 | 1 00 | 1 50 | 1 25 | 1 50 | 1 00 |
| Fairmount | 3 37½ | 1 25 | 1 00 | 1 00 | | | | |
| Florida | 5 00 | 2 50 | 3 00 | 2 00 | 3 00 | 2 75 | 3 00 | 2 00 |
| Gould and Curry | 44 00 | 20 00 | 24 00 | 19 00 | 35 00 | 18 25 | 32 25 | 23 00 |
| Globe | 2 50 | 1 50 | 2 75 | 1 00 | 2 12½ | 1 12½ | 2 50 | 1 00 |
| Golden Chariot | 23 50 | 18 50 | 21 50 | 18 00 | 19 00 | 14 50 | 19 00 | 13 00 |
| Gold Hill Quartz | 4 50 | 4 00 | | | | | 2 00 | 1 50 |
| Greene | 3 25 | 1 00 | | | | | | |
| General Lee | | | | | 1 25 | 30 | 75 | 75 |
| Hale & Norcross | 92 00 | 52 00 | 65 50 | 46 00 | 58 00 | 38 00 | 59 00 | 40 00 |
| Huhn & Hunt | 2 12½ | 1 25 | 1 87½ | 1 12½ | 1 75 | 1 25 | 2 00 | 87½ |
| Hayes | 1 50 | 75 | 87½ | 37½ | 1 25 | 50 | 75 | 62½ |
| Hermes | | | | | 3 00 | 1 00 | | |
| Hartford | | | | | | | | |
| Imperial | 10 00 | 6 00 | 7 00 | 5 62½ | 7 87½ | 5 87½ | 7 00 | 5 87½ |
| Ingomar | 60 | 40 | 40 | 25 | 35 | 25 | 25 | 25 |
| Independent | 2 25 | 1 00 | 1 00 | 1 00 | 1 50 | 87½ | 1 50 | 75 |
| Ida Elmore | 7 00 | 1 25 | 2 75 | 1 25 | 2 87½ | 1 25 | 3 25 | 2 50 |
| Independent and Omega | 2 50 | 2 00 | 2 50 | 2 50 | | | | |
| Idaho, (Grass Valley) | | | | | | | | |
| Indus | | | | | | | | |
| Julia | 11 50 | 6 50 | 7 00 | 2 50 | 6 00 | 2 75 | 5 62½ | 3 25 |
| Justice | 17 00 | 9 00 | 15 50 | 8 00 | 12 00 | 8 50 | 11 25 | 7 50 |
| Jackson | 45 | 20 | 35 | 30 | 33½ | 30 | 30 | 25 |
| Jacob Little | 5 00 | 3 50 | 4 25 | 3 00 | 3 00 | 2 00 | | |
| Josephine Consolidated | 2 50 | 2 00 | | | | | | |
| Kentuck | 36 00 | 19 50 | 29 00 | 18 00 | 25 50 | 18 00 | 25 00 | 14 25 |
| Knickerbocker | 9 00 | 5 00 | 6 25 | 3 50 | 7 37½ | 3 25 | 6 75 | 5 00 |
| Kentucky | 60 | 31½ | 50 | 50 | 30 | 20 | | |
| K. K. Consolidated | 5 00 | 5 00 | | | | | | |
| Kossuth | 14 00 | 6 50 | 7 00 | 6 00 | 7 00 | 5 00 | 7 00 | 5 00 |
| Lady Bryan | 3 00 | 37½ | 50 | 25 | 75 | 50 | 62½ | 10 |
| Leo | 1 00 | 65 | 85 | 80 | 2 75 | 60 | 1 37½ | 87½ |
| Louise | 40 | 40 | | | | | | |
| Lady Washington | 5 75 | 4 50 | 5 00 | 4 00 | 4 87½ | 3 50 | 4 00 | 3 00 |

*Highest and lowest price of mining-stocks for* 1874—Continued.

| Companies. | January. | | February. | | March. | | April. | |
|---|---|---|---|---|---|---|---|---|
| | Highest price. | Lowest price. | Highest price. | Lowest price. | Highest price. | Lowest price. | Highest price. | Lowest price. |
| Lower Comstock | $2 00 | $1 50 | | | | | | |
| Meadow Valley | 15 50 | 10 25 | $12 25 | $10 25 | $12 50 | $10 25 | $11 50 | $8 87 |
| Monitor Belmont | 8 00 | 3 75 | 5 25 | 4 50 | 6 00 | 4 50 | 4 87½ | 4 00½ |
| McMahon | 30 | 30 | ... | ... | ... | ... | ... | ... |
| Mammoth | 50 | 25 | 30 | 25 | 35 | 20 | 25 | 25 |
| Mahogany | 9 50 | 4 50 | 8 00 | 2 25 | 3 00 | 1 50 | 6 50 | 3 25 |
| Mint | 60 | 35 | 50 | 15 | 50 | 20 | 50 | 30 |
| Midas | 5 00 | 5 00 | ... | ... | ... | ... | ... | ... |
| Magenta | ... | ... | ... | ... | ... | ... | ... | ... |
| Mansfield | ... | ... | ... | ... | ... | ... | ... | ... |
| Mexican | ... | ... | ... | ... | ... | ... | ... | ... |
| Newark | 4 00 | 3 00 | 4 00 | 1 25 | 3 00 | 1 50 | 2 87½ | 1 12½ |
| New York Consolidated | 5 50 | 3 62½ | 4 50 | 2 00 | 4 25 | 2 25 | 4 50 | 2 75 |
| Nevada | 2 75 | 1 50 | 1 50 | 1 12½ | 2 25 | 1 00 | 1 75 | 1 25 |
| North Belmont | 2 00 | 1 25 | 75 | 25 | 75 | 20 | 40 | 25 |
| North Carson | ... | ... | ... | ... | ... | ... | ... | ... |
| North Utah | ... | ... | ... | ... | ... | ... | ... | ... |
| Ophir | $315 00<br>50 00 | 155 00<br>32 00 | 40 50 | 26 50 | 29 00 | 19 25 | 26 37½ | 16 50 |
| Overmann | 144 00 | 72 50 | 114 50 | 43 00 | 60 00 | 40 50 | 65 00 | 46 00 |
| Original Hidden Treasure | 4 50 | 3 00 | 6 00 | 6 00 | 9 00 | 3 50 | 9 00 | 4 50 |
| Occidental | 6 50 | 3 00 | 3 50 | 3 00 | 4 50 | 3 50 | 4 50 | 3 00 |
| Original Gold Hill | ... | ... | ... | ... | ... | ... | ... | ... |
| Pioche, (west) | 87½ | 75 | ... | ... | ... | ... | ... | ... |
| Peavine | 50 | 10 | 50 | 50 | 30 | 20 | ... | ... |
| Pioche | 9 00 | 5 50 | 8 00 | 5 00 | 6 75 | 5 00 | 6 50 | 4 00 |
| Phœnix | 80 | 30 | 10 | 10 | 7 | 7 | 25 | 25 |
| Page & Panaca | 3 00 | 1 25 | 3 12½ | 1 50 | 4 00 | 1 50 | 3 87½ | 1 00 |
| Portland | ... | ... | 1 50 | 50 | 2 50 | 1 00 | 1 87½ | 20 |
| Picton | 2 00 | 50 | 62½ | 50 | ... | ... | 30 | 25 |
| Prussian | 3 25 | 3 00 | 3 00 | 3 00 | 3 00 | 3 00 | 3 00 | 3 00 |
| Patten | ... | ... | 2 00 | 2 00 | ... | ... | 1 50 | 1 50 |
| Poorman | ... | ... | ... | ... | ... | ... | ... | ... |
| Pacific | ... | ... | ... | ... | ... | ... | ... | ... |
| Phil. Sheridan | ... | ... | ... | ... | ... | ... | ... | ... |
| Quintero | 2 00 | 1 50 | 25 | 25 | 25 | 25 | 95 | 25 |
| Raymond & Ely | 57 00 | 25 00 | 36 00 | 29 00 | 33 00 | 25 00 | 28 00 | 20 00 |
| Red Jacket, (Idaho) | 3 00 | 1 25 | 2 75 | 2 00 | 2 50 | 1 75 | 2 00 | 1 50 |
| Rye Patch | 10 00 | 6 50 | 7 00 | 4 00 | 4 50 | 2 12½ | 3 75 | 2 50 |
| Rock Island | 4 75 | 1 75 | 2 75 | 2 00 | 1 75 | 1 25 | 1 62½ | 1 62½ |
| Red Jacket, (Gold Hill) | 40 | 10 | 25 | 25 | ... | ... | 50 | 50 |
| Silver Cloud | 3 00 | 1 00 | ... | ... | ... | ... | ... | ... |
| Savage | 149 00 | 78 00 | 120 50 | 80 00 | 97 00 | 71 00 | 91 00 | 58 00 |
| Segregated Belcher | 190 00 | 100 00 | 123 00 | 80 00 | 110 00 | 81 00 | 107 00 | 83 00 |
| Succor | 4 50 | 2 00 | 4 00 | 2 87½ | 3 25 | 2 75 | 3 00 | 2 00 |
| Sierra Nevada | 34 00 | 20 00 | 24 50 | 17 00 | 26 25 | 18 50 | 26 00 | 18 00 |
| Silver Peak | 1 50 | 50 | 50 | 25 | 87½ | 75 | 1 62½ | 87½ |
| Silver Hill | 15 50 | 9 75 | 12 00 | 9 25 | 12 00 | 9 00 | 10 50 | 8 50 |
| Standard | 50 | 50 | ... | ... | ... | ... | ... | ... |
| South Chariot | 16 00 | 12 00 | 27 00 | 14 00 | 25 00 | 22 50 | 31 00 | 24 50 |
| South Overmann | 1 75 | 75 | 1 00 | 50 | 1 00 | 40 | 50 | 50 |
| Spring Valley Water | ... | ... | 84 50 | 84 00 | 86 00 | 86 00 | 90 00 | 87 00 |
| San Francisco Gas-Light | 68 00 | 68 00 | 68 00 | 68 00 | 70 00 | 69 50 | 73 50 | 70 00 |
| Senator | 3 25 | 1 75 | 1 50 | 1 50 | 1 50 | 1 25 | 75 | 62½ |
| South Comstock | 2 50 | 2 50 | ... | ... | ... | ... | ... | ... |
| Segregated Rock Island | 1 62½ | 65 | 1 25 | 80 | 1 12½ | 87½ | 1 00 | 62½ |
| Saint Patrick | 7 75 | 6 00 | 6 50 | 4 00 | 4 00 | 2 25 | 2 87½ | 1 50 |
| Sutro | 6 00 | 4 00 | 4 50 | 3 50 | 3 00 | 3 00 | 3 00 | 50 |
| South Star | 3 75 | 2 25 | 3 00 | 2 75 | ... | ... | 3 00 | 3 00 |
| Silver Cord | ... | ... | ... | ... | ... | ... | ... | ... |
| Segregated Caledonia | ... | ... | ... | ... | ... | ... | ... | ... |
| Trench | 9 25 | 5 00 | 4 50 | 4 00 | 5 50 | 3 00 | 5 00 | 5 00 |
| Tyler | 2 25 | 1 00 | 1 00 | 75 | 1 25 | 80 | 1 25 | 75 |
| Union Consolidated | 30 50 | 14 50 | 22 50 | 12 50 | 19 00 | 13 50 | 19 50 | 15 00 |
| Utah | 12 00 | 4 50 | 9 00 | 5 75 | 6 50 | 5 50 | 6 00 | 2 .5 |
| Virtue | 25 | 25 | ... | ... | ... | ... | ... | ... |
| Woodville | 3 00 | 1 50 | 2 50 | 1 12½ | 2 00 | 1 50 | 3 00 | 1 62½ |
| Washington and Creole | 4 87½ | 6 50 | 4 00 | 1 25 | 4 00 | 1 62½ | 4 62½ | 2 00 |
| War Eagle | 3 50 | 2 00 | 4 50 | 2 00 | 4 00 | 3 25 | 2 50 | 1 75 |
| Whitman | 3 00 | 2 00 | 2 00 | 1 00 | 2 00 | 95 | ... | ... |
| Watson | 5 75 | 5 00 | 5 00 | 4 50 | 3 00 | 2 00 | 4 00 | 3 00 |
| Wells, Fargo & Co. Mining Co | ... | ... | ... | ... | ... | ... | ... | ... |
| Ward | ... | ... | ... | ... | ... | ... | ... | ... |
| Yellow Jacket | 132 00 | 68 00 | 78 00 | 58 00 | 86 00 | 57 00 | 85 00 | 75 00 |

*Highest and lowest price of mining-stocks for* 1874—Continued.

| Companies. | May. | | June. | | July. | | August. | |
|---|---|---|---|---|---|---|---|---|
| | Highest price. | Lowest price. | Highest price. | Lowest price. | Highest price. | Lowest price. | Highest price. | Lowest price. |
| Alpha | $13 00 | $8 00 | $19 00 | $10 75 | $13 25 | $11 00 | $15 00 | $11 00 |
| Alps | 25 | 20 | 30 | 25 | ...... | ...... | ...... | ...... |
| American Flag | 4 00 | 3 00 | 3 50 | 3 00 | 3 50 | 1 50 | 2 75 | 1 50 |
| Arizona and Utah | ...... | ...... | ...... | ...... | ...... | ...... | ...... | ...... |
| Adams Hill | ...... | ...... | ...... | ...... | ...... | ...... | ...... | ...... |
| Alamo | 5 00 | 5 00 | ...... | ...... | ...... | ...... | ...... | ...... |
| American Flat | 6 00 | 5 50 | 8 25 | 6 00 | 8 00 | 6 50 | 8 50 | 6 25 |
| American Consolidated | ...... | ...... | ...... | ...... | ...... | ...... | 5 00 | 1 00 |
| Andes | 1 25 | 87½ | 1 00 | 62½ | 75 | 25 | 1 00 | 25 |
| Alta | ...... | ...... | ...... | ...... | ...... | ...... | ...... | ...... |
| American Flat, (south) | ...... | ...... | ...... | ...... | ...... | ...... | ...... | ...... |
| Belcher | 87 00 | 75 50 | 85 00 | 77 50 | 85 75 | 62 00 | 71 75 | 61 25 |
| Baltimore Consolidated | 6 87½ | 5 00 | 8 00 | 5 00 | 8 50 | 6 87½ | 7 75 | 6 25 |
| Belmont | 7 00 | 3 62½ | 5 87½ | 4 00 | 5 00 | 4 00 | 9 00 | 4 00 |
| Best and Belcher | 27 00 | 21 00 | 29 50 | 23 00 | 26 75 | 22 00 | 35 00 | 22 00 |
| Buckeye | 1 00 | 1 00 | 1 00 | 1 00 | ...... | ...... | 2 75 | 50 |
| Bacon | 2 50 | 2 50 | 6 00 | 4 00 | 5 00 | 3 50 | 5 00 | 3 00 |
| Bowery | 30 | 25 | 37½ | 20 | 40 | 20 | 20 | 20 |
| Bullion | 2 00 | 1 50 | 11 00 | 6 75 | 8 50 | 7 50 | 9 00 | 7 00 |
| Baltic Consolidated | ...... | ...... | ...... | ...... | ...... | ...... | ...... | ...... |
| Bellevue | 2 87½ | 1 00 | 2 37½ | 1 50 | 2 50 | 1 12½ | 1 50 | 25 |
| Bowers | ...... | ...... | ...... | ...... | 4 00 | 4 00 | ...... | ...... |
| Consolidated Virginia | 85 00 | 66 50 | 84 75 | 78 00 | 85 00 | 75 00 | 84 00 | 73 00 |
| Chollar Potosi | 65 00 | 52 00 | 92 00 | 59 00 | 73 50 | 46 75 | 50 00 | 46 75 |
| Crown Point | 89 00 | 77 00 | 86 50 | 79 00 | 83 00 | 65 50 | 79 00 | 67 50 |
| Confidence | 8 00 | 6 50 | 13 25 | 7 00 | 13 37½ | 10 00 | 14 62½ | 10 00 |
| Caledonia | 22 00 | 14 50 | 24 00 | 16 00 | 19 75 | 15 00 | 24 00 | 15 00 |
| Cederberg | 3 00 | 2 50 | 3 00 | 1 00 | 2 25 | 2 00 | 2 00 | 1 00 |
| Chief of the Hill | 75 | 50 | ...... | ...... | ...... | ...... | 20 | 5 |
| Charter Oak | 1 50 | 87½ | 75 | 75 | 50 | 25 | ...... | ...... |
| Chief, (east extension) | ...... | ...... | ...... | ...... | ...... | ...... | ...... | ...... |
| Central | ...... | ...... | ...... | ...... | ...... | ...... | ...... | ...... |
| Central No. 2 | ...... | ...... | ...... | ...... | ...... | ...... | ...... | ...... |
| Caroline | ...... | ...... | ...... | ...... | ...... | ...... | ...... | ...... |
| Chariot Mill | 5 50 | 4 50 | 6 75 | 5 50 | 8 87½ | 6 50 | 10 25 | 7 50 |
| Challenge | ...... | ...... | ...... | ...... | 3 00 | 2 00 | 3 00 | 3 00 |
| Consolidated Washoe | ...... | ...... | ...... | ...... | ...... | ...... | ...... | ...... |
| California | 43 25 | 33 00 | 40 00 | 35 00 | 39 75 | 35 00 | 40 00 | 35 00 |
| Chapman | ...... | ...... | ...... | ...... | ...... | ...... | ...... | ...... |
| Consolidated Amador | ...... | ...... | ...... | ...... | 4 50 | 4 00 | ...... | ...... |
| Condor | 25 | 25 | ...... | ...... | ...... | ...... | ...... | ...... |
| Cherry Creek | ...... | ...... | 1 87½ | 1 12½ | 3 25 | 1 25 | 3 00 | 1 50 |
| Crown Point Ravine | ...... | ...... | ...... | ...... | 1 00 | 50 | 1 12½ | 87½ |
| Calaveras | ...... | ...... | ...... | ...... | 1 00 | 1 00 | 1 00 | 1 00 |
| Calistoga | ...... | ...... | ...... | ...... | ...... | ...... | ...... | ...... |
| Dayton | 6 50 | 5 00 | 6 00 | 4 25 | 5 50 | 4 75 | 6 00 | 5 00 |
| Dardanelles | ...... | ...... | 6 00 | 3 50 | 12 00 | 5 00 | 13 00 | 11 00 |
| Dexter | ...... | ...... | ...... | ...... | ...... | ...... | ...... | ...... |
| Daney | 2 37½ | 1 00 | 2 75 | 2 00 | 2 25 | 2 00 | 2 00 | 1 50 |
| Empire Mill | 3 87½ | 3 25 | 8 87½ | 3 50 | 7 75 | 6 12½ | 9 12½ | 7 00 |
| Eureka Consolidated | 21 25 | 17 00 | 20 62½ | 16 00 | 18 50 | 10 00 | 13 50 | 10 00 |
| Exchequer | 35 00 | 20 00 | 80 00 | 34 00 | 66 00 | 58 00 | 73 00 | 61 00 |
| Eclipse | 4 87½ | 4 00 | 8 50 | 4 50 | 7 25 | 5 00 | 8 00 | 5 50 |
| Eureka | 10 00 | 8 50 | 9 50 | 8 00 | 8 00 | 5 00 | 5 50 | 5 50 |
| Empire | 15 00 | 6 00 | 14 75 | 9 00 | 8 00 | 1 50 | 2 50 | 1 25 |
| El Dorado (south) Consolidated | 8 00 | 5 37½ | 6 00 | 4 00 | 4 50 | 1 75 | 2 50 | 1 62½ |
| El Dorado (north) Consolidated | 2 75 | 2 50 | 2 75 | 2 37½ | 2 50 | 2 25 | 2 25 | 1 00 |
| Europa | ...... | ...... | ...... | ...... | ...... | ...... | ...... | ...... |
| Empress | ...... | ...... | ...... | ...... | ...... | ...... | ...... | ...... |
| Franklin | 1 25 | 1 25 | 2 50 | 2 00 | 2 50 | 1 50 | 2 00 | 1 50 |
| Fairmount | ...... | ...... | ...... | ...... | ...... | ...... | ...... | ...... |
| Florida | 3 00 | 3 00 | ...... | ...... | ...... | ...... | ...... | ...... |
| Gould & Curry | 23 00 | 16 00 | 28 50 | 18 50 | 23 50 | 17 25 | 24 00 | 17 25 |
| Globe | 2 25 | 1 00 | 2 00 | 1 00 | 2 75 | 75 | 1 75 | 1 37½ |
| Golden Chariot | 17 25 | 14 50 | 16 00 | 12 50 | 11 00 | 5 00 | 9 25 | 5 00 |
| Gold Hill Quartz | 2 50 | 1 50 | 4 00 | 1 00 | 4 50 | 2 50 | 3 75 | 2 75 |
| Greene | ...... | ...... | ...... | ...... | ...... | ...... | ...... | ...... |
| General Lee | ...... | ...... | ...... | ...... | ...... | ...... | ...... | ...... |
| Hale & Norcross | 44 00 | 31 50 | 61 00 | 37 50 | 56 00 | 38 50 | 47 00 | 38 50 |
| Huhn & Hunt | 1 00 | 50 | 50 | 50 | 60 | 50 | 25 | 25 |
| Hayes | ...... | ...... | ...... | ...... | ...... | ...... | ...... | ...... |
| Hermes | ...... | ...... | 1 00 | 1 00 | ...... | ...... | ...... | ...... |
| Hartford | ...... | ...... | ...... | ...... | ...... | ...... | ...... | ...... |

*Highest and lowest price of mining-stocks for* **1874—Continued.**

| Companies. | May. | | June. | | July. | | August. | |
|---|---|---|---|---|---|---|---|---|
| | Highest price. | Lowest price. | Highest price. | Lowest price. | Highest price. | Lowest price. | Highest price. | Lowest price. |
| Imperial | $5 25 | $4 00 | $9 37½ | $4 37½ | $8 25 | $6 75 | $9 00 | $7 25 |
| Ingomar | 25 | 10 | 35 | 25 | 25 | 20 | 25 | 20 |
| Independent | 70 | 50 | 30 | 25 | 75 | 75 | 87½ | 62½ |
| Ida Elmore | 2 87½ | 1 25 | 3 25 | 2 50 | 2 75 | 75 | 2 50 | 75 |
| Independent and Omega | | | | | | | 1 50 | 1 00 |
| Idaho, (Grass Valley) | | | | | 700 00 | 700 00 | | |
| Indus | | | | | | | | |
| Julia | 3 50 | 1 75 | 4 25 | 1 75 | 3 75 | 2 00 | 2 62½ | 1 75 |
| Justice | 7 75 | 6 25 | 8 50 | 7 00 | 8 00 | 5 12½ | 12 00 | 6 75 |
| Jackson | | | | | | | | |
| Jacob Little | | | | | | | | |
| Josephine Consolidated | | | | | | | | |
| Kentuck | 15 75 | 11 00 | 19 00 | 13 50 | 16 00 | 12 00 | 16 50 | 12 75 |
| Knickerbocker | 5 00 | 3 87½ | 6 75 | 4 00 | 6 00 | 2 75 | 6 25 | 4 25 |
| Kentucky | | | | | | | | |
| K. K. Consolidated | 5 50 | 2 00 | 5 00 | 3 50 | 3 75 | 3 00 | 3 00 | 2 50 |
| Kossuth | 5 75 | 4 00 | 1 50 | 1 25 | 1 12½ | 75 | 1 00 | 75 |
| Lady Bryan | 50 | 50 | 75 | 37½ | 75 | 75 | 1 00 | 1 00 |
| Leo | 1 12½ | 1 12½ | 1 00 | 75 | 85 | 62½ | 1 50 | 85 |
| Louise | | | | | | | | |
| Lady Washington | | | | | | | 2 50 | 2 50 |
| Lower Comstock | | | | | | | | |
| Meadow Valley | 11 50 | 9 00 | 9 50 | 7 75 | 8 75 | 4 50 | 7 25 | 4 50 |
| Monitor Belmont | 4 00 | 2 50 | 3 00 | 2 87½ | 2 50 | 1 00 | 3 00 | 1 00 |
| McMahon | | | | | | | | |
| Mammoth | 25 | 20 | 15 | 15 | 17 | 10 | 15 | 12½ |
| Mahogany | 5 50 | 4 00 | 7 50 | 5 25 | 5 00 | 3 00 | 9 25 | 3 00 |
| Mint | 25 | 25 | 35 | 25 | 25 | 25 | 12½ | 12½ |
| Midas | | | | | | | | |
| Magenta | | | | | 2 00 | 2 00 | 2 00 | 1 50 |
| Mansfield | | | | | | | | |
| Mexican | | | | | | | | |
| Newark | 3 25 | 70 | 3 12½ | 2 00 | 2 00 | 95 | 1 37½ | 37½ |
| New York Consolidated | 2 75 | 1 75 | 3 87½ | 1 25 | 3 50 | 1 75 | 4 25 | 3 12½ |
| Nevada | 1 25 | 1 25 | 1 62½ | 1 25 | | | 1 25 | 1 25 |
| North Belmont | 25 | 25 | | | 10 | 10 | 25 | 10 |
| North Carson | | | | | | | | |
| North Utah | | | | | | | | |
| Ophir | 18 00 | 8 87½ | 21 00 | 10 00 | 22 00 | 16 00 | 24 50 | 17 25 |
| Overmann | 46 50 | 23 00 | 45 00 | 22 75 | 36 00 | 25 00 | 51 50 | 27 50 |
| Original Hidden Treasure | 2 00 | 2 00 | 4 00 | | | | | |
| Occidental | 3 00 | 2 50 | 5 00 | 2 50 | 4 00 | 3 00 | 4 00 | 2 75 |
| Original Gold Hill | | | | | | | | |
| Pioche, (west) | | | | | | | | |
| Peavine | | | | | | | | |
| Pioche | 7 25 | 4 00 | 10 00 | 5 50 | 7 75 | 4 00 | 8 00 | 4 00 |
| Phœnix | 75 | 25 | 50 | 25 | | | | |
| Page & Panaca | 1 50 | 25 | 1 75 | 1 75 | 1 00 | 50 | 1 75 | 25 |
| Portland | 50 | 25 | 50 | 50 | | | | |
| Picton | 30 | 12½ | 63 | 25 | 40 | 25 | 60 | 25 |
| Prussian | | | 2 00 | 2 00 | 2 50 | 1 62½ | 3 00 | 2 50 |
| Patten | 75 | 50 | 1 50 | 75 | 1 50 | 1 50 | 50 | 50 |
| Poorman | | | 3 00 | 1 00 | 1 50 | 1 00 | 1 00 | 1 00 |
| Pacific | | | | | | | | |
| Phil. Sheridan | | | | | | | | |
| Quintero | 55 | 20 | 35 | 25 | 10 | 5 | 40 | 5 |
| Raymond & Ely | 30 75 | 20 00 | 24 00 | 20 00 | 20 50 | 9 25 | 15 00 | 9 25 |
| Red Jacket, (Idaho) | 3 00 | 2 00 | 3 75 | 2 25 | 3 00 | 1 50 | 1 50 | 1 25 |
| Rye Patch | 2 62½ | 1 50 | 1 87½ | 1 50 | 2 00 | 75 | 2 50 | 87½ |
| Rock Island | 1 50 | 1 00 | 2 25 | 1 75 | 3 50 | 2 00 | 4 37½ | 2 00 |
| Red Jacket, (Gold Hill) | | | | | | | | |
| Silver Cloud | | | | | | | 1 75 | 50 |
| Savage | 80 00 | 60 00 | 100 00 | 67 50 | 90 00 | 57 00 | 69 00 | 48 00 |
| Segregated Belcher | 87 00 | 64 50 | 95 00 | 67 00 | 89 00 | 68 00 | 117 00 | 71 00 |
| Succor | 2 25 | 1 00 | 2 50 | 1 25 | 2 00 | 1 00 | 1 87½ | 1 50 |
| Sierra Nevada | 17 00 | 15 00 | 28 50 | 15 25 | 25 50<br>4 75 | 20 50<br>4 00 | 4 87½ | 3 37½ |
| Silver Peak | 1 12½ | 1 00 | 1 00 | 75 | 25 | 25 | 30 | 25 |
| Silver Hill | 9 00 | 6 00 | 9 25 | 6 00 | 7 25 | 4 50 | 8 75 | 6 00 |
| Standard | 3 50 | 2 50 | 5 50 | 4 00 | | | | |
| South Chariot | 31 00<br>13 00 | 30 00<br>11 00 | 13 00 | 11 00 | 12 50 | 9 00 | 14 00 | 8 75 |
| South Overmann | | | | | 60 | 50 | 75 | 50 |

*Highest and lowest price of mining-stocks for 1874*—Continued.

| Companies. | May. | | June. | | July. | | August. | |
|---|---|---|---|---|---|---|---|---|
| | Highest price. | Lowest price. | Highest price. | Lowest price. | Highest price. | Lowest price. | Highest price. | Lowest price. |
| Spring Valley Water | $92 50 | $90 00 | $95 50 | $94 50 | $93 00 | $92 50 | $92 50 | $90 50 |
| San Francisco Gas-Light | 75 00 | 75 00 | 75 00 | 75 00 | 78 50 | 78 50 | 83 50 | 80 00 |
| Senator | | | | | 1 00 | 25 | 1 00 | 62½ |
| South Comstock | | | | | | | 62½ | 62½ |
| Segregated Rock Island | 87½ | 50 | 90 | 50 | 75 | 62½ | 87½ | 50 |
| Saint Patrick | 2 75 | 1 50 | | | | | 75 | 40 |
| Sutro | | | 3 00 | 2 25 | 2 50 | 1 87½ | 2 00 | 1 00 |
| South Star | | | | | | | | |
| Silver Cord | | | 2 50 | 1 75 | 1 50 | 50 | 1 75 | 50 |
| Segregated Caledonia | | | | | | | | |
| Trench | | | 10 00 | 6 50 | 9 00 | 4 00 | | |
| Tyler | 95 | 35 | 1 00 | 40 | 1 00 | 50 | 70 | 50 |
| Union Consolidated | 15 00 | 10 00 | 14 75 | 10 00 | 13 00 | 11 00 | 18 50 | 10 00 |
| Utah | 3 50 | 1 50 | 6 00 | 2 50 | 6 00 | 2 50 | 5 00 | 3 00 |
| Virtue | | | | | | | | |
| Woodville | 2 75 | 1 00 | 3 50 | 62½ | 3 62½ | 2 50 | 8 50 | 2 75 |
| Washington and Creole | 2 00 | 1 25 | 2 25 | 50 | 2 00 | 75 | 1 50 | 35 |
| War Eagle | 4 75 | 2 00 | 4 25 | 2 50 | 3 00 | 2 00 | 2 62½ | 1 25 |
| Whitman | | | | | | | 1 50 | 1 50 |
| Watson | 6 25 | 4 50 | 5 00 | 4 00 | 5 00 | 5 00 | | |
| Wells, Fargo & Co. Mining Co | | | | | | | | |
| Ward | | | | | | | | |
| Yellow Jacket | 79 00 | 64 00 | 100 00 | 67 50 | 93 00 | 66 00 | 83 00 | 66 00 |

| Companies. | September. | | October. | | November. | | December. | |
|---|---|---|---|---|---|---|---|---|
| | Highest price. | Lowest price. | Highest price. | Lowest price. | Highest price. | Lowest price. | Highest price. | Lowest price. |
| Alpha | $25 00 | $13 25 | $23 00 | $14 50 | $25 00 | $15 75 | $45 00 | $20 00 |
| Alps | | | 1 50 | 25 | | | | |
| American Flag | 3 50 | 1 25 | 5 00 | 2 50 | 2 50 | 2 00 | 3 00 | 2 00 |
| Arizona and Utah | | | | | | | | |
| Adams Hill | | | | | | | | |
| Alamo | | | | | | | | |
| American Flat | 17 50 | 7 00 | 13 75 | 8 25 | 10 00 | 7 50 | 9 00 | 6 00 |
| American Consolidated | 20 00 | 5 00 | 20 00 | 9 50 | | | | |
| Andes | 2 12½ | 90 | 2 00 | 1 00 | 6 50 | 1 12½ | 6 00 | 3 00 |
| Alta | 3 00 | 3 00 | 3 00 | 1 00 | | | 1 50 | 1 00 |
| American Flat, (south) | | | 1 00 | 1 00 | | | | |
| Belcher | 72 50 | 64 00 | 79 50 | 55 50 | 58 00 | 42 50 | 61 00 | 44 00 |
| Baltimore Consolidated | 17 00 | 7 25 | 14 00 | 7 50 | 10 00 | 8 00 | 9 75 | 7 00 |
| Belmont | 9 37½ | 7 00 | 8 50 | 7 25 | 9 87½ | 7 50 | 18 25 | 9 00 |
| Best and Belcher | 38 00 | 31 00 | 45 00 | 29 00 | { 32 00<br>{ 25 75 | 32 00<br>14 00 | } 83 50 | 24 50 |
| Buckeye | 4 00 | 2 50 | 3 75 | 3 50 | | | | |
| Bacon | 7 00 | 4 00 | 7 00 | 5 00 | 7 00 | 5 00 | 18 00 | 5 50 |
| Bowery | 20 | 20 | | | | | | |
| Bullion | 24 50 | 8 50 | 23 50 | 14 00 | 21 50 | 15 00 | 60 00 | 16 00 |
| Baltic Consolidated | | | | | | | | |
| Bellevue | 50 | 20 | | | 50 | 50 | | |
| Bowers | | | | | | | | |
| Consolidated Virginia | 94 00 | 80 50 | 120 00 | 91 00 | 188 00 | 121 00 | 590 00 | 135 00 |
| Chollar Potosi | 84 00 | 54 50 | 96 00 | 53 00 | 69 00 | 53 00 | 95 00 | 63 00 |
| Crown Point | 70 00 | 55 75 | 70 00 | 54 00 | 67 00 | 44 00 | 83 00 | 42 00 |
| Confidence | 30 00 | 13 25 | 29 00 | 19 00 | 39 00 | 20 00 | 52 00 | 34 00 |
| Caledonia | 33 00 | 20 00 | 34 50 | 17 00 | 25 00 | 18 00 | 34 00 | 22 00 |
| Cederberg | 1 00 | 75 | 1 00 | 1 00 | 50 | 50 | | |
| Chief of the Hill | 35 | 10 | 50 | 40 | 20 | 15 | | |
| Charter Oak | 50 | 25 | | | | | | |
| Chief, (east extension) | | | | | | | | |
| Central | | | | | | | | |
| Central No. 2 | | | | | | | | |
| Caroline | | | | | | | | |

*Highest and lowest prices of mining-stocks for* 1874—Continued.

| Companies. | September. | | October. | | November. | | December. | |
|---|---|---|---|---|---|---|---|---|
| | Highest price. | Lowest price. | Highest price. | Lowest price. | Highest price. | Lowest price. | Highest price. | Lowest price. |
| Chariot Mill | $12 50 | $9 25 | $17 00 | $10 00 | $9 50 | $5 50 | $5 50 | $3 00 |
| Challenge | 10 50 | 6 50 | 9 50 | 5 50 | 13 00 | 6 00 | 18 00 | 9 87½ |
| Consolidated Washoe | | | | | | | | |
| California | 63 50 | 37 00 | 65 00 | 52 00 | 126 00 | 59 75 | 540 00 | 135 50 |
| Chapman | | | | | | | | |
| Consolidated Amador | | | | | | | | |
| Condor | | | | | | | | |
| Cherry Creek | 1 50 | 50 | 1 00 | 50 | 50 | 50 | 3 00 | 25 |
| Crown Point Ravine | 2 25 | 87½ | 1 50 | 75 | 1 00 | 50 | 1 25 | 50 |
| Calaveras | | | | | | | | |
| Calistoga | 11 12½ | 10 00 | 11 37½ | 10 00 | 11 25 | 10 00 | 10 25 | 10 25 |
| Dayton | 8 00 | 5 00 | 7 75 | 5 00 | 6 00 | 5 25 | 5 87½ | 3 50 |
| Dardanelles | 14 00 | 12 50 | 18 00 | 15 00 | 18 00 | 17 50 | 25 00 | 18 00 |
| Dexter | | | | | | | | |
| Daney | 4 75 | 1 25 | 4 50 | 1 25 | 2 50 | 1 25 | 2 50 | 1 50 |
| Empire Mill | 11 00 | 8 25 | 11 00 | 8 87½ | 17 50 | 9 25 | 17 75 | 13 00 |
| Eureka Consolidated | 12 00 | 10 75 | 15 00 | 11 50 | 14 00 | 12 25 | 17 00 | 12 50 |
| Exchequer | 230 00 | 70 00 | 220 00 | 170 00 | 275 00 | 126 00 | 250 00 | 185 00 |
| Eclipse | 12 75 | 6 75 | 10 00 | 6 75 | 12 50 | 7 50 | 16 00 | 9 50 |
| Eureka | 10 00 | 7 00 | 9 00 | 8 00 | 8 50 | 8 00 | 8 50 | 8 00 |
| Empire | 1 37½ | 50 | 2 00 | 1 00 | 1 00 | 25 | 1 25 | 25 |
| El Dorado (south) Consolidated | 3 00 | 1 75 | 2 50 | 1 50 | 1 75 | 1 00 | 3 25 | 1 25 |
| El Dorado (north) Consolidated | 1 87½ | 1 12½ | 2 00 | 1 50 | 1 37½ | 1 00 | 1 50 | 1 00 |
| Europa | | | | | | | | |
| Empress | | | 5 00 | 1 50 | | | | |
| Franklin | 3 50 | 2 00 | 3 00 | 50 | 1 75 | 1 50 | 1 50 | 87½ |
| Fairmount | 2 00 | 2 00 | 2 00 | 2 00 | | | | |
| Florida | | | | | | | | |
| Gould & Curry | 29 00 | 20 00 | 35 00 | 25 00 | 30 50 / 26 00 | 29 00 / 14 25 | 63 00 | 24 00 |
| Globe | 3 87½ | 1 25 | 3 25 | 1 25 | 1 75 | 1 25 | 1 75 | 75 |
| Golden Chariot | 7 25 | 4 25 | 6 50 | 2 50 | 5 00 | 2 00 | 4 50 | 1 75 |
| Gold Hill Quartz | 8 00 | 3 75 | 6 25 | 4 50 | 8 00 | 6 00 | 9 00 | 5 00 |
| Greene | 1 75 | 1 50 | | | | | | |
| General Lee | | | | | | | | |
| Hale & Norcross | 60 00 | 41 00 | 65 00 | 41 00 | 62 00 | 43 00 | 74 00 | 55 00 |
| Huhn & Hunt | | | | | | | | |
| Hayes | | | | | | | | |
| Hermes | | | | | | | | |
| Hartfod | 1 75 | 1 50 | | | | | | |
| Imperial | 11 87½ | 8 12½ | 11 50 | 9 00 | 16 00 | 10 12½ | 21 00 | 14 00 |
| Ingomar | 20 | 20 | | | | | | |
| Independent | 50 | 25 | 50 | 50 | | | | |
| Ida Elmore | 2 75 | 75 | 2 50 | 75 | 2 50 | 1 50 | 2 50 | 1 50 |
| Independent and Omega | 1 50 | 1 00 | | | | | | |
| Idaho, (Grass Valley) | | | | | | | | |
| Indus | | | 2 75 | 87½ | 4 00 | 10 | 3 00 | 10 |
| Julia | 7 00 | 2 00 | 9 50 | 2 62½ | 7 37½ | 2 75 | 12 00 | 5 25 |
| Justice | 38 00 | 9 25 | 60 00 | 25 00 | 48 50 | 28 00 | 140 00 | 41 00 |
| Jackson | | | 30 | 10 | | | | |
| Jacob Little | | | | | | | | |
| Josephine Consolidated | | | | | | | | |
| Kentuck | 26 00 | 15 00 | 28 00 | 14 75 | 23 00 | 15 50 | 26 00 | 19 00 |
| Knickerbocker | 9 25 | 5 00 | 9 00 | 4 00 | 7 25 | 2 75 | 7 00 | 5 00 |
| Kentucky | | | | | | | | |
| K. K. Consolidated | 3 50 | 3 00 | 4 00 | 2 00 | 2 25 | 2 00 | | |
| Kossuth | 3 25 | 75 | 3 00 | 1 62½ | 4 00 | 2 25 | 3 50 | 2 37½ |
| Lady Bryan | 1 50 | 50 | 3 00 | 75 | 4 25 | 2 25 | 7 00 | 3 00 |
| Leo | 3 00 | 1 25 | 3 25 | 75 | 1 50 | 87½ | 2 12½ | 1 00 |
| Louise | 75 | 75 | 20 | 20 | | | | |
| Lady Washington | 5 00 | 2 50 | 4 50 | 1 50 | 2 25 | 2 00 | 4 50 | 1 37½ |
| Lower Comstock | | | | | | | | |
| Meadow Valley | 6 25 | 5 00 | 10 75 | 6 00 | 6 25 | 5 37½ | 7 00 | 5 25 |
| Monitor Belmont | 2 50 | 2 00 | 2 50 | 1 50 | 1 75 | 50 | 3 00 | 1 25 |
| McMahon | | | | | | | | |
| Mammoth | 35 | 15 | 40 | 20 | 25 | 25 | 40 | 40 |
| Mahogany | 5 75 | 2 75 | 4 00 | 3 25 | 5 62½ | 1 37½ | 4 50 | 3 00 |
| Mint | 50 | 15 | 45 | 10 | 35 | 20 | 50 | 25 |
| Midas | | | 13 75 | 12 00 | | | | |
| Magenta | | | | | 2 00 | 2 00 | 2 00 | 2 00 |
| Mansfield | 5 12½ | 5 00 | 6 75 | 5 50 | 3 00 | 3 00 | | |
| Mexican | | | | | | | 50 00 | 15 00 |
| Newark | 1 50 | 20 | 2 00 | 50 | 62½ | 37½ | 1 37½ | 50 |

*Highest and lowest prices of mining-stocks for* 1874—Continued.

| Companies. | September. |  | October. |  | November. |  | December. |  |
|---|---|---|---|---|---|---|---|---|
|  | Highest price. | Lowest price. | Highest price. | Lowest price. | Highest price. | Lowest price. | Highest price. | Lowest price. |
| New York Consolidated | $7 00 | $3 00 | $6 50 | $2 50 | $4 50 | $2 50 | $5 00 | $3 00 |
| Nevada | 2 00 | 1 50 | 2 25 | 1 50 |  |  |  |  |
| North Belmont | 20 | 15 | 15 | 10 | 15 | 12½ | 20 | 15 |
| North Carson |  |  | 2 50 | 2 00 | 3 00 | 2 12½ | 3 00 | 3 00 |
| North Utah |  |  | 1 62½ | 1 00 | 1 50 | 1 00 | 2 00 | 1 00 |
| Ophir | 52 50 | 21 62½ | 64 50 | 43 00 | 100 00 | 60 25 | 212 00 | 95 00 |
| Overmann | 85 00 | 39 50 | 110 00 | 49 00 | 70 00 | 51 50 | 100 00 | 66 00 |
| Original Hidden Treasure |  |  | 3 00 | 3 00 |  |  |  |  |
| Occidental | 4 50 | 2 50 | 4 50 | 2 50 | 4 00 | 2 75 | 8 00 | 3 87½ |
| Original Gold Hill | 5 00 | 2 75 | 3 25 | 1 00 | 2 00 | 1 12½ | 2 00 | 1 37½ |
| Pioche, (west) |  |  |  |  |  |  |  |  |
| Peavine |  |  |  |  |  |  |  |  |
| Pioche | 8 25 | 6 00 | 9 00 | 5 00 | 6 00 | 4 00 | 5 75 | 3 00 |
| Phœnix |  |  |  |  | 25 | 25 |  |  |
| Page & Panaca | 1 75 | 25 | 1 37½ | 10 |  |  |  |  |
| Portland |  |  |  |  |  |  |  |  |
| Picton | 1 00 | 45 | 1 00 | 30 | 50 | 12½ | 1 50 | 50 |
| Prussian | 5 00 | 2 50 | 5 00 | 3 00 | 4 75 | 4 00 | 4 25 | 3 00 |
| Patten | 1 25 | 50 |  |  |  |  |  |  |
| Poorman |  |  | 2 00 | 50 | 50 | 50 |  |  |
| Pacific | 1 62½ | 50 | 1 50 | 50 | 1 00 | 75 | 1 00 | 50 |
| Phil. Sheridan |  |  | 62½ | 37½ |  |  | 4 75 | 50 |
| Quintero |  |  | 15 | 10 | 15 | 12½ | 25 | 25 |
| Raymond & Ely | 17 00 | 9 50 | 23 00 | 15 00 | 17 75 | 14 50 | 27 00 | 14 75 |
| Red Jacket, (Idaho) | 1 00 | 1 00 |  |  |  |  |  |  |
| Rye Patch | 2 00 | 1 00 | 3 87½ | 1 50 | 6 00 | 3 25 | 6 00 | 3 00 |
| Rock Island | 7 00 | 3 50 | 7 00 | 4 50 | 4 50 | 3 37½ | 9 00 | 3 50 |
| Red Jacket, (Gold Hill) |  |  |  |  |  |  |  |  |
| Silver Cloud | 1 50 | 75 |  |  |  |  |  |  |
| Savage | 81 00 | 51 50 | 113 00 | 66 00 | 30 00 | 74 00 | 151 00 | 87 00 |
| Segregated Belcher | 175 00 | 93 00 | 255 00 | 100 00 | 140 00 | 110 00 | 177 50 | 118 00 |
| Succor | 6 00 | 1 50 | 4 87½ | 2 75 | 3 75 | 2 75 | 7 00 | 3 75 |
| Sierra Nevada | 10 50 | 4 50 | 9 87½ | 7 00 | 13 50 | 8 62½ | 28 00 | 13 50 |
| Silver Peak |  |  |  |  |  |  |  |  |
| Silver Hill | 18 00 | 6 75 | 13 50 | 6 00 | 10 37½ | 6 25 | 15 00 | 9 00 |
| Standard |  |  | 2 75 | 2 50 |  |  |  |  |
| South Chariot | 18 50 | 8 00 |  |  | 1 25 | 75 | 2 00 | 1 23 |
| South Overmann | 1 00 | 62½ | 1 00 | 1 00 |  |  |  |  |
| Spring Valley Water | 91 50 | 91 50 | 94 00 | 92 00 |  |  | 99 00 | 99 00 |
| San Francisco Gas-Light | 86 50 | 86 50 |  |  |  |  |  |  |
| Senator | 2 25 | 87½ | 1 75 | 1 00 | 1 37½ | 75 | 1 87½ | 1 00 |
| South Comstock |  |  |  |  | 2 25 | 2 00 | 3 00 | 2 00 |
| Segregated Rock Island | 2 00 | 75 | 2 00 | 1 00 | 1 50 | 1 00 | 2 75 | 50 |
| Saint Patrick |  |  | 1 25 | 1 25 | 1 87½ | 1 00 | 1 75 | 1 00 |
| Sutro | 1 00 | 75 | 2 00 | 1 00 | 1 75 | 1 00 | 4 00 | 2 25 |
| South Star |  |  |  |  |  |  |  |  |
| Silver Cord | 1 50 | 87½ | 1 50 | 50 | 1 25 | 50 | 87½ | 87½ |
| Segregated Caledonia |  |  |  |  |  |  | 3 00 | 1 25 |
| Trench | 10 00 | 8 00 | 7 00 | 5 00 | 9 00 | 6 00 | 16 00 | 9 00 |
| Tyler | 1 50 | 60 | 1 50 | 50 | 87½ | 37½ | 1 00 | 15 |
| Union Consolidated | 27 50 | 15 00 | 25 00 | 17 00 | 46 00 | 19 50 | 95 00 | 34 00 |
| Utah | 5 75 | 3 00 | 4 75 | 3 00 | 6 75 | 3 25 | 13 00 | 5 00 |
| Virtue |  |  |  |  |  |  |  |  |
| Woodville | 13 00 | 6 75 | 10 00<br>4 00 | 5 00<br>1 50 | 2 62½ | 1 25 | 4 00 | 1 50 |
| Washington and Creole | 1 75 | 1 00 | 2 50 | 1 25 | 1 25 | 75 | 1 12½ | 75 |
| War Eagle | 2 00 | 1 00 | 2 50 | 2 00 | 3 50 | 2 00 | 4 00 | 1 25 |
| Whitman | 1 37½ | 1 37½ |  |  |  |  |  |  |
| Watson |  |  |  |  | 3 00 | 2 50 |  |  |
| Wells, Fargo & Co. Mining Co |  |  | 1 00 | 1 00 |  |  |  |  |
| Ward |  |  |  |  |  |  | 3 25 | 3 00 |
| Yellow Jacket | 125 00 | 78 00 | 129 00 | 84 00 | 133 00 | 87 00 | 180 00 | 124 50 |
| Total | 22, 394, 805 |  | 30, 760, 910 |  | 26, 969, 020 |  | 50, 062, 145 |  |

NOTE.—The following companies have increased their capital stock, as indicated by the two prices given in months when changes were made: Ophir, in January; Alpha, in February; Bullion, in February; South Chariot, in May; Sierra Nevada, in July; Woodville, in October; Best & Belcher, in November, and Gould & Curry, in November.

# INDEX OF MINES, MILLS, WORKS, ETC.

[NOTE.—This index contains the names of such individual enterprises as are alluded to in the foregoing report. In most cases the counties in which the mines or works are situated are also given. Reduction-works belonging to mines may be sought also under the names of the mines.]

## A.

C.

D.

F.



G.

## H.

I.

N.

## O.

P.

## W.

## Y.



## Z.

# INDEX OF COUNTIES, DISTRICTS, ETC.

A.



B.

C.

D.



E.



F.



G.

H.



I.



J.



K.

## Q.



## R.



## S.

T.



U.

# INDEX OF SUBJECTS.

## A.



## B.



## C.

D.



F.



G.



H.

I.



K.



L.



M.



N.



O.



P.



Q.

R.



S.



T.



U.



V.



W.



Z.

CPSIA information can be obtained
at www.ICGtesting.com
Printed in the USA
LVHW112036150720
660524LV00014B/527